ANTIMICROBIAL POLYMERS

ANTIMICROBIAL POLYMERS

Edited by

JOSÉ M. LAGARÓN
Institute of Agrochemistry and Food Technology (IATA)
Spanish Council for Scientific Research (CSIC)

MARÍA J. OCIO
Institute of Agrochemistry and Food Technology (IATA)
Spanish Council for Scientific Research (CSIC), and
Faculty of Pharmacy, University of Valencia

AMPARO LÓPEZ-RUBIO
Institute of Agrochemistry and Food Technology (IATA)
Spanish Council for Scientific Research (CSIC)

WILEY

A JOHN WILEY & SONS, INC., PUBLICATION

Published by John Wiley & Sons, Inc., Hoboken, New Jersey
Published simultaneously in Canada

Library of Congress Cataloging-in-Publication Data:

Antimicrobial polymers / edited by José M. Lagarón, María J. Ocio, Amparo López-Rubio.
p. cm.
Includes bibliographical references and index.
ISBN 978-0-470-59822-1 (cloth)
1. Antimicrobial polymers. 2. Materials–Microbiology. 3. Plastics–Additives.
I. Lagaron, Jose Maria. II. Ocio, Maria Jose. III. Rubio, Amparo Lopez.
TP1180.A58A68 2011
668.4—dc23

2011022704

10 9 8 7 6 5 4 3 2 1

CONTENTS

PREFACE vii

CONTRIBUTORS ix

1 **ANTIMICROBIAL PACKAGING POLYMERS. A GENERAL INTRODUCTION** 1
José M. Lagarón, María J. Ocio, and Amparo López-Rubio

2 **BACTERIAL RESISTANCE AND CHALLENGES OF BIOCIDE PLASTICS** 23
Nibras A.A.M. Ahmed

3 **"CLICK CHEMISTRY" TO DERIVED ANTIMICROBIAL POLYMERS** 51
P. Lecomte, R. Riva, and C. Jérôme

4 **CHITOSAN AND CHITOSAN BLENDS AS ANTIMICROBIALS** 71
Patricia Fernandez-Saiz

5 **THYMOL IN NANOCOMPOSITES. A CASE STUDY** 101
Maria Dolores Sanchez-Garcia

6 **BACTERIOCINS IN PLASTICS** 117
Gianluigi Mauriello and Francesco Villani

7 **ANTIMICROBIAL ENZYMES AND NATURAL EXTRACTS IN PLASTICS** 159
Marcella Mastromatteo, Marianna Mastromatteo, Amalia Conte, and Matteo Alessandro Del Nobile

8 **ANTIMICROBIAL PEPTIDES** 195
Jose F. Marcos and Paloma Manzanares

9 **RECOMBINANT ANTIMICROBIAL PEPTIDES** 227
J. Carlos Rodríguez-Cabello, Carmen García-Arévalo, Alessandra Girotti, Laura Martín, and Mercedes Santos

10 NOVEL ANTIMICROBIALS OBTAINED BY ELECTROSPINNING METHODS 261
Sergio Torres-Giner

11 SILVER- AND NANOSILVER-BASED PLASTIC TECHNOLOGIES 287
Antonio Martínez-Abad

12 ANTIMICROBIAL PLASTICS BASED ON METAL-CONTAINING NANOLAYERED CLAYS 317
José M. Lagarón and Maria A. Busolo

13 NANOMETALS AS ANTIMICROBIALS 327
R.P. Allaker, M.A. Vargas-Reus, and G.G. Ren

14 TITANIUM DIOXIDE-BASED PLASTIC TECHNOLOGIES 351
Marta Fernández-García, María L. Cerrada, Anna Kubacka, and Marcos Fernández-García

15 TISSUE-IMPLANT ANTIMICROBIAL INTERFACES 379
E. Marsich, A. Travan, I. Donati, G. Turco, F. Bellomo, and S. Paoletti

16 CHARACTERIZING THE INTERACTIONS BETWEEN CELL MEMBRANES AND ANTIMICROBIALS VIA SUM-FREQUENCY GENERATION VIBRATIONAL SPECTROSCOPY 429
Christopher W. Avery and Zhan Chen

17 GAS-BASED ANTIMICROBIALS IN ACTIVE PACKAGING 459
Siriyupa Netramai, Maria Rubino, and Loong-Tak Lim

18 CURRENT LEGISLATION IN ANTIMICROBIALS 489
Javier Gómez

19 HUMAN SAFETY AND ENVIRONMENTAL CONCERNS ASSOCIATED WITH THE USE OF BIOCIDES 527
Orla Condell, Carol Iversen, Karen Power, and Séamus Fanning

INDEX 561

PREFACE

Antimicrobial technologies are rapidly spreading to many industrial applications thanks to increased consumer awareness in the safe handling of everyday appliances and surfaces. As a result, there is also a concomitant rise in the number of research activities in biocides from both fundamental and applied perspectives. Biocides are currently applied in many application areas such as health care and medicine, building and construction, furniture and office accessories, shoes and apparel, as well as food packaging and coatings. This is the result of the increasing value perception of hygiene and cleanliness among the public, perhaps borne out of the rapidly spread outbreak of infections, so loudly reported by the media during the last several years. Most of these outbreaks are associated with globalization and increased internationalization of individuals' business and leisure activities. In a global context, people move across continents within one day and get potentially exposed to unknown (to their immune system) threats and infections. The discussion of whether this is completely within a rational or historic perspective is beyond the scope of this book, but what is certainly true is that the highly competitive environment in which we currently liaise makes it very "inconvenient" for professional people to be absent from work as a result of illness. Moreover, the growing world population and increasing common shared spaces facilitate microbial contamination and spread in locations such as schools, supermarkets, airports, shopping malls, convention centers, and hospitals. Additionally, the rushing activity resulting from complying with frenzied schedules reduces the time to approach hygiene more carefully within homes and other children's habitats.

Nevertheless, it is in the food and medical fields where microbial contamination can cause special seriousness and grave consequences to people in general. In the food area, this is lately more the case because current trends toward mildly preserved foods to ensure product quality and freshness by making use of alternative hurdle preservation technologies may eventually reduce food safety aspects. In this context, antimicrobials have thus become increasingly important in food handling areas, in food itself, and in "active" packaging strategies. Of course, biocides cannot be considered substitutes for hygiene, and in reckoning this, the food industry operations will become a lot safer by appropriate usage of biocides in combination with currently employed good hygiene practices.

In the medical area, antimicrobials have always been of concern, but novel antimicrobial technologies and nanotechnologies are being implemented in more applications such as implant interphases, coatings, and others to avoid postoperatory infections.

In agreement with this new interest in biocides and with the current trend of giving value to natural and renewable resources as well as to avoid (1) the inherent collateral toxicity and side effects of biocide synthetic chemicals, (2) legislation, and (3) application barriers, naturally-derived biocides are rapidly growing in various applications areas, particularly in food packaging and in biomedical applications. As the use of plastics and bioplastics is currently very widespread and many articles and surfaces are made of plastics, these materials currently offer considerable interest as matrices for the incorporation of highly functional biocides. Even traditional and novel ceramic-based matrices and various other finishes and coatings of other materials are made of plastics, from which biocides can be fixated or control released. Moreover, some bioplastics and modern fabrication and nanofabrication of bioplastics can also lead to very efficient antimicrobial plastics.

This pioneering book gathers, by reviewing appropriate scientific literature and technological developments in the field, up-to-date scientific and technological understanding to generate and describe antimicrobial plastics, putting special emphasis on natural biocides and on the nanofabrication of biocides. After a general introductory chapter by the editors and others related to bacterial resistance, several chapters deal with the design, phenomenology, and understanding of novel antimicrobial technologies to be implemented in plastics or made of plastics, discussing nanotech strategies such as various nanometals and electrospun fiber-based biopolymeric biocides. Additionally, biocide implant interphases, chitosan-based biocides, advanced "click" chemistry to render antimicrobial plastics, peptides and recombinant peptides, bacteriocins, antimicrobial and antioxidant natural extracts, and gas-based biocides incorporated into plastics or bioplastics are presented and discussed. The book ends with two chapters related to both global legislation in the area and human safety and environmental concerns associated with the use of biocides and plastic-based biocides.

We believe this book to be of general interest to biologists, microbiologists, biotechnologists, medical doctors, pharmacists, polymer scientists, food scientists and technologists, and other academic and industrial professionals with interest in antimicrobials. We seriously hope to have offered a balanced, interesting, and innovative perspective in the area not only for advanced readers but also for industrial decision makers and for those approaching the field with limited knowledge.

The Editors

CONTRIBUTORS

Nibras A.A.M. Ahmed, Department of Oral Biology, University of Oslo, Norway

R.P. Allaker, Queen Mary University of London, Barts and The London School of Medicine and Dentistry, Institute of Dentistry, London, U.K.

Christopher W. Avery, University of Michigan, Ann Arbor, Michigan, USA

F. Bellomo, Department of Life Sciences, University of Trieste, Trieste, Italy

Maria A. Busolo, Instituto de Agroquímica y Tecnología de Alimentos (IATA), NanoBioMatters R&D S.L., Paterna, (Valencia), Spain

María L. Cerrada, Instituto de Ciencia y Tecnología de Polímeros (ICTP-CSIC), Madrid, Spain

Zhan Chen, University of Michigan, Ann Arbor, Michigan, USA

Orla Condell, University College Dublin, Belfield, Dublin, Ireland

Amalia Conte, University of Foggia, Foggia, Italy

Matteo Alessandro Del Nobile, University of Foggia, Foggia, Italy

I. Donati, Department of Life Sciences, University of Trieste, Trieste, Italy

Séamus Fanning, University College Dublin, Belfield, Dublin, Ireland

Marcos Fernández-García, Instituto de Catálisis y Petroleoquímica, (ICP-CSIC), Madrid, Spain

Marta Fernandez-García, Instituto de Ciencia y Tecnología de Polímeros (ICTP-CSIC), Madrid, Spain

Patricia Fernandez-Saiz, Institute of Agrochemistry and Food Technology (IATA), CSIC, Burjassot, Spain

Carmen García-Arévalo, GIR Bioforge, University of Valladolid, Valladolid, Spain

Alessandra Girotti, GIR Bioforge, University of Valladolid, Valladolid, Spain

Javier Gómez, Treqtop Services, S.L., Valencia, Spain

Carol Iversen, Nestlé Research Center, Lausanne, Switzerland

C. Jérôme, Center for Education and Research on Macromolecules (CERM), University of Liège, Liège, Belgium

Anna Kubacka, Instituto de Catálisis y Petroleoquímica, (ICP-CSIC), Madrid, Spain

José M. Lagarón, Instituto de Agroquímica y Tecnología de Alimentos (IATA), CSIC, Valencia, Spain

P. Lecomte, Center for Education and Research on Macromolecules (CERM), University of Liège, Liège, Belgium

Loong-Tak Lim, Department of Food Science, University of Guelph, Guelph, Ontario, Canada

Amparo López-Rubio, Instituto de Agroquímica y Tecnología de Alimentos (IATA), CSIC, Valencia, Spain

Paloma Manzanares, Departamento de Biotecnología de Alimentos, Instituto de Agroquímica y Tecnología de Alimentos (IATA), Consejo Superior de Investigaciones Científicas (CSIC), Burjassot, Valencia, Spain

Jose F. Marcos, Departamento de Ciencia de los Alimentos, Consejo Superior de Investigaciones Científicas (CSIC), Burjassot, Valencia, Spain

E. Marsich, Department of Life Sciences, University of Trieste, Trieste, Italy

Laura Martín, GIR Bioforge, University of Valladolid, Valladolid, Spain

Antonio Martínez-Abad, Instituto de Agroquímica y Tecnología de Alimentos (IATA), CSIC Paterna, Valencia, Spain

Marcella Mastromatteo, University of Foggia, Foggia, Italy

Marianna Mastromatteo, University of Foggia, Foggia, Italy

Gianluigi Mauriello, Department of Food Science, University of Naples Federico II, Portici, Naples, Italy

Siriyupa Netramai, School of Packaging, Michigan State University, East Lansing, Michigan and Mahidol University, Karnjanaburi, Thailand

María J. Ocio, Instituto de Agroquímica y Tecnología de Alimentos (IATA), CSIC, Valencia, Spain

S. Paoletti, Department of Life Sciences, University of Trieste, Trieste, Italy

Karen Power, University College Dublin, Belfield, Dublin, Ireland

G.G. Ren, School of Engineering and Technology, University of Hertfordshire, Hatfield, U.K.

R. Riva, Center for Education and Research on Macromolecules (CERM), University of Liège, Liège, Belgium

J. Carlos Rodríguez-Cabello, GIR Bioforge, University of Valladolid, Valladolid, Spain

Maria Rubino, School of Packaging, Michigan State University, East Lansing, Michigan

Maria Dolores Sanchez-Garcia, Instituto de Agroquímica y Tecnología de Alimentos (IATA), CSIC, Valencia, Spain

Mercedes Santos, GIR Bioforge, University of Valladolid, Valladolid, Spain

Sergio Torres-Giner, Instituto de Agroquímica y Tecnología de Alimentos (IATA), CSIC, Paterna (Valencia), Spain

A. Travan, Department of Life Sciences, University of Trieste, Trieste, Italy

G. Turco, Department of Life Sciences, University of Trieste, Trieste, Italy

M.A. Vargas-Reus, Queen Mary University of London, Barts and The London School of Medicine and Dentistry, Institute of Dentistry, London, U.K.

Francesco Villani, Department of Food Science, University of Naples Federico II, Portici, Naples, Italy

CHAPTER 1

ANTIMICROBIAL PACKAGING POLYMERS. A GENERAL INTRODUCTION

JOSÉ M. LAGARÓN, MARÍA J. OCIO, and AMPARO LÓPEZ-RUBIO
Instituto de Agroquímica y Tecnología de Alimentos (IATA), CSIC, Valencia, Spain

CONTENTS

1.1 Pathogens in Food: Public Health Importance 2
1.2 Primary Contamination and Its Causes 2
  1.2.1 *Salmonella* spp. 5
  1.2.2 *L. monocytogenes* 5
  1.2.3 *S. aureus* 5
  1.2.4 *C. jejuni* 6
  1.2.5 *E. coli* O157:H7 6
1.3 Prevention and Control 7
1.4 Antimicrobial Agents for Food Preservation 8
1.5 Methods to Determine the Antimicrobial Activity 9
1.6 Plastics and Bioplastics in Packaging 10
1.7 Antimicrobials in Polymers 12
1.8 Conclusions 16
References 16

Antimicrobial Polymers, First Edition.
Edited by José M. Lagarón, María J. Ocio, and Amparo López-Rubio.

This chapter introduces in a general and brief fashion the subjects discussed across the book. Albeit the concepts can be generally considered, it has an application perspective that relates more intensely to the food area. This is because antimicrobials, thanks to the recent publication of the European Commission regulation on active and intelligent materials and articles intended to come into contact with food (EC 450/2009) as well as to the increasing number of submissions to the U.S. Food and Drug Administration (FDA), are attracting a lot of new academic and industrial interest for implementation into plastic packaging materials.

1.1 PATHOGENS IN FOOD: PUBLIC HEALTH IMPORTANCE

Without a doubt, the most relevant sectors regarding the seriousness of microbial contamination are hospitals and medical equipment and foods. Hence, antimicrobials have traditionally been of great relevance to these areas. Foods are perhaps attracting even greater general attention in particular nowadays because they constitute a permanent part of our daily life and are increasingly making use of plastic packaging for their presentation to the consumer. Infections and intoxications associated with consumption of foods are a growing concern worldwide. In this regard, the World Health Organization (WHO) and the European Food Safety Authority (EFSA) report annually on the main agents causing food-borne toxic infections. The results suggest that in recent years there has been a slight increase in food-borne diseases in many parts of the world and that the emergence of new or newly recognized food-borne problems have been identified and associated with consumption of foods. According to the WHO [1, 2], one of the main reasons for this increase is that the microbial population have adapted through natural selection leading to the development of antibiotic resistance, acquisition of new virulence factors, or changes in the ability to survive in adverse environmental conditions. In addition, the change of population dietary habits produced by the growing demand for prepared foods and minimally processed foods has contributed significantly in increasing the number of outbreaks of food-borne illnesses.

1.2 PRIMARY CONTAMINATION AND ITS CAUSES

Most pathogenic bacteria associated with food-borne diseases are zoonotic (animal origin), and their carriers are usually healthy animals from which are transmitted to a wide variety of foods such as *Salmonella* spp., *Escherichia coli*, or *Campylobacter jejuni*. Other pathogens such as *Listeria monocytogenes* are widely distributed in the environment or are part of the natural microbiota of humans as *Staphylococcus aureus*. In these two last cases, food contamination occurs during processing as a result of failures of hygienic practices in the food chain.

Farm animals usually acquire microbial hazards as a result of horizontal transmission from their environment. The principal sources are other infected animals, contaminated water, and wildlife such as birds or rodents. This horizontal transmission can be exacerbated by intensive husbandry, which promotes overcrowding and interferes with the maintenance of adequate hygiene to which animals are subjected in many farms [3, 4]. Finally, the products derived from these infected animals can reach the consumer at some point.

The rising incidence of microbial food-borne disease has focused attention on the sources of contamination. Because animal products have been directly responsible for more than the 50% of the total food-borne outbreaks in the 1990s, emphasis has been paid to these types of products such as meat, poultry, eggs, and milk [5]. Nevertheless, in recent years, the demand for fresh fruits and vegetables has increased in the industrial countries as a consequence of the awareness of the health benefits associated with eating fresh produce [6]. Nowadays, therefore, outbreaks of food-borne diseases have been increased probably because crops in the field can be contaminated with pathogens carried by farm animals and human beings. The main risk factors include proximity to irrigation wells and surface waterways exposed to feces from cattle and wildlife, exposure in fields to wild animals and their waste materials, and improperly composted animal manure used as fertilizer [7]. The public health implications are especially serious when the products affected are those usually consumed without cooking such as salad fruits and vegetables [8]. Although the frequency of food-borne outbreaks in gastrointestinal illness associated with fruit and vegetables seems to be low compared with products of animal origin, ready-to-eat fruit and vegetables requiring minimal or no further processing prior to consumption have been implicated as vehicles for transmission of infectious microorganisms. Even more, food-borne illnesses associated with fruit and vegetables seem to be increasing in many countries because of mainly the increase in global food distribution.

The last global report available from the European Union is from 2006. In this report, the number of food-borne outbreaks, causative agents, and number of affected people by these gastrointestinal illnesses were shown. For instance, in this year, 5,705 outbreaks involving a total of 53,568 people, resulting in 5,525 hospitalizations (10.3%) and 50 deaths (0.1%), were reported. Table 1.1 summarizes the number of reported food-borne outbreaks in the Europe Union (EU) in 2006. These results show that *Salmonella* was the food-borne pathogen most frequently recorded by the EU, although there has been a slight decline in recent years. *Campylobacter* infections are the second most common zoonotic agent, affecting 1,304 people (6.9%). *S. aureus* was the etiological agent of 4.1% of the outbreaks (236) and affected 2,057 people, causing two deaths. Verotoxic-producing *E. coli* was responsible for 0.8% of the outbreaks, and although it is one of the most harmful pathogens, it only caused one death. On average, *Listeria* was the most severe pathogen causing 9 outbreaks that affected 120 people, of which 74.2% (89) was hospitalized and 17 died. The most common food vehicle was eggs and egg products, which were responsible for 17.8% of

TABLE 1.1 Causative agents responsible for foodborne outbreaks, 2006 (all countries)

	Outbreaks				Human cases		
Causative agent	No.	% of total	General	Household	No.	No. admitted to hospital	No. of deaths
Salmonella	3,131	53.9	1,520	1,611	22,705	3,185	23
Unknown	952	16.4	610	342	9,437	947	2
Viruses	587	10.2	373	214	13,345	553	3
Campylobacter	400	6.9	116	284	1,304	65	0
Staphylococcus	236	4.1	157	79	2,057	277	2
Toxins	86	1.5	20	66	834	261	3
Clostridium	81	1.4	55	26	1,651	44	2
Bacillus	78	1.3	66	12	964	34	0
Histamine	71	1.2	62	9	370	41	0
Pathog. *E. coli*	48	0.8	25	23	750	103	1
Shigella	33	0.6	19	14	138	22	0
Yersinia	26	0.4	11	15	604	15	2
Giardia	18	0.3	13	5	44	—	0
Trichinella	18	0.3	5	13	202	113	0
Listeria	9	0.2	5	4	120	89	17
Other	9	0.2	5	4	31	2	0
Cryptosporidium	7	0.1	4	3	59	0	0
Brucella	6	0.1	3	3	43	3	0
Flavivirus	6	0.1	2	4	26	25	0
Klebsiella	3	0.1	2	1	109	1	0
Streptococcus	2	<0.1	2	0	236	—	—
EU Total	5,705	98.2	3,000	2,706	53,546	5,523	50
Total	5,807	100.0	3,075	2,732	55,029	5,780	55

Source: The EFSA Journal, 2007;130:252–352 [9].

these outbreaks, whereas unspecified meat was reported as the causative source in 10.3% of the outbreaks. Fish and fish products were the source of 4.6% and dairy products of 3.2% of the outbreaks [9].

Food-borne disease caused by microbiological hazards is a large and growing public health problem. Most countries with systems for reporting cases of food-borne diseases have documented significant increases over the past few decades in the incidence of diseases caused by microorganisms in foods, including *Salmonella* spp., *S. aureus*, *C. jejuni*, *L. monocytogenes*, or *E. coli* O157 among others.

1.2.1 *Salmonella* spp.

Salmonella spp. is a heterotrophic, mesophile, gram-negative bacteria that is worldwide one of the major infections transmitted via food ingestion, causing gastroenteritis, diarrhea, vomiting, abdominal cramping, enteric fever, septicemia, and in severe cases even death [10]. In August 2005, 2,138 cases of salmonella gastroenteritis were reported to the National Centre for Epidemiology (CNE) in Spain. The reported cases were epidemiologically and microbiologically linked to a single brand of precooked, vacuum-packed roast chicken that was commercially distributed throughout Spain. Although it did not report any human deaths, this produced a substantial economic loss to the producing company.

1.2.2 *L. monocytogenes*

Another more harmful microorganism is *L. monocytogenes*, a food-borne pathogen of particular concern in ready-to-eat (RTE) products because of its ability to survive and grow at refrigeration temperatures and of its capacity to tolerate high heat and high concentrations of salt [11]. Even after cleaning, a prevalence of *L. monocytogenes* of 10% was detected in surface samples of the investigated equipment of small Spanish processing plants of traditional fermented products [12]. *L. monocytogenes* causes food-borne listeriosis, a disease that occurs largely in pregnant woman and the elderly leading to illness, miscarriages, and death [13]. Recently, a *Listeria* outbreak linked to a meat product plant in Toronto has expanded to 26 cases, and 12 people have died, although it is not yet clear how many of the deaths were directly from the illness.

1.2.3 *S. aureus*

S. aureus is a gram-positive bacterium able to produce sufficient enterotoxins to cause illness from an inoculum size of ca. 10^5 CFU/mL. Although death from staphylococcal food poisoning is rare (0.03% of cases), it presents several symptoms such as vomiting, diarrhea, or abdominal cramping. A wide range of foods is involved as sources of staphylococci food poisoning in restaurants where meals are previously prepared in a central chicken and subsequently transported. As *S. aureus* lives in the nasal membranes, skin, gastrointestinal tract, and so on, carriers or infected food handles may easily transmit these organisms to food. If food is contaminated and temperature abused before cooking, heating will destroy the bacteria, but heat-stable enterotoxin may remain and cause illness. A recent study conducted on determination of histamine in swordfish filets implicated in an incident of food-borne poisoning that caused illness in 43 victims in December 2004 in central Taiwan revealed that *S. aureus* seemed to be the histamine former strain [14].

1.2.4 *C. jejuni*

Campylobacter infection is estimated to be the leading cause of bacterial food-borne illness in the industrial countries, and food-borne transmission accounts for approximately 80% of all infections [15]. More than 90% of all human *Campylobacter* infections are caused by *C. jejuni* and *C. coli*. Natural reservoirs are wild birds, and chickens become colonized shortly after birth and are the most important source for human infection. *C. jejuni* is a gram-negative microaerophile, and its optimal grown conditions are a neutral pH and 41°C to 42°C of temperature. It has been frequently isolated from the gastrointestinal tract of wild birds, humans, and other mammals [16]. Human infections are normally attributed to consumption of contaminated uncooked poultry and cross-contamination from this source as well as to contact with cattle including consumption of beef and milk. Symptoms and signs usually include fever, abdominal cramping, and diarrhea as a mild illness but occasionally severe, leading to meningitis, pneumonia, miscarriage, and Guillain–Barré syndrome [17]. Deaths attributable to *C. jejuni* infection have been reported but rarely occur [18].

1.2.5 *E. coli* O157:H7

E. coli infection is transmitted to humans mainly through consumption of contaminated foods such as raw or undercooked meat and milk. Fecal contamination of water and other foods, as well as cross-contamination during handling, are also important causes of infection. The main symptoms that occur after infection are abdominal pain and diarrhea that may progress to bloody diarrhea. Recovery is usually carried out in 10 days, but when it affects the elderly or children, it can lead to hemolytic uremic syndrome characterized by acute renal failure and hemolytic anemia [19]. Most confirmed human *E. coli* O157:H7 outbreaks have been associated with the consumption of undercooked ground beef and, less frequently, unpasteurized milk. *E. coli* O157:H7 is in a close relation to modern approaches to food preservation such as minimally processed foods. Although the incidence of outbreaks produced by this organism is low, in recent years, it has been slightly increased. The latest data available from the EFSA provided by the EU countries for 2006 collected a total of 4,916 cases of toxic infections associated with *E. coli* O157:H7, of which 13 occurred in Spain. An important virulence factor of this microorganism is its acid resistance.

The bacteria possess at least three acid-resistance systems that account for their well-known ability to tolerate acid environments [20]. Some studies have revealed that *E. coli* O157:H7 isolates are resistant to antibiotics [21].

1.3 PREVENTION AND CONTROL

Hazards can be introduced at any point from farm to table. To control food-borne diseases, careful attention has to be paid in the following major factor groups: inappropriate temperature, use of inadequate raw materials, environmental factors, and inadequate handling [1]. Even though heating of foods kills potentially hazardous pathogens, and refrigeration prevents their multiplication, an inappropriate use of temperature such as inadequate refrigeration and inadequate cooking could be resulting in human outbreaks. Use of microbiological contaminated raw materials or contained contaminated ingredients was reported in 20% of the outbreaks in Europe in the 1990s. Inadequate handling such as cross-contamination, inadequate processing, insufficient hygiene, and reusing leftovers are also high-risk factors resulting in food-borne diseases. Among environmental factors, contamination by personnel was the most frequently reported contributing factor, followed by contaminated equipment and use of inadequate rooms.

Therefore, preventing and controlling food-borne infections requires constant efforts along the entire chain of production [22]. The above detailed epidemiologic investigations of how contamination occurs in outbreaks settings can identify such points of contamination and, thus, can target the development of improved control strategies. In this sense, an extensive list of options is available for improving the prevention and control of food-borne diseases. Among them the most relevant was the implementation of the Hazard Analysis Critical Control Point (HACCP) process. All of these programs require the food industries to identify points in food production where the contamination may occur and target resources toward processes that may reduce or eliminate food-borne hazards [23]. To provide a basis for educating food handlers and consumers is also crucial.

New prevention technologies will be critical to food safety in the future. To ensure the food safety and to extend the shelf life of food products, processing, such as freezing, drying, pasteurization, or sterilization, is often necessary [24]. The growing demand for slightly processed products with the same guarantees of innocuousness than those treated by traditional methods of preservation has urged researchers to focus most of their efforts on studying new ways of ensuring the food safety and extending the shelf life of food products [25]. Nowadays, diverse methods such as mild heating, irradiation, high pressure, and modified atmosphere or antimicrobial packaging are applied, often combined, for food preservation. The use of mild treatments combined with refrigeration has led to a continuous increase in the incidence of emergent pathogens [26].

One way to prevent the growth of pathogens in food is to use antimicrobial agents [27]. Prudent use of antimicrobials may prolong their effectiveness by preventing serious public health problems such as the antibiotic acquired and cross-resistances in some pathogenic bacterial strains [28].

1.4 ANTIMICROBIAL AGENTS FOR FOOD PRESERVATION

Antimicrobial agents are playing a great role in effective control of food-borne pathogens, food manufacturers, and others within the food industry [29]. Food antimicrobials are compounds employed to control microbial contamination by reducing the growth rate and maximum growth population, extending the lag phase, or inactivating microorganisms in the food, while maintaining sensorial and nutritional quality and freshness. Traditionally, the main use of food antimicrobials has been to extend shelf life and preserve the quality by inactivation of spoilage microorganisms [30]. And only a few antimicrobials have been used exclusively to control the growth of specific food-borne pathogens [31]. The most relevant cases are nitrates to inhibit the growth of *Clostridium botulinum* in cured meats, nisin and lysozyme to inactivate *C. botulinum* in pasteurized processed cheese, and lactate and diacetate against *L. monocytogenes* among others [32]. Nowadays, however, antimicrobials have been used increasingly to inhibit the outgrowth of pathogens in foods because the increase in consumer demand for minimally processed foods is causing the appearance of food-borne pathogen outbreaks. As many antimicrobial agents are bacteriostatic or fungistatic, they are often applied in combination with other food preservation procedures in order to prolong the shelf life of the food products. The combination of several microbial controls is sometimes called "hurdle technology" [33].

An antimicrobial agent is a chemical or biological agent that either destroys or inhibits the growth of pathogenic and spoilage microorganisms. Antimicrobials can be classified as traditional or naturally occurring [34, 35].

These chemicals may include synthetic compounds, which are added intentionally to the foods or natural occurring, biologically derived substances, the so-called naturally occurring antimicrobials [36]. Examples of synthetic additives include chemical antimicrobials such as formic and propionic acid [37].

Natural antimicrobials may exhibit antimicrobial activity as additives in foods [38] and include those that are present or derived from plant or animal tissues and those produced by microorganisms.

Different chemicals such as propionates, sorbates, and benzoates have been successfully used as reliable antimicrobial substances to control several microbial hazards. However, increasing concern of the suspected carcinogenic nature of synthetic additives, such as butylated hydroxytoluene (BHT) and butylated hydroxyanisole (BHA) [39], has resulted in a consumer increase in avoiding foods prepared with preservatives of chemical origin [40]. Legislation has

TABLE 1.2 Relevant antimicrobial systems: Chemical synthetic *versus* naturally occurring compounds

Chemically synthetic antimicrobials	Naturally occurring antimicrobials
• Organic and inorganic acids	• Lactoferrin
• Thiosulphinates	• Enzymes
• Antibiotics	• Bacteriocins
• Fungicides	• Plant essential oils
• Metal salts	• Spice essential oils
• Isothiocyanates	• Chitosan
• Chelating agents	• Porphyrins
	• Metals

restricted the use and permitted levels of some currently accepted preservatives in different foods. This has created problems for the industry because the susceptibility of some microorganisms to most currently used preservatives is diminishing [41]. Although the use of chemical preservatives is still essential in food processing, the replacement of chemicals by more natural alternatives is appropriate when the chemicals are no longer acceptable.

Although the use of chemical preservatives is still essential in food processing, the replacement of chemicals by more natural alternatives is appropriate when the chemicals are no longer acceptable or questioned because of toxicological side effects. Natural alternatives are therefore needed to achieve a sufficiently long shelf life of foods and a high degree of safety with respect to food-borne pathogenic microorganisms [40]. Attention is therefore shifting toward the use of natural preservatives [42]. However, it is evident that these natural alternatives are not always as effective as the traditionally used synthetic chemicals. The major antimicrobial compounds included in plastics are shown in Table 1.2.

1.5 METHODS TO DETERMINE THE ANTIMICROBIAL ACTIVITY

There have been nearly as many methods used for determining antimicrobial efficacy as there are compounds [43]. This makes it difficult to compare results from different laboratories and even more complex to determine the potential success of an antimicrobial in food because of methods that are inappropriate or lack significance [44]. Despite the large number of methods available, the agar diffusion method has almost certainly been the most widely used method for determining the antimicrobial activity or for screening antimicrobial substances [44, 45] and has often been referred to as the disk assay [43]. Figure 1.1 shows the major methods used to evaluate the efficacy of food antimicrobial agents.

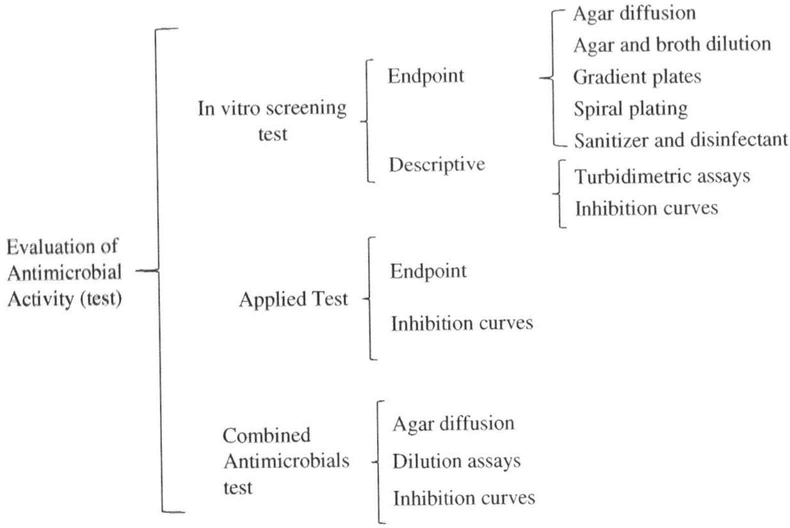

FIGURE 1.1 Methods of evaluation of antimicrobial activities.
Source: Davidson and Parish (1989) and López-Malo et al. (2000a)

1.6 PLASTICS AND BIOPLASTICS IN PACKAGING

Plastics are a family of materials whose use has been increased in many applications since the late 1930s. The most rapidly adopting area of application for plastics has been packaging. To the extent that nowadays, plastic packaging is the largest application for plastics (ca. 37% in Europe), and within the packaging niche, food packaging amounts to the largest plastics demanding application. Plastics bring in enormous advantages, such as thermoweldability, flexibility in thermal and mechanical properties, lightness, integrated projects (integrating forming, filling, and sealing), and low price [46–48]. Nevertheless, plastics also have some limitations when compared with more traditional materials like metals, alloys, or ceramics. One of the most relevant characteristics of plastics, often regarded as a disadvantage, is their permeability to the transport of low-molecular-weight components in the form of permeability, migration (of polymer residues and/or additives), or scalping (sorption of aroma compounds). Transport or barrier properties are determined by permeability ($P = S \times D$), diffusion (D), solubility (S), and partition (P) coefficients. Other limitations of plastics are a comparatively low thermal resistance and a strong interdependence between thermal and mechanical properties and barrier properties. Despite that plastic materials continue to expand and replace the conventional use of paperboard, tinplate cans, and glass, because of the above-mentioned positive characteristics and the development of multilayer systems, which can include metalized layers for the higher barrier and ultraviolet (UV) protection demanding applications.

More recently there has been a current trend to substitute petroleum-based materials by renewable biobased-derived plastics, which will reduce the oil dependence, facilitate the afterlife of the packaging by for instance composting, and reduce the carbon footprint of the food packaging industry. Regarding biodegradable (renewable and nonrenewable) materials, three families are usually considered: polymers directly extracted from biomass such as the polysaccharides chitosan, starch, and cellulose and proteins such as gluten and zein. A second family makes use of oil-based monomers or of biomass-derived monomers but uses classic chemical synthetic routes to obtain the final biodegradable polymer; this is the case of, for instance, polycaprolactones (PCLs), polyvinyl-alcohols (PVOHs), and copolymers (EVOHs) and for the case of sustainable monomers of polylactic acid (PLA) [46–48]. The third family makes use of polymers produced by natural or genetically modified microorganisms such as polyhydroxyalcanoates (PHAs) and polypeptides [49]. The latter materials can be engineered to exert antimicrobial performance.

The bioplastics more commercially viable at the moment are some biodegradable polyesters, which can be processed by conventional processing equipment and are being used in several monolayers and multilayer applications already, particularly in the food packaging and biomedical field. The most widely researched thermoplastic sustainable biopolymers for monolayer packaging applications are starch, PHA, and PLA. From these, starch and PLA biopolymers are without a doubt the most interesting families of biodegradable materials because they have become commercially available (by, for instance, companies such as Novamont and Natureworks, respectively), are produced on a large industrial scale, and present an interesting balance of properties. Of particular interest in food packaging is the case of PLA because of its excellent transparency and relatively good water resistance. Nevertheless, these materials suffer from shortages such as barrier and thermal properties when compared with, for instance, PET, and therefore, it is of great industrial interest to enhance the barrier properties of these materials while maintaining its inherently good properties such as transparency and biodegradability [50–56].

Despite this, other materials also are extracted from biomass resources such as proteins (e.g., zein), polysaccharides (e.g., chitosan), and lipids (e.g., waxes) with excellent potential as carriers of antimicrobial systems. The main drawbacks of these families of materials is their inherently high rigidity, difficult processability using conventional processing equipment, and for the cases of proteins and polysaccharides, the very strong water sensitivity originating from their hydrophobic character, which leads to a strong plasticization of many properties including the excellent oxygen barrier as relative humidity and water sorption increase in the material. The low water resistance of proteins and polysaccharides strongly handicap to some extent its use. Nevertheless, chitosan and zein biopolymers exhibit two very interesting characteristics, one is that the chitosan displays antimicrobial properties [57, 58] and the other is that zein shows an unusually high water resistance compared with other similar biomaterials [59]. Furthermore, the zein in a resin form can also be heat processed.

In general, polymers and particular biomass-derived polymers, because of their intrinsic mass transport properties discussed above, constitute ideal carriers for antimicrobials because of the advantage of being tuneable in terms of control release and the possibility of combining several polymers through blending or multilayer extrusion, to tailor the performance.

Traditional plastic use in, for instance, packaging has been defined as a passive barrier to delay the adverse effect of the environment over the contained product. Nevertheless, the current tendencies include the development of plastic technologies that interact with the environment and with the product, playing an active role in its preservation. Moreover, plastics, when applied to, for instance, food packaging, can also be designed to impact the safety of the consumer by integration of functional active ingredients in the packaging structure. These new food packaging systems have been developed as a response to trends in consumer preferences toward mildly preserved, fresh, tasty, healthy, and convenient food products with prolonged shelf life and can be used to compensate for shortages in the packaging design. In addition, changes in retail practices, such as globalization of markets resulting in longer distribution distances, present major challenges to the packaging industry that finally act as driving forces for the development of new and improved packaging concepts that extend shelf life while maintaining the safety, quality, and health aspects of the packaged foods.

The use of new active packaging strategies compatible with minimal processing technologies has attracted much attention from the food industry because it could have a significant impact on shelf life extension without compromising the food safety. Antimicrobial packaging has been identified as one of the most promising forms of active food packaging technologies. Therefore, research into this area has been increased significantly during the past 10 years. Basically, antimicrobial compounds can be incorporated into or coated onto plastic packaging to control microbial contamination by reducing the growth rate and maximum growth population, extending the lag-phase, or inactivating microorganisms present on the food surface through contact [60].

1.7 ANTIMICROBIALS IN POLYMERS

The demand for antimicrobial packaging is forecast to expand rapidly from a very low base over the next decade. The main objective of antimicrobial plastics is to control undesirable microorganisms by means of the incorporation of antimicrobial substances into or coated onto the packaging materials [61, 62]. Additionally, biocides are included into plastics to stabilize or fixate them, to promote antistaining and self-cleaning performance, and for the control release of these. The principles of action of antimicrobial packaging technologies are based on (1) intended migration (control release, see Figure 1.2) of biocide species either to the liquid phase and/or to the head space, (2) absorption of essential components for microbial growth, and (3) by direct contact

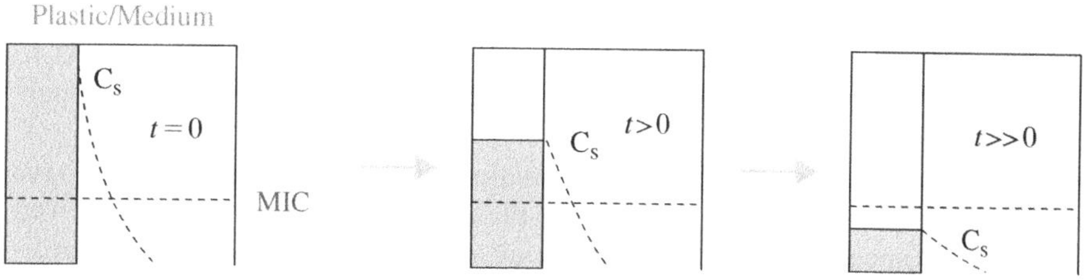

FIGURE 1.2 Schematics showing a control release process from an antimicrobial plastic. Sufficient uptake above the minimum inhibitory (MIC) concentration is essential in antimicrobial plastic packaging.

(nonintended migration). The basic requirements for the selection of biocides to constitute antimicrobial plastics are product stability during shelf life and with the polymer matrix, sufficient solubility in the polymer to avoid biocide concentration depletion below the minimum inhibitory concentration during service (see Figure 1.2), efficacy at low concentrations, biocidal against a broad spectrum of pathogenic microorganisms, thermally stable during polymer processing, does not alter the quality of the product by providing odor or taste, and cost effective.

Different chemicals such as organic and inorganic acids, chlorine dioxide, silver ions and nanoparticles, zinc oxide, magnesium oxide, chitosan, alcohols, ammonium compounds, or amines have been successfully incorporated as antimicrobial substances into plastic materials. However, the current trend nowadays is the preference for natural over synthetic chemistries; therefore, much attention is being paid to the use of bacterial starter cultures, biopreservatives, and plant extracts as antimicrobial hurdles as they present a perceived lower risk to the consumers. Thus, these natural antimicrobial additives are expected to be of increased interest because degradation into harmful products or by-products is thought to be lower than for synthetic chemicals. The biopreservatives suggested as antimicrobials include bacteriocins such as nisin and pediocin and antimicrobial enzymes such as lysozyme, lactoperoxidase, chitinase, and glucose oxidase.

Different systems have been developed that have been widely reported in the previous literature. Ag-substituted zeolite is commercialized in Japan and other countries and exhibits strong and broad antimicrobial attributes [63]. Nevertheless, the real effectivity of this system is not well understood, the requisite migration from polymers is minimal, and the silver ions antimicrobial effects are weakened by sulfur-containing amino acids in many food products [64].

Commercial examples of Ag-based zeolites are Zeomic (Shinanen New Ceramics Co. Ltd., Tokyo, Japan), AgIon (AgIon Technologies Inc., Wakefield, MA), and Apacider (Sangi Group America, Los Angeles, CA). EU and FDA food-contact–approved nanotechnology systems based on biocide metals, natural extracts, and other principles and to be used within plastics and ceramic substrates are also commercially available by a nanotech company under the general trade name of Bactiblock (NanoBioMatters S.L., Valencia, Spain) [65–67]. Nanodispersion of these systems in plastics and bioplastics leads to synergies in other properties (enhanced barrier, etc.) while retaining transparency and other good properties of the polymeric matrix.

Volatile substances, like SO_2, chlorine dioxide, or allyl isothiocyanate, have also been studied to be incorporated into plastics [68]. Chlorine dioxide received FDA acceptance as a packaging material antimicrobial agent. It is an antimicrobial gas released from a basic chlorine containing chemical upon exposure to moisture. Its main advantage is that it functions away from the plastic, and thus, it is one of the few packaging antimicrobials that do not require direct contact with the food. The antimicrobial properties of dermaseptin S4 derivative [69] have also been discussed in plastic packaging.

Other possible antimicrobial substances are food preservatives such as sorbates, benzoates, propionates, and parabens, all of them covered by FDA regulations [70]. Sorbate-releasing plastic films are a good example of successful research and development of antimicrobial packaging. The plastic resin and the antimicrobial agents were mixed, extruded, and pelletized to produce masterbatch resins. Films containing sodium propionate have also been proved to be useful in prolonging the shelf life of bread by retarding microbial growth [71].

An interesting commercial development is the marketing of food-contact–approved kitchen products such as chopping boards, tabletops, dish cloths, and so on, which contain triclosan, an antimicrobial aromatic chloro-organic compound, which is also used in soaps, shampoos, and so on [72]. The use of triclosan for food-contact applications was allowed in EU countries by the SCF (Scientific Committee for Food) in the 10th additional list of monomers and additives for food contact materials (SCF, 2000), with a quantitative restriction on migration of 5 mg/kg of food. Nevertheless, the future use of triclosan is being currently questioned because of potential toxicological side effects and is being replaced by metal-based and natural-products–based antimicrobials.

In this respect, the EU has launched a new regulation [73] that legislates the use of antimicrobial products to come into contact with foods, which will trigger the implementation of antimicrobial plastics in packaging within Europe. A range of different antimicrobial plastics are being currently assessed by the EFSA for commercial implementation into food packaging applications in the EU area and in countries of influence. Nevertheless, although the use of antimicrobials in plastics in the food area is expected to grow over the next decade, the U.S. FDA regulations limiting the use of antimicrobials as additives in packaging with direct food and drug contact will restrict broader usage. Concerns regarding the migration of these substances into food will prevent regulatory approval for their use in packaging. Another drawback of many condensed antimicrobials is that they are effective only at the surface, and significant microbial activity often takes place below the surface. However, the significant media emphasis on health and hygiene matters and their role in the spread of infections is creating great opportunities for antimicrobial additives in packaging for personal care and pharmaceutical products.

In the academic literature, other examples of antimicrobial polymers are based on the following principles: thymol [74] and thymol in nanocomposites [75], chitosan and chitosan derivatives [76, 77], nisin [78], 2-nonanona [79], natamycin [80], cinnamon, rosemary, garlic essential oils, oregano and thyme [81–83], clove oils [84], carvacrol [85], lysozyme [86], and many others that can be found in previous work and recent literature reviews [87–90].

The incorporation of plant extracts and essential oils in different plastic systems, like adhesive layers, is without a doubt one of the most interesting areas of current research [88]. Some of these compounds also have antioxidant properties, which makes them very suitable for designing multifunctional plastic materials with greater stability for the contact products.

Of particular interest is also the biobased polymer chitosan. This system has been widely researched, but it has been only recently that the phenomenology of the polysaccharide and its biocide properties have been more clearly understood and adequate characterization methods put in place to optimize its biocide capacity in food packaging and coating applications [76, 91, 92] (see chapter 4).

Recent developments and optimization in nanofabrication by electrospinning of this biopolymer and others indicate that nanostructured fiber mats of biopolymers have very strong biocide properties [93].

Direct application of antimicrobial compounds onto the food surface can be inefficient because of their rapid diffusion within the bulk of food [94]. The incorporation of the above and other antimicrobial agents into the packaging

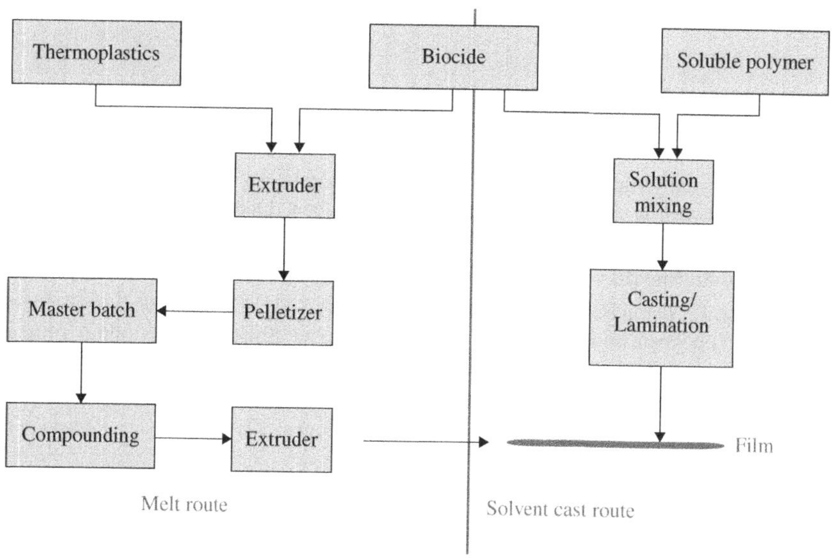

FIGURE 1.3 Schematics for the typical production of antimicrobial plastics via compounding or coating.

material, leading to antimicrobial films, could improve their activity maintaining an optimal effect during the whole storage period. According to Cooksey [61], there are three basic categories to form antimicrobial films: (1) the incorporation of volatile biocide substances into a sachet connected to the package, (2) the direct incorporation of the antimicrobial agent into the packaging film by, for instance, melt compounding routes (see Figure 1.3), and (3) coating of the packaging with a matrix that acts as a carrier for the antimicrobial agent (see Figure 1.3). For nonvolatile compounds that have an increased risk of losing their properties during the thermal polymer processing methods such as extrusion and injection moulding, the latter option seems to be the most appropriate method.

1.8 CONCLUSIONS

The incorporation of antimicrobials within plastic appliances and surfaces has attracted over the last few years a great deal of research and development attention because of widespread concerns at all levels in safety and hygiene. The reasons for incorporating biocides in plastics are to make plastic objects aseptic from a microbial contamination viewpoint, to render them antistaining or antifouling, to stabilize or fixate the biocides, and because of the particular mass transport properties of plastics, to procure control release of the biocides to tailor antimicrobial performance. It is clear that the future will bring a lot of new developments in the area with special focus on the use of bioplastics as matrixes, in the development of natural and biologically derived biocides, and in the application of nanotechnology tools for increased efficiency and specificity.

REFERENCES

1. WHO Regional Office for Europe. (1999). Food safety. In: *Overview of the Environment and Health in Europe in the 1990s*. Copenhagen, Denmark, pp. 51–52 (http://www.euro.who.int/document/e66792.pdf; accessed June 7, 2011).
2. WHO Foodborne diseases, emerging. (2002). Fact Sheet No. 124.
3. Berndtson, E., Danielsson-Tham, M. L., Engvall, A. (1996). *Campylobacter* incidence on a chicken farm and the spread of *Campylobacter* during the slaughter process. *International Journal of Food Microbiology*, *32*, 35–47.
4. Davies, R., Breslin, M. (2001). Environmental contamination and detection of *Salmonella enterica* serovar *enteriditis* in laying flocks. *The Veterinary Record*, *149*, 699–704.
5. WHO Regional Publications European Series No. 96. (2004). *Food and Health in Europe: A New Basis for Action*. Eds.: Robertson, A., Tirado, C., Lobstein, T., Jermini, M., Knai, C., Jensen, J. H., Ferro-Luzzi, A., James, W. P. T.
6. Johnston, L. M., Jaykus, L. A., Moll, D., Martinez, M. C., Anciso, J., Mora, B., Moe, C. L. (2005). A field study of the microbiological quality of fresh produce. *Journal of Food Protection*, *68*, 1840–1847.

7. Doyle, M. P., Erickson, M. C. (2008). Summer meeting 2007—the problems with the fresh produce: An overview. *Journal of Applied Microbiology, 105*, 317–330.
8. Little, C. L., Gillespie, I. A. (2008). Prepared salads and public health. *Journal of Applied Microbiology, 105*(6), 1729–1743.
9. EFSA. (2007). The community summary report on trends and sources of zoonoses, zoonotic agents, antimicrobial resistance and foodborne outbreaks in the European Union in 2006. *The EFSA Journal, 130*, 2–352.
10. Marques, A., Encarnaçao, S., Pedro, S., Nunes, M. L. (2008). In vitro antimicrobial activity of garlic, oregano and chitosan against *Salmonella enterica. World Journal of Microbiology and Biotechnology, 24*, 2357–2360.
11. Ye, M., Neetoo, H., Chen, H. (2008). Control of *Listeria monocytogenes* on ham steaks by antimicrobials incorporated into chitosan-coated plastic films. *Food Microbiology, 25*, 260–268.
12. Garriga, M., Fadda, S., Aymerich, M. T., Hugas, M. (2004). Inhibition of the hygienic status of traditional dry sausage workshops in Catalonia. *Proceedings of the 2nd International Conference on Food Factory of the Future*, Laval, France.
13. Nguyena, V. T., Gidleyb, M. J., Dykesc, G. A. (2008). Potential of a nisin-containing bacterial cellulose film to inhibit *Listeria monocytogenes* on processed meats. *Food Microbiology, 25*, 471–478.
14. Pantazi, D., Papavergou, A., Pournis, N., Kontominas, M. G., Savvaidis, I. N. (2008). Shelf-life of chilled fresh Mediterranean swordfish (*Xiphias gladius*) stored under various packaging conditions: Microbiological, biochemical and sensory attributes. *Food Microbiology, 25*, 136–143.
15. Mead, P. S., Slutsker, L., Dietz, V., McCaig, L. F., Bresee, J. S., Shapiro, C., Griffin, P. M., Tauxe, R. V. (1999). Food-related illness and death in the United States. *Emerging Infectious Diseases, 5*, 607–625.
16. Alter, T., Scherer, K. (2006). Stress response of *Campylobacter* spp. and its role in food processing. *Journal of Veterinary Medicine B, 53*(8), 351–357.
17. Levin, R. E. (2007). *Campylobacter jejuni*: A review of its characteristics, pathogenicity, ecology, distribution, subspecies characterization and molecular methods of detection. *Food Biotechnology, 21*(3–4), 271–347.
18. Nachamkin, I. (1997). *Campylobacter jejuni*. In: *Food Microbiology: Fundamentals and Frontiers*. Eds.: Doyle, M. P., Beuchat, L. R., Montville, T. J. pp. 159–170. ASM Press, Washington, D. C.
19. Donnenberg, M. S., Nataro, J. P. (2000). The molecular pathogenesis of *Escherichia coli* infections. In: *Microbial Foodborne Diseases: Mechanisms of Pathogenesis and Toxin Synthesis*. Eds.: Cary, J. W., Linz, J. E., Bhatnagar, D. pp. 87–130. Technomic Publishing, Lancaster, U.K.
20. Lin, C. M., Moon, S. S., Doyle, M. P., McWatters, K. H. (2002). Inactivation of Escherichia coli O157:H7, Salmonella enteric serotype Enteritidis, and Listeria monocytogenes on lettuce by hydrogen peroxide and lactic acid and by hydrogen peroxide with mild heat. *Journal of Food Protection, 65*, 1215–1220.
21. Ateba, C. N., Bezuidenhout, C. C. (2008). Characterisation of Escherichia coli O157 strains from humans, cattle and pigs in the North-West Province, South Africa. *International Journal of Food Microbiology, 128*, 181–188.
22. Tauxe, R. V. (2002). Emerging foodborne pathogens. *International Journal of Food Microbiology, 78*, 31–41.

23. Alterkruse, S. F., Cohen, M. L., Swerdlow, D. L. (1997). Emerging food diseases. *Perspectives*, *3*(3), 285–293.

24. Matser, A. A., Krebbers, B., van den Berg, R. W., Bartels, P. V. (2004). Advantages of high pressure sterilization on quality of food products. *Trends in Food Science & Technology*, *15*(2), 79–85.

25. Soliva-Fortuny, R. C., Martin-Belloso, O. (2003). New advances in extending the shelf-life of fresh-cut fruits: A review. *Trends in Food Science & Technology*, *14*(9), 5774–5779.

26. Ross, A. I. V., Griffiths, M. W., Mittal, G. S., Deeth, H. C. (2003). Combining nonthermal technologies to control foodborne microorganisms. *Trends in Food Science & Technology*, *89*, 125–138.

27. Tiwari, B. K., Valdramidis, V. P., Donnell, C. P. O., Muthukumarappan, K., Bourke, P., Cullen, P. J. (2009). Application of natural antimicrobials for food preservation. *Journal of Agricultural and Food Chemistry*, *57*, 5987–6000.

28. Mor-Mur, M., Yuste, J. (2010). Emerging bacterial pathogens in meat and poultry: An overview. *Food and Bioprocess Technology*, 3-24-35.

29. Smid, E. J., Gorris, L. G. M. (1999). Natural antimicrobials for food preservation. In: *Handbook of Food Preservation*. Eds.: Rahman, M. S., Marcel Dekker, New York.

30. Davidson, P. M., Harrison, M. A. (2002). Resistance and adaptation to food antimicrobials, sanitizers, and other process controls. *Food Technology*, *56*(11), 69–78.

31. Davidson, P. M., Zivanovic, S. (2003). The use of natural antimicrobials. In: *Food Preservation Techniques*. Eds.: Zeuthen, P., Bogh-Sorensen, L., Woodhead, Cambridge, U.K.

32. FDA. (2000). Sodium diacetate, sodium lactate and potassium lactate; use as food additives. Food and Drug Administration, Washington, D.C., Fed. Reg. 65, pp. 17128–17129.

33. Leistner, L. (2000). Basic aspects of food preservation by hurdle technology. *International Journal of Food Microbiology*, *55*, 181–186.

34. Davidson, P. M. (2001). Chemical preservatives and natural antimicrobial compounds. In: *Food Microbiology: Fundamentals and Frontiers*. Eds.: Doyle, M. P., Beuchat, L.R., Montville, T. J. ASM Press, Washington, D. C.

35. Davidson, P. M., Branen, A. L. (2005). Food antimicrobials—an introduction. In: *Antimicrobials in Food*. Eds.: Davidson, P. M., Sofos, J. N., Branen, A. L. CRC Taylor & Francis, Boca Raton, FL.

36. Sofos, J. N., Beuchat, L. R., Davidson, P. M., Johnson, E. A. (1998). Naturally occurring antimicrobials in food. Task Force Report No. 132, Council for Agricultural Science and Technology, Ames, IA, p. 103.

37. Duffy, C. F., Powe, R. F. (2001). Antioxidant and antimicrobial properties of some Chinese plant extracts. *International Journal of Antimicrobial Agents*, *17*(6), 527–529.

38. Naidu, A. S., Bidlack, W. R., Clemens, P. A. (1999). Probiotic spectra of lactic acid bacteria (LAB). *Critical Reviews in Food Science and Nutrition*, *39*(1), 13–126.

39. Jayaprakasha, G. K., Selvi, T., Sakariah, K. K. (2003). Antibacterial and antioxidant activities of grape (Vitis vinifera) seed extracts. *Food Research International*, *36*(2), 117–122.

40. Rauha, J. P., Remes, S., Heinonen, M., Hopia, A., Kähkönen, M., Kujala, T., Pihlaja, K., Vuorela, H., Vuorela, P. (2000). Antimicrobial effects of Finnish plant extracts containing flavonoids and other phenolic compounds. *International Journal of Food Microbiology*, *56*(1), 3–12.

41. Brul, S., Coote, P. (1999). Preservative agents in foods-Mode of action and microbial resistance mechanisms. *International Journal of Food Microbiology*, *50*(1–2), 1–17.

42. Menon, K. V., Garg, S. R. (2001). Inhibitory effect of clove oil on Listeria monocytogenes in meat and cheese. *Food Microbiology*, *18*(6), 647–650.

43. Davidson, P. M., Parish, M. E. (1989). Methods for testing the efficacy of food antimicrobials. *Food Technology*, *43*(1), 148–155.

44. Lopez-Malo, A., Palou, E., Parish, M. E., Davidson, P. M. (2005). Methods for activity assay and evaluation of results. In: *Antimicrobials in Food*. Eds.: Davidson, P. M., Sofos, J. N., Branen, A. L. CRC Taylor & Francis, Boca Raton, FL.

45. Tsai, C. E., Kondo, F. (2001). Improved agar diffusion method for detecting residual antimicrobial agents. *Journal of Food Protection*, *64*(3), 361–366.

46. Petersen, K., Nielsen, P. K., Bertelsen, G., et al. (1999). Potential of biobased materials for food packaging. *Trends in Food Science & Technology*, *10*, 52–68.

47. Haugaard, V. K., Udsen, A. M., Mortensen, G., Hoegh, L., Petersen, K., Monahan, F. (2001). Food biopackaging. In: *Biobased Packaging Materials for the Food Industry—Status and Perspectives*. Ed.: Weber, C. J. A report from the EU concerted action project: Production and application of biobased packaging materials for the food industry (Food Biopack), funded by DG12 under the contract PL98 4046 http://www.biodeg.net/fichiers/Book%20on%20biopolymers%20(Eng).pdf

48. Arvanitoyannis, I., Psomiadou, E., Biliaderis, C. G., Ogawa, H., Kawasaki, H., Nakayama, O. (1997). Biodegradable films made from low density polyethylene (LDPE), ethylene acrylic acid (EAA), polycaprolactone (PLC) and wheat starch for food packaging applications: Part 3. *Starch-Starke*, *49*, 306–322.

49. Reguera, J., Lagaron, J. M., Alonso, M., Reboto, V., Calvo, B., Rodriguez-Cabello, J. C. (2003). Thermal behavior and kinetic analysis of the chain unfolding and refolding and of the concomitant nonpolar solvation and desolvation of two elastin-like polymers. *Macromolecules*, *36*, 8470–8476.

50. Koening, M. F., Huang S. J. (1995). Biodegradable blends and composites of polycaprolactone and starch derivatives. *Polymer*, *36*, 1877–1882.

51. Bastiolo, C., Bellotti, V., Del Tredici, G. F., Lombi, R., Montino, A., Ponti, R. (1992). Biodegradable polymeric compositions based on starch and thermoplastic polymers. World Patent WO 9219680, filed May 4, 1992, and issued Nov. 12, 1992.

52. Park, E. S., Kim, M. N., Yoon, J. S. (2002). Grafting of polycaprolactone onto poly (ethyleneco-vinyl alcohol) and application to polyethylene-based biodegradable blends. *Journal of Polymer Science Part B: Polymer Physics*, *40*, 2561–2569.

53. Tsuji, H., Yamada, T. (2003). Blends of aliphatic polyesters. VIII. Effects of poly(l-lactideco-e-caprolactone) on enzymatic hydrolysis of poly(l-lactide), poly(e-caprolactone), and their blend films. *Journal of Applied Polymer Science*, *87*, 412.

54. Jacobsen, S., Fritz, H. G. (1996). Filling of poly(lactic acid) with native starch. *Polymer Engineering & Science*, *36*, 2799–2804.

55. Jacobsen, S., Degée, P. H., Fritz, H. G., Dubois, P. H., Jérome, R. (1999). Polylactide (PLA)—a new way of production. *Polymer Engineering & Science, 39*, 1311–3119.
56. Chen, C. C., Chueh, J. Y., Tseng, H., Huang, H. M., Lee, S. Y. (2003). Preparation and characterization of biodegradable PLA polymeric blends. *Biomaterials, 24*, 1167–1173.
57. Möller, H., Grelier, S., Pardon, P., Coma, V. (2004). Antimicrobial and physicochemical properties of chitosan-HPMC-based films. *Journal of Agricultural and Food Chemistry, 52*, 6585–6591.
58. Dong, S. C., Manjeet, S. C. (2004). Biopolymer-based antimicrobial packaging: A review. *Critical Reviews in Food Science and Nutrition, 44*, 223–237.
59. Wang, Y., Padua, G. W. (2004). Water sorption properties of extruded zein films. *Journal of Agricultural and Food Chemistry, 52*, 3100–3105.
60. Lagaron, J. M., Gimenez, E., Sanchez-Garcia, M. D. (2008). Thermoplastic nanobiocomposites for rigid and flexible food packaging applications. In: *Environmentally Compatible Food Packaging*. Ed.: Chiellini, E. Woodhead, Cambridge, U.K.
61. Cooksey, K. (2001). Antimicrobial food packaging materials. *Additives for Polymers, 8*, 6–10.
62. Han, J. H. (2000). Antimicrobial food packaging. *Food Technology, 54*, 56–65.
63. Vermeiren, L., Devlieghere, F., van Beest, M., de Kruijf, N., Debevere, J. (1999). Developments in the active packaging of foods. *Trends in Food Science & Technology, 10*, 77–86.
64. Brody, A. L. (2001). What's active in active packaging? *Food Technology, 55*, 104–106.
65. Lagaron, J. M. (2006). Novel antimicrobial and bioactive packaging technologies based on biomaterials and nanotechnology. Paper presented at the Pira International Conference: "Future Markets for Active and Antimicrobial Packaging," London, U.K.
66. Lagaron, J. M. (2007). Application of nanotechnology to active and bioactive compounds. Paper presented at the 5th FunctionalFoodNet (FFNet) Congress, Valencia, Spain.
67. Sanchez-Garcia, M. D., Gimenez, E., Ocio, M. J., Lagaron, J. M. (2008). Novel nanobiocomposites with antimicrobial and barrier properties of interest in active packaging applications. Paper presented at the ANTEC Conference 2008, Milwaukee, WI.
68. Appendini, P., Hotchkiss, J. H. (2002). Review of antimicrobial food packaging. *Innovative Food Science & Emerging Technologies, 3*, 113–126.
69. Miltz, J., Rydlo, T., Mor, A., Polyakov, V. (2006). Potential evaluation of a dermaseptin S4 derivative for antimicrobial food packaging applications. *Packaging Technology and Science, 19*, 345–354.
70. Floros, J. D., Dock, L. L., Han, J. H. (1997). Active packaging technologies and applications. *Food Cosmetics Drug Packaging, 20*, 10–17.
71. Soares, N. F. F., Rutishauser, D. M., Melo, N., Cruz, R. S., Andrade, N. J. (2002). Inhibition of microbial growth in bread through active packaging. *Packaging Technology and Science, 15*, 129–132.

72. Chung, D., Papadakis, S. E., Yam K. L. (2003). Evaluation of a polymer coating containing triclosan as the antimicrobial layer for packaging materials. *International Journal of Food Science and Technology*, *38*, 165–169.

73. Regulation (EC) No. 450/2009. (2009). European Food Safety Authority (EFSA).

74. Del Nobile, M. A., Conte, A., Incoronato, A. L., Panza, O. (2008). Antimicrobial efficacy and release kinetics of thymol form zein films. *Journal of Food Engineering*, *89*, 57–63.

75. Sanchez-Garcia, M.D., Gimenez, E., Ocio, M. J., Lagaron, J. M. (2008). Novel polycaprolactone nanocomposites containing thymol of interest in antimicrobial film and coating applications. *Journal of Plastic Film and Sheeting*, *24*, 239–251.

76. Fernandez-Saiz, P., Lagarón, J. M., Hernández-Muñoz, P., Ocio, M. J. (2008). Characterization of antimicrobial properties on the growth of S. aureus of novel renewable blends of gliadins and chitosan of interest in food packaging and coating applications. *International Journal of Food Microbiology*, *124*, 13–20.

77. Belalia, R., Grelier, S., Benaissa, M., Coma, V. (2008). New bioactive biomaterials based on quartenized chitosan. *Journal of Agricultural and Food Chemistry*, *56*, 1582–1588.

78. Nguyen, V. T., Gidley, M. J., Dykes, G. A. (2008). Potential of a nisin-containing bacterial cellulose film to inhibit Listeria monocytogenes on processed meats. *Food Microbiology*, *25*, 471–478.

79. Almenar, E., Del-Valle, V., Hernández-Muñoz, P., Lagarón, J. M., Catalá, R., Gavara, R. (2007). Equilibrium modified atmosphere packaging of wild strawberries. *Journal of the Science of Food and Agriculture*, *87*, 1931–1939.

80. Oliveira, T. M., Soares, N. D. F., Pereira, R. M., Fraga, K. D. (2007). Development and evaluation of antimicrobial natamycin-incorporated film in Gorgonzola cheese conservation. *Packaging Technology and Science*, *20*, 147–153.

81. Becerril, R., Gómez-Lus, R., Goni, P., López, P., Nerin, C. (2007). Combination of analytical and microbiological techniques to study the antimicrobial activity of a new active food packaging containing cinnamon or oregano against E-coli and S-aureus. *Analytical and Bioanalytical Chemistry*, *388*, 1003–1011.

82. López, P., Sánchez, C., Batlle, R., Nerin, C. (2007). Vapor-phase activities of cinnamon, thyme, and oregano essential oils and key constituents against foodborne microorganisms. *Journal of Agricultural and Food Chemistry*, *55*, 4348–4356.

83. Seydim, A. C., Sarikus, G. (2006). Antimicrobial activity of whey protein based edible films incorporated with oregano, rosemary and garlic essential oils. *Food Research International*, *39*, 639–644.

84. Matan, N., Rimkeeree, H., Mawson, A. J., Chompreeda, P., Haruthainthanasan, V., Parker, M. (2006). Antimicrobial activity of cinnamon and clove oils under modified atmosphere conditions. *International Journal of Food Microbiology*, *107*, 180–185.

85. Charlier, P., Ben Arfa, A., Preziosi-Belloy, L., Gontard, N. (2007). Carvacrol losses from soy protein coated papers as a function of drying conditions. *Journal of Applied Polymer Science*, *106*, 611–620.

86. Buonocore, G. G., Conte, A., Corbo, M. R., Sinigaglia, M., Del Nobile, M. A. (2005). Mono and multilayer active films containing lysozyme as antimicrobial agent. *Innovative Food Science & Emerging Technologies*, *6*, 459–464.

87. López-Rubio, A., Almenar, E., Hernández-Muñoz, P., Lagaron, J. M., Catalá, R., Gavara, R. (2004). Overview of active polymer-based packaging technologies for food applications. *Food Reviews International*, *20*, 357–386.
88. Rodríguez, A., Batlle, R., Nerin, C. (2007). The use of natural essential oils as antimicrobial solutions in paper packaging. Part II. *Progress in Organic Coatings*, *60*, 33–38.
89. Kenawy, E. R., Worley, S. D. M., Broughton, R. (2007). The chemistry and applications of antimicrobial polymers: A state-of-the-art-review. *Biomacromolecules*, *8*, 1359–1384.
90. Rhim, J. W. (2007). Potential use of biopolymer-based nanocomposite films in food packaging applications. *Journal of Food Science and Biotechnology*, *16*, 691–709.
91. Lagaron, J. M., Fernandez-Saiz, P., Ocio, M. J. (2007). Using ATR-FTIR spectroscopy to design active antimicrobial food packaging structures based on high molecular weight chitosan polysaccharide. *Journal of Agricultural and Food Chemistry*, *55*, 2554–2562.
92. Fernandez-Saiz, P., Lagaron, J. M., Ocio, M. J. (2009). Optimization of the biocide properties of chitosan for its application in the design of active films of interest in the food area. *Food Hydrocolloids*, *23*, 913–921.
93. Torres-Giner, S., Ocio, M. J., Lagaron, J. M. (2008). Development of active antimicrobial fiber based chitosan polysaccharide nanostructures using electrospinning. *Engineering in Life Science*, *8*, 303–314.
94. Siragusa, G. R., Dickson, J. S. (1992). Inhibition of Listeria monocytogenes on beef tissue by application of organic acids immobilized in a calcium alginate gel. *Journal of Food Science*, *57*, 293–296.
95. SCF. (2000). Opinion of the Scientific Committee on Food on the 10th addional list of monomers and additives for food contact materials.

CHAPTER 2

BACTERIAL RESISTANCE AND CHALLENGES OF BIOCIDE PLASTICS

NIBRAS A.A.M. AHMED
Department of Oral Biology, University of Oslo, Norway

CONTENTS

2.1 Preface 24
2.2 Resistance to Antibiotics 25
2.3 Mechanisms of Resistance 25
  2.3.1 Innate Resistance to Antibiotics 26
  2.3.2 Acquired Resistance, Mutations, and Selective Pressure 26
  2.3.3 Physiological Resistance 29
2.4 Consequences of Bacterial Resistance 31
  2.4.1 Clinical Resistance 31
2.5 Biocide Plastics 33
2.6 Challenges of Biocide Plastics 34
  2.6.1 Resistance to Biocides 34
  2.6.2 Cross Resistance to Antibiotics 37
  2.6.3 Efficacy 38
  2.6.4 Sustainability 40
2.7 Conclusions and Future Trends 41
References 41

Antimicrobial Polymers, First Edition.
Edited by José M. Lagarón, María J. Ocio, and Amparo López-Rubio.

2.1 PREFACE

Microbes constitute an integral part of our environment and dwell within our bodies in a delicate symbiotic balance. Yet, the resilience and flexibility of these single-celled beings have enabled them to survive for millions of years in complex and hostile conditions, challenging their evolution to sophisticated virulence levels. The disruption in human ecological balance by pathogenic microbes has plagued and threatened the existence of humans throughout history.

The invention of Leeuwenhoek's microscopic lens (1677) ignited a revolution of discoveries in microbiology. However, the culprit role of the microbes in disease and the significance of disinfection in health remained largely unknown until Ignaz Semmelweis first introduced the chlorinated lime hand washing policy more than 150 years ago (1850). Subsequently, the presentation of the germ theory by Louis Pasteur (1862) and the practice of antiseptic principle by Joseph Lister (1867) initiated the discipline of microbial medicine, sterilization, and disinfection.

During the 20th century, successful isolation and application of antibacterial molecules from *Penicillium* fungi succeeded in eradicating several pathogens, marking the beginning of the antibiotic revolution. Subsequently, synthetic classes of antibiotics specifically targeting intercellular microbial structures and selectively avoiding toxicity to human cells were introduced. Concomitantly, introduction of biocide disinfectants and implementation of sterilization guidelines contributed to a significant reduction in microbial contamination. As a result, the spread of communicable disease was reduced and lifespan in the developed countries more than doubled.

The hopeful antibiotic era was, however, shortly lived with the inevitable realization that with the aid of antibiotics, we can win certain battles in the short run against pathogens. However, with continuous bacterial evolution, we will inevitably lose the ultimate antibacterial war. The decrease in bacterial infections brought a concomitant steady rise in bacterial resistance.

> There is probably no chemotherapeutic drug to which in suitable circumstances the bacteria cannot react by in some way acquiring 'fastness' [resistance].
>
> —Alexander Fleming, 1946 [1]

Despite the dramatic change in chemotherapeutic approaches, widespread antibiotic resistance and dissemination of communicable diseases remain the leading causes of morbidity, mortality, and rising health costs. Fatal nosocomial and community acquired infections together with the emergence and spread of untreatable multidrug-resistant pathogens threaten the future of medicine in the absence of adjunctive methods to sterilize surfaces and contain the spread of these deadly pathogens.

In our modern era, hygienic practices, disinfection, and biocides played a crucial role in reducing spread of infections in hospitals, industrial areas, and agricultural fields. Yet, environmental and synthetic surfaces remain potential microbial reservoirs. Recently, the possibilities of incorporating biocides into materials, particularly in plastics, have opened novel opportunities to inhibit microbial adhesion and limit fomite transmission.

This chapter reviews the consequence and challenges associated with widespread use of antibiotics and biocides. The advantages and future challenges of biocide plastics in controlling infections are also discussed.

2.2 RESISTANCE TO ANTIBIOTICS

Resistance to any antimicrobial agent is an ongoing process that may originate from spontaneous mutations or disseminate from resistant genes contributed by other bacteria. Although bacterial mutation occurs naturally and under various environmental conditions, mutations can increase frequency under stress conditions and at subinhibitory antibiotic concentrations [2, 3]. Bacteria have an effective ability to adapt to stress through natural selection of new mutations [3]. Ultimately, the presence of antimicrobials in the environment will lead to selection of the fittest bacterial mutants that possess an advantageous modification against the antibiotic action. Antibiotics may even act as mutation and resistance promoters [4]. The resistance to antibiotics is further accentuated by global widespread prophylactic administration of antibiotics in both medical, agricultural and industrials fields. Most human pathogens have developed a resistance to a certain antibiotic and several other pathogens have even acquired multi-drug resistance (MDR).

Perhaps one of the most prominent and grave occurrences of acquired resistance to antibiotics appeared in the Methicillin-resistant *Staphylococcus aureus* (MRSA). Methicillin, which is a narrow-spectrum β-lactam antibiotic, was used to treat gram-positive infections in the 1960s but was shortly discarded after the development of MRSA. Subsequently, MRSA outbreaks have spread globally from immune-compromised hospital patients to various sectors of the community threatening low-risk individuals [5]. The dilemma of MRSA becomes more critical with the development of increasingly virulent strains as antibiotic resistance to more potent and novel β-lactams (vancomycin) emerge, forecasting the development of pandemic untreatable infections. MRSA infections have become resistant to all the β-lactam antibiotics, and most infections are treated with "last-resort" intravenous vancomycin. It is estimated that in 2005, approximately 18,650 patients died from MRSA infections in the USA alone [6].

2.3 MECHANISMS OF RESISTANCE

Antibiotics generally target specific intracellular structures including cell-wall synthesis (e.g., β-lactams), protein synthesis (e.g., macrolides and tetracyclines), and DNA replication (e.g., fluoroquinolone) [7]. Thus, resistance to antibiotics, whether in clinical or environmental settings, can result from various mechanisms depending on the bacterial organism and the mode of action of the antibiotic. Bacterial survival strategies to combat antibiotics, therefore, may be categorized into three principle mechanisms, listed as follows.

2.3.1 Innate Resistance to Antibiotics

Innate resistance occurs naturally in certain species to certain antibiotics in the absence of selection pressure [8]. In intrinsic resistance, all strains of the bacterial species are equally resistant to the same antimicrobial class.

In the ecological environment, several bacteria naturally produce antibiotics to inhibit microbial competitors [8, 9]. Antibiotics at low concentrations may also serve as signaling functions in bacteria [8]. In certain cases, bacteria may even survive on antibiotics as a sole carbon source in the soil [10]. Thus, several bacteria have the natural ability to survive their antibiotic effect and may even integrate resistance mechanisms into their metabolic and virulence functions [8, 11].

Innate resistance can result from the following:

1. Efflux pumps: The presence of efflux pumps confer natural resistance against harmful by products, antimicrobials, and toxic environmental substances by constantly expelling them from the cell and decreasing their intracellular concentration [8, 11, 12].
2. Decreased permeability: The cell envelope (cell membranes and peptidoglycan) may display poor permeability to certain types of molecules, including antibiotics. In gram-negative bacteria, spores, and mycobacteria, the intrinsic envelope naturally limits antibiotic penetration [12].

Pseudomonas aeruginosa, which is a serious and debilitating nosocomial pathogen, can colonize and thrive on a wide range of environmental and artificial surfaces. *P. aeruginosa* infections target immune-compromised patients, particularly those suffering from cystic fibrosis, burns, and HIV. The inherent ability of *P. aeruginosa* to resist a wide range of antibiotics because of its active efflux pumps and cell envelope impermeability remains the most challenging factor in treatment of its persistent infections [13].

2.3.2 Acquired Resistance, Mutations, and Selective Pressure

Acquired resistance refers to the attainment of resistant traits to previously susceptible species. Acquired bacterial resistance to various environmental toxins and antibiotics had developed long before the antibiotic era, albeit at a significantly slower pace [8, 14]. The widespread use of antibiotics in medical, veterinary, industrial, and agricultural settings has been the major driving force for rapid resistance spread. In fact, resistance has emerged to every antibiotic class introduced. The association between antibiotic use and resistance can clearly be documented by a sudden rise in resistance to a given antibiotic shortly after its administration [5]. Acquired resistance results from the selection of fittest mutants during the course of evolution. Spontaneous mutation in bacterial cultures occur at a rate of 10^{-8} per base pair per generation with beneficial mutations that confer selective advantage to resistance and virulence occurring as high as 1% in *Escherichia coli* [15]. Alternatively, bacteria may acquire the mutant genes from other species through horizontal gene transfer. Thus, although antibiotics may initially kill the susceptible strains, inevitably a portion of the mutant strains may not only be immune to the antibiotic effect but in addition harbor resistant genes for vertical and horizontal transfer. Subinhibitory concentrations of antibiotics provoke a range of transcriptional changes in bacteria including stress alert genes (SOS) and the DNA-repair process [16]. As a result, mutation frequencies increase under subinhibitory concentrations that do not inhibit bacterial growth, allowing sufficient opportunities for survival of the fittest [2]. Subinhibitory concentrations are likely to be encountered intermittently between antibiotic administration in various body compartments and when excreted outside the host in environmental residues [17]. The high resistance to pathogens in hospitals is most likely not initiated by mobile resistant genes but rather by selection of the fittest mutant pathogens that survive the extensive antibiotic therapy [8].

Resistant mutations and horizontally acquired resistant genes may contribute to the following [7] (Figure 2.1):

1. Alterations in target proteins and drug binding sites: For example, alteration in penicillin binding proteins confers resistance to β-lactam antibiotic and alterations in 50S ribosomal proteins results in resistance to macrolides. Bacteria may also develop an alternative route for metabolism to bypass the antibiotic effect and confer resistance to sulfonamides and trimethoprim.
2. Altered enzyme production and antibiotic inactivating: the most prevalent examples present in β-lactamase enzymes that are produced by gram-negative bacteria and confer resistance to β-lactam antibiotics.
3. Altered outer membrane proteins and reduction in cell permeability. For example, changes in the number and structure of porins confer resistance to aminoglycosides in both gram-positive and gram-negative bacteria.
4. Upregulation of efflux pumps: Efflux pumps may be prompted to expel antibiotics out of the cell and confer resistance to tetracyclines and macrolides in gram-positive bacteria.

2.3.2.1 Spread of Resistance. Once bacteria develop a resistance, the genetic spread of these mechanisms is only a matter of time. Bacteria may acquire resistant genes through the following mechanisms.

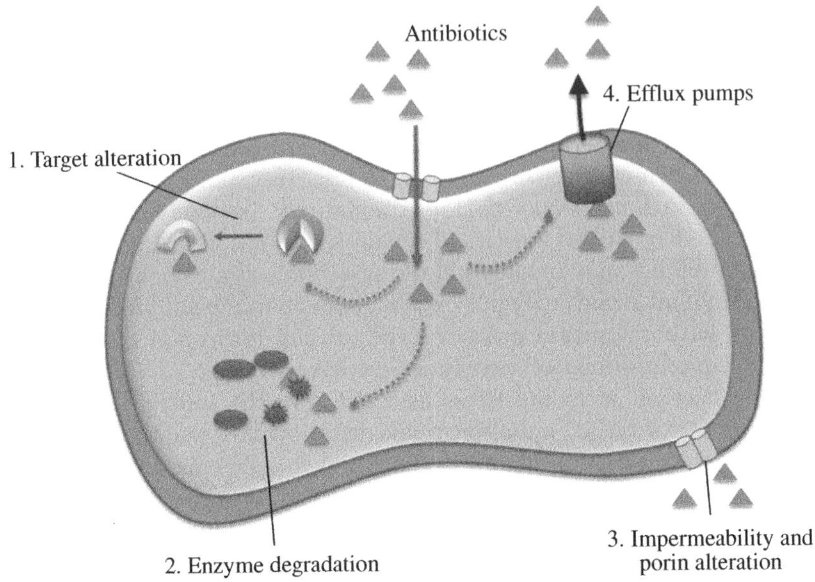

FIGURE 2.1 Acquired resistance mechanisms in bacteria.

2.3.2.1.1 Horizontal Gene Transfer. Is the most efficient and common resistance mechanism for dissemination of resistance in which nonparental transfer of genetic material occurs between different species or genera. The mechanism of genetic transfer occurs through passing mobile genetic elements either by conjugation (plasmid exchange), transduction (bacteriophages), or transformation (environmental DNA after cell lysis) [9]. However, the exchange of mobile genetic elements and transfer of resistance usually requires close proximity between microorganisms in areas with high microbial load and low antibiotic concentration. Thus, resistance development may be encountered in clinical environments, in host tissue, and in the soil [8, 9, 18].

2.3.2.1.2 Vertical Gene Transfer. Mutant resistant cells pass on their resistant phenotype to subsequent generations during the replication process.

2.3.2.2 Multidrug Resistance. Multi-drug resistance refers to the ability of a microorganism to resist multiple classes of antimicrobials (at least three) by multiple strategies. The spread of MDR in bacteria results from the accumulation of resistance genes on plasmids [19, 20]. The challenge with MDR is not only in that these microorganisms are immune to a wide spectrum of antimicrobials including antibiotics and biocides but also in the possibility of rapid spread of their MDR traits.

Generally, MDR is attributed to acquisition of mobile genetic elements of resistance that influence innate nonspecific resistance, particularly efflux pumps and membrane impermeability. Several intrinsic efflux pumps in gram-negative organisms can pump out toxic environmental molecules from the cytoplasm and across the cell envelope preventing their accumulation and, subsequently, their antibiotic effect. Another mechanism of MDR is reduced membrane permeability through cell wall thickening and porin channel alterations [1, 12].

Examples of multi-drug organisms are as follows:

- MRSA
- Multiresistant enterococci
- Multiresistant pneumococci
- *E. coli* and *Klebsiella pneumoniae*
- *P. aeruginosa*
- Extensively drug-resistant *Mycobacterium tuberculosis*

2.3.3 Physiological Resistance

When examined under planktonic growth, bacteria initially appear as single organisms in a static state. Yet bacteria, in nature, rarely exist in their free-floating planktonic mode but instead appear as a cluster of dynamic, active biofilm communities capable of communication and collaboration. The ability to form a biofilm is a vital survival strategy, especially under challenging environments. In fact, biofilm appears to be the default mode of growth that is necessary to achieve optimum microbial defense, colonization and beneficial collaboration [21]. Bacteria have a predilection for biofilm formation at interfaces between solid, air, and liquid surfaces [22]. These interfaces offer bacteria the advantage of acquiring their nutritional, colonial, and atmospheric requirements.

Once formed, biofilms present a challenge to all antimicrobials and particularly those targeting surface infections. Biofilms are a potential hazard for chronic medical infections (e.g., cystic fibrosis, catheters infections, and prosthetic implants failure), industrial challenges (e.g., food spoilage and marine biofouling), and environmental considerations (e.g., wastewater treatment).

The persistence of biofilms and difficulty in their eradication are principally caused by the following:

1. Resistance to antimicrobials: Biofilms confer a significantly higher resistance to antimicrobial treatment compared with planktonic cells [23].
2. Resistance to mechanical removal: Bacterial attachment is highly resistant to mechanical debridement and may also function as focal spots for spread of infection to other surfaces.
3. Resistance to innate immunity: Biofilms shield microorganisms from inflammatory and immune cells [24].
4. Close proximity between microorganisms facilitates genetic exchange and increases potential risk for development of resistance and cross resistance.

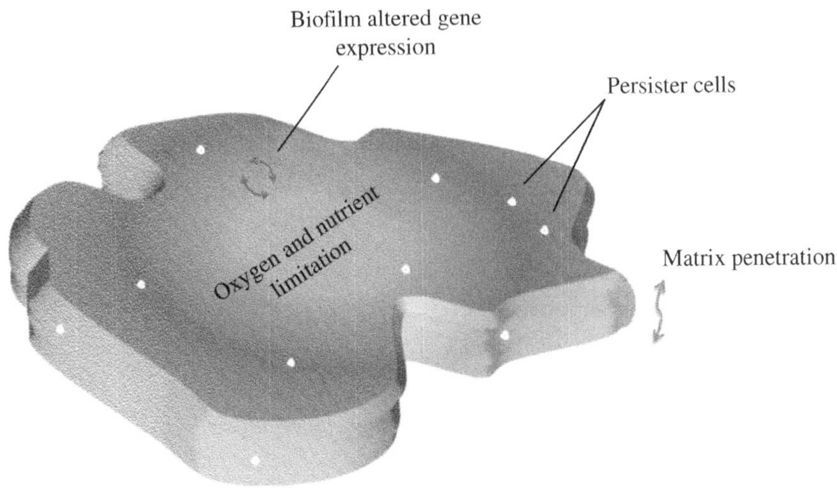

FIGURE 2.2 Major resistance mechanisms in bacterial biofilms.

Resistance mechanisms in the biofilm microbial community differ from that of planktonic bacteria. Whereas resistance in individual planktonic cells results from the acquisition of resistant traits via mutations and horizontal resistance spread, in biofilms, resistance is mediated by multicellular mechanisms [25].

Several characteristics of biofilms define their natural ability to suppress antimicrobials [26, 27] (Figure 2.2):

1. Matrix penetration: Shortly after initial attachment, biofilm bacteria excrete and embed themselves in a protective architectural framework referred to as the biofilm matrix. The matrix constitutes of water, microbial cells, exopolysaccharides, proteins, and extracellular DNA [28]. However, microbial cells constitute only 2% to 5% of the total biofilm matrix [28]. The biofilm matrix is thus the foremost protective shield against environmental and antimicrobial substances.

> From whence I conclude, that the Vinegar with which I washt my Teeth, kill'd only those Animals which were on the outside of the scurf, but did not pass thro the whole substance of it.
>
> —Antonie van Leeuwenhoek, 1683 [29]

The matrix may play a significant role in increasing resistance by acting as a diffusion barrier thus reducing the antibiotic concentrations that reach the deeper biofilm cells [26]. Slow matrix penetration increases the enzymatic inactivation of seeping antimicrobials by biofilm bacteria [26].

As the antibiotics penetrate the matrix, the therapeutic concentrations may be diluted to subinhibitory levels. At subinhibitory concentrations, several antibiotics have been shown to stimulate biofilm formation by either increasing extrapolysaccharide production or increasing the biofilm biomass [30, 31]. Intracellular signaling, quorum sensing, and genetic regulation have also been shown to play a role in increasing biofilm formation at subinhibitory antibiotic concentrations [32].

2. Altered microenvironments: Within biofilms, oxygen and nutrient gradients may affect metabolic activity and slow growth [26]. Oxygen tension and metabolites may be limited to superficial layers, whereas deeper layers maintain low metabolic activity in a dormant state [33]. Because antibiotics function on actively growing bacteria, dormant bacterial states or non-growing bacteria may escape the antimicrobial effect [34]. Thus, bacteria at the edge of a biofilm will be more susceptible to antimicrobial treatment than slow-growing organisms in deeper layers. Furthermore, the altered metabolic activity of starved bacteria may activate stress response genes in biofilms rendering them more resistant [34].
3. The biofilm phenotype possesses specific regulatory pathways that increase up-regulation of the biomass, matrix productions, and the induction of MDR operons and efflux pumps [30].
4. Persister cells are spore-like cells in biofilm cultures that persistently can resist antibiotic treatment because of their dormant, nongrowing state and stress-induced resistance mechanisms [35].

2.4 CONSEQUENCES OF BACTERIAL RESISTANCE

2.4.1 Clinical Resistance

2.4.1.1 Nosocomial Infections. In modern-day medical services, the threat of acquiring infections from hospitals has grown concomitantly with the rise of antibiotic resistance. Immune-compromised patients risk lethal consequences from bacterial and viral infections during their hospital stays. Increase in multiresistant pathogens warrants the emergence of extremely virulent and highly resistant pathogens. Most complications arise from difficulty in controlling surface biocontamination and as a consequence of contact with health care workers, fomite transmission, and shared hospital surfaces. Nosocomial bacteria, fungi, and viruses can persist on dry and inanimate surfaces for long durations lasting from days to months in the absence of routine disinfection [36, 37]. In contrast, routine disinfection of common hospital surfaces may not have a significant effect on the incidence of nosocomial infections because of a short duration of action and the ability of microorganisms to recolonize these surfaces within a few hours of their elimination [38, 39].

2.4.1.2 Biomaterials in Medicine. One of the major advances in medicine has been the introductions of artificial surfaces and devices into medical care including catheters, contact lenses, dental implants, heart valves, and synthetic joint replacement implants. However, increased dependence on artificial implant replacement technology has contributed new attachment opportunities to opportunistic as well as to previously non-pathogenic resident flora. For example, *Staphylococcus epidermidis* is primarily a human commensal bacterium of the skin and mucosal surfaces with low pathogenic potential. With the introduction of implants, *S. epidermidis* became one of the most significant pathogens in biomaterial-associated infections [40]. Persistent *S. epidermidis* infections result from their ability to form dense, cohesive biofilms on polymers and metal surfaces [41]. Implant biofilm infections impede successful tissue integration, increase antibiotic resistance, and evade host immunity [40].

2.4.1.3 Environmental Antibiotic Pollution. The widespread use of antibiotics in medical, agricultural, and industrial areas has drawn concerns for rising antibiotic pollution in aquatic and soil environments [42]. Antibiotic pollutants originate from extensive antibiotic treatment used in animal farms and hospitals. Indeed, several antibiotics are excreted from the host without change shortly after antibiotic administration [17]. The biggest risk for antibiotic environmental contamination is spread of resistant genes that are harbored within the soil [9, 10]. Figure 2.3 displays a model for horizontal spread of MDR genes in *E. coli.*

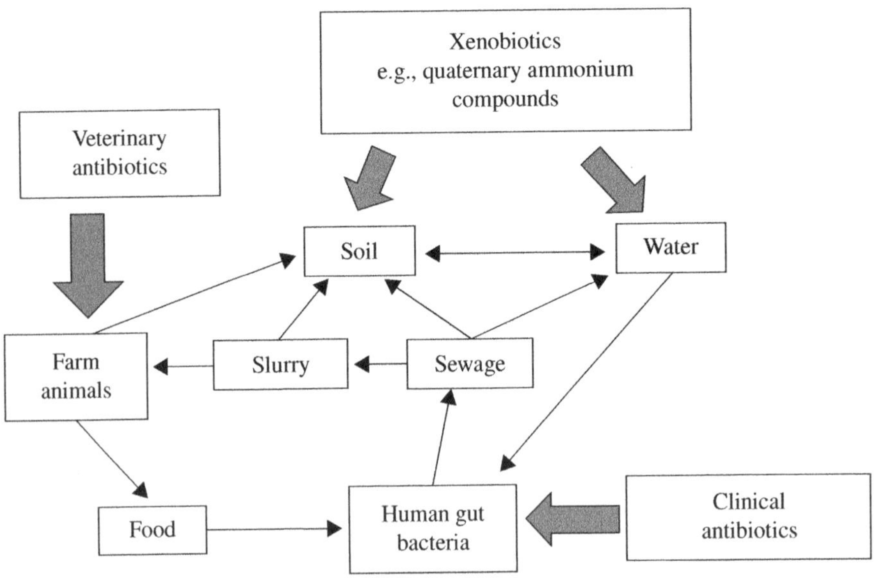

FIGURE 2.3 Spread of resistance genes in *E. coli*. Thick arrows display major selective pressures on antibiotic resistance genes. Thin arrows show the significant direction of gene flow.

Source: Adapted from Reference 5 with permission from Oxford University Press.

2.4.1.4 Antibiotic Resistance in the Food Industry, Agriculture, and Veterinary Medicine. Antibiotic resistance in food-borne pathogens has become a global concern [43]. The prophylactic treatment of farm animals, antibiotic treatment in veterinary medicine and in agriculture resistance has contributed to resistance spread via food-borne pathogens. Animals that harbor resistant strains have the potential to spread resistance to the human host via food products, direct contact, or by an environmental route [44]. In addition, biofilm formation remains a major obstacle in food spoilage, particularly on packaging materials [21]. The buildup of microbial films may originate in food processing plants rendering antimicrobials ineffective in preventing infections and increased infection persistence in factory environments [45].

2.4.1.5 The Role of Biofilms in Biofouling and Biocorrosion. The buildup of environmental organic material (abiotic) and subsequently the attachment of microbes on metal surfaces can lead to biological induced deterioration (biocorrosion) [22]. Biocorrosion is caused by microbial use of oxygen, hydrogen, breakdown of protective coatings, microbial acid production, volatile compounds release, and ions concentrations [22]. The subsequent detachment of microbes (biofouling) can be a potential source of bacterial contamination in water systems [22].

Traditional biocides have displayed limited efficacy against biocorroding and biofouling biofilms in aquatic and terrestrial environments. Environmental pollution from antifouling biocides are another obstacle in liquid biocide use [46]. Therefore, alternative strategies with more effective outcomes and minimum environmental impact are needed.

2.5 BIOCIDE PLASTICS

Initially, biocides were used for the purpose of topical sterilization of inanimate objects, hospital surfaces, and intact skin. However, the spread of resistant pathogens together with the increased dependency on polymers in medicine and industry call for new approaches to targeting bacteria. The concept of manufacturing antimicrobial surfaces and materials that prevent bacterial attachment has been introduced more than a decade ago. Recently, the possibility of incorporation biocides either in surface coatings or within the plastic matrix has attracted considerable global attention [47, 48]. Novel delivery systems and modified surfaces represent a promising solutions for persistent surface contamination and bacterial spread [49, 50]. The application of plastic biocides has been extended to a wide variety of industrial appliances including water treatment plants and textiles. In addition, it has found useful applications in the food industry, preventing the growth of potentially lethal food-borne bacteria and fungi [37, 47]. In the medical field, antimicrobial polymers have been applied to catheters, wound dressings, implants, hospital floors, hospital garments, and medical devices [37, 51, 52].

When compared with liquid biocides and systemic antibiotics, novel biocide plastics technologies contribute several novel advantages. First, surface antimicrobials display a broad and effective action against surface-borne, air-borne, and water-borne pathogens, in addition to broad antibacterial, antifungal and antiviral action [53, 54].

Sterilization guidelines with liquid biocides in hospitals require routine disinfection at high concentrations for long durations of time [37]. These concentrations may cause allergic adverse effects to the medical staff and patients. Thus, the presence of bound antimicrobials within the material may assist in reducing the need for vigorous disinfection. Biocide plastics may also maintain sterilization on inaccessible surfaces or those that may deteriorate under consistent cleaning. Furthermore, bound antimicrobial agents minimize the threat to environmental biocide pollution by reducing residual release of agents, increasing their efficiency and prolonging their antimicrobial effect [47]. Rapid biocidal action in plastics reduces the survival time and size of microbial load on fomite surfaces. Reductions in surface contamination may reduce nosocomial infections especially because hospital surfaces can harbor dangerous bacteria, viruses, and fungi for long durations of time [37]. The constant battle with treating biofilms infections is aimed at preventing biofilm formation, dispersing established biofilms, or penetrating the matrix barrier in an attempt to eliminate persistent infection. However, biofilm formation may be reduced or eliminated on biocide plastics by targeting pathogens immediately upon contact. Last, biocide plastics may reduce the need for prophylactic and systemic antibiotic doses, especially for immune compromised patients thus reducing the concern for resistance dissemination and systemic complications [55].

2.6 CHALLENGES OF BIOCIDE PLASTICS

2.6.1 Resistance to Biocides

2.6.1.1 Innate Resistance to Biocides. Biocides target multiple microbial structures concurrently. Thus, unlike antibiotic resistance, resistance to biocides is predominantly mediated by nonspecific innate resistance [56]. In nature, several microorganisms may produce biocides to disrupt the growth of other rival microbes in their microenvironment, allowing them to proliferate without competition. For example, several oral streptococci produce hydrogen peroxide to eliminate competition over substrate availability and attachment sites [57, 58]. Bacteriocins inhibit the growth of closely related bacterial species and possess broad spectrum activity against a wide variety of gram-positive bacteria [59]. Thus, microorganisms that produce and survive these toxic substances evolved an innate mechanism to counteract their biocidal effect [57, 59].

The following Mechanisms of innate resistance to biocides are similar to antibiotic innate resistance:

1. **Permeability barrier:** Changes in cell permeability is naturally encountered in spores [60], mycobacteria [61, 62], and gram-negative bacteria [12]. The waxy cell envelope in mycobacteria and external coats on bacteria spores makes them the most resistant type of cells to biocide treatment [60, 61]. Gram-negative bacteria are inherently more resistant to biocides than gram-positive bacteria because of outer membrane barrier and multiple efflux pumps [63, 64].
2. **Efflux pumps:** Microbial cells can expel a wide range of toxic molecules by employing efflux pumps, and preventing intracellular accumulation of biocides [11, 63].
3. **Biofilms and physiological resistance:** In nature, bacterial cells exist predominantly within a matrix of exopolymers collectively referred to as biofilms [28]. Biofilms present a physiological form of resistance that shields bacteria from unfavorable environmental factors. As mentioned previously, bacterial survival within biofilms can be attributed to several factors including limited matrix penetration that renders biofilm bacteria resistant to several antimicrobials (pp. 30–31, Reference 25).

2.6.1.2 Acquired Bacterial Resistance to Biocides. Bacterial resistance to biocide has been reported since the 1950s [65]. In practice, the resistance to biocides is more difficult to attain than resistance to conventional antibiotics [66]. This may be attributed to the fact that biocides are used at high concentrations, and therefore, the possibilities of reaching subinhibitory concentrations are less likely to be encountered in practice [67]. However, the danger in biocides resides in their increasing demand and their improper use at low concentrations, escalating the possibility of emerging resistance and antibiotic cross resistance [37, 66].

Several studies have purposed that unrestricted biocides use may promote a rise in bacterial resistance. For example, the wide scale incorporation of quaternary ammonium compounds and triclosan in cosmetic and hygiene products has contributed already to increased resistance and antibiotic cross-resistant development [68, 69]. The acquired mechanisms are often a result of multiple exposures to low biocide concentrations [70]. For instance, low-level resistance to phenolic biocides have been observed *in vitro* laboratory experiments as a result of membrane protein alteration [71]. The spread of resistance genes may also contribute to increase resistance to biocides. For example, resistance to hydrogen peroxide and metals ions may result from extracellular genetic stress transfer from nonviable cultures [72].

The acquired resistance to biocides occurs through the following mechanisms:

1. Overexpression of efflux pumps [70]
2. Membrane permeability alterations [66, 73]

3. Target overproduction, target alteration, and enzymatic biocide degradation [74–76]
4. Increased mutation frequency, altered growth rate, altered protein expression, and phenotypic changes leading to general resistance to unrelated compounds [70, 71, 77, 78]

The incorporation of biocides into polymers may limit the possibility of releasing subinhibitory concentrations and thus reduce acquired resistance. However, the possibility remains that several microorganisms may be inherently resistant to the biocide plastics or that they may evolve to acquire resistance determinants with time. In one study, after several weeks of exposure to peroxide sanitizer, biofilm formation by *Listeria monocytogenes* was initially reduced. However, with continuous and repeated exposure, biofilm formation eventually increased because of the development of stress resistance [78]. It is thus vital to survey bacterial behavior on biocide plastics *in vivo* to investigate whether bacteria will evolve novel adaptation and resistant mechanisms to biocidal polymers.

Examples of resistance development to surface-incorporated biocides

1. **Bacteriocins:** Bacteriocins are natural antimicrobial proteins produced by certain bacteria to combat other closely related bacteria in the vicinity [59]. For example, the bacteriocin nisin, produced by *Lactobacillus lacti*, targets cell wall synthesis of food-associated pathogens, especially *L. monocytogenes*, and therefore, it has been widely used for natural biopreservation in the food industry [59]. The incorporation of nisin into antimicrobial food packaging films has proved to be an effective method for controlling food preservation and inhibition of bacterial spoilage [79, 80].

   Increased resistance to nisin has been associated with a gradual increase in minimum inhibitory concentrations and changes in gene expression including the overexpression of binding proteins [76]. *L. monocytogenes* mutants have developed resistant after exposure to nisin by alteration in the cell envelope [74, 75]. Of a greater concern is that resistance to nisin may promote cross resistance to antibiotics [81]. However, most studies that have investigated nisin resistance were performed in laboratory settings and subsequent investigations on food trails are needed [82].
2. **Silver:** Silver has historically been used as a natural broad spectrum antimicrobial agent that is effective against both gram-positive and gram-negative bacteria, as well as viruses and yeast infections [56, 83–85]. Silver has a long history of use as an antimicrobial, with applications in water storage dating to the 5th century BC. Silver ions have multiple antimicrobial modes of action ranging from cell membrane damage, protein denaturation, and interference with DNA replication and metabolism [83, 85, 86]. Silver salts have been used in health care to treat wound infections, diabetic traumas and as antibacterial components in dental

amalgam, catheter coatings, and in heart valves [83, 87]. Silver salts are particularly useful for combating deadly wound infections with *P. aeruginosa* and other burn-related infections [84]. Silver-resistant bacteria have been detected since the 1970s and are most frequently detected in burn hospital departments where silver salts promote the selection of high resistant strains [85, 88]. Resistance is predominately mediated by efflux pumps and reduce membrane permeability [88, 89]. The spread of silver resistance is mediated by plasmids that may also confer resistance to several other heavy metals and antibiotics [56, 88].

2.6.2 Cross Resistance to Antibiotics

The second major concern with widespread biocide application has revolved around the possibility of emerging cross resistance between antibiotics and subinhibitory biocide concentrations. Common resistance mechanisms between biocides and antibiotics develop principally from bacterial stress adaptation to the presence of antimicrobials leading to increased efflux activity and porin proteins alteration [90, 91]. For example, Randall et al. [70] demonstrated that the exposure of *Salmonella entericaserovar typhymurium* to subinhibitory concentrations of biocides can stimulate resistance to different antibiotics. Walsh et al. [71] showed that *in vitro* subinhibitory exposure to phenols may cause antibiotic cross-resistance in *E. coli* and *P. aeruginosa.*

Common resistance mechanisms between biocides and antibiotics may involve the following mechanisms:

1. Common innate resistance as a result of nonspecific reaction against toxic molecules. Innate resistance may be caused by impermeability after changes in cell envelope, reductions in porins or lipopolysaccharide alterations [12, 64, 66] and multidrug efflux pumps [66, 70].
2. Common acquired resistance induced by stress related gene expression and horizontal spread of plasmid-mediated resistance [92].
3. Coresistance: genetically linked resistance genes [93].
4. Physiological resistance in biofilms.

However, the evidence linking common resistance and cross resistance between biocides and antibiotics has several potential pitfalls. First, most studies are based on *in vitro* laboratory experiments [37, 66]. These studies have focused on resistant development after biocidal exposure at subinhibitory values that may not reflect the concentrations of biocides used in practice [66, 67].

Second, when comparing biocides to antibiotics, few similarities may be summarized, as follows:

1. Biocides and antibiotics are similarly taken up by passive diffusion through the cell envelope [66, 73, 94].

2. Biocides and antibiotics are similarly processed by bacterial metabolic systems [94].
3. Intrinsic and physiologic (biofilm) resistance mechanisms may overlap between biocides and antibiotics [66].

However, despite these few similarities, significant differences exist between biocial and antibacterial agents that would make resistance development to biocides less discernible (Table 2.1). Therefore, the methods used to assess antibiotic resistance can lead to inappropriate conclusions when applied to biocides [66].

2.6.3 Efficacy

Russell and colleagues reported that liquid biocides are not equally potent against all microorganisms (Figure 2.4). In addition to the susceptibility of the micro-organism, efficacy of liquid biocides depend on a range of factors including time, concentration, temperature and pH [37, 66, 94, 97, 98]. Efficacy of biocide plastics may require additional factors to be taken into perspective. To engage in biocidal action, the active surface of biocide plastics must initially contact the bacterial surfaces, especially for biocide plastics that primarily target bacterial cell wall [95]. Bacteria may display differences in adhesion to polymeric surfaces as a consequence of differences in their bacterial cell wall structure [99]. Biocide plastics may thus succeed in impeding certain micro-organisms; however, others may flourish under a selective advantage. The incorporation of biocides into polymers and surfaces cannot be accountable for reductions in biofilm formation alone as several other surface factors participate in biofilm formation including the conditioning film, surface roughness, polarity, and hydrophobicity [26, 100, 101]. Prior to bacterial attachment, the surfaces are initially conditioned with water, electrolytes, and organic substances derived from the environment. The organic conditioning layer is the result of protein adsorption and is a prerequisite for bacterial adhesion and biofilm formation. For example, medical devices are coated with a layer of blood proteins shortly after implantation [99, 102, 103]. Organic layers are deposited onto stainless steel surfaces derived from aquatic environments [104]. The formation of these environmental conditioning layers occurs naturally irrespective of the bacteriocidal action and is principally the result of surface hydrophobicity and charge. The concern with organic deposited layers may not be only in that they act as a physical barrier between the microorganism and the sterilizing surface, but it may in addition neutralize and compromise biocidal efficacy [94, 98, 105, 106]. In certain instances, the formation of the conditioning layer may be desirable when using medical implants to enhance cell attraction, improve tissue integration, and promote healing [55]. Thus, the conditioning layer and buildup of environmental residues have to be taken into account when considering efficacy of biocide plastics [66, 107].

TABLE 2.1 Major differences between biocides and antibiotics

Biocide	Antibiotic	References
Antimicrobial action		
Effective against bacteria, viruses, and fungi	Effective against bacteria	[51, 73, 118]
Nonspecific, multiple antimicrobial targets	Specific intracellular bacterial targets	[7, 51, 66, 95, 96]
Used in high bacteriocidal concentrations for disinfection and low bacteriostatic for food preservation	Mechanism of action of the antibiotic determines the bacteriostatic or bacteriocidal action	[7, 51, 66, 67, 95, 118]
Cannot be administered systematically, may be toxic to mammalian cells at bacteriocidal concentrations	Used in systemic and topical antibacterial chemotherapy administration, nontoxic to mammalian cells at bacteriocidal concentrations	[51, 66]
Resistance		
Lower potential for resistant development	Clinical and environmental resistance development is rapid and clinically evident	[7, 66, 90, 95]
Resistance less evident in clinical practice and involves *in vitro* studies		
Predominately innate resistance	Predominately acquired resistance	[51, 66]
Biocide activity is best assessed by testing bacteriocidal (lethality) activity	Antibiotic activity is assesses by testing minimum inhibition concentration (MIC)	[66, 94]

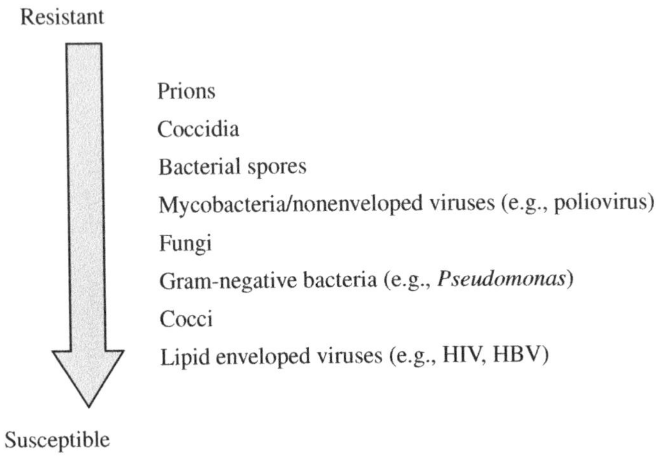

FIGURE 2.4 Relative suscetpibilites of enities (Prions and Coccidia) and microorganisms to biocides.
Source: Adapted from Reference 73 with permission from Oxford University Press.

Biocide plastics efficiency may be influenced also by interfering substances that inhibit the bacteriocidal action. In addition to conditioning layer, debris, pus, and blood may accumulate on medical surfaces. Dead microbial cells that accumulate on antimicrobial surfaces may generate debris that neutralize the antimicrobial effect and attract subsequent bacterial attachment [108, 109]. For example, despite efficient bacteriocidal action, both viable and nonviable bacteria accumulate and adhere on silver-coated fabrics and quaternary ammonium silane-coated biomedical devices [110, 111]. In the food industry, food residue and debris may inactivate the biocidal action [105]. It is thus imperative that the antimicrobial action of biocide polymers extends beyond the surface and accumulating debris. The antibacterial action of biocide plastics must also be maintained by following routine and stringent cleaning practices that remove the dead deposited cells.

Limited biofilm penetration remains a significant obstacle in targeting nosocomial infections and hospital pathogen reservoirs [23]. Biocidal surfaces may reduce the number of adhering bacteria but may not eradicate them completely [53, 112]. In active growing environments, the reduction in adhesion alone cannot reduce biofilm formation [21]. Biofilms can thus occur on antimicrobial surfaces if a conditioning layer forms in the presence of high bacterial load [113]. Thus, biocide polymers that can inhibit initial bacteria attachment or kill bacteria on contact must also be effective in highly contaminated environments [114, 115].

2.6.4 Sustainability

Biocidal polymers, whether in medical or industrial application, must sustain their effect for a sufficient duration of time. Biocidal polymers and antimicrobial surface coatings are generally either bound or slowly released into the environment [55]. The disadvantage of several antimicrobial releasing materials is their short duration of action and limited effectiveness that may eventually be depleted [54, 55]. Leaching antimicrobials may also cause toxicity to surrounding tissues and environmental contamination [55]. Released antimicrobial agents into the environment and loss of antimicrobial agents from surface (leaching) can result in subinhibitory concentrations that may promote antimicrobial resistance [35, 54, 55].

Permanent, nonleaching biocide plastics are more desirable because they reduce environmental contamination, prevent tissue toxicity, limit antimicrobial resistance development, and provide long-term effectiveness [54, 55]. However, immobilizing antimicrobial molecules may undermine their bacteriocidal activity and limit their access to microbes in the surrounding environment [54, 55].

2.7 CONCLUSIONS AND FUTURE TRENDS

Studies investigating the role of biocide plastics in the development of microbial resistance remain very limited and random. Inoculum size and duration of experiment may affect the long-term estimation of bacterial adhesion and bacteriocidal activity [78, 96, 116, 117]. Most studies are performed *in vitro* under controlled laboratory settings and for relatively short durations of time on single or few microorganisms of interest neglecting other microorganisms that may still act as resistant reservoirs in the environment. Studies conducted in laboratory settings also lack essential factors that may be present *in vivo* including inflammation, compatibility, and environmental changes [55]. Hence, there is a need to investigate thoroughly the effect of biocide plastics on bacterial ecology, biofilms formation, resistance development, and cross-resistance to antibiotics in circumstances that simulate practical environments.

Multi-drug resistance and global spread of resistant determinants forecast the emergence of untreatable infections. Biocide polymers represent an attractive advantage over conventional antibiotics and liquid biocides in containing the spread of lethal pathogens. Biocide plastics may minimize surface microbial loads, prevent biofilm formation, reduce environmental pollution, and prevent subinhibitory antimicrobial concentrations that contribute to an acquired bacterial resistance.

However, several potential and undetermined challenges remain that may emerge after widespread application of biocide plastics. Therefore, it is essential to monitor the effect of plastic biocides on microbial ecology, bacterial resistance, and antibiotic cross-resistance.

REFERENCES

1. Alekshun, M. N., Levy, S. B. (2007). Molecular mechanisms of antibacterial multidrug resistance. *Cell*, *128*, 1037–1050.

2. Henderson-Begg, S. K., Livermore, D. M., Hall, L. M. (2006). Effect of subinhibitory concentrations of antibiotics on mutation frequency in *Streptococcus pneumoniae*. *Journal of Antimicrobial Chemotherapy*, *57*, 849–854.
3. Bjedov, I., Tenaillon, O., Gerard, B., Souza, V., Denamur, E., Radman, M., Taddei, F., Matic, I. (2003). Stress-induced mutagenesis in bacteria. *Science*, *300*, 1404–1409.
4. Blazquez, J. (2003). Hypermutation as a factor contributing to the acquisition of antimicrobial resistance. *Clinical Infectious Diseases*, *37*, 1201–1209.
5. Hawkey, P. M., Jones, A. M. (2009). The changing epidemiology of resistance. *Journal of Antimicrobial Chemotherapy*, *64*(Suppl 1), i3–10.
6. Klevens, R. M., Morrison, M. A., Nadle, J., Petit, S., Gershman, K., Ray, S., Harrison, L. H., Lynfield, R., Dumyati, G., Townes, J. M., Craig, A. S., Zell, E. R., Fosheim, G. E., McDougal, L. K., Carey, R. B., Fridkin, S. K. (2007). Invasive methicillin-resistant *Staphylococcus aureus* infections in the United States. *Journal of the American Medical Association*, *298*, 1763–1771.
7. Walsh, C. (2000). Molecular mechanisms that confer antibacterial drug resistance. *Nature*, *406*, 775–781.
8. Martinez, J. L. (2009). The role of natural environments in the evolution of resistance traits in pathogenic bacteria. *Proceedings of the Royal Society B: Biological Sciences*, *276*, 2521–2530.
9. D'Costa, V. M., Griffiths, E., Wright, G. D. (2007). Expanding the soil antibiotic resistome: Exploring environmental diversity. *Current Opinion in Microbiology*, *10*, 481–489.
10. Dantas, G., Sommer, M. O., Oluwasegun, R. D., Church, G. M. (2008). Bacteria subsisting on antibiotics. *Science*, *320*, 100–103.
11. Piddock, L. J. (2006). Multidrug-resistance efflux pumps—not just for resistance. *Nature Reviews Microbiology*, *4*, 629–636.
12. Nikaido, H. (2001). Preventing drug access to targets: Cell surface permeability barriers and active efflux in bacteria. *Seminars in Cell and Developmental Biology*, *12*, 215–223.
13. Wagner, V. E., Iglewski, B. H. (2008). *P. aeruginosa* Biofilms in CF Infection. *Clinical Reviews in Allergy and Immunology*, *35*, 124–134.
14. Hughes, V. M., Datta, N. (1983). Conjugative plasmids in bacteria of the 'pre-antibiotic' era. *Nature*, *302*, 725–726.
15. Perfeito, L., Fernandes, L., Mota, C., Gordo, I. (2007). Adaptive mutations in bacteria: High rate and small effects. *Science*, *317*, 813–815.
16. Mesak, L. R., Miao, V., Davies, J. (2008). Effects of subinhibitory concentrations of antibiotics on SOS and DNA repair gene expression in *Staphylococcus aureus*. *Antimicrobial Agents and Chemotherapy*, *52*, 3394–3397.
17. Hirsch, R., Ternes, T., Haberer, K., Kratz, K. L. (1999). Occurrence of antibiotics in the aquatic environment. *Science of the Total Environment*, *225*, 109–118.
18. Jernberg, C., Lofmark, S., Edlund, C., Jansson, J. K. (2007). Long-term ecological impacts of antibiotic administration on the human intestinal microbiota. *International Society for Microbial Ecology Journal*, *1*, 56–66.

19. Guerra, B., Soto, S. M., Arguelles, J. M., Mendoza, M. C. (2001). Multidrug resistance is mediated by large plasmids carrying a class 1 integron in the emergent *Salmonella enterica* serotype [4,5,12:i:-]. *Antimicrobial Agents and Chemotherapy*, *45*, 1305–1308.

20. Oppegaard, H., Steinum, T. M., Wasteson, Y. (2001). Horizontal transfer of a multidrug resistance plasmid between coliform bacteria of human and bovine origin in a farm environment. *Applied and Environmental Microbiology*, *67*, 3732–3734.

21. Brooks, J. D., Flint, S. H. (2008). Biofilms in the food industry: Problems and potential solutions. *International Journal of Food Science and Technology*, *43*, 2163–2176.

22. Coetser, S. E., Cloete, T. E. (2005). Biofouling and biocorrosion in industrial water systems. *Critical Reviews in Microbiology*, *31*, 213–232.

23. Smith, K., Hunter, I. S. (2008). Efficacy of common hospital biocides with biofilms of multi-drug resistant clinical isolates. *Journal of Medical Microbiology*, *57*, 966–973.

24. Fedtke, I., Gotz, F., Peschel, A. (2004). Bacterial evasion of innate host defenses—the *Staphylococcus aureus* lesson. *International Journal of Medical Microbiology*, *294*, 189–194.

25. Anderson, G. G., O'Toole, G. A. (2008). Innate and induced resistance mechanisms of bacterial biofilms. *Current Topics in Microbiology and Immunology*, *322*, 85–105.

26. Donlan, R. M., Costerton, J. W. (2002). Biofilms: Survival mechanisms of clinically relevant microorganisms. *Clinical Microbiology Review*, *15*, 167–193.

27. Patel, R. (2005). Biofilms and antimicrobial resistance. *Clinical Orthopaedics and Related Research*, 41–47.

28. Sutherland, I. W. (2001). The biofilm matrix—an immobilized but dynamic microbial environment. *Trends in Microbiology*, *9*, 222–227.

29. van Leeuwenhoek, A. (1684). Letter. *Philosophical Transactions of the Royal Society of London*, *14*, 568–574.

30. Bagge, N., Schuster, M., Hentzer, M., Ciofu, O., Givskov, M., Greenberg, E. P., Hoiby, N. (2004). *Pseudomonas aeruginosa* biofilms exposed to imipenem exhibit changes in global gene expression and beta-lactamase and alginate production. *Antimicrobial Agents and Chemotherapy*, *48*, 1175–1187.

31. Rachid, S., Ohlsen, K., Witte, W., Hacker, J., Ziebuhr, W. (2000). Effect of subinhibitory antibiotic concentrations on polysaccharide intercellular adhesin expression in biofilm-forming *Staphylococcus epidermidis*. *Antimicrobial Agents and Chemotherapy*, *44*, 3357–3363.

32. Ahmed, N. A., Petersen, F. C., Scheie, A. A. (2009). AI-2/LuxS is involved in increased biofilm formation by *Streptococcus intermedius* in the presence of antibiotics. *Antimicrobial Agents and Chemotherapy*, *53*, 4258–4263.

33. Walters, M. C., 3rd, Roe, F., Bugnicourt, A., Franklin, M. J., Stewart, P. S. (2003). Contributions of antibiotic penetration, oxygen limitation, and low metabolic activity to tolerance of *Pseudomonas aeruginosa* biofilms to ciprofloxacin and tobramycin. *Antimicrobial Agents and Chemotherapy*, *47*, 317–323.

34. Anderl, J. N., Zahller, J., Roe, F., Stewart, P. S. (2003). Role of nutrient limitation and stationary-phase existence in *Klebsiella pneumoniae* biofilm resistance to ampicillin and ciprofloxacin. *Antimicrobial Agents and Chemotherapy*, *47*, 1251–1256.

35. Lewis, K. (2008). Multidrug tolerance of biofilms and persister cells. *Current Topics in Microbiology and Immunology*, *322*, 107–131.
36. Kramer, A., Schwebke, I., Kampf, G. (2006). How long do nosocomial pathogens persist on inanimate surfaces? A systematic review. *BMC Infectious Diseases*, *6*, 130.
37. Maillard, J. Y. (2005). Antimicrobial biocides in the healthcare environment: Efficacy, usage, policies, and perceived problems. *Therapeutics and Clinical Risk Management*, *1*, 307–320.
38. Dettenkofer, M., Spencer, R. C. (2007). Importance of environmental decontamination—a critical view. *Journal of Hospital Infection*, *65*(Suppl 2), 55–57.
39. Dharan, S., Mourouga, P., Copin, P., Bessmer, G., Tschanz, B., Pittet, D. (1999). Routine disinfection of patients' environmental surfaces. Myth or reality? *Journal of Hospital Infection*, *42*, 113–117.
40. Schierholz, J. M., Beuth, J. (2001). Implant infections: A haven for opportunistic bacteria. *Journal of Hospital Infection*, *49*, 87–93.
41. Ziebuhr, W., Hennig, S., Eckart, M., Kranzler, H., Batzilla, C., Kozitskaya, S. (2006). Nosocomial infections by *Staphylococcus epidermidis*: how a commensal bacterium turns into a pathogen. *International Journal of Antimicrobial Agents*, *28*(Suppl 1), S14–20.
42. Martinez, J. L. (2009). Environmental pollution by antibiotics and by antibiotic resistance determinants. *Environmental Pollution*, *157*, 2893–2902.
43. Threlfall, E. J. (2002). Antimicrobial drug resistance in Salmonella: Problems and perspectives in food- and water-borne infections. *FEMS Microbiology Reviews*, *26*, 141–148.
44. Teuber, M. (2001). Veterinary use and antibiotic resistance. *Current Opinion in Microbiology*, *4*, 493–499.
45. Vestby, L. K., Moretro, T., Langsrud, S., Heir, E., Nesse, L. L. (2009). Biofilm forming abilities of Salmonella are correlated with persistence in fish meal- and feed factories. *BMC Veterinary Research*, *5*, 20.
46. Cloete, T. E. (2003). Biofouling control in industrial water systems:What we know and what we need to know. *Materials and Corrosion*, *54*, 520–526.
47. Kenawy el, R., Worley, S. D., Broughton, R. (2007). The chemistry and applications of antimicrobial polymers: A state-of-the-art review. *Biomacromolecules*, *8*, 1359–1384.
48. D'Arcy, N. (2001). Antimicrobials in plastics: a global review. *Plastics, Additives and Compounding*, *3*, 12–15.
49. Smith, A. W. (2005). Biofilms and antibiotic therapy: Is there a role for combating bacterial resistance by the use of novel drug delivery systems? *Advanced Drug Delivery Reviews*, *57*, 1539–1550.
50. Wood, P., Jones, M., Bhakoo, M., Gilbert, P. (1996). A novel strategy for control of microbial biofilms through generation of biocide at the biofilm-surface interface. *Applied Environmental Microbiology*, *62*, 2598–2602.
51. White, D. G., McDermott, P. F. (2001). Biocides, drug resistance and microbial evolution. *Current Opinion in Microbiology*, *4*, 313–317.
52. Hart, E., Azzopardi, K., Taing, H., Graichen, F., Jeffery, J., Mayadunne, R., Wickramaratna, M., O'Shea, M., Nijagal, B., Watkinson, R., O'Leary, S., Finnin, B., Tait, R., Robins-Browne, R. (2010). Efficacy of antimicrobial polymer coatings

in an animal model of bacterial infection associated with foreign body implants. *Journal of Antimicrobial Chemotherapy*, *65*, 974–980.

53. Chen, Z., Luo, J., Sun, Y. (2007). Biocidal efficacy, biofilm-controlling function, and controlled release effect of chloromelamine-based bioresponsive fibrous materials. *Biomaterials*, *28*, 1597–1609.
54. Lewis, K., Klibanov, A. M. (2005). Surpassing nature: Rational design of sterile-surface materials. *Trends in Biotechnology*, *23*, 343–348.
55. Vasilev, K., Cook, J., Griesser, H. J. (2009). Antibacterial surfaces for biomedical devices. *Expert Review of Medical Devices*, *6*, 553–567.
56. McDonnell, G., Russell, A. D. (1999). Antiseptics and disinfectants: Activity, action, and resistance. *Clinical Microbiology Reviews*, *12*, 147–179.
57. Pericone, C. D., Park, S., Imlay, J. A., Weiser, J. N. (2003). Factors contributing to hydrogen peroxide resistance in *Streptococcus pneumoniae* include pyruvate oxidase (SpxB) and avoidance of the toxic effects of the fenton reaction. *Journal of Bacteriology*, *185*, 6815–6825.
58. Kreth, J., Zhang, Y., Herzberg, M. C. (2008). Streptococcal antagonism in oral biofilms: *Streptococcus sanguinis* and *Streptococcus gordonii* interference with *Streptococcus mutans*. *Journal of Bacteriology*, *190*, 4632–4640.
59. Cintas, L. M., Casaus, M. P., Herranz, C., Nes, I. F., Hernández, P. E. (2001). Review: Bacteriocins of Lactic Acid Bacteria. *Food Science and Technology International*, *7*, 281–305.
60. Russell, A. D. (2001). Mechanisms of bacterial insusceptibility to biocides. *American Journal of Infection Control*, *29*, 259–261.
61. Russell, A. D. (1996). Activity of biocides against mycobacteria. *Society for Applied Bacteriology Symposium Series*, *25*, 87S–101S.
62. Lambert, P. A. (2002). Cellular impermeability and uptake of biocides and antibiotics in gram-positive bacteria and mycobacteria. *Symposium Series (Society for Applied Microbiology)*, 46S–54S.
63. Nikaido, H. (1996). Multidrug efflux pumps of gram-negative bacteria. *Journal of Bacteriology*, *178*, 5853–5859.
64. Nikaido, H. (1994). Prevention of drug access to bacterial targets: permeability barriers and active efflux. *Science*, *264*, 382–388.
65. Russell, A. D. (2004). Bacterial adaptation and resistance to antiseptics, disinfectants and preservatives is not a new phenomenon. *Journal of Hospital Infection*, *57*, 97–104.
66. Russell, A. D. (2003). Biocide use and antibiotic resistance: The relevance of laboratory findings to clinical and environmental situations. *Lancet Infectious Disease*, *3*, 794–803.
67. Weber, D. J., Rutala, W. A. (2006). Use of germicides in the home and the healthcare setting: Is there a relationship between germicide use and antibiotic resistance? *Infection Control and Hospital Epidemiology*, *27*, 1107–1119.
68. Hegstad, K., Langsrud, S., Lunestad, B. T., Scheie, A. A., Sunde, M., Yazdankhah, S. P. (2010). Does the wide use of quaternary ammonium compounds enhance the selection and spread of antimicrobial resistance and thus threaten our health? *Microbial Drug Resistance*, *16*, 91–104.

69. Yazdankhah, S. P., Scheie, A. A., Hoiby, E. A., Lunestad, B. T., Heir, E., Fotland, T. O., Naterstad, K., Kruse, H. (2006). Triclosan and antimicrobial resistance in bacteria: An overview. *Microbial Drug Resistance*, *12*, 83–90.

70. Randall, L. P., Cooles, S. W., Coldham, N. G., Penuela, E. G., Mott, A. C., Woodward, M. J., Piddock, L. J., Webber, M. A. (2007). Commonly used farm disinfectants can select for mutant *Salmonella enterica* serovar *Typhimurium* with decreased susceptibility to biocides and antibiotics without compromising virulence. *Journal of Antimicrobial Chemotherapy*, *60*, 1273–1280.

71. Walsh, S. E., Maillard, J. Y., Russell, A. D., Catrenich, C. E., Charbonneau, D. L., Bartolo, R. G. (2003). Development of bacterial resistance to several biocides and effects on antibiotic susceptibility. *Journal of Hospital Infection*, *55*, 98–107.

72. Rowbury, R. J. (2003). How killed enterobacterial cultures can activate living organisms to resist lethal agents or conditions. *Science Progress*, *86*, 157–178.

73. Russell, A. D. (2003). Similarities and differences in the responses of microorganisms to biocides. *Journal of Antimicrobial Chemotherapy*, *52*, 750–763.

74. Crandall, A. D., Montville, T. J. (1998). Nisin resistance in *Listeria monocytogenes* ATCC 700302 is a complex phenotype. *Applied and Environmental Microbiology*, *64*, 231–237.

75. Davies, E. A., Adams, M. R. (1994). Resistance of *Listeria monocytogenes* to the bacteriocin nisin. *International Journal of Food Microbiology*, *21*, 341–347.

76. Gravesen, A., Sorensen, K., Aarestrup, F. M., Knochel, S. (2001). Spontaneous nisin-resistant *Listeria monocytogenes* mutants with increased expression of a putative penicillin-binding protein and their sensitivity to various antibiotics. *Microbial Drug Resistance*, *7*, 127–135.

77. Mols, M., van Kranenburg, R., Tempelaars, M. H., van Schaik, W., Moezelaar, R., Abee, T. Comparative analysis of transcriptional and physiological responses of *Bacillus cereus* to organic and inorganic acid shocks. *International Journal of Food Microbiology*, *137*, 13–21.

78. Pan, Y., Breidt, F., Jr., Kathariou, S. (2006). Resistance of *Listeria monocytogenes* biofilms to sanitizing agents in a simulated food processing environment. *Applied and Environmental Microbiology*, *72*, 7711–7717.

79. Neetoo, H., Ye, M., Chen, H. (2007). Effectiveness and stability of plastic films coated with nisin for inhibition of *Listeria monocytogenes*. *Journal of Food Protection*, *70*, 1267–1271.

80. Cooksey, K. (2005). Effectiveness of antimicrobial food packaging materials. *Food Additives and Contaminants*, *22*, 980–987.

81. Naghmouchi, K., Kheadr, E., Lacroix, C., Fliss, I. (2007). Class I/Class IIa bacteriocin cross-resistance phenomenon in *Listeria monocytogenes*. *Food Microbiology*, *24*, 718–727.

82. Gravesen, A., Jydegaard Axelsen, A. M., Mendes da Silva, J., Hansen, T. B., Knochel, S. (2002). Frequency of bacteriocin resistance development and associated fitness costs in *Listeria monocytogenes*. *Applied and Environmental Microbiology*, *68*, 756–764.

83. Percival, S. L., Bowler, P. G., Russell, D. (2005). Bacterial resistance to silver in wound care. *Journal of Hospital Infection*, *60*, 1–7.

84. Silver, S., Phung le, T., Silver, G. (2006). Silver as biocides in burn and wound dressings and bacterial resistance to silver compounds. *Journal of Industrial Microbiology and Biotechnology*, *33*, 627–634.

85. Clement, J. L., Jarrett, P. S. (1994). Antibacterial silver. *Metal Based Drugs*, *1*, 467–482.

86. Edwards-Jones, V. (2009). The benefits of silver in hygiene, personal care and healthcare. *Letters in Applied Microbiology*, *49*, 147–152.

87. Silver, S. (2003). Bacterial silver resistance: Molecular biology and uses and misuses of silver compounds. *FEMS Microbiology Reviews*, *27*, 341–353.

88. McHugh, G. L., Moellering, R. C., Hopkins, C. C., Swartz, M. N. (1975). *Salmonella typhimurium* resistant to silver nitrate, chloramphenicol, and ampicillin. *Lancet*, *1*, 235–240.

89. Li, X. Z., Nikaido, H., Williams, K. E. (1997). Silver-resistant mutants of *Escherichia coli* display active efflux of $Ag+$ and are deficient in porins. *Journal of Bacteriology*, *179*, 6127–6132.

90. Maillard, J. Y. (2007). Bacterial resistance to biocides in the healthcare environment: Should it be of genuine concern? *Journal of Hospital Infection*, *65*(Suppl 2), 60–72.

91. Langsrud, S., Sundheim, G., Holck, A. L. (2004). Cross-resistance to antibiotics of *Escherichia coli* adapted to benzalkonium chloride or exposed to stress-inducers. *Journal of Applied Microbiology*, *96*, 201–208.

92. Beaber, J. W., Hochhut, B., Waldor, M. K. (2004). SOS response promotes horizontal dissemination of antibiotic resistance genes. *Nature*, *427*, 72–74.

93. Sidhu, M. S., Heir, E., Leegaard, T., Wiger, K., Holck, A. (2002). Frequency of disinfectant resistance genes and genetic linkage with beta-lactamase transposon Tn552 among clinical staphylococci. *Antimicrobial Agents and Chemotherapy*, *46*, 2797–2803.

94. SCENIHR (Scientific Committee on Emerging and Newly Identified Health Risks). (2009). Assessment of the antibiotic resistance effects of biocides. Brussels, Belgium, January 19.

95. Walsh, S. E., Maillard, J. Y., Russell, A. D., Catrenich, C. E., Charbonneau, D. L., Bartolo, R. G. (2003). Activity and mechanisms of action of selected biocidal agents on Gram-positive and -negative bacteria. *Journal of Applied Microbiology*, *94*, 240–247.

96. Bayston, R., Vera, L., Mills, A., Ashraf, W., Stevenson, O., Howdle, S. M. (2009). *In vitro* antimicrobial activity of silver-processed catheters for neurosurgery. *Journal of Antimicrobial Chemotherapy*, *65*, 258–265.

97. Sagripanti, J. L., Bonifacino, A. (1996). Comparative sporicidal effects of liquid chemical agents. *Applied Environmental Microbiology*, *62*, 545–551.

98. Spotts W. E. A., Beatty, M. E., Taylor, T. H., Jr., Weyant, R., Sobel, J., Arduino, M. J., Ashford, D. A. (2003). Inactivation of *Bacillus anthracis* spores. *Emerging Infectious Diseases*, *9*, 623–627.

99. Terada, A., Yuasa, A., Kushimoto, T., Tsuneda, S., Katakai, A., Tamada, M. (2006). Bacterial adhesion to and viability on positively charged polymer surfaces. *Microbiology*, *152*, 3575–3583.

100. Patel, J. D., Ebert, M., Ward, R., Anderson, J. M. (2007). *Staphylococcus epidermidis* biofilm formation: Effects of biomaterial surface chemistry and serum proteins. *Journal of Biomedical Materials Research Part A*, *80*, 742–751.

101. Ratner, B. D., Bryant, S. J. (2004). Biomaterials: Where we have been and where we are going. *Annual Review of Biomedical Engineering*, *6*, 41–75.

102. Francolini, I., Donelli, G. (2010). Prevention and control of biofilm-based medical-device-related infections. *FEMS Immunology and Medical Microbiology*, *59*, 227–238.

103. Francolini, I., Norris, P., Piozzi, A., Donelli, G., Stoodley, P. (2004). Usnic acid, a natural antimicrobial agent able to inhibit bacterial biofilm formation on polymer surfaces. *Antimicrobial Agents and Chemotherapy*, *48*, 4360–4365.

104. Compère, C., Bellon-Fontaine, M.-N., Bertrand, P., Costa, D., Marcus, P., Poleunis, C., Pradier, C.-M., Rondot, B., Walls, M. G. (2001). Kinetics of conditioning layer formation on stainless steel. *Biofouling*, *17*, 129–145.

105. Hilgren, J., Swanson, K. M., Diez-Gonzalez, F., Cords, B. (2007). Inactivation of *Bacillus anthracis* spores by liquid biocides in the presence of food residue. *Applied and Environmental Microbiology*, *73*, 6370–6377.

106. Nakanishi, K., Sakiyama, T., Imamura, K. (2001). On the adsorption of proteins on solid surfaces, a common but very complicated phenomenon. *Journal of Bioscience and Bioengineering*, *91*, 233–244.

107. Bayston, R., Ashraf, W., Bhundia, C. (2004). Mode of action of an antimicrobial biomaterial for use in hydrocephalus shunts. *Journal of Antimicrobial Chemotherapy*, *53*, 778–782.

108. Sambhy, V., MacBride, M. M., Peterson, B. R., Sen, A. (2006). Silver bromide nanoparticle/polymer composites: Dual action tunable antimicrobial materials. *Journal of the American Chemical Society*, *128*, 9798–9808.

109. Simoes, M., Simoes, L. C., Cleto, S., Pereira, M. O., Vieira, M. J. (2008). The effects of a biocide and a surfactant on the detachment of *Pseudomonas fluorescens* from glass surfaces. *International Journal of Food Microbiology*, *121*, 335–341.

110. Klueh, U., Wagner, V., Kelly, S., Johnson, A., Bryers, J. D. (2000). Efficacy of silver-coated fabric to prevent bacterial colonization and subsequent device-based biofilm formation. *Journal of Biomedical Materials Research*, *53*, 621–631.

111. Gottenbos, B., van der Mei, H. C., Klatter, F., Nieuwenhuis, P., Busscher, H. J. (2002). *In vitro* and *in vivo* antimicrobial activity of covalently coupled quaternary ammonium silane coatings on silicone rubber. *Biomaterials*, *23*, 1417–1423.

112. Seyfriedsberger, G., Rametsteiner, K., Kern, W. (2006). Polyethylene compounds with antimicrobial surface properties. *European Polymer Journal*, *42*, 3383–3389.

113. Chiang, W. C., Schroll, C., Hilbert, L. R., Moller, P., Tolker-Nielsen, T. (2009). Silver-palladium surfaces inhibit biofilm formation. *Applied and Environmental Microbiology*, *75*, 1674–1678.

114. Fernandes, J. C., Eaton, P., Gomes, A. M., Pintado, M. E., Xavier Malcata, F. (2009). Study of the antibacterial effects of chitosans on *Bacillus cereus* (and its spores) by atomic force microscopy imaging and nanoindentation. *Ultramicroscopy*, *109*, 854–860.

115. Hill, K. E., Malic, S., McKee, R., Rennison, T., Harding, K. G., Williams, D. W., Thomas, D. W. An *in vitro* model of chronic wound biofilms to test wound

dressings and assess antimicrobial susceptibilities. *Journal of Antimicrobial Chemotherapy*, *65*, 1195–1206.

116. Vera, L., Ashraf, W., Mills, A., Stephenson, O., Bayston, R. (2009). An *in vitro* investigation of the antimicrobial activity of silver-processed catheters for external ventricular drainage. *Cerebrospinal Fluid Research*, *6*(Suppl 2), S18.
117. Junker, L. M., Hay, A. G. (2004). Effects of triclosan incorporation into ABS plastic on biofilm communities. *Journal of Antimicrobial Chemotherapy*, *53*, 989–996.

CHAPTER 3

"CLICK CHEMISTRY" TO DERIVED ANTIMICROBIAL POLYMERS

P. LECOMTE, R. RIVA, and C. JÉRÔME
Center for Education and Research on Macromolecules (CERM), University of Liège, Liège, Belgium

CONTENTS

3.1 Introduction . . . 51
3.2 Synthesis and Polymerization of Monomers Functionalized by Cationic Salts . . . 52
3.3 Functionalization of Polymer Chains . . . 55
3.4 Click Chemistry . . . 57
3.5 The "Click" Copper-Mediated Azide/Alkyne Cycloaddition Reaction . . . 58
3.6 Antibacterial Polypropylene by Click Chemistry . . . 58
3.7 Antibacterial Aliphatic Polyesters by Click Chemistry . . . 60
3.8 Outlook . . . 65
Acknowledgments . . . 66
References . . . 66

3.1 INTRODUCTION

Infection by bacteria is an important issue in many fields such as packaging and hospital furniture because these pathogen microorganisms are responsible for many diseases [1]. Biocides are chemicals that inhibit the growth of microorganisms, and some of them have been made available on the market place, as illustrated by alcohols [2], biguanides [3], and halogen-releasing agents [4]. In addition, quaternary ammonium and phosphonium salts have proved to be

Antimicrobial Polymers, First Edition.
Edited by José M. Lagarón, María J. Ocio, and Amparo López-Rubio.

potential antiseptics and disinfectants for clinical purposes [5]. Salton et al. proposed a five-step mechanism for the antimicrobial action of quaternary ammonium salts [6]: (1) adsorption and penetration of the agent into the cell wall; (2) reaction with the cytoplasmic membrane followed by its disorganization; (3) leaking of intracellular low-molecular-weight material; (4) degradation of the proteins and nucleic acids; and (5) wall lysis caused by autolytic enzymes. As a result of the temporary activity of all these compounds, repeated applications are required for a long-term biocide effect. In the field of polymers, two main types of materials can be distinguished, depending on whether the additive is temporarily trapped within the polymer [7, 8] or permanently attached to the chains [3, 9]. Dispersion of a low-molecular-weight biocide, such as a heavy metal [7] or silver [8], within a polymer matrix is an example of the first family of materials. The major issue is the possible migration and release of the antimicrobial agent. This undesired effect can be avoided by chemically bonding the active molecule to the matrix. The permanency of the effect depends then on the stability of the chemical bond between the biocide and the polymer. Typical examples are polymers substituted by quaternary ammonium salts [10, 11], phosphonium salts [12, 13], and pyridinium cations [14, 15]. As a rule, these cationic biocides interact with the negatively charged membrane of the bacteria, which is accordingly disrupted by the long alkyl substituent of the cationic salt and finally disintegrated.

Two strategies allowing synthesizing polymers bearing cationic salts have to be distinguished. The first strategy relies on the preparation of a functionalized monomer followed by its direct polymerization. The second strategy relies on the grafting of the cationic salt onto the preformed polymer chains.

3.2 SYNTHESIS AND POLYMERIZATION OF MONOMERS FUNCTIONALIZED BY CATIONIC SALTS

Kanazawa et al. synthesized aliphatic polyesters by polycondensation of a mixture of ethane-1,2-diol (ethylene glycol), dimethyl benzene-1,4-dicarboxylate (dimethyl terephthalate), and various alkyl tributylphosphonium 3,5-bis (methoxycarbonyl) benzenesulfonate (Figure 3.1) [16]. The length of the alkyl group was varied to assess its impact on the antimicrobial properties.

The surface antibacterial activity of the polyester films against *Staphylococcus aureus* (*S. aureus*) and *Escherichia coli* (*E. coli*) was assessed. These films exhibited a high surface antibacterial activity against *S. aureus* and *E. coli*. The activity depends on the structure and the compositional ratio of the phosphonium salts. The amount of the released phosphonium salts was very small, so that the release of the phosphonium biocides can be expected to occur over a long period. The surface antibacterial activity of the polyester films was more bacteriostatic than bactericidal because no morphological change was observed in the strain cells.

Radical polymerization was implemented by Ikeda as shown by the synthesis of a polyacrylate bearing pendent biguanide groups (Figure 3.2). The polymerization

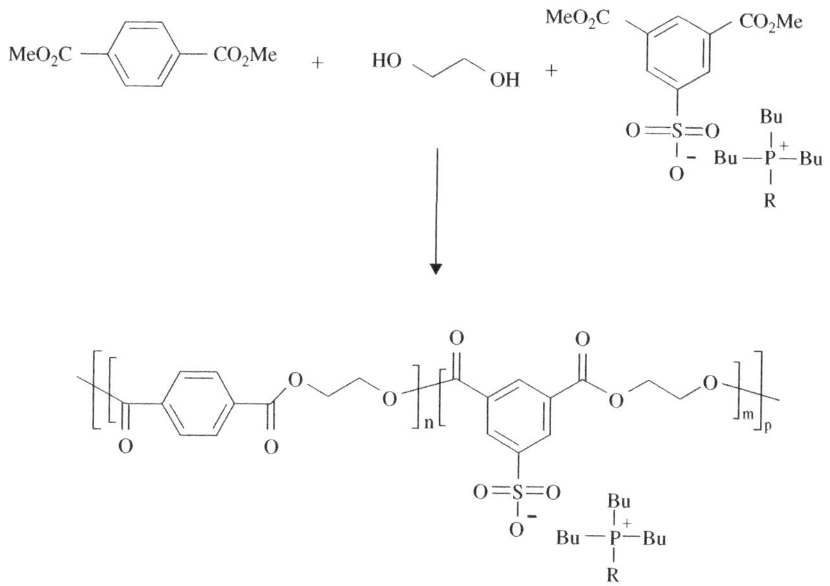

FIGURE 3.1 Synthesis of aliphatic polyesters bearing pendant phosphonium salts according to Kanazawa [16].

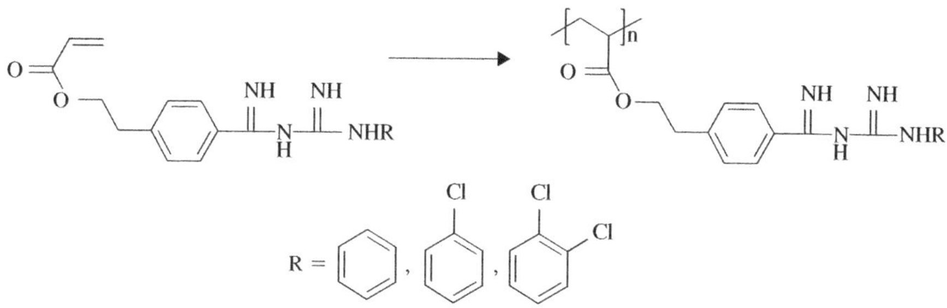

FIGURE 3.2 Radical polymerization of an acrylate substituted by biguanide groups according to Ikeda [3].

was carried out at 60 °C in *N*,*N*-dimethylformamide (DMF) with 2,2′-azo(2)-amidinopropane dihydrochloride as an initiator [3].

Radical polymerization was also used to synthesize polystyrenes bearing pendant phosphonium groups (Figure 3.3) [17, 18]. The antimicrobial activity of this class of polymers was proved against *S. aureus* and *E. coli*. The highest antimicrobial activity of polystyrene bearing pendant octyl phosphonium salts suggests that 8 carbon atom is the optimum chain length.

Various poly(trialkylvinylbenzylammonium chloride)s were synthesized by radical polymerization of the corresponding monomers (Figure 3.4) [18].

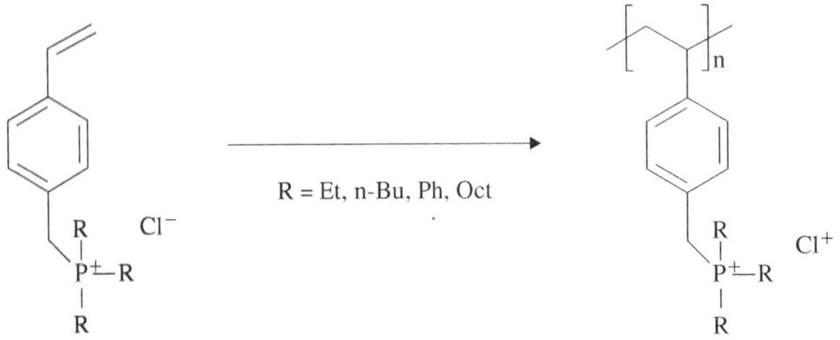

FIGURE 3.3 Radical polymerization of styrenes substituted by phosphonium salts [17].

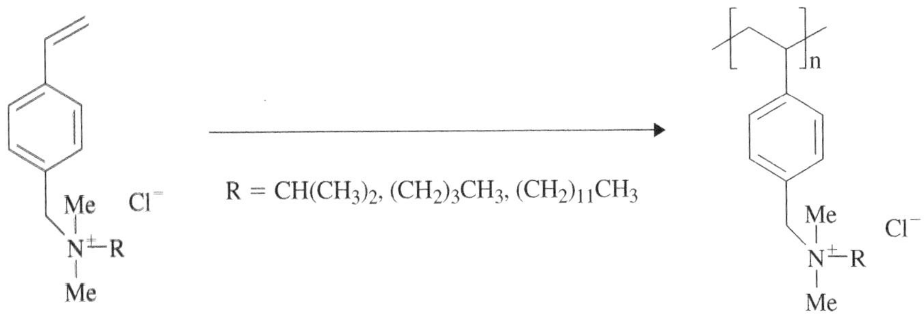

FIGURE 3.4 Radical polymerization of styrenes substituted by ammonium salts [18].

These polymers turned out to be more active against gram-positive bacteria such as *Bacillus subtilis* and *S. aureus* than against gram-negative bacteria such as *E. coli*, *Aerobacter aerogenes*, and *Pseudomonas aeruginosa*. Interestingly enough, compounds with the longest alkyl chain studied (dodecyl) were found to exhibit a high activity. Moreover, the antimicrobial activity of polymers was higher compared with the monomer except for the dodecyl quaternized polymer.

The polymerization of monomers substituted by cationic salts has thus been widely used to synthesize antibacterial polymers. Nevertheless, the number of polymerization mechanism used for that sake remains limited because the cationic salt has to be tolerated by the propagating species of the polymerization. In the examples shown above, polycondensation and radical polymerization were successfully used. In addition, as far as the polymerization is carried out in solution, the choice of the solvent is also critical because it is necessary to use an efficient solvent depending on the mechanism of polymerization and, of course, a solvent able to dissolve the monomer. A complementary approach was thus considered and was based on the direct functionalization of polymer chains as it will be detailed in the next section.

3.3 FUNCTIONALIZATION OF POLYMER CHAINS

Many works reported on the synthesis of antibacterial polymers by quaternization of duly prefunctionalized polymer chains. For instance, Ikeda et al. synthesized polyesters bearing pendant pyridinium salts by a two step-procedure (Figure 3.5) [19]. First, a polyester bearing pendant pyridines was synthesized by the polycondensation of diethyl (pyridin-4-ylmethyl)propanedioate and 1,4-butanediol. Finally, pendant pyridines were quaternized with ethyl bromide, butyl bromide, or octyl bromide. The polyester with the longer chain shows the highest antibacterial activity.

Kawabata and Nishiguchi synthesized poly(*N*-benzyl-4-vinylpyridinium salt) by quaternization of poly(4-vinylpyridine) by, for example, benzyl chloride, benzyl bromide, and butyl bromide (Figure 3.6) [20]. This polymer was found to exhibit a strong antibacterial activity against gram-positive bacteria, whereas it was less active against gram-negative bacteria. The antibacterial activity of this cationic, polymeric disinfectant was considerably greater than that of the corresponding monomeric compound and was approximately equal to that of conventional disinfectants such as benzalkonium chloride.

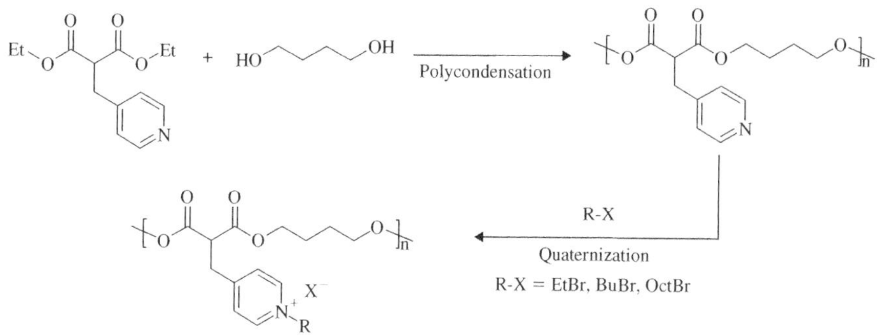

FIGURE 3.5 Synthesis of antibacterial aliphatic polyesters by quaternization of pendant pyridines [18].

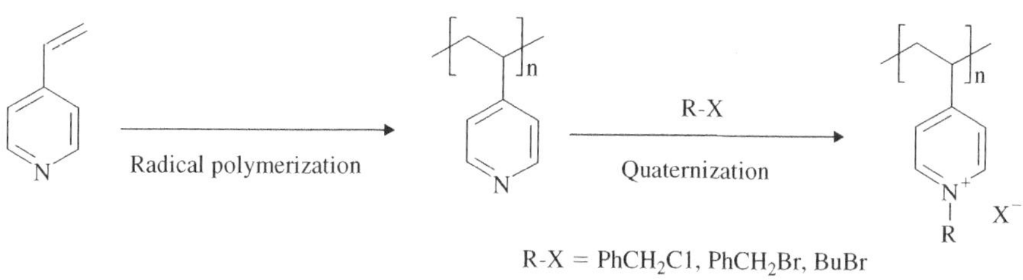

FIGURE 3.6 Synthesis of antibacterial polystyrene by quaternization of pendant pyridines [18].

Another example is provided by Lenoir et al. who quaternized the amino groups of poly(ethylene-*co*-butylene)-*b*-poly[2-(dimethamino) ethyl methacrylate] (PEB-*b*-PDMAEMA) diblock copolymers by octyl bromide (Figure 3.7) [21]. The quaternization was not quantitative (conversion: 64%). The diblock copolymer was synthesized by the atom transfer radical polymerization (ATRP) of 2-(dimethylamino) ethyl methacrylate (DMAEMA) initiated by a bromo-end-capped poly(ethylene-*co*-butylene) (PEB-Br).

Nevertheless, the efficiency of quaternization reactions is often limited. It is indeed difficult to obtain quantitative quaternization reactions, especially when a high content of pendant ammonium is targeted. This issue stems from the change of the conformation of the chain during the quaternization reaction depending on the solvent. As long as the efficiency of the solvent decreases with the conversion, the sites to be quaternized are less available because of steric hindrance. It is then more difficult for the quaternization agent to diffuse toward the reaction site and the reaction is then kinetically disfavored. It is thus highly desirable to find very efficient reactions allowing for grafting cationic

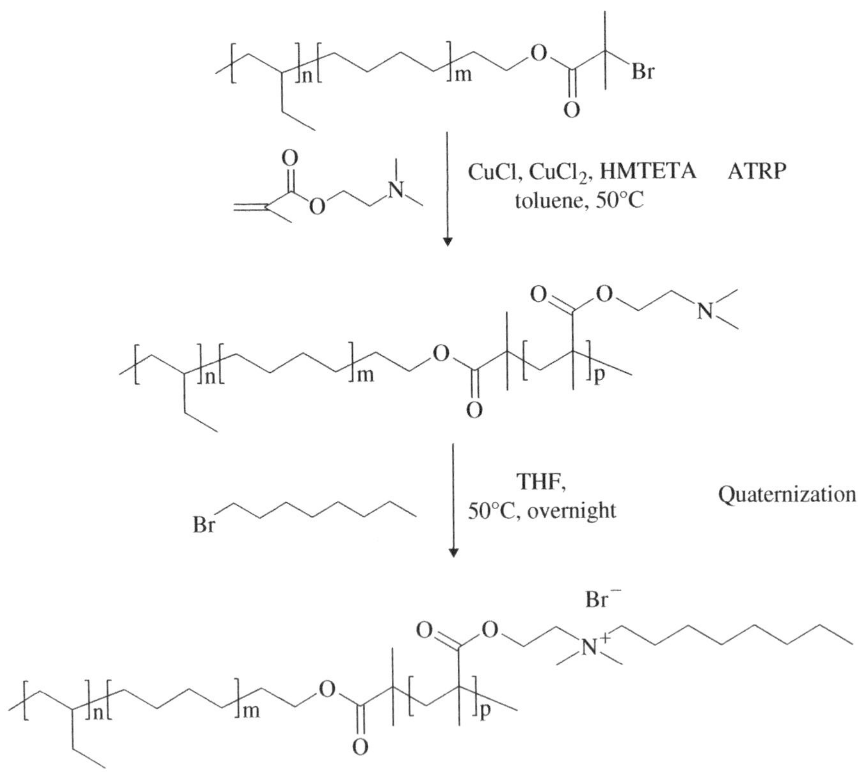

FIGURE 3.7 Synthesis of an antibacterial diblock copolymer by quaternization of the poly(2-(dimethylamino) ethyl methacrylate) sequence [21].

salts onto polymers in order to bring about antibacterial properties. Toward this end, click reactions have turned out to be very promising.

3.4 CLICK CHEMISTRY

The concept of click reactions was defined by Sharpless et al. [22]. To be considered a click reaction, a reaction has to fulfill a set of stringent criteria. The reaction must be modular, wide in scope, provide very high yields, and generate only inoffensive by-products that can be removed by nonchromatographic methods. The reaction has to be stereospecific but not necessarily enantioselective. The reaction conditions have to be simple, and ideally, the reaction should be insensitive to oxygen and water. The starting materials and reagents have to be very easily available. It is mandatory to perform the reaction either in bulk, with a benign solvent such as water, or with a solvent easily removed. The products must be purified by nonchromatographic methods, such as crystallization or distillation, and the product must be stable under physiological conditions. Click reactions achieve their required characteristics by having a high thermodynamic driving force, usually greater than 20 kcal/mol. Such processes proceed rapidly to completion and tend to be highly selective for a single product.

Several reactions are reported to fulfill these criteria:

1. Cycloaddition of unsaturated species, especially 1,3-dipolar cycloaddition reactions and Diels–Alder reactions.
2. Nucleophilic substitution reactions, particularly ring-opening reactions of strained heterocyclic electrophiles such as epoxides, aziridines, aziridinium ions, and episulfonium ions.
3. Carbonyl chemistry of the non-aldol type, such as formation of ureas, thioureas, aromatic heterocycles, oxime ethers, hydrazones, and amides.
4. Additions to carbon–carbon multiple bonds, especially oxidative cases such as epoxidation, dihydroxylation, aziridination, and sulfenyl halide addition, but also Michael additions of nucleophilic reactants.

It seemed very promising to use these click reactions to synthesize polymers bearing pendant cationic salts. For instance, Lou et al. implemented a strategy based on the ring-opening reaction of epoxides by thiols (Figure 3.8) [23]. They synthesized an aliphatic polyester bearing epoxides by using ring-opening polymerization rather than polycondensation. Interestingly enough, this technique allows for better control of the molecular weight than polycondensation and allows for obtaining polyesters with a narrow molecular weight distribution. The pendant epoxides were thereafter reacted with 2-(diethylamino) ethanethiol. The pendant amines were finally quaternized with benzyl bromide. Nevertheless, this approach is complicated because too many steps are required to graft ammonium salts. Finally, a quaternization reaction remained involved in the last step of the process.

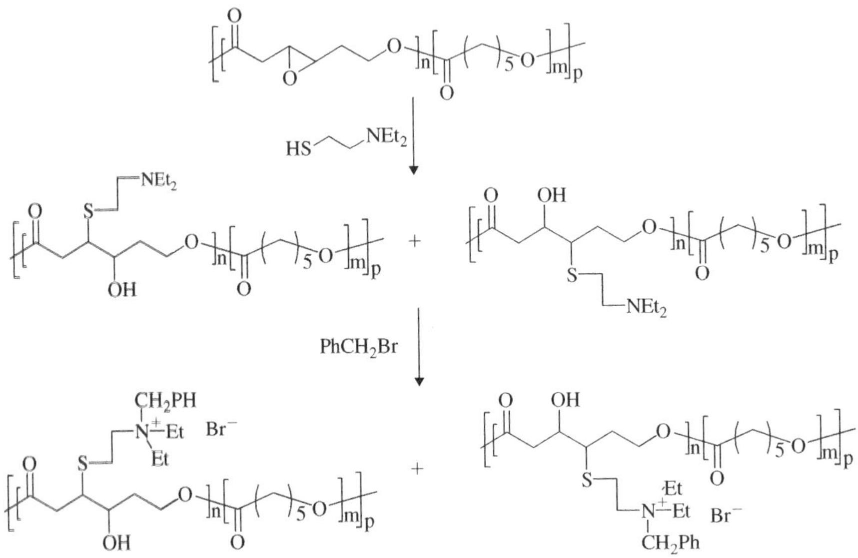

FIGURE 3.8 Synthesis of an aliphatic polyester by implementing a click epoxide ring-opening reaction [23].

3.5 THE "CLICK" COPPER-MEDIATED AZIDE/ALKYNE CYCLOADDITION REACTION

Among "click" reactions, the copper-mediated azide/alkyne cycloaddition reaction (the CuAAc reaction) remains at the time being particularly efficient. Indeed, many researchers consider abusively this reaction as the "click" reaction. The cycloaddition reaction of azides and alkynes is very old but is not very efficient and selective. Meldal [24] and Sharpless et al. [25] found out that the efficiency and the selectivity increases in the presence of copper (I) catalysts. Interestingly enough, quantitative conversions were observed under very mild conditions. Interestingly enough, this reaction turned out to be very efficient to functionalize not only low-molecular-weight molecules but also polymers. Nowadays, this reaction is widely used in the frame of macromolecular engineering as pointed out by several recent reviews [26–33].

3.6 ANTIBACTERIAL POLYPROPYLENE BY CLICK CHEMISTRY

Thomassin et al. used the CuAAC reaction to synthesize a polymer material, which combines in a single material the good mechanical properties of polypropylene and the antibacterial activity of a nonquaternized polymeric biocide, i.e., poly[2-(*tert*-butylamino)ethyl methacrylate] (PTBAEMA) [34]. First, α-azido, ω-bromo-PTBAEMA was obtained by the ATRP of 2-(*tert*-butylamino) ethyl methacrylate initiated by 3-azidopropyl bromoisobutyrate (Figure 3.9). Second, the bromo end-group was reduced into a hydrogen by reaction with tributyltin hydride (Figure 3.9). Third, the azido end-group was converted into an amino end-group by the CuAAC reaction of prop-2-yn-1-amine (propargyl amine) and the azido end-group (Figure 3.9). It is worth pointing out that the reduction step was necessary because the coupling of the two homopolymers by the CuAAC reaction was not efficient in the presence of the bromo end-group because of an undesired coupling reaction shown in Figure 3.10. Finally, the high immiscibility of the polypropylene and PTBAEMA was tackled by reactive compatibilization. For that sake, maleic anhydride-grafted polypropylene was

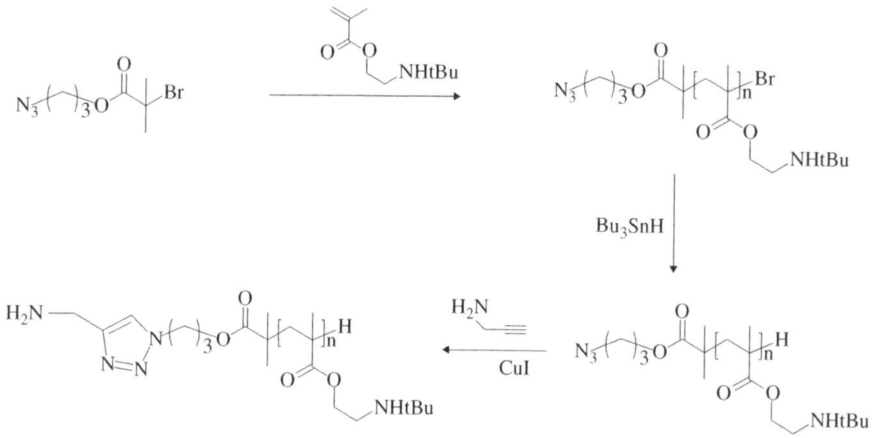

FIGURE 3.9 Synthesis of α-amino poly[2-(*tert*-butylamino)ethyl methacrylate] [34].

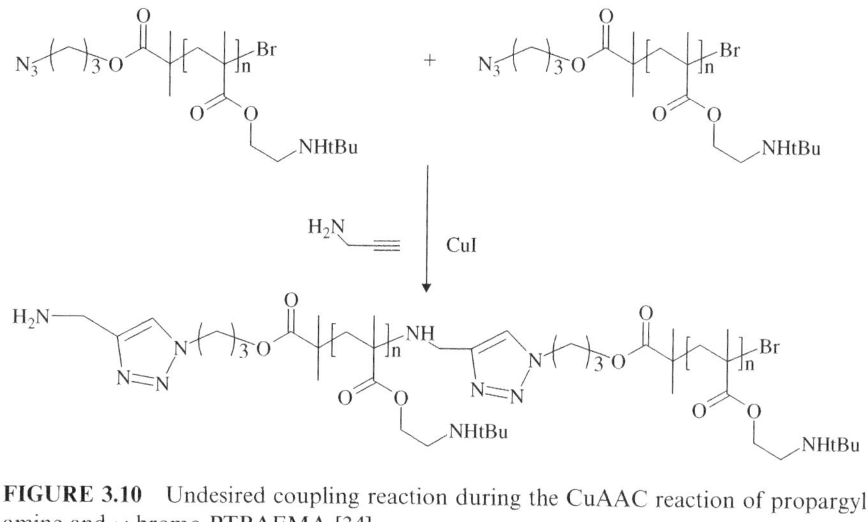

FIGURE 3.10 Undesired coupling reaction during the CuAAC reaction of propargyl amine and ω-bromo-PTBAEMA [34].

reacted with primary amine-end-capped PTBAEMA in the melt to obtain a PP-*g*-PTBAEMA graft copolymer. This copolymer was then melt dispersed within neat PP and processed as fibers, whose antimicrobial properties were assessed by the viable cell counting method against *E. coli*. The anchoring of the PTBAEMA chains onto PP prevented it from being released from the surface of the fibers, which resulted in a long-lasting activity.

3.7 ANTIBACTERIAL ALIPHATIC POLYESTERS BY CLICK CHEMISTRY

As far as polypropylene is concerned, this polymer is not biodegradable. To synthesize polymers exhibiting both properties of biodegradability and antimicrobial activity, an appealing approach relies on the use of the CuAAC reaction to graft antimicrobial agents onto aliphatic polyesters. Two approaches were reported by Emrick et al. [35] and Riva et al. [36, 37], for the implementation of the CuAAC reaction for the functionalization of aliphatic polyesters. To synthesize a polyester bearing pendant alkynes, Emrick et al. synthesized a copolyester of polycaprolactone (PCL) by copolymerization of εCL and 3-(prop-2-yn-1-yl)tetrahydro-2*H*-pyran-2-one [38]. Azides were thereafter grafted by the CuAAC reaction (Figure 3.11) [35]. Conversely, Riva et al.

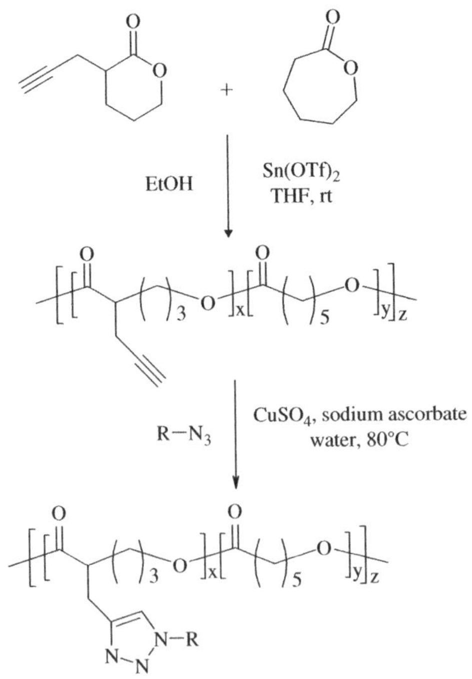

FIGURE 3.11 Coupling of azides onto aliphatic polyesters bearing pendant alkynes [35].

used the reverse approach and grafted alkynes onto aliphatic polyesters bearing pendant azides (Figure 3.12) [39]. First, 2-chlorocyclohexanone was reacted with meta-chloroperbenzoic acid into 3-chlorooxepan-2-one (αClεCL or α-chloro-ε-caprolactone). Second, αClεCL and oxepan-2-one (εCL or ε-caprolactone) were copolymerized by using 2,2-dibutyl-2-stanna-1,3-dioxepane (DSDOP) as an initiator [40]. The content of pendant functional groups could be tailored at will just by changing the molar ratio of the two co-monomers. The molecular weight of the polyester was predetermined by the co-monomers-to-initiator molar ratio. Third, pendant chlorides were reacted with sodium azide to be converted into pendant azides. The reaction was quantitative, and no degradation of the polyester was observed. Finally, pendant azides were reacted with a set of alkynes, and more particularly with *N*,*N*,*N*-triethylprop-2-yn-1-ammonium bromide, which allows for grafting a cationic salt (Figure 3.12) [32, 36].

It is worth recalling that aliphatic polyesters are biodegradable polymers. The degradation takes place by transesterification and hydrolysis of ester bonds. The functionalization of aliphatic polyesters is thus challenging because it is necessary to use a very efficient reaction under conditions where no degradation is observed. The number of reactions meeting these requirements is very limited. To this respect, the CuAAC turned out to be particularly successful. Emrick et al. reported that the CuAAC reaction is very efficient and no degradation was observed during the functionalization as far as PCL is concerned as an aliphatic polyester. Nevertheless, in their work, the CuAAC reaction was carried out at 80 °C for several hours in water and in the presence of a copper catalyst [35]. Unfortunately, these conditions are not general because many aliphatic polyesters are too sensitive and degrade under these conditions. Later on, Riva et al. carried out the CuAAC reaction in DMF or in THF, thus in the absence of water as a solvent, in order to disfavor degradation by hydrolysis of ester bonds [36, 37]. The CuAAC reaction was then catalyzed

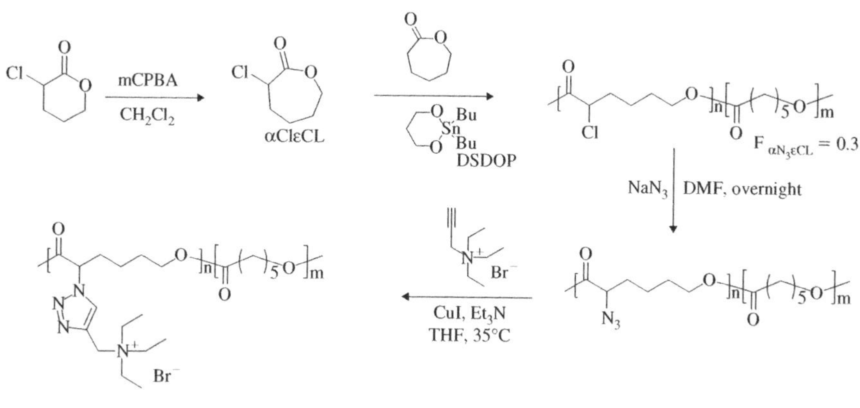

FIGURE 3.12 Coupling of alkynes onto aliphatic polyesters bearing pendant azides [36].

by copper iodide in the presence of an amine as a base. It turned out these conditions are the most efficient for functionalizing a wide range of aliphatic polyesters by the CuAAC reaction with minimized degradation. The extent of degradation depends on the sensitivity of the aliphatic polyester, the functionalization agent, the nature of the solvent, and the temperature.

Riva et al. extended their process shown in Figure 3.12 to the grafting of *N*,*N*-dimethyl-*N*-prop-2-yn-1-yloctan-1-ammonium bromide in order to impart an antibacterial behavior to the aliphatic polyester [39]. The main goal of this work was to develop a polymer that exhibits a permanent biocide activity during use and degrades afterward. Furthermore, PCL is also very well known for its thermodynamic miscibility with other (co)polymers such as acrylonitrile butadiene styrene (ABS) copolymers and poly(vinyl chloride) (PVC). These (co)polymers could thus be made antibacterial by blending or coating with PCL functionalized by a suitable biocide agent. In the work of Riva et al. [39], CuBr was used rather than CuI as a catalyst for the CuAAC reaction, with CuI being known as a biocide, which could interfere during the determination of the antibacterial behavior of the polymer. The biocidal effect of the so-obtained PCL was assessed by the shake flask method [41]. It was checked that the antibacterial activity resulted from the contact of bacteria with the copolyester and not from the ammonium salts that could be released by any hydrolysis of the polyester chain. It was also shown that catalytic copper remnants could not be at the origin of the observed antibacterial activity. Finally, copolyesters with nonquaternized amines also exhibited an antibacterial activity, which was even higher compared with quaternized amines. This difference was ascribed to a better dispersion of the noncationic polymer in water, resulting in a much larger surface area in contact with the bacteria rather than the intrinsic higher antimicrobial activity of polyesters bearing pendant amines instead of ammonium salts.

Interestingly enough, a two-step strategy was also investigated to graft ammonium salts onto aliphatic polyesters bearing pendant azide groups by the CuAAC reaction [39]. First, *N*,*N*-dimethylprop-2-yn-1-amine was coupled onto poly($\alpha N_3\varepsilon$CL-*co*-εCL) ($\alpha N_3\varepsilon$CL stands for α-azido-ε-caprolactone or 3-azidooxepan-2-one). Finally pendant amines were quaternized by reaction with 1-bromooctane (Figure 3.13). Nevertheless, it turned out that the quaternization reaction was not quantitative (76%). Conversely, the direct coupling of *N*,*N*-dimethyl-*N*-prop-2-yn-1-yloctan-1-ammonium bromide was quantitative (Figure 3.13) as a new example that highlights the remarkable efficiency of the CuAAC reaction.

It is worth noting that the sequence of the CuAAC reaction and the ring-opening polymerization can be reversed. Indeed, Riva et al. reported on the synthesis of *N*,*N*-dimethyl-*N*-[1-(7-oxooxepan-4-yl)-1H-1,2,3-triazol-4-yl] methyloctan-1-ammonium bromide by the CuAAC reaction of 5-azidooxepan-2-one ($\gamma N_3\varepsilon$CL or γ-azido-ε-caprolactone) and *N*,*N*-dimethyl-*N*-prop-2-yn-1-yloctan-1-ammonium bromide (Figure 3.14) [42]. 5-azidooxepan-2-one was obtained by the reaction of sodium azide and 5-bromooxepan-2-one ($\gamma Br\varepsilon$CL),

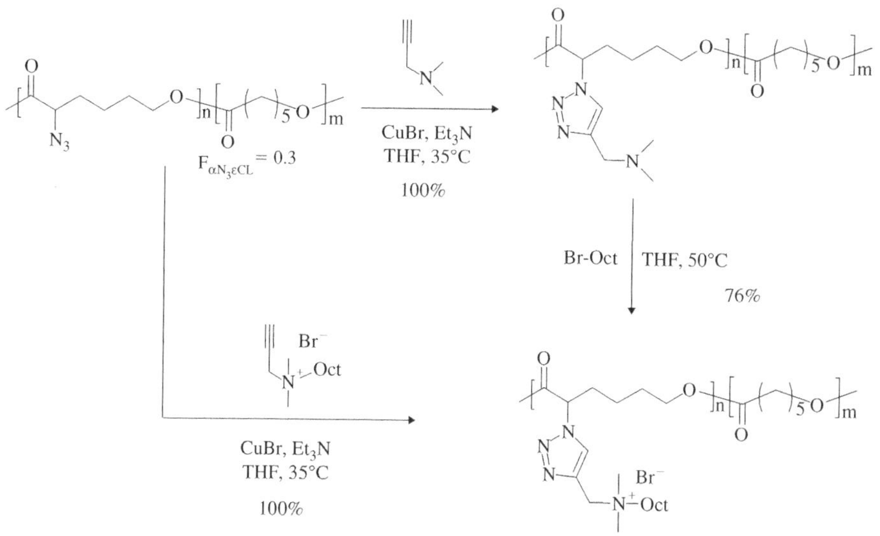

FIGURE 3.13 Synthesis of aliphatic polyesters bearing pendant ammonium salts by the CuAAC reaction [39].

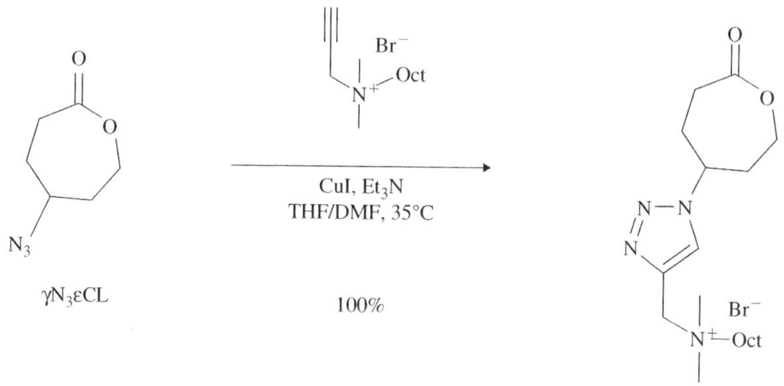

FIGURE 3.14 Synthesis of ε-caprolactone γ-substituted by an ammonium salt [42].

which was synthesized according to a previously reported procedure [43] as shown in Figure 3.15.

It is worth pointing out that it was necessary to use $\gamma N_3\varepsilon CL$ rather than the 3-azidooxepan-2-one ($\alpha N_3\varepsilon CL$ or α-azido-ε-caprolactone) to synthesize ε-caprolactone γ-substituted by an ammonium salt (Figure 3.14) even though the synthesis of this lactone was more complicated and required several steps (Figure 3.15). Indeed, it was shown that the CuAAC reaction of 3-azidooxepan-2-one and alkynes brought about an undesired ring opening of the lactone. Obviously, the lactone was not stable enough because of the presence of the

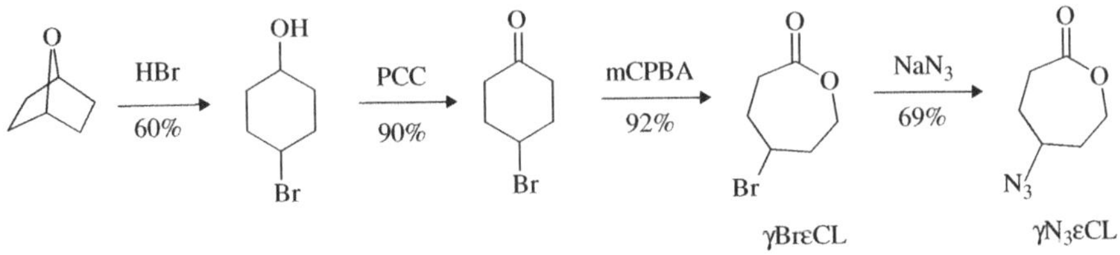

FIGURE 3.15 Synthesis of γ-azido-ε-caprolactone [43].

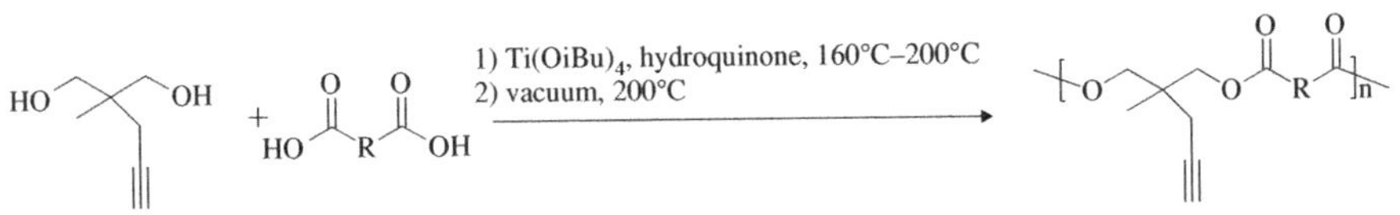

FIGURE 3.16 Synthesis of aliphatic polyesters bearing pendant alkynes by polycondensation [44].

functional group at the α-position [42]. As a result of the more difficult synthesis of the lactone, this approach was not investigated further and no ring-opening polymerization of *N*,*N*-dimethyl-*N*-prop-2-yn-1-yloctan-1-ammonium bromide was reported.

In the work of Riva et al. [36, 37] and Emrick et al. [35], functionalized aliphatic polyesters were synthesized by ring-opening polymerization (Figures 3.11 and 3.12). The main advantage relies on the living character of this polymerization, which allows for the synthesis of aliphatic polyester with a predetermined molecular weight and a narrow polydispersity index. Nevertheless, as long as such a control is not required for the application, polycondensation should also be considered as a valuable alternative for the synthesis of aliphatic polyesters functionalizable by the CuAAC reaction. Toward this end, Du Prez et al. synthesized a copolyester bearing pendant alkynes group by the polycondensation of 2-methyl-2-(prop-2-yn-1-yl)propane-1,3-diol and diacarboxylic acids (Figure 3.16) [44]. Nevertheless, no result dealing with the grafting of ammonium salts to impart an antimicrobial activity was reported until now.

The process of Riva et al. shown in Figure 3.12 could easily be extended to the synthesis of networks just by substituting diacetylic compounds as cross-linking agents for monoacetylenic compounds. Indeed, this approach was implemented by Zednik et al. who reported a process where aliphatic polyesters bearing pendant azide groups were easily functionalized and cross-linked by a one-pot process using the CuAAC reaction [45]. The strategy of Zednik et al. was extended to the synthesis of antimicrobial networks as shown in Figure 3.17 [46]. A polyester bearing pendant azide groups was synthesized by the ring-opening polymerization of 5-bromooxepan-2-one initiated by aluminium alkoxide followed by the conversion of pendant bromide atoms into azide functions by the reaction of sodium azide. The grafting of pendant ammonium salts was carried

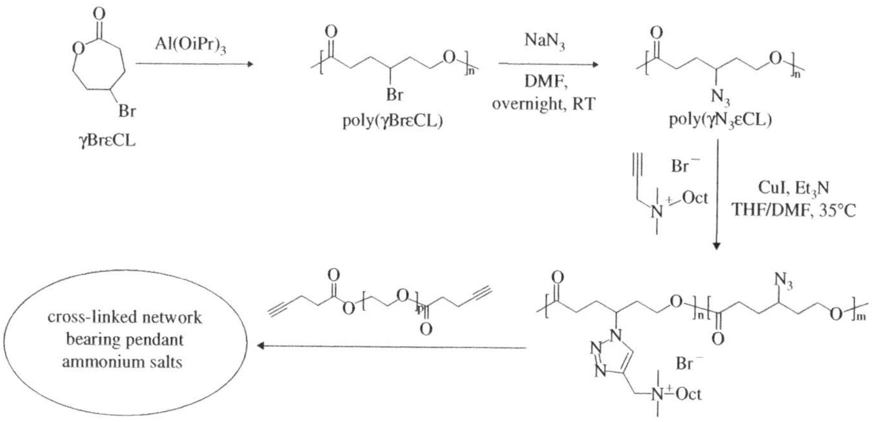

FIGURE 3.17 Synthesis of antimicrobial hydrogels [46].

out by the CuAAC reaction of 50 mol% of *N*,*N*-dimethyl-*N*-prop-2-yn-1-yloctan-1-ammonium bromide with respect to pendant azide groups. Finally, the polyester bearing a mixture of ammonium salts and azide groups was not isolated and was immediately reacted with telechelic acetylenic poly(ethylene oxide) (PEO) to obtain the corresponding amphiphilic network. It is worth recalling that aliphatic polyesters are hydrophobic and that the PEO-based cross-linker is soluble in water, and thus, the networks form hydrogels in water. The antimicrobial behavior of the gel was proved by the "shake flask" method [41]. In the frame of this synthesis, CuI was used to catalyze the CuAAC reaction. Nevertheless, it was checked that the low amount of CuI used for this reaction was not sufficient to impart an antibacterial behavior to the gel.

It is worth noting that $\gamma N_3\varepsilon CL$ had again to be used rather than $\alpha N_3\varepsilon CL$ as a monomer. Indeed, whenever poly($\alpha N_3\varepsilon CL$) was synthesized by the reaction of poly($\alpha Cl\varepsilon CL$) and sodium azide, a slight degradation was systematically observed, which was not the case as far as poly($\alpha N_3\varepsilon CL$-*co*-εCL) copolyesters were concerned. The poly($\alpha N_3\varepsilon CL$) homopolymer then turned out to be slightly more sensitive than the corresponding poly($\alpha N_3\varepsilon CL$-*co*-εCL) copolyesters. The lower sensitivity of poly($\gamma N_3\varepsilon CL$) compared with poly($\alpha N_3\varepsilon CL$) was accounted for by the γ position of the functional group, thus, further away from the ester.

3.8 OUTLOOK

The grafting of judiciously chosen ammonium or phosphonium salts is a tool for imparting an antibacterial behavior to polymers. Among many reactions that can be used to graft ammonium and phosphonium salts onto polymers, the CuAAC deserves particular attention because this reaction is quantitative under very mild conditions. Until now, the CuAAC was implemented to graft ammonium salts onto aliphatic polyesters with the aim of making available biodegradable antibacterial polymers. Nevertheless, as long as biodegradability is not required for the targeted application, the CuAAC remains also very attractive to functionalize nonbiodegradable polymers as shown, for example, in the work of Thomassin et al. [34].

Although the CuAAC reaction is very efficient, some severe drawbacks limit future developments. The CuAAC implies the use of azides, which are known to explode very violently, especially as far as low-molecular-weight azides are concerned. To limit the risks as much as possible, it is thus better to graft azide rather than alkyne groups onto polymers. Low-molecular-weight azides do not have to be handled in the laboratory. In addition, Riva et al. developed a one-pot procedure that avoids the need to handle and store azido-functionalized polymers in the laboratory [37]. These azido-containing polymers were synthesized by reaction of sodium azide with the corresponding halogen-functionalized polymers and were immediately reacted with alkynes under appropriate conditions without being isolated in the laboratory. Even though this procedure is safe at the laboratory level, it remains that the use of azides limits the development of future commercial applications. Moreover, catalytic remnants are very difficult to extract from the polymer even though some procedures are reported in the literature [47, 48]. The samples obtained after CuAAC are thus many times colorful and the metal is prone to interfere with the antimicrobial properties. The metal-free azide-alkyne cycloaddition was reported to tackle these issues. For that sake, Bertozzi et al. reported on the metal-free, strain-promoted Huisgen's cycloaddition carried out by the reaction of azides and fluorocyclooctynes [49]. Nevertheless, the synthesis of these alkynes is demanding, which limits the practical use of this approach [50]. At the time being, an intense research effort is devoted to the development of metal-free click reactions [51]. To this respect, the thiol-ene reaction is very popular [52, 53], but one can also mention the thiol-yne reaction [54] nitrile oxide-azide reaction [55]. The success of the CuAAC reaction for the synthesis of "clicked" antibacterial polymers paves the way for future developments by using these metal-free click reactions.

ACKNOWLEDGMENTS

The authors are indebted to the "Politique scientifique fédérale" for general support to CERM in the frame of the "PAI 6/27: Functional Supramolecular Systems". P.L. IS "Chercheur Qualifié" by the "National Fund for Scientific Research" (FNRS).

REFERENCES

1. Takai, K., Ohtsuka, T., Senda, Y., Nakao, M., Yamamoto, K., Matsuoka, J., Hirai, Y. (2002). Antibacterial properties of antimicrobial-finished textile products. *Microbiology and Immunology*, *46*, 75–81.

2. Ali, Y., Dolan, M. J., Fendler, E. J., Larson, E. L. (2001). *Desinfection, Sterilization, and Preservation*, 5th Edition. Ed.: Block, S. S. pp. 229–253. Lippincott Williams & Wilkins, Philadelphia, PA.
3. Ikeda, T., Yamaguchi, H., Tazuke, S. (1984). New polymeric biocides: Synthesis and antibacterial activities of polycations with pendant biguanide groups. *Antimicrobial Agents and Chemotherapy*, *26*, 139–144.
4. Bloomfield, S. F. (1996). *Handbook of Disinfectants and Antiseptics*, Ed.: Ascenzi, J. M. pp. 133–158. Marcel Dekker, New York, NY.
5. Frier, M. (1971). *Inhibition and Destruction of the Microbial Cell*, Ed.: Hugo, W. B. pp. 107–120. Academic Press, Ltd., London, U.K.
6. Salton, M. R. J. (1968). Lytic agents, cell permeability, and monolayer penetrability. *The Journal of General Physiology*, *52*, 227–252.
7. Nonaka, T., Uemera, Y., Enishi, K., Kurihara, S. (1996). Antibacterial activity of resin-containing triethylenetetramine side chains and/or thiol groups-metal complexes. *Journal of Applied Polymer Science*, *62*, 1651–1657.
8. Ignatova, M., Labaye, D. E., Lenoir, S., Strivay, D., Jérôme, R., Jérôme, C. (2003). Immobilization of silver in polypyrrole/polyanion composite coatings: Preparation, characterization, and antibacterial activity. *Langmuir*, *19*, 8971–8979.
9. Broxton, P., Woodcock, P. M., Gilbert, P. (1983). A study of the antibacterial activity of some polyhexamethylene biguanides towards Escherichia coli ATCC 8739. *Journal of Applied Bacteriology*, *54*, 345–350.
10. Gerba, C. P., Janauer, G. E., Costello, M. (1984). Removal of poliovirus and rotavirus from tapwater by a quaternary ammonium resin. *Water Research*, *18*, 17–19.
11. Hugues, C., Bessy, C., Bartolomeo, P., Mairgaillan, A. (2003). Complexation of an acrylic resin by tertiary amines: Synthesis and characterisation of new binders for antifouling paints. *European Polymer Journal*, *39*, 319–326.
12. Kanazawa, A., Ikeda, T. (2000). Multifunctional tetracoordinate phosphorus species with high self-organizing ability. *Coordination Chemistry Reviews*, *198*, 117–131.
13. Kenawy, E. R., Adbdel-Hay, F. I., El-Raheem, A., El-Sanshoury, R., El-Newehy, M. H. (1998). Biologically active polymers: Synthesis and antimicrobial activity of modified glycidyl methacrylate polymers having a quaternary ammonium and phosphonium groups. *Journal of Controlled Release*, *50*, 145–152.
14. Li, G. J., Shen, J. R., Zhu, Y. L. (2000). A study of pyridinium-type functional polymers. III. Preparation and characterization of insoluble pyridinium-type polymers. *Journal of Applied Polymer Science*, *78*, 668–675.
15. Li, G. J., Shen, J. R. (2001). Preparation and antibacterial activity of pyridinium-carrying polypropylene particles. *Polymer Preprints (American Chemical Society Division of Polymer Chemistry)*, *42*(2), 360–361.
16. Kanazawa, A., Ikeda, T., Endo, T. (1993). Polymeric phosphonium salts as a novel class of cationic biocides. V. Synthesis and antibacterial activity of polyesters releasing phosphonium biocides. *Journal of Polymer Science, Part A: Polymer Chemistry*, *31*, 3003–3011.
17. Kanazawa, A., Ikeda, T., Endo, T. J. (1993). Novel polycationic biocides: Synthesis and antibacterial activity of polymeric phosphonium salts. *Journal of Polymer Science, Part A: Polymer Chemistry*, *31*, 335–343.

18. Kanazawa, A., Ikeda, T., Endo, T. J. (1994). Polymeric phosphonium salts as a novel class of cationic biocides. 7. Synthesis and antibacterial activity of polymeric phosphonium salts and their model compounds containing long alkyl chains. *Journal of Applied Polymer Science*, *53*, 1237–1244.
19. Ikeda, T., Hirayama, H., Suzuki, K., Yamaguchi, H., Tazuke, S. (1986). Biologically active polycations, 6. Polymeric pyridinium salts with well-defined main chain structure. *Makromolekulare Chemie*, *187*, 333–340.
20. Kawabata, N., Nishiguchi, M. (1988). Antibacterial activity of soluble pyridinium-type polymers. *Applied and Environmental Microbiology*, *54*, 2532–2535.
21. Lenoir, S., Pagnoulle, C., Detrembleur, C., Galleni, M., Jérôme, R. (2006). New antimicrobial cationic surfactants prepared by Atom Transfer Radical Polymerization. *Journal of Polymer Science, Part A: Polymer Chemistry*, *44*(3), 1214–1224.
22. Kolb, H. C., Finn, K. M. G., Sharpless, K. B. (2001). Click chemistry: Diverse chemical function from a few good reactions. *Angewandte Chemie International Edition*, *40*, 2004–2021.
23. Lou, X., Detrembleur, C., Lecomte, P., Jérôme, R. (2002). Controlled synthesis and chemical modification of unsaturated aliphatic (co)polyesters based on 6,7-dihydro-2(3H)-oxepinon. *Journal of Polymer Science, Part A: Polymer Chemistry*, *40*, 2286–2297.
24. Tornøe, C. W., Christensen, C., Meldal, M. (2002). Peptidotriazoles on solid phase: [1,2,3]-triazoles by regiospecific copper(I)-catalyzed 1,3-dipolar cycloadditions of terminal alkynes to azides. *Journal of Organic Chemistry*, *67*, 3057–3064.
25. Rostovtsev, V. V., Green, L. G., Fokin, V. V., Sharpless, K. B. (2002). A stepwise Huisgen cycloaddition process: Copper(I)-catalyzed regioselective ligation of azides and terminal alkynes. *Angewandte Chemie International Edition*, *41*, 2596–2599.
26. Lecomte, P., Riva, R., Jérôme, C., Jérôme, R. (2008). Macromolecular engineering of biodegradable polyesters by ring-opening polymerization and 'click' chemistry. *Macromolecular Rapid Communications*, *29*, 982–997.
27. Lecomte, P., Riva, R., Jérôme, C., Jérôme, R. (2008). Macromolecular engineering of biodegradable polyesters by ring-opening polymerization and 'click' chemistry. *Macromolecular Rapid Communications*, *29*, 982–997.
28. Meldal, M. (2008). Polymer "clicking" by CuAAC reactions. *Macromolecular Rapid Communications*, *29*, 1016–1051.
29. Fournier, D., Hoogenboom, R., Schubert, U. S. (2007). Clicking polymers: A straightforward approach to novel macromolecular Architectures. *Chemical Society Reviews*, *36*, 1369–1380.
30. Binder, W. H., Sachsenhofer, R. (2007). 'Click' chemistry in polymer and materials science. *Macromolecular Rapid Communications*, *28*, 15–54.
31. Lutz, J.-F. (2007). 1,3-dipolar cycloadditions of azides and alkynes: A universal ligation tool in polymer and materials science. *Angewandte Chemie International Edition*, *46*, 1018–1025.
32. Evans, R. A. (2007). The rise of azide–alkyne 1,3-dipolar 'click' cycloaddition and its application to polymer science and surface modification. *Australian Journal of Chemistry*, *60*, 384–395.
33. Iha, R. K., Wooley, K. L., Nystrom, A. M., Burke, D. J., Kade, M. J., Hawker, C. J. (2009). Applications of orthogonal "click" chemistries in the synthesis of functional soft materials. *Chemical Reviews*, *109*, 5620–5686.

34. Thomassin, J.-M., Lenoir, S., Riga, J., Jérôme, R., Detrembleur, C. (2007). Grafting of poly[2-(tert-butylamino)ethyl methacrylate] onto polypropylene by reactive blending and antibacterial activity of the copolymer. *Biomacromolecules, 8*, 1171–1177.
35. Parrish, B., Breitenkamp, R., Emrick, T. (2005). PEG- and peptide-grafted aliphatic polyesters by click chemistry. *Journal of the American Chemical Society, 127*, 7404–7410.
36. Riva, R., Schmeits, S., Stoffelbach, F., Jérôme, C., Jérôme, R., Lecomte, P. (2005). Combination of ring-opening polymerization and "click chemistry" towards functionalization of aliphatic polyesters. *Chemical Communications*, 5334–5336.
37. Riva, R., Jérôme, C., Jérôme, R., Lecomte, P. (2007). Combination of ring-opening polymerization and "click chemistry" towards functionalization and grafting of poly(ε-caprolactone). *Macromolecules, 40*, 796–803.
38. Parrish, B., Emrick, T. (2002). Functional polyesters prepared by polymerization of α-allyl(valerolactone) and its copolymerization with ε-caprolactone and δ-valerolactone. *Journal of Polymer Science, Part A: Polymer Chemistry, 40*, 1983–1990.
39. Riva, R., Lussis, P., Lenoir, S., Jérôme, C., Jérôme, R., Lecomte, P. (2008). Contribution of "click chemistry" to the synthesis of antimicrobial aliphatic copolyester. *Polymer, 49*, 2023–2028.
40. Lenoir, S., Riva, R., Lou, X., Detrembleur, C., Jérôme, R., Lecomte, P. (2004). Ring-opening polymerization of α-chloro-ε-caprolactone and chemical modification of poly(α-chloro-ε-caprolactone) by atom transfer radical processes. *Macromolecules, 37*, 4055–1061.
41. Charlot, A., Sciannaméa, V., Lenoir, S., Faure, E., Jérôme, R., Jérôme, C., Van De Weerdt, C., Martiall, J., Archambeau, C., Willet, N., Duwez, A.-S., Fustin, C.-A., Detrembleur, C. (2009). All-in-one strategy for the fabrication of antimicrobial biomimetic films on stainless steel. *Journal of Materials Chemistry, 19*, 4119–4125.
42. Riva, R., Chafaqi, L., Jérôme, R., Lecomte, P. (2007). Synthesis of new substituted lactones by "click" chemistry. *Arkivoc, 10*, 292–306.
43. Detrembleur, C., Mazza, M., Halleux, O., Lecomte, P., Mecerreyes, D., Hedrick, J.L., Jérôme, R. (2000). Ring-Opening Polymerization of γ-bromo-ε-caprolactone: A novel route to functionalized aliphatic polyesters. *Macromolecules, 33*, 14–18.
44. Billiet, L., Fournier, D., Du Prez, F. (2008). Combining "click" chemistry and step-growth polymerization for the generation of highly functionalized polyesters. *Journal of Polymer Science, Part A: Polymer Chemistry, 46*, 6552–6554.
45. Zednik, J., Riva, R., Lussis, L., Jérôme, C., Jérôme, R., Lecomte, P. (2008). H-responsive amphiphilic networks. *Polymer, 49*, 697–702.
46. Lussis, P. (2006). Synthèse de nouveaux polyesters antibactériens et biodégradables par cycloaddition de Huisgen. Master Thesis, University of Liège, Wallonia, Belgium.
47. Grignard, B., Schmeits, S., Riva, R., Detrembleur, C., Lecomte, P., Jérôme, C. (2009). First example of click copper(I) catalyzed azide-alkyne cycloaddition in supercritical carbon dioxide: application to the functionalization of aliphatic polyesters. *Green Chemistry, 11*, 1525–1529.
48. Bonami, L., Van Camp, W., Van Rijckegem, D., Du Prez, F. E. (2009). Facile access to an efficient solid-supported click catalyst system based on poly(ethyleneimine). *Macromolecular Rapid Communications, 30*, 34–38.
49. Agard, N. J., Baskin, J. M., Prescher, J. A., Lo, A., Bertozzi, C. R. (2006). *ACS Chemical Biology, 1*, 644–648.

50. Lutz, J.-F. (2008). Copper-free azide-alkyne cycloadditions: new insights and perspectives. *Angewandte Chemie International Edition*, *47*, 2182–2184.
51. Becer, C. R., Hoogenboom, R., Schubert, U. S. (2009). Click chemistry beyond metal-catalyzed cycloaddition. *Angewandte Chemie International Edition*, *48*, 4900–4908.
52. Lutz, J.v-F., Schlaad, H. (2008). Modular chemical tools for advanced macromolecular engineering. *Polymer*, *49*, 817–824.
53. Lowe, A.vB. (2010). Thiol-ene "click" reactions and recent applications in polymer and materials synthesis. *Polymer Chemistry*, *1*, 17–36.
54. Hoogenboom, R. (2010). Thiol-yne chemistry: a powerful tool for creating highly functional materials. *Angewandte Chemie International Edition*, *49*, 3415–3417.
55. Singh, I., Zarafshani, Z., Lutz, J.-F., Heaney, F. (2009). Metal-free "click" chemistry: efficient polymer modification via 1,3-dipolar cycloaddition of nitrile oxides and alkynes. *Macromolecules*, *42*, 5411–5413.

CHAPTER 4

CHITOSAN AND CHITOSAN BLENDS AS ANTIMICROBIALS

PATRICIA FERNANDEZ-SAIZ
Institute of Agrochemistry and Food Technology (IATA), CSIC, Burjassot, Spain

CONTENTS

4.1 General Properties of Chitosan . . . 72
4.1.1 Waste/Effluent Water Purification . . . 72
4.1.2 Cosmetics . . . 73
4.1.3 Fat Trapper Agent . . . 73
4.1.4 Pharmaceutical and Biomedical Applications: Controlled Drug Release, Tissue Engineering . . . 73
4.1.5 Agriculture . . . 73
4.1.6 Biosensors—Industrial Membrane Bioreactors . . . 74
4.1.7 Control of Enzymatic Browning in Fruits . . . 74
4.1.8 Clarification and Deacidification of Fruit Juices . . . 74
4.1.9 Improvement of the Emulsification in Mayonnaise Preparation . . . 75
4.1.10 Antioxidant Activity in Meat and Fish . . . 75
4.2 Limitations of Chitosan . . . 75
4.3 Understanding the Biocide Properties of Chitosan Films . . . 77
4.4 Designing Water-Resistant Chitosan-Based Films . . . 78
4.5 Active Antimicrobial Packaging Based on Chitosan . . . 81
4.5.1 On the Effect of Chitosan Formulation . . . 81
4.5.2 On the Effect of the Cells Parameters . . . 83
4.5.3 On the Effect of Culture Conditions . . . 85

Antimicrobial Polymers, First Edition.
Edited by José M. Lagarón, María J. Ocio, and Amparo López-Rubio.

4.5.4 On the Effect of Film-Forming and Storage Conditions of Chitosan Films . . . 86
4.5.5 On the Effect of Chitosan onto Real Food Systems . . . 87
4.6 Chitosan-Based Systems in Biomedical Applications . . . 88
4.6.1 Tissue Engineering . . . 89
4.6.2 Wound Dressing . . . 89
4.6.3 Cancer Diagnosis . . . 90
4.7 Future Perspectives . . . 90
References . . . 91

4.1 GENERAL PROPERTIES OF CHITOSAN

Chitosan (β-(1,4)-2amino-2-deoxy-D-glucose) is a biodegradable, biocompatible, nontoxic aminopolysaccharide obtained by deacetylation of chitin, which is one of the most abundant biopolymer in the world. This biopolymer shows excellent film- and coating-forming properties when cast from organic acidic water solutions [1–10]. Moreover, it presents an inherent antimicrobial character against the growth of pathogen microorganisms, which has already been widely demonstrated in a general range of foods such as bread, strawberries, juices, mayonnaise, milk, rice cake, etc. [11–16]. Although the exact mechanism by which chitosan exerts its antimicrobial activity is currently unknown, it has been suggested that the polycationic nature of this biopolymer that forms from acidic solutions below pH 6.5 is a crucial factor. Thus, it has been proposed that the positively charged amino groups of the glucosamine units interact with negatively charged components in microbial cell membranes altering their barrier properties, thereby preventing the entry of nutrients or causing the leakage of intracellular contents [17–21]. Another reported mechanism involves the penetration of low-molecular-weight chitosan in the cell, the binding to DNA, and the subsequent inhibition of RNA and protein synthesis [22]. Chitosan has been shown also to activate several defense processes in plant tissues, and it inhibits the production of toxins and microbial growth because of its ability to chelate metal ions [11, 23].

Because of its diversified range of applications, chitosan is a biomolecule with great potential. Although the previously mentioned antimicrobial capacity of chitosan has been the most interesting application in recent years, several possibilities may be realized for value addition to this basic raw material by suitable chemical or enzymatic modification, polymer grafting, or selective depolymerization. Some of the actual uses of chitosan are reported as follows.

4.1.1 Waste/Effluent Water Purification

Treatment of wastewaters is one the major current applications of chitin/chitosan-based formulations because of their excellent coagulating, flocculating and metal chelating properties. An awareness of the ecological and health problems associated with heavy metals and pesticides and their accumulation through the food chain prompted the need of rigorous purification of industrial water [24, 25]. The ability of the free NH_2 groups of chitosan to form coordinate/covalent bonds with metal ions is of great interest [26]. Thus, chitosan powder and/or chitosan-dried films are of considerable use in metal ion complexing because chitosan presents most of its free amino groups above the pKa value ($\approx$6.5). The use of chitosan for potable water purification has been approved by the United States Environmental Protection agency up to a maximum level of 10 ppm [24]. Chitosan, carboxymethyl chitosan, and cross-linked chitosan have been shown to be effective in removing Cd^{+2}, Cu^{+2}, Hg^{+2}, Ni^{+2}, and Zn^{+2} from wastewater and industrial effluent [27, 28].

4.1.2 Cosmetics

Because chitosan becomes viscous on being solubilized in acid, it is often used in creams, lotions, and nail lacquers. Depolymerized chitosan is being used as active ingredient in products such as hair shampoo and conditioner because its aqueous solutions are viscous, retain moisture, and import softness to hair and skin.

4.1.3 Fat Trapper Agent

Because of the ability of forming ionic bonds at low pH, chitosan can bind *in vitro* to different types of anions such as bile acids or free fatty acids [29]. Large proportions of these bound lipids are thus excreted. Inside the digestive tract, chitosan forms micelles with cholesterol in the alkaline fluids in the upper part of the intestine, resulting in the depression of absorption of dietary cholesterol and circulation of cholic acid to the liver. Because of the formation of cholic acid from the blood cholesterol in the liver, it tends to decrease blood cholesterol concentration. Deuchi et al. [30] proposed that chitosan is solubilized in the stomach to form an emulsion with intragastric oil droplets and begins to precipitate in the small intestine at pH 6–6.5. With the aggregation of polysaccharide chains, the oil droplets are entrapped in their matrices thereby passing through the lumen and emptying into feces.

4.1.4 Pharmaceutical and Biomedical Applications: Controlled Drug Release, Tissue Engineering

Chitosan has been used in the pharmaceutical industry for a variety of biomedical applications. These particular uses are discussed deeply in a subsequent section.

4.1.5 Agriculture

Chitosan could also be employed as a coating of fertilizers, pesticides, herbicides, nematocides, and insecticides for their controlled release to soil to reduce environmental damage caused by excessive use of these agrochemicals. Coating the seeds and leaves with chitosan films could prevent microbial infections and induce genes in plants that help them to resist diseases.

4.1.6 Biosensors—Industrial Membrane Bioreactors

Enzyme immobilization is a method to keep enzyme molecules confined in a distinct phase separated from the bulk phase while allowing the exchange between these two phases. The immobilization of penicillin G acylase on different physical forms of chitosan, namely beads, particles, and powder was studied by Braun et al. [31], who observed activity retention of 40%, 93%, and 100%, respectively. Another study by Siso et al. [32] demonstrated that microencapsulation in chitosan beads was an effective enzyme immobilization method for invertase and α-amylase. In general, chitosan gels as enzyme immobilization supports could offer great potential for a multiplicity of applications ranging from biosensors for both *in situ* measurements of environmental pollutants and metabolite control in artificial organs. In addition, chitosan has been used in wine, sugar, and fish industries to remove organic contaminants from wastewaters.

4.1.7 Control of Enzymatic Browning in Fruits

Coating fruits with semipermeable films has generally been shown to retard ripening by modifying the endogenous CO_2, O_2, and ethylene levels of fruits. Chitosan coating is likely to modify the internal atmosphere without causing anaerobic respiration because chitosan films are more selectively permeable to O_2, than to CO_2. Therefore, chitosan coating with its ability to modify internal atmosphere in the tissue and fungistatic property has a potential to prolong storage life and control decay of fruits. Furthermore, edible coatings can be used as a vehicle for incorporating functional ingredients such as antioxidants, flavors, colors, antimicrobial agents, and nutraceuticals. Thus, several workers have endeavored to incorporate calcium, vitamin E, or oleic acid into chitosan film formulation to prolong the shelf life and to improve the nutritional value of fruits [33–36].

4.1.8 Clarification and Deacidification of Fruit Juices

The processing of clarified fruit juices commonly involves the use of clarifying aids, including gelatin, bentonite, silica sol, tannins, polyvinylpyrrolidone, or combinations of these compounds [37]. Chitosan with a partial positive charge has been shown to possess acid-binding properties [38] and to be effective in aiding the separation of colloidal and dispersed particles from food processing wastes. Clarification of fruit juices with chitosan has been attempted by Soto-Peralta et al. [37], Chatterjee et al. [39], and Rungsardthong et al. [40]. Chitosan (2% dissolved in water) was more effective in reducing the turbidity of juices than bentonite and gelatin. The appearance and acceptability of the juices, after chitosan treatment, could be significantly increased and the turbidity reduced. Chitosan may also be used to control acidity in fruit juices because of its acid-binding properties. The effects of chitosan treatment on the reduction of titratable acidity provide a potential for acidity control in other food systems. Enzymatic browning in apple and pear juices can be prevented by chitosan treatments probably because of the ability of the positively charged polymer to coagulate suspended solids to which polyphenol oxidase is bound.

4.1.9 Improvement of the Emulsification in Mayonnaise Preparation

Unlike other polysaccharides, chitosan possesses a positive ionic charge and has both reactive amino and hydroxyl groups, which give it the ability to bond chemically with negatively charged protein. When pH is less than 6.5, chitosan solution carries a positive charge along its backbone. Because of its polar groups, chitosan also provides additional stabilization from hydration forces [41]. According to Filar and Wirick [42], chitosan functions only in acid systems to show possible utility as a thickener and stabilizer.

4.1.10 Antioxidant Activity in Meat and Fish

Meat and meat products are highly susceptible to lipid oxidation, which leads to the rapid development of rancid of rancid or warmed-over flavor. Chitosan possesses antioxidant and antibacterial capacity [43] and may retard the lipid oxidation and inhibit the growth of spoilage bacteria in meat during storage. Furthermore, Kanatt et al. [44] investigated the use of irradiated chitosan as a natural antioxidant to minimize lipid peroxidation of radiation-processed lamb meat. Irradiation of chitosan at 25-kGy dose of gamma radiation resulted in a 6-fold increase in its antioxidant activity as compared with the nonirradiated chitosan, as measured by 1,1-diphenyl-2-picrylhydrazyl scavenging activity.

Seafood products are highly susceptible to quality deterioration because of the lipid oxidation of unsaturated fatty acids, catalyzed by the presence of high concentrations of hematin compounds and metal ions in the fish muscle [45]. Furthermore, seafood quality is highly influenced by autolysis, contamination by the growth of microorganisms, and loss of protein functionality [46]. Several studies indicate the antioxidant capacity of chitosan added to the fish muscle, which depends on its molecular weight and concentration. Chitosan may retard lipid oxidation by chelating ferrous ions present in the fish model system, thus eliminating prooxidant activity of ferrous ions or preventing their conversion to ferric ion. Also, coating with chitosan was effective in reducing about 50% moisture loss of salmon fillets compared with the control noncoated fillets, and in delaying lipid oxidation [47].

4.2 LIMITATIONS OF CHITOSAN

Although there is no doubt that chitosan can be effectively used in a wide variety of applications as mentioned in the previous section, individual users employ chitosan with varying physicochemical properties and perhaps from various sources. Thus, the question arises as to how to produce chitosan globally with consistent properties. Each batch of chitosan produced from the same manufacturer may differ in its quality. For proper quality control in the chitosan production, there is a critical need to establish less expensive and reliable analytical methods, especially for the evaluation of the molecular weight and degree of deacetylation [48].

Furthermore, the traditional chitosan production involves deproteinization, demineralization, decolorization, and deacetylation. The physicochemical characteristics of chitosan can be variously affected by the production methods and the crustacean species (that is, molecular weight, viscosity, degree of deacetylation, hydrophilicity, water, and fat absorption capacities, and so on). Consequently, these final physicochemical characteristics may affect effectiveness of chitosan as, for instance, food preservative or edible coating materials [48].

Moreover, the classic chitosan production process is costly, thus limiting its applications. Simplification of chitosan production, for example, by the elimination of deproteinization and/or demineralization, or by the reduction of reaction time required for deproteinization and demineralization, would considerably reduce production cost because of a reduction in chemical usage, process time, and voluminous wastewater discharge. Recently, No et al. [49] prepared various chitosan types by using a simplified process and used them as coating materials effectively to extend the shelf life of eggs.

Regarding food applications, the characteristic astringent/bitterness taste of chitosan currently limits its use, to some extent, as a food additive or preservative. The incorporation of L-arginine and adenosine monophosphate, both considered as generally recognized as safe in the United States, may be used to mask or minimize this effect, and should be further investigated [48].

In addition, the numerous research works mentioned in this chapter have been carried out in a laboratory scale. As a consequence, additional research on, for instance, the quality and shelf-life extension of foods coated or packaged with chitosan-based materials should be conducted on a pilot plant and on a industrial scale to ascertain the commercial value and the scaling-up difficulties of the use of this interesting biopolymer or of its nanobiocomposites. In any case, in Europe, chitosan is still not permitted as food contact material, although this may change in the near future. Nevertheless, chitosan does surely have significant opportunities for implementation as a natural antimicrobial packaging material under the new EU regulation for active packaging (Commission Regulation (EC) No. 450/2009).

One major drawback of chitosan film is its high sensitivity to humidity; thus, it may not be appropriate for use when it is in direct contact with foods and/or for direct handling. In addition, the effects of chitosan production protocols, film-casting solvents, and plasticizer contents on sorption behavior of chitosan films should be investigated. Some published works, which will be mentioned in a subsequent section, have been focused to develop antimicrobial chitosan films that are less sensitive to humidity.

Similarly and regarding biomedical applications, chitosan lacks the mechanical properties required for regeneration of many tissues. For example, the tensile strength of an articular cartilage is approximately 27 MPa [50], whereas chitosan has shown a strength of only 5–7 MPa in the wet state [51]. To overcome these limitations, chitosan has been combined with several synthetic polymers that could extend mechanical integrity to the matrix. For example, the addition of poly(vinylalcohol) increased the tensile strength of dry chitosan membranes from 70 to 220 MPa and their extension from 4% to 11% [52]. In making hydrogels for drug delivery, the primary concern is to influence the network structure, pH-dependent swelling behavior (synonymous with hydrophilicity), and degradation kinetics of chitosan gels [53]. In wound dressings, enhancing the mechanical properties and antimicrobial nature of chitosan films may also be the main objective. In tissue engineering, the control of biomechanical and degradation properties of chitosan is important because of the wide range of properties exhibited by natural tissues. For example, soft tissues such as skin have a strength of 7 MPa, whereas a cortical bone has strength of 133 MPa, measured in the longitudinal direction [50]. Similarly, the spreading and proliferation kinetics of fibroblasts are different from that of chondrocytes.

Mechanical, barrier, and degradation properties of chitosan could also be enhanced through the incorporation of nanoclays. This issue has also been discussed in a subsequent section.

4.3 UNDERSTANDING THE BIOCIDE PROPERTIES OF CHITOSAN FILMS

While there is no doubt that chitosan has shown a tremendous potential as antimicrobial component in a number of cases, a clear understanding between the phenomenology of the chitosan biopolymer as a biocide agent and its molecular structure, functional chemistry and physical state for optimum performance is of paramount importance from an applied view point. attenuated total reflection Fourier transformed infrared (ATR-FTIR) spectroscopy has become one of the most versatile techniques for the quick identification and characterization of substances. The mechanism of antimicrobial action of chitosan has been studied by this technique through the intensity of the carboxylate biocide groups ($-NH_3^+ \; ^-OOCH$) at 1546 and at 1405 cm^{-1} in various chitosan samples [4,5]. The authors demonstrated that high-molecular-weight chitosan seems to act only in its acetate (salt) form and as carrier of the activated protonated species, which by passing from the film to the microbial solution lead to the inactivation of certain microorganisms. Moreover, a novel methodology based on the use of a normalized infrared band centered at 1405 cm^{-1} was proposed that can be correlated with the antimicrobial character of the biopolymer.

In relation to the feasible mechanisms by which chitosonium acetate decreases the number of "active" protonated groups, Lagaron et al. [5] argued two feasible mechanisms. First, the biocide carboxylate groups that form when chitosan is cast from acetic acid solutions are being continuously evaporated from the formed film in the form of acetic acid (mechanism I) in the presence of environmental humidity, rendering weak biocide film systems. Furthermore, after direct contact of the cast chitosonium acetate film with liquid water, water solutions or the high-moisture tryptone soy agar hydrogel, a positive rapid migration, with diffusion coefficient faster than $3.7\ 10^{-12}\,m^2/s$, of protonated glucosamine water soluble molecular fractions (mechanisms II) takes place from the film into the liquid phase, yielding strong antimicrobial performance and leaving in the remaining cast film only the non-water-soluble chitosan fractions.

In a more recent work from the same research group, it has been showed that only dissolved polysaccharide molecules from the gliadin-chitosonium acetate blends acted as the antimicrobial agent [6]. This finding is in accordance with the work performed by Zhai et al. [54], who evaluated the antimicrobial capacity of starch/chitosan blends. In this particular case, the films presented significant biocide properties even with a low content of chitosan but uniquely when formed under the action of irradiation. The degradation of chitosan by the irradiation treatment improved its dissolution into the nutrient broth, enabling a detectable bactericidal effect. These results seem to contradict, however, a previous study performed by Vartiainen et al. [55] on the antimicrobial capacity of chitosan immobilized in a bioriented polypropylene matrix, where chitosan was reported to preserve a strong bactericidal activity without any notable leaching. This finding was probably obtained because bacteria were directly exposed over the film, and therefore, no migration was required for the microbes to be exposed. Tanabe et al. [56] published that the antimicrobial capacity of chitosan was probably a result of the irreversible adsorption of bacteria into the film. Despite the above arguments, there is a large body of experimental evidence that suggests that chitosan acts more efficiently from its solution form by precipitation onto, or even by penetration through, the microbial membrane [4, 19, 57, 58].

4.4 DESIGNING WATER-RESISTANT CHITOSAN-BASED FILMS

In the design of a chitosan-based packaging system for food packaging applications, the presence of the high levels of humidity typically found in many foods has to be taken into account. In this sense, it has been found that the development of a chitosan active formulation in the food contact layer as a coating leads to both a rapid release of the biocide groups onto the food surface and the film partial or total dissolution. By means of physical or chemical treatments, film disintegration in water could be prevented, but at the same time, dissolution of the protonated glucosamine groups in water would be blocked and deactivated, hence reducing the biocide character of the formulation. In this sense, the incorporation of cross-linking agents that form covalent bonds between the chains of chitosan preventing polysaccharide–water interactions could be a means to hold the film structure. Previous studies in relation to this matter have already been performed by the incorporation of compounds such as glutaraldehyde, glyoxal, or epichlorohydrin [59–62]. Of additional concern is, of course, the inherent toxicity of these cross-linking agents that can restrict their use. Polymer surface modifications, such as plasma treatments, have been also exploited to improve the adhesion between chitosan and other matrices like polypropylene [55, 63–66]. Another strategy to overcome the drawback of film dissolution is to blend chitosan with a more moisture-resistant polymer to obtain water-resistant chitosan-based antimicrobial films. Some researchers have included chitosan into other biopolymer matrices such as cellulose, starch, or Konjac glucomannan to improve the mechanical and water-swelling properties of antimicrobial chitosan films [54, 67–73]. Nevertheless, many published studies on this matter generated biocomposite materials with optimum antimicrobial activity but poor water barrier properties.

A recent work performed by Fernandez-Saiz et al. [6] showed an inhibitory effect of the growth of *Staphylococcus aureus* when a composite matrix gliadin/chitosonium acetate (60/40% wt/wt) was tested, maintaining the film integrity in the nutrient broth. Nevertheless, although film disintegration was prevented, some turbidity in the nutrient medium was detected. Indeed, the gliadins protein network must exert a blocking effect against the dissolution of biocide chitosonium acetate chains, albeit some of the latter chains do still migrate to exert their required antimicrobial role. Similar findings were observed by Tanabe et al. [56], who studied keratin/chitosan films and observed a significant reduction of the swelling ratio as well as a significant reduction of *Escherichia coli* (61%) for the ratio 5:1 (keratin:chitosan). Other published works studied the water barrier and the antimicrobial activity of high- and low-molecular-weight chitosonium-acetate-based solvent-cast blends with ethylene-vinyl alcohol (EVOH) copolymers against *S. aureus* and *Salmonella spp* [9]. These samples showed excellent antimicrobial activity as well as enhanced water barrier when low-molecular-weight chitosan was used as the dispersed phase in the blend, specifically 80/20% (wt/wt) EVOH/chitosan.

In view of the results of the mentioned works, the most appropriate ratio between chitosan and the water-resistant polymer will depend on their compatibility. Also, it could be concluded that chitosan-based blends are shown to be a suitable means to solve the excessive hydrophilic character of chitosan while retaining its biocide performance.

Another possible approach to modify biopolymer film properties is to make hybrid films with organic polymers and nanosized clay minerals such as layered silicates, which are known as nanocomposite films [74–77]. Nanocomposite films consisting of inorganic nanolayers of layered silicate, such as montmorillonite (MMT) clay and organic polymers have recently evoked intense research interest in the material and polymer science areas [77–80]. Usually, polymer/clay nanocomposites comprise an organic/inorganic hybrid polymer matrix containing platelet-shaped clay particles that have sizes in the order of a few nanometers thick and several hundred nanometers long. Partly because of their high aspect ratios and high surface area, the clay particles, if properly dispersed in the polymer matrix at a loading level of 1–5 wt%, impart unique combinations of physical and chemical properties that make these nanocomposites attractive for making films and coatings for a variety of industrial applications, such as food packaging. Examples of such property enhancements include decreased permeability to gases and liquids, better resistance to solvents, increased thermal stability, and improved mechanical properties [77, 78]. Moreover, biodegradability is retained; that is, after final degradation, only inorganic, natural minerals (actually soil) will be left [81–83].

Because chitosan is a polycation in acid conditions, it can be easily adsorbed on the MMT-Na surface. This property has been extensively used to elaborate chitosan/MMTNa hybrid materials by the solvent route. Solutions were prepared by chitosan addition into 1% or 2% (v/v) of acetic acid solution. MMT-Na was dispersed into water to obtain a 2 wt% clay suspension. To avoid any structural alteration of the phyllosilicate structure, the polysaccharide solution was adjusted with NaOH to pH 4.9 and then was slowly added to the clay suspension at ambient temperature. This mixture was stirred and finally washed with purified water to remove acetate and then casted [84–88]. Darder et al. [84] demonstrated chitosan intercalation thanks to a shift of the MMT-Na diffraction peak to lower angles. Moreover, a broadening and intensity decrease in the diffraction peak was observed, indicating a disordered intercalated/exfoliated structure [85, 86]. The interlayer chitosan structure was studied by Darder et al. [84, 85] evidencing the adsorption of two chitosan layers on the clay surface and even inside the interlayer spacing. The first chitosan layer was mainly adsorbed thanks to electrostatic interactions between the chitosan $-NH3+$ groups and the MMT negative charges. The second-layer adsorption was promoted by hydrogen bonds established between the chitosan amino and –OH groups and the clay substrate. At low MMT content, several authors have even shown the formation of an exfoliated nanostructure [86–90]. At higher MMT content, namely more than 5 wt%, the formation of intercalated/flocculated structure was observed [86].

The thermal transitions of these nanobiocomposites were investigated and related to the material dispersion state. Günister et al. [88] measured the effect of the nanofiller addition on the Tg (by differential scanning calorimetry) and observed an increase directly linked to the ionic interactions established between the chitosan and the nanofiller, which reduced the chains mobility.

Furthermore, the increases in the tensile strength correlated to a small decrease in the strain at break were observed in the different chitosan nanobiocomposite [87]. These raises were induced by the nanofillers/chitosan interactions, which enhance the stress transfer at the interface. The strain at break decrease was related to the morphology of the chitosan/MMT hybrid materials, which displayed, in the best case, an intercalated/exfoliated structure. Such a stiffness increase is already well reported into the literature and is correlated to the clay rigidity and dispersion state [91].This property has been extensively used to elaborate chitosan/MMTNa hybrid materials by solvent route [84–88].

In a work performed by Rhim et al. [92], four different types of chitosan-based nanocomposite films were developed, and their film properties and antimicrobial functions were tested. By compositing with nanoparticles (such as unmodified and organically modified montmorillonites, nanosilver, and silver zeolite), mechanical and water vapor barrier properties of chitosan films were increased significantly ($P < 0.05$) compared with those of control chitosan films, and various degrees of antimicrobial activity were observed depending on the nanoparticles used.

In another work performed by Lagaron and Fendler [83], two commercially available clay additives were included in methyl cellulose and chitosan matrixes by a solution casting process. In this study, the water barrier properties of the nanobiocomposites were found to be enhanced to a significant extent compared with the pure matrix materials without losing biodegradability as mentioned previously. The nanoclay was also found to reduce the water solubility significantly in the polymer hence resulting in increased water resistance.

4.5 ACTIVE ANTIMICROBIAL PACKAGING BASED ON CHITOSAN

4.5.1 On the Effect of Chitosan Formulation

Several works have been published regarding the influence of the molecular weight of chitosan on its antimicrobial properties [93–99]. Some of them have demonstrated that chitooligosaccharides (COS), which are soluble in water, were the least effective in terms of biocide properties [97–99]. Moreover, in a more recent work carried out by Qin et al. [58] on the evaluation of chitosan solutions against the growth of *Candida albicans*, *E. coli*, and *S. aureus*, it has been shown that only water insoluble chitosan in organic acidic solutions, i.e., chitosonium salts, exhibit efficient biocide properties. In contrast, a research performed by Fernandes et al. [100] showed that the growth of *E. coli* was markedly inhibited by COS, and this inhibition decreased slightly as molecular weight increased. In another work performed by Fernandez-Saiz et al. [7], changes in molecular weight of the chitosan materials tested, i.e., 310–375 KDa and 50–190 KDa, did not lead to significant variations in biocide properties. Therefore, despite the many reports on the topic, the reported conclusions are divergent probably because the methodology employed for the susceptibility tests varies from work to work. Thus, Jeon et al. [93] and Liu et al. [96] showed that the biocide effects increased with molecular weight whereas Liu et al. [95] found the contrary trend. In these works, the bacterial growth was evaluated by optical density. By this method, important differences on turbidity do not necessarily correspond to significant differences in bacterial counts and real effects may be masked. Besides, published works are not all comparable as individual researchers used different types of chitosan with varying physicochemical properties (affected by production methods) and perhaps from various sources [48].

Concerning the degree of deacetylation (DD), several works consider this feature and the antimicrobial properties of chitosan undoubtedly increased with this variable. The antimicrobial performance tends to increase during an increase in the DD of chitosan, which is related with an increase in the positive charge of the polymer [47, 101–103].

Although several studies report the influence of the organic solvent employed to obtain chitosan films on their physical or antimicrobial properties [3, 104, 105], the effect of the solvent concentration to obtain chitosan dispersions has been scarcely studied in the literature. A recent work compared the biocide properties of films of chitosonium acetate obtained from 0.5 wt% and 1 wt% of acetic acid in the solution [7]. Although no statistical differences were obtained between samples, a certain trend could also be observed in which chitosan dispersions prepared from an acid-water solution with 1% acetic acid led to matrices with superior antimicrobial properties. A feasible explanation may be the fact that solubilization of chitosan is influenced by acetic acid concentration in the solution. In this sense, a previous work performed by Rinaudo et al. [106] demonstrated through potentiometric assays that the complete solubilization of chitosan in acidic water caused by a maximum degree of protonation occurs with a stochiometry $[AcOH]/[Chit\text{-}NH_2] = 0.6$. Such a ratio corresponds to the one used in the mentioned work, i.e., for chitosan dispersions of 1.5 wt% polymer and 1 wt% acetic acid.

The antimicrobial performance of chitosan films after a neutralization process has also been studied by some authors. For this purpose, chitosan films were, for instance, washed in a 0.1M NaOH solution for 30 min, subsequently rinsed thoroughly with distilled water, and redried at 37 °C prior to the susceptibility test. A considerable drop in the intensity of the active antimicrobial bands after the neutralization treatment was demonstrated by ATR-FTIR [4]. Indeed, the molecular structure of the neutralized film was similar than that of chitosan as received. Moreover, neutralization results in the loss of biopolymer solubility in aqueous media. As expected, when the biocide properties of a neutralized chitosan film were studied no antimicrobial activity was observed, indicating that amine groups are no longer activated by titration with an alkaline component [4]. These preceding findings are also in conformity with the results obtained by Ouattara et al. [107], who showed the lack of biocide properties of neutralized chitosan films when applied onto the surface of processed meat. In apparent opposition to these results, Tang et al. [59] obtained a bactericidal effect toward *S. aureus* and *E. coli* when biocide properties for suspensions of both neutralized chitosan films and neutralized crosslinked chitosan films were measured. Among some feasible reasons for this behavior could be either the incomplete neutralization process of the tested films or a protonation reaction during the preparation of the studied film suspensions because the solvent composition to obtain the suspensions was not specified in the work.

For chitosan matrices designed to be applied as food contact materials, in many cases, the humid heat sterilization process becomes necessary. Even though the sterilization of the chitosan solution showed no significant impact on the biocide properties of the cast material, the antimicrobial activity of just formed autoclaved films (with and without steam contact) demonstrated that the sterilized films lost their biocide properties by this treatment [7]. Both dry and humid heat sterilization processes of a cast film did cause an increase in yellowness or even an important browning effect, respectively, probably because of the formation of double bonds or colored complexes [108, 109]. These chemical changes are thought to be responsible for the insoluble character of the annealed samples.

Fernandez-Saiz et al. [7] analyzed the influence of the thickness of the chitosan matrix in its biocide properties and showed no significant differences in the films ranging from 25 to 100 μm. When the ATR-FTIR spectra were obtained for these particular samples, only minor differences were obtained between the different samples in terms of the intensity of the active carboxylate active bands. These results suggest that the amount of protonated groups in the final film does not seem to depend on the thickness for the range evaluated in this particular work. Thus, even in thicker chitosan films, because of their hydrophilic character, water diffuses very rapidly into the film and an immediate release of a significant fraction of the carboxylate groups is produced.

4.5.2 On the Effect of the Cells Parameters

An important issue is the lack of proper standardized methods to determine the effectiveness of antimicrobial polymers. According to the standard method described by the *National Committee for Clinical Laboratory Standards* [110], minimal inhibitory concentration is the lowest concentration of an antimicrobial agent that inhibits visible growth, after a predetermine incubation period (usually 18 to 24 h). When Fernandez-Saiz et al. [7] studied the biocide properties of chitosonium acetate solutions or films, some turbidity was detected in the test tubes containing the nutrient broth even before bacterial inoculation. This additional turbidity causes an overestimation of bacterial final concentrations when calculated by optical density. Furthermore, important differences on turbidity do not necessarily correspond to significant differences in bacterial counts. For that reason, some previous published works on the antimicrobial capacity of chitosan films showed alterations or even a lack performance when evaluated by optical density [93, 95, 96, 100]. In other published works, the biocide properties of chitosan films [111, 112] and of chitosan solutions [113] were studied by the agar diffusion method with no satisfactory results, probably because no dissolution of chitosan took place through the agar. In contrast, several reports studying this biomaterial showed considerable antimicrobial capacity when tested by means of other techniques, such as macrodilution method and agar plate counts or microcalorimetry [4, 58, 99].

When the effect of the cell age on the antibacterial properties of chitosan was evaluated, it has been showed that the sensitivity of *S. aureus* seems to be greater when bacteria were inoculated in mid-log phase [7]. Liu et al. [95] obtained comparable results when testing chitosonium acetate solutions at different growth stages of *E. coli*. However, not all the published works agree with the mentioned examples. For instance, the preceding results do not agree with those obtained by Chen and Chou [114], who studied the effect of water-soluble lactose chitosan derivative against *S. aureus* growth in different physiological states, obtaining a greater susceptibility when late-log phase cells were tested. According to Tsai and Su [115], the susceptibility of *E. coli* cells to chitosan may depend on the cells' surface electronegativity, which progressively increases from the early exponential phase to the late exponential phase. At the beginning of the stationary phase, this parameter starts to decrease. Nevertheless, another study performed by the same authors [116] about the action of low-molecular-weight chitosan against the growth of *Bacillus cereus* showed that cells in stationary phase were less sensitive, suggesting that changes in surface electronegativity seem to be different depending on bacteria type.

When the influence of bacteria type was studied, the published results demonstrate that gram-positive bacteria are generally more sensitive to the polymer than gram-negative bacteria (see Figure 4.1). It is worth noting that whereas some antimicrobial agents (for instance, bacteriocins) seem not to act against gram-negative bacteria because of the resistance conferred by the outer membrane [117], chitosan has demonstrated to have a significant antimicrobial effect against the growth of gram-negative bacteria such as *Salmonella spp.* These results are in accordance with the hypothesis of an electrostatic

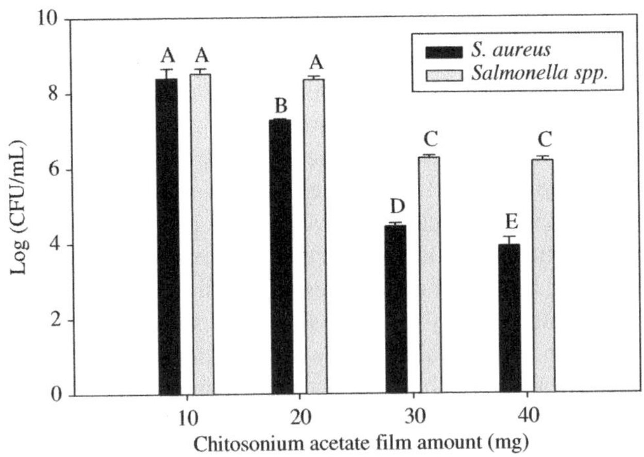

FIGURE 4.1 The antimicrobial capacity of chitosonium acetate films against *S. aureus* or *Salmonella spp* after an incubation for 24 h at 37 °C. The cells in the midexponential position with an initial inoculum size of approximately 10^5 CFU/mL were employed in all cases. Different letters indicate significantly different groups determined by Tukey's test ($P < 0.05$). The control samples without film showed a final bacterial concentration of approximately 8 $\log_{10}$ (CFU/mL).

interaction between chitosan and the cell wall. In this sense, gram-positive microorganisms should be more susceptible than gram negative because their wall is composed by a thick peptidoglycan layer and by polymers called teichoic acids. This teichoic acid backbone is highly charged by phosphate groups with negative charge, which could establish electrostatic interactions with cationic antimicrobial compounds such as the chitosan salts. In contrast, Chung et al.'s work [57] supported the major susceptibility of the gram-negative bacteria to chitosan on the reasoning that gram-negative bacteria possesses higher negative charge values on the cell surface.

4.5.3 On the Effect of Culture Conditions

The pH of the starting culture presented a large effect on the antimicrobial effectiveness of chitosan [7, 99]. Thus, the tests carried out at lower pHs (≤ 6.2) showed less bacterial counts than the performed at pH = 7.4, indicating a stronger biocide effect. This finding is related to the particular pK_a of this biopolymer (that is, 6.4). At this pH, the amount of positively charged amino groups (active groups) is close to 75%, whereas at pH of 7.4, this quantity decreases until approximately 10% [118]. Additionally, it has been published that at lower pH values the physiology of the cell is more prone to suffer damage. Thus, at lower pH conditions, the carboxyl and phosphate groups of the bacterial surface are anionic and offer potential sites for electrostatic binding of chitosan [18].

The influence of the incubation temperature has been also investigated to a large extent. Thus, in a recent study on the antimicrobial properties of medium-molecular-weight chitosan against *Salmonella enterica*, the antimicrobial agent presented strong antimicrobial properties at 10 °C, whereas the inhibition was almost inexistent at 20 °C [119]. In the same way, the addition of low-molecular-weight chitosan in cooked rice samples inhibited or retarded increases in total aerobic counts and *B. cereus;* this effect was more significant at a low temperature (i.e., 18 °C) [116]. Thus, at low temperatures, even if bacterial cells presented an optimal growth, they could be injured somehow, and these damages hampered their repair mechanisms in presence of chitosan, showing a higher susceptibility to the polymer at this condition. Nevertheless, the mentioned findings do not agree with a study performed by Chen and Chou [114], who observed that as the temperature was increased from 5 °C to 37 °C, the population of *S. aureus* in the chitosan derivative-containing deionized water decreased. Furthermore, It has been observed that *Listeria monocytogenes* showed to be more resistant to the biocide action of chitosan at 12 °C than at other studied temperatures (4 and 37 °C) [10]. The reason of such behavior could be that this microorganism is known to adapt to lower temperatures by producing phospholipids with shorter and more branched fatty acids. Also, the high susceptibility at 37 °C could be attributed to exhaustion caused by efforts to multiply at a growth promoting temperature [120]. Similarly, Briandet et al. [121] found that *L. monocytogenes* grown at 8 °C had significantly less surface electronegativity than cells grown at 15 °C, 20 °C, and 37 °C. They speculated that the reduced charge at 8 °C was the result of the synthesis of acclimation proteins and fewer carboxylic groups in the cell wall composition. In view of the mentioned results, the temperature of incubation seems to be an important factor that could define the antimicrobial performance of chitosan because bacterial cells are more or less susceptible to the polymer depending on this factor.

From a practical point of view, the antimicrobial effectiveness of a chitosan-based packaging in terms of microbiological safety will depend on the storage temperature of the specific food product (that is, 4 °C for refrigerated foods or 12 °C to 25 °C for nonrefrigerated foods). Thus, whereas chitosan could not be effective at 37 °C, it should present excellent biocide properties at 4 °C.

4.5.4 On the Effect of Film-Forming and Storage Conditions of Chitosan Films

From an application viewpoint, the effectiveness of an antimicrobial agent when applied in or coated over foods or packaging materials may deteriorate during film forming, distribution, and storage. Hence, the chemical stability of an incorporated substance is likely to be affected by the forming process or the storage conditions until its use. Consequently, to obtain highly efficient biocides, all these parameters should be assessed and optimized for the application. The properties of chitosan films obtained under specific forming and storage conditions in terms of mechanical and barrier properties have been investigated to some extent [108, 122–125]. Moreover, several published works have examined the chemical changes of chitosan salts during film forming and even at different storage conditions using infrared spectroscopy [4, 5, 108, 124, 126, 127].

The optimal procedure to preserve the full biocide capacity of chitosan-based films with minimum losses of the activated antimicrobial species in the materials has been thoroughly investigated [8]. Thus, it has been demonstrated that just-formed low-molecular-weight chitosonium acetate films presented significant biocide properties against *S. aureus* and *Salmonella* when cast at 37 or 80 °C, whereas this capacity was reduced when cast at 120 °C. In addition, the films preserved significant biocide properties when maintained at low temperatures and dry conditions (4, 23 °C, 0% RH) during the storage periods studied. In contrast, when the same films were maintained at high-relative-humidity conditions (i.e., 75%) or higher temperatures (i.e., 37 °C), the samples presented a progressive yellow coloration and a gradual loss of their antimicrobial capacity because of water resistance induced by chemical and/o physical alterations in the films. Larena and Caceres [128] also studied the color changes of chitosan films after a storage period and obtained a decrease in the membrane transmittance with temperature, suggesting nonenzymatic browning reactions, which led to the presence of conjugated double bonds in the structure of the polymer. Similar findings were reported by Srinivasa et al. [125] and by Kam et al. [124] after certain thermal treatments of the chitosan matrices. In good agreement with this behavior, it has been observed by ATR-FTIR spectroscopy that chitosonium acetate films cast at 120 °C or stored at 75% RH preserved a significant fraction of the biocide carboxylate chemistry after contact with water because of a strong reduction in cast film solubility [8].

4.5.5 On the Effect of Chitosan onto Real Food Systems

The effectiveness of this compound against food spoilage microorganisms has been tested in recent years to assess the potential of using chitosan as a natural preservative when applied as an edible coating. For instance, Tsai et al. [47] studied its application on fish preservation obtaining an increase of the shelf life of the product from 5 to 9 days. Other research performed by Sagoo et al. [129] evaluated the effect of treatments with chitosan solutions and reached an increase of the shelf life of raw sausages stored at chill temperatures from 7 to 15 days. A work on the effect of chitosan-gelatin blends as a coating for fish patties showed a difference of around two log cycles between the control and the coated batches for total bacterial counts, *Pseudomonas*, and enterobacteria at 8–11 days of storage [130]. The effects of the addition of low-molecular-weight chitosan on raw rice before steam cooking were also analyzed by Tsai et al. [116] who found an effective inhibition of the total anaerobic bacteria and *B. cereus* in cooked rice stored at 37 and 18 °C for 48 and 72 h, respectively. Other work performed by Juneja et al. [131] demonstrated that the incorporation of chitosan glutamate into beef or turkey may reduce the potential risk of *Clostridium perfringens* spore germination and development during abusive cooling from 54.4 °C to 7.2 °C in 12, 15, or 18 h. Vasconez et al. [132] recently demonstrated that a salmon slice coating assay showed that the chitosan solution was the best coating because aerobic mesophillic and psycrophilic cell counts were reduced, and pH and weight loss remained acceptable throughout refrigerated storage, extending global quality to 6 days. Nevertheless, in all previously mentioned works, chitosan was always applied as a food coating. Ye et al. [133] studied the control of *L. monocytogenes* on ham steaks by the use of chitosan-coated plastic films. In this research, the antimicrobial properties of chitosan became negligible when this polysaccharide was tested in the form of insoluble films because they seem to be incapable of diffusing through a solid medium such as agar. In contrast, Ouattara et al. [107] obtained satisfactory results when studying the inhibition of surface spoilage bacteria in processed meats by the application of chitosan-based films. A recent work demonstrated the biocide properties of chitosonium acetate film in fish soup [10]. In this study, the effects of the microorganism, the temperature of incubation, and the food substrate were analyzed. The results showed an increase in the lag phase of *L. monocytogenes* of approximately 7 days when 20-mg chitosan film were added into 10 mL of fish soup at 4 °C (see Figure 4.2).

Of additional concern is the loss of the antimicrobial capacity of chitosan or chitosan films when tested on real food systems, probably because of their irreversible binding by some of the negatively charged compounds in the food. This finding was also observed by Rhoades and Roller [134] when evaluated chitosan effects against the natural microflora in an apple-elderflower juice and in hummus. Fernandez-Saiz et al. [10] also observed a significant reduction of

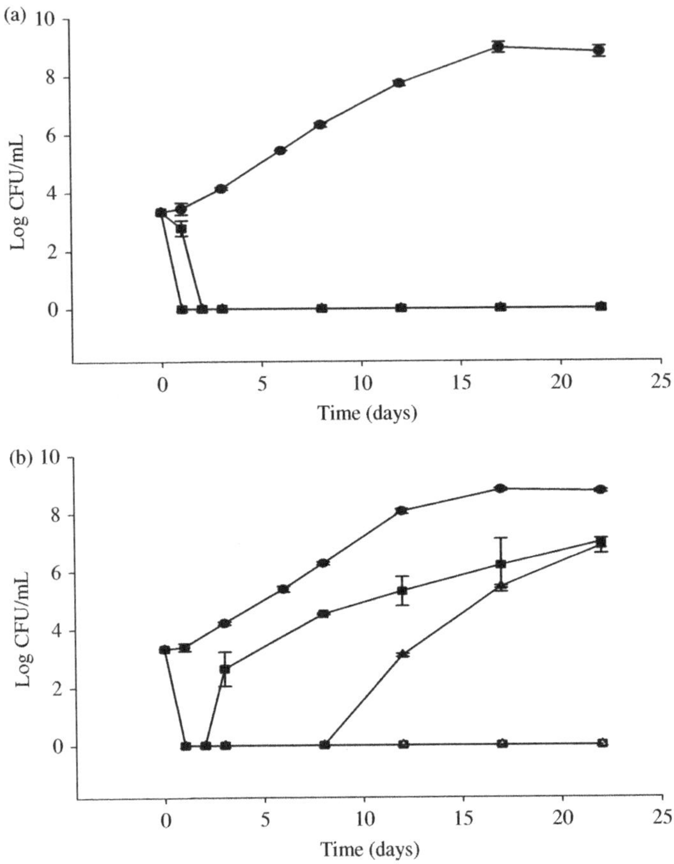

FIGURE 4.2 The effect of chitosonium acetate films on the growth of *L. monocytogenes* in TSB (a) and fish soup (b) after incubation at 4 °C. Film amount: 80 mg (γ), 40 mg (△), 20 mg (▲), 10 mg (■), and control without film (•).

the biocide properties of chitosan films in tests performed in the fish soup compared with the laboratory nutrient medium (i.e., triptone soy broth). These results also agree with a work performed by Devlieghere et al. [135] who studied the influence of food components on the biocide properties of chitosan against *Candida lambica*, and they found a protective effect for bacteria when starch, wheat proteins, and NaCl were added to the nutrient medium. Oil, conversely, had no influence in this particular work.

4.6 CHITOSAN-BASED SYSTEMS IN BIOMEDICAL APPLICATIONS

As mentioned previously, chitin and chitosan are biocompatible, biodegradable, and nontoxic polymers. These properties make chitosan useful for several biomedical applications in tissue engineering, wound healing, drug delivery, or cancer diagnosis.

4.6.1 Tissue Engineering

Tissue engineering involves the use of living cells that are manipulated through their extracellular environment or genetically to develop biological substitutes for implantation into the body. There are few basic requirements for designing polymer scaffolds: high porosity, high surface area, biodegradability (degradation rate matches the rate of neotissue formation), structural integrity, and appropriate mechanical properties to prevent collapsing during neotissue formation. Also, the scaffold should be nontoxic to cells and be biocompatible, positively interacting with the cells to promote cell adhesion, proliferation, migration, and differentiated cell function. In that sense, electrospun nanofiber materials are similar to human native extracellular matrix [136], and hence, it can be a promising scaffolding material for cell culture and tissue engineering applications. Because chitosan inherently have poor mechanical properties, other biomaterials like hydroxyapatite or bioactive glass ceramic (BGC) should be added. Research on chitin/nanoBGC and chitosan/nanoBGC composite scaffolds for tissue engineering developed by lyophilization technique showed adequate porosity when the nanoBGC were homogenously distributed [137]. Furthermore, chitin or chitosan/nanoBGC composite scaffolds showed adequate swelling and degradation properties as well as bioactivity. Cytocompatability of the chitin/nanoBGC and chitosan/nanoBGC scaffolds was assessed by direct contact tests and cell attachment studies; no signs of toxicity were detected. Also, the cells were found to be attached to the pore walls offered by the scaffolds. In another work, the composite scaffolds of chitosan (CS)-gelatin with nanoBGC were prepared by blending chitosan and gelatin with nanoBGC [138]. The results showed macroporous internal morphology in the scaffold with pore size ranging from 150 to 300 μm. Degradation and swelling behavior of the nanocomposite scaffolds decreased, whereas protein adsorption increased with the addition of nanoBGC. Direct contact test and cell attachment studies indicated that the nanocomposite scaffolds are efficient for cell attachment and spreading.

4.6.2 Wound Dressing

Dev et al. [139] prepared poly(lactic acid) (PLA)/CS nanoparticles by the emulsion method for anti-HIV drug delivery applications. For this work, the hydrophilic antiretroviral drug lamivudine was loaded into PLA/CS nanoparticles. The encapsulation efficiency, the *in vitro* drug release behavior, and the cytotoxicity were studied using absorption spectrophotometry. The results showed that the drug release rate from PLA/CS nanoparticles was lower in the acidic pH when compared with the alkaline pH, probably because of the repulsion between H+ ions and cationic groups present in the polimeric nanoparticles.

4.6.3 Cancer Diagnosis

Many of the quantum dots investigated for bioimaging of tissues and cancerous cells (cadmium sulphide, calcium selenide, etc.) are cytotoxic because of their heavy metal composition. As an alternative, in a study performed by Sasidharan et al. [140], a system based on chitosan encapsulated mannosylated zinc sulphide showed low toxicity but an active targeting of cancer cells exhibiting strong fluorescence and long stability. Fluorescence microscopic observations revealed the targeting specificity of mannosylated ZnS nanocrystals toward the mannose-bearing KB cells (a cell line derived from a human carcinoma), with no specific attachment on the normal cells.

4.7 FUTURE PERSPECTIVES

Antimicrobial packaging is a promising system for the future improvement of food quality and preservation during processing and storage. Antimicrobial packaging can also be helpful in extending the food shelf life. Chitosan offered itself as a versatile and promising biodegradable polymer for food packaging. In addition, it possesses immense potential as an antimicrobial packaging material because of its antimicrobial activity and nontoxicity. However, individual researchers used chitosan with varying physicochemical properties and perhaps from various sources. Thus, the question arises as to how to produce chitosan standard materials globally with consistent and reproducible properties. For proper quality control in the chitosan production, there is a critical need to establish less expensive and reliable analytical methods, especially for the evaluation of molecular weight and degree of deacetylation [48]. One major drawback of chitosan film is its high sensitivity to humidity, and thus, it may not be appropriate for use when it is in direct contact with foods and/or for direct handling unless blending or other water-resistant strategies are used. In this sense, several works have been successful in improving the functional properties of chitosan films by blending this material with other more water-resistant film-forming materials. Another possible approach is to make hybrid films with chitosan and nanosized-clay minerals such as layered silicates. By making composites with nanoparticles, mechanical and water barrier properties of chitosan were observed to increase significantly, without altering their antimicrobial properties and biodegradability.

In any case, the numerous research works mentioned in this chapter have been carried out in a laboratory scale. As a consequence, subsequent research on the quality and shelf life extension of foods coated or packaged with chitosan-based materials should be conducted on a pilot plant and industrial scale to ascertain the commercial value and the scaling-up difficulties of the use of this interesting biopolymer or of its nanobiocomposites. In any case, in Europe, chitosan is still not permitted as food contact material, although this may change in the near future. Nonetheless, chitosan does have significant opportunities for implementation as a natural antimicrobial packaging material under the new EU regulation for active packaging (Commission Regulation (EC) No. 450/2009).

In the same way, biomedical applications are still at the laboratory level. Additional studies are necessary before we can expect clinical applications and commercialization of chitin and chitosan-based nanofibers.

REFERENCES

1. Yingyuad, S., Ruamsin, S., Reekprkhon, D., Douglas, S., Pongamphai, S., Siripatrawan, U. (2006). Effect of chitosan coating and vacuum packaging on the quality of refrigerated grilled pork. *Packaging Technology and Science*, *19*, 149–157.
2. Caner, C., Vergano, P. J., Wiles, J. L. (1998). Chitosan film mechanical and permeation properties as affected by acid, plasticizer and storage. *Journal of Food Science*, *63*, 1049–1053.
3. Kim, K. M., Son, J. H., Kim, S. K., Weller, C., Hanna, M. A. (2006). Properties of chitosan films as function of pH and solvent type. *Journal of Food Science*, *71*, 119–124.
4. Fernandez-Saiz, P., Ocio, M. J., Lagaron, J. M. (2006). Film forming and biocide assessment of high molecular weight chitosan as determined by combined ATR-FTIR spectroscopy and antimicrobial assays. *Biopolymers*, *83*, 577–583.
5. Lagaron, J. M., Fernandez-Saiz, P., Ocio, M. J. (2007). Using ATR-FTIR spectroscopy to design active antimicrobial food packaging structures based on high molecular weight chitosan polysaccharide. *Journal of Agriculture and Food Chemistry*, *55*, 2554–2562.
6. Fernandez-Saiz, P., Lagaron, J. M., Hernandez-Muñoz, P., Ocio, M. J. (2008). Characterization of the antimicrobial properties against *S. aureus* of novel renewable blends chitosan and gliadins of interest in active food packaging and coating applications. *International Journal Food Microbiology*, *124*, 13–20.
7. Fernandez-Saiz, P., Lagaron, J. M., Ocio, M. J. (2009). Optimization of the biocide properties of chitosan for its application in the design of active films of interest in the food area. *Food Hydrocolloids*, *23*, 913–921.
8. Fernández-Saiz, P., Lagarón, J. M., Ocio, M. J. (2009). Optimization of the film-forming and storage conditions of chitosan as an antimicrobial agent. *Journal Agriculture and Food Chemistry*, *57*, 3298–3307.
9. Fernandez-Saiz, P., Ocio, M. J., Lagaron, J. M. (2010). Antibacterial chitosan based blends with Ethylene-vinyl alcohol copolymer. *Carbohydrate Polymers*, *80*(3), 874–884.
10. Fernandez-Saiz, P., Soler, C., Lagaron, J. M., Ocio, M. J. (2010). Effects of chitosan films on the growth of *Listeria monocytogenes*, *Staphylococcus aureus* and *Salmonella spp.* in laboratory media and in fish soup. *International Journal of Food Microbiology*, *137*(2–3), 287–294.
11. El Ghaouth, A., Arul, J., Ponnampalam, R., Boulet, M. (1991). Chitosan coating effect on storability and quality of fresh strawberries. *Journal of Food Science*, *56*(6), 1618–1620.
12. Lee, J. W., Lee, H. H., Rhim, J. W. (2000). Shelf life extension of white rice cake and wet noodle by the treatment with chitosan. *Korean Journal Food Science Technology*, *32*(4), 828–833.

13. Roller, S., Covill, N. (1999). The antifungal properties of chitosan in laboratory media and apple juice. *International Journal Food Microbiology*, *47*, 67–77.
14. Roller, S., Covill, N. (2000). The antimicrobial properties of chitosan in mayonnaise and mayonnaise-based shrimp salads. *Journal Food Protection*, *63*(2), 202–209.
15. Ha, T. J., Lee, S. H. (2001). Utilization of chitosan to improve the quality of processed milk. *Journal Korean Society. Food Science and Nutrition*, *30*(4), 630–634.
16. Lee, H. Y., Kim, S. M., Kim, J. Y., Youn, S. K., Choi, J. S., Park, S. M., Ahn, D. H. (2002). Effect of addition of chitosan on improvement for shelf-life of bread. *Journal Korean Society. Food Science and Nutrition*, *31*(3), 445–450.
17. Ralston, G. B., Racey, M. V., Wrench, P. M. (1964). Inhibition of fermentation in baker's yeast by chitosan. *Biochimica et Biophysica Acta*, *93*, 652–655.
18. Helander, I. M., Nurmiaho-Lassila, E. L., Ahvenainen, R., Rhoades, J., Roller, S. (2001). Chitosan disrupts the barrier properties of the outer membrane of Gram-negative bacteria. *International Journal of Food Microbiology*, *71*, 235–244.
19. Liu, H., Du, Y., Xiaohui Wang, X., Sun, L. (2004). Chitosan kills bacteria through cell membrane damage. *International Journal Food Microbiology*, *95*, 147–155.
20. Mayachiew, P., Devahastin, S., Mackey, B. M., Niranjan, K. (2010). Effects of drying methods and conditions on antimicrobial activity of edible chitosan films enriched with galangal extract. *Food Research International*, *43*(1), 125–132.
21. Ganan, M., Carrascosa, V., Martinez-Rodriguez, A. J. (2009). Antimicrobial activity of chitosan against Campylobacter spp. and other microorganism and its mechanism of action. *Journal of Food Protection*, *8*, 1735–1738.
22. Hadwiger, L. A., Kendra, D. F., Fristensky, B. W., Wagoner, W. (1985). *Chitosan both activates genes in plants and inhibits RNA synthesis in fungi*. In: *Chitin in Nature and Technology*. Eds.: Muzzarelli, R. A. A., Jeuniaux, C., Gooday, G. W. pp. 209–222. Plenumx Press, New York, NY.
23. Cuero, R. G., Osuji, G., Washington, A. (1991). N-carboxymethyl chitosan inhibition of aflatoxin production: Role of zinc. *Biotechnology Letters*, *13*, 441–444.
24. Knorr, D. (1984). Use of chitinous polymers in food – a challenge for food research and development. *Food Technology*, *38*: 85–97.
25. Jeuniaux, C. (1986). Chitosan as a tool for the purification of water. In: *Chitin in Nature and Technology*. Eds.: Muzzarelli, R. A. A., Jeuniaux, C., Gooday, G. W. pp. 551–567. Plenum Press, New York, NY.
26. Immizi, S. A., Iqbal, J., Isa, M. (1996). Colletion of metal ions present in waste water samples of different sites of Pakistan using biopolymer chitosan. *Journal of Chemical Society of Pakistan*, *18*, 312–315.
27. Muzzarelli, R. A. A. (Ed.). (1977). *Chitin*. Pergamon Press, Oxford, U.K.
28. McKay, H. S., Blair, J. R., Gardner, M. (1982). Absorption of dyes on chitin. I. Equilibrium studies. *Journal of Applied Polymer Science*, *27*, 3043–3046.
29. Muzzarelli, R. A. A. (1996). Chitosan-based dietary foods. *Carbohydrate Polymers*, *29*, 309–316.
30. Deuchi, K., Kanauchi, O., Imasato, Y., Kobayashi, E. (1994). Decreasing effect of chitosan on the apparent fat digestibility by rats fed on high-fat diet. *Bioscience, Biotechnology, and Biochemistry*, *58*(9), 1613–1616.

31. Braun, J., Le Chanu, P., Le Goffic, F. (1989). The immobilization of penicillin G acylase on chitosan. *Biotechnology and Bioengineering*, *33*, 242–246.
32. Siso, M. I. G., Lang, E., Carreno-Gomez, B., Becerra, M., Espinar, F. O., Mendez, J. B. (1997). Enzyme encapsulation on chitosan microbeads. *Process Biochemistry*, *32*(3), 211–216.
33. Han, C., Zhao, Y., Leonard, S. W., Traber, M. G. (2004). Edible coatings to improve storability and enhance nutritional value of fresh and frozen strawberries (*Fragaria* × *ananassa*) and raspberries (*Rubus ideaus*). *Postharvest Biology and Technology*, *33*, 67–78.
34. Hernandez-Munoz, P., Almenar, E., Ocio, M. J., Gavara, R. (2006). Effect of calcium dips and chitosan coatings on postharvest life of strawberries (*Fragaria* × *ananassa*). *Postharvest Biology and Technology*, *39*, 247–253.
35. Park, S. I., Stan, S. D., Daeschel, M. A., Zhao, Y. (2005). Antifungal coatings on fresh strawberries (*Fragaria* × *ananassa*) to control mold growth during cold storage. *Journal of Food Science*, *70*(4), M202–M207.
36. Vargas, M., Albors, A., Chiralt, A., Gonzalez-Martinez, C. (2006). Quality of cold-stored strawberries as affected by chitosan-oleic acid edible coatings. *Postharvest Biology and Technology*, *41*, 164–171.
37. Soto-Peralta, N. V., Moller, H., Knorr, D. (1989). Effects of chitosan treatments on the clarity and color of apple juice. *Journal of Food Science*, *54*(2), 495–496.
38. Imeri, A. G., Knorr, D. (1988). Effect of chitosan on yield and compositional data of carrot and apple juice. *Journal of Food Science*, *53*(6), 1707–1709.
39. Chatterjee, S., Chatterjee, S., Chatterjee, B. P., Guha, A. K. (2004). Clarification of fruit juice with chitosan. *Process Biochemistry*, *39*, 2229–2232.
40. Rungsardthong, V., Wongvuttanakul, N., Kongpien, N., Chotiwaranon, P. (2006). Application of fungal chitosan for clarification of apple juice. *Process Biochemistry*, *41*, 589–593.
41. Del Blanco, L. F., Rodriguez, M. S., Schulz, P. C., Agullo, E. (1999). Influence of the deacetylation degree on chitosan emulsification properties. *Colloid and Polymer Science*, *77*, 1087–1092.
42. Filar, L. J., Wirick, M. G. (1978). Bulk and solution properties of chitosan. In: *Proceedings of the First International Conference on Chitin/Chitosan*. Eds. Muzzarelli, R. A. A., Pariser, E. R. pp. 169–181. MIT Sea Grant Program, Cambridge, MA.
43. Kamil, J. Y. V. A., Jeon, Y. J., Shahidi, F. (2002). Antioxidative activity of chitosans of different viscosity in cooked comminuted flesh of herring (*Clupea harengus*). *Food Chemistry*, *79*, 69–77.
44. Kanatt, S. R., Chander, R., Sharma, A. (2004). Effect of irradiated chitosan on the rancidity of radiation-processed lamb meat. *International Journal of Food Science and Technology*, *39*, 997–1003.
45. Decker, E. A., Hultin, H. O. (1992). Lipid oxidation in muscle foods via redox iron. In: *Lipid Oxidation in Food*. Ed.: Angelo, A. J. pp. 33–54. American Chemical Society, Washington, D.C.
46. Jeon, Y. J., Kamil, J. Y. V. A., Shahidi, F. (2002). Chitosan as an edible invisible film for quality preservation of herring and Atlantic cod. *Journal of Agricultural and Food Chemistry*, *50*, 5167–5178.

47. Tsai, G. J., Su, W. H., Chen, H. C., Pan, C. L. (2002). Antimicrobial activity of shrimp chitin and chitosan from different treatments and applications of fish preservation. *Fisheries Science*, *68*, 170–177.
48. No, H. K., Meyers, S. P., Prinyawiwatkul, W., Xu, Z. (2007). Applications of chitosan for improvement of quality and shelf life of foods: A review. *Journal of Food Science*, *72*, 87–100.
49. No, H. K., Prinyawiwatkul, W., Meyers, S. P. (2005). Comparison of shelf life of eggs coated with chitosans prepared under various deproteinization and demineralization times. *Journal of Food Science*, *70*(6), S377–S382.
50. Black, J., Hastings, G. (1998). *Handbook of Biomaterials Properties*. Chapman and Hall, London, U.K.
51. Cheng, M. Y., Deng, J. U., Yang, F., Gong, Y. D., Zhao, N. M., Zhang, X. F. (2003). Study on physical properties and nerve cell affinity of composite films from chitosan and gelatin solutions. *Biomaterials*, *24*(17), 2871–2880.
52. Park, S. Y., Jun, S. T., Marsh, K. S. (2001). Physical properties of PVOH/chitosan blended films cast from different solvents. *Food Hydrocolloids*, *15*(4–6), 499–502.
53. Lee, S. J., Kim, S. S., Lee, Y. M. (2000). Interpenetrating polymer network hydrogels based on poly(ethylene glycol) macromer and chitosan. *Carbohydrate Polymers*, *41*(2), 197–205.
54. Zhai, M., Zhao, L., Yoshii, F., Kume, T. (2004). Study on antibacterial starch/chitosan blend film formed under the action of irradiation. *Carbohydrate Polymers*, *57*, 83–88.
55. Vartiainen, J., Rättöa, M., Tapperb, U., Paulussenc, S., Hurmea, E. (2005). Surface modification of atmospheric plasma activated BOPP by immobilizing chitosan. *Polymer Bulletin*, *54*, 343–352.
56. Tanabe, T., Okitsu, N., Tachinaba, A., Yamauchi, K. (2002). Preparation and characterization of keratin-chitosan composite film. *Biomaterials*, *23*, 817–825.
57. Chung, Y., Su, Y., Chen, C., Jia, G., Wang, H., Wu, J. C. G., Lin, J. (2004). Relationship between antibacterial activity of chitosan and surface characteristics of cell wall. *Acta Pharmacology Sinica*, *25*, 932–936.
58. Qin, C., Li, H., Xiao, Q., Liu, Y., Zhu, J., Du, Y. (2006). Water-solubility of chitosan and its antimicrobial activity. *Carbohydrate Polymers*, *63*, 367–374.
59. Tang, R., Du, Y., Fan, L. (2003). Dialdehyde starch-crosslinked chitosan films and their antimicrobial effects. *Journal of Polymer Science*, *41*, 993–997.
60. Suto, S., Ui, N. (1996). Chemical crosslinking of hydroxypropyl cellulose and chitosan blends. *Journal of Applied Polymer Science*, *61*, 2273–2278.
61. Tual, C., Espuche, E., Escoubes, M., Domard, A. (2000). Transport properties of chitosan membranes: Influence of crosslinking. *Journal of Polymer Science*, *38*, 1521–1529.
62. Zheng, H., Du, Y. M., Yu, J. H., Xiao, L. (2000). The properties and preparation of crosslinked chitosan films. *Chemical Journal of Chinese Universities*, *21*, 809–812.
63. Elsabee, M. Z., Abdou, E. S., Nagy, K. S. A., Eweis, M. (2008). Surface modification of polypropylene films by chitosan and chitosan/pectin multilayer. *Carbohydrate Polymers*, *71*, 187–195.
64. Abdou, E. S., Elkholy, S. S., Elsabee, M. Z., Mohamed, E. (2008). Improved antimicrobial activity of polypropylene and cotton nonwoven fabrics by surface

treatment and modification with chitosan. *Journal of Applied Polymer Science, 108*, 2290–2296.

65. Yang, J. M., Lin, H. T., Wu, T. H., Chen, C. C. (2003). Wettability and antibacterial assessment of chitosan containing radiation-induced graft nonwoven fabric of polypropylene-g-acrylic acid. *Journal of Applied Polymer Science*, *90*(5), 1331–1136.
66. Hu, S. G., Jou, C. H., Yang, M. C. (2002). Surface grafting of polyester fiber with chitosan and the antibacterial activity of pathogenic bacteria. *Journal of Applied Polymer Science*, *86*(12), 2977–2983.
67. Pelissari, F. M., Grossmann, M. V. E., Yamashita, F., Pined, E. A. G. (2009). Antimicrobial, mechanical, and barrier properties of cassava starch-chitosan films incorporated with oregano essential oil. *Journal of Agricultural and Food Chemistry*, *57*(16), 7499–7504.
68. Li, B, Peng, J., Yie, X., Xie, B. (2006). Enhancing physical properties and antimicrobial activity of konjac glucomannan edible films by incorporating chitosan and nisin. *Journal of Food Science*, *71*, 174–178.
69. Sebti, I., Chollet, E., Degraeve, C. N., Peyrol, E. (2007). Water sensitivity, antimicrobial, and Physicochemical analyses of edible films based on HPMC and/or Chitosan. *Journal of Agriculture and Food Chemistry*, *55*, 693–699.
70. Park, S. I., Daeschel, M. A., Zhao, Y. (2004). Functional properties of antimicrobial lysozyme-chitosan composite films. *Journal of Food Science*, *69*(8), M215–M221.
71. Moller, H., Grelier, S., Pardon, P., Coma, V. (2004). Antimicrobial and physico-chemical properties of chitosan-HPMC-based films. *Journal of Agriculture and Food Chemistry*, *52*, 6585–6591.
72. Wu, Y. B., Yu, S. H., Mi, F. L., Wu, C. W., Shyu, S. S., Peng, C. K., Chao, A. C. (2004). Preparation and characterization on mechanical and antibacterial properties of chitosan/cellulose blends. *Carbohydrate Polymers*, *57*, 435–440.
73. Shen, X. L., Wu, J. M., Chen, Y., Zhao, G. (2010). Antimicrobial and physical properties of sweet potato starch films incorporated with potassium sorbate or chitosan. *Food Hydrocolloids*, *24*(4), 285–290.
74. Giannelis, E. P. (1996). Polymer layered silicate nanocomposites. *Advanced Materials*, *8*(1), 29–35.
75. Lagaly, G. (1999). Introduction: From clay mineral-polymer interactions to clay mineral-polymer nanocomposites. *Applied Clay Science*, *15*, 1–9.
76. Fischer, S., Vlieger, J., Batenburg, L., Fischer, H. Kock, T. (2000). 'Green' nano-composite materials- new possibilities for bioplastics. *Materialen*, *16*(3–7), 12.
77. Alexandre, M., Dubois, P. (2000). Polymer-layered silicate nanocomposites: Preparation, properties and use of a new class of materials. *Material Science Engineering*, *28*, 1–63.
78. Sinha, R. S., Okamoto, M. (2003). Polymer/layered silicate nanocomposites: A review from preparation to processing. *Progress in Polymer Science*, *28*(11), 1539–1641.
79. Pandey, J. K., Kumar, A. P., Misra, M., Mohanty, A. K., Drzal, L. T., Singh, R. P. (2005). Recent advances in biodegradable nanocomposites. *Journal of Nanoscience and Nanotechnology*, *5*(4), 497–526.
80. Sinha, S. R., Bousmina, M. (2005). Biodegradable polymers and their layered silicate nanocomposites: In greening the 21st century materials world. *Progress in Materials Science*, *50*, 962–1079.

81. Schmidt, D. (2002). New advances in polymer/layered silicate nanocomposites. *Current Opinion in Solid State and Materials Science*, *6*(3), 205–212.

82. Yu, Y. H. (2003). Preparation and properties of poly(vinyl alcohol)-clay nanocomposite materials. *Polymer*, *44*, 3553–3560.

83. Lagaron, J. M., Fendler, A. (2009). High water barrier nanobiocomposites of methyl cellulose and chitosan for film and coating applications. *Journal of Plastic Film Sheet*, *25*(1), 47–59.

84. Darder, M., Colilla, M., Ruiz-Hitzky, E. (2003). Biopolymer-clay nanocomposites based on chitosan intercalated in montmorillonite. *Chemistry of Materials*, *15*, 3774–3780.

85. Darder, M., Colilla, M., Ruiz-Hitzky, E. (2005). Chitosan–clay nanocomposites: Application as electrochemical sensors. *Applied Clay Science*, *28*, 199–208.

86. Wang, S. F., Shen, L., Tong, Y. J., Chen, L., Phang, I. Y., Lim, P. Q., Liu, T. X. (2005). Biopolymer chitosan/montmorillonite nanocomposites: Preparation and characterization. *Polymer Degradation and Stability*, *90*, 123–131.

87. Xu, Y., Ren, X., Hanna, M. A. (2006). Chitosan/clay nanocomposite film preparation and characterization. *Journal of Applied Polymer Science*, *99*, 1684–1691.

88. Günister, E., Pestreli, D., Unlu, C. H., Atici, O., Gungor, N. (2007). Synthesis and characterization of chitosan-MMT biocomposite systems. *Carbohydrate Polymers*, *67*, 358–365.

89. Wang, X., Du, Y., Yang, J., W. X., Shi, X., Hu, Y. (2006). Preparation, characterization and antimicrobial activity of chitosan/layered silicate nanocomposites. *Polymer*, *47*, 6738–6744.

90. Lin, K. F., Hsu, C. Y., Huang, T. S., Chiu, W. Y., Lee, Y. H., Young, T. H. (2005). A novel method to prepare chitosan/montmorillonite nanocomposites. *Journal of Applied Polymer Science*, *98*, 2042–2047.

91. Luo, J. J., Daniel, I. M. (2003). Characterization and modeling of mechanical behavior of polymer/clay nanocomposites. *Composite Science and Technology*, *63*(11), 1607–1616.

92. Rhim, J., Hong, S., Park, H., Perry, K. W. (2006). Preparation and characterization of chitosan-based nanocomposite films with antimicrobial activity. *Journal of Agricultural and Food Chemistry*, *54*, 5814–5822.

93. Jeon, Y. J., Park, P. J., Kim, S. K. (2001). Antimicrobial effect of chitooligosaccharides produced by bioreactor. *Carbohydrate Polymers*, *44*, 71–76.

94. Gerasimenko, D. V., Avdienko, I. D., Bannikova, G. E., Zueva, O. Y., Varlamov, V. P. (2004). Antibacterial effects of water-soluble low-molecular-weight chitosan on different microorganisms. *Applied Biochemistry and Microbiology*, *40*, 253–257.

95. Liu, N., Chen, X. G., Park, H. J., Liu, C. G., Liu, C. S., Meng, X. H., Yu, L. J. (2006). Effect of MW and concentration of chitosan on antibacterial activity of *Escherichia Coli*. *Carbohydrate Polymers*, *64*, 60–65.

96. Liu, X. F., Guan, Y. L., Yang, D. Z., Li, Z., Yao, K. D. (2001). Antibacterial action of chitosan and carboxymethylated chitosan. *Journal of Applied Polymer Science*, *79*, 1324–1335.

97. Zivanovic, S., Basurto, C. C., Chi, S., Davidson, P. M., Weiss, J. (2004). Molecular weight of chitosan influences antimicrobial activity in oil-in-water emulsions. *Journal of Food Protection*, *67*, 952–959.

98. Uchida, Y., Izume, M., Ohtakara, A. (1989). Preparation of chitosan oligomers with purified chitosanase and its application. In: *Chitin and Chitosan: Sources, Chemistry, Biochemistry, Physical Properties and Applications*. Eds.: Skjåk-Bræk, G., Anthonsen, T., Sandford, P. pp. 373–382. Elsevier Applied Science, London, U.K.

99. No, H. K, Park, N. Y., Lee, S. H., Meyers, S. P. (2002). Antibacterial activity of chitosan and chitosan oligomers with different molecular weights. *International Journal Food Microbiology*, *74*, 65–72.

100. Fernandes, J. C., Tavaria, F. K., Soares, J. C., Ramos, O. S., Monteiro, M. J., Pintado, M. E., Malcata, F. X. (2008). Antimicrobial effects of chitosan and chitooligosaccharides, upon Staphylococcus aureus and Escherichia coli, in food model systems. *Food Microbiology*, *25*, 922–928.

101. Takahashi, T., Imai, M., Suzuki, I., Jun, S. (2008). Growth inhibitory effect on bacteria of chitosan membranes regulated with deacetylation degree. *Biochemical Engineering Journal*, *40*, 485–491.

102. Chen, Y. M., Chung, Y. C., Wang, L. W., Chen, K. T., Li, S. Y. (2002). Antibacterial properties of chitosan in waterborne pathogen. *Journal of Environmental Science and Health, Part A*, *37*, 1379–1390.

103. Hongpattarakere, T., Riyaphan, O. (2008). Effect of deacetylation conditions on antimicrobial activity of chitosans prepared from carapace of black tiger shrimp (penaeus monodon). *Songklanakarin Journal of Science and Technology*, *30*(1), 1–9.

104. Caner, C. (2005). The effect of edible eggshell coatings on egg quality and consumer perception. *Journal of the Science of Food and Agriculture*, *85*, 1897–1902.

105. Park, S. Y., Marsh, K. S., Rhim, J. W. (2002). Characteristics of different molecular weight chitosan films affected by the type of organic solvents. *Journal of Food Science*, *67*, 194–197.

106. Rinaudo, M., Pavlov, G., Desbrieres, J. (1999). Influence of acetic acid concentration on the solubilisation of chitosan. *Polymer*, *40*, 7029–7032.

107. Ouattara, B., Simard, R. E., Piette, G., Bégin, A., Holley, R. A. (2000). Inhibition of surface spoilage bacteria in processed meats by application of antimicrobial films prepared with chitosan. *International Journal Food Microbiology*, *62*, 139–148.

108. Zotkin, M. A., Vikhoreva, G. A., Smotrina, T. V., Derbenev, M. A. (2004). Thermal modification and study of the structure of chitosan films. *Fibre Chemistry*, *36*, 16–20.

109. Marreco, P. R., Moreira, P. L., Genari, S. C., Moraes, A. (2005). Effects of different sterilization methods on the morphology, mechanical properties, and citotoxicity of chitosan membranes used as wound dressings. *Journal of Biomedical Materials Research*, *71*, 268–277.

110. NCCLS. (1999). Methods for determining bactericidal activity of antimicrobial agents; approved guideline, M26-A, Volume *19*, Number 18. NCCLS, Wayne, PA.

111. Pranoto, Y., Rakshit, S. K., Salokhe, V. M. (2005). Enhancing antimicrobial activity of chitosan films by incorporating garlic oil, potassium sorbate and nisin. *LWT- Food Science and Technology*, *38*, 859–865.

112. Zivanovic, S., Chi, S., Draughon, A. F. (2005). Antimicrobial activity of chitosan films enriched with essential oils. *Journal of Food Science*, *70*, 45–51.

113. Coma, V., Martial-Gros, A., Garreau, S., Copinet, A., Salin, F., Deschamps, A. (2002). Edible antimicrobial films based on chitosan matrix. *Journal of Food Science*, *67*, 1162–1169.

114. Chen, Y. L., Chou, C. C. (2005). Factors affecting the susceptibility of *Staphylococcus aureus* CCRC 12657 to water soluble lactose chitosan derivative. *Food Microbiology*, *22*, 29–35.
115. Tsai, G. J., Su, W. H. (1999). Antibacterial activity of shrimp chitosan against Escherichia coli. *Journal of Food Protection*, *62*(3), 239–243.
116. Tsai, G. J., Tsai, M. T., Lee, J. M., Zhong, M. Z. (2006). Effects of Chitosan and low-molecular-weight chitosan on Bacillus cereus and application in the preservation of cooked rice. *Journal of Food Protection*, *69*, 2168–2175.
117. Rodriguez, E., Calzada, J., Arqués, J. L., Rodriguez, J. M., Nuñez, M., Medina, M. (2005). Antimicrobial activity of pediocin-producing *Lactococcus lactis* on *Listeria monocytogenes*, *Staphylococcus aureus* and *Escherichia coli* O157:H7 in cheese. *International Dairy Journal*, *15*, 51–57.
118. Igarashi, K., Nakano, Y. (2003). Dissociation properties of amino groups in the chitosan gel particles prepared by the suspension evaporation method. *Journal of Chemical Engineering of Japan*, *36*, 716–719.
119. Marques, A., Encarnaçao, S., Pedro, S., Nunes, M. L. (2008). In vitro antimicrobial activity of garlic, oregano and chitosan against *Salmonella enterica*. *World Journal Microbiology Biotechnology*, *24*, 2375–2360.
120. Simpson, C. A., Geornaras, I., Yoon, Y., Scanga, J. A., Kendall, P. A., Sofos, J. N. (2008). Effect of inoculum preparation procedure and storage time and temperature on the fate of *Listeria monocytogenes* on inoculated salami. *Journal Food Protection*, *71*, 494–501.
121. Briandet, R., Meylheuc, T., Maher, C., Bellon-Fontaine, M. N. (1999). *Listeria monocytogenes* Scott A: Cell surface charge, hydrophobicity, and electron donor and acceptor characteristics under different environmental growth conditions. *Applied Environmental Microbiology*, *65*, 5328–5333.
122. Butler, B. L., Vergano, P. J., Testin, R. F., Bunn, J. M., Wiles, J. L. (1996). Mechanical and barrier properties of edible chitosan films as affected by composition and storage. *Journal Food Science*, *61*, 953–955.
123. Ritthidej, G. C., Phaechamud, T., Koizumi, T. (2002). Moist heat treatment on physicochemical change of chitosan salt films. *International Journal Pharmaceutics*, *232*, 11–22.
124. Kam, H. M., Khor, E., Lim, L. Y. (1999). Storage of partially deacetylated chitosan films. *Journal of Biomedical Materials Research*, *48*(6), 881–888.
125. Srinivasa, P. C., Ramesh, M. N., Kumar, K. R., Tharanathan, R. N. (2004). Properties of chitosan films prepared under different drying conditions. *Journal Food Engineering*, *63*, 79–85.
126. Demarger-Andre, S., Domard, A. (1994). Chitosan carboxylic acid salts in solution and in the solid state. *Carbohydrate Polymers*, *23*, 211–219.
127. Osman, Z., Arof, A. K. (2003). FTIR studies of chitosan acetate based polymer electrolytes. *Electrochimica Acta*, *48*, 993–999.
128. Larena, A., Caceres, D. A. (2004). Variability between chitosan membrane surface characteristics as function of its composition and environmental conditions. *Applied Surface Science*, *238*, 273–277.
129. Sagoo, S., Board, R., Roller, S. (2002). Chitosan inhibits growth of spoilage microorganisms in chilled pork products. *Food Microbiology*, *19*, 175–182.

130. Lopez-Caballero, M. E., Gómez-Guillén, M. C., Pérez-Mateos, M., Montero, P. (2005). A chitosan-gelatin blend as a coating for fish patties. *Food Hydrocolloids*, *19*, 303–311.

131. Juneja, V. K., Thippareddi, H., Bari, L., Inatsu, Y., Kawamoto, S., Friedman, M. (2006). Chitosan protects cooked ground beef and turkey against *Clostridium perfringens* spores during chilling. *Journal of Food Science*, *71*, 236–240.

132. Vasconez, M. B., Flores, S. K., Campos, C. A., Alvarado, J., Gerschenson, L. N. (2009). Antimicrobial activity and physical properties of chitosan-tapioca starch based edible films and coatings. *Food Research International*, *42*(7), 762–769.

133. Ye, M., Neetoo, H., Chen, H. (2008). Control of *Listeria monocytogenes* on ham steaks by antimicrobials incorporated into chitosan-coated plastic films. *Food Microbiology*, *25*, 260–268.

134. Rhoades, J., Roller, S. (2000). Antimicrobial actions of degraded and native chitosan against spoilage organisms in laboratory media and foods. *Applied Environmental Microbiology*, *66*, 80–86.

135. Devlieghere, F., Vermeulen, A., Debevere, J. (2004). Chitosan: Antimicrobial activity, interactions with food components and applicability as a coating on fruits and vegetables. *Food Microbiology*, *21*, 703–714.

136. Jayakumar, R., Prabaharan, M., Nair, S. V., Tamura, H. (2010). Novel chitin and chitosan nanofibers in biomedical applications. *Biochechnology Advances*, *28*, 142–150.

137. Peter, M., Binulol, N., Soumya, S., Nair, S. V., Tamura, H., Jayakumar, R. (2010). Nanocomposite scaffolds of bioactive glass ceramic nanoparticles disseminated chitosan matrix for tissue engineering applications. *Carbohydrate Polymers*, *79*, 284–289.

138. Peter, M., Binulol, N., Nair, S. V., Selvamurugan, N., Tamura, H., Jayakumar, R. (2010). Novel biodegradable chitosan-gelatin/nano-bioactive glass ceramic composite scaffolds for alveolar bone tissue engineering. *Chemical Engineering Journal*, *158*, 353–361.

139. Dev, A., Binulal, N., Anitha, A., Nair, S. V., Furuike, T., Tamura, H. (2010). Preparation of novel poly(lactic acid)/chitosan nanoparticles for anti HIV drug delivery applications. *Carbohydrate Polymers*, *80*, 833–838.

140. Sasidharan, S., Aswathy, J., Manzoor, K., Nair, S. V., Menon, D. (2011). Mannosylated zinc sulphide nanocrystals for targeted cancer diagnosis. *Carbohydrate Polymers*, *85*(1), 37–43.

CHAPTER 5

THYMOL IN NANOCOMPOSITES. A CASE STUDY

MARIA DOLORES SANCHEZ-GARCIA
Instituto de Agroquímica y Tecnología de Alimentos (IATA), CSIC, Valencia, Spain

CONTENTS

5.1 Understanding Thymol and its Biocide Efficiency 101
5.2 Generating Thymol-Based Films by Casting Methods 104
5.3 Using Nanoclays to Control Solubility and Diffusion: Modeling Control Release 106
5.3.1 Release of Thymol from PCL-Nanoclay Systems 107
5.3.2 PCL vs. Clay-Containing Nanocomposite System 107
5.3.3 Thymol Release as a Function of Clay Concentration 107
5.4 Summary and Future Trends 111
References 113

5.1 UNDERSTANDING THYMOL AND ITS BIOCIDE EFFICIENCY

With the increase in consumer awareness for food safety and health standards, there is a general concern regarding the use of chemical preservatives in the food chain [1]. In response to this, biopreservatives and naturally derived antimicrobial (AM) additives are becoming more important as they represent a perceived lower risk to consumers [2]. More extensive attempts are being made in the search for alternative AM compounds based on plant extracts [3]. For example, the AM effect of essential oils and their active constituents against many foodborne pathogenic bacteria, including *Salmonella enterica*, *Campylobacter jejuni*

Antimicrobial Polymers, First Edition.
Edited by José M. Lagarón, María J. Ocio, and Amparo López-Rubio.

[4], *Staphylococcus aureus*, and *Vibrio parahaemolyticus* [5], have been studied. The use of natural extracts is often preferred because of less complex regulation processes and consumer preference when compared with chemical AM agents [6]. Plant extracts, in particular, such as grapefruit seed, cinnamon, horseradish, and cloves, have been added to packaging systems to demonstrate effective AM activity against spoilage and pathogenic bacteria [7–9].

The essential oils of various biologically active plant species have become popular in recent years. There are several research studies about the essential oils and their AM activity in liquid as well as in vapor media [4–25]. Friedman et al. [4] analyzed a broad variety of naturally occurring and potentially food-compatible, plant-derived oils and oil compounds for AM activity. The extracts showed promising AM activity against several species of bacterial food-borne pathogens, including *C. jejuni, Escherichia coli, Listeria monocytogenes*, and *S. enterica*. The use of natural AM compounds is not only important in the control of human and plant diseases of microbial origin but also in the preservation of packaged food products [6]. Fyfe et al. [10] studied the inhibition of *L. monocytogenes* and *S. enteritis* by combinations of plant essential oils with either benzoic acid or methyl-paraben. This work highlighted the fact that the essential oil of basil was a potent inhibitor of both microbial species. Koga et al. [11] studied the bactericidal activity of basil and sage essential oils against a range of bacteria, and their findings showed that gram-positive bacteria showed higher resistance to basil essential oil than gram-negative bacteria. In addition to AM activity, basil oil is often used as a flavorant in tomato-based products that have high acidity and that are prone to spoilage by acid-tolerant food microflora [12–13].

Deans and Ritchie [14] screened the AM spectrum of 50 plant essential oils against 25 genera of bacteria, and all bacteria showed a reasonably broad sensitivity to the oils tested. The AM and antifungal properties of essential oils of different species of *Ocimum* have been predominantly associated with the main constituent linalool [15], and there have been synergistic effects attributed to the combination of linalool with other essential oils against *Rhizopus nigrans* [16]. Couladis et al. [17] demonstrated the antifungal activity of thymol and proved that it was a potent inhibitor of molds, thus confirming its potential use in food preservation. Prasad et al. [18] studied the AM activity of essential oils of *O. basilicum* and reported that these oils were more effective against the gram-positive than the gram-negative bacteria. All gram-positive bacteria including *Bacillus sacharolyticus, B. stearothermophilus, B. subtilis, B. thurengiensis, Micrococcus glutamicus*, and *Sarcina lutea* were inhibited by each of these basil essential oils. Only the gram-negative strain *Salmonella weltevreden*, however, was not suppressed by the oils. Lahariya and Rao [19] studied the AM effectiveness of the essential oil of *O. basilicum* tested *in vitro* against 10 different microorganisms. They founded that the essential oil was more active in inhibiting the growth of *Bacillus pumilus* and had less activity against the fungi [19]. Sweet basil (*Ocimum basilicum L.*) is a popular culinary herb that has been widely used as a food ingredient [20]. Sweet basil has also been used for many years as a food flavorant and as an ingredient in dental and oral health-care products [21]. Additionally, basil essential oils have been reported to possess AM activity against a spectrum of gram-positive and gram-negative bacteria as well as against important food-borne pathogens [10], molds [22], and yeasts [23]. Coating of low-density polyethylene (LDPE) films or blending LDPE with basil extracts prior to extrusion are some of the techniques used for obtaining AM films [24].

Suppakul et al. [25] recently published research articles focusing on the potential of basil extract in the field of AM food packaging. The investigation, evaluation, efficacy, and feasibility effect of basil AM agents when incorporated into LDPE films against a wide selection of microorganisms including *Staphylococcus aureus*, *Listeria innocua*, *Escherichia coli*, and *Saccharomyces cerevisiae* were discussed. The resulting LDPE films proved to be promising as an active AM packaging material.

Monoterpene phenols are created by plants and are major components of essential oils. They are organic (carbon-based) molecules that are made up by two connected rings of carbon. Each of these rings, called isoprene, has five carbon molecules. (Isoprene is the building block for a wide variety of other substances, including other terpene molecules, cholesterol, steroid hormones, vitamin A, and natural rubber.)

In addition to the two rings of isoprene, monoterpene phenols contain a hydroxyl group (OH) as well as a variety of different side chains of atoms. These side chains are what make each different monoterpene phenol molecule unique.

Thymol (also known as 2-isopropyl-5-methylphenol, IPMP) is a natural monoterpene phenol derivative of cymene, $C_{10}H_{14}O$, which presents the molecule structure depicted in Figure 5.1.

Sefidkon et al. [26] extracted 19.6% thymol by hydrodistillation from aerial parts of *Thymus eriocalyx jalas* growing in various locations in central Iran with the major component observed to be linalool (1.8–60.4%). Kalvandi et al. [27] extracted 42.8% to 43.1% thymol from essential oils obtained from the *Thymus eriocalyx* (Ronniger) species. Couladis et al. [28, 29] obtained 59% thymol from

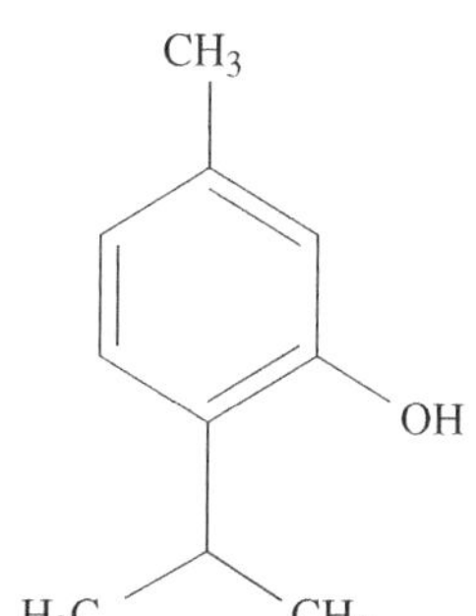

FIGURE 5.1 Chemical structure of thymol.

essential oils extracted from *Thymus striatus* collected from the Orjen Mountains. Thymol was observed to be the major constituent in *T. kotschyanus* (19.6%), *T. carnosus* (36.6%), *T. pubescisus* (27.1%), and *T. serpullum* (18.7%). The hydrodistillation of essential oils from the Saturja species in Turkey contained 17.5–43.5% thymol. These examples illustrate the diversity of locations and variation in extractable quantities of naturally occurring thymol.

Thymol has received considerable attention as an antimicrobial agent with very high antifungal activity and very low minimal inhibitory concentration (MIC) values and as a possible food antioxidant.

Thymol is effective against *E. coli* [30–33] and is significantly effective against *Staphylococcus aureus* [34]. Thymol seems to harm bacterial cell membrane, as well as enter the bacteria and change cell processes. These effects may account for their antibacterial properties [34, 35].

Toxic effects on membrane structure and function have been generally used to explain the antimicrobial action of essential oils and their monoterpenoid components [35, 36]. In fact, as a result of their lipophilic character, monoterpenes will preferentially partition from an aqueous phase into membrane structures [36, 37]. This results in membrane expansion, increased membrane fluidity and permeability, disturbance of membrane-embedded proteins, inhibition of respiration, and alteration of ion transport processes. Helander et al. [38] have described the effects of selected essential oil components on outer membrane permeability in gram-negative bacteria, which provides evidence that monoterpene uptake is also determined by the permeability of the outer envelope of the target microorganism.

Recent studies have revealed that the essential oils of oregano and thyme are active against strains of *E. coli* [39–43]. The MIC for thymol presents a range of values in the literature, which reflects the differences in media composition, methodology, and strains of bacteria used. The MIC of thymol is around 1.2–3.0 mmol/L reported previously [44–46]. In another study, the MIC for thymol was between 0.66 and 3.3 mmol/L [47]. This activity could lead to opportunities for the use of these essential oils or their components in improving the safety of certain foods.

5.2 GENERATING THYMOL-BASED FILMS BY CASTING METHODS

Active packaging has been defined as a system in which the product, the package, and the environment interact in a positive way to extend shelf life or to achieve some characteristics that cannot be obtained otherwise [48].

The main aim of active packaging is to change the condition of packaged food in order to extend the shelf life [49]. This practice can improve food safety and sensory properties, while maintaining the quality of packaged food. Active packaging techniques for conservation and improving quality and safety of foods can be divided into three classes: (1) absorbing systems; (2) releasing systems; and (3) other speciality systems for temperature, ultraviolet light, and microwave control systems (although this latter category is sometimes referred to as "intelligent packaging" [50]). Active packaging materials that can release active compounds for enhancing the quality and safety of a wide range of foods during extended storage are particularly important.

Therefore, the active packaging consists of adding an active substance in the packaging material improving the functionality of the packaging [51]. For example, the active substances in the polymer allow for absorbing the oxygen, controlling the concentration of carbon dioxide or ethylene, releasing ethanol or antioxidant substances or antimicrobial substances, as well as controlling the humidity and the growth of microorganisms [52].

Antimicrobial activity can be attained by adding AM agents to a packaging system during manufacturing or by using AM polymeric materials [3]. There are three typical packaging systems with AM agent activity: (1) those in which the AM substance is absorbed in the packaging material; (2) those that contain the AM agent immobilized; and (3) release systems. The absorption system removes the essential factors for microbial growth from the food systems, thus inhibiting their growth. The immobilization system does not release AM agents but controls the growth of microorganisms through direct contact surface. The release system allows for the migration of the AM agent (solute or gas) into the food or the headspace inside the package to inhibit the growth of microorganisms. Whereas a gaseous AM agent can penetrate through any space, a solute AM agent cannot migrate through the air space between the food and the packaging material. The release kinetics of a packaging system is studied by measuring the release rate of the AM agent into a food simulant or by measuring the effectiveness in inhibiting microbial growth and extending the shelf life of food.

Controlled release packaging is a new generation of packaging materials that can release active compounds at different controlled rates suitable for enhancing the quality and safety of foods during extended storage. The substances that can be included in the release packaging are nutrients, antimicrobials, antioxidants, enzymes, flavors, and nutraceuticals. The antimicrobial substances in the release packaging permit a gradual migration to the food during the storage and distribution of this packaged food product. The antimicrobial packaging is effective for minimizing the superficial contamination in foods, and for that reason, the application of this antimicrobial packaging is very attractive in food products like meat, fruits, and vegetables. The antimicrobial substances used in the food packaging that can migrate to the food are considered food additives, thus having to comply with valid legislation [53].

Development of thymol-based films by casting methods can be considered an active packaging. Sanchez-García et al. [54] developed antimicrobial biodegradable films based on thymol and polycaprolactone (PCL) prepared by casting method.

Seydim and Sarikus [55] developed whey protein-based edible films with the incorporation of oregano, rosemary, and garlic essential oils by casting method to release antimicrobial constituents in food packaging.

Mastromatteo et al. [56] reported the development of active packaging materials based on zein-thymol films prepared by casting. The release rate of thymol from the developed monolayer and multilayer films, as affected by layers thickness, was evaluated in this work.

5.3 USING NANOCLAYS TO CONTROL SOLUBILITY AND DIFFUSION: MODELING CONTROL RELEASE

Nanocomposites can be designed to control the release of, for instance, antimicrobial natural components from packaging materials. The addition of antimicrobial substances in the nanocomposite biodegradable packaging is a way to improve the release of the natural antimicrobial agents such as thymol. The addition of nanoclays in the release packaging system allows for the absorption of the antimicrobial substance and the control release of this substance. Nanocomposite of PCL/layered organoclay with volatile antimicrobial substance system thus represents a new concept in developing drug delivery systems, which can also have antimicrobial effects. The presence of the nanoadditives can be tailored to enhance the solubility of the antimicrobial component in the matrix and to decrease or increase its kinetics of release. The kinetics of thymol release depends on the type of clay and the organic modification. According to the requirements of the packaged product, the adequate clay can be chosen as well as the chemical modification of the clay for the specific needs of the product. These new nanobiocomposites with antimicrobial agents are environmentally friendly, specifically designed for enhancing properties (barrier, thermal, and mechanical properties, blocking food-altering ultraviolet [UV]-visible spectroscopy [VIS] radiation) [57–59] of specific materials while maintaining clarity and that make use only of food-contact–approved components and precursors. These characteristics make them safe, bio-complying and readily applicable in current packaging applications without the need for changes in legislation.

In summary, the addition of food-contact–approved layered silicates to biodegradable polymers through innovative technology is a formidable tool for improving the properties of these biodegradable polymers and, therefore, to enhance packaged food quality aspects. In this context, functional nanoadditives, such as nanoclays, with tailor-made and food-contact–approved surface modifications can have a significant potential to enhance mechanical and barrier properties of these materials, for controlled release of active functional components (antimicrobial, antioxidant, etc.) and to block food-altering UV-VIS radiation. As an example of these novel active nanobiocomposite systems, in this chapter, we will focus on the results for thymol-containing nanobiocomposites obtained by the author via solution casting. A new family of food-contact–approved organoclays (Nanoter, Nanobiomatters S.L., Paterna, Spain) was used to enhance the physical barriers of biodegradable packaging materials to enhance food quality aspects in passive and active food packaging applications.

Sánchez-García et al. [60] studied the modeling of the thymol release: the release of the thymol in the PCL, the influence of the different clays in the thymol release, and the concentration of the clays.

5.3.1 Release of Thymol from PCL-Nanoclay Systems

Infrared (IR) spectroscopy can be used to follow the release of thymol from polymeric systems. The IR spectrum of the thymol type is very similar to that of the corresponding chemotype of basil, recently described elsewhere [61]. The most intense peaks at 738 and 807 cm^{-1} are assigned to ring vibrations of the thymol chemistry [62]. Systematic release studies were performed for thymol incorporated in the nanocomposite of the biodegradable material for a time period of up to 2 weeks. Figure 5.2a shows the release of thymol from the formed film as followed by the decrease of the band at 807 cm^{-1} over time. Figure 5.2b shows the evolution of the intensity of the bands of PCL, thymol, and chloroform as a function of time. Figure 5.2b shows that the chloroform was evaporated in a few hours. At this time (time cero), the release of thymol can be studied. The figure also shows a concomitant increase of the PCL signal in the ATR evanescent field as a result of polymer deswelling (thickness reduction) as thymol is released from the matrix and the polymer reduces its free volume.

5.3.2 PCL vs. Clay-Containing Nanocomposite System

The influence of different nanoclays at 5 wt% clay content on thymol release has been studied by Sanchez-Garcia et al. [60] in PCL nanocomposites.

Figure 5.3 shows, as an example, how the solubility of a natural antimicrobial agent, in this case thymol, can be enhanced and the release delayed in PCL by the presence of specific nanocarriers in the polymer matrix.

Curiously, the thymol diffusion coefficient is observed to decrease more with the nanoadditives in these biocomposite. In Table 5.1, the thymol diffusion and solubility coefficients for the samples of PCL with different organoclays are summarized. From the results, a reduction in thymol diffusion coefficient between 58% and 65% is observed with respect to the unfilled material (see Figure 5.4a). This is likely to be the result of the larger tortuosity effect imposed to the diffusion of the biocide by the dispersed nanoclay.

Figure 5.4b shows that thymol relative uptake increases with the addition of 5% clays. These results confirm that it is feasible to control the uptake and release of thymol by incorporation of laminar nanoclays in bioplastics such as PCL.

5.3.3 Thymol Release as a Function of Clay Concentration

Thymol release was also studied as a function of the concentration of the clays. Figure 5.5 shows the influence of the concentration of clay in the thymol release.

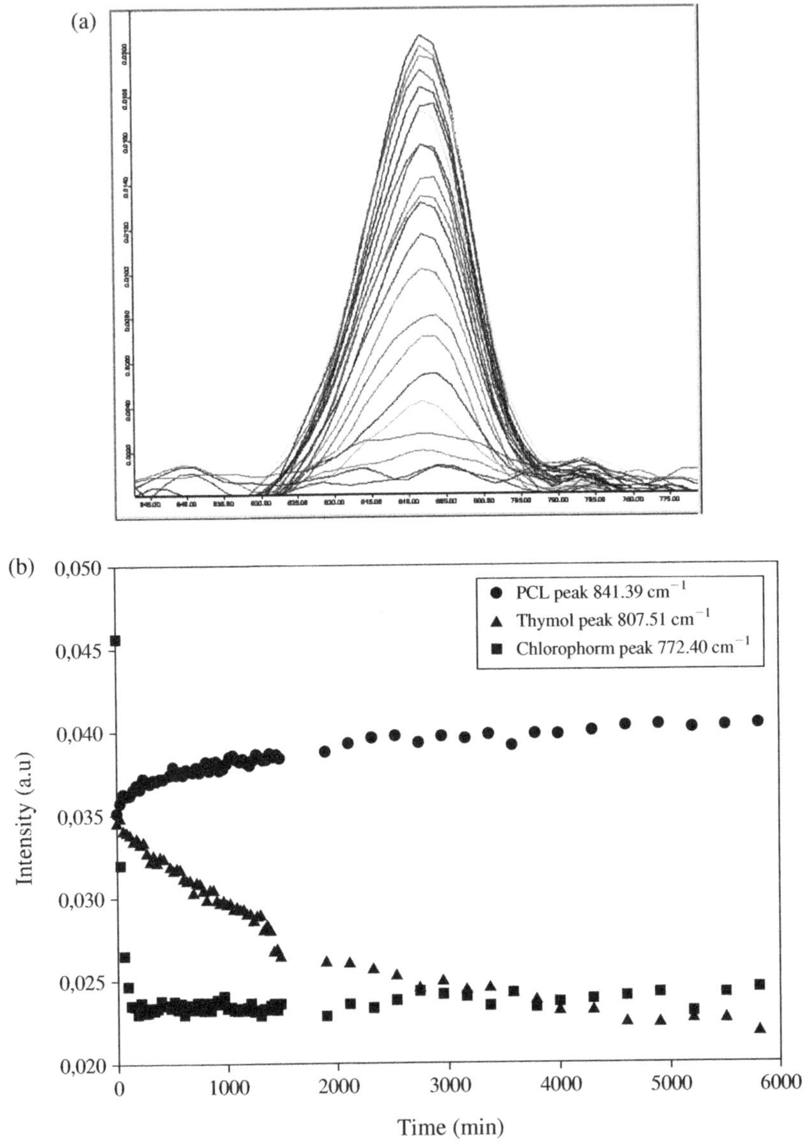

FIGURE 5.2 (a) ATR-FTIR spectra of the thymol release from neat PCL. (b) Intensity of the bands of PCL, thymol, and chloroform as a function of time.

It also shows that with the increase of clay content, the release is more gradual in time. As expected, with the increase of clay content, the tortuosity effect increases and the thymol release decreases with the presence of the dispersed clays.

The thymol diffusion coefficient (see Figure 5.6a) is shown to decrease with the increase of content of the nanoadditive in the biocomposite as expected from

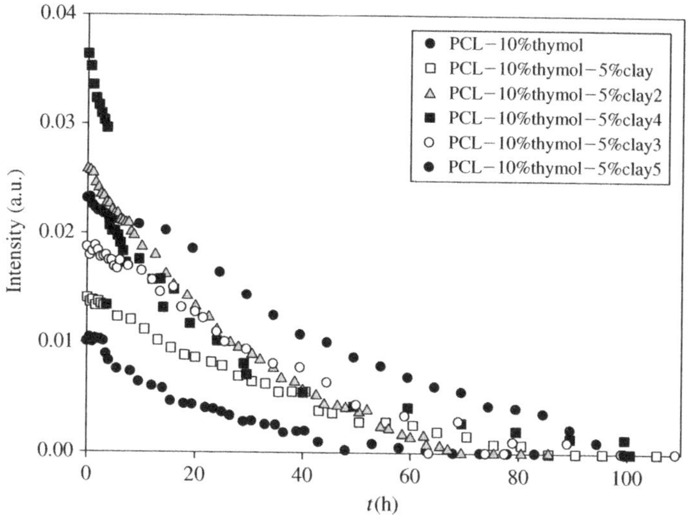

FIGURE 5.3 ATR desorption kinetics of release of thymol in the PCL matrix and their nanobiocomposites.

TABLE 5.1 Diffusion and solubility coefficient of release of thymol for the PCL films and their nanocomposites containing 5 wt% of different clays

	D(m^2/s)	Solubility (g/m^3Pa)
PCL−10 wt% thymol	7.34^{-14}	$1.21e^{-4}$
PCL−10 wt% thymol−5% clay	$1.57e^{-14}$	$2.70e^{-4}$
PCL−10 wt% thymol−5% clay2	$0.73 \pm 0.15e^{-14}$	$2.50e^{-4}$
PCL−10 wt% thymol−5% clay3	$0.39 \pm 0.07e^{-14}$	$2.69 \pm 0.19e^{-4}$
PCL−10 wt% thymol−5%clay4	$1.54e^{-14}$	$2.86e^{-4}$
PCL−10 wt% thymol−5%clay5	$0.22 \pm 0.03e^{-14}$	$3.31 \pm 0.60e^{-4}$

observation of Figure 5.5. Figure 5.6b also shows that thymol relative uptake increases with the increase of clay content.

Table 5.2 summarizes the diffusion and solubility coefficients for the thymol release in the nanocomposites containing 1, 5, 10, and 20 wt% of a specific type of clay (referred to as "clay5"). From the results, a reduction in thymol diffusion coefficient of ca. 42%, 59%, 75%, and 92% can be observed with the addition of 1, 5, 10, and 20 wt% of clay5, respectively, with regard to the unfilled material. Furthermore, the solubility coefficient increases of ca. 156%, 173%, 431%, and 391% with the addition of 1, 5, 10, and 20 wt% of clay5, respectively.

As a summary, the concentration of the clay is another factor that influences thymol release. So, the tortuosity effect of the clays dispersed in the matrix increases with increase the clay content.

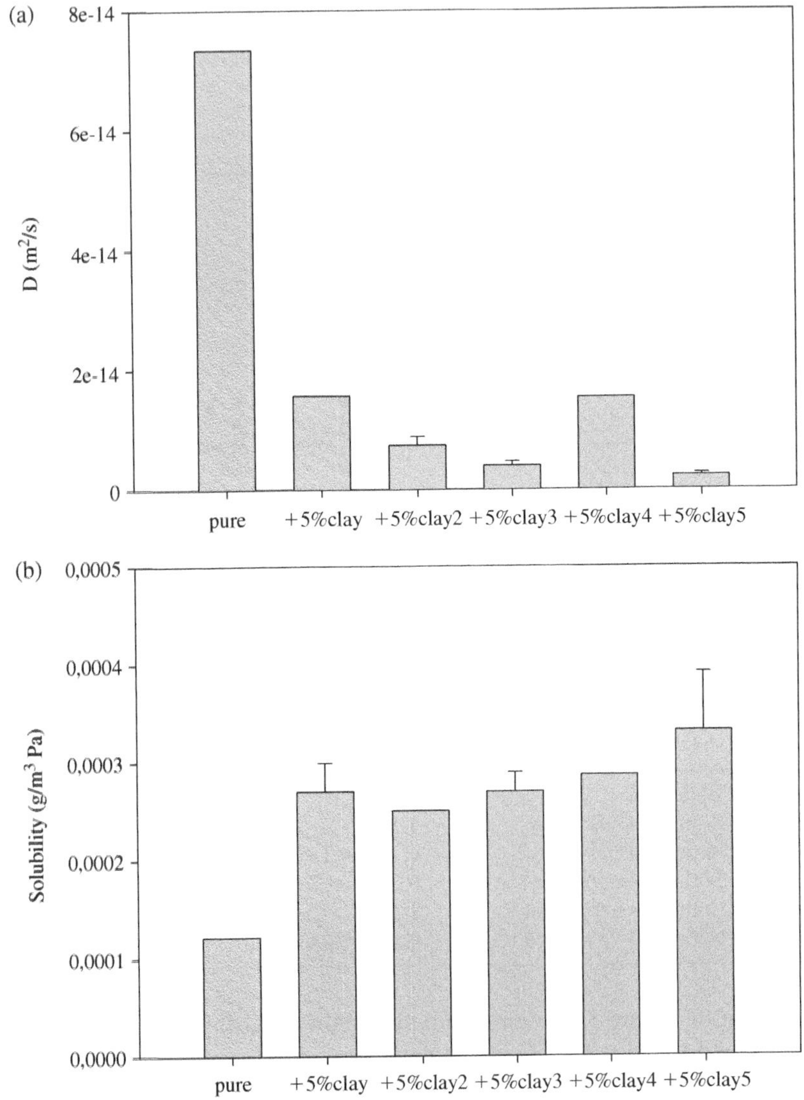

FIGURE 5.4 (a) Diffusion coefficient (m^2/s) for the release of thymol from the PCL matrix and the PCL nanocomposite with different clays. (b) Solubility coefficient (g/m^3 Pa) of thymol as determined by ATR-FTIR spectroscopy in neat PCL matrix and in the nanocomposite with different clays.

Mastromatteo et al. [63] also reported the kinetics release of thymol from zein-based films by monitoring with an HPLC, chromatographic method. The developed active films were put into a container with distilled water at room temperature. And the active compound concentration in the surrounding

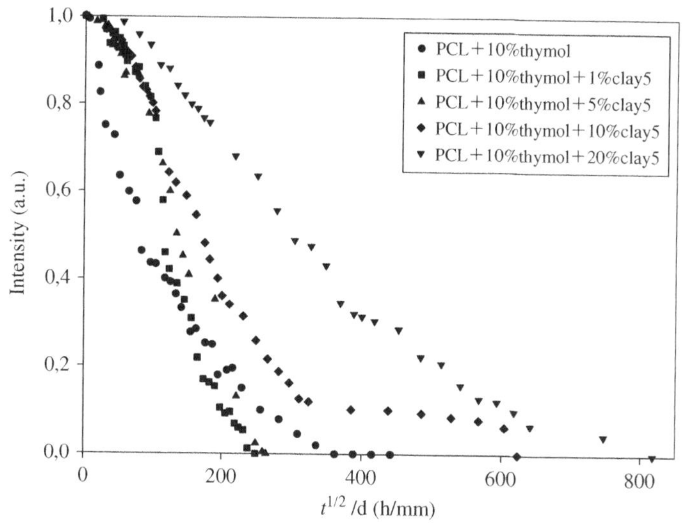

FIGURE 5.5 ATR desorption release of thymol in the PCL matrix and their nanobiocomposites in function of the concentration of the clay.

solution, until reaching an equilibrium value, was monitored by HPLC. Mastromatteo et al. reported a diffusion coefficient of thymol in the zein film around of ca. 6.52 $e^{-12} m^2/s$ by chromatographic method. [63]. However, the diffusion coefficient of thymol in the PCL film determined by ATR-spectroscopic method is around 7.34 $e^{-14} m^2/s$; in this case, the diffusion is lower than the diffusion from zein films and determined by chromatographic method.

5.4 SUMMARY AND FUTURE TRENDS

The combination of active technologies such as antimicrobials and nanotechnologies such as nanocomposites can synergistically lead to bioplastic formulations with balanced properties and functionalities for their implementation in packaging applications. As an example of bioactive packaging, the formulation of novel antimicrobial nanocomposites of PCL has been presented as a way to control solubility and diffusion of the natural biocide agent thymol. The results presented in this chapter confirm that it is feasible to control the uptake and release of thymol through incorporation of laminar nanoclays into bioplastics such as PCL. Therefore, incorporation of these nanoadditives can also be used not only to enhance the solubility of natural biocides into biopolymeric matrices but also to control the release of natural antimicrobials with interest in the design of novel active antimicrobial film and coating systems.

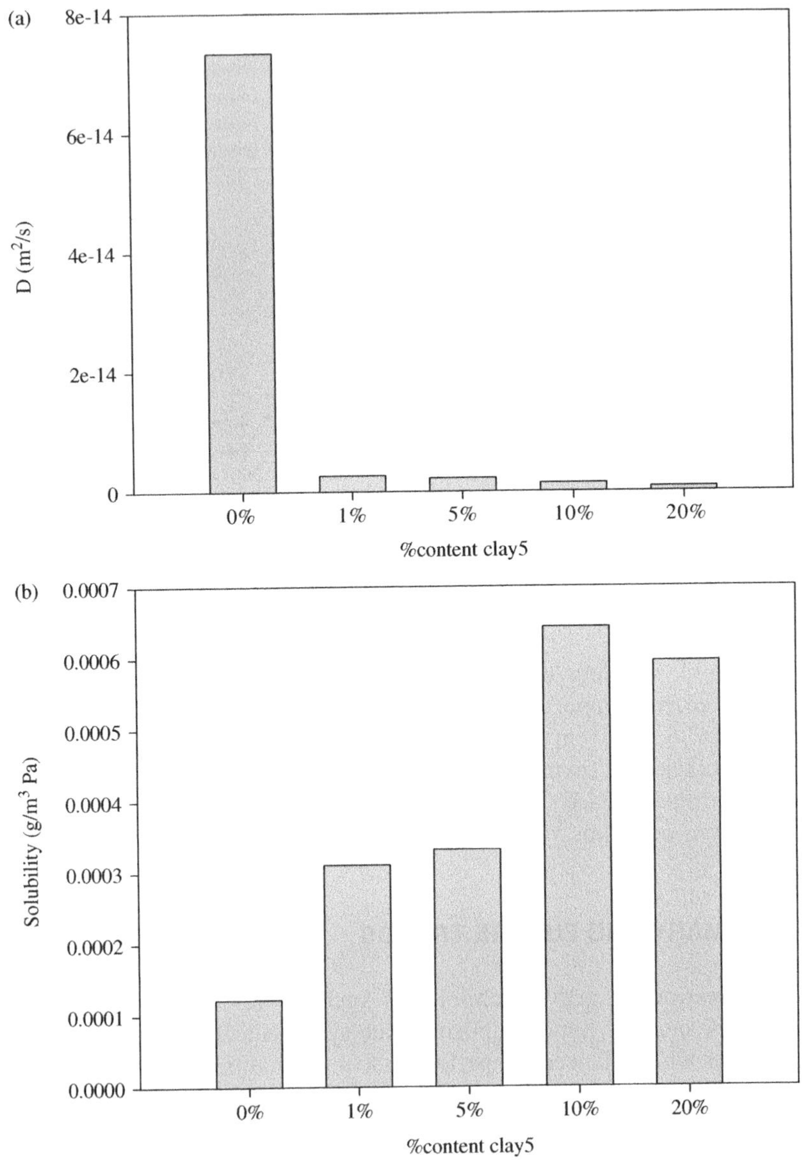

FIGURE 5.6 (a) Diffusion coefficient (m^2/s) for the release of thymol from the PCL matrix and the PCL nanocomposite with different content of clay5. (b) Solubility coefficient (g/m^3Pa) of thymol as determined by ATR-FTIR spectroscopy in neat PCL matrix and in the nanocomposite with different content of clay5.

TABLE 5.2 Diffusion and solubility coefficient of release of thymol for the PCL films and their nanocomposites containing 1, 5, 10, and 20 wt% of clay5

	D(m^2/s)	Solubility (g/m^3Pa)
PCL-10 wt% thymol	7.34^{-14}	$1.21e^{-4}$
PCL-10 wt% thymol-1% clay5	$0.25e^{-14}$	$3.10e^{-4}$
PCL-10 wt% thymol-5% clay5	$0.22 \pm 0.04e^{-14}$	$3.31 \pm 0.60e^{-4}$
PCL-10 wt% thymol-10% clay5	$0.13 \pm 0.04e^{-14}$	$6.43e^{-4}$
PCL-10 wt% thymol-20% clay5	$0.07e^{-14}$	$5.95e^{-4}$

REFERENCES

1. Azaz, A. D., Kurkcuoglu, M., Satil, F., Baser, H. C., Tumen, G. (2005). In vitro antimicrobial activity and chemical composition of some Satureja essential oils. *Flavour and Fragrance Journal*, *20*, 587–591.
2. Nicholson, M. D. (1998). The role of natural antimicrobials in food/packaging biopreservation. *Journal of Plastic Film and Sheeting*, *14*, 234–241.
3. Hotchkiss, J. H. (1997). Food packaging interactions influencing quality and safety. *Food Additives and Contaminants*, *14*, 601–607.
4. Friedman, M., Henika, P. R., Mandrell, R. E. (2002). Bacterial activity of plants essential oils and some of their isolated constituents against Campylobacter jejuni, Escherichia coli, Listeria monocytogenes and Salmonella enterica. *Journal of Food Protection*, *65*, 1545–1560.
5. Juven, B. J., Kanner, J., Sched, F., Weisslowicz, H. (1994). Factors that interact with the antibacterial action of thyme oil and its active constituents. *Journal of Applied Bacteriology*, *76*, 626–631.
6. Baratta, T. M., Dorman, D. J. H., Denes, G. S., Figueiredo, C. A., Barroso, G. J., Ruberto, G. (1998). Antimicrobials and antioxidant properties of some commercial oils. *Flavour and Fragrance Journal*, *13*, 235–244.
7. Ha, J. U., Kim, Y. M., Lee, D. S. (2001). Multilayered antimicrobial polyethylene films applied to the packaging of ground beef. *Packaging Technology and Science*, *14*, 55–62.
8. Lee, D. S., Hwang, Y., Cho, S. H. (1998). Developing antimicrobial packaging film for curled lettuce and soyabean sprouts. *Food Science and Biotechnology*, *7*, 117–121.
9. Hong, S., Park, J., Him, D. (2000). Antimicrobial and physical properties of food packaging films incorporated with some natural compounds. *Food Science and Biotechnology*, *9*, 38–42.
10. Fyfe, L., Armstrong, F., Stewart, J. (1998). Inhibition of Listeria monocytogenes and Salmonella enteridities by combinations of plant oils and derivatives of benzoic acid: The development of synergistic antimicrobial combinations. *International Journal of Antimicrobial Agents*, *9*, 130–136.
11. Koga, T., Hirota, N., Takumi, K. (1999). Bactericidal activities of essential oils of basil and sage against a range of bacteria and the effects of these essential oils on Vibrio parahaemolyticus. *Microbiology Reserve*, *154*, 267–273.
12. Dziezak, J. D. (1989). Spices. *Food Technology*, *43*, 102–115.

13. Frierheller, M. C. (1991). Modified atmosphere packaging of miscellaneous products. In: *Modified Atmosphere Packaging of Food*. Eds.: Ooraikul, B., Stile, M. E. Ellis Horwood Limited, Chichester, U.K.
14. Deans, J. G., Ritchie, G. (1987). Antimicrobial properties of plant essential oils. *International Journal of Food Microbiology*, *5*, 165–180.
15. Sinha, G. K., Gulathi, B. C. (1990). Antimicrobial and antifungal study of some essential oils and some of their constituents. *Indian Perfumer*, *34*, 126–129.
16. Reuveni, R., Fleischer, A., Putievsk, E. (1984). Fungistatic activity of essential oils from ocimum basilicum. *Journal of Essential Oil*, *110*, 20–22.
17. Couladis, M., Tzakou, O., Kujundzic, S., Sokovic, M., Mimica-Dukic, N. (2004). Chemical analysis and antifungal activity of Thymus striatus. *Phytotherapy Research*, *18*, 40–42.
18. Prasad, G., Kumar, A., Singh, A. K., Bhattacharya, A. K., Singh, K., Sharma, V. D. (1986). Antimicrobial activity of essential oils of some Ocicum species and clove oil. *Fitoterapia*, *57*, 429–432.
19. Lahariya, A. K., Rao, J. T. (1979). In Vitro antimicrobial studies of the essential oils of Cyprus scariosus and Ocimum basilicum. *Indian Drugs*, 1916, 150–152.
20. Dziezak, J. D. (1989). Spices. *Food Technology*, *43*, 102–115.
21. Guenther, E. (1952). *The Essential Oils*, Van Nostrad Co. Inc., New York, NY.
22. Arora, R., Pandey, G. N. (1977). The application of essential oils and their isolates for blue mould decay control in Citus reticulata Blanco. *Journal of Food Science and Technology*, *14*, 137–146.
23. Conner, D. E., Beuchat, L. R. (1984). Effects of essential oils from plants on growth of food spoilage yeasts. *Journal of Food Science*, *49*, 429–434.
24. Han, C. D., Kim, Y. W., Chen, S. J. (1975). Melt behaviour and blending of polymers. *Journal of Applied Polymer Science*, *19*, 28–31.
25. Suppakul, P., Miltz, J., Sonneveld, K., Bigger, S. W. (2003a). Active packaging technologies with an emphasis on antimicrobial packaging and its applications. *Journal of Food Science, Concise Reviews and Hypotheses*, *68*, 408–420.
26. Sefidkon, F., Kalvandi, R., Arti, M., Barazandeh, M. M. (2005). Essential oil variability of Thymus eriocalyx (Ronniger) Jalas. *Flavour and Fragrance Journal*, *20*, 521–524.
27. Kalvandi, R., Sefidkon, F., Atri, M., Mirza, M. (2005). Analysis of the essential oil of Thymus eriocalyx from Iran. *Flavour and Fragrance Journal*, *19*, 341–343.
28. Couladis, M., Tzakou, O., Kujundzic, S., Sokovic, M., Mimica-Dukic, N. (2004). Chemical analysis and antifungal activity of Thymus striatus. *Phytotherapy Research*, *18*, 40–42.
29. Sefidkon, F., Kalvandi, R., Arti, M., Barazandeh, M. M. (2005). Essential oil variability of Thymus eriocalyx (Ronniger) Jalas. *Flavour and Fragrance Journal*, *20*, 521–524.
30. Xu, Z., Ji, P., Burt, S. (2004). Essential oils: Their antibacterial properties and potential applications in foods—A review. *International Journal of Food Microbiology*, *94*: 223–253.
31. Dorman, H. J. D., Deans, S. G. (2000). Antimicrobial agents from plants: antibacterial activity of plant volatile oils. *Journal of Applied Microbiology*, *88*: 308–316.
32. Hammer, K. A., Carson, C. F., Riley, T. V. (1999). Antimicrobial activity of essential oils and other plant extracts. *Journal of Applied Microbiology*, *86*, 985–990.

33. Smith-Palmer, A., Stewart, J., Fyfe, L. (1998). Antimicrobial properties of plant essential oils and essences against five important food-borne pathogens. *Letters in Food Microbiology*, *26*, 118–122.
34. Cristani, M., D'Arrigo, M., Mandalari, G., Castelli, F., Sarpietro, M. G., Micieli, D., Trombetta, D. (2007). Interaction of four monoterpenes contained in essential oils with model membranes: Implications for their antibacterial activity. *Journal of Agricultural and Food Chemistry*, *55*(15), 6300–6308.
35. Xu, J., Zhou, F., Ji, B. P., Pei, R. S., Xu, N. (2008). The antibacterial mechanism of carvacrol and thymol against Escherichia coli. *Letters in Applied Microbiology*, *47*(3), 174–179.
36. Trombetta, D., Castelli, F., Sarpietro, M. G., Venuti, V., Cristani, M., Daniele, C., Saija, A., Mazzanti, G., Bisignano, G. (2005). Mechanisms of antibacterial action of three monoterpenes. *Antimicrobial Agents and Chemotherapy*, *49*(6), 2474–2478.
37. a. Sikkema, J., de Bont, J. A. M., Poolman, B. (1994). Interactions of cyclic hydrocarbons with biological membranes. *Journal of Biological Chemistry*, *269*, 8022–8028; b. Sikkema, J., de Bont, J. A. M., Poolman, B. (1995). Mechanisms of membrane toxicity of hydrocarbons. *Microbiological Reviews*, *59*, 201–222.
38. Helander, I. M., Alakomi, H.-L., Latva-Kala, K., Mattila-Sandholm, T., Pol, I., Smid, E. J., Gorris, L. G. M., von Wright, A. (1998). Characterization of the action of selected essential oil components on gram-negative bacteria. *Journal of Agricultural and Food Chemistry*, *46*, 3590–3595.
39. Burt, S. (2004). Essential oils: Their antibacterial properties and potential applications in foods–A review. *International Journal of Food Microbiology*, *94*, 223–253.
40. Burt, S. A., Reinders, R. D. (2003). Antibacterial activity of selected plant essential oils against Escherichia coli O157:H7 *Letters in Applied Microbiology*, *36*, 162–167.
41. Dorman, H. J. D., Deans, S. G. (2000). Antimicrobial agents from plants: Antibacterial activity of plant volatile oils. *Journal of Applied Microbiology*, *88*, 308–316.
42. Hammer, K. A., Carson, C. F., Riley, T. V. (1999). Antimicrobial activity of essential oils and other plant extracts. *Journal of Applied Microbiology*, *86*, 985–990.
43. Smith-Palmer, A., Stewart, J., Fyfe, L. (1998). Antimicrobial properties of plant essential oils and essences against five important food-borne pathogens. *Letters in Food Microbiology*, *26*, 118–122.
44. Cosentino, S., Tuberoso, C. I. G., Pisano, B., Satta, M., Mascia, V., Arzedi, E., Palmas, F. (1999). In vitro antimicrobial activity and chemical composition of Sardinian Thymus essential oils. *Letters in Applied Microbiology*, *29*, 130–135.
45. Helander, I. M., Alakomi, H.-L., Latva-Kala, K., Mattila-Sandholm, T., Pol, I., Smid, E. J., Gorris, L. G. M., Von Wright, A. (1998). Characterization of the action of selected essential oil components on gram-negative bacteria. *Journal of Agricultural and Food Chemistry*, *46*, 3590–3595.
46. Olasupo, N. A., Fitzgerald, D. J., Gasson, M. J., Narbad, A. (2003). Activity of natural antimicrobial compounds against Escherichia coli and Salmonella enterica serovar Typhimurium. *Letters in Applied Microbiology*, *36*, 448–451.
47. Walsh, S. E., Maillard, J.-Y., Russell, A. D., Catrenich, C. E., Charbonneau, D. L., Bartolo, R. G. (2003). Activity and mechanisms of action of selected biocidal agents on gram-positive and -negative bacteria. *Journal of Applied Microbiology*, *94*, 240–247.

48. Miltz, J., Passy, N., Mannhwim, C. H. (1995). Trends and applications of active packaging systems. In: *Food and Packaging Materials – Chemical Interaction.*, Eds.: Ackerman, P., Jägerstad, M., Ohlsson, P. pp. 201–210. The Royal Society of Chemistry, London, U.K.
49. Ahvenainen, R. (2003). Active and intelligent packaging. In: *Novel Food Packaging Techniques*. Ed.: Ahvenainen, R. Woodhead Publishing Limited, Cambridge, U.K.
50. Han, J. H. (2003). *Novel food packaging techniques*. In: *Antimicrobial Food Packaging*. Ed.: Ahvenainen, R. Woodhead Publishing Limited, Cambridge, U.K.
51. Appendini, P., Hotchkiss, J. H. (2002). Review of antimicrobial food packaging. *Innovative Food Science & EmergingTechnologies*, *3*, 113–126.
52. Gennadios, A., Hanna, M. A., Kurth, L. B. (1997). Application of edible coatings on meats, poultry and seafoods: A review. *Lebensmittel-Wissenschaft und Technology*, *30*, 337–350.
53. Real Decreto 142/2002, de 1 de febrero, por el que se aprueba la lista positiva de aditivos distintos de colorantes y edulcorantes para su uso en la elaboración de productos alimenticios, así como sus condiciones de utilización. BOE núm. 44 de 20 de febrero.
54. Sanchez-Garcia, M. D., Gimenez, E., Ocio, M. J., Lagaron, J. M. (2008). Novel polycaprolactone nanocomposites containing thymol of interest in antimicrobial film and coating applications. *Journal of Plastic Film and Sheeting*, *24*, 239–251.
55. Seydim, A. C., Sarikus, G. (2006). Antimicrobial activity of whey protein based edible films incorporated with oregano, rosemary and garlic essential oils. *Food Research International*, *39*(5), 639–644
56. Mastromatteo, M., Barbuzzi, G., Conte, A., Del Nobile, M. A. (2009). Controlled release of thymol from zein based film. *Innovative Food Science and Emerging Technologies*, *10*, 222–227
57. Sanchez-Garcia, M. D., Gimenez, E., Lagaron, J. M. (2008). Development and characterization of novel nanobiocomposites of bacterial poly(3-hydroxybutirate), layered silicates and poly(caprolactone). *Journal of Applied Polymer Science*, *108*(8), 2787–2801.
58. Sanchez-Garcia, M. D., Lagaron, J. M. (2010). Novel clay based nanobiocomposites of biopolyesters with synergistic barrier to UV light. *Journal of Applied Polymer Science*, *118*(1), 188–199.
59. Sanchez-Garcia, M. D., Hilliou, L., Lagaron, J. M. (2010). Nanobiocomposites of carrageenan, zein and mica of interest in food packaging and coating applications. *Journal of Agricultural Food and Chemistry*, *58*(11), 6884–6894.
60. Sanchez-Garcia, M. D., Ocio, M. J., Gimenez, E., Lagaron, J. M. (2008). Novel polycaprolactone nanocomposites containing thymol of interest in antimicrobial film and coating applications. *Journal of Plastic Film and Sheeting*, *24*(3–4), 239–250.
61. Schulz, H., Schrader, B., Quilitzsch, R., Pfeffer, S., Krüger, H. (2003). *Journal of Agricultural and Food Chemistry*, *51*, 2475.
62. Schulz, H., Quilitzsch, R, Krüger, H. (2003). Rapid evaluation and quantitative analysis of thyme, origano and chamomile essential oils by ATR-IR and NIR spectroscopy. *Journal of Molecular Structure*, 661–662, 299–306
63. Mastromatteo, G., Barbuzzi, A., Conte, M. A., Nobile, D. (2009). Controlled release of thymol from zein based film. *Innovative Food Science and Emerging Technologies*, *10*, 222–227

CHAPTER 6

BACTERIOCINS IN PLASTICS

GIANLUIGI MAURIELLO and FRANCESCO VILLANI
Department of Food Science, University of Naples Federico II, Portici, Naples, Italy

CONTENTS

6.1 Bacteriocins 117
6.1.1 Generalities 117
6.1.2 Types and Characteristics of LAB Bacteriocins 119
6.1.3 Sources of LAB Bacteriocins 124
6.1.4 Advantages and Limitations 125
6.2 Designing Films and Surfaces Containing Active Bacteriocins 126
6.2.1 Introduction 126
6.2.2 Factors to Take Into Account for the Production of a Bacteriocin-Based Antimicrobial Packaging 130
6.2.3 *In Vitro* Antimicrobial Activity of Films Containing Bacteriocins 142
6.3 Applications in Food Packaging 143
6.4 Future Trends 147
References 147

6.1 BACTERIOCINS

6.1.1 Generalities

Food preservation aims to reduce or eliminate the presence of spoilage and pathogenic organisms in food. Modern trends in consumer demand and food production, distribution, and legislation have led to the need to broaden the strategies of food preservation. Among the many and various existing or

Antimicrobial Polymers, First Edition.
Edited by José M. Lagarón, María J. Ocio, and Amparo López-Rubio.

proposed technologies to assure safety and stability of food, those based on the use of lactic acid bacteria (LAB) and/or their antimicrobial metabolites have been used for centuries [1, 2].

LAB are a heterogeneous group of Gram-positive bacteria that share many physiological features. They play an essential natural role in the production of fermented food because of their ability to assure desirable changes in the taste, texture, flavor, and nutritional value as well as inhibit undesirable microorganisms. Moreover, a wide variety of species and strains is employed as starter cultures in manufacture of dairy, meat, vegetable, and bakery products [3, 4], and LAB are also part of intestinal microbiota with beneficial health effects in humans [5]. The long history of use in food suggests that LABs do not cause any risk to human health and are designed as generally recognized as safe (GRAS) and are considered as "food-grade" organisms [6, 7].

Inhibitory compounds produced by LAB during food fermentations (organic acids, hydrogen peroxide, carbon dioxide, diacetyl, low-molecular-weight antimicrobial substances, and bacteriocins) are extremely important both in preventing the growth of spoilage and pathogenic bacteria [8, 9] and in influencing the composition of mixed starter cultures [10–12]. Among these substances, antimicrobial proteinaceous compounds produced by bacteria that are active against other bacteria, despite varying greatly in chemical nature, mode of action, and host specificity, have been known traditionally as bacteriocins [9].

Bacteriocin-producing bacteria may occur over the whole range of Gram-negative and Gram-positive species and have been isolated from a wide variety of habitats [13]. Bacteriocins and lantibiotics produced by LAB have been the subject of extensive studies because of their potential use as novel, natural food preservatives [2, 14]. Moreover, the study of the biochemistry and genetics of bacteriocins and their use as phenotypic markers has also provided much valuable information on the genetics and physiology of LAB [15, 16].

Bacteriocin-producing species have now been identified among all the genera that comprise the LAB, including *Lactococcus*, *Streptococcus*, *Lactobacillus*, *Leuconostoc*, *Pediococcus*, and *Carnobacterium*, as well as several *Enterococcus* spp. and *Bifidobacterium* spp. [9]. Nevertheless, the only bacteriocin approved for use as food additive is the nisin that is produced by certain strains of *Lactococcus lactis* subsp. *lactis*. It inhibits the growth of several Gram-positive microorganisms, including other LAB strains and some food-borne pathogenic bacteria [17].

The industrial interest of LAB bacteriocins has led to a vast increase in research on these compounds, which has generated a wealth of information on biochemical and genetic characteristics, spectrum of activity, mode of action, and applications [6, 18–20].

In general, bacteriocins are small cationic peptides that display hydrophobic or amphiphilic properties, and their mode of action, in most cases, targets bacterial membrane by forming pores resulting in efflux of ions and sometimes adenosine triphosphate (ATP) [21].

Some strains of LAB can synthesize multiple bacteriocins. In some cases, the structural diversity of the bacteriocins is the result of different peptides, whereas in other cases, the differences are caused by hydrolysis or oxidation of the bacteriocin. This ability should provide to the bacteriocin producer a wider inhibitory capacity and, therefore, an enhanced ability to compete with other bacteria. Moreover, the inhibitory activity of some bacteriocins depends on the complementary action of two heterogeneous peptides [11]. However, bioavailability and efficiency of antibacterial action and, hence, stability of bacteriocin activity are strongly determined by the bacteriocin environment (nutrients, pH, temperature, etc.) [22]. The genes responsible for production, regulation and immunity are generally arranged in operons [9].

The bacteriocins produced by LAB can have narrow or broad bactericidal spectrum of inhibitory activity, mostly confined to Gram-positive bacteria [23, 24]. It has been observed that Gram-negative bacteria become sensitive toward bacteriocin action only if the outer membrane is destabilized by physical or chemical stresses [25–27].

The bacteriocins produced by LAB are considered as potent food preservatives. Their use in the food industry is justified by the fact that they fulfill several requisites, such as they (1) are nontoxic to humans, (2) do not alter the nutritional and organoleptic properties of food, (3) are effective at low concentrations, (4) are sufficiently stable during storage, and (5) are accepted by recognized authorities for regulatory approval. Furthermore, there is an increased awareness by consumers of the risks arising from the use of artificial chemical preservatives used to control food-borne pathogens. This trend leads consumers to prefer naturally-produced preservatives such as bacteriocins [23]. According to Deegan et al. [6], the highly promising results of the study of existing bacteriocins and discovery of new bacteriocins looks promising for application in the food industry to improve the quality and safety of food.

Bacteriocins can be incorporated into food through different ways. In addition to the traditional use of bacteriocin-producing starter strains (bacteriocins can be produced *in situ*) in food fermentations, bacteriocins can be added directly to food as purified or semipurified preparations or by incorporating an ingredient previously fermented with a bacteriocin-producing strain [6, 18, 28].

Currently, new techniques in bacteriocin application are being directed toward their binding to polymeric packaging. This technology, which is referred to as active packaging, will be treated in this chapter.

6.1.2 Types and Characteristics of LAB Bacteriocins

The term "bacteriocin(e)" was used by Jacob et al. [29] as a general term for highly specific antibacterial proteins. More appropriately, bacteriocins produced by LAB can be defined as ribosomally synthesized peptides or proteins produced by different bacteria that inhibit strains and species that are usually, but not always, closely related to the producing bacteria. The producer has a specific immunity mechanism against its own bacteriocin(s) [22].

The first report on LAB bacteriocins started in 1928, when it was reported that some *Streptococcus* (*Lactococcus*) strains, which were used in cheese making, inhibited the growth of other LAB strains [30]. A few years later, in 1933, a proteinaceous antimicrobial substance produced by some strains of the species *Lactococcus lactis* subsp. *Lactis* was described [31] and later named nisin.

A large number of new bacteriocins produced by LAB has been identified and characterized, and several classification schemes have been proposed over the years. A first attempt of classification of LAB bacteriocins was developed by Klaenhammer in 1993 [15], on the basis of chemical structure, heat stability, molecular mass, enzymatic sensitivity, presence of modified amino acids, and mode of action of these antimicrobial compounds. According to this scheme, four classes of LAB bacteriocins are distinguished:

Class I (lantibiotics): These are small (<5 kDa) ribosomally synthesized peptides characterized by the presence of several unusual thioether amino acids (residues of lanthionine, Lan; β-methyllanthionine, MeLan; dehydroalanine, and dehydrobutyrine), which are generated through posttranslational modification. The lantibiotics (term derived from "lanthionine containing antibiotics") are further subdivided into types A (linear subtypes) and B (globular subtypes) on the basis of chemical structure and antimicrobial mode of action [32].

Class II (non-lantibiotics): Class II LAB are small (<10 kDa), heat-stable, membrane-active, non-lanthionine-containing peptides. They, unlike lantibiotics, undergo only minor posttranslational modifications of the N-terminal leader peptide. This class was divided into the following three groups or subclasses: (1) Class IIa is represented by pediocin-like *Listeria* active peptides, which contain the N-terminal located consensus sequence YGNGCXV where X is any amino acid; (2) class IIb consists of bacteriocins requiring two different peptides for activity, and (3) class IIc contains peptides whose translocation across the cytoplasmic membrane is dependent on the general (*sec*-dependent) secretory pathway [33].

Class III: Class III LAB consist of large (>30 kD) heat-labile, cell wall-active bacteriocins. Little information is available on this group of bacteriocins [34].

Class IV: Class IV LAB include complex bacteriocins that contain glycoproteins or lipoproteins that require nonproteinaceous moieties for activity. However, the definition of this class still requires additional information [35].

An additional class is added to the classification scheme, the class V, which was proposed by Kemperman et al. [36], comprising ribosomally synthesized, nonmodified cyclic antibacterial peptides.

Cotter et al. [18] proposed a simplification of the Klaenhammer classification scheme for LAB bacteriocins. They divide the bacteriocins produced by Gram-positive bacteria into two groups. Class I contains lantibiotics containing modified amino acids such as lanthionine and β-methyllanthionine, as well as several dehydrated amino acids. Class II contains non-lanthionine–containing bacteriocins, which are subdivided into subgroups IIa (listeria-active peptides) and IIb (two-peptide bacteriocins). The large heat-labile, nonlantibiotics of class III are renamed bacteriolysins, whereas the complex bacteriocins of class IV have been removed. It was suggested that the cyclic LAB bacteriocins (Kemperman's class V) are reassigned to the non-lanthionine–containing class IIc.

Another modification to the bacteriocins classification schemes has been proposed by Heng et al. [22]. In agreement with Cotter et al. [18], they proposed that classes IIc (thiol-activated peptides) and IV (chemically complex bacteriocins) were eliminated. Moreover, Klaenhammer's class III is subdivided into IIIa (bacteriolysins) and IIIb (nonlytic proteins), the cyclic LAB bacteriocins reclassified in a newly defined class IV, and class IIc was proposed to be a repository for all unmodified class II inhibitors other than the listeria-active (type IIa) and multicomponent bacteriocins (type IIb). However, according to Ennahar et al. [33], the classification of bacteriocins requires constant updating as the knowledge on the characteristics of these antimicrobials rapidly increases, making it difficult to formulate a classification scheme that is permanent and uniquely able to allocate all kinds of existing and novel bacteriocins.

Here, we will focus on classes I and II that include only the bacteriocins that have found widespread application in food, inviting the interested reader to consult the many valuable and excellent reviews published on the subject [18, 22].

Class I bacteriocins comprise linear (type A) and globular (type B) polycyclic peptides that are posttranslationally modified and contain the unusual amino acids dehydroalanine, lanthionine, and β-methyllanthionine. After the gene–encoded, prelantibiotic formation, the posttranslational modification reaction begins with the enzymatic dehydration of hydroxyl amino acids, serine and threonine, to yield 2,3-didehydroalanine (Dha) and 2,3-didehydrobutyrine (Dhb), respectively. Subsequently, Dha or Dhb condenses with the sulphydryl group of a neighboring cysteine (Cys) to form thioether bridges, yielding either lantionine (Lan) or β-methyllanthionine (MeLan) residues, respectively. The covalently closed rings within the formerly linear peptide confer both structure and functionality to the bacteriocin. The two enzymes LanB and LanC, or in some cases only one enzyme, LanM, are involved in the formation of lanthionine and methyllanthionine [18, 32]. The modified prelantibiotics undergo proteolytic processing to release the leader peptide for lantibiotic activation. For type A lantibiotics, the leader peptides are cleaved by a serine protease Lan P. This event could occur before or after the peptide is exported from the producing cell via a dedicated ABC transporter LanT, depending on the intracellular or extracellular location of the protease LanP. For type B lantibiotics, the leader peptides are cleaved concomitantly with export through a hybrid ABC transporter [37].

Most class I bacteriocins exhibit a broad inhibitory spectrum. They inhibit both closely related bacteria, such as species from the genera *Enterococcus*, *Lactobacillus*, *Lactococcus*, *Leuconostoc*, *Pediococcus*, and *Streptococcus*, as well as several less closely related Gram-positive bacteria, such as *Listeria monocytogenes*, *Staphylococcus aureus*, *Bacillus cereus*, and *Clostridium botulinum*.

Bacteriocins generally exert their antimicrobial action by interfering with the cell wall or the membrane of target organisms either by inhibiting cell wall biosynthesis or causing pore formation, which subsequently results in death. Type A lantibiotics kill the target cell through the formation of pores, leading to the dissipation of membrane potential and the efflux of small metabolites from sensitive cells [18, 38]. It is also known that some bacteriocins bind to certain docking molecules, e.g., lipid II or mannose permease. The binding of some lantibiotics, such as nisin, to lipid II facilitates a dual mechanism of action involving the prevention of peptidoglycan synthesis and pore formation. It is also thought that the concentration of docking molecule on the cell surface determines the susceptibility of the target cell [39]. The antimicrobial mechanism of type B lantibiotics is by the inhibition of specific enzyme functions [15].

Among the type A lantibiotics, the most prominent member is nisin. Nisin is the best-known and best-studied Gram-positive bacteriocin with the longest history of safe use in the food industry [40]. It was marketed first in England in 1953, whereas in 1969 it was assessed to be safe for food use by the Joint Food and Agriculture Organization/World Health Organization Expert Committee of Food Additives [41]. It is a thermoresistant, small (34 residues, with a molecular mass of 3510 Da) polycyclic peptide formed by dehydrated residues (dehydroalanine and dehydrobutyrine) and five rings of lanthionine and/or β-methyllanthionine that contain disulfur bridges [8, 18, 42, 43]. Two naturally occurring nisin variants have been produced by strains of *L. lactis* subsp. *lactis*: nisin A, being the most active type, and nisin Z, which differs in a single amino acid residue (histidine for asparagine) at position 27 [42].

In the European Economic Community, nisin was added to the food additive list as number E234 [44]. In 1988, it was approved by the Food and Drug Administration (FDA) for use in pasteurized, processed cheese spreads [45]. Nisin is a permitted food additive in more than 50 countries, and it is commercially made in a partially purified form and marketed as Nisaplin (Gillco Products, Inc., San Marcos, CA), which is a preparation that contains 2.5% nisin with NaCl (77.5%) and nonfat dried milk (12% protein and 6% carbohydrate). The formulation Nisaplin ND (nondairy nisin) does not contain milk proteins, which facilitates its solubility in aqueous solutions, and it contains nisin (6.25%), NaCl (90%), carbohydrates (2%), and proteins (approximately 2%) [41]. The stability of nisin powder is high if it is preserved from moisture and is stored at temperatures not exceeding 25°C. Nisin solubility and stability in solution are affected by pH. The solubility of nisin at pH 2.2 is approximately 56 mg/mL, dropping to 1.5 mg/mL at pH 6.0 and to 0.25 mg/mL at pH 8.5 [46]. In any case, because of the low concentrations in which nisin is used in foods, which rarely exceed 0.025 mg/mL, its solubility does not raise any problems. Heinemann et al. [47] reported that in both low-acid (pH values of 6.1 to 6.9) and highly acid foods (pH values of 3.3 to 4.5), the heat treatments for 3 min at 121°C destroyed 25% to 50% of the added nisin. It is reported that optimum heat stability of nisin occurs at pH 3.0 to 3.5, although it is suggested that food components may protect to some extent the antimicrobial substance [41].

Nisin has a broad antimicrobial activity spectrum, inhibiting the growth of several Gram-positive microorganisms, including LAB, *Brochothrix thermosphacta*, Gram-positive endospores and pathogenic bacteria such as *S. aureus*, *L. monocytogenes*, and *C. botulinum* [48]. In normal circumstances, nisin does not inhibit yeasts, molds, or Gram-negative bacteria because of their lantibiotic impermeable outer cell membrane. Nisin sensitivity could vary among strains even of the same species [17]. Nisin is suitable for use in a wide range of foods, such as dairy (especially cheeses), meat products, liquid egg products, fruit juices, and canned and packaged foods [17].

Class II of non-lanthionine–containing bacteriocins includes small (30–100 amino acids), heat-stable, and commonly not posttranslationally modified peptides. They are synthesized as prepeptide mostly containing a conserved N-terminal leader peptide and a characteristic double glycine proteolytic processing site. This leader peptide is recognized and cleaved by a dedicated ABC transporter, which results in the traslocation of the active bacteriocin into the medium [33]. Although several class II bacteriocins have been shown to be effective in the biopreservation of foods, none is licensed or marketed as a food additive in a partially purified form. Different subclasses (IIa to IIf) have been suggested for class II bacteriocins [49], although only two subclasses are common to all classification systems: class IIa pediocin-like *or Listeria*-active peptides with a consensus sequence in the N-terminus and class IIb, poration complexes formed by oligomers of two different proteinaceous peptides [18]. Class II bacteriocins are produced by a variety of LAB belonging to the genera *Pediococcus, Enterococcus, Carnobacterium, Lactobacillus*, and *Leuconostoc.*

Class IIa is the largest and most extensively studied group. It has a conserved N-terminal amino acid sequence YGNGVXCXXXXCXV or "pediocin box" with the two cysteine residues forming a disulfide bridge [50]. They show a high specific activity against the food-borne pathogen *L. monocytogenes*. Most pediocin-like bacteriocins have a narrow spectrum of activity and inhibit closely related Gram-positive bacteria. The prototype of this group is pediocin A (PA-1; formerlynamed AcH), a 4.6-kDa peptide containing 44 amino acids. It is produced by some strains of the species *Pediococcus acidilactici* (meat isolates), by two strains of *Pediococcus parvulus* (vegetable origin), and by one strain of *Lactobacillus plantarum* isolated from cheese [50]. The most outstanding feature of pediocin is its high activity against *Listeria*. This characteristic confers great technological interest. Pediocin PA-1 demonstrates broad inhibitory spectra and could inhibit certain less closely related Gram-positive bacteria, such as *S. aureus* as well as vegetative cells of *Clostridium* and *Bacillus* species [50]. In general, pediocins are relatively resistant to heat, pH, organic solvents, and hydrostatic pressure. Currently, pediocin PA-1 is commercially available as LAB fermentate that contains the bacteriocin named Alta 2341 and is used as a food preservative for ready-to-eat meat products, salads, and sauces [51, 52].

The bacteriocins of class II are bacteriocins whose activity requires the complementary activity of two peptides [53] that individually tested have little or no activity. It has been proposed that the two separate peptides could be subdivided into type E (enhancing), in which one peptide has antibacterial activity that is enhanced by the other (i.e., thermophilin 13, lactacin F, and plantaracin S), and type S (synergistic), in which the peptides have little or no activity alone (i.e., lactococcin G, plantaricin A) [49, 54, 55].

6.1.3 Sources of LAB Bacteriocins

Bacteriocin-producing species of LAB are commonly isolated from different food products in which they play an important role as starter cultures, co-cultures, or bioprotective cultures for improving and assuring their safety and stability. Bacteriocinogenic strains have been identified in the genera *Lactococcus*, *Streptococcus*, *Lactobacillus*, *Leuconostoc*, *Pediococcus*, and *Carnobacterium* as well as several *Enterococcus* spp. and *Bifidobacterium* spp. [9]. Bacteriocin production may play a role in the regulation of mixed natural population dynamics within a fermenting ecosystem [13, 56, 57]. Tagg [58] reported that all freshly bacteria isolated from their natural ecosystems, are probably equipped with the capability of expressing bacteriocins and failure in the isolation of bacteriocin producers is only because of the inability to create *in vitro* the conditions to show that the bacterial isolate can produce bacteriocins.

Although several bacteriocins have been purified and characterized, there is still a considerable interest to isolate new and more effective bacteriocin-producing strains from LAB.

Commonly, the search for new bacterial isolates producing bacteriocins involves the following steps: (1) isolation of bacteria from food products, (2) screening for bacteriocin activity against selected pathogenic or spoilage indicators, (3) characterizsation of bacteriocins and of the bacteriocin producer strain, and (4) applications of bacteriocin and/or producers in food models and in *situ* experiments.

To detect *in vitro* the antimicrobial activity of potential bacteriocin-producing bacteria, the most commonly used techniques are based on the "spot-on-lawn" approach. Simultaneous and deferred antagonism can be applied [59, 60]. In the direct or simultaneous antagonism, the indicator bacteria is added to tempered, molten agar used to prepare pour plates. Successively, the potential producing strain is spot inoculated onto the hardened agar and the cultures are concurrently incubated. The plates are then examined for zones of inhibition around the growth of producing strains. In the deferred antagonism test, the producing and indicator strains are grown separately on optimal media and at optimal incubation parameters (time, temperature, and atmosphere). The producing culture is spot inoculated on the surface of the agar medium and allowed to grow. A soft agar overlay seeded with the indicator is poured on the plate where growth of the producers occurred. After reincubation, inhibition is considered positive in the presence of a detectable clearing zone around the colony of the producer strain.

The agar-well diffusion method and critical dilution assay can be used for the detection and quantitative measurement of bacteriocins produced in liquid media. Aliquots of serial two-fold dilution of cell-free supernatant fluids of the potential producer strains, adjusted to neutral pH, are added to wells cut in agar plates previously inoculated with the indicator strain and allowed to diffuse. After incubation, the plates are examined for zones of inhibition. Bacteriocin activity is expressed as activity (arbitrary) units per milliliter (AU/mL). One arbitrary unit (AU) was defined as the reciprocal of the highest dilution yielding a zone of growth inhibition on the indicator lawn [60–62].

To verify the proteinaceous nature of a newly potential bacteriocin, hydrolysis by various proteolytic enzymes (trypsin, α-chymotrypsin, pepsin, papain, and proteinase K) can be applied. In the deferred antagonism test, inhibition from proteinaceous substances could be confirmed by spotting aliquots of proteolytic enzymes solutions near the spotted producer strain, which resulted in the disappearance of the inhibition zone near the enzyme spot. Proteolytic enzyme solutions could also be added to supernatant fluids in which the bacteriocin producer is grown. After incubation at 37°C for 1 h, the residual bacteriocin activity could be assayed by the agar-well diffusion assay [63, 60]. New and more rapid methods of bacteriocin detection have been proposed. Matrix-assisted laser desorption/ionization time-of-flight mass spectrometry, was developed for rapid examination of culture supernatants for the presence of some bacteriocins [64]. Furthermore, polymerase chain reaction, dot-blot, and colony hybridization methods have been used to detect the genes responsible for bacteriocin production and regulation in bacterial cultures [65–67].

6.1.4 Advantages and Limitations

The bacteriocins produced by food-grade LAB attracted attention for their potential use in food control of spoilage and pathogenic microorganisms, as well as for the general tendency to decrease the use of chemical additives. Their use as ideal food biopreservative is justified by the several fulfilled requisites for a food preservative [68].

Bacteriocins have proven nontoxic to humans. They have been consumed for centuries as products of LAB but their use in food as purified substances requires legal approval by the regulatory agencies of different countries. Currently, nisin is the only purified bacteriocin approved for food use in the United States and in Europe. Its nontoxicity has been established on data collected in approximately 20 years of research regarding its safety. Claypool et al. [69] compared the residual antimicrobial activity of nisin and penicillin in saliva after the consumption of chocolate milk containing the two antimicrobial substances. It was found that 1 min after the consumption, only 1/40 of the activity of the original nisin concentration could be detected in the saliva. In contrast, it was found that when the chocolate milk contained penicillin, the saliva showed antibacterial activity for a greater length of time. Hara et al. [70] showed that trypsin inactivated the nisin. This study supports the conclusion that ingested nisin would not have an effect on the gut microbiota. In 1969, the FAO/WHO Expert Committee indicated that a level of 3.3×10^6 units/kg body weight had no adverse effect, allowing an acceptable daily intake of 3.3×10^4 units/kg body weight [71]. Nisin was found to have no effect on animals and the oral LD50 in mice is 6950 mg/kg body weight [70]. In 1983 nisin was added as number E234 to the food additive list of European Community [44], whereas in 1988 it was approved by the FDA as GRAS for use as a direct ingredient in human food [45].

The use of bacteriocins in a food system has many advantages over traditional preservatives. They are compatible with most existing food technologies and are sufficiently stable during storage. Moreover, they do not affect the organoleptic and sensory properties of foods, do not alter their nutritional properties, and are active under refrigerated storage and effective at low concentrations [2, 72, 73].

Nevertheless, the use of bacteriocins in food can present some drawbacks. One main limitation is related to the fact that the chemical composition and physical conditions of food can negatively influence the bioavailability and stability of the molecules and, thus, the *in situ* antimicrobial efficacy of the bacteriocins. This reduced efficacy may be attributed to a range of factors including a limited diffusion of the bacteriocin throughout the product resulting in an uneven distribution, the binding of the antimicrobial substance to fat and/or protein food components, or the inactivation by food natural proteases [25, 72, 74–76]. Another important limitation is that most bacteriocins are generally active against Gram-positive and not against Gram-negative bacteria, yeasts, or molds. Treatments such as hydrostatic pressure, heat, freezing and thawing, and chelating agents such as ethylenediaminetetraacetic acid (EDTA) that alter the outer membrane barrier functions could potentiate bacteriocins' activity in Gram-negative bacteria [26, 27, 76]. These findings suggest a more effective use of bacteriocins in hurdle technology for food preservation [6].

Other limitations to the use of bacteriocins can be derived from the fact that their preparation in purified form requires many technical skills and that there is a need to overcome regulatory barriers to their use. These facts increase the costs.

6.2 DESIGNING FILMS AND SURFACES CONTAINING ACTIVE BACTERIOCINS

6.2.1 Introduction

Modern food packaging is believed to have begun in the 19th century with the invention of canning by Nicholas Appert [77]. However, although they were discovered in first half of the 19th century, synthetic plastics have been used in the food packaging only in the second part of 20th century, after the World War II. In fact, plastic is the newest packaging material in comparison with metal, glass, and paper. During the early 20th century, plastic jars, tubs, and films from polyolefins, polyvinyl, polyethylene, vinylidene, vinyl chloride, surlyn, and nylon became the packaging of choice for food manufacturers as well as consumers [78]. Recently, materials derived from renewable biomass sources have found large interest in the field of food packaging. Unfortunately, excluding of cellophane, which is a cellulose-based material already extensively used in food packaging for many years, no bioplastic material has currently found a large application in the food packaging industry.

According to Brandsch and Piringer [79], plastics can be classified according to whether they are made from converted natural products or from completely synthetic products. In all cases, plastics are polymers of high molecular mass. Among the polymers arising from fossil-based raw materials, polyethylene is the most widely used, mass-produced plastic. Beer et al. [80] estimated the worldwide capacity of polyethylene (PE) in 2007 to 79 million tons per year. Moreover, the Global Business Intelligence research report on the global market of low-density polyethylene (LDPE) refers to a global LDPE demand in 2009 close to 18.4 million tons, whereas the industrial market research institution Ceresana Research refers in "Market Study: Polyethylene—LDPE" that in 2009 the global LDPE market had a volume of approximately $22.2 billion (15.9 billion Euros). However, polypropylene (PP) and poly(vinyl chloride) (PVC) are extensively used plastics. In fact, films from PE, PP, and PVC are recognized to have optimal technological features for food packaging applications. A list of polymers used for food packaging purposes and belonging to the plastics and bioplastics group is provided in Table 6.1. It is important in this chapter to

TABLE 6.1 Common plastics and bioplastics used in food packaging

Plastics		Bioplastics	
Material	Abbreviation	Material	Abbreviation
Low-density polyethylene	LDPE	Protein- (wheat, soy, corn zein, and whey) based polymers	—
Linear low-density polyethylene	LLDPE	Starch-based polymers	—
High-density polyethylene	HDPE	Gluten-based polymers	—
Polypropylene	PP	Celulose-based polymers	—
Polybutene	PB	Hydroxypropyl cellulose[a]	HPC
Polymethylpentene	PMP	Polylactic acid	PLA
Polyethylene terephtalate	PET	Polycaprolactone	PCL
Polystyrene	PS	Polyhydroxyalkanoates	PHAs
Polyvinyl alcohol	PVOH	Polyhydroxybutyrate[b]	PHB
Polyvinyl chloride	PVC	Chitosan	—
Polytetrafluoroethylene	PTFE		
Ethylene vinyl alcohol	EVOH		
Ethylene vinyl acetate	EVA		

[a]One of the most used cellulose-based polymers.
[b]One of the most used polyhydroxyalkanoates.

underline the differences among biobased, biodegradable, and edible bioplastics. Notably, biobased and biodegradable are not equal terms. The biobased materials may have biodegradability as one of their properties, whereas biodegradable materials not necessarily are biobased [81]. Finally, biobased materials are not necessarily edible.

According to the now no longer new concepts, food-packaging materials do not provide only barrier and protective functions. Various kinds of active substances can now be incorporated into the packaging material to improve its functionality and give it new or extra functions [82]. Labuza and Breen [83] definitively introduced the concept of "active packaging," which was defined as "an intelligent or smart system that involves interactions between package or package components and food or internal gas atmosphere and complies with consumer demands for high quality, fresh-like and safe products." Within the active packaging family, which comprises a large variety of packaging types, the antimicrobial packaging could be a potential alternative to control the growth of spoilage and pathogenic bacteria in food. In fact, antimicrobial food packaging acts to reduce, inhibit, or retard the growth of microorganisms that may be present in the packed food or packaging material itself [84]. Many antimicrobial substances, both organic and inorganic, have been used to confer antimicrobial activity to both synthetic and bio-based polymers for food-packaging applications. The incorporation of antimicrobials into the packaging can reduce the addition of larger quantities of antimicrobials that are usually added to food. The gradual release of an antimicrobial from a packaging film to the food may have an advantage over dipping and spraying. In these processes, the antimicrobial activity may be reduced by food components or dilution below active concentration because of migration into the food matrix [84]. This scenario seems particularly true for proteinaceous antimicrobials such as the bacteriocins, which have shown to confer effective antimicrobial activity to food packaging materials.

The use of bacteriocins to generate active plastic materials has received a considerable attention in the last ten years. Daeschel et al. [85] published one of the first contributions in this field in which they reported the use of nisin to activate hydrophilic and hydrophobic silicon surfaces, which can be in contact with food. Since then many articles describing the application of bacteriocins for the development of antimicrobial food packaging solutions have been published. Bacteriocins have been used for the functionalization of many different polymers, both synthetic and natural. In the latter case polymers such as cellulose [86–90], chitosan [91, 92], polylactic acid [93, 94], and pullulan [95] were used. In contrast, synthetic polymers and copolymers such as PE, LDPE, linear LDPE (LLDPE), ethylene vinyl acetate (EVA), PVC, poly(vinyl alcohol) (PVOH), biaxially oriented polypropylene, and nylon have been handled to produce antimicrobial food packaging by using bacteriocins or other proteinaceous antimicrobial compounds. All these polymers are usually known as plastics and they are also used to produce flexible plastic films for food packaging application. Usually, an extrusion process is carried out to obtain the

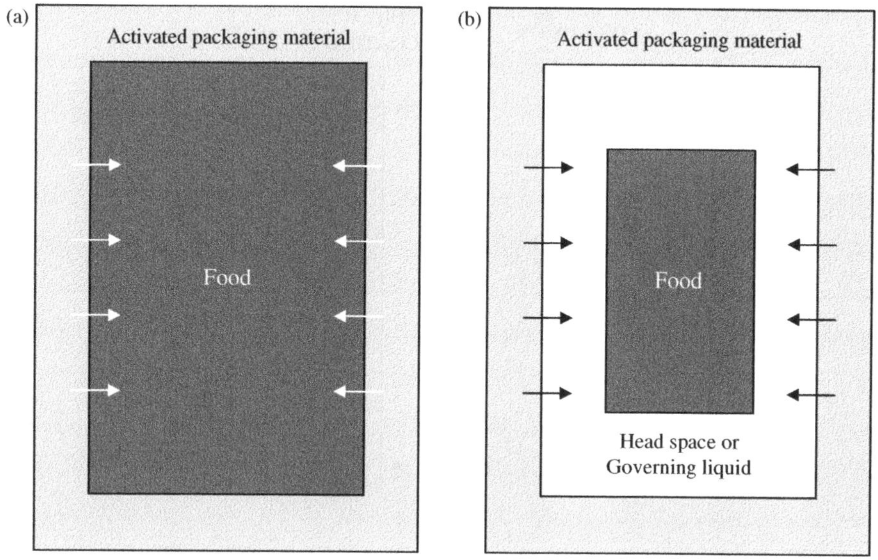

FIGURE 6.1 Action of an antimicrobial packaging by direct contact (a) or through the release of the antimicrobial into the headspace (b).

flexible plastic films. The flexible packaging industry registered in the last years a big growth in all the sectors and specifically in the food packaging, mainly because of the lowest production cost and reduced storage space needed compared with other packaging solutions.

The action of an antimicrobial packaging can be performed by direct contact of the packaging with the food or through the release of the antimicrobial into the headspace of the packed food (Figure 6.1). According to the classification of Appendini and Hotchkiss [84], the bacteriocin incorporation in the packaging materials may be carried out through (1) incorporation of the antimicrobial agents directly into polymers and (2) coating or adsorbing antimicrobials onto polymer surfaces.

In the case of active antimicrobial packaging containing bacteriocins, the antimicrobial performance is achieved by direct contact of the food with the active packaging material. In fact, bacteriocins are nonvolatile substances and they can act, as previously described, after the contact with the target cell. The incorporation into the packaging material can be carried out by including bacteriocins within the polymer matrix or by surface coating (Figure 6.2). Overall, binding of antimicrobials to polymeric surfaces has been achieved by different means, which range from simply spreading antimicrobial solutions onto the polymer surface, or by more sophisticated means, such as combining the antimicrobial with binders [96]. The bacteriocin inclusion in the plastic material can be achieved by blending protein with the polymer granules before the extrusion process. In the case of some biobased materials, bacteriocins can be incorporated into the polymer matrix by a casting process.

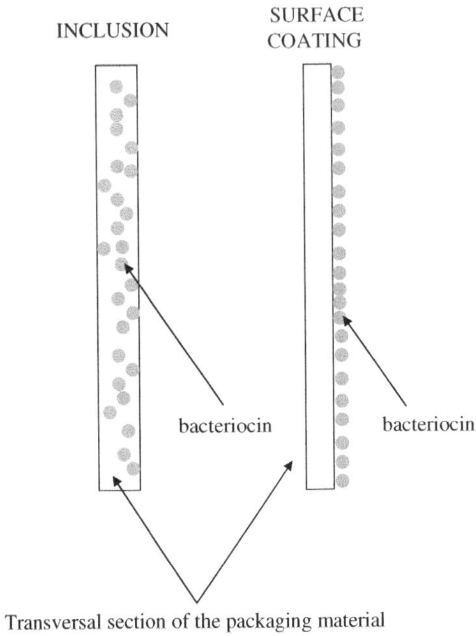

FIGURE 6.2 Schematic representation of a plastic film containing a bacteriocin into the polymer matrix (inclusion) or onto the surface (surface coating).

6.2.2 Factors to Take Into Account for the Production of a Bacteriocin-Based Antimicrobial Packaging

According to Han [82], several factors affect the design or modeling of an antimicrobial film or package. This is true also when a bacteriocin-based antimicrobial packaging has to be designed.

6.2.2.1 Bacteriocin Inactivation. First of all, the possible inactivation of the antimicrobial during the process and the storage of packaging must be taken into account. Bacteriocins are proteins and like proteins, they are susceptible to partial or total thermal inactivation. Furthermore, a prolonged exposition to nonrefrigerated temperatures may lead to a modification of the protein chemical structure with a consequent loss/reduction of the antimicrobial activity. As reported previously, bacteriocins are heat-resistant molecules. However, the temperatures used during the extrusion process, which are set according to the melting point of the polymer used, are probably too high to preserve the antimicrobial activity of bacteriocins totally. Nevertheless, Siragusa et al. [97] and Cutter et al. [98] demonstrated the retention of nisin activity when incorporated into LDPE-based films by an extrusion process carried out at 120°C. Actually, this temperature is not common; generally, temperatures are higher for an industrial PE extrusion process aimed to get a transparent flexible film. On the contrary, a PE film obtained by the extrusion of a master batch containing the antilisterial bacteriocin 32Y did not show antimicrobial activity when placed in contact with the target microorganism [99]. Beside these few experiences of bacteriocin inclusion in plastic materials by an extrusion method, many works describe the incorporation of bacteriocins into bioplastic materials by casting methods. Overall, the inclusion of bacteriocins in biopolymers by solvent-casting methods did not lead to a reduction of the antimicrobial activity. In fact, the antimicrobial activity of bacteriocins is not generally affected by either organic solvent contact or the temperature used for the drying of biopolymer solutions. About the reduction/loss of the antimicrobial activity of bacteriocins during the storage of the active packaging, some contributors take into account this relevant aspect in the design of an active packaging. Scannel et al. [86] investigated the antimicrobial stability of cellulose-based material impregnated with either lacticin 3147 or nisin solution during the storage at ambient and refrigerated temperatures. They found an apparent decrease in activity for both adsorbed bacteriocins in the first week of storage; the activity thereafter was constant for 4 months. Likewise, Mauriello et al. [100] demonstrated that the activity of a bacteriocin-coated antilisterial PE film was still stable after 4 months of storage at room temperature.

6.2.2.2 Evaluation of the Residual Bacteriocin Activity. It could be interesting, according to some authors, to know the "residual antimicrobial activity" of the packaging in terms of activity of the material with respect to the antimicrobial substance activity before the process. As reported previously, the antimicrobial activity of the bacteriocins is usually expressed as AU/mL, and it is referred to a specific bacterial strain (Figure 6.3). Only for the nisin, the activity can be also expressed in international units per mL (IU/mL) [101]. In any case, the bacteriocins are tested in solution. The antimicrobial activity of a plastic material may be qualitatively or quantitatively measured. In the first case, the packaging is usually judged as "active" or "nonactive" (sometimes "slightly active") after an experiment in which the material is placed in contact with a bacterial population grown in an agar medium (Figure 6.4). It is judged "active" when a growth inhibition zone appears underneath and around the active plastic sample. In some cases, this method can be inappropriately adapted as a quantitative evaluation method by measuring the diameter/surface of the growth inhibition zone. In contrast, a quantitative measurement can be expressed as the capability to reduce the cell load, generally as a function of time, of a particular medium (food or non-food, solid or liquid) (Figure 6.5). In this context, it is not an easy task to compare the antimicrobial activity of the packaging with the antimicrobial activity of bacteriocin before incorporation into the plastic to evaluate the "residual antimicrobial activity." An alternative quantitative method to evaluate the activity of a plastic film coated with a bacteriocin solution has been proposed by Ercolini et al. [102], who carried out a fluorescence viable staining of the bacteriocin susceptible cells directly on the surface of the film. The advantage of this method is to evaluate rapidly

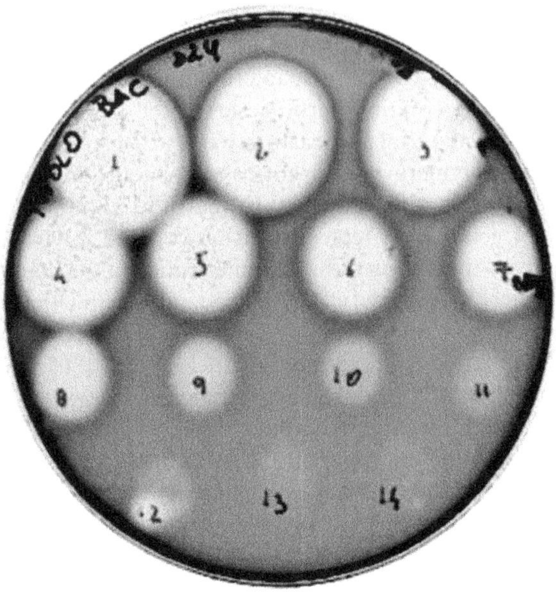

FIGURE 6.3 Evaluation of the arbitrary units per ml (AU/mL) of a bacteriocin solution.

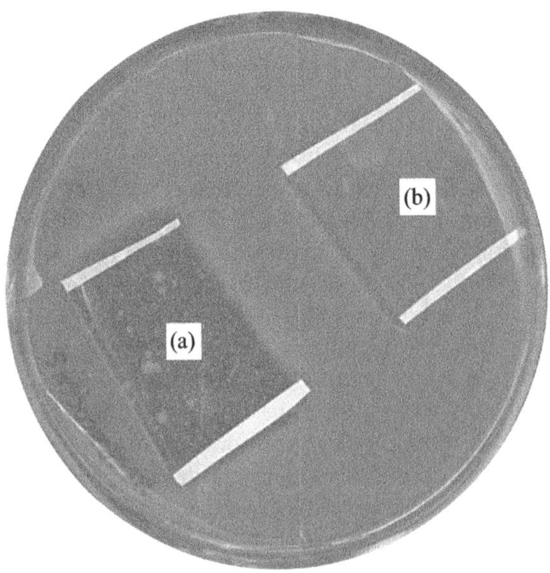

FIGURE 6.4 Evaluation of the antimicrobial activity of a plastic film treated with a bacteriocin. (a) Treated film. (b) Control film.

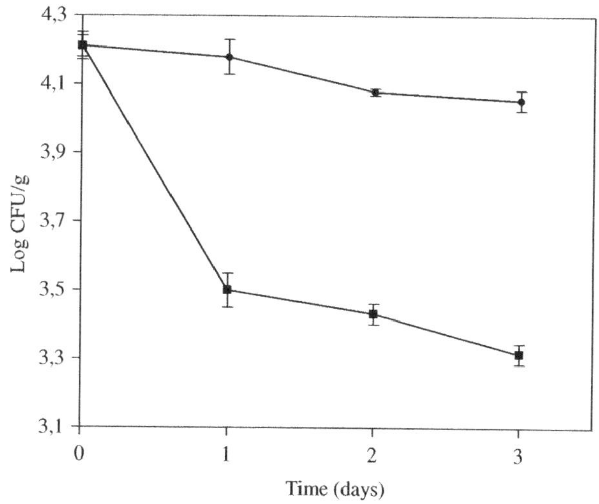

FIGURE 6.5 A quantitative evaluation of the antimicrobial activity of a PE film after a bacteriocin coating. Viable counts of listeria population during the storage at 4°C of an artificially contaminated solid food. (■) Food packed with the bacteriocin activated PE film. (•) Food packed with untreated film.

and efficiently the antimicrobial activity of the antimicrobial packaging. Appendini and Hotchkiss [84] reported the determination of the minimum inhibitory concentration as a method to assess the activity of antimicrobial packaging. However, to the best of our knowledge, no author applied this method for bacteriocin-containing active packaging. It could be interesting to highlight that ISO 22196:2007 specifies a method of evaluating the antibacterial activity of antibacterial-containing plastic products (including intermediate products), based on the film contact method implemented previously in Japan by the Society of Industrial Technology for Antimicrobials Articles [103]. Finally, Leung et al. [104] described an immunoassay method for determining the surface concentration of nisin on ethylene acrylic acid copolymer films activated by immersion into nisin solutions.

6.2.2.3 Microbial Target. It is crucial to consider the microbial target of the food packaging in manufacturing of antimicrobial plastic films because of the relatively narrow activity spectrum of bacteriocins. Bacteriocins show antimicrobial activity only against Gram-positive bacteria. We will observe later that it is possible to extend the antimicrobial activity also against Gram-negatives by combining bacteriocins with other antimicrobial substances. Food-related bacteria are usually divided in three groups as follows: (1) spoilage bacteria, (2) pathogenic bacteria, and (3) useful bacteria (related to food fermentations). Other food-related microorganisms, which are not included in the mentioned groups, do not have a known role in affecting the properties of foods. Moreover, as well known, the scenario is complicated by the possible overlapping between the spoilage and useful bacteria. A bacteriocin-based antimicrobial food packaging can be designed to be active against one or more of these groups of bacteria. Of course, the target bacteria will be mainly correlated to the food type (e.g., origin, fermented or not, shelf life expected, storage condition, and bacterial hazard generally associated). Definitively, based on the bacteriocin used, the packaging will perform antimicrobial activity against a wide or narrow spectrum of bacterial species and will be applied to control the growth of pathogenic and/or spoilage bacterial species. In a recent review, Joerger [96] analyzed quantitative data on the food application of antimicrobial films and reported that nisin is the antimicrobial most frequently used either as the only antimicrobial or in conjunction with other antimicrobials. Accordingly, nisin is also the bacteriocin most frequently used in the active packaging area. The choice of nisin could be suggested for several reasons, the first of which is without doubt the easy commercial availability. Moreover, nisin is the only bacteriocin licensed as a food additive; its spectrum of activity comprises numerous pathogenic and spoilage bacterial species, the combination with other antimicrobial substances has proved to extend the activity against Gram-negative bacteria. Other than nisin, other bacteriocins have been used by several authors to obtain an effective antimicrobial packaging. In Table 6.2 [105–166], a list of proposed antimicrobial packaging containing bacteriocins both alone and in combination with other antimicrobials is presented.

6.2.2.4 Bacteriocin Migration from the Packaging Material. The antimicrobial effectiveness of an active packaging is strictly related to the mechanism of action of the antimicrobial principle incorporated in the plastic material. The mechanism of action of bacteriocins or, at least, what is the most likely mechanism of action, involves the contact with bacterial target, as was explained previously. Consequently, the bacteriocin migration is crucial, and its kinetic, from the plastic material. On the other hand, most of the developed antimicrobial packaging contemplates the diffusion of the antimicrobial substance from the plastic material into the food matrix. Migration is generally considered as a diffusion process; therefore, the diffusion coefficient of bacteriocins in the packaging layers could be used for the evaluation of the ability of the substance to be released. Some authors have developed mathematical models to describe this phenomenon, particularly in biobased plastic films. Among them, in two recent works, Guiga et al. [154, 162] reported a "desorption coefficient" of nisin from hydroxypropyl methylcellulose (HPMC), whereas other authors calculate the best known "diffusion coefficient" [115, 119, 157, 159, 163, 164]. Chollet et al. [153], who reported both coefficients, monitored the nisin desorption from an active multilayer PE-based film with a coating of bacteriocin-loaded HPMC and the nisin diffusion in agarose gels supplemented with two different fat concentrations. They found a strict correlation between nisin mass transfer from the film to the artificial food system and both fat

TABLE 6.2 Bacteriocins, plastic materials and microbial targets used in each work describing the antimicrobial properties of plastics containing bacteriocins

Bacteriocin	Plastic material	Target microorganism	References
Nisin	Silicon	*Listeria monocytogenes*	105
Bacteriocins from *Lactococcus lactis* 7962 and *Pediococcus acidilactici* K	Cellulose	*Listeria monocytogenes*	106
Nisin; Nisin +	Soy protein; corn zein	*Lactobacillus plantarum*; *Escherichia coli*	107
Nisin	LDPE	*B. thermosphacta*; *Lactobacillus helveticus*	97
Nisin; Lacticin NK24	Polyamide (coated LDPE)	*Micrococcus flavus*	108
Nisin +	PVC, LLDPE, nylon	*Salmonella* Thyphimurium	109
Nisin +	Zein	*Lactobacillus plantarum*	110
Nisin; Lacticin 3147	Cellulose-based paper; PE:PA	*Lactococcus lactis*; *Listeria innocua*; *Staphylococcus aureus*	86
Nisin	HPMC/PEG	*Listeria innocua*; *Staphylococcus aureus*; *Micrococcus luteus*	87
Nisin; Nisin +	PE; PE/PEO	*Brochothrix thermosphacta*	98
Nisin; Nisin +	Corn zein	*Listeria monocytogenes*; *Salmonella* Enteritidis	111
Nisin	Whey protein; Soy protein; Egg albumen; Wheat gluten	*Listeria monocytogenes*	112
Nisin +	Soy protein	*Listeria monocytogenes*	113
Nisin; Nisin +	Zein	*Listeria monocytogenes*	115
Nisin	EVA (coated paper); Acrylic polymer (coated paper)	*Micrococcus flavus*	115
Nisin; Lacticin NK24	Polyamide (coated LDPE)	Total aerobic bacteria and coliform bacteria of fresh oysters and ground beef	116
Nisin	HPMC	*Listeria innocua*; *Staphylococcus aureus*	117
Nisin	HPMC	*Listeria monocytogenes*; *Staphylococcus aureus*	88

(*Continued*)

TABLE 6.2 (Continued)

Bacteriocin	Plastic material	Target microorganism	References
Nisin	Corn-zein proteins; Wheat gluten	*Lactobacillus plantarum*	118
Nisin	Crosslinked PVOH	Antimicrobial activity was not determined	119
Nisin	Corn-zein proteins; Wheat gluten	*Lactobacillus plantarum*	120
Nisin	Cellulose (coated PE)	*Listeria monocytogenes*	121
Nisin	MC/HPMC (coated PE)	*Listeria monocytogenes*	122
Nisin +	EVA (coated paper)	*Micrococcus flavus*; *Escherichia coli* O157:H7; *Listeria monocytogenes*	123
Nisin	LDPE; LLDPE; Primacor 3440 (Dow Chemical, Freeport, TX); Primacor 3340; EVOH	*Lactobacillus leichmanii*	124
Nisin	HPMC	*Micrococcus luteus*	125
Nisin +	Soy protein	*Listeria monocytogenes*; *Escherichia coli* O157:H7; *Salmonella gaminara*	126
Nisin	MC/HPMC (undeclared coated film)	*Listeria monocytogenes*	127
Nisin	MC/HPMC/PEG (coated LDPE)	*Listeria monocytogenes*	128
Nisin	MC/HPMC/PEG (coated LDPE)	*Listeria monocytogenes*	129
Nisin; Nisin +	EVA (coated papepr)	Total aerobic bacteria of pasteurized milk, yeasts of raw orange juice	89
Nisin; Nisin +	EVA (coated papepr)	Total aerobic bacteria of pasteurized milk cream	90
Nisin	Cellulose	*Listeria monocytogenes*	130

Bacteriocin from *Lactobacillus curvatus* 32Y	PE/OPA	*Listeria monocytogenes*	100
Nisin	Cellophane	*Enterococcus hirae*; total mesophilic bacteria	131
Nisin	K-carrageenan; dextrin, chitosan; MC; HPMC	*Lactobacillus plantarum*	132
Nisin +	Zein	*Listeria monocytogenes*	133
Nisin	LDPE	*Micrococcus luteus*	134
Nisin	Wheat gluten	*Listeria monocytogenes*; *Salmonella* Thyphimurium	135
Nisin +	Chitosan	*Escherichia coli*; *Staphylococcus aureus*; *Salmonella* Thyphimurium; *Listeria monocytogenes*; *Bacillus cereus*	91
Bacteriocin from *Pediococcus parvulus* VKMX133	Zein-whey protein	*Listeria innocua*	136
Bacteriocin from *Lactobacillus curvatus* 32Y	PE-OPA	*Listeria monocytogenes*	102
Nisin +	Konjac glucomannan	*Escherichia coli*; *Staphylococcus aureus*; *Listeria monocytogenes*; *Bacillus cereus*	137
Nisin +	Konjac glucomannan	*Escherichia coli*; *Staphylococcus aureus*; *Listeria monocytogenes*; *Bacillus cereus*	138
Nisin	Starch	*Listeria innocua*	139
Nisin +	Soy protein	*Listeria monocytogenes*	140
Enterocins A and B from *Enterococcus faecium* CTC492; Sakacin K from *Lactobacillus sakei* CTC494; Nisin; Nisin +	three-layer film: perforated PP/PA/non perforated PP	*Listeria monocytogenes*	141
Nisin	PLA; Pectin/PLA	*Lactobacillus plantarum*	142
Enterocins A and B from *Enterococcus faecium* CTC492	Arginate; Zein; Polyvinyl alcohol	*Listeria monocytogenes*	143

(*Continued*)

TABLE 6.2 (Continued)

Bacteriocin	Plastic material	Target microorganism	References
Nisin	MC/HPMC/PEG (coated LDPE; EVA; Nucrel® 0403; Nucrel® 0903)	*Lactococcus lactis* subsp. *Lactis*; *Listeria monocytogenes*	144
Lactocins 705 and AL705 from *Lactobacillus curvatus* CRL705	LLDPE/EVA	*Lactobacillus plantarum*; *Listeria innocua*	145
Nisin +	Calcium alginate	*Listeria monocytogenes*; *Salmonella* Anatum	146
Nisin; Nisin +	Whey protein isolate	*Listeria monocytogenes*; *Escherichia coli* O157: H7; *Salmonella* Thyphimurium	147
Enterocin 416K1 from *Enterococcus casseliflavus* IM 416K1	PE/PEG (coated LDPE)	*Listeria monocytogenes*	148
Nisin	PLA	*Listeria monocytogenes*; *Escherichia coli* O157: H7; *Salmonella* Enteritidis	93
Nisin	Sodium caseinate	*Listeria monocytogenes*	149
Nisin	Bacterial cellulose	*Listeria monocytogenes*	150
Nisin; Nisin +	Cellulose-derivative polymer	*Staphylococcus aureus*; *Listeria monocytogenes*; *Penicillium* sp.; *Geotrichum* sp.	151
Nisin; Nisin +	Soy protein	*Listeria monocytogenes*; *Escherichia coli* O157: H7; *Salmonella* Thyphimurium	152
Nisin	HPMC (coated PE/PA/PE three-layer film)	*Kocuria rizophila*	153
Nisin	HPMC (coated PE/PA/PE three-layer film)	*Kocuria rizophila*	105
Nisin	Pectin/PLA	*Listeria monocytogenes*	155
Nisin +	Alginate	Total mesophilic and psychrophilic microflora of fresh northern snakehead (*Channa argus*) fillets	156
Nisin	Whey protein isolate	*Micrococcus luteus*; *Brochothrix thermosphacta*	157
Pediocin	Cellulose	*Listeria innocua*; *Salmonella* sp.	158

Nisin	Polybutylene adipate-co-terephthalate	Antimicrobial activity was not determined	159
Nisin	Sodium caseinate	*Listeria innocua*	160
Nisin +	Bilayer film: co-extruded VDC-EVA/LDPE	Total mesophilic microflora, lactic acid bacteria, *Brochothrix thermosphacta*, *Carnobacterium* spp. and *Enterobacteriaceae* of beef	161
Nisin	Three-layer film: ethylcellulose/HPMC plus nisin/ethylcellulose	*Kocuria rizophila*	162
Nisin	PVdC (coated LDPE)	*Lactobacillus helveticus*	163
Nisin +	PA/LDPE	*Penicillium expansum*; *Penicillium culmorum*; *Lactobacillus helveticus*; *Listeria ivanovii*; *Listeria monocytogenes*; aerobic sporeforming bacteria and *Bacillus cereus* of Blatcke zlato cheese	164
Nisin +	PLA	*Escherichia coli* O157:H7	94
Nisin +	Whey protein	*Listeria monocytogenes*; *Pseudomonas aeruginosa*; *Yarrowia lipolytica*; *Penecillium roqueforti*; *Penicillium chrysogenum*; *Penicillium commune*	165
Sakacin A from *Lactobacllus sakei* DSMZ 6333	Pullulan	*Listeria monocytogenes*	95
Enterocin EJ97 from *Enterococcus faecalis* EJ97	LDPE	*Bacillus coagulans*	166

Nisi +: nisin supplemented with other substances.

content and nisin purity. Guiga et al. [162] illustrated the nisin desorption kinetics from multilayer cellulose-based films into saline solution by applying mass balance equations. They found the total nisin desorption from two-layer films within 30 min, whereas a complete desorption occurred in approximately 20 h when three-layer films (nisin entrapped in the inner layer) were used. In addition, desorption modeling showed that the use of three-layer films obtains desorption coefficients 118 times lower than with two-layer films. According to Buonocore et al. [119] the release kinetics of an active compound from a highly hydrophilic polymer placed in contact with water depends on (1) water diffusion, (2) macromolecular matrix relaxation kinetic, and (3) diffusion of the active compound through the swollen polymeric network. These authors prepared a cross-linked PVOH-based film with nisin and developed a mathematical model taking into account the previously mentioned involved phenomena to describe the antimicrobial agent release kinetic from the active material into distilled water. Hanusova et al. [163] calculated the bacteriocin diffusion coefficient from PE films coated with polyvinyldichloride varnish supplemented with different concentrations of nisin into acidified physiological saline solution. These authors found that the diffusion coefficient was slightly related to the coating thickness and, surprisingly, that only less than 10% of the nisin added to the films was released. Although the different authors present similar findings both in terms of desorption and diffusion, the release of the bacteriocins overall from the materials is mainly related to film thickness, polymer matrix type, and temperature. Of course, other important factors investigated infrequently, such as food composition, may affect the release of bacteriocin. It should be noted that the bacteriocin release kinetics can be evaluated in the following two different ways: (1) monitoring the quantity of the bacteriocin released into the contact solution (food simulant) as a function of time or (2) monitoring the quantity of the bacteriocin remaining within the polymer material during time. Some authors investigated the bacteriocin release kinetics from plastic films without the need to develop prediction models. For example, Grower et al. [128] measured the diameter of growth inhibition halos of *L. monocytogenes* cultured in contact with peptone water (10-μL drops) in which a nisin-activated film (LDPE coated with MC/HPMC/polyethylene glycol (PEG)/nisin solution) sample was been immersed for 1 min, 30 min, 60 min, 8 h, 24 h, 4 days, and 8 days. The plastic films were activated with three different concentrations of nisin. The authors found an uncontrolled release of nisin; in fact, the film with the lower concentration of nisin did not release bacteriocin between 60 min and 4 days; the other two films did not release nisin between 24 h and 4 days. However, all the films released nisin after 8 days of contact, in such quantities to determine the larger growth inhibition halos of *L. monocytogenes* with respect to previous times. Similar findings were described by Mauriello et al. [134], which monitored the release of nisin from a LDPE film by measuring the antimicrobial activity of two simulants (water and phosphate-buffered saline at pH 3.5) after 1 to 7, 24, 30, 48, and 72 h of contact with the active plastic film. During this period, an alternation of low and high values of antimicrobial activity of simulating solutions was found, suggesting a mechanism of release and back-absorption of nisin from/to the film. Overall, most studies on the bacteriocins released from plastic materials report an uncontrolled release from the polymers, and that the release is mainly concentrated in the first period of contact with food simulants.

6.2.2.5 Mechanical Properties of Plastic Films. Some mechanical properties of plastic films are particularly relevant to their use in food packaging application. In fact, these properties may have a key role in the technological evaluation of a plastic film. They are as follows: heat resistance, cold resistance, water vapor permeability, gas permeability, stiffness, impact resistance, clarity (transparency), thickness, tensile strength, ultimate elongation, Young modulus, weldability, and self-adhesion. The presence of a bacteriocin into or onto a plastic film may affect its mechanical properties. Therefore, in an active packaging design, it is essential to foresee the evaluation of the structure plastic film modification after the activation process. Unfortunately, this aspect is taken into account little by the researchers even though it is crucial for the industrial application of antimicrobial packaging. However, it should also be noted that the authors who investigated this aspect did not find differences between standard and bacteriocin-treated plastic films in their mechanical properties. Grower et al. [129] evaluated the thickness, tensile strength, and vapor transmission of LDPE films coated with 2,500 and 10,000 IU/cm^2 of nisin, which was diluted in a cellulose-based coating solution. They reported that the bacteriocin coating led to a significant progressive increase of both film thickness and tensile strength. Instead, vapor transmission decreased significantly after the coating, but no differences were shown between the films treated with the two levels of nisin. Finally, the authors concluded that the modification of mechanical properties did not adversely affect the effective application of the film for food packaging. Similarly, Guiga et al. [154] and Chollet et al. [153] did not find notable differences between PE plastic films coated with HPMC solutions with or without nisin in the mechanical properties investigated. In addition, the mechanical properties of edible films have been investigated before and after the treatment with bacteriocin solutions for the manufacture of antimicrobial packaging. Generally, in these films, the mechanical properties are mainly related to the type and quantity of plasticizers used. So, the presence of bacteriocins in the film composition slightly affects their properties. Actually, 30% of glycerol triacetate in the composition of a polylactic acid film deeply affected its mechanical properties [94]. On the contrary, the presence of 5% of EDTA and 5% of Nisaplin in a polylactic acid film containing 30% of glycerol triacetate slightly affected its mechanical properties [94]. Recently, Pintado et al. [165] measured thickness, water vapor permeability, puncture strength and percentage of elongation of whey protein films plasticized with glycerol or sorbitol and containing malic acid, nisin and natamycin. The authors reported that the films showed to be effective carriers of antimicrobials and that the mechanical properties were not significantly compromised.

6.2.3 *In Vitro* Antimicrobial Activity of Films Containing Bacteriocins

In Table 6.2, a list of the microbial targets used to evaluate the antimicrobial activity of developed packaging containing bacteriocins is presented. Although in Applications in Food Packaging we report the use of bacteriocin-treated packaging in real food systems, here we provide a summary of works in which antimicrobial packaging solutions were tested *in vitro*. *L. monocytogenes* (or alternately *Listeria innocua*) is without doubt the most used microbial target to investigate the antimicrobial performance of active food packaging containing bacteriocins from both synthetic and biobased polymers. One of the first contributions on plastic materials treated with bacteriocins was published by Bower et al. [105], who prepared nisin-adsorbed, silanized silicon surfaces at different hydrophobicity, proving their effectiveness of reducing the cellular adhesion of *L. monocytogenes* and to be lethal to *L. monocytogenes* cells that do attach. However, this work aimed to develop antimicrobial food contact surfaces and not antimicrobial food packaging materials. Instead, other studies provided evidence that biobased polymer films incorporating bacteriocins developed with food packaging intents controlled the growth of listeria and/or other microbial populations in solid or liquid media. Ko et al. [112] registered the reduction of *L. monocytogenes* viable cells of 1–2 Log CFU/mL when a bacterial suspension of approximately 10^5 CFU/mL was placed in contact with edible films containing nisin. Moreover, several authors [87, 88, 117, 122, 128, 129, 144] proved the antimicrobial effectiveness of HPMC films containing nisin against *L. monocytogenes* or *L. innocua* cultures. In contrast, little antilisterial activity was shown by a polylactic acid film coated with a nisin solution, whereas it was enhanced by the presence of pectin in the film composition, which probably favors bacteriocin release [155]. Bacteriocins different from nisin were also used in similar studies. For example, two bacteriocins produced by the strain *Lactobacillus curvatus* CRL705 were used to treat LLDPE films, which inhibited the growth of *L. innocua* in agar media [145]. Iseppi et al. [148] developed an antimicrobial packaging using the antilisterial bacteriocin produced by the strain *Enterococcus casseliflavus* IM 416K1. The antimicrobial packaging film, which inhibited the growth of *L. monocytogenes* NCTC 10888 in agar medium, was constituted by an LDPE coated with a PE/PEG solution containing the bacteriocin. Another potential pathogen, *S. aureus*, was used as microbial target to prove the efficacy of plastic films containing bacteriocins to control the microbial growth. Coma et al. [87] reported the antimicrobial effectiveness of HPMC films containing nisin against *S. aureus*, *L. innocua*, and *Micrococcus luteus*. Sebti and Coma [117] and Sebti et al. [88] reported that similar antimicrobial polymers demonstrated an inhibitory activity toward *S. aureus* and *L. innocua*, and *S. aureus* and *L. monocytogenes*, respectively.

Other Gram-positive bacteria were used to show the antimicrobial effectiveness of polymers treated with bacteriocins. Since *M. luteus* (syn. *Micrococcus flavus*) was used as test organism in the first studies on the nisin [167], the plastic polymers treated with nisin have been tested against this bacterium in several works [87, 108, 115, 157, 134]. Despite the different origin of the polymers and the different conditions (nisin concentration, temperature, time of contact, etc.) adopted by these authors, the materials always controlled the growth of *M. luteus*.

As reported previously, bacteriocins show their antimicrobial activity only against Gram-positive bacteria. However, several studies demonstrated that the activity could be extended to Gram-negative bacterial species when the nisin is combined with other substances [26]. In the application of antimicrobial packaging, the bacteriocins, mostly nisin, have been coupled to different compounds to enhance and extend the antimicrobial activity. The most used substance for this purpose is EDTA, which is a food-grade chelating agent that prevents oxidation reactions catalyzed by metal ions. It is well known that EDTA disrupts the lipopolysaccharide in the outer membrane of Gram-negative bacteria by chelating Ca^{2+} and Mg^{2+} ions, which confer stability to this structure [168]. For example, heat press and casting method were used by Padgett et al. [107] to produce biodegradable corn zein or soy protein films. The films treated with nisin inhibited the growth of *L. plantarum* cells. The same film inhibited also *Escherichia coli* when EDTA was added to the film formulation. Corn zein was also used by Hoffman et al. [111], who reported that the presence of nisin, EDTA, and lauric acid in film composition controlled the growth of a 10^4 CFU/mL *Salmonella* Enteritidis starting population, which reached approximately 10^9 CFU/mL after 48 ore in a laboratory liquid medium. Edible soy protein films were prepared including nisin, grape seed extract, EDTA, and their combinations. After 1 h of contact in liquid medium, the film with all the components reduced the population of *E. coli* and *Salmonella* Thyphimurium of approximately 2 and 0.5 log cycles, respectively. In contrast, the film with nisin alone did not show significant differences with the control film [111]. EDTA was used also to enhance the activity of the enterocin EJ97 against the spore-forming microorganism *Bacillus coagulans* [166]. An interesting association was found by Hanusova et al. [164], who combined nisin with a naturally occurring antifungal agent produced by *Streptomyces natalensis*, the natamycin. The authors prepared a nisin/natamicin solution coated with PA/LDPE film, which showed antimicrobial activity against both Gram-positive bacteria and filamentous fungi grown in agar media. Similar findings were reported by Pintado et al. [165], who included nisin, natamycin, and malic acid into films made with whey protein isolate. Finally, the growth of the pathogenic microorganism *E. coli* O157:H7 was controlled by a PLA film containing nisin and EDTA [94].

6.3 APPLICATIONS IN FOOD PACKAGING

Nisin is the antimicrobial found most frequently in film tests either as the only antimicrobial or in conjunction with other antimicrobials [96]. However, many other works describe the use of bacteriocins different from nisin to functionalize plastic films for food packaging applications. In this part of the chapter, we review the literature on the application of antimicrobial food packaging containing bacteriocins on real food systems. In a pioneering work, Siragusa et al. [97] described the development of an active plastic film by extruding a blend of powdered LDPE with 0.1% (w/w) of nisin. The film obtained showed antimicrobial activity against *B. thermosphacta* ATCC 11509 used to contaminate a beef carcass surface tissue artificially, which was prewrapped with the antimicrobial packaging and then stored under vacuum at 4°C in a standard vacuum-packaging bag. In contrast, Daeschel and McGuire [169] deposited in 1995 the invention entitled "Bactericidal surfaces and articles with attached bacteriocin" (http://www.freepatentsonline.com/5451369.pdf).

The experiments on real food systems are crucial to validate an antimicrobial packaging for functioning in food packaging applications. The antimicrobial effect of plastic films containing bacteriocins can be performed only in the case of a close contact with the food, both in the case of solid or liquid foods. In some cases, the bacteriocin can be released in the governing liquid, which works as an interface between the packaging material and food. The antimicrobial effectiveness of a plastic film in a food system is usually investigated by an artificial contamination of food with pathogen or spoilage microorganisms, and then the fate of microbial population during the time of contact between packaging and food is monitored. This type of experiment is indicated as "challenge test," in which a microbial target is "challenged" to overcome a hurdle, which is the bacteriocin in this case.

The food system most involved in challenge tests with antimicrobial packaging containing bacteriocins is fresh or processed meat, even though other foods have been subjected to similar investigations. Of the numerous works describing the activity of these packaging in *in vitro* experiments, some investigations have been carried out in foods artificially contaminated with *Listeria* strains. A group of researchers described the use of a food packaging material treated with a pediocin to pack turkey breast, ham, and beef samples previously contaminated with *L. monocytogenes* ATCC 19115 [106]. Unfortunately, in this work, the authors described too vaguely the material activation process. Moreover, we did not find clarifying information on the origin of the material used and it is not illustrated well in the work, the ClearTite from Viskase Corporation (Chicago, IL).

In another work, coupled polyethylene-oriented polyamide (PE-OPA) plastic films were coated with a bacteriocin produced by *L. curvatus* 32Y by different methods [100]. In particular, an industrial coating procedure was used to produce an antimicrobial plastic film utilized for packing pork steaks and hamburgers artificially contaminated with a culture of *L. monocytogenes* V7. Viable counts of Listeria were lower by approximately 1 log CFU/g during 3 days of storage when both pork steaks and hamburgers were packed with the active film compared with the meat samples unpacked or packed with the traditional plastic film. The same active packaging was used by Ercolini et al. [102] to package single frankfurters surface contaminated with *L. monocytogenes* V7. The authors verified the effectiveness of the package in reducing the live population of Listeria during storage at 4°C for 24 h. Furthermore, just after 15 min of contact with the packaging film led to a significant live cells reduction as determined using the LIVE/DEAD Bacterial Viability Kit (*Bac*Light; Molecular Probes Inc., Eugene, OR). Scannel et al. [86] registered an effective antimicrobial activity of a PE/PA nisin-coated film used to vacuum pack cheddar cheese samples surface inoculated with *L. innocua* or *S. aureus* cultures.

In addition, bioplastic materials were treated with bacteriocins to produce antimicrobial packaging for controlling Listeria cells in intentionally contaminated foods. Nisin-impregnated, soy-based films reduced cell numbers on turkey bologna contaminated with *L. monocytogenes* from 10^6 to 10^5 CFU/g after 21 days [113]. Alginate-, zein-, and polyvinyl alcohol-based films, all added with the enterocins A e B produced by *Enterococcus faecium* CTC492, were used to air pack or vacuum pack cooked ham slices inoculated with a three-strain cocktail of *L. monocytogenes* [143]. The results of this study confirmed the greater effectiveness of bacteriocin-containing antimicrobial films when maintained in direct contact with the food, as well as in the case of vacuum packaging. Another study was conducted by Nguyen et al. [150], who described the packaging with a nisin-containing cellulose film of frankfurters inoculated with *L. monocytogenes* Scott A. In this case, the pathogen population was maintained at a level of 2 log CFU/g lower than the control samples just after 2 days of storage at 4°C, remaining constant during all the 14 days of the experiment. Recently, a sakacin A-containing pullulan film was tested in a challenge study to control the growth of two strains of *L. monocytogenes* experimentally spread onto the surface of turkey deli meat in two separate experiments [95]. In this case, the difference between treated and control samples was of 1–3 log CFU/g according to the strain used as target.

Some authors described experiments in which antimicrobial packaging was used with the aim to extend the shelf life of the food product. In this context, many authors developed antimicrobial packaging containing bacteriocins as a tool to control the growth of *B. thermosphacta*, which is recognized as one of the most important meat-spoiling agent. Cutter et al. [98] tested the antimicrobial properties of plastic films obtained by extruding powdered PE or PE+ poly (ethylene oxide) (PEO) (70% PE and 30% PEO) with 0.1% of nisin in dried milk (Nisaplin), against *B. thermosphacta* ATCC 11509 inoculated onto the surface of postrigor beef carcass sections. The viable cells of *B. thermosphacta* ATCC 11509 in the samples vacuum packed with PE+nisin along 21 days of storage at 4°C was 1–2 log cycles lower than the control samples (packed with PE or PEO alone). Interestingly, in the samples packed with PE+PEO+nisin, the cell load of *B. thermosphacta* ATCC 11509 was approximately 5 log cycles lower than the control samples. Similar antimicrobial performance was shown also by the formulation including EDTA (PE+nisin+EDTA), which did not show significant differences with PE+PEO+nisin at 3, 7, and 14 days of storage. Instead, after 21 days of storage, the cell load of *B. thermosphacta* ATCC 11509 was approximately 4 log cycles higher than the samples packed with PE+PEO+nisin polymer film. The authors reported that PEO is water soluble and hygroscopic, thus facilitating the release of nisin from the films made from this polymer. In addition, other microbial species were used in challenge tests to study the effectiveness of bacteriocin-treated packaging materials to slow the spoilage of food products. An LDPE film coated with a nisin solution was used in a challenge study in which raw, pasteurized, and ultrahigh temperature milk were contaminated with *M. luteus* ATCC 10240 and stored at 4°C for 7 days in a package manufactured with the antimicrobial film. The package retarded the microbial growth in all milk types and the pH decreased in raw milk, suggesting a detrimental effect on the acidifying microflora [134].

To validate packaging for controlling the spoilage phenomenon, many authors monitored the growth of autochthonous microorganisms, single species, or groups whose cell load is usually considered as index of spoilage. For example, Kim et al. [116] observed slowed-down growth of total aerobic bacteria and coliform bacteria populations of fresh oysters and ground beef by storing the products in LDPE film coated with a nisin solution, thus prolonging the shelf life of these food products. Guerra et al. [131] monitored the total bacterial population of fresh veal meat packed into cellophane film impregnated with nisin to assess the effectiveness of this material of extending the shelf life of fresh meat. Furthermore, fresh orange juice, pasteurized milk, and pasteurized milk cream were placed in contact with a nisin-coated paperboard, and a total aerobic bacterial count was performed during 12 days [89, 90]. The results of these works showed that the paper containing nisin possessed the potential for controlling the growth of autochthonous spoilage microflora of these foods. Finally, in a recent work, Ercolini et al. [161] reported that beef samples packed under vacuum with plastic (coextruded of copolymer vinylidene chloride-EVA and LDPE) bags surface treated with a nisin-EDTA solution showed a reduction of viable counts of meat spoilage Gram-positive bacteria (carnobacteria, lactic acid bacteria, and *B. thermosphacta*) and *Enterobacteriaceae* compared with control samples. Also in other works, the combination of the bacteriocins with other substances extended the activity of the antimicrobial food packaging toward Gram-negative bacteria. In fact, the cell load of *S.* Thyphimurium in intentionally contaminated broiler drumstick samples was reduced by 1–3 log CFU/g (according to the different packaging conditions) when packaged with different materials (PVC, LLDPE, and nylon) treated with different combinations of nisin, EDTA, citric acid, and Tween 80 [109]. Lee et al. [123] found that paperboard coated with a combination of nisin and chitosan not only slowed the growth of natural aerobic bacteria and yeast population of fresh orange juice and pasteurized milk, but also it reduced the population of *E. coli* O157:H7 both in distilled water and microbial liquid medium better than paperboard coated with nisin or chitosan alone. Another cellulose-based film containing pediocin and EDTA was used to package sliced ham contaminated with a single strain of *L. monocytogenes* or *Salmonella* sp. Although the film showed effective antimicrobial activity against Listeria, it presented only a slight reduction of *Salmonella* sp. population during the storage period [158]. In some works, edible polymers containing bacteriocins were used as coatings of frankfurters, which are particularly suitable for this type of experiment [130, 133, 140, 147]. Cellulose, zein, soy protein, and whey protein coatings containing nisin alone or combined with other substances (i.e., potassium sorbate, grape seed extract, green tea extract, malic acid, and EDTA) were developed and challenge tests were carried out against *L. monocytogenes* [130, 133, 140, 147], *E. coli* O157:H7, and *S.* Thyphimurium [147].

6.4 FUTURE TRENDS

Active packaging is a deep-rooted concept both in the scientific community and in industrial food production. Within this field, antimicrobial packaging is met with large interest. As reported in previous sections of this chapter, the use of bacteriocins to develop antimicrobial food packaging materials has been investigated. However, the development of new plastic materials in particular bioplastics, as well as the impact that nanotechnology is having on food science, could be an opportunity for extending the possibilities to combine bacteriocins with food packaging materials. Recent studies described the antimicrobial activity of phosphatidylcoline nanovesicles containing bacteriocins [170, 171], whereas nanoparticles were used to vehicle an antimicrobial essential oil in plastic film for food packaging purposes [172]. These techniques could improve the controlled release of a bacteriocin from a plastic film, therefore helping to design an antimicrobial packaging containing bacteriocins. Another important topic is the combination of bacteriocins with other natural food preservatives. Although many authors described the use of bacteriocins with substances such as EDTA and organic acids, and as many authors reported the treatment of plastic materials with essential oils, little attention has been given to combine bacteriocins with plant extracts (e.g., essential oils) in plastic films for use in food packaging applications. Finally, an attractive idea is to incorporate viable bacteriocin-producer cells directly into plastic materials. This strategy could be a solution to the uncontrolled release of bacteriocins from plastic films. Furthermore, the continuous production of bacteriocins by the entrapped cells could solve the problem related to the depletion of the antimicrobial during the time of contact with the food products.

REFERENCES

1. Caplice, E., Fitzgerald, G. F. (1999). Food fermentations: role of microorganisms in food production and preservation. *International Journal of Food Microbiology*, *50*, 131–149.
2. Gálvez A., Abriouel, H., Benomar, N., Lucas, R. (2010). Microbial antagonists to food-borne pathogens and biocontrol. *Current Opinion in Biotechnology*, *21*, 14–148.
3. Stiles, M. E., Holzapfel, W. H. (1997). Lactici acid bacteria of foods and their current taxonomy. *International Journal of Food Microbiology*, *36*, 1–29.

4. Leroy, F., De Vuyst, L. (2004). Lactic acid bacteria as functional starter cultures for the food fermentation industry. *Trends in Food Science & Technology, 15*, 67–78.
5. Schroeter, J., Klaenhammer, T. (2009). Genomics of lactic acid bacteria. *FEMS Microbiology Letters, 292*, 1–6.
6. Deegan, L. H., Cotter, P. D., Hill, C., Ross, P. (2006). Bacteriocins: biological tools for bio-preservation and shelf-life extension. *International Dairy Journal, 16*, 1058–1071.
7. De Vuyst, L., Leroy, F. (2007). Bacteriocins from lactic acid bacteria: production, purification, and food applications. *Journal of Molecular Microbiology and Biotechnology, 13*, 194–199.
8. De Vuyst, L., Vandamme, E. J. (1994). *Bacteriocins of lactic acid bacteria, Blackie Academic and Professional*, Glasgow, Scotland.
9. Jack, R. W., Tagg, J. R., Ray, B. (1995). Bacteriocins of Gram-positive bacteria. *Microbiological Review, 59*, 171–200.
10. Lindgren, S. W., Dobrogosz, W. J. (1990). Antagonistic activities of lactic acid bacteria in food and feed fermentations. *FEMS Microbiology Review, 87*, 149–164.
11. Nes, I. F., Diep, D. B., Håvarstein, L. S., Brurberg, M. B., Eijsink V., Holo, H. (1996). Biosynthesis of bacteriocins in lactic acid bacteria. *Antonie Van Leeuwenhoek, 70*, 113–128.
12. Thunnel, R. K., Sandine, W. E. (1985). Types of starter cultures. In: *Bacterial Starter Cultures for Foods*. Ed.: Gilliland, S. E. CRC Press, Boca Raton, FL.
13. Dykes, G. A. (1995). Bacteriocins: ecological and evolutionary significance. *Trends in Ecology and Evolution, 10*, 186–189.
14. Nettles, C. G., Barefoot, S. F. (1993). Biochemical and genetic characteristics of bacteriocins of food-associated lactic acid bacteria. *Journal of Food Protection, 56*, 338–356.
15. Klaenhammer, T. R. (1993). Genetics of bacteriocins produced by lactic acid bacteria. *FEMS Microbiology Review, 12*, 39–86.
16. Nes, I. F., Johnsborg, O. (2004). Exploration of antimicrobial potential in LAB by genomics. *Current Opinion in Biotechnology, 15*, 100–104.
17. Sobrino-López, A., Martın-Belloso, O. (2008). Use of nisin and other bacteriocins for preservation of dairy products. *International Dairy Journal, 18*, 329–343.
18. Cotter, P. D., Hill, C., Ross, R. P. (2005). Bacteriocins: developing innate immunity for food. *Nature Review Microbiology, 3*, 777–788.
19. Breukink, E. (2006). A lesson in efficient killing from two-component lantibiotics. *Molecular Microbiology, 61*, 271–273.
20. Drider, D., Fimland, G., Héchard, Y., McMullen, L. M., Prévost, H. (2006). The continuing story of class IIa bacteriocins. *Microbiology and Molecular Biology Reviews, 70*, 564–582.
21. Abee, T. (1995). Pore-forming bacteriocins of Gram-positive bacteria and self-protection mechanisms of producer organisms. *FEMS Microbiology Letters, 129*, 1–10.
22. Heng, N. C. K., Wescombe, P. A., Burton, J. P., Jack, R. W., Tagg J. R. (2007). The diversity of bacteriocins in Gram-positive bacteria. In: *Bacteriocins: Ecology and Evolution*. Eds.: Riley, M. A., Chavan, M. A. pp. 45–92. Springer-Verlag, Berlin, Germany.

23. Abee, T., Krockel, L., Hill, C. (1995). Bacteriocins: modes of action and potentials in food preservation and control of food *poisoning*. *International Journal of Food Microbiology, 28*, 169–185.
24. Ryan, M. P., Rea, M. C., Hill, C., Ross, R. P. (1996). An application in cheddar cheese manufacture for a strain of *Lactococcus lactis* producing a novel broad-spectrum bacteriocin, lacticin 3147. *Applied and Environmental Microbiology, 62*, 612–619.
25. Gänzle, M. G., Weber, S., Hammes, W. P. (1999). Effect of ecological factors on the inhibitory spectrum and activity of bacteriocins. *International Journal of Food Microbiology, 46*, 207–217.
26. Gálvez, A., Abriouel, H., Lucas, R., Ben Omar, N. (2007). Bacteriocin-based strategies for food preservation. *International Journal of Food Microbiology, 120*, 51–70.
27. Gálvez, A., Lopez, R. L., Abriouel, H., Valdivia, E., Omar, N. B. (2008). Application of bacteriocins in the control of foodborne pathogenic and spoilage bacteria. *Critical Reviews in Biotechnology, 28*, 125–152.
28. Chen, H., Hoover, D. G. (2003). Bacteriocins and their food applications. *Comprehensive Reviews in Food Science and Food Safety, 2*, 82–100.
29. Jacob, F., Lwoff, A., Siminovitch, L., Wallman, E. (1953). Definition de quelques termes relatifs à la Pysogenie. *Annales de Institut Pasteur, 84*, 222–224.
30. Rogers, L. A., Whittier, E. D. (1928). Limiting factors in lactic fermentation. *Journal of Bacteriology, 16*, 211–229.
31. Mattick, A. T.R., Hirsch, A. (1947). Further observations on an inhibitory substance (nisin) from lactic streptococci. *Lancet, 2*, 5–7.
32. McAuliffe, O., Ross, R. P., Hill, C. (2001). Lantibiotics: structure, biosynthesis and mode of action. *FEMS Microbialogical Review, 25*, 285–308.
33. Ennahar, S., Sashihara, T., Sonomoto, K., Ishizaki, A. (2000). Class IIa bacteriocins: biosynthesis, structure and activity. *FEMS Microbiology Reviews, 24*, 85–106.
34. Joerger, M. C., Klaenhammer, T. R. (1986). Characterization and purification of helveticin J and evidence for a chromosomally determined bacteriocin produced by *Lactobacillus helveticus* 481. *Journal of Bacteriology, 167*, 439–446.
35. Oscariz, J. C., Pisabarro, A. G. (2001). Classification and mode of action of membrane-active bacteriocins produced by Gram-positive bacteria. *International Microbiology, 4*, 13–19.
36. Kemperman, R., Kuipers, A., Karsens, H., Nauta, A., Kuipers, O., Kok, J. (2003). Identification and characterization of two novel clostridial bacteriocins, circularin A and closticin 574. *Applied and Environmental Microbiology, 69*, 1589–1597.
37. Nes, I. F., Tagg, J. R. (1996). Novel lantibiotics and their pre-peptides. *Antonie Van Leeuwenhoek, 69*, 89–97.
38. Jack, R. W., Bierbaum, G., Hiedrich, C., Sahl, H.-G. 1995. The genetics of lantibiotic biosynthesis, *Bioassays, 17*, 793–802.
39. Wiedemann, I., Breukink, E., van Kraaij, C., Kuipers, O. P., Bierbaum, G., de Kruijff, B., Sahl, H. -G. (2001). Specific binding of nisin to the peptidoglycan precursor lipid II combines pore formation and inhibition of cell wall biosynthesis for potent antibiotic activity. *Journal of Biology Chemistry, 276*, 1772–1779.

40. Hurst, A. (1983). Nisin and other inhibitory substances from lactic acid bacteria. In: *Antimicrobials in Foods*. Eds.: Branen, A. L., Davidson, P. M. pp. 327–335. Marcel Dekker, New York,
41. Delves-Broughton, J., Blackburn, P., Evans, R. J., Hugenholtz, J. (1996). Applications of the bacteriocin nisin. *Antonie van Leuwenhoek*, *69*, 193–202.
42. Mulders, J. W. M., Boerrigter, I. J., Rollema, H. S., Siezen, R. J., de Vos, W. M. (1991). Identification and characterization of the lantibiotic nisin Z, a natural nisin variant. *European Journal of Biochemistry*, *201*, 581–584.
43. Kupke, T., Götz, F. (1996). Post-translational modifications of lantibiotics. *Antonie van Leeuwenhoek*, *69*, 139–150.
44. European Economic Community. (1983). European Economic Community Commission Directive. *Official Journal of European Union 255*, 1–6 *83/463/EEC*.
45. Federal Register. (1988). Nisin preparation: affirmation of GRAS status as a direct human food ingredient. *Federal Register*, *54*, 11247–11251.
46. Liu, W., Hansen, J. N. (1990). Some chemical and physical properties of nisin, a small-protein antibiotic produced by *Lactococcus lactis*. *Applied and Environmental Microbiology*, *56*, 2551–2558.
47. Heinemann, B., Voris, L., Stumbo, C. R. (1965). Use of nisin in processing food products. *Food Technology*, *19*, 592–596.
48. O'Sullivan, L., Ross, R. P., Hill, C. (2002). Potential of bacteriocin producing lactic acid bacteria for improvements in food safety and quality. *Biochimie*, *84*, 593–604.
49. van Belkum, M. J., Stiles, M. E. (2000). Non-lantibiotic antibacterial peptides from lactic acid bacteria, *Natural Product Reports*, *17*, 323–335.
50. Eijsink, V. G. H., Skeie, M., Middelhoven, P. H., Brurberg, M. B., Nes, I. F. (1998). Comparative studies of class IIa bacteriocins of lactic acid bacteria. *Applied and Environmental Microbiology*, *64*, 3275–3281.
51. Szabo, E. A., Cahill, M. E. (1998). Nisin and ALTA tm 2341 inhibit the growth of *Listeria monocytogenes* on smoked salmon packaged under vacuum or 100% CO_2. *Letters in Applied Microbiology*, *28*, 373–377.
52. Rodriguez, J. M., Martinez, M. I., Kok, J. (2002). Pediocin PA-1, a wide-spectrum bacteriocin from lactic acid bacteria. *Critical Review in Food Science and Nutrition*, *42*, 91–121.
53. Nissen-Meyer, J., Holo, H., Hvarstein, L. S., Sletten, K., Nes, I. F. (1992). A novel lactococcal bacteriocin whose activity depends on the complementary action of two peptides. *Journal of Bacteriology*, *174*, 5686–5692.
54. Allison, G. E., Fremaux, C., Klaenhammer, T. R. (1994). Expansion of bacteriocin activity and host range upon complementation of 2 peptides encoded within the lactacin F operon. *Journal of Bacteriology*, *176*, 2235–2241.
55. Marciset, O., Jeronimus-Stratingh, M. C., Mollet, B., Poolman, B. (1997). Thermophilin 13, a nontypical antilisterial poration complex bacteriocin that functions without a receptor. *Journal of Biological Chemistry*, *272*, 14277–14284.
56. Collins, E. B. (1961). Domination among strains of lactic acid streptococci with attention to antibiotic production. *Applied Microbiology*, *9*, 200–205.
57. Klaennammer, T. R. (1988). Bacteriocins of lactic acid bacteria. *Biochimie*, *70*, 337–349.

58. Tagg, J. R. (1992). Bacteriocins of Gram-positive bacteria: an opinion regarding their nature, nomenclature and numbers. In: *Bacteriocins, Microcins and Lantibiotics*. NATO ASI Series, Vol. H65. Eds.: James, R., Lazdunski C., Pattus, F. pp. 33–35. Springer, Berlin, Germany.

59. Tagg, J. R., Dajani, A. S., Wannamaker, L. W. (1976). Bacteriocins of Gram positive bacteria. *Bacteriology Review*, *40*, 722–756.

60. Parente, E., Brienza, C., Moles, M., Ricciardi, A. (1995). A comparison of methods for the measurement of bacteriocin activity. *Journal of Microbiological Methods*, *22*, 95–108.

61. Villani, F., Salzano, G., Sorrentino, E., Pepe, O., Marino, P., Coppola, S. (1993). Enterocin 226 NWC, a bacteriocin produced by *Enterococcus faecalis* 226 active against *Listeria monocytogenes*. *Journal of Applied Bacteriology*, *74*, 380–387.

62. Villani, F., Mauriello, G., Aponte, M., Blaiotta, G., Pepe, O., Moschetti, G. (2001). Detection and characterization of a bacteriocin, garviecin L1–5, produced by *Lactococcus garvieae* isolated from raw cow's milk. *Journal of Applied Microbiology*, *90*, 430–439.

63. Villani, F., Pepe, O., Moschetti, G., Salzano, G., Parente, E., Coppola, S. (1995). Antagonistic activity of lactic acid bacteria isolated from natural whey starters for water-buffalo mozzarella cheese manufacture. *Italian Journal of Food Science*, *3*, 221–234.

64. Rose, N. L., Sporns, P., McMullen, L. M. (1999). Detection of bacteriocins by matrix assisted laser desorption/ionization time-of-flight mass spectrometry. *Applied and Environmental Microbiology*, *65*, 2238–2242.

65. Rodriguez, J. M., Cintas, L. M., Casaus, P., Suarez, A., Hernandez, P. E. (1995). PCR detection of the lactocin S structural gene in bacteriocin-producing lactobacilli from meat. *Applied and Environmental Microbiology*, *61*, 2802–2808.

66. Rodriguez, E., Martinez, M. I., Medina, M., Hernandez, P. E., Rodriguez, J. M. (1998). Detection of enterocin AS-48-producing dairy enterococci by dot-blot and colony hybridization. *Journal of Dairy Research*, *65*, 143–148.

67. Garde, S., Rodriguez, E., Gaya, P., Medina, M., Nunez, M. (2001). PCR detection of the structural genes of nisin Z and lacticin 481 in *Lactococcus lactis* subsp. *lactis* INIA 415, a strain isolated from raw milk Manchego cheese. *Biotechnology Letters*, *23*, 85–89.

68. Daeschel, M. A. (1989). Antimicrobial substances from lactic acid bacteria for use as food preservatives. *Food Technology*, *43*,164–167.

69. Claypool, L., Heinemann, B., Voris, L., Stumbo, C. R., (1966). Residence time of nisin in the oral cavity following consumption of chocolate milk containing nisin. *Journal of Dairy Science*, *49*, 314–316.

70. Hara, S., Yakazo, K., Nakakawaji, K., Takeuchi, T., Kobayasi, T., Sata, M., Imai, Z., Shibuya, T. (1962). An investigation of toxicity of nisin with a particular reference to experimental studies of its oral administration and influences by digestive enzymes. *Journal of Tokyo Medical University*, *20*, 176–207.

71. FAO/WHO Expert Committee on Food Additives. (1969). Specifications for identity and purity of some antibiotics. 12th Report. WHO Technical Report Series, No. 430. World Health Organization, Geneva, Switzerland.

72. O'Keeffe, T., Hill, C. (2000). *Bacteriocins*. In: *Encyclopaedia of Food Microbiology*. Eds.: Robinson, R. K., Batt, C. A., Patel, P. D. pp. 183–197. Academic Press, London, U.K.

73. Ross, R. P., Galvin, M., McAuliffe, O., Morgan, S. M., Maire P., Ryan M. P., Twomey, D. P., Meaney, W. J., Hill C. (1999). *Antonie van Leeuwenhoek*, *76*, 337–346.
74. Hugas, M., Garriga, M., Aymerich, M. T., Montfort, J. M. (1995). Inhibition of *Listeria* in dry fermented sausages by the bacteriocinogenic *Lactobacillus sake* CTC494. *Journal of Applied Bacterialogy*, *79*, 322–330.
75. Schillinger, U., Geisen, R., Holzapfel, W. H. (1996). Potential of antagonistic microorganisms and bacteriocins for the biological preservation of foods. *Trends in Food Science and Technology*, *7*, 158–164.
76. Leroy, F., De Vuyst, L., (1999). The presence of salt and a curing agent reduces bacteriocin production of *Lactobacillus sakei* CTC 494, a potential starter culture for sausage fermentation. *Applied and Environmental Microbiology*, *65*, 5350–5356.
77. Brody, A. L., Bugusu, B., Han, J. H., Sand, C. K., McHugh, T. H. (2008). Innovative food packaging solutions. *Journal of Food Science*, *73*, R107–R116.
78. Cutter, C. N. (2002). Incorporation of antimicrobials into packaging materials. *Proceedings of the 55th Reciprocal Meat Conference*, pp. 83–87.
79. Brandsch, J., Piringer, O. (2008). Characteristics of plastic materials. In: *Plastic Packaging*, 2nd ed. Eds.: Piringer, O. G., Baner, A. L. pp. 15–60. Wiley-VCH, Weinheim, Germany.
80. Beer, G., Wiesecke, J., Mannebach, G., Auffermann, J., Vogt, H., Bisson, P., Marczinke, B. (2005). Low-density polyethylene. *Kunststoffe-Plast Europe*, *95*, 54–59.
81. Weber, C. J., Haugaard, V., Festersen, R., Bertelsen, G. (2002). Production and application of biobased packaging materials for the food. *Food Additives and Contaminants*, *19*, 172S–177S.
82. Han, J. H. (2000). Antimicrobial food packaging. *Food Technology*, *54*, 56–65.
83. Labuza, T. P., Breene, W. M. (1989). Applications of "active packaging" for improvement of shelf-life and nutritional quality of fresh and extended shelf-life foods. *Journal of Food Processing and Preservation*, *13*, 1–69.
84. Appendini, P., Hotchkiss, J. H. (2002). Review of antimicrobial food packaging. *Innovative Food Science and Emerging Technologies*, *3*, 113–126.
85. Daeschel, M. A., McGuire, J., Al-Makhlafi, H. (1992). Antimicrobial activity of nisin adsorbed to hydrophilic and hydrophobic silicon surfaces. *Food Contamination and Toxicology*, *55*, 731–735.
86. Scannel, A. G. M., Hill, C., Ross, R. P., Marx, S., Hartmeier, W., Arendt, E. K. (2000). Development of bioactive food packaging materials using immobilised bacteriocins Lacticin 3147 and Nisaplin®. *International Journal of Food Microbiology*, *60*, 241–249.
87. Coma, V., Sebti, I., Pardon, P., Deschamps, A., Pichavant, F. H. (2001). Antimicrobial edible packaging based on cellulose ethers, fatty acids, and nisin incorporation to inhibit *Listeria innocua* and *Staphylococcus aureus*. *Journal of Food Protection*, *64*, 470–475.
88. Sebti, I., Pichavant, F. H., Coma, V. (2002). Edible bioactive fatty acid-cellulosic derivative composites used in food-packaging applications. *Journal of Agricultural and Food Chemistry*, *50*, 4290–4294.
89. Lee, C. H., An, D. S., Lee, S. C., Park, H. J. Lee, D. S. (2004). A coating for use as an antimicrobial and antioxidative packaging material incorporating nisin and α-tocopherol. *Journal of Food Engineering*, *62*, 323–329.

90. Lee, C. H., Park, H. J., Lee, D. S. (2004). Influence of antimicrobial packaging on kinetics of spoilage microbial growth in milk and orange juice. *Journal of Food Engineering*, *65*, 527–531.

91. Pranoto, Y., Rakshit, S. K., Salokhe, V. M. (2005). Enhancing antimicrobial activity of chitosan films by incorporating garlic oil, potassium sorbate and nisin. *LWT*, *38*, 859–865.

92. Benabbou, R., Zihler, A., Desbiens, M., Khedar, E., Subirade, M., Fliss, I. (2009). Inhibition of *Listeria monocytogenes* by a combination of chitosan and divergicin M35. *Canadian Journal of Microbiology*, *55*, 347–355.

93. Jin, T., Zhang, H. (2008). Biodegradable polylactic acid polymer with nisin for use in antimicrobial food packaging. *Journal of Food Science*, *73*, M127–M134.

94. Liu, L. S., Jin, T., Coffin, D. R., Liu, C.-K., Hicks, K. B. (2010). Poly(lactic acid). membranes containing bacteriocins and EDTA for inhibition of the surface growth of Gram-negative bacteria. *Journal of Applied Polymer Science*, *117*, 486–492.

95. Trinetta, V., Floros, J. D., Cutter, C. N. (2010). Sakacin A-containing pullulan film: an active packaging system to control epidemic clones of *Listeria monocytogenes* in ready-to-eat foods. *Journal of Food Safety*, *30*, 366–381.

96. Joerger, R. D. (2007). Antimicrobial films for food applications: a quantitative analysis of their effectiveness. *Packaging Technology and Science*, *20*, 231–273.

97. Siragusa, G. R., Cutter, C. N., Willett, J. L. (1999). Incorporation of bacteriocin in plastic retains activity and inhibits surface growth of bacteria on meat. *Food Microbiology*, *16*, 229–235.

98. Cutter, C. N., Willett, J. L., Siragusa, G. R. (2001). Improved antimicrobial activity of nisin-incorporated polymer films by formulation change and addition of food grade chelator. *Letters in Applied Microbiology*, *33*, 325–328.

99. Mauriello, G., Casaburi, A., Villani, F. (2003). Development of polyethylene films for food packaging activated with a food-grade antlisterial bacteriocin. *1st FEMS Congress of European Microbiologists*, Ljubljana, Slovenia. Abstract in *FEMS Microbiology Letters*, *222*, 30S.

100. Mauriello, G., Ercolini, D., La Storia, A., Casaburi, A., Villani, F. (2004). Development of polyethylene filmsfor food packaging activated with an antlisterial bacteriocin from *Lactobacillus curvatus* 32Y. *Journal of Applied Microbiology*, *97*, 314–322.

101. *FAO Nutrition Meetings Report Series N°45A*. (1969). Specifications for Identity and Purity of Some Antibiotics: Nisin. 12th Report. Edited by Joint FAO/WHO Expert Committee on Food Additives, pp. 53–67.

102. Ercolini, D., La Storia, A., Villani, F., Mauriello, G. (2006). Effect of a bacteriocin-activated polythene film on *Listeria monocytogenes* as evaluated by viable staining and epifluorescence microscopy. *Journal of Applied Microbiology*, *100*, 765–772.

103. SIAA. (1998). Test methods for assessing the antimicrobial strength of products processed with antimicrobial agents. *Society of Industrial Technology for Antimicrobial Articles*, Japan.

104. Leung, P. P. Khadre, M., Shellhammer, T. H., Yousef, A. E. (2002). Immunoassay method for quantitative determination of nisin in solution and on polymeric films. *Letters in Applied Microbiology*, *34*, 199–204.

105. Bower, C. K., McGuire, J., Daeschel, M. A. (1995). Suppression of *Listeria monocytogenes* colonization following adsorption of nisin onto silica surfaces. *Applied and Environmental Microbiology*, *61*, 992–997.
106. Ming, X., Weber, G. H., Ayres, J. W., Sandine, W. E. (1997). Bacteriocins applied to food packaging material to inhibit *Listeria monocytogenes* on meats. *Journal of Food Science*, *62*, 413–415.
107. Padgett, T., Han, I. Y., Dawson, P. L. (1998). Incorporation of food-grade antimicrobial compounds into biodegradable packaging films. *Journal of Food Protection*, *61*, 1330–1335.
108. Kim, Y. -M., Lee, N.-K., Paik, H.-D., Lee, D.-S. (2000). Migration of bacteriocin from bacterion-coated film and its antimicrobial activity. *Food Science and Biotechnology*, *9*, 325–329.
109. Natrajan, N., Sheldon, B. W. (2000). Efficacy of nisin-coated polymer films to inactivate *Salmonella* Typhimurium on fresh broiler skin. *Journal of Food Protection*, *63*, 1189–1196.
110. Padgett, T., Han, I. Y., Dawson, P. L. (2000). Effect of lauric acid addition on the antimicrobial efficacy and water permeability of corn zein films containing nisin. *Journal of Food Processing and Preservation*, *24*, 423–432.
111. Hoffman, K. L., Han, I. Y., Dawson, P. L. (2001). Antimicrobial effects of corn zein films impregnated with nisin, lauric acid, and EDTA. *Journal of Food Protection*, *64*, 885–889.
112. Ko, S., Janes, M. E., Hettiarachchy, N. S., Johnson, M. G. (2001). Physical and chemical properties of edible films containing nisin and their action against *Listeria monocytogenes*. *Food Science*, *66*, 1006–1011.
113. Dawson, P. L., Carl, G. D., Acton, J. C., Han, I. Y. (2002). Effect of lauric acid and nisin-impregnated soy-based films on the growth of Listeria monocytogenes on turkey bologna. *Poultry Science*, *81*, 721–726.
114. Janes, M. E., Kooshesh, S., Johnson, M. G. (2002). Control of *Listeria monocytogenes* on the surface of refrigerated, ready-to-eat chicken coated with edible zein film coatings containing nisin and/or calcium propionate. *Journal of Food Science*, *67*, 2754–2757.
115. Kim, Y. -M., An, D. -S., Park, H. -J., Park, J. -M., Lee, D. S. (2002). Properties of nisin-incorporated polymer coatings as antimicrobial packaging materials. *Packaging Technology and Science*, *15*, 247–254.
116. Kim, Y.-M., Paik, H.-D., Lee, D.-S. (2002). Shelf-life characteristics of fresh oysters and ground beef as affected by bacteriocin-coated plastic packaging film. *Journal of the Science of Food and Agriculture*, *82*, 998–1002.
117. Sebti, I., Coma, V. (2002). Active edible polysaccharide coating and interactions between solution coating compounds. *Carbohydrate Polymers*, *49*, 139–144.
118. Teerakarn, A., Hirt, D. E., Acton, J. C., Rieck, J. R., Dawson, P. L. (2002). Nisin diffusion in protein films: effects of film type and temperature. *Journal of Food Science*, *67*, 3019–3025.
119. Buonocore, G. G., Del Nobile, M. A., Panizza, A., Corbo, M. R., Nicolais, L. (2003). A general approach to describe the antimicrobial agent release from highly swellable films intended for food packaging applications. *Journal of Controlled Release*, *90*, 97–107.

120. Dawson, P. L., Hirt, D. E., Rieck, J. R., Acton, J. C., Sotthibandhu, A. (2003). Nisin release from films is affected by both protein type and film-forming method. *Food Research International, 36*, 959–968.

121. Cha, D. S., Chen, J., Park, H. J., Chinnan, M. S. (2003). Inhibition of *Listeria monocytogenes* in tofu by use of a polyethylene film coated with a cellulosic solution containing nisin. *International Journal of Food Science and Technology, 38*, 499–503.

122. Cha, D. S., Cooksey, K., Chinnan, M. S., Park, H. J. (2003). Release of nisin from various heat-pressed and cast film. *Lebensmittel-Wissenschaft und-Technologie, 36*, 209–213.

123. Lee, C. H., An, D. S., Park, H. J., Lee, D. S. (2003). Wide-spectrum antimicrobial packaging materials incorporating nisin and chitosan in the coating. *Packaging Technology and Science, 16*, 99–106.

124. Leung, P. P., Yousef, A. E., Shellhammer, T. H. (2003). Antimicrobial properties of nisin-coated polymeric films as influenced by film type and coating conditions. *Journal of Food Safety, 23*, 1–12.

125. Sebti, I., Delves-Broughton, J., Coma, V. (2003). Physicochemical properties and bioactivity of nisin-containing cross-linked hydroxypropylmethylcellulose films. *Journal of Agricultural and Food Chemistry, 51*, 6468–6474.

126. Eswaranandam, S., Hettiarachchy, N. S., Johnson, M. G. (2004). Antimicrobial activity of citric, lactic, malic, or tartaric acids and nisin-incorporated soy protein film against *Listeria monocytogenes, Escherichia coli O157:H7, and Salmonella Gaminara. Journal of Food Science, 69*, M79–M84.

127. Franklin, N. B., Cooksey, K. D., Getty, K. J. K. (2004). Inhibition of *Listeria monocytogenes* on the surface of individually packaged hot dogs with a packaging film coating containing nisin. *Journal of Food Protection, 67*, 480–485.

128. Grower, J. L., Cooksey, K., Getty, J. K. (2004). Release of nisin from methylcellulose-hydroxypropyl methylcellulose film formed on low-density polyethylene film. *Journal of Food Science, 69*, FMS107–FMS111.

129. Grower, J. L., Cooksey, K., Getty, J. K. (2004). Development and characterization of an antimicrobial packaging film coating containing nisin for inhibition of *Listeria monocytogenes. Journal of Food Protection, 67*, 475–479.

130. Luchansky, J. B., Call, J. E. (2004). Evaluation *of nisin-coated cellulose casings for the con*trol of *Listeria monocytogenes* inoculated onto the surface of commercially *prepared frankfurt*ers. *Journal of Food Protection, 67*, 1017–1021.

131. Guerra, N. P., Macías, C. L., Agrasar, A. T., Castro, L. P. (2005). Development of a bioactive packaging cellophane using Nisaplin® as biopreservative agent. *Letters in Applied Microbiology, 40*, 106–110.

132. Hong, S.-I., Lee, J.-W., Son, S.-M. (2005). Properties of polysaccharide-coated polypropylene films as affected by biopolymer and plasticizer types. *Packaging Technologies and Science, 18*, 1–9.

133. Lungu, B., Johnson, M. G. (2005). Potassium sorbate does not increase control of *Listeria monocytogenes* when added to zein coatings with nisin on the surface of full fat turkey frankfurter pieces in a model system at 4°C. *Journal of Food Science, 70*, M95–M99.

134. Mauriello, G., De Luca, E., LaStoria, E., Villani, F., Ercolini, D. (2005). Antimicrobial activity of a nisin-activated plastic film for food packaging. *Letters in Applied Microbiology*, *41*, 464–469.

135. McCormick, K., Han, I. Y., Acton, J., Sheldon, B. W., Dawson, P. L. (2005). In-package pasteurization combined with biocide-impregnated films to inhibit *Listeria monocytogenes* and *Salmonella* Typhimurium in turkey bologna. *Journal of Food Science*, *70*, M52–M57.

136. Quintero-Salazar, B., Vernon-Carter, E. J., Guerrero-Legarreta, I., Ponce-Alquicira, E. (2005). Incorporation of the antilisterial bacteriocin-like inhibitory substance from *Pediococcus parvulus* VKMX133 into film-forming protein matrices with different hydrophobicity. *Journal of Food Science*, *70*, M398–M403.

137. Li, B., Peng, J., Yie, X., Xie, B. (2006). Enhancing physical properties and antimicrobial activity of konjac glucomannan edible films by incorporating chitosan and nisin. *Journal of Food Science*, *71*, C174–C178.

138. Li, B., Kennedy, J. F., Peng, J. L., Yie, X., Xie, B. J. (2006). Preparation and performance evaluation of glucomannan-chitosan-nisin ternary antimicrobial blend film. *Carbohydrate Polymers*, *65*, 488–494.

139. Sanjurjo, K., Flores, S., Gerschenson, L., Jagus, R. (2006). Study of the performance of nisin supported in edible films. *Food Research International*, *39*, 749–754.

140. Theivendran, S., Hettiarachchy N. S., Johnson, M. G. (2006). Inhibition of *Listeria monocytogenes* by nisin combined with grape seed extract or green tea extract in soy protein film coated on turkey frankfurters. *Journal of Food Science*, *71*, M39–M44.

141. Jofré, A., Garriga, M., Aymerich, T. (2007). Inhibition of *Listeria monocytogenes* in cooked ham through active packaging with natural antimicrobials and high-pressure processing. *Journal of Food Protection*, *70*, 2498–2502.

142. Liu, L. S., Finkenstadt, V. L., Liu, C. K., Jin, T., Fishmanm M. L., Hicks, K. B. (2007). Preparation of poly(lactic acid) and pectin composite films intended for applications in antimicrobial packaging. *Journal of Applied Polymer Science*, *106*, 801–810.

143. Marcos, B., Aymerich T., Monfort, J. M., Garriga, M. (2007). Use of antimicrobial biodegradable packaging to control Listeria monocytogenes during storage of *cooked ham. International Journal of Food Microbiology*, *120*, 152–158.

144. Neetoo, H., Ye, M., Chen, H. (2007). Effectiveness and stability of plastic films coated with nisin for inhibition of *Listeria monocytogenes*. *Journal of Food Protection*, *70*, 1267–1271.

145. Blanco Massani, M., Fernandez, M. R., Ariosti, A., Vignolo, G. (2007). Development and characterization of an active polyethylene film containing Lactobacillus curvatus CRL705 bacteriocins. *Food Additives and Contaminants*, *25*, 1424–1430.

146. Datta, S., Janes, M. E., Xue, Q.-G., Losso, J., La Peyre, J. F. (2008). Control of *Listeria monocytogenes* and *Salmonella* Anatum on the surface of smoked salmon coated with calcium alginate coating containing oyster lysozyme and nisin. *Journal of Food Science*, *73*, M67–M71.

147. Gadang, V. P., Hettiarachchy N. S., Johnson, M. G., Owens, C. (2008). Evaluation of antibacterial activity of whey protein isolate coating incorporated with nisin, grape seed extract, malic acid, and EDTA on a turkey frankfurter system. *Journal of Food Science*, *73*, M389–M394.

148. Iseppi, R., Pilati, F., Marini, M., Toselli, M., de Niederhäusern, S., Guerrieri, E., Messi, P., Sabia, C., Manicardi, G., Anacarso, I., Bondi, M. (2008). Anti-listerial activity of a polymeric film coated with hybrid coatings doped with Enterocin 416K1 for use as bioactive food packaging. *International Journal of Food Microbiology*, *123*, 281–287.

149. Kristo, E., Koutsoumanis, P., Biliaderis, C. G. (2008). Thermal, mechanical and water vapour barrier properties of sodium caseinate films containing antimicrobials and their inhibitory action on *Listeria monocytogenes*. *Food Hydrocolloids*, *22*, 373–386.

150. Nguyen, V. T., Gidley, M. J., Dykes, G. A. (2008). Potential of a nisin-containing bacterial cellulose film to inhibit *Listeria monocytogenes* on processed meats. *Food Microbiology*, *25*, 471–478.

151. dos Santos Pires, A. C., de Fátima F. S., de Andrade, N. J., da Silva, L. H. M., Camilloto, G. P., Bernardes, P. C. (2008). Development and evaluation of active packaging for sliced mozzarella preservation. *Packaging Technology and Science*, *21*, 375–383.

152. Sivarooban, T., Hettiarachchy, N. S., Johnson, M. G. (2008). Physical and antimicrobial properties of grape seed extract, nisin, and EDTA incorporated soy protein films. *Food Research International*, *41*, 781–785.

153. Chollet, E., Swesi, Y., Degraeve, P., Sebti, I. (2009). Monitoring nisin desorption from a multi-layer polyethylene-based film coated with nisin loaded HPMC film and diffusion in agarose gel by an immunoassay (ELISA). method and a numerical modelling. *Innovative Food and Emerging Technologies*, *10*, 208–214.

154. Guiga, W., Galland, S., Peyrol, E., Degraeve, P., Carnet-Pantiez, A., Sebti, I. (2009). Antimicrobial plastic film: physico-chemical characterization and nisin desorption modelling. *Innovative Food Science Emerging Technologies*, *10*, 203–207.

155. Jin, T., Liu, L. S., Zhang, H., Hicks, K. (2009). Antimicrobial activity of nisin incorporated in pectin and polylactic acid composite films against *Listeria monocytogenes*. *International Journal of Food Science and Technology*, *44*, 322–329.

156. Lu, F., Liu, D., Ye, X., Wei, Y., Liu, F. (2009). Alginate-calcium coating incorporating nisin and EDTA maintains the quality of fresh northern snakehead (*Channa argus*) fillets stored at 4°C. *Journal of Science of Food and Agriculture*, *89*, 848–854.

157. Rossi-Márquez, G., Han, J. H., García-Almendárez, B., Castaño-Tostado, E., Regalado-González, C. (2009). Effect of temperature, pH and film thickness on nisin release from antimicrobial whey protein isolate edible films. *Journal of the Science of Food and Agriculture*, *89*, 2492–2497.

158. Santiago-Silva, P., Soares, N. F. F., Nóbrega, J. E., Júnior, A. W., Barbosa, K. B. F., Volp, A. C. P., Zerdas, E. R. M. A., Würlitzer, N. J. (2009). Antimicrobial efficiency of film incorporated with pediocin (Alta® 2351) on preservation of sliced ham. *Food Control*, *20*, 85–89.

159. Bastarrachea, L., Dhawan, S., Sablani, S. S., Powers, J. (2010). Release kinetics of nisin from biodegradable poly(butylenes adipate-co-terephthalate). films into water. *Journal of Food Engineering*, *100*, 93–101.

160. Cao-Hoang, L., Grégoire, L., Chaine, A., Waché, Y. (2010). Importance and efficiency of in-depth antimicrobial activity for the control of listeria development with nisin-incorporated sodium caseinate films. *Food Control*, *21*, 1227–1233.

161. Ercolini, D., Ferrocino, I., La Storia, A., Mauriello, G., Gigli, S., Masi, P., Villani, F. (2010). Development of spoilage microbiota in beef stored in nisin activated packaging. *Food Microbiology*, *27*, 137–143.
162. Guiga, W., Swesi, Y., Galland, S., Peyrol, E., Degraeve, P., Sebti, I. (2010). Innovative multilayer antimicrobial films made with Nisaplin® or nisin and cellulosic ethers: physico-chemical characterization, bioactivity and nisin desorption kinetics. *Innovative Food Science Emerging Technologies*, *11*, 352–360.
163. Hanusova, K., Klaudisova, K., Milatova, E., Svirakova, E., Dobias, J., Voldrich, M., Koza, V., Marek, M. (2010). Release of nisin from polyvinyldichloride lacquer coated on a polyethylene film. *Journal of Food and Nutrition Research*, *49*, 21–29.
164. Hanusova, K., Stastna, M., Votavova, L., Klaudisova, K., Dobias, J., Voldrich, M., Koza, V., Marek, M. (2010). Polymer films releasing nisin and/or natamycin from polyvinyldichloride lacquer coating: nisin and natamycin migration, efficiency in cheese packaging. *Journal of Food Engineering*, *99*, 491–496.
165. Pintado, C. M. B. S., Ferreira, M. A. S. S., Sousa, I. (2010). Control of pathogenic and spoilage microorganisms from cheese surface by whey protein containing malic acid, nisin and natamycin. *Food Control*, *21*, 240–246.
166. Viedma, P. M., Ercolini, D., Ferrocino, I., Abriouel, H., Omar, N. B., López, R. L., Gálvez, A. (2010). Effect of polythene film activated with enterocin EJ97 in combination with EDTA against *Bacillus coagulans*. *LWT-Food Science and Technology*, *43*, 514–518.
167. Tramer, J., Fowler, G. (1964). Estimation of nisin in foods. *Journal of Science of Food and Agriculture*, *15*, 522–528.
168. Lambert, R. J. W., Hanlon, G. W., Denyer, S. P. (2004). The synergistic effect of EDTA/antimicrobial combinations on *Pseudomonas aeruginosa*. *Journal of Applied Microbiology*, *96*, 244–253.
169. Daeschel, M. A., McGuire, J. (1995). Bactericidal surfaces and articles with attached bacteriocin. United States Patent 5,451,369, issued Sept. 19, 1995.
170. Teixeira, M. L., dos Santos, J., Silveira, N. P., Brandelli, A. (2008). Phospholipid nanovesicles containing a bacteriocin-like substance for control of *Listeria monocytogenes*. *Innovative Food Science and Emerging Technologies*, *9*, 49–53.
171. da Silva Malheiros, P., Micheletto, Y. M. S., da Silveira, N. P., Brandelli, A. (2010). Development and characterization of phosphatidylcholine nanovesicles containing the antimicrobial peptide nisin. *Food Research International*, *43*, 1198–1203.
172. Persico, P., Ambrogi, V., Carfagna, C., Cerruti, P., Ferrocino, I., Mauriello, G. (2009). Nanocomposite polymer films containing carvacrol for antimicrobial active packaging. *Polymer Engineering and Science*, *49*, 1447–1455.

CHAPTER 7

ANTIMICROBIAL ENZYMES AND NATURAL EXTRACTS IN PLASTICS

MARCELLA MASTROMATTEO, MARIANNA MASTROMATTEO,
AMALIA CONTE, and MATTEO ALESSANDRO DEL NOBILE
University of Foggia, Foggia, Italy

CONTENTS

7.1 Introduction . . . 160
7.2 Enzymes and Natural Extracts . . . 160
7.2.1 Generalities, Advantages and Limitations of Enzymes . . . 160
7.2.2 The Particular Case of Glucose-Oxidase . . . 162
7.2.3 Types and Sources of Natural Extracts . . . 166
7.3 Designing Films and Surface-Containing Active Enzymes and/or Natural Extracts . . . 167
7.3.1 Chemical Binding of Enzymes . . . 167
7.3.2 Physical Retention of Enzymes . . . 170
7.3.3 Nanoscaled Systems with Enzymes . . . 171
7.3.4 Applications of Entrapped Enzymes to Foodstuffs . . . 172
7.3.5 Physical Retention of Natural Extracts . . . 174
7.3.6 Natural Extracts in Microencapsulated Systems . . . 176
7.4 Modeling and Delivery Systems for Controlled Release of Active Agents . . 177
7.4.1 Diffusion-Limited Release: Reservoir and Matrix Systems . . . 178
7.4.2 Swelling-Controlled Delivery Systems . . . 181
7.5 Future Trends . . . 184
References . . . 185

Antimicrobial Polymers, First Edition.
Edited by José M. Lagarón, María J. Ocio, and Amparo López-Rubio.

7.1 INTRODUCTION

In agreement with the current trend of giving value to natural and renewable resources and to avoid the inherent collateral toxicity of biocide synthetic chemicals, natural biocides are rapidly growing in various areas, particularly in food packaging and in biomedical applications. In the same industrial applications, the direct use of natural active compounds can have disadvantages because of their sensitivity to processing conditions and to trace levels of substances that can actuate as inhibitors, resulting in a short operational life or inactivation. Therefore, the use of plastic and bioplastic is currently widespread, as these materials offer considerable interest as matrixes for the incorporation of highly functional biocides. This chapter will gather the understanding of the coupling of plastics and biocides, taking into account enzymes and natural extracts as active compounds. The factors influencing the choice of the immobilization method and of the biomaterial support will be analyzed. The research activity aimed to develop active systems intended for pharmaceutical or bioengineering fields, and upcoming possibilities to incorporate biocatalysts and natural extracts in polymers for food packaging applications, will be discussed. Moreover, the fundamental mechanisms involved in active compound release and most diffuse mathematical approaches to describe controlled-release systems will be presented, along with recent advances in delivery devices development.

7.2 ENZYMES AND NATURAL EXTRACTS

7.2.1 Generalities, Advantages and Limitations of Enzymes

The term *enzyme*, which comes from Greek *ενζυμον*, "in leaven," was used for the first time at the end of the 19th century by the physiologist Wilhelm Kühne. Later, Eduard Buchner began to study the ability of yeast extracts that lacked any living yeast cells to ferment sugar and named the enzyme that brought about the fermentation of sucrose "zymase." Buchner received the Nobel Prize in Chemistry for his biochemical research and his discovery of cell-free fermentation. Following Buchner's example, enzymes are usually named according to the reaction they carry out. Typically, to generate the name of an enzyme, the suffix *-ase* is added to the name of its substrate or the type of reaction. Enzymes are proteins that catalyze chemical reactions by converting the substrates into different molecules, called products. Because enzymes are selective for their substrates and speed up only a few reactions from among many possibilities, the set of enzymes made in a cell determines which metabolic pathways occur in that cell. As all catalysts, enzymes are not consumed by the reactions they catalyze, nor do they alter the equilibrium of these reactions. However, enzymes do differ from most other catalysts by being much more specific. Enzyme activity can be affected by other molecules; inhibitors are molecules that decrease enzyme activity, whereas activators are molecules that increase activity. Some enzymes are used commercially in the synthesis of antibiotics or to speed up biochemical reactions [1]. Activity is also affected by temperature, chemical environment (*e.g.*, pH), and concentration of substrate. The proof that enzymes are pure proteins was definitively provided by Northrop and Stanley (1930), who worked on the digestive enzymes pepsin, trypsin, and chymotrypsin. The lysozyme structure was solved by X-ray crystallography. This high-resolution structure of lysozyme marked the beginning of the field of structural biology and the effort to understand how enzymes work at an atomic level of detail. Lysozyme, which is a lytic enzyme found in foods such as milk and eggs, is a muramidase that hydrolyses β1-4 linkages between *N*-acetylmuramic acid and *N*-acetylglucosamine. It is known to inhibit some gram-positive bacteria, but alone it is ineffective against gram-negative bacteria. Commercially, lysozyme has been used primarily to prevent late blowing in semihard cheeses, which is caused by the fermentation of lactate by butyric acid bacteria, primarily *Clostridium tyrobutyricum*. Enzymes are generally globular proteins and range from just 62 amino acid residues in size to more than 2,500 residues in the animal fatty acid synthase. Like all proteins, enzymes are made as long, linear chains of amino acids that fold to produce a three-dimensional product. Each unique amino acid sequence produces a specific structure, which has unique properties. Individual protein chains may sometimes group together to form a protein complex. Most enzymes can be denatured—that is, unfolded and inactivated—by heating or chemical denaturants, which disrupt the three-dimensional structure of the protein. Depending on the enzyme, denaturation may be reversible or irreversible [2].

Contrary to chemical compounds, enzymes present the advantage of being highly specific catalysts not consumed during processing. For this reason, they received considerable attention in many biotechnological processes in the food industry, as described in the extensive review of Fernandez et al. [3]. Enzymes have been also widely used to develop films of biomacromolecules in the thickness range of 10–100 nm for sensors as well as for patterning of organic systems [4, 5] to be applied in the pharmaceutical, technical, and bioengineering fields. In fact, thin lysozyme films of thickness up to more than 100 nm have been produced in a dry environment by matrix-assisted pulsed laser evaporation from a water ice matrix, thus demonstrating that a significant part of the lysozyme molecules is transferred to the substrate without decomposition and that the protein activity is preserved [6]. Because of the increasing requirement for reducing pollution in textile production, greater attention has been also given to enzymatic processes of textiles as effective alternatives to conventional chemical treatments because of the nontoxic and eco-friendly characteristics of enzymes [7]. A new functionalization method for wool fabrics based on immobilization of lysozyme was recently investigated with success by Wang et al. [8] by using glutaraldehyde as cross-linking agent.

Marine environment and in particular actinomycetes as marine microorganisms, are potential source for industrially important enzymes [9]. Actinomycetes comprise a wide range of distinct microorganisms that are not present in the terrestrial environment and are considered highly valuable as they produce various antibiotics and other therapeutically useful compounds with diverse biological activities, and probably, they are the most important group of organisms studied extensively for the discovery of drugs and other bioactive metabolites [10].

Even though European Directive 95/2 EC classifies some enzymes as food additives, thus allowing their use as food ingredients, in some industrial applications the direct use of biocatalysts can have disadvantages because of their sensitivity to processing conditions and to trace levels of substances that can actuate as inhibitors, resulting in short operational life or inactivation [11]. Over the use of free enzymes, the immobilization of enzymes in materials was applied with numerous technological advantages such as reusability, improved stability to temperature, resistance to denaturing compounds, and improved activity [12]. To the specific aim of catalyzing a reaction, which is considered beneficial from a nutritional point of view, i.e., decreasing the concentration of a nondesired food constituent, and/or producing a food substance beneficial to the health of the consumer, many enzymes are currently being used in several food transformation processes or they are being incorporated in packaging materials. The choice of the immobilization method and of the biomaterial support for the manufacturing of enzymatic packages depends on many factors, such as the nature of the biocatalyst (e.g., whole cells or purified enzymes, from fungal or bacterial origin, either native or genetically modified), the storage conditions, and the specific food application of the biocatalyst [13]. More details dealing with incorporation of enzymes in materials are widely presented and discussed subsequently.

7.2.2 The Particular Case of Glucose-Oxidase

The glucose oxidase enzyme is an oxido-reductase that catalyses the oxidation of glucose to hydrogen peroxide and *D*-glucono-δ-lactone. In cells, it aids in breaking the sugar down into its metabolites. Glucose oxidase is widely used for the determination of free glucose in body fluids (diagnostics), in vegetal raw material, and in the food industry. It also has many applications in biotechnology, typically in enzyme assays for biochemistry including biosensors used in nanotechnologies [14].

It is often extracted from *Aspergillus niger* and *Penicillium* spp.. The antimicrobial activity is a result of the cytotoxicity of the H_2O_2 formed, although the lowering of pH by the production of *D*-gluconic acid may also influence the growth of some microorganisms [15]. It is used in the food industry to remove small amounts of oxygen from food products or glucose from diabetic drinks. Glucose oxidase is found in honey and acts as a natural preservative; on the surface of the honey, it reduces atmospheric oxygen to hydrogen peroxide, which acts as an antimicrobial barrier [16].

Although biocatalysts display high catalytic efficiency and exquisite selectivity, they are often unstable under harsh environmental and industrial operational conditions. Immobilization constitutes a strategy to overcome this stability problem [17]. Polymers have been employed as matrices for their attractive thermomechanical and chemical properties. Synthetic polymers are usually the material of choice for enzyme immobilization [18], although various examples involving the use of natural polymers such as cellulose are described in the literature [19]. Traditional enzyme immobilization can be achieved by entrapment within the polymer during polymerization, by selective or non-selective adsorption on a polymeric material, or by covalent attachment to a support [20].

In a study of Amitai et al. [21], medical grade polyurethane ChronoFlex AR (CF) was electrospun together with glucose oxidase (GOX) and horseradish peroxidase (HRP). The electrospun fibers were collected as a uniform, water insoluble, flexible elastomeric matrix with an average fiber diameter of 1 ± 0.2 mm. Biocidal activity of CF/enzyme fibers resulted in >6-log unit reduction of both *Escherichia coli* and *Staphylococcus aureus* challenges. A time-course of biocidal activity displayed a 3–4 log reduction of *E. coli* and *S. aureus* within the first 5 min and complete kill (>6 logs) within 15 min. A dose–response study of fiber weight (0.5–30 mg/mL) exhibited complete kill of *E. coli* (>6 logs) and at least 99.99% *S. aureus* kill (>4 logs) with as little as 1 mg fiber. The fibers were reusable with slightly less activity on the second use and significant activity after continuous soaking in buffer for up to 7 days. Electrospun CF/GOX/HRP fibers adhered to a thin film with embedded sodium iodide and glucose caused a complete kill of *E. coli* (>7-log units) and *Staphylococcus aureus* (6-log unit reduction) within 1 h at 37 °C.

Murtinho et al. [22] used cellulosic derivatives (cellulose acetate, cellulose propionate, and cellulose acetate-butyrate) as membranes for the immobilization of enzymes. These were then characterized by differential scanning calorimetry (DSC), scanning electron microscopy (SEM) and contact angle evaluation. Subsequently, catalase ($H_2O_2 : H_2O_2$ oxireductase; EC 1.11.1.6), alcohol oxidase (alcohol : oxygen oxireductase; EC 1.1.3.13) and glucose oxidase (β-D-glucose : oxygen 1-oxireductase; EC 1.1.3.4) were covalently linked to these membranes. The catalytic activity and stability of these enzymes, when immobilized, were examined. The results obtained showed that characteristics such as crystallinity, hydrophilicity, and permeability can be correlated with the activity and stability of enzymes that are immobilized in membranes. The cellulose acetate membrane, which was prepared by evaporation, gave the more active conjugate support enzyme. The membranes prepared by the immersion technique were more crystalline and therefore less suitable for enzyme immobilization. The highly hydrophobic membranes, which were obtained from the propionate and the butyrate esters of cellulose reduced the activities but gave better storage stability. The cellulose-derivative membranes used by Murtinho et al. [22] are a good model for the study of the factors that influence the behavior of immobilized enzyme systems. Fusion enthalpies, water sorption values, and diffusion coefficients are useful in helping to establish practical parameters for the creation of effective supports for the immobilization of enzymes.

Tirkes et al. [23] examined the immobilization of GOX in several conductive matrices including graft copolymers of pyrrole with different block chain lengths, namely: polypyrrole (PPy); poly(pyrrole-graft-polytetrahydrofuran) (PPy/1), and (PPy/3), with molecular weights of 92000 and 14000, respectively; and poly(pyrrole-graft polystyrene/polytetrahydrofuran) (PPy/2). The electrolysis conditions and parameters like enzyme activity, temperature, K_m and V_{max} (affinity of enzyme towards substrate and maximum reaction rate, respectively), were investigated. Concerning the activity and the stability of GOX, better conditions were obtained with the matrices having soft segments (polytetrahydrofuran) with a suitable molecular weight. Tirkes et al. [23] estimated that with the high-molecular-weight polymers the grafting of pyrrole groups decreases, causing long pyrrole chains. Hence, the matrix shows the characteristics of pure PPy. The presence of cauliflower structures in PPy/(2) matrix and also its long operational stability were rather unusual when compared with the results reported by Erginer et al. [24]. This reflects that the type of enzyme has a significant effect on the morphology and the operational stability of enzyme electrodes.

In their study, Isık et al. [25] examined the immobilizations of invertase and glucose oxidase by entrapment within polypyrrole and poly 2-methylbutyl-2-(3-thienyl) acetate)/polypyrrole, PMBTA/PPy, copolymer matrix via electrochemical method. The synthesis and characterization of both PMBTA and PMBTA/PPy copolymer was previously described by Levent et al. [26]. Optimum temperature and kinetic parameters (V_{max}, the maximum reaction rate and K_m the affinity of enzyme toward substrate) for the immobilized enzymes were examined. Furthermore, the surface morphologies of enzyme-entrapped films were investigated. The operational and storage stabilities of the enzyme electrodes were determined. Isık et al. [25] showed that PMBTA/PPy copolymer can be used as enzyme immobilization matrix for invertase and glucose oxidase. Sodium dodecyl sulfate was found to be the best supporting electrolyte for the entrapments. Both enzymes were dispersed homogeneously in the electrode because they show appreciable activity with respect to free enzyme activity. The immobilization of invertase in PMBTA/PPy matrices is more feasible than in PPy matrices with regard to their optimum temperature and storage stability. The immobilization of glucose oxidase for both PMBTA/PPy and PPy matrices reveals almost the same stability. As for the stability, in PMBTA/PPy matrices, invertase immobilization is more feasible than that of glucose oxidase.

In the study by Gursel et al. [27], the immobilization of enzymes, invertase, and glucose oxidase were achieved in conducting copolymers of *N*-pyrrolyl terminated polydimethylsiloxane/polypyrrole (PDMS/PPy) matrices via electrochemical polymerization. The kinetic parameters, V_{max} and K_m, of both free and immobilized enzymes were determined. The effect of supporting electrolytes, p-toluene sulfonic acid and sodium dodecyl sulfate, on enzyme activity and film morphologies was examined. The optimum temperatures and operational stabilities of immobilized enzymes were determined. PDMS/PPy copolymer matrix was found to exhibit significantly enhanced properties compared with pristine polypyrrole in terms of relative enzyme activities, kinetic parameters, and operational stabilities.

In a work of Vasileva and Godjevargova [28], the microfiltration membranes of polyamide-6 (PA-6) were modified preliminarily by hydrochloric acid (HCl) and activated subsequently by 1,2-diaminoethane (DAE). They were used as matrix for covalent immobilization of GOX by using glutaraldehyde. Some characteristics of the initial and modified membranes were studied: amino groups content, degree of hydrophilicity, water transmission rate, water permeability coefficient, and average pore radius. The amount of bound protein and relative activity of GOX immobilized onto the modified membrane matrix were determined. In the course of studying the properties of unmodified and modified PA-6 membranes, Vasileva and Godjevargova [28] proved that the highest degree of hydrophilicity and amino groups content and the highest decrease of water transmission rate, coefficient of permeability, and average pore radius was found for the membrane modified with 6 wt% HCl and 10 wt% DAE. The amount of bound protein and the relative activity of the immobilized GOX was correlated to the amount of amino groups and to the degree of hydrophilicity. The highest relative activity (81.4%) was measured for GOX immobilized onto membrane modified with 6 wt% HCl and 10 wt% DAE. The relative activity of GOX immobilized onto the modified membrane matrix was identical with activity of immobilized glucose oxidase on acrylonitrile-vinylimidazol copolymer membrane, described previously by Godjevargova et al. [29], but the performances in the presented membrane were higher. Studying the properties of native and immobilized onto modified PA-6 membrane (with 6 wt% HCl and 10 wt% DAE) GOX, the authors proved that the immobilized one shows greater pH, as well as thermal and storage stability.

In a work of Vartiainen et al. [30], GOX was covalently immobilized onto amino- and carboxyl-plasma-activated biorientated polypropylene (BOPP) films via glutaraldehyde and carbodiimide chemistries. GOX immobilized onto amino-activated BOPP films using 2.5% glutaraldehyde produced higher enzymatic activities than GOX immobilized by 0.4% carbodiimide.

GOX covalently immobilized onto BOPP films inhibited completely the growth of *E. coli* and substantially inhibited the growth of *Bacillus subtilis*; thus, they may have great potential to be exploited in various antimicrobial packaging film applications.

A new procedure for preparing organic-inorganic hybrid gel of chitosan–SiO_2 was developed by Yang et al. [31]. Novel porous gels were prepared by coupling chitosan with tetramethoxy silicane through the sol–gel approach at ambient conditions. The characteristics of the gel were evaluated using Fourier transform infrared (FT-IR) spectroscopy and SEM. GOX was selected as a model enzyme to assess its potential applicability for the enzyme immobilization purpose. The immobilization procedure stabilized the enzyme substantially, as 86% and 56% of the immobilized GOX activity could still be discerned after being kept at 30 °C for 10 and 15 days, respectively, whereas most of the free enzyme turned inactive after a 5-day standing in the same manner. A SEM scrutiny of the chitosan–SiO_2 gel revealed that it possessed a highly porous microstructure. A subsequent attention to its potential application for the enzyme immobilization showed that the newly prepared gel was an excellent enzyme carrier for GOX. Specifically, under the optimized condition (formaldehyde 20 mg/mL, GOX 2 mg/mL and pH 6.5), the enzyme immobilization yield (IY%) was 97% with the immobilized GOX activity of 1585 U/g. The Michaelis constant (K_m) for immobilized GOX was disclosed to be 36.5 mmol/L at 25 °C, and the highest enzyme activity could be expected at pH 5.6. The immobilized enzyme was more stable than the free one. The results described herein are of considerable interest in the possible application of the immobilized enzymes.

7.2.3 Types and Sources of Natural Extracts

Spices and herbs are generally used in foodstuffs for enhancing the flavor or color attributes. Moreover, these materials have antimicrobial and antioxidant activity [32–34]. The most common synthetic antioxidants used in the food industry are butylatedhydroxytoluene (BHT), butylatedhydroxyanisole (BHA) and *tert*-butylhydroquinone (TBHQ). However, these synthetic antioxidants are banned in several countries because of their carcinogenic risk [35, 36]. Recently, natural antioxidants have become a part of the diet in human nutrition with the aim of decreasing the risk of diseases such as coronary heart disease, cancer [34, 37, 38], and diabetes [39]. Moreover, natural antioxidants are reported to be more powerful than the synthetic antioxidants, especially rosemary, sage, and green tea extracts [34–36, 38]. The literature is replete with reports of extracts obtained from edible and nonedible plant materials, which possess antioxidant compounds [40]. Extracts from herbs and spices were found to be effective for controlling the development of rancidity in fats and oils [41, 42].

Plant essential oils, constituted mainly of terpenoides, have been studied for their antimicrobial activity against many microorganisms including several pathogens [43, 44]. In particular, the activity of oils from *Labiatae* [45, 46] and citrus fruits [47] have been investigated by several authors. In addition, the action of single constituents of these oils has been exploited to improve the understanding of the cell targets of the molecules [48, 49].

Green tea leaf extracts are becoming increasingly important as a functional food in the diet because of their high polyphenol content [50]. Its polyphenol content can increase up to 36% (dry basis) based on climate, season, or variety [36]. The tea catechins and other polyphenols are free radical scavengers, metal chelators, inhibitors of transcription factors, and enzymes. Therefore, green tea extracts have been used as natural antioxidants, antibacterial agents, and antiviral agents [35, 37, 50]. Green tea extract has been used in canola oil [51], marine oil [36], rapeseed oil [38], soybean oil, and corn oil [34]. *Thymbra spicata* (a member of Lamiaceae family) is used in meat, fish, and foodstuffs as a spice and used as a herbal tea in some eastern Mediterranean countries [52]. The major essential oil components derived from *T. spicata* are carvacrol (about 86%), thymol (about 4%), *E*-3-caren-2-ol (about 3%), γ-terpinene (about 1.9%) [52]. The essential oil has antioxidant, antibacterial, and antifungal activities [32, 53, 54].

Gutierrez et al. [55] evaluated the efficacy of plant essential oils, oregano, and thyme for controlling the natural spoilage microflora on ready-to-eat lettuce and carrots and also considering their impact on sensory properties.

Natural extracts from microorganisms (e.g., bacteriocins and organic acids) and animals (e.g., lysozyme from eggs, chitosan from crab and shrimp shell wastes, and transferrins from milk) were used as natural food preservatives [56]. Bari et al. [57] studied the combined efficacy of bacteriocins such as nisin and pediocin with organic acids in reducing the *Listeria monocytogenes* population of inoculated fresh-cut produce. Chitosan has also been documented to possess antimicrobial activity, thus increasing the shelf life of perishable foods. The use of chitosan in fruits, vegetables, and other products, both as an ingredient or as edible coating, has been widely reported [58].

7.3 DESIGNING FILMS AND SURFACE-CONTAINING ACTIVE ENZYMES AND/OR NATURAL EXTRACTS

What we call the preservation role is a fundamental requirement of food packaging because it is directly related with the safety of the consumer. Extraenhanced functions have been sought by the food packaging sector for meeting consumer demand: packaging acting to improve the service of a product, packaging acting to communicate and simplify the new uses of a good, and packaging to simplify the offer of complex solutions [59]. The efforts to improve the performances of packaging solutions and control fresh foods can be directed toward many working areas. Active packaging is the most relevant innovative idea applied for consumer satisfaction [12, 60]. Active packaging has been defined as a system in which the product, package, and environment interact in a positive way to extend shelf life of product or to achieve some characteristics that cannot be obtained otherwise [61]. An overview of applications on bioactive food packaging is listed in Table 7.1. Recently, an additional step forward was made by developing active packaging with antimicrobial properties that contains natural instead of synthetic additives [62]. Traditional packaging materials derived from petroleum oil are neither readily recyclable nor environmentally sustainable and impose several health risks associated with, for instance, the migration of harmful additives. Thus, it would be highly desirable to develop biodegradable (including edible) and sustainable matrixes capable of safe and long shelf-life integration of bioactive substances [63]. Among the functional substances, enzymes and natural extracts are thought to be very suitable for their incorporation in the package wall. The reasons for this are outlined subsequently.

7.3.1 Chemical Binding of Enzymes

The choice of both the immobilization method and the material support for the manufacturing of enzymatic packages depends on the nature of the biocatalyst,

TABLE 7.1 Applications of bioactive food packaging

Active component	Packaging material	Food	References
Batteriocins/Vitamins			
nisin	LDPE	Culture media, milk	[60]
Nisin, natamycin	Cellulose-derivative	Culture media, sliced mozzarella cheese	[61]
Nisin	Tapioca starch	Culture media	[62]
Nisin, α-tocopherol	VAE	Culture media, milk cream	[63]
Nisin	Alginate	Fresh beef	[64]
Nisin	Galactomannans	Ricotta cheese	[65]
Essential oils			
Lemongrass essential oil	PVA	Culture media	[66]
Thymol	Corn zein	Culture media	[67]
Oregano, rosemary and garlic essential oils	Whey protein isolate	Culture media	[68]
Garlic oil	Chitosan	Culture media	[52]
Grapefruit seed extract	PE	Minced beef	[69]
Oregano essential oils	Milk protein	Beef muscle	[70]
Enzymes			
Lysozyme	Corn zein	Culture media	[71]
Lysozyme	CA	Culture media	[72]
Lysozyme	PVA	Culture media	[73], [74]
Lysozyme	Corn zein	Culture media	[75]
Other release agents			
Potassium sorbate	Chitosan	Fresh strawberries	[76]
Ascorbic acid, citric acid	Carrageenan and whey protein	Apple slices	[77]
Ascorbic acid, sorbic acid	MC	Cut-pear wedges	[78]
α-tocopheryl	Chitosan	Fresh and frozen strawberries	[79]
Sorbic acid	Beeswax, wheat gluten	Culture media	[80]
Ag^+	PEO	Culture media, apple juice	[81]
Calcium, vitamin E	Xanthan gum	Carrots	[82]
Calcium, zinc, vitamin E	Chitosan	Fresh fruits and vegetables	[83]

CA = cellulose acetate; LDPE = low–density polyethylene; MC = methyl cellulose; PE = polyethylene; PVA = poly vinyl alcohol; VAE = vinyl acetate ethylene; PEO = polyethylenoxide.

the envisaged storage conditions, the type of food to be packed, and the specific application of the biocatalyst [12].

Generally speaking, the two main targeted benefits of enzyme immobilization are easy separation of enzyme from the product and reuse of the enzyme. Easy separation of the enzyme from the product simplifies enzyme applications and permits reliable and efficient reaction technology. Enzyme reuse provides a number of cost advantages, which are often an essential prerequisite for establishing an economically viable enzyme-catalyzed process. The properties of immobilized enzyme preparations are governed by the properties of both the enzyme and the carrier material. The interaction between the two provides an immobilized enzyme with specific chemical, biochemical, mechanical, and kinetic properties [64].

Immobilization can be achieved in many ways, but it always affects enzyme activity to some extent. Various techniques have been developed for enzyme immobilization, including adsorption to insoluble materials, entrapment in polymeric gels, encapsulation in membranes, cross-linking with reagents, or covalent linking onto soluble carriers. The most common method for immobilization of enzymes is the encapsulation [65]. The sol-gel encapsulation method makes use of inorganic gels (silica gels) that can easily be tailored to a large range of porous textures, network structures, surface functionalities, and processing conditions; however, even though the methodology will be useful for developing novel functional materials, additional research is needed to tailor the technology to the specific requirements of encapsulation [66]. Despite the industrial feasibility of most techniques, often the enzyme immobilization must be consistently improved. For example, even with large interest in the galactosidase immobilization for continuous hydrolysis of lactose on dairy products, it is difficult to obtain a catalyst showing simultaneously high activity, high stability, and optimal mechanical properties. Therefore, free enzyme is still widely used in dairy industry around the world [67].

Covalent attachment and cross-linking are techniques that involve either the modification of the biopolymer surface or the use of toxic chemical compounds, usually unsuited for food applications.

Via covalent attachment to the sole histidine aminoacid, the immoblization of hen egg white lysozyme to inert soluble polystyrene beads was achieved. This single point immobilization may minimize the steric interference between the support and the immobilized enzyme [68]. As extensively reviewed by Goddard and Hotchkiss [69], the covalent union between the polymer and the biocatalyst is based on the chemical activation of a surface and the attachment of nucleophilic groups of the proteins. This union transforms the polymer in a stable carrier that, in the best case, does not leach the protein to the surroundings, thus avoiding the arrival of undesirable compounds to a potential food matrix put in contact with the activated polymer. Lactase and glucose oxidase were immobilized, respectively, by covalent attachment to different polymers displaying high activity yields [70, 71]. Lysozyme, which is active against gram-positive bacteria, and chitinase, which is active against bacteria and fungi, have been covalently immobilized by Hotchkiss [72] and Wang and Chio [73], respectively. However, their activity after immobilization was too low for practical applications.

To overcome the problem of toxic compounds to activate the polymer surface, mineral supports, such as inorganic nanocarriers, could be added to biopolymer materials to promote enzyme binding by cation exchange, physical adsorption, or ionic binding [74]. Alpha-amylase, glucoamylase, and invertase were immobilized through adsorption and grafting onto clays generally used as nanofillers in polymers to reinforce mechanical and barrier properties to low-molecular-weight compounds of packaging films [75]. Among the biomaterials already studied for the immobilization of enzymes and with potential interest in functional bioactive packaging, carrageenan [76], chitosan [77–79], gelatin [80], alginate [81], polylactic acid, and polyglycolic acid [82] are very promising materials because of their natural, biodegradable, biocompatible, nontoxic and readily available character.

7.3.2 Physical Retention of Enzymes

A simple and flexible tool to realize physical retention of enzymes is through entrapment. To this aim, porous solids, biodegradable matrices, gels, synthetic fibers, beads, or membranes have been widely developed [83–88]. Andersson et al. [89] developed an entrapment method to make an enzyme-based oxygen scavenger laminate. The enzyme solution, containing various additives, was applied on a paper carrier, which was placed between two polyethylene films. The laminate was then heated under pressure and the system proved to be successful, and no significant efficiency loss was observed. This easy method of introducing glucose oxidase in the package wall can also be applied to other enzymes with functional bioactivities, but it also becomes clear that to transform food components the product must be able to get into direct contact with the immobilized enzymes or whole cells.

The most common examples of enzyme entrapment in polymeric matrices to provide antimicrobial properties have been developed with lysozyme. The first example of lysozyme retained in different polymers was provided by Appendini and Hotchkiss [90]. Lysozyme was immobilized on polyvinyl alcohol (PVOH) beads, nylon 6,6 pellets, and cellulose triacetate (CTA) films, exerting the better results on *M. lysodeikticus* when it was bound to CTA. PVOH was also used by Buonocore et al. [91, 92], who reported the suitability of highly swellable polymers being able to modulate the release kinetics of lysozyme and nisin by changing the degree of the chemically induced cross-linking. The same research group also proposed multilayer structures consisting of two external control layers and an inner layer containing lysozyme, showing that a satisfactory control of lysozyme release might be achieved after cross-linking with glyoxal [93]. In contrast, LaCoste et al. [94] proposed the use of smart blending for developing novel controlled-release packaging materials with active properties. Interesting results on controlled release of lysozyme were also obtained by a Turkish research group that worked on the feasibility in controlling the release rate of partially purified lysozyme incorporated in zein films. The authors showed that partially purified lysozyme presented good storage stability in lyophilized form and in zein films containing salts of ethylenediaminetetraacetic acid (EDTA). The only disadvantage of using partially purified lysozyme is the nonhomogenous distribution of the hydrophilic enzyme in hydrophobic zein films at high concentrations. Research results suggested that additional edible ingredients may also be employed to minimize structural changes and obtain more uniform antimicrobial films [87]. Hence, in a subsequent step, the authors evaluated the efficacy of the partially purified lysozyme with EDTA incorporated into zein films, in combination with chickpea albumin extracts and bovine serum albumin against *B. subtilis* and *E. coli*, demonstrating the benefits of using functional protein extracts to control lysozyme release rate and, consequently, its activity [95]. A valid polymeric structure with a significant controlled release of lysozyme, intended for internal food packaging applications, was also represented by the asymmetric porous cellulose acetate-based film containing lysozyme, developed by Gemili et al. [96]. Different research groups showed the advantage of using asymmetric-membrane capsules and asymmetric coatings on drug tablets to control the release rate of drugs [97]; however, before the work of Gemili et al. [96], the potential use of these structures in preparing controlled-release materials for food packaging had not been investigated. The results of the work carried out by these authors clearly demonstrated that the porous films can be used as novel internal food packaging systems with controlled-release properties and could be tailored by changing the composition of the initial casting solution. Subsequent studies are still needed to test the effectiveness of these active structures on real food systems.

7.3.3 Nanoscaled Systems with Enzymes

Enzyme nanoscale immobilization has attracted great interest in biotechnological processes, thus providing good opportunities also in the area of food packaging [98]. Rudra et al. [99] carried out an easy application of layer-by-layer assemblies to create a multilayer film with a thickness in the nanoscale range. The system was made up of poly(L-glutamic acid) and hen egg white lysozyme by sequential adsorption of oppositely charged polyelectrolytes on a solid support. Because of its effectiveness against *Micrococcus luteus* in liquid medium, it could represent an interesting strategy for food preservation. Another technique that can also provide deposition of enzyme layers in the nanometer range is represented by the matrix-assisted pulsed laser evaporation, which was tested with success using lysozyme [6]. Electrospinning is a promising technology that can produce polymeric ultrafine fibers. The method consists in the generation of an electric field around a droplet of a polymer that provokes its deformation to form nanobeads, nanofibers, or microfibers [100]. Electrospun fibers are reasonably much thinner in diameter and thus higher in surface-to-volume ratio than fibers fabricated using conventional mechanical extrusion or spinning processes because the elongation can be accomplished via a contactless scheme. The mats composed of electrospun fibers are excellent candidates for applications in filtration, drug delivery, food texture alterations, encapsulation of food additives, active and bioactive packaging elements, reinforcement in composite materials, and biomaterial for wound dressings and scaffolds for tissue engineering [101, 102]. The luciferase-loaded polyvinyl alcohol nanofibers obtained by electrospinning by Zeng et al. [103] represents one of the first attempts to tailor drug delivery systems with a controlled release of active compounds.

7.3.4 Applications of Entrapped Enzymes to Foodstuffs

The direct use of antimicrobial proteins can have disadvantages based on their sensitivity to processing conditions and to trace levels of substances that can act as inhibitors, resulting in short operational life or inactivation. The efficacy of many enzymes may be reduced by certain food components. Glass and Johnson [104] demonstrated that fat may reduce the efficacy of some antimicrobials added to or found naturally in foods, probably because many food preservatives interact with the phospholipids within the bacterial cell membrane, interfering with normal transport of substrate into the cells. The fat found in foods may compete for the lipophilic molecule, effectively reducing the concentration available for antimicrobial activity. However, their application in various appropriately tailored carriers usually provides the bioactive proteins with additional stability under tough working conditions of pH and temperature and an adequate environment either for their repeated use or for their controlled release [105]. The research has already achieved good results in the final set up of devices based on enzymes and/or natural extracts incorporated in polymeric materials. However, few applications in real working conditions are reported in the literature. In the subsequent paragraphs, the main relevant applications are presented and discussed.

The potential use of lysozyme as a food preservative has invoked considerable interest, particularly in Japan [106]. The antimicrobial spectrum of lysozyme could be enhanced when it is used with other substances, such as EDTA [107], disodium pyrophosphate, pentasodium tripolyphosphate [108], caffeic acid, and cinnamic acid [109], but the biocatalyst combined to Na_2-EDTA is generally known to compromise the appearance of the product. Lysozyme is a lytic enzyme found in many natural systems, and it is used in cheese manufacturing to prevent the growth of lactate-fermenting and gas-forming *Clostridia* spp. [110].

Sinigaglia et al. [111] demonstrated that it is possible to prolong the shelf life of mozzarella cheese by dissolving lysozyme and Na_2-EDTA in the product brine. However, no information dealing with sensorial properties of cheese has been reported. To achieve good antimicrobial effects without compromising product acceptability, Conte et al. [112] also applied lysozyme for preserving mozzarella cheese, but the authors incorporated lysozyme and Na_2-EDTA in an edible coating based on sodium alginate, thus replacing the traditional brine with an active hydrogel. The active coating controlled the microbial proliferation of cheese without compromising the sensorial properties of the product. The same active system was also applied with success in combination to modified atmospheric conditions to enhance the efficacy of the packaging system in prolonging product shelf life. Edible coatings have been widely applied to food for centuries, and they are intended to provide some level of protection, prevent mass transfer from one food component to another, and enhance the appearance of fruits and vegetables [113].

Gill and Holley [114] used lysozyme to limit bacterial growth in meat after demonstrating that a combination treatment of lysozyme, nisin, and EDTA might be effective in controlling the growth of spoilage and pathogenic bacteria in cured meat products. However, according to Han [115], in most cases, the lysozyme would not be effective against gram-negative bacteria. Güçbilmez et al. [95] produced zein-based films effective against *E. coli*. The authors also reported that gram-negative bacteria are protected by their outer membrane surrounding the peptidoglycan. Indeed, lyzozyme exhibits antimicrobial activity by splitting the bonds between *N*-acetylmuramic acid and *N*-acetylglucosamine of the peptidoglycan in the cell wall of bacteria. The destabilization of the outer membrane by EDTA, potentially used in combination, could increase the activity on gram-negative bacteria.

Lysozyme immobilized on PVOH surface was also used against *Alicyclobacillus acidoterrestris* in apple juice [116]. To obtain the active film, a solution of lysozyme (2%) dissolved in distilled water and an amount of glutaraldehyde were sprayed on a PVOH wet film to distribute the antimicrobial compound and the bonding agent uniformly. Glacial acetic acid was also added as reaction catalyst. To test the antimicrobial effects of films, microbiological tests were conducted with success on a single strain and on a culture cocktail of *A. acidoterrestris* at 44 °C. By monitoring the viable cell concentration of the juice put in contact with the active system, it was possible to demonstrate that the antimicrobial film maintained its efficacy also at different surface-to-volume ratios. The same microbial tests were also conducted on viable spores of *A. acidoterrestris*, thus demonstrating the efficacy of developed active film in inhibiting the viable spores more easily than the cells.

The use of enzymes in the juice industry has contributed to increasing the yield and production of various types of juices. For example, pectinases are added to degrade the pectic substances and minimize the impacts of these compounds on the color, turbidity, and viscosity of the final product. Moreover, enzymes that can remove bitterness of citrus juice and extract pigments, among other applications, are of great interest in the juice industry [117].

With regard to naringinase, different works have been reported for the use of the enzyme to reduce juice bitterness by hydrolysis of naringin. In particular, whereas Soares and Hotchikiss [118] immobilized naringinase in plastic packages, Munish et al. [119] studied the applicability of alginate-entrapped naringinase to reduce the bitterness of kinnow juice. The application of the active system with naringinase achieved a 60% decrease in bitterness. This result suggested the desirability of ultrafiltration to minimize product inhibition in order to maximize naringin hydrolysis and subsequent fortification by the permeate at the end of debittering to formulate sweetened juice with nourishing and natural characteristics. PVOH cross-linked with glutaraldehyde to entrap naringinase was used by Del Nobile et al. [120]. The authors demonstrated that the enzyme only hydrolyzed the naringin diffusing into the film, thus compromising effective applications for the food industry.

As mentioned previously, another application of enzymes in fruit production is that aimed to depectinization. The process is currently carried out using two pectic enzymes in sequence: pectinesterase (PE) and polygalacturonase (PG), which involve the deesterification and depolymerization of the pectin molecules. Some disadvantages of using these enzymes are the release of methanol and the formation of colloidal precipitates in the system between the deesterified pectin and the endogenous Ca^{2+}. Pectin lyase is an enzyme that is subject to growing interest for the substitution of the previously mentioned two pectic enzymes. The latter enzyme was immobilized by physical adsorption or by the formation of covalent bonds on organic (cellulose and its derivatives, and X-ray anomalous diffraction-amberlites) and inorganic (sulfides, g-alumina, and bentonite) supports. The supports that permitted an effective immobilization with good activity levels and sufficient stability under the operational conditions were found to be the acrylic resin activated with trichlorotriazine and, in particular, bentonite activated with glutaraldehyde [121, 122].

Finally, the incorporation in the packaging material of certain enzymes, such as glucose oxidase, represents an active packaging solution that can eliminate residual oxygen in the headspace and/or prevent the contact of food with atmospheric oxygen entering the package. This type of oxygen scavenger packaging solution is also receiving a great deal of attention because the contact of small amounts of oxygen with sensitive foodstuffs might provoke their oxidation, the development of rancidity, color changes, off flavors, and of course, the growth of aerobic microorganisms. The Bioka Ltd. Company is already commercializing a biocatalyst-based technology using glucose oxidase in an individual sachet, which is offered as an oxygen-scavenging technology intended for food quality preservation.

7.3.5 Physical Retention of Natural Extracts

Spices are rich in phenolic compounds, such as flavonoids and phenolic acids, which exhibit a wide range of biological effects, including antioxidant and antimicrobial properties [62, 123–125]. The direct incorporation of essential oils to food will result in the immediate reduction of the bacterial population but may alter the sensorial characteristics of the food. In contrast, the incorporation of essential oils into edible films may be particularly interesting. Seydim and Sarikus [126] showed that oregano and garlic oil were, for example, effective in whey protein-based films against *S. aureus*, *Salmonella enteritidis*, *Listeria monocytogenes*, *E. coli*, and *Lactobacillus plantarum*. Basil (*Ocimum basilicum*) is a popular culinary herb, and its essential oils have been used extensively for many years in the flavoring of confectionary and baked goods, condiments, sausages and meats, and so on [62]. According to these authors, basil essential oil has a potential use in food preservation, especially in conjunction with technologies of antimicrobial packages for food products.

Ponce et al. [127] found that the use of a chitosan coating enriched with oleoresins applied to butternut squash did not produce a significant antimicrobial effect. However, it improved the antioxidant protection of the minimally processed squash offering a great advantage in the prevention of browning reactions, which typically result in quality loss in fruits and vegetables.

Rojas-Grau et al. [128] found that the inclusion of lemongrass (1 and 1.5% w/w), oregano oil (0.1 and 0.5% w/w) and vanillin (0.3 and 0.6% w/w) in apple puree-alginate edible coatings significantly inhibited the growth of psychrophilic aerobes, yeasts and moulds of fresh-cut Fuji apples stored at 4 °C. Lemongrass- and oregano oil-containing coatings exhibited the strongest antimicrobial activity against *Listeria innocua* (4-log reduction). Moreover, a significant reduction in the rates of O_2 depletion and CO_2 production was observed in samples containing high concentrations of essential oils.

Vanillin incorporated into chitosan/methyl cellulose films provided an inhibitory effect against *E. coli* bacteria and *Saccharomyces cerevisiae* yeast on fresh-cut cantaloupe and pineapple stored at 10 °C [129]. Fitzgerald et al. [130], based on the studies conducted using *E. coli*, *L. plantarum*, and *L. innocua*, found that the inhibitory activity of vanillin resided primarily in its ability to affect the integrity of the cytoplasmic membrane detrimentally, with the resultant loss of ion gradient, loss of pH homeostasis, and inhibition of respiratory activity.

Raybaudi-Massilia et al. [131] investigated the effects of alginate-based edible coatings incorporated with malic acid and essential oils of cinnamon, palmarosa, and lemongrass (0.3% and 0.7%) and their main active compounds (eugenol, geraniol, and citral, 0.5%) on the microbiological, physicochemical shelf life, and safety of fresh-cut Piel de Sapo melon (*Cucumis melo* L.). The incorporation of essential oils or their active compounds into the edible coatings prolonged the microbiological shelf life by more than 21 days. Significant reductions in *Salmonella enteritidis* population in inoculated coated fresh-cut melon were achieved. In particular, palmarosa oil incorporated at 0.3% into the coating seems to be a promising preservation alternative for fresh-cut melon because it had a good acceptance by panelists, maintained the fruit quality, inhibited the native flora growth, and reduced *S. enteritidis* population.

Bentayeb et al. [132] developed a new antioxidant film that incorporated a natural extract of rosemary; a fast method for the direct determination of carnosic acid in the packaging material, to evaluate the antioxidant capacity of the new active plastic, was developed and optimized. The rosemary extract was screened, and carnosic acid and carnosol were determined as the major antioxidant components that were responsible for the antioxidant properties of the whole extract. This method was used to characterize the antioxidant activity of the film throughout the shelf life of the packaging material itself, which corresponded to different degrees of carnosic acid degradation. The analytical method investigated by Bentayeb et al. [132] could be applied to various natural extracts that contain phenolic acids as their main antioxidants, which would expand the potential applications of the new active material.

Pranoto et al. [133] showed that the incorporation of garlic oil in chitosan films improved the antimicrobial efficacy of chitosan against food pathogenic bacteria. Some authors showed that oregano, rosemary extracts, and garlic oil exerted different effectiveness against various food-borne spoilage bacteria if incorporated in whey protein films [126], whereas Del Nobile et al. [134] investigated the antimicrobial properties of corn zein films containing different concentrations of thymol against various spoilage cells and spores. Most scientific investigations available in the literature on the release systems of essential oils intended for food applications demonstrate that to date, a few attempts have been made to develop real controlled-release systems. One of the few examples is represented by the work of Mastromatteo et al. [135] that developed zein-based mono and multilayer films loaded with spelt bran and thymol (35% w/w) to obtain edible composite materials to control thymol release. To this aim, the thickness of film layers and the amount of biodegradable fibres were varied. The results highlighted that release rate decreased with the increase of film thickness for both monolayer and multilayer films, without spelt bran addition. Conversely, a significant increase of thymol release rate with the increase of the bran concentration was recorded for both monolayer and multilayer films, probably because the addition of spelt bran promoted the formation of microchannels that interconnected the thymol phase, which by-passed the zein matrix and led to an increase in the active compound release rate.

7.3.6 Natural Extracts in Microencapsulated Systems

Another effort to deliver essential oils has been performed by microencapsulation processes. In truth, microencapsulation is a feasible alternative to increase the stability of these compounds, but it also influences release characteristics. Generally, essential oils undergo oxidative degradation; hence, there is a need for protection of these compounds against environmental factors, which contribute to their deterioration, e.g., oxygen, light, and moisture. Microencapsulation using carbohydrates such as hydrolyzed starches, emulsifying starches, and gums (especially gum acacia) as the most common carrier materials [136–139], is the most common approach to tackle this problem. During the last decade, several essential oils have been microencapsulated in various chemical structures and by different processes [140]. In the work of Leimann et al. [140], the microencapsulation of lemongrass essential oil by the simple coacervation method was investigated, using cross-linked poly(vinyl alcohol) as the wall-forming material. Antimicrobial activity and the antimicrobial properties of the encapsulated oil were determined, demonstrating that the process of microencapsulation did not deteriorate the encapsulated essential oil.

Baranauskiene et al. [141] investigated the properties of milk proteins for encapsulation of natural flavorings such as oregano (*Origanum vulgare L.*), citronella (*Cymbopogon nardus G.*) and marjoram (*Majorana hortensis L.*). The authors determined the retention of volatile flavor compounds and evaluate the efficiency of microencapsulation; they assessed the changes in the composition of flavors taking place during processing and measured the release of volatiles from microencapsulated products.

Krishnan et al. [142] studied the microencapsulation of cardamom oleoresin by spray drying using gum arabic, maltodextrin, and a commercially available modified starch as wall materials. The microcapsules were evaluated for the content and stability of volatiles, nonvolatiles, entrapped 1,8-cineole, and entrapped α-terpinyl acetate for 6 weeks. Gum arabic was found to be a better wall material for the encapsulation of cardamom oleoresin compared with maltodextrins and modified starch. Ramos [143] studied the therapeutic efficiency of copaiba oil encapsulated with gum arabic, showing that the efficiency of the essential oil was not affected by the encapsulation process.

Shu et al. [144] developed a spray drying method for preparation of lycopene microcapsules using gelatin and sucrose as wall materials. The resulting microcapsules were characterized in terms of lycopene isomerization, storage stability, particle size, particle size distribution, and outer and inner structures. The optimal conditions to obtain the microencapsulated lycopene with good storage stability were determined. These conditions were as follows: ratio of gelatin/sucrose of 3/7, ratio of core and wall material of 1/4, feed temperature of 55 °C, inlet temperature of 190 °C, homogenization pressure of 40 MPa, and lycopene purity of not less than 52%. Under these conditions, the microencapsulated lycopene showed some isomerization but good storage stability.

The work of Ersus and Yurdagel [145] was focused on the production of spray-dried anthocyanin extracts from black carrot to determine the effects of different spray-drying temperatures on the anthocyanin content of the powders to evaluate the effects of maltodextrins with different dextrose equivalent on the properties of spray-dried powders and their storage stability. The results showed that for spray drying of black carrot anthocyanins, higher air inlet temperatures (>160 °C to 180 °C) caused more anthocyanin losses and the dextrose equivalent affected the anthocyanin content powder at the end of the drying process. Storage at 4 °C increased the half-life of spray dried anthocyanin pigments three times with respect to 25 °C storage temperature.

Spray-dry microencapsulation has been used to improve the stability of carotenoids in carrot pulp and paprika oleoresin [146].

Heinzelmann and Franke [147] compared the application of a freeze-drying technique with respect to the production of dried microencapsulated fish oil. They concluded that the production of dried microencapsulated fish oil by freezing and subsequent freeze drying offers an opportunity to achieve a product with good oxidation stability.

7.4 MODELING AND DELIVERY SYSTEMS FOR CONTROLLED RELEASE OF ACTIVE AGENTS

For many years, the major focus of active material-related research has been focused on the discovery and the synthesis of new molecules. In the last decades, increasing attention was devoted to the manner in which these compounds are delivered. There has been considerable interest in developing controlled-release systems in a wide range of applications such as drug delivery, crop and seed protection in agriculture, animal nutrition, food additives, chemical reagents, curing agents, and catalysts.

Mathematical modeling plays an important role in facilitating the polymeric network design by identifying key parameters and molecule release mechanisms. Therefore, the fundamental mechanisms involved in release phenomena and the mathematical approaches in polymeric network design aimed at describing controlled-release systems are widely discussed in the subsequent sections.

7.4.1 Diffusion-Limited Release: Reservoir and Matrix Systems

In the diffusion-limited release of active agents, a substance is released from a device by diffusion from its interior to the surrounding. There are two main types of diffusion-controlled systems: the reservoir system and the matrix system [148]. In a reservoir system, the active agent is enclosed by an inert outer membrane, whereas in a matrix system the drug is dissolved or dispersed uniformly throughout the rate-controlling polymeric matrix. Barriers may be microporous, macroporous, or nonporous. The most commonly used barriers are the nonporous homogeneous films. The release rate from a reservoir system depends on the thickness, area, and permeability of the barrier. In a reservoir containing an excess of active agent, the release rate follows zero-order kinetics (i.e., the release rate is constant). According to Pothakamury and Barbosa-Cánovas [149], the principal steps in the release of an active ingredient from a reservoir system are (1) diffusion of the active agent within the reservoir, (2) dissolution or partitioning of the active agent between the reservoir carrier fluid and the barrier, (3) diffusion through the barrier and partitioning between the barrier and the elution medium (i.e., the surrounding food), and (4) transport away from the barrier surface into the food. The rate-limiting step in the release of the active ingredient is the diffusion through the polymeric barrier. Several excellent reviews dealing with mathematical equations to describe reservoir systems have been published, and the most relevant one is examined hereinafter.

Manca and Rovaglio [150] proposed a general approach for the release of active principles from microencapsulated drugs, based on the "first principles" model, which combines both phenomenological insights and theoretical equations. As a first hypothesis and for the sake of simplicity, the authors assumed a microcapsule system characterized by spherical particles with the same diameter and the same coating thickness. Consequently, it is possible to focus the attention on a single particle of given core diameter f and coating thickness s_{coat}. During the experimental release, the water present in the container penetrates the particle's coating and starts to dissolve the core. A liquid solution is formed inside the particle, and it coexists with the remaining solid core that shrinks along time. The spherical solid nucleus goes on dissolving into the internal liquid solution up to the singularity condition ($f = 0$). At the same time, the internal liquid solution becomes richer in active principle and its concentration gets higher than the external agitated solution. Consequently, because of the positive gradient, the drug is transported from the inside to the outside of the particle. A diffusive release phenomenon is therefore originated. The solid core dissolves, saturating the internal liquid solution, up to a final equilibrium between the internal and external liquid concentrations of the dissolved compound. The mass balance of the solid core nucleus within the particle can be characterized as follows:

$$\frac{dM_{\text{core}}}{dt} = \rho_{\text{core}} \cdot \frac{\pi}{6} \cdot 3\phi^2 \frac{d\phi}{dt} = -\pi \cdot \phi^2 \cdot k_c \cdot (\omega_s - \omega_i) \tag{1}$$

where M_{core} is the core mass of a single particle, ρ_{core} is the core density, ϕ is the core diameter, ω_s is the saturation concentration of active principle, ω_i is the active principle concentration within the internal solution, and k_c is the mass transfer coefficient within the internal particle solution.

The core dissolution described in Eq. (1) is favored by the concentration gradient between the core-solution interface and the internal solution bulk. Manca and Rovaglio [150] assumed that the active principle is immediately dissolved at the interface between solid and liquid. Consequently, the saturation concentration ω_s is reached within such interface. The ω_i concentration is lower than ω_s, and at the same time, it continues to grow until the solid core vanishes. When the solid core is dissolved completely, ω_i inverts its trend, decreasing up to the asymptotic equilibrium condition with the external liquid solution. As regard the active principle within the internal solution, it will pass through the polymer coating before leaving it for the external aqueous solution. The description of such a step requires a second term of the mass balance referred to the compound flux between the internal liquid solution and the polymer coating. Subsequently, the authors described another mass balance equation for the active principle within the polymer coating. A last equation is necessary to complete the mathematical model proposed by Manca and Rovaglio [150] relative to the mass balance of active principle outside the microcapsules. The use of the penetration theory to describe this step is innovative with respect to conventional theories found in the literature and provides a clear explanation of the release dependency from the square root of time [151]. The presence of a second term in the mass balance equation brings to the following formula:

$$\frac{dM_i}{dt} = \pi \cdot \phi^2 \cdot k_c \cdot (\omega_s - \omega_i) - k_l \cdot A_i \cdot (\omega_i - \omega_{\text{coat}}) \tag{2}$$

where M_i is the active principle mass within the internal solution, A_i is the internal coating area, ω_{coat} is the active principle concentration within the polymer coating, and k_l describes the overall mass transport coefficient that considers both the resistance given by the boundary layer near the internal coating surface and the resistance of the polymer coating, which is evaluated by means of the penetration theory. That parameter is a function of the compound diffusivity into the polymer coating (D_{coat}). The mass balance equation for the active principle within the polymer coating is the following:

$$\frac{dM_{coat}}{dt} = k_i \cdot A_i \cdot (\omega_i - \omega_{coat}) - k_e \cdot A_o \cdot (\omega_{coat} - \omega_{bulk}) \tag{3}$$

where M_{coat} is the active principle mass within the polymer coating, ω_{bulk} is the active principle concentration within the external liquid bulk, A_o is the external coating area, and k_e is the overall mass transfer coefficient through the coating, which lumps together the resistance of both coating and external layer. Another equation is also necessary to complete the mathematical model proposed by Manca and Rovaglio [150] relative to the mass balance of the active principle outside the microcapsule. With NP as the total number of microcapsules present in the solution, one has:

$$\frac{dM_{bulk}}{dt} = k_o \cdot A_o \cdot (\omega_{coat} - \omega_{bulk}) \cdot NP \tag{4}$$

where M_{bulk} is the active principle mass within the external solution and k_o is the mass transfer coefficient within the external liquid solution. Under the assumption of a single particle size, NP can be evaluated as:

$$NP = \frac{M_{tot}}{M_{part}} = \frac{M_{tot}}{M_{core} + M_{pol}} \tag{5}$$

where M_{tot} is the particles total mass, and M_{part} is the total mass of a particle constituted by the mass of the active principle M_{core} and the mass of the polymer coating M_{pol}. The numerical integration of the resulting ordinary differential equation system, with initial value conditions, was performed by the authors referring to two distinct active principles, sodium salicylate and potassium chloride, both microencapsulated within ethylcellulose. The release depends essentially on two parameters: the coating thickness (s_{coat}) and the compound diffusivity into the polymeric coating (D_{coat}). The nonuniform characteristics of the polymer coating, such as porosity, tortuosity, thickness, and eccentricity, can be lumped together into the D_{coat} parameter. For this reason, D_{coat} becomes a tuning parameter of the physical model, and it can be renamed as apparent diffusivity.

Regardless of the detailed modeling approach, the results of the release curves obtained by Manca and Rovaglio [150] are not satisfactory when compared with the available experimental data. The main hypothesis characterizing the mathematical model of drug release was based on the statement that the active compound could be described by only one particle diameter. For the sake of simplicity, a single diameter value was assumed for all the sample particles. Conversely, a particle size distribution should be considered to increase the detail of description. Manca and Rovaglio [150] found a good accordance between experimental data and simulated values when a granulometric distribution was taken into account.

The principal steps involved in the release of the active agent from a matrix system depend on whether the active agent is dissolved or dispersed in the matrix. The steps involved in the release of an active agent that is dissolved in the matrix are [152] (1) diffusion of the active agent to the surface of the matrix, (2) partition of the active agent between the matrix and the elution medium (i.e., the surrounding food), and (3) transport away from the matrix surface. The desorption of the active agent from matrix systems was first described by Crank [153]. The one-dimensional mass diffusion equation for a sphere of specified diameter d and radius R is expressed by Fick's second law as:

$$\frac{\partial C}{\partial t} = \frac{1}{r^2}\left\{\frac{\partial}{\partial_r}\left(D(t)r^2\frac{\partial}{\partial r}\right)\right\} \tag{6}$$

where C is the concentration of the active agent, r is the radial location inside the sphere, and D is the drug diffusion coefficient in the polymeric matrix. Solving the previous equation with the suitable boundary and initial conditions ($C = 0$, $r = R$, $t > 0$; $C = C_1$, r, $t = 0$, where C_1 is the initial active agent concentration) the following equation for the total amount of diffusing active agent leaving a sphere of diameter d [153] is given:

$$\frac{M_{t,d}}{M_{\infty,d}} = \left(1 - \frac{6}{\pi^2}\sum_{j=1}^{\infty}\frac{1}{j^2}e^{(-j^2\pi^2 T/R^2)}\right); \; T = \int_0^t D(t)dt \tag{7}$$

where $M_{t,d}$ and $M_{\infty,d}$ represent the mass of active agent released from a sphere of diameter d, at time t and $t = \infty$, respectively.

Regarding the release of an active agent that is dispersed in the matrix the steps are [152]: (1) dissolution into the matrix, (2) diffusion to the surface, and (3) transport away from the surface of the matrix. Assuming pseudo steady-state release kinetics, the amount of active agent released at the surface of the polymer is given by [154]:

$$M_t = A[D_m C_{as}(2C_{ao} - C_{as})t]^{1/2} \tag{8}$$

where C_{as} is the solubility of the active agent in the polymer, C_{ao} is the concentration of the active agent in the polymer, A is the barrier area, and D_m is the diffusion coefficient of the active agent through the barrier.

7.4.2 Swelling-Controlled Delivery Systems

In swelling-controlled systems, the active agent, dissolved or dispersed in a polymeric matrix, cannot diffuse to any significant extent within the matrix because of its low diffusion coefficient. When the polymer matrix is placed in a thermodynamically compatible medium, the polymer swells because of the absorption of fluid (penetrant) from the medium. The active agent diffusion coefficient in the swollen part of the matrix increases then it diffuses into the outer medium. In a diffusion-controlled system, the matrix is assumed to be unaffected during the release process, whereas in swelling-controlled systems, the membrane undergoes a transition from a glassy to a rubbery state after interaction with the penetrant. The polymer chains in the rubbery state, being more mobile than in the glassy state, allow the active agent to diffuse out of the matrix more rapidly. The release rate is determined by the glass-to-rubbery transition process. The quantity of active agent released at any time t, M_t, is given by:

$$\frac{M_t}{M_\infty} = k_a t^n \tag{9}$$

where M_∞ is the initial active-agent loading of the polymer, and k_a and n are system parameters that depend on both the nature of the polymer-penetrant-active agent interactions and on the geometry of the release device. The value of the parameter n indicates the type of active agent transport mechanism. When $n = 0.5$, the active compound is released by simple Fickian diffusion (stochastic phenomenon), whereas when $n = 1.0$, diffusion is described as "case II diffusion." In the latter situation, the rate of solvent uptake by the polymer is largely determined by the rate of swelling and relaxation of the polymer chains. "Super case II transport" occurs when $n > 1.0$. In the $0.5 < n < 1.0$ region, diffusion process is a combination of Fickian and non-Fickian diffusion, and it is known as "anomalous diffusion."

Several approaches are reported in the literature to take into account quantitatively the stochastic diffusion (related to Brownian motion) and the relaxation phenomenon involved in mass diffusion [155–157].

Flores et al. [158] proposed a mathematical model that can describe the release of sorbate from tapioca starch edible film. A linear superimposition of the previously described mechanisms is one of the simplest ways to combine them:

$$M(t) = M_F(t) + M_R(t) \tag{10}$$

where $M(t)$ is the total amount of active compound released at time t, $M_F(t)$ is the contribution of stochastic phenomena to the amount of active compound released at time t, and $M_R(t)$ is the contribution of polymer relaxation phenomena to the amount of active compound released at time t.

Mass transport related to Brownian motions is generally described by means of the Fick's model. In the specific case of diffusion through a plane sheet with constant boundary conditions, the solution of the Fick's second law is:

$$M_F(t) = M_F^{\text{eq.}} \cdot \left\{ 1 - \frac{8}{\pi^2} \cdot \sum_{n=0}^{n=\infty} \frac{1}{(2 \cdot n + 1)^2} \cdot \exp\left[-D \cdot (2 \cdot n + 1)^2 \cdot \pi^2 \cdot \frac{t}{\ell^2} \right] \right\} \tag{11}$$

where $M_F^{eq.}$ is the amount of low-molecular-weight compound released at equilibrium as a consequence of stochastic phenomena, D is the diffusion coefficient through the swollen polymeric matrix, and ℓ is the film thickness.

Polymer relaxation is driven by the distance of the system from the equilibrium; it can be quantitatively described through a first-order kinetic type equation:

$$M_R(t) = M_R^{eq.}\left[1 - \exp\left(-\frac{t}{\tau}\right)\right] \tag{12}$$

where $M_R^{eq.}$ is the amount of low-molecular-weight compound released at equilibrium as a consequence of polymer relaxation, where τ is the relaxation time associated to polymer relaxation.

Substituting Eqs. (11) and (12) in Eq. (10), the following equation is obtained:

$$M(t) = M_R^{eq.} \cdot \left\{1 - \frac{8}{\pi^2} \cdot \sum_{n=0}^{n=\infty} \frac{1}{(2 \cdot n + 1)^2} \cdot \exp\left[-D \cdot (2 \cdot n + 1)^2 \cdot \pi^2 \cdot \frac{t}{\ell^2}\right]\right\} + M_R^{eq.} \cdot \left[1 - \exp\left(-\frac{t}{\tau}\right)\right] \tag{13}$$

By defining X_F as the ratio between $M_F^{eq.}$ and $M^{eq.}$, where $M^{eq.} - M_F^{eq.} + M_R^{eq.}$, the preceding equation can be rearranged in the following form:

$$M(t) = M^{eq} \cdot \left\{X_F \cdot \left\{1 - \frac{8}{\pi^2} \cdot \sum_{n=0}^{n=\infty} \frac{1}{(2 \cdot n + 1)^2} \cdot exp\left[-D \cdot (2 \cdot n + 1)^2 \cdot \pi^2 \cdot \frac{t}{\ell^2}\right]\right\} + (1 - X_F) \cdot \left[1 - \exp\left(-\frac{t}{\tau}\right)\right]\right\} \tag{14}$$

The advantage of rearranging Eq. (13) into Eq. (14) is that X_F is a measure of the deviation of the transport mechanism from the ideal Fickian behavior. In fact, by definition, X_F ranges from 0 to 1; for X_F equal to 1, the preceding equation is the solution of the Fick's second law, whereas for X_F equal to 0, anomalous diffusion is obtained.

Flores et al. [158] found that the proposed model satisfactorily fits the experimental data, suggesting that despite the complexity of the phenomena involved during the release of an active compound from the polymeric matrix, the adopted model can be used to obtain useful information on the release mechanism of low-molecular-weight compounds from films.

More complex models were described by Buonocore et al. [91, 155]. In particular, Buonocore et al. [91] developed a mathematical model to predict lysozyme release kinetic from cross-linked PVOH into aqueous solution. The model was developed taking into account the diffusion of water molecules into the polymeric film and the counterdiffusion of the incorporated antimicrobial agent from the film to the aqueous solution. In particular, it was considered that the lysozyme release from a swelling polymer could be regarded as "anomalous diffusion" with moving boundary conditions. The study was conducted developing two models, one describing the water uptake kinetic (Eq. 15) and the other describing the enzyme release (Eq. 16):

$$\frac{d}{dt} \cdot \int_{v(t)} (\rho \cdot C_W) \cdot dV = \int_{S(t)} \left[D_F^W \cdot \frac{\partial}{\partial x} (\rho \cdot C_W) \cdot \vec{n} \right] \cdot dS \tag{15}$$

$$\frac{d}{dt} \cdot \int_{V(t)} (\rho \cdot C_L) \cdot dV = \int_{S(t)} \left[D_F^L \cdot \frac{\partial}{\partial x} (\rho \cdot C_L) \cdot \bullet \vec{n} \right] \cdot dS \tag{16}$$

where D_F^W is the water diffusion coefficient, D_F^L the lysozyme diffusion coefficient, C_W is the local water concentration, C_L is the lysozyme concentration, ρ is the density of the polymeric matrix, $S(t)$ is the surface of the volume $V(t)$, $\vec{n}$ is the versor normal to the surface, and x is axial coordinate. The assumption made by the authors is that the elements have a volume that changes during hydration (the volume of each element changes in such a way that it contains always the same amount of macromolecular matrix according to the Lagrangian approach) instead of considering elements with a fixed volume (Eulerian approach). The first model was fitted to the water sorption data and the obtained parameters were used into the second model to fit the lysozyme release data. The mass balance for the penetrant and the active substance were numerically solved by means of the finite elements method. The model proposed by Buonocore et al. [91] also takes into account the dependence of the water diffusion coefficient on the degree of cross-link.

The same model was used again [155] to describe the release kinetics of several active compounds (i.e., lysozyme, nisin, and sodium benzoate) from cross-linked PVOH. As in the previous study, the model takes into account the dependence of water diffusion coefficient on the degree of cross-link by varying the boundary conditions.

7.5 FUTURE TRENDS

Nowadays, the inclusion of active compounds in biobased films is considered a tool of key interest to extend product shelf life or to induce controlled transformations in the foodstuffs. Therefore, the incorporation of biocatalysts can result in more attractive and active applications for edible polymers in the near future. The use of antimicrobial enzymes in active food packaging develops the hurdle concept in food processing. Moreover, certain bioconversions might be conveniently implemented in prepacked foodstuffs to transform foods for special purposes during storage. However, despite the great industrial interest, active or bioactive packages where transformations take place by the action of some fixed or released enzymes have not been implemented to date. The existence of many regulatory constrains seems to limit the expansion of active packaging technologies, especially in Europe, where the Directive on this technology is currently under preparation. Starting from the environmentally sustainable solutions, the way found through innovation goes up to the safety and to the implementation of material characteristics. The next aims are the development of active eco-friendly packaging solutions. Taking into consideration the already existing approaches and the performances of new biopolymers, it would be expected that this research field will expand, achieving commercial applications in parallel with the developments in edible films and coatings. Research opportunities also exist to integrate carriers with antimicrobial properties and package indicators into the same device. Combining intelligent and active packaging offers many intriguing possibilities, thus allowing the development of more sophisticated packaging systems. In the event that the actual levels need to be surpassed to achieve the desired biofunctionality of the product, the industry would have to seek approval for the strategies to convey their products in ways that comply with current or future legislations.

REFERENCES

1. Hult, K., Berglund, P. (2003). Engineered enzymes for improved organic synthesis. *Current Opinion in Biotechnology*, *14*, 395–400.
2. Dunaway-Mariano, D. (2008). Enzyme function discovery. *Structure*, *16*, 1599–1600.
3. Fernandez, A., Cava, D., Ocio, M. J., Lagaron, J. M. (2008). Perspective for biocatalysts in food packaging. *Trends in Food Science & Technology*, *19*, 198–206.
4. Jawaheer, S., White, S. F., Rughooputh, S. D. D. V., Cullen, D. C. (2003). Development of a common biosensor format for an enzyme based biosensor array to monitor fruit quality. *Biosensors and Bioelectronics*, *18*, 1429–1437.
5. Ren, G. L., Xu, X. H., Liu, Q., Cheng, J., Yuan, X. Y., Wu, L. L. (2006). Electrospun poly(vinyl alcohol)/glucose oxidase biocomposite membranes for biosensor applications. *Reactive & Functional Polymers*, *66*, 1559–1564.
6. Purice, A., Schou, J., Kingshott, P., Pryds, N., Dinescu, M. (2007). Characterization of lysozyme films produced by matrix assisted pulsed laser evaporation (MAPLE). *Applied Surface Science*, *253*, 6451–6455.
7. Chen, J., Wang, Q., Hua, Z. Z., Dua, G. C. (2006). Research and application of biotechnology in textile industries in China. *Enzyme Microbiology & Technology*, *40*, 1651–1655.
8. Wang, Q., Fan, X., Hu, Y., Yuan J., Cui, L., Wang, P. (2008). Antibacterial functionalization of wool fabric via immobilizing lysozymes. *Bioprocess and Biosystems Engineering*, *32*, 633–639.
9. Ramesh, S., Mathivanan, N. (2009). Screening of marine actinomycetes isolated from the Bay of Bengal, India for antimicrobial activity and industrial enzymes. *World Journal of Microbiology and Biotechnology*, *25*, 2103–2111.

10. Zhang, L., An, R., Wang, J., Sun, N., Zhang, S., Hu, J., Kuai, J. (2005). Exploring novel bioactive compounds from marine microbes. *Current Opinion in Microbiology*, *8*, 276–281.
11. Olempska-Beer, Z. S., Merker, R. I., Ditto, M. D., DiNovi, M. J. (2006). Food-processing enzymes from recombinant microorganisms: A review. *Regulatory Toxicology and Pharmacology*, *45*, 144–158.
12. Lopez-Rubio, A., Almenar, E., Hernandez-Munoz, P., Lagaron, J. M., Catala, R., Gavara, R. (2006). Overview of active polymer-based packaging technologies for food applications. *Food Reviews International*, *20*, 357–387.
13. Krajewska, B. (2004). Application of chitin- and chitosan-based materials for enzyme immobilizations: A review. *Enzyme and Microbial Technology*, *35*, 126–139.
14. Raba, J., Mottola, H. A. (1995). Glucose Oxidase as an analytical reagent. *Critical Reviews in Analytical Chemistry*, *25*, 1–42.
15. Fuglsang, C. C., Johansen, C., Christgau, S., Adler-Nissen, J. (1995). Antimicrobial enzymes: Applications and future potential in the food industry. *Trends in Food Science & Technology*, *6*, 390–396.
16. Schepartz, A. I., Subers, M. H. (1964). The glucose oxidase of honey I. Purification and some general properties of the enzyme. *Biochimica et Biophysica Acta (BBA) — Specialized Section on Enzymological Subjects*, *85*, 2228–2237.
17. Katchalski-Katzir, E. (1993). Immobilized enzymes – learning from past successes and failures. *Trends Biotechnology*, *11*, 471–478.
18. Dias, S. F., Vilas-Boas, L., Cabral, J. M. S., Fonseca, M. M. R. (1991). Production of ethyl butyrate by Candida rugosa lipase immobilized in polyurethane. *Biocatalysis 5*, 21–34.
19. Chiaratini, L., Magnani, M. (1997). Immobilization of enzymes and proteins on red blood cells. *In: Immobilization of Enzymes and Cells*. Ed.: Bickerstaff, G. F., pp. 143,152. Humana, Totowa, NJ.
20. Bickerstaff, G. F. (1997). Immobilization of enzymes and cells: Some practical considerations. *In: Immobilization of Enzymes and Cells*. Ed.: Bickerstaff, G. F., pp. 1,11. Humana, Totowa, NJ.
21. Amitai, G., Andersen, J., Wargo, S., Asche, G., Chir, J., Koepsel, R., Russell, A. J. (2009). Polyurethane-based leukocyte-inspired biocidal materials. *Biomaterials*, *30*, 6522–6529.
22. Murtinho, D., Lagoa, A. R., Garcia, F. A. P., Gil, M. H. (1998). Cellulose derivatives membranes as supports for immobilization of enzymes. *Cellulose*, *5*, 299–308.
23. Tirkes, S., Toppare, L., Alkan, S., Bakir, U., Önen, A., Yağcı;, Y. (2002). Immobilization of glucose oxidase in polypyrrole/polytetrahydrofuran graft copolymers. *International Journal of Biological Macromolecules*, *30*, 81–87.
24. Erginer, R., Toppare, L., Alkan, S., Bakir, U. (2000). Immobilization of invertase in functionalized copolymer matrices. *Reactive and Functional Polymers*, *45*, 227–233.
25. Isık, S., Alkan, S., Toppare, L., Cianga, I., Yagcı, Y. (2003). Immobilization of invertase and glucose oxidase in poly 2-methylbutyl-2-(3-thienyl) acetate/ polypyrrole matrices. *European Polymer Journal*, *39*, 2375–2381.
26. Levent, A., Toppare, L., Cianga I., Yagcı, Y. (2003). Synthesis and characterization of conducting copolymers of (S)-2-Methylbutyl-2-(3-thienyl)acetate with Pyrrole and Thiophene. *Macromolecular Chemistry and Physics*, *204*, 1118–1122.

27. Gursel, A., Alkan, S., Toppare, L., Yagcı, Y. (2003). Immobilization of invertase and glucose oxidase in conducting H-type polysiloxane/polypyrrole block copolymers. *Reactive and Functional Polymers, 57*, 57–65.
28. Vasileva, N., Godjevargova, T. (2004). Study on the behaviour of glucose oxidase immobilized on microfiltration polyamide membrane. *Journal of Membrane Science, 239*, 157–161.
29. Godjevargova, T., Konsulov, V., Dimov, A., Vasileva, N. (2000). Behavior of glucose oxidase immobilized on ultrafiltration membranes obtained by copolymerising acrylonitrile and N-vinylimidazol. *Journal of Membrane Science, 172*, 279–285.
30. Vartiainen, J., Rättö, M., Paulussen, S. (2005). Antimicrobial activity of glucose oxidase-immobilized plasma-activated polypropylene films. *Packaging Technology and Science, 18*, 243–251.
31. Yang, Y. M., Wang, J. W., Tan, R. X. (2004). Immobilization of glucose oxidase on chitosan–SiO2 gel. *Enzyme and Microbial Technology, 34*, 126–131.
32. Baydar, H., Sağdıc, O., Gülcan Ö., Karadoğan, T. (2004). Antibacterial activity and composition of essential oils from Origanum, Thymbra, and Satureja species with commercial importance in Turkey. *Food Control, 15*, 169–172.
33. Sagdıc, O., Özcan, M. (2003). Antibacterial activity of Turkish spice hydrolysis. *Food Control, 14*, 141–143.
34. Yanishlieva, N. V., Marinova, E. M. (2001). Stabilisation of edible oils with natural antioxidants. *European Journal of Lipid Science and Technology, 103*, 752–767.
35. Tang, S., Kerry, J. P., Sheehan, D., Buckley, D. J., Morrissey, P. A. (2001). Antioxidative effect of added tea catechins on susceptibility of cooked red meat, poultry and fish patties to lipid oxidation. *Food Research International, 34*, 651–657.
36. Wanasundara, U. N., Shahidi, F. (1998). Antioxidant and pro-oxidant activity of green tea extracts in marine oils. *Food Chemistry, 63*, 335–342.
37. Higdon, J. V., Frei, B. (2003). Tea catechins and polyphenols: Health effects, metabolism, and antioxidant functions. *Critical Reviews in Food Science and Nutrition, 43*, 89–143.
38. Zandi, P., Gondon, M. H. (1999). Antioxidant activity of extracts from old tea leaves. *Food Chemistry, 64*, 285–288.
39. McCune, L. M., Johns, T. (2002). Antioxidant activity in medicinal plants associated with the symptoms of diabetes mellitus used by the Indigenous Peoples of the North American boreal forest. *Journal of Ethnopharmacology, 82*, 197–205.
40. Alexander, P., Ritsuko, M., Miechal, S., Bat-Sheva, C., Fostik-Magyar, C., Dubinsley, Z. (1998). Natural antioxidatnt activity in some microalgal spices. *International Journal of Plant Science, 46*, 169–176.
41. Farag, R. S., Badel, A. Z. M. A., El Baraty, G. S. A. (1989). Influence of thyme and clove essential oils on cotton seed oil oxidation. *Journal of the American Oil Chemists Society, 66*, 800–804.
42. Kohchi, Y. (1995). Antioxidative activity of spices and herbs. *Food Ingredients Journal of Japan, 163*, 44–55.
43. Delaquis, P. J., Stanich, K., Girard, B., Mazza, G. (2002). Antimicrobial activity of individual and mixed fractions of dill, cilantro, coriander and eucalyptus essential oils. *International Journal of Food Microbiology, 74*, 101–109.

44. Dorman, H. J. D., Deans, S. G. (2000). Antimicrobial against from plants: Antimicrobial activity of plant volatile oils. *Journal of Applied Microbiology*, *88*, 308–316.

45. Lambert, R. J. W., Skandamis, P. N., Coote, P. J., Nychas, G. J. E. (2001). A study of the minimum inhibitory concentration and mode of action of oregano essential oil, thymol, and carvacrol. *Journal of Applied Microbiology*, *91*, 453–462.

46. Tassou, C. C., Koutsoumanis, K., Nychas, G. J. E. (2000). Inhibition of *Salmonella enteritidis* and *Staphylococcus aureus* in nutrient broth by mint essential oil. *Food Research International*, *33*, 273–280.

47. Ben-Yehoshua, S., Rodov, V., Peretz, J. (1998). Constitutive and induced resistance of citrus fruits against pathogens. In: Disease Resistance in Fruits. Eds.: Johonson, G. I., Highly, E., Joyce, D. C. pp. 78–89. Australian Centre for International Agricultural Research. ACIAR Proceeding No. 80. Canberra, Australia.

48. Karatzas, A. K., Bennik, M. H. J., Smid, E. J., Kets, E. P. W. (2000). Combined action of S-carvone and mild heat treatment on *Listeria monocytogenes* Scott A. *Journal of Applied Bacteriology*, *89*, 296–301.

49. Vàzquez, B. I., Fente, C., Franco, C. M., Vazquez, M. J., Cepeda, A. (2001). Inhibitory effects of eugenol and thymol on *Penicillium citrinum* strains in culture media and cheese. *International Journal of Food Microbiology*, *67*, 157–163.

50. Manzocco, L., Anese, M., Nicoli, M. C. (1998). Antioxidant properties of green tea extracts as affected by processing. *Lebensmittel-Wissenschaft und Technologie*, *31*, 694–698.

51. Chen, Z. Y., Chan, P. T., Ma, H. M., Fung, K. P., Wang, J. (1996). Antioxidative effect of ethanol green tea extracts on oxidation of canola oil. *Journal of the American Oil Chemists' Society*, *73*, 375–380.

52. Özel, M. Z., Göğüs, F., Lewis, A. C. (2003). Subcritical water extraction of essential oil from Thymbra spicata. *Food Chemistry*, *82*, 381–386.

53. Dorman, H. J. D., Peltoketo, A., Hiltunen, R., Tikkanen, M. J. (2003). Characterisation of the antioxidant properties of de-odourised aqueous extracts from selected Lamiaceae herbs. *Food Chemistry*, *83*, 255–262.

54. Hanc, S., Sahin, S., Yilmaz, L. (2003). Isolation of volatile oil from thyme (Thymbra spicata) by steam distillation. *Nahrung/Food*, *47*, 252–255.

55. Gutierrez, J., Bourke, P, Lonchamp, J., Barry-Ryan, C. (2009). Impact of plant essential oils on microbiological, organoleptic and quality markers of minimally processed vegetables. *Innovative Food Science and Emerging Technologies*, *10*, 195–202.

56. Meyer, A. S., Suhr, K. I., Nielsen, P. (2002). Natural food preservatives. *In: Minimal Processing Technologies in the Food Industries*. Eds.: Ohlsson, T. Bengtsson, N. Woodhead Publishing Ltd. Cambridge, U.K.

57. Bari, M. L., Ukuku, D. O., Kawasaki, T., Inatsu, Y., Isshiki, K., Kawamoto. S. (2005). Combined efficacy of nisin and pediocin with sodium lactate, citric acid, phytic acid, and potassium sorbate and EDTA in reducing the Listeria monocytogenes population of inoculated fresh-cut produce. *Journal of Food Protection*, *68*, 1381–1387.

58. No, H. K., Meyers, S. P., Prinyawiwatkul, W., Xu, Z. (2007). Application of chitosan for improvement of quality and shelf life of foods–A review. *Journal of Food Science*, *72*, 87–100.

59. Lee, D. S., Yam, K. L., Piergiovanni, L. (2008). *Food Packaging Science and Technology*, CRC Press, Boca Raton, FL.

60. Appendini, P., Hotchkiss, J. H. (2002). Review of antimicrobial food packaging. *Innovative Food Science and Emerging Technologies*, *3*, 113–126.

61. Cutter, C. N. (2002). Microbial control by packaging: A review. *Critical Reviews in Food Science and Nutrition*, *42*, 151–161.

62. Suppakul, P., Miltz, J., Sonneveld, K., Bigger, S. W. (2003). Active packaging technologies with an emphasis on antimicrobial packaging and its applications. *Journal of Food Science*, *68*, 408–420.

63. Siracusa, V., Rocculi, P., Romani, S., Dalla Rosa, M. (2008). Biodegradable polymers for food packaging: A review. *Trends in Food Science & Technology*, *19*, 634–643.

64. Tischer, W., Kasche, V. (1999). Immobilized enzymes: Crystals or carriers? *Trends in Biotechnology*, *17*, 325–335.

65. Brody, A. L., Budny, J. A. (1995). Enzymes as active packaging agents. *In: Active Food Packaging*. Ed.: Rooney Glasgow, M. L.: pp. 174,192. Blackie Academic Professional. London, U.K.

66. Jürgen-Lohmann, D. L., Nacke, C., Legge, R. L., Simon, L. C. (2009). Preparation and metholdology for chemical mapping of sol-gel thin films containing lysozyme. *Journal Sol Gel Science Technology*, *50*, 77–86.

67. Di Serio, M., Maturo, C., De Alteriis, E., Parascandola, P., Tesser, R., Santacesaria, E. (2003). Lactose hydrolysis by immobilized β-galactosidase: The effect of the supports and the kinetics. *Catalysis Today*, 79–80, 333–339.

68. Wu, Y., Daeschel, M. A. (2007). Lytic antimicrobial activity of hen egg white lysozyme immobilized to polystyrene beads. *Food Microbiology and Safety*, *72*, 369–374.

69. Goddard, J. M., Hotchkiss, J. H. (2007). Polymer surface modification for the attachment of bioactive compounds. *Progress in Polymer Science*, *32*, 698–725.

70. Goddard, J. M., Talbert, J. N., Hotchkiss, J. H. (2007). Covalent attachment of lactase to low-density polyethylene films. *Journal of Food Science*, *72*, 6–41.

71. Vartiainen, J., Ratto, M., Paulussen, S. (2005). Antimicrobial activity of glucose oxidase-immobilized plasma-activated polypropylene films. *Packaging Technology and Science*, *18*, 243–251.

72. Hotchkiss, J. H. (1997). Food packaging interactions influencing quality and safety. *Food Additives & Contaminants*, *14*, 601–607.

73. Wang, C., Chio, S. (1998). Reversible immobilization of chitinase via coupling to reversible soluble polymer. *Enzyme and Microbial Technology*, *22*, 636–640.

74. Sarkar, J. M., Leonowicz, A., Bollag, J. M. (1989). Immobilization of enzymes on clays and soils. *Soil Biology and Biochemistry*, *21*, 223–230.

75. Gopinath, S., Sugunan, S. (2007). Enzymes immobilized on montmorillonite K10: Effect of adsorption and grafting on the surface properties and the enzyme activity. *Applied Clay Science*, *35*, 67–75.

76. Van de Velde, F., Lourenço, N. D., Pinheiro, H. M., Bakker, M. (2002). Carrageenan: A food-grade and biocompatible support for immobilisation techniques. *Advanced Synthesis and Catalysis*, *344*, 815–835.

77. Agullò, E., Rodriguez, M. S., Ramos, V., Albertengo, L. (2003). Present and future role of chitin and chitosan in food. *Macromolecular Bioscience*, *3*, 521–530.
78. Chang, M. Y., Juang, R. S. (2005). Activities, stabilities, and reaction kinetics of three free and chitosaneclay composite immobilized enzymes. *Enzyme and Microbial Technology*, *36*, 75–82.
79. Hoven, V. P., Tangpasuthadol, V., Angkitpaiboon, Y., Vallapa, N., Kiatkamjornwong, S. (2007). Surface-charged chitosan: Preparation and protein adsorption. *Carbohydrate Polymers*, *68*, 44–53.
80. Nagatomo, H., Matsushita, Y., Sugamoto, K., Matsui, T. (2005). Preparation and properties of gelatin-immobilized b-glucosidase from Pirococcus furiosus. *Bioscience, Biotechnology, and Biochemistry*, *69*, 128–136.
81. Roy, I., Gupta, M. N. (2004). Hydrolysis of starch by a mixture of glucoamylase and pullulanase entrapped individually in calcium alginate beads. *Enzyme and Microbial Technology*, *34*, 26–32.
82. Lazzeri, L., Cascone, M. G., Quiriconi, S., Morabito, L., Giusti, P. (2005). Biodegradable hollow microfibres to produce bioactive scaffolds. *Polymer International*, *54*, 101–107.
83. Bower, C. K., Avena-Bustillos, R. J., Olsen, C. W., McHugh, T. H., Bechtel, P. J. (2006). Characterization of fish-skin gelatin gels and films containing the antimicrobial enzyme lysozyme. *Journal of Food Science*, *71*, 141–145.
84. Ertan, F., Yagar, H., Balkan, B. (2007). Optimization of alpha-amylase immobilization in calcium alginate beads. *Preparative Biochemistry & Biotechnology*, *37*, 195–204.
85. Haider, T., Husain, Q. (2007). Calcium alginate entrapped preparations of Aspergillus orza beta galactosidase: Its stability and applications in the hydrolysis of lactose. *International Journal of biological macromolecules*, *41*, 72–80.
86. Padgett, T., Han, I. Y., Dawson, P. L. (1998). Incorporation of food-grade antimcirobial compounds into biodegradable packaging films. *Journal of Food protection*, *61*, 1330–1335.
87. Mecitoglu, C., Yemenicioglu, A., Arslanoglu, A., Elmaci, Z. S. Korel, F., Cetin, A. E. (2006). Incorporation of partially purified hen egg white lysozyme into zein films for antimcrobial food packaging. *Food Research Intenational*, *39*, 12–21.
88. Park, S. I., Daeschel, M. A., Zhao, Y. (2004). Functional properties of antimicrobial lysozymeechitosan composite films. *Journal of Food Science*, *69*(8), 215–221.
89. Andersson, M., Andersson, T., Adlercreutz, P., Nielsen, T., Hörnsten, E. G. (2002). Towards an enzyme-based oxygen scavenging laminate. Influence of industrial lamination conditions on the performance of glucose oxidase. *Biotechnology and Bioengineering*, *79*, 37–42.
90. Appendini, P., Hotchkiss, J. H. (1997). Immobilization of lysozyme on food contact polymers as a potential antimicrobial films. *Packaging Technology and Science*, *10*, 271–279.
91. Buonocore, G. G., Del Nobile, M. A., Panizza, A., Bove, S., Battaglia, G., Nicolais, L. (2003). Modeling the lysozyme release kinetics from antimicrobial films intended for food packaging applications. *Journal of Food Science*, *68*, 1365–1370.
92. Buonocore, G. G., Sinigaglia, M., Corbo, M. R., Bevilacqua, A., La Notte, E., Del Nobile, M. A. (2004). Controlled release of antimicrobial compounds from highly swellable polymers. *Journal of Food Protection*, *67*, 1190–1194.

93. Buonocore, G., Conte, A., Corbo, M. R., Sinigaglia, M., Del Nobile, M. A. (2005). Mono-and multilayer active films containing lysozyme as antimicrobial agent. *Innovative Food Science and Emerging Technologies*, *6*, 459–464.

94. LaCoste, A., Schaich, K. M., Zumbrunnen, D., Yam, K. L. (2005). Advancing controlled release packaging through smart blending. *Packaging Technology and Science*, *18*, 77–87.

95. Gŋbilmez, Ç. M., Yemenicioğlu, A., Arslanoğlu, A. (2007). Antimicrobial and antioxidant activity of edible zein films incorporated with lysozyme, albumin proteins and disodium EDTA. *Food Research International*, *40*, 80–91.

96. Gemili, S., Yemenicioğlu, A., Altınkaya, S. A. (2009). Development of cellulose acetate based antimicrobial food packaging materials for controlled release of lysozyme. *Journal of Food Engineering*, *90*, 453–462.

97. Altinkaya, S. A., Yenal, H. (2006). In vitro drug release rates from asymmetric–membrane tablet coatings: Prediction of phase-inversion dynamics. *Biochemical Engineering Journal*, *28*, 131–139.

98. Desimone, M. F., Matiacevich, S. B., del Pilar Buera, M., Diaz, L. E. (2008). Effects of relative humidity on enzyme activity immobilized in sol-gel-derived silica nanocomposites. *Enzyme and Microbial Technology*, *42*, 583–588.

99. Rudra, J. S., Dave, K., Haynie, D. T. (2006). Antimicrobial polypeptide multilayer nanocoatings. *Journal of Biomaterials Science*, *17*, 1301–1315.

100. Torres-Giner, S., Gimenez, E., Lagaron, J. M. (2008). Characterization of a morphology and thermal properties of zein, prolamine nanostructured obtained by electrospinning. *Food Hydrocolloids*, *22*, 601–614.

101. Ramakrishna, S., Fujihara, K., Teo, W. E., Yong, T., Ma, Z., Ramakrishna, R. (2006). Electrospun nanofibers: Solving global issues. *Materials Today*, *1*, 9–3.

102. Zeng, J., Xu, X., Chen, X., Liang, Q., Bian, X., Yang, L., Jing, X. (2003). Biodegradable electrospun fibers for drug delivery. *Journal of Controlled Release*, *92*, 227–231.

103. Zeng, J., Aigner, A., Czubayko, F., Kissel, T., Wendorff, J. H., Greiner, A. (2005). Poly(vinyl alcohol) nanofibers by electrospinning as a protein delivery system and the retardation of enzyme release by additional polymer coatings. *Biomacromolecules*, *6*, 1484–1488.

104. Glass, K. A., Johnson, E. A. (2004). Antagonistic effect of fat on the antibotulinal activity of food preservatives and fatty acids. *Food Microbiology*, *21*, 675–682.

105. Kandimalla, V. B., Tripathi, V. S., Ju, H. Y. (2006). Immobilization of biomolecules in sol-gels: Biological and analytical applications. *Critical Reviews in Analytical Chemistry*, *36*, 73–106.

106. Cunningham, F. E., Proctor, V. A., Goetsch, S. J. (1991). Egg-white lysozyme as a food preservative: An overview. *World's Poultry Science Journal*, *47*, 141–163.

107. Branen, J. K., Davidson, P. M. (2004). Enhancement of nisin, lysozyme, and monolaurin antimicrobial activities by ethylenediaminetetraacetic acid and lactoferrin. *International Journal of Food Microbiology*, *90*, 63–74.

108. Boland, J. S., Davidson, P. M., Weiss, J. (2003). Enhanced inhibition of Escherichia coli O157:H7 by lysozyme and chelators. *Journal of Food Protection*, *66*, 1783–1789.

109. Masschalck, B., Michiels, C. W. (2003). Antimicrobial properties of lysozyme in relation to foodborne vegetative bacteria. *Critical Reviews in Microbiology*, *29*, 191–214.

110. Crapisi, A., Lante, A., Pasini, G., Spettoli, P. (1993). Enhanced microbial cell lysis by the use of lysozyme immobilized on different carrier. *Process Biochemistry*, *28*, 17–21.

111. Sinigaglia, M., Bevilacqua, A., Corbo, M. R., Pati, S., Del Nobile, M. A. (2008). Use of active compounds for prolonging the shelf life of Mozzarella cheese. *International Dairy Journal 18*, 624–630.

112. Conte, A., Gammariello, D., Di Giulio, S., Attanasio, M., Del Nobile, M. A. (2009). Active coating and modified atmosphere packaging to extend the shelf life of Fior di latte cheese. *Journal of Dairy Science*, *92*, 887–894.

113. Brody, A. L. (2005). Edible packaging. *Food Technology*, *56*, 65–66.

114. Gill, A. O., Holley, R. A. (2000). Inhibition of bacterial growth on ham and bologna by lysozyme, nisin and EDTA. *Food Research International*, *33*, 83–90.

115. Han, J. H. (2005). Antimicrobial packaging systems. *In: Innovations in Food Packaging*. Ed.: Han, J. H. pp. 80,107. Elsevier Academic Press, Amsterdam, the Netherlands.

116. Conte, A., Sinigaglia, M., Del Nobile, M. A. (2006). Antimicrobial effectiveness of lysozyme immobilized on polyvinylalcohol-based film against *Alicyclobacillus acidoterrestris*. *Journal of Food Protection*, *4*, 861–865.

117. Ribeiro, D. S., Henrique, S. M. B., Oliveira, L. S., Macedo, G. A., Fleuri, L. F. (2010). Enzymes in juice processing: a review. *International Journal of Food Science and Technology*, *4*, 5635–641.

118. Soares, N. F., Hotchikiss, J. H. (1998). Naringinase immobilisation in packaging films for reducing naringin concentration in grapefruit juice. *Journal of Food Science*, *63*, 61–65.

119. Munish, P., Marwaha, S. S., Kothari, R. M. (1996). Studies on the applicability of alginate-entrapped naringinase for the debittering of kinnow juice. *Enzyme and Microbial Technology*, *8*, 281–285.

120. Del Nobile, M. A., Piergiovanni, L., Buonocore, G. G., Fava, P., Puglisi, M. L., Nicolais, L. (2003). Naringinase immobilization in polymeric films intended for food packaging applications. *Journal of Food Science*, *68*, 2046–2049.

121. Spagna, G., Pifferi, P. G., Gilioli, E. (1995). Immobilization of a pectinlyase from Aspergillus niger for application in food technology. *Enzyme and Microbial Technology*, *17*, 729–738.

122. Spagna, G., Pifferi, P. G., Tramontini, M. (1995). Immobilization and stabilization of pectinlyase on synthetic polymers for application in the beverage industry. *Journal of Molecular Catalysis A: Chemical*, *101*, 99–105.

123. Oussalah, M., Caillet, S., Salmiéri, S., Saucier L., Lacroix, M. (2004). Antimicrobial and antioxidant effects of milk protein-based film containing essential oils for the preservation of whole beef muscle. *Journal of Agricultural and Food Chemistry*, *52*, 5598–5605.

124. Matan, N., Rimkeeree, H., Mawson, A. J., Chompreeda, P., Haruthaithanasan, V., Parker, M. (2006). Antimicrobial activity of cinnamon and clove oils under modified atmosphere conditions. *International Journal of Food Microbiology*, *107*, 180–185.

125. Han, X., Shen, T., Lou, H. (2007). Dietary polyphenols and their biological significance. *International Journal of Molecular Science*, *8*, 950–988.

126. Seydim, A. C., Sarikus, G. (2006). Antimicrobial activity of whey protein based edible films incorporated with oregano, rosemary and garlic essential oils. *Food Research International*, *39*, 639–644.

127. Ponce, A. G., Roura, S. I., del Valle, C. E., Moreira, M. R. (2008). Antimicrobial and antioxidant activities of edible coatings enriched with natural plant extracts: *In vitro* and *in vivo* studies. *Postharvest Biology and Technology*, *49*, 294–300.

128. Rojas-Grau, M. A., Raybaudi-Massilia, R. M., Soliva-Fortuny, R. C., Avena-Bustillos, R. J., McHugh, T. H., Martìn-Belloso, O. (2007). Apple puree-alginate edible coating as carrier of antimicrobial agents to prolong shelf-life of fresh-cut apples. *Postharvest Biology and Technology*, *45*, 254–264.

129. Sangsuwan, J., Rattanapanone, N., Rachtanapun, P. (2008). Effect of chitosan/methyl cellulose films on microbial and quality characteristics of freshcut cantaloupe and pineapple. *Postharvest Biology and Technology*, *49*, 403–410.

130. Fitzgerald, D. J., Stratford, M., Gasson, M. J., Ueckert, J., Bos, A., Narbad, A. (2004). Mode of antimicrobial action of vanillin against Escherichia coli, Lactobacillus and Listeria innocuau. *Journal of Applied Microbiology*, *97*, 104–113.

131. Raybaudi-Massilia, R. M., Mosqueda-Melgar, J., Martìn-Belloso, O. (2008). Edible alginate-based coating as carrier of antimicrobials to improve shelf-life and safety of fresh-cut melon. *International Journal of Food Microbiology*, *121*, 313–327.

132. Bentayeb, K., Rubio, C., Batlle, R., Nerín, C. (2007). Direct determination of carnosic acid in a new active packaging based on natural extract of rosemary. *Analytical and Bioanalytical Chemistry*, *389*, 1989–1996.

133. Pranoto, Y., Rakshit, S. K., Salokhe, V. M. (2005). Enhancing antimicrobial activity of chitosan films by incorporating garlic oil, potassium sorbate and nisin. *LWT-Food Science and Technology*, *38*, 859–865.

134. Del Nobile, M. A., Conte, A., Incoronato, A. L., Panza, O. (2008). Antimicrobial efficacy and release kinetics of thymol from zein films. *Journal of Food Engineering*, *89*, 57–63.

135. Mastromatteo, M., Barbuzzi, G., Conte, A., Del Nobile, M. A. (2009). Controlled release of thymol from zein based film. *Innovative Food Science & Emerging Technologies*, *10*, 222–227.

136. Reineccius, G. A. (1988). Spray drying of food flavours. *In: Flavour Encapsulation*. Eds.: Reineccius, G. A., Risch, S. J. pp. 55, 66. American Chemical Society, Washington, D.C.

137. Reineccius, G. A. (1989). Flavour encapsulation. *Food Reviews International*, *5*, 448.

138. Rosenberg, M., Kopelman, I. J., Talmon, Y. (1990). Factors affecting retention in spray-drying microencapsulation of volatile materials. *Journal of Agricultural and Food Chemistry*, *38*, 1288–1294.

139. Versic, R. J. (1988). Flavour encapsulation—An overview. *In: Flavour Encapsulation*. Eds.: Reineccius, G. A., Risch, S. J. pp. 1, 6. American Chemical Society, Washington, D.C.

140. Leimann, F. V., Gonçalves, O. H., Machado, R. A. F., Bolzan, A. (2009). Antimicrobial activity of microencapsulated lemongrass essential oil and the effect of experimental parameters on microcapsules size and morphology. *Materials Science and Engineering*, *C29*, 430–436.

141. Baranauskiene, R., Venskutonis, R. P., Dewettinck, K., Verhé, R. (2005). Properties of oregano (*Origanum vulgare L.*), citronella (*Cymbopogon nardus G.*) and marjoram (*Majorana hortensis L.*) flavors encapsulated into milk protein-based matrices. *Food Research International*, *39*, 413–425.

142. Krishnan, S., Bhosale, R.. Singhal, R. S. (2005). Microencapsulation of cardamom oleoresin: Evaluation of blends of gum arabic, maltodextrin and a modified starch as wall materials. *Carbohydrate Polymers*, *61*, 95–102.
143. Ramos, M. F. de S. (2006). Ph.D. Dissertation, São Paulo University, São Paulo, Brazil.
144. Shu, B., Yu, W., Zhao, Y., Liu., X. (2006). Study on microencapsulation of lycopene by spray-drying. *Journal of Food Engineering*, *76*, 64–669.
145. Ersus, S., Yurdagel, U. (2007) Microencapsulation of anthocyanin pigments of black carrot (*Daucuscarota L.*) by spray drier. *Journal of Food Engineering*, *80*, 805–812.
146. Leach, G., Oliveria, G., Morias, R. (1986). Spray drying of Dunaliella salina to produce a beta-carotene rich powder. *Journal of Industrial Microbiology and Biotechnology*, *20*, 82–85.
147. Heinzelmann, K., Franke, K. (1999). Using freezing and drying techniques of emulsions for the microencapsulation of fish oil to improve oxidation stability. *Colloids and Surfaces B: Biointerfaces*, *12*, 223–229.
148. Baker, R. W. (1987). *Controlled Release of Biologically Active Agents*. Wiley, New York, NY.
149. Pothakamury, U. R., Barbosa-Cánovas, G. V. (1995). Fundamental aspects of controlled release in foods. *Trends in Food Science & Technology*, *6*, 397–406.
150. Manca, D., Rovaglio, M. (2003). Modeling the controlled release of microencapsulated drugs: Theory and experimental validation. *Chemical Engineering Science*, *58*, 1337–1351.
151. Bird, R. B., Stewart, W. E., Lightfoot, E. N. (1960). *Transport Phenomena*. Wiley, New York, NY.
152. Fan, L. T., Singh, S. K. (1989). *Controlled release: A quantitative treatment*. Springer-Verlag, New York, NY.
153. Crank, J. (1956). *The Mathematics of Diffusion*, Oxford University Press, London, U.K.
154. Langer, R., Peppas, N. A. (1983). Chemical and physical structure of polymers as carriers for controlled release of bioactive agents: A review. *Macromolecular Chemistry and Physics*, *C23*, 61–126.
155. Buonocore, G. G., Del Nobile, M. A., Panizza, A., Corbo, M. R., Nicolais, L. (2003). A general approach to describe the antimicrobial agent release from highly swellable films intended for food packaging applications. *Journal of Controlled Release*, *90*, 97–107.
156. Del Nobile, M. A., Massera, M. (2000). Modeling of water sorption kinetic in spaghetti during overcooking. *Cereal Chemistry*, *77*, 615–619.
157. Del Nobile, M. A., Mensitieri, G., Manfredi, C., Arpaia, A., Nicolais, L. (1996). Low molecular weight molecules diffusion in advanced polymers for food packaging applications. *Polymers for Advanced Technologies*, *7*, 409–417.
158. Flores, S., Conte, A., Campos, C., Gerschenson, L., Del Nobile, M. A. (2007). Mass transport properties of tapioca-based active edible films. *Journal of Food Engineering*, *81*, 580–586.

CHAPTER 8

ANTIMICROBIAL PEPTIDES

JOSE F. MARCOS[1] and PALOMA MANZANARES[2]
[1] Departamento de Ciencia de los Alimentos, Consejo Superior de Investigaciones Científicas (CSIC), Burjassot, Valencia, Spain
[2] Departamento de Biotecnología de Alimentos, Instituto de Agroquímica y Tecnología de Alimentos (IATA), Consejo Superior de Investigaciones Científicas (CSIC), Burjassot, Valencia, Spain

CONTENTS

8.1 Summary . . . 195
8.2 Antimicrobial Peptides: Definition and Properties . . . 196
8.3 Examples of Natural Antimicrobial Peptides . . . 197
8.4 Antimicrobial Peptides by Design . . . 201
8.5 Mode of Action of Antimicrobial Peptides . . . 203
8.6 Production of Antimicrobial Peptides . . . 205
8.7 Incorporation of Antimicrobial Peptides Into Polymers . . . 206
8.8 Antimicrobial Peptide Encapsulation by Liposomes or Microparticles . . . 209
8.9 Nanostructure-Forming Antimicrobial Peptides . . . 210
8.10 Future Trends . . . 211
Acknowledgments . . . 211
References . . . 212

8.1 SUMMARY

Antimicrobial peptides are widespread in nature and show great potential as novel antimicrobials for the treatment of infectious diseases, crop protection, and the food industry. The aim of this chapter is to provide basic knowledge on

Antimicrobial Polymers, First Edition.
Edited by José M. Lagarón, María J. Ocio, and Amparo López-Rubio.

the properties of antimicrobial peptides and to introduce their potential use as components of antimicrobial polymers.

8.2 ANTIMICROBIAL PEPTIDES: DEFINITION AND PROPERTIES

Antimicrobial peptides (AMPs) and peptide-related molecules are widespread in nature in organisms all along the phylogenetic scale, including bacteria, fungi, plants, and animals [1, 2]. In microorganisms, their biological significance is mainly related to competition for nutrients. In higher multicellular organisms, AMPs are considered part of an ancestral system of defense against microbial pathogen infection [3]. Descriptions of antimicrobial activity also exist among peptides/proteins from animals that are toxic weapons against predators or preys. AMPs are peptides and small-sized proteins that have direct killing or inhibitory activity against microorganisms. Antimicrobial activity resulting from enzymatic activity, for instance, as in the case of lysozimes or glucanases, is not considered in this class. In addition, some AMPs also display an array of additional properties such as antiviral activity [4–7], modulation of the innate immune response [2, 8], cytotoxicity against cancer cells [9], or intrinsic capability of cell penetration [10]. AMPs have been extensively studied in recent years because they show great potential as novel antimicrobials in view of the decreasing efficacy of conventional antibiotics in clinic [2, 3], alternatives to the massive use of fungicides in plant protection [11, 12], or as food preservatives [13, 14]. The objective of this chapter is to provide basic knowledge and references on the properties of AMP to introduce their potential use as components of antimicrobial polymers.

Significant benefits for the use of AMPs as antimicrobial agents include their broad spectrum of activity that, in some cases, expands not only gram-positive and -negative bacteria but also fungi. Such a wide activity is often associated with undesirable toxicity against nontarget cells/organisms including human cells, a negative property that modification of peptides aims to counteract (see subsequent sections). Cytotoxicity is mostly caused by interaction and disruption of biological membranes followed by cell lysis, which occurs because of the amphipathic and cationic properties of many AMPs. However, as explained in detail subsequently, cell disruption is not the only antimicrobial mechanism for an increasing number of AMPs that also affect intracellular functions and are, therefore, considered as multitarget drugs. Multiple mechanisms of action within a single molecule make the selection of microbial resistance difficult, and only a limited number of reports describe the isolation of resistant strains under laboratory conditions.

Additional advantages of AMPs include that they act at low concentrations, usually micromolar and in some classes as bacteriocins to the nanomolar range. Because of their peptidic nature, most AMPs are also highly biocompatible and susceptible to biodegradation; in fact, increasing the stability of many short antimicrobial peptides by reducing their susceptibility to proteolysis is a major goal of AMP improvement, which would render benefits both for their practical application and their production through biotechnology.

8.3 EXAMPLES OF NATURAL ANTIMICROBIAL PEPTIDES

The amino acid sequence for more than 1,000 natural AMPs are currently known, and the number is growing continuously [15, 16]. Despite this enormous diversity, most AMPs share common biophysical properties. They are small, from just 5–6 amino acid residues in some synthetic peptides to approximately 50 in natural antimicrobial proteins; most of them are cationic and have a positive charge at physiological conditions because of the presence of arginine and lysine residues. In addition, they also have a high proportion (up to 50%) of hydrophobic residues and fold or arrange into a variety of amphipathic structures and conformations. Cationic charge and amphipathic arrangement base their propensity of *in vitro* interaction with anionic lipid bilayers (see section 8.5).

The purpose of this chapter is not to provide a detailed survey of the extensive variety of AMP, which has been addressed in recent reviews [1, 17, 18]. Only significant examples are summarized in this chapter to illustrate the distinct properties of the major classes of AMPs. Amino acid sequences of some examples are detailed in Table 8.1 [19–35]. Short, linear AMPs can fold into α-helixes (Figure 8.1a) [28, 31, 36, 37] or adopt nonconventional amphipatic conformations (Figure 8.1b) on membrane interaction or in the presence of membrane-mimicking conditions. Longer peptides/proteins are surprisingly conserved in terms of structures, which frequently are small, compact, and maintained by disulfide bridges (Figure 8.1d). The dodecapeptide bactenecin is a minimum example of this compact structure maintained by a single disulfide bridge [19] (Table 8.1).

Magainins and cecropins (Table 8.1) are representatives of AMPs unstructured in solution that adopt amphipathic α-helical structures during interaction with membranes [38]. Magainins (Figure 8.1a) were originally isolated from the skin of the frog *Xenopus laevis*, and since the demonstration of the activity of the corresponding synthetic peptides [39], they have been extensively targeted for amino acid sequence modifications. Cecropins are a family of AMPs originally isolated from the hemolymph of insects [40, 41]. Both magainins and cecropins are well known for their antibacterial properties; additionally, antifungal activities against human and plant pathogens have been described. They are among peptides with strong nonspecific toxicity that make their practical application difficult and, therefore, have been extensively used in terms of sequence modification, structure activity studies, and as modules to design peptide fusions with the double objective of improving antimicrobial activity while reducing toxicity against plant or animal cells (see next section, and also peptides CAMEL and MSI-99 in Table 8.1) [42–45].

TABLE 8.1 Examples of short AMPs

Peptide	Amino acid sequence[a]	Source	Notes	References
Bactenecin	RLCRIVVIRVCR	Bovine		[19]
BM0	RFWWFRRR	Designed	Combinatorial screening	[20]
Buforin II	TRSSRAGLQFPVGRVHRLLRK	Toad		[21]
CAMEL	KWKLFKKIGAVLKVL	Designed	Hybrid of cecropin A and melittin	[22]
Cecropin A	KWKLFKKIEKVGQNIRDGIIKAGPAVAVVGQATQIAK	Insect		[23]
D4E1	FKLRAKIKVRLRAKIKL	Designed	*de novo design*	[24]
E14LKK	LKKLLKLLKKLLKL	Designed	*de novo design*	[25]
HHC-10	KRWWKWIRW	Designed	Virtual *in silico* screening	[26]
Histatin 5	DSHAKRHHGYKRKFHEKHHSHRGY	Human		[27]
Indolicidin	ILPWKWPWWPWRR	Bovine		[28]
$\text{LfcinB}_{4\text{-}14}$	RRWQWRMKKLG	Bovine	Fragment from Lactoferrin	[29]
LL-37	LLGDFFRKSKEKIGKEFKRIVQRIKDFLRNLVPRTES	Human		[30]
Magainin 2	GIGKFLHSAKKFGKAFVGEIMNS-NH2	Frog		[31]
Melittin	GIGAVLKVLTTGLPALISWIKRKRQQ	Insect		[32]
MSI-99	GIGKFLKSAKKFGKAFVKILNS	Designed	Sequence analog of Magainin 2	[33]
PAF26	RKKWFW	Designed	Combinatorial screening	[34]
Tat (47-58)	GRKKRRQRRRPPQ	Viral	Fragment from HIV-1 Tat	[35]

[a] One letter amino acid code is shown to indicate peptide sequences.

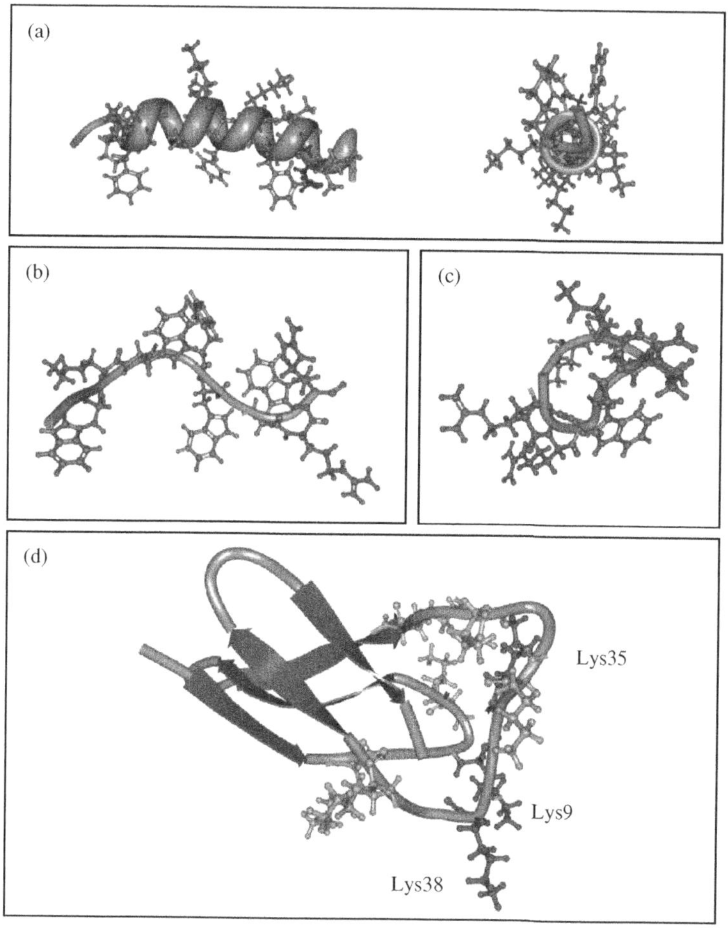

FIGURE 8.1 Selected molecular structures of antimicrobial peptides determined by solution nuclear magnetic resonance methods. Peptide 3D structures and identifications (PDB ID) were obtained from Protein Data Bank (PDB, www.rcsb.org/pdb/) and visualized with PDB ProteinWorkshop v3.9 viewer [36]. **(a)** α-helix folding of the 23-residue peptide magainin 2 in dodecylphosphocholine micelles (PDB ID: 2MAG) [31]. Side view (left) and top view from the N-terminus (right) are shown. The side chains of hydrophobic isoleucine, leucine and phenylalanine residues are shown, as well as cationic arginine and lysine, to visualize the separation and amphipathicity of the peptide. **(b)** Structure of the 13-residue indolicidin bound to dodecylphosphocholine micelles (PDB ID: 1G89) [28]. **(c)** Structure of the cyclic synthetic derivative of bovine lactoferrin LfcinB4-14 in SDS micelles (PDB ID: 1Y58) [29]. **(d)** The structure of the antifungal protein from *Penicillium chrysogenum* [37]. The five antiparallel β-strands forming two orthogonal β-sheets are shown. The cationic side chains exposed in the two largest loops at the right are shown, including the three lysine residues (labeled with their residue number) that were demonstrated to enhance toxicity by site-directed mutagenesis. (See color insert.)

Mammalian epithelial surfaces produce antimicrobial cathelicidins from longer inactive precursors [46, 47]. The biosynthesis of the antimicrobial motif fused to predomains and/or prodomains that target the product to specific locations and require specific proteolytic cleavage for activation is common in higher organisms. Contrary to other species, humans encode a unique member of the cathelicidin family named LL-37 (Table 8.1), which can be naturally processed into multiple peptides with distinct antimicrobial and immunomodulatory functionalities [48].

Among the linear peptides, notorious examples of AMP are enriched in certain amino acids. Indolicidin is a small peptide rich in tryptophan residues isolated from bovine neutrophils [49, 50]. It has an extended and flexible structure that adopts distinct unusual conformations depending on the assay conditions [28, 50] (Figure 8.1b). Different tryptophan-rich AMPs have been identified, and a role of this amino acid in the interaction with biological membrane mimetics has been postulated [51]. Histatins are a group of sequence-related peptides present in human saliva that are rich in a specific residue, in this case histidine (Table 8.1). The biological significance for the enrichment in this particular residue is unclear.

Disulfide bridge-containing small proteins are a common structural scaffold among AMPs. Included in this class are mammalian defensins [52], insect defensins, antimicrobial proteins from plants such as thionins, lipid transfer proteins and plant defensins [53, 54], fungal antifungal proteins [37, 55], and the antimicrobial plectasin [56, 57]. Broadly, they contain 2–4 disulfide bonds that conform a compact structure with antiparallel β strands that is stable against adverse biophysical and biochemical conditions, including resistance against high temperatures and protease degradation (Figure 8.1d). Selected members such as insect defensins and plectasin contain α-helixes in their structure.

Bacteriocins are ribosomally synthesized peptides produced by specific bacteria, mostly food-related lactic acid bacteria, which are active against other bacterial species. Producer strains have immunity mechanisms to protect themselves from peptide killing. They have been extensively characterized in relation to their use as food preservatives [14, 58]. A major group within this class is the so-called lantibiotics [59], which are peptides subjected to posttranslational modifications by specific enzymes that result in nonproteinaceous amino acids and an unusual cyclic bond, the lanthionine ring. These modifications confer structural constraints and high stability. Nisin and mersacidin are two well-known examples in this class. Nisin is a 3.5-kDa cationic polypeptide produced by specific strains of *Lactoccoccus lactis* strains, which inhibits a variety of gram-positive bacteria. It is one of the most studied bacteriocins and has been approved as a food preservative in more than 40 countries [60].

Non-ribosomally AMP are a heterogeneous class produced also by microorganisms that code for specific enzymes called nonribosomal peptide synthetases (NRPS) [61]. Their structure may contain non-natural amino acids, cyclic rings, or covalently linked glycan or acyl groups. These modifications from the canonical peptidic primary structure have guided *de novo* designs to improve

AMP activity (see subsequent sections). They include polymixins and gramicidins, which are two families of cyclic lipopeptides that include distinct but highly homologous sequences. Gramicidin S is a well-known model of peptides with pore-forming properties in lipid bilayer membranes [62].

An unexpected source of AMP is the proteolysis products of larger proteins, which do not necessarily have antimicrobial activity. A representative example is that of lactoferricins (Lfcin) and lactoferrampins (LFampin), which are peptides obtained from the milk protein lactoferrin (LF) [63–65]. Both in bovine and human LF, pepsin digestion renders smaller peptidic fragments that have direct antimicrobial activity qualitatively distinct from the parental protein. A continuously increasing number of synthetic peptides derived from LF has been generated and shown to have antibacterial and/or antifungal properties. Also, several different approaches have been taken to the biotechnological production of LF peptides that include bacterial and yeast producers as well as plants been used as biofactories. Similarly, AMPs have been obtained from the sequence of other well-known proteins such as lysozyme [66] or hemoglobin [67].

8.4 ANTIMICROBIAL PEPTIDES BY DESIGN

The development of synthetic peptide chemistry, the increased number of biotech companies with competitive peptide synthesis facilities, and high-throughput methodologies to synthesize and screen large collections and libraries of peptides have increased substantially the number and diversity of non-natural synthetic peptides with antimicrobial activity [12, 68]. Synthetic AMPs have been designed *de novo* based on the properties of natural AMPs or identified using combinatorial and nonbiased approaches. Examples of these approaches are the peptides D4E1, BM0, and PAF26 (Table 8.1). Lead peptides have been optimized in their properties with the use of rational design. These studies have contributed substantially to increase the number and diversity of known AMPs.

The three main properties to be modulated by peptide modification/design are increased activity against microbial cells, reduced toxicity to nontarget cells (human, animal, or plant), and increased stability against (proteolityc) degradation. The two former properties are reflected in an increase of cell selectivity (microbial versus host cells) or the therapeutic index in the case of biomedical applications [69].

Peptide analogs of natural AMP have been synthesized with substituted, deleted, or extended amino acids. Numerous examples exist of such rational designs. Synthetic analogs of cecropins [43], magainins [70], melittin [71], indolicidin [72], or lactoferricins [73, 74] have been produced and studied, through the modification of amino acid sequence, shortening to determine minimal antimicrobial motifs, or enlargement of peptide length, even by fusion of fragments from different peptides [75] (Table 8.1). Of course, all these approaches can be combined and applied in an iterative way to increase optimization [76], or they can be applied to *de novo* non-natural AMP sequences

[77, 78]. They have been used to determine structure–function relationships of AMP [79, 80], which are instrumental in the knowledge of peptide mode of action (see next section) and the design of novel AMP [81].

Combinatorial peptide libraries [82] have contributed to the nonbiased discovery of novel AMP sequences of synthetic origin [83, 84]. Initially, combinatorial libraries were random, synthesized with D-enantiomers of amino acids, and representive of all the potential diversity of a peptide of given (and usually short) length. The D- or L-amino acid versions of these short AMPs do not usually present significant differences in antimicrobial potency [20, 85]. Once lead compounds have been defined, the combinatorial approach can be applied to the extension of AMP sequences [86, 87] or the optimization of conformationally restricted or sequence-restricted peptides [20, 88, 89]. Examples of peptides identified after combinatorial approaches are shown in Table 8.1.

Synthetic libraries have limitations regarding the total number of molecules that can be screened. To overcome these limitations, bioinformatic or chemoinformatic approaches are being developed to test in a virtual way the activity of an even higher number of sequence variations of longer peptides [90–92]. *In silico* predictions of these virtual libraries are subsequently tested through the synthesis of the identified peptides and inhibition experiments, and in some reports, predictive models have been shown to be significantly powerful as in the case of HHC-10 (Table 8.1) [26].

Non-natural modifications such as the introduction of D-, unusual, or substituted amino acid residues also have great influence on peptide selectivity and stability [69]. Other examples of modifications that increase peptide activity are found in nature, as is the case of the cyclic peptide bactenecin (Table 8.1). Peptide cyclation greatly increases stability because of the absence of free peptide ends, and also has the potential to improve the activity and selectivity as a consequence of the restriction of the peptide structure in a preferred conformation [93, 94], such as the better separation of cationic and hydrophobic residues of a cyclic synthetic analog of the lactoferricin fragment $LfcinB_{4\text{-}14}$ (Figure 8.1c) [29].

Because antimicrobial lipopeptides as polymyxins posses fatty acid moieties that are essential to their activities, acylation of peptides to form synthetic lipopeptides has been reported to promote antimicrobial activity, in some cases of otherwise nonactive peptide motifs of very short length (i.e., four residues) [95–97]. This increased activity is related to an increased interaction with lipid bilayers.

Another non-natural modification with a great impact on stability is the covalent multimerization of AMP monomers in the form of branched arrays or dendrimers [98]. Dendrimeric antimicrobial peptides have been developed that are selective for microbial surfaces [99].

In a broad sense, peptidomimetics are oligomeric molecules that mimic peptide backbone and biophysical properties, while avoiding at the same time undesirable properties of natural peptides such as proteolytic susceptibility [100, 101]. Among the peptoids, the peptidic side chains are attached to the amide nitrogen rather than to the α-carbon. Several examples of antimicrobial molecules exist for these side chains [102]. However, a potential drawback of all these non-natural modifications of AMP is that they restrict the production of peptides through biotechnological approaches (see section 8.6), and thus, the production of the antimicrobial is limited to synthetic procedures, which have a wide range of cost-effectiveness depending on the particular modification/procedure.

8.5 MODE OF ACTION OF ANTIMICROBIAL PEPTIDES

Although this chapter does not discuss in detail the mechanism of antimicrobial action of peptides, some brief ideas and the reference to relevant reviews must be mentioned. Mechanistic characteristics of AMP should be considered to develop its mode of application/delivery as an antimicrobial agent.

Initial studies concluded that antimicrobial activity was related to the capacity of cationic amphipathic peptides to interact and disrupt biological membranes, thus resulting in direct killing through permeabilization [103, 104]. The treatment of microorganisms with AMP concentrations above minimal inhibitory resulted in microbe permeabilization that correlated with microbicidal potency. As consequence, many contributions dealing with AMP mechanisms relied on models of peptide disruption of biological membranes. Although this holds true for most AMPs, it is becoming increasingly evident that alternative mechanisms exist, even among peptides known to be membrane disrupting [105–108]. Significant examples reproduced in more than one AMP include the binding to membrane components [57, 109, 110], the interaction with chaperone-like proteins [111, 112], or the induction of DNA damage and apoptosis [113–115]. Treatment with distinct AMP either at inhibitory [116–122] or subinhibitory concentrations [123] induced specific transcriptomic changes not related with permeabilization. As result, AMPs are viewed as multitarget drugs with overlapping actions that should difficult the raise of microbial resistance [124].

Different stages of peptide action are envisioned [108] (Figure 8.2). First, AMP must be attracted and interact with microbial cell envelopes. Not surprisingly, cell envelope/membrane charge is a major determinant of susceptibility to cationic AMPs [3,125, 126], and in some examples, the cell wall is the target itself of peptide action [105]. Also, the induction of genes related to cell wall reinforcement is a response common to functionally distinct AMPs [123, 127]. The diffusion of peptides across bacterial outer layers to allow their access to bacterial membranes also determines peptide action [128]. Even an intact fungal cell wall has been shown to be necessary for the proper activity of a plant defense [129].

A significant example is the antibacterial peptide nisin; it is established that the membrane-bound peptidoglycan precursor lipid II acts as a binding target to attract the peptide to the cell membrane and promote cell lysis [109, 130, 131]. This binding also inhibits peptidoglycan-containing cell wall biosynthesis. Nisin is active at the nanomolar range against bacteria containing lipid II, roughly

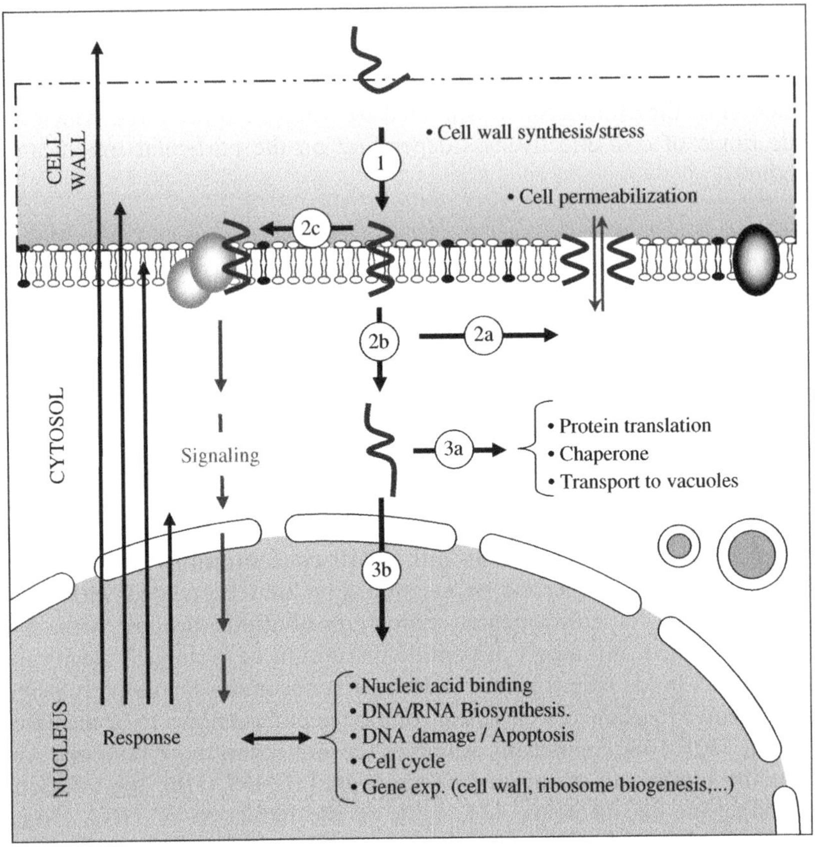

FIGURE 8.2 General model of the antimicrobial mechanism of AMP (modified from Reference 108). The figure shows a schematic view of a eukaryotic cell. A major difference with a bacterial cell relies on potential peptide import to nucleus, which has been demonstrated for specific AMPs in fungal cells (see text for details). Three major steps are depicted in the antimicrobial mechanism of AMP (represented by a thick blue line). (1) Interaction with the outer microbial structures, being the electrostatic attraction the initial driving force of the interaction, and diffusion to the inner cell membrane. There are examples of AMP that affect cell wall synthesis and/or cause cell wall stress. (2) Interaction with cell membrane components that results in disruption of lipid bilayer through diverse mechanism and consequently cell permeation (2a), internalization/translocation to cell interior (2b), and/or signaling of peptide exposure (2c). (3) Intracellular targets have been demonstrated in distinct examples as related to peptide activity (see text for details). AMP folding (thick line) may change on each step. It is also known that microorganism have sensing mechanisms to detect and signal peptide presence, a process that is required to change gene expression and protein activities to counteract peptide action at the cytosol, membrane, cell envelope, or cell environment.

three orders of magnitude more active than peptides that act only through permeabilization. Nisin is not active against fungi; however, yeast spheroplasts devoid of a cell wall are lysed in the presence of nisin at concentrations that do not affect intact cells as consequence of the intrinsic lytic capacity of nisin [132]. Therefore, even in the case of membrane-perturbing peptides, additional factors such as their propensity to interact with specific compounds might modulate (and enhance) their activity. Recently, it has been shown that lipid II is also the target of the amphipathic antifungal protein plectasin [57].

As mentioned previously, once access is gained to the cell membrane, disruption and permeation mechanisms are still considered a major property and determinant of activity of AMP. However, cell-penetrating properties are demonstrated for an increasing number of peptides such as buforin II [133], PAF26 [134], or histatin 5 [112] (Table 8.1). AMPs that have nondisruptive cell translocation activity share biophysical properties with the so-called cell penetrating peptides, which questioned the differences between antimicrobial and cell penetrating peptides [10, 12]. Peptides derived from HIV-1 Tat (Table 8.1), which initially were known as cell penetrating, have antimicrobial effects on distinct microorganisms. This activity correlated with internalization into microbial cells [35, 135]. In any case, it is assumed that peptides acting in this way should be used in solution to exert their mechanism onto microbes effectively.

AMPs that penetrate microbial cells might disturb cell homeostasis in different ways determined by their intrinsic properties as well as their targeting/interaction with cellular organelles. Becuase of their cationic nature, most AMPs readily bind nucleic acids (DNA and RNA) *in vitro*, which result in a broad inhibition of DNA synthesis, transcription, and/or messenger RNA translation inside cells [136–140]. In addition to the chaperon binding and apoptosis mentioned previously, more specific intracellular mechanisms have been demonstrated. Sorting by intracellular vacuoles seems to limit the antifungal toxicity of histatin 5 [141]. The pea defensin PSD1 locates at the fungal nucleus, interacts with nuclear proteins, and affects normal cell cycle progression [142]. The peanut RNase AhPR-10 has been shown to locate inside hyphae and kill susceptible fungi [143]. A mutation devoid of RNase activity internalizes into hyphae but does not affect fungal growth or membrane permeability, thus separating penetration from RNase activity and also linking the latter with the antimicrobiosis. Additional examples exist in which cell penetration is separated from antimicrobial activity [138], thus confirming that penetration and killing can be separated steps of the antimicrobial mechanism of AMPs.

8.6 PRODUCTION OF ANTIMICROBIAL PEPTIDES

Although examples such as nisin exist in which recovery of AMP from the producer (micro)organism is feasible [144], a major general disadvantage of AMP use is their poor bioavailability, which makes peptide production critical for commercial development. Cost-effective production of short peptides is not easy to reach and will strongly depend on the final application of the product and its added value. A first decision point is the selection of synthetic versus biotechnological approaches, or even a combination of them. The synthetic approach is generally straightforward but is more expensive than heterologous production using recombinant DNA technology. Peptide synthesis services are offered by several biotech companies and will not be discussed in this chapter. Most of the synthetic chemical modifications of AMP that have been shown to improve activity (see preceding section) are usually not suitable for recombinant production, such as cyclation, inclusion of D-enantiomers, or dendrimer structures. Remarkably, engineering the specific synthetic genes has allowed the heterologous expression of lipopeptides [145–147].

Numerous reports describe the heterologous production of linear AMPs in microorganisms, mostly in the model bacteria *Escherichia coli* and the yeast *Pichia pastoris* [148–151]. Key aspects to be considered are the intrinsic toxicity of AMP to microbes and the poor stability of short peptide sequences; both of these properties make the accumulation to high yields difficult. Solutions include the use of fusions to carrier proteins to increase stability and/or reduce toxicity [152, 153], fusion to inactivating sequences that decrease/eliminate toxicity [154, 155], targeting to inclusion bodies that also facilitate recovery [153], concatemer enlargement of multiple units of the peptide [156, 157], secretion from the producer organism [158], or promoter regulation of gene expression. Subsequent purification of peptides is facilitated by the addition of recovery tags such as the six histidines motif.

8.7 INCORPORATION OF ANTIMICROBIAL PEPTIDES INTO POLYMERS

Several recent works report the inclusion or attachment of AMPs to a variety of materials and demonstrated the antimicrobial properties of the resulting structures. A focus of the studies was to determine how restricted mobility affects the accessibility of the peptides to microbial cells and their antimicrobial activity. Two main approaches are used depending on the importance of peptide mobility, and these strategies relate to whether the peptide is embedded within a matrix to allow the controlled release or is covalently immobilized on a surface. In this latter case, the stability of the immobilization may also allow the partial release of the peptide to the solution. In some applications, it is undesirable to have the antimicrobial released into the surrounding system (i.e., food packaging), whereas the controlled release of AMPs can be an advantageous in some biomedical applications. An additional benefit of the incorporation of AMPs into polymers or surfaces is that to some degree, it protects peptides from the environment and limits their proteolytic degradation.

These approaches have a broad range of potential applications. In this section, we briefly describe several recent studies including mainly biomedical devices and food packaging. Among biomedical applications, the development of antibiofilm surfaces is one of the main areas of research. Bacteria have a strong ability to attach to solid surfaces, forming colonies and, subsequently, biofilms [159, 160]. Biofilms act as reservoirs of microbial pathogens, which cause infectious diseases [160, 161]. Among strategies that can be adopted to prevent microbial biofilm growth on solid surfaces, the most common one relies on the coating of the surface with materials that have antimicrobial properties. AMPs are increasingly studied as sources of antimicrobial activity for very diverse coatings and materials. Magainin I was covalently bound to a mixed thiol self-assembled monolayer on gold surfaces [162]. The antibacterial activity of the resulting surface was tested against three gram-positive bacteria, and the results revealed a reduction of the adhesion of bacteria to the surface by more than 50%, together with the killing of bacteria that adhered to the surface. No release of the peptide was observed upon contact with the bacterial suspension and the antimicrobial activity was kept up to six months after the first use. Three different peptides including nisin were also covalently immobilized on another metal surface, stainless steel, also resulting in a very stable coating [163].

Various types of polymeric materials have been used. A synthetic AMP named E14LKK (Table 8.1) has been attached onto a poly(ethylene) (PE) film, and the antimicrobial activity of the modified film against *E. coli* was demonstrated [25]. To avoid that the hydrophobic nature of the PE could disrupt the peptide structure and prevent antimicrobial activity, a hydrophilic spacer molecule, poly(ethyleneglycol) (PEG), was first attached to the PE and the peptide then attached to the free terminus of the PEG. The PEG spacer also provides mobility to the peptide, which is beneficial for activity.

The layer-by-layer (LbL) deposition of polyelectrolytes based on the alternate deposition of polycations and polyanions onto solid substrates is a simple and robust method to coat surfaces. AMP can be easily embedded in the multilayer architecture to prepare biocidal films. Nondegradable and noneluting films were the focus of initial studies in which LbL films containing antibacterial and antifungal AMP were assembled. These films were found to have strong action against various gram-positive and gram-negative bacteria and fungi [164, 165]. Recent studies with nondegradable film systems containing gramicidin A, which is a hydrophobic AMP acting against gram-positive microorganisms, showed that the peptide was partially released into the solution surrounding the film, which contributed to the strong antibacterial activity exhibited [166]. Therefore, the introduction of film coatings that can release AMP in a controlled way is of interest for biomedical coating and wound-healing applications. In this context, delivery of AMPs from degradable LbL films has been demonstrated with ponericin G1, an AMP found in the venom of predatory ants, with a strong activity against a wide range of bacteria, including *Staphylococcus aureus*, and with low hemolytic activity [167]. The use of different film architectures allowed to obtain various peptide loadings and release profiles, including burst as well as linear release. Film-released ponericin did not suffer any loss of activity against *S. aureus* and inhibited bacterial attachment, which is a necessary step in preventing biofilm formation [168]. This kind of film provides the level of control over peptide loading and release kinetics required in medically relevant applications, including coatings for implant materials and bandages, while reducing susceptibility to bacterial resistance.

The prevention of biofilm formation was also the aim of adsorption of Histatin 5 on poly(methyl metacrylate) (PMMA) denture base surfaces [169]. Histatin 5 has fungicidal and fungistatic effects on *Candida albicans* cells. The authors reported the adsorption of histatin 5 onto a PMMA surface modified to provide functional groups and demonstrated the effectiveness for reducing *C. albicans* biofilm formation. The amount of *C. albicans* colonization on histatin-adsorbed surface was significantly less than the control. The authors suggested the application to PMMA dentures to reduced denture-induced stomatitis. With the same objective, histatin 5 and two synthetic variants were covalently bound to polydimethlisiloxane (PDMS), using 4-azido-2,3,5,6-tetrafuoro-benzoic acid as an interlinkage molecule. The PDMS functionalized with one of the histatin 5 variant yielded the highest reduction of the number of *C. albicans* biofilm cells. The results indicated that the reductions are peptide dependent [170].

Silk-based biopolymers represent an important option because of their impressive mechanical properties, environmental stability, biocompatibility, and biodegradability. In this context, silk fibroin films modified by linkage to a cecropin B antimicrobial peptide have been suggested as a new option to engineer the surface-modified implants and prevent biomaterial-derived infection [171]. The antimicrobial peptide was covalently bound to the film surface by carbodiimide chemistry method. It was found that *E. coli* and *S. aureus* reductions of the surface-modified films were approximately of 90%, and the films showed antimicrobial durability after 144 h.

Another material that has opened new opportunities in the field of implant applications is the mesostructured silica [172]. Novel mesoporous silica systems, using a one-pot evaporation induced self-assembly method, have been developed for the incorporation of the antimicrobial peptide LL-37 [172]. Mesoporous materials may offer opportunities as implantable biomaterials or surface coatings of such biomaterials. The functionalization with 3-mercaptopropyltrimethoxysilane provided a hydrophobic character to pore walls, increasing the interaction with the peptide and resulting in an effective decrease of the release kinetics of LL-37. Mesoporous silica containing LL-37 displayed potent bactericidal properties against both gram-positive and gram-negative bacteria, and they showed low toxicity.

The development of biosensors for detection of pathogens is another area of application of immobilized AMP as part of bionanostructures. Dermaseptin S1 was immobilized within nanostructured layered films in conjunction with nickel tetrasulfonated phthalocyanines [173]. This linear peptide isolated from the skin secretion of frogs folds into an amphipathic helix, and it is active against the infective form of *Leishmania*, which is a protozoan parasite that causes various serious diseases threatening millions of people worldwide. Leishmanicidal activity of the nanostructured film material was confirmed. Unexpectedly, inmobilized dermaseptin on top of a conductive electrode displayed electroactivity as a function of the number of *Leishmania* cells in the electrolytic solution, suggesting its potential application in detection and diagnosis.

Another major area of application of antimicrobial peptide-based materials is food packaging. Current developments in food preservation include active packaging. The incorporation of antimicrobial agents directly into polymeric packaging allows combining the preservative functions of antimicrobials with the protective functions of preexisting packaging concepts. For commercialization, antimicrobials films would be made and stored before they are used for packaging food products; thus, these films must maintain their antimicrobial activity during long periods of transportation and storage.

Bacteriocins have been widely proposed as natural preservatives of food [58]. Nisin and other lesser known bacteriocins have been incorporated or adsorbed to food packaging materials to control pathogenic and food spoilage microorganisms [174–176].

Nisin can be adsorbed to packaging materials, which would be ideally in direct contact with a food surface, such as cellulose-based inserts, to be used concomitantly with modified atmosphere packaging between portions of sliced food products [176]. Nisin-adsorbed material maintained its activity for several months, both at room temperature and under refrigeration. The antimicrobial packaging reduced the population of lactic acid bacteria and extended the shelf life of sliced cheese and ham stored under modified atmospheres and at refrigeration temperatures. Nisin-adsorbed bioactive inserts reduced levels of microbial load of *Listeria innocua* and *S. aureus* in both products.

Nisin-coated films with different chemical composition and surface properties have been developed for inhibition of *Listeria monocytogenes* [177]. Different film types (low-density polyethylene, ethylene-vinyl acetate copolymer, and three types of ethylene-methacrylic acid copolymers) did not have any significant effect on the antimicrobial activity, and all maintained stable activity at room temperature and at refrigeration. Also, nisin has been incorporated in polyvinylidene copolymer films which exhibited inhibitory effects on *L. monocytogenes*, although in this case, the barrier and mechanical properties of the films decreased [178].

In comparison with nisin, little information is available on the use of nonbacteriocin AMP for antimicrobial food packaging. A derivative of dermaseptin S4 called K4K20-S4 has been tested as a wrapping film coated with the peptide or as a coating containing the peptide applied directly onto the surface of the foodstuff [179]. It was concluded that the latter showed a greater efficiency against molds and aerobic bacteria, even at lower surface concentrations of AMP in the coating.

8.8 ANTIMICROBIAL PEPTIDE ENCAPSULATION BY LIPOSOMES OR MICROPARTICLES

The successful delivery/application of AMPs can be achieved by their inclusion into particles or liposomes that separate and protect them from the environment, enhancing their stability and efficacy as well as ensuring constant delivery [180].

Liposomes are microscopic vesicles composed of lipid membranes surrounding discrete aqueous compartments. The encapsulation of bioactive peptides in liposomes has many industrial applications including pharmaceuticals, therapeutics, cosmetics, anticancer drugs, and gene-therapy drugs. In the food industry, liposomes have been used to deliver food flavors and nutrients, and more recently, liposomes are being investigated to incorporate food antimicrobials, mainly nisin and lysozyme (reviewed in References 181 and 182).

Nisin has been encapsulated in liposomes that were used as antimicrobials in milk-related products [183, 184], demonstrating maintenance of nisin activity over a 6-month period and strong reductions of bacterial counts. Nisin encapsulation in liposomes with different lipid composition affect the release of the antimicrobial, which was minimized in phosphatidylcholine/cholesterol liposomes [183, 185]. The antimicrobial activity of encapsulated nisin is dependent of the methodology used to prepare the liposomes [186].

Recently, a near-complete inhibition of *E. coli* 0157:H7 with liposomal nisin at concentrations below those reported necessary for unencapsulated nisin has been described [187] suggesting that liposomes may represent a powerful technology for the encapsulation of antimicrobials and the control of food-borne pathogens.

Other types of microcapsules formed by emulsions from different mixtures can been used to entrap and deliver AMPs, which is exemplified with lactoferrin [188]. An additional advantage in this latter example is that delivery could be temperature regulated.

8.9 NANOSTRUCTURE-FORMING ANTIMICROBIAL PEPTIDES

Certain peptides can spontaneously associate to form different types of nanostructures [189–191]. This is the reason why peptides have recently begun to be exploited for their potential use as biomaterials, such as filaments and fibrils, surfactants, and peptide hybrids. For this purpose, peptides offer attractive features, such as their biocompatibility and rich chemical, structural, and functional diversity, in conjunction with their easiness of production. Not just synthetic production, because examples already exist of biotechnological production of potentially nanostructure-forming peptides [192–194].

The cationic and amphipathic properties of some AMP can allow their self-association in higher order structures. An additional refinement is thus the possibility that the AMP themselves can form the biomaterials to be used for their delivery. Examples are still scarce. In the initial reports, the self-assembling into nanostructures properties of true AMPs resulted from covalent binding to other chemical moieties, as fatty acids in the case of short lipopeptides [97]. Authors reported the assembly of distinct lipopeptides into nanostructures of different morphology, which could explain differences in bioactivity, thus suggesting that the nanostructure itself had a major role in the microbicidal activity. Furthermore, a direct relation has been proposed between the propensity of nanoaggregation and the capacity of lipid bilayer (biomembrane) destabilization that results in microbicidal activity, for an alanine-rich cationic peptide [195]. The underlying idea that deserves additional attention is that the antimicrobial activity itself is a consequence of peptide aggregation.

Amphiphilicity and cationicity are shared by proteins that can catalyze biomineralization reactions. Following this idea, a specific, cationic AMP catalyzed self-biomineralization within different inorganic matrices [196]. Silica antimicrobial nanoparticles were obtained that protect the peptide from degradation and facilitated its continuous release to the media.

A cationic and penetrating type peptide was fused to a cholesterol molecule to promote self-assembly and improve membrane permeability [197]. The resulting self-assembled nanoparticles were reported to have broad spectrum of antibiosis against gram-positive bacteria, yeast, and fungi. These particles were more potent than the unassembled peptide counterpart. Their *in vivo* efficacy and absence of toxicity in animal models [197, 198] make the approach promising for its testing in other environments.

8.10 FUTURE TRENDS

The interest and potential application of AMPs is supported by a continuously growing number of scientific reports. However, their practical use and commercial development need yet to be confirmed in a significant number of examples. Problems encountered that affect the efficacy of peptides in the real world relate to cost-effective production, stability, unspecific toxicity, and delivery to desired target/location. This scenario is common to other bioactive peptides. The inclusion of peptides into, or delivery by, novel polymers or structures could represent not only a tailored use of AMP but also part of the solution to these more general practical problems. The mechanism of antimicrobial action of peptides in solution and as part of such structures is still a major area of research that must be addressed for each peptide of interest. Materials scientists, peptide chemists, and applied microbiologists should interact to develop these novel antimicrobial polymers effectively.

ACKNOWLEDGMENTS

We apologize to all the investigators whose research could not be appropriately cited because of space limitations. Our work on antimicrobial peptides was supported by grants VIN03 007 C2-1, BIO2006-09523, Consolider Ingenio 2010 Fun-C-Food CSD2007-00063, and BIO2009-12919 from the Spanish Government.

REFERENCES

1. Zasloff, M. (2002). Antimicrobial peptides of multicellular organisms. *Nature*, *415*, 389–395.
2. Hancock, R. E. W., Sahl, H. G. (2006). Antimicrobial and host-defense peptides as new anti-infective therapeutic strategies. *Nature Biotechnology*, *24*, 1551–1557.
3. Peschel, A., Sahl, H. G. (2006). The co-evolution of host cationic antimicrobial peptides and microbial resistance. *Nature Reviews Microbiology*, *4*, 529–536.
4. Marcos, J. F., Beachy, R. N., Houghten, R. A., Blondelle, S. E., Pérez-Payá, E. (1995). Inhibition of a plant virus infection by analogs of melittin. *Proceedings of the National Academy of Sciences of the USA*, *92*, 12466–12469.
5. Wachinger, M., Kleinschmidt, A., Winder, D., Von Pechmann, N., Ludvigsen, A., Neumann, M., Holle, R., Salmons, B., Erfle, V., Brack-Werner, R. (1998). Antimicrobial peptides melittin and cecropin inhibit replication of human immunodeficiency virus 1 by suppressing viral gene expression. *Journal of General Virology*, *79*, 731–740.
6. Edward, R., Jr., McDougall, B., Tran, D., Selsted, M. E. (1998). Anti-HIV-1 activity of indolicidin, an antimicrobial peptide from neutrophils. *Journal of Leukocyte Biology*, *63*, 94–100.
7. Klotman, M. E., Chang, T. L. (2006). Defensins in innate antiviral immunity. *Nature Reviews Immunology*, *6*, 447–456.
8. Lai, Y., Gallo, R. L. (2009). AMPed up immunity: How antimicrobial peptides have multiple roles in immune defense. *Trends in Immunology*, *30*, 131–141.
9. Hoskin, D. W., Ramamoorthy, A. (2008). Studies on anticancer activities of antimicrobial peptides. *Biochimica et Biophysica Acta—Biomembranes*, 1778, 357–375.
10. Henriques, S. T., Melo, M. N., Castanho, M. A. R. B. (2006). Cell-penetrating peptides and antimicrobial peptides: How different are they? *Biochemical Journal*, *399*, 1–7.
11. Montesinos, E. (2007). Antimicrobial peptides and plant disease control. *FEMS Microbiology Letters*, *270*, 1–11.
12. Marcos, J. F., Muñoz, A., Pérez-Payá, E., Misra, S., Lopez-García, B. (2008). Identification and rational design of novel antimicrobial peptides for plant protection. *Annual Review of Phytopathology*, *46*, 273–301.
13. Rydlo, T., Miltz, J., Mor, A. (2006). Eukaryotic antimicrobial peptides: Promises and premises in food safety. *Journal of Food Science*, *71*, 125–135.
14. Cotter, P. D., Hill, C., Ross, R. P. (2005). Bacteriocins: Developing innate immunity for food. *Nature Reviews Microbiology*, *3*, 777–788.
15. Wang, G., Li, X., Wang, Z. (2009). APD2: The updated antimicrobial peptide database and its application in peptide design. *Nucleic Acids Research*, *37*, D933–D937.
16. Fjell, C. D., Hancock, R. E. W., Cherkasov, A. (2007). AMPer: A database and an automated discovery tool for antimicrobial peptides. *Bioinformatics*, *23*, 1148–1155.
17. Giuliani, A., Pirri, G., Nicoletto, S. (2007). Antimicrobial peptides: An overview of a promising class of therapeutics. *Central European Journal of Biology*, *2*, 1–33.
18. Papagianni, M. (2003). Ribosomally synthesized peptides with antimicrobial properties: Biosynthesis, structure, function, and applications. *Biotechnology Advances*, *21*, 465–499.

19. Romeo, D., Skerlavaj, B., Bolognesi, M., Gennaro, R. (1988). Structure and bactericidal activity of an antibiotic dodecapeptide purified from bovine neutrophils. *Journal of Biological Chemistry*, *263*, 9573–9575.
20. Monk, B. C., Niimi, K., Lin, S., Knight, A., Kardos, T. B., Cannon, R. D., Parshot, R., King, A., Lun, D., Harding, D. R. K. (2005). Surface-active fungicidal D-peptide inhibitors of the plasma membrane proton pump that block azole resistance. *Antimicrobial Agents and Chemotherapy*, *49*, 57–70.
21. Park, C. B., Kim, M. S., Kim, S. C. (1996). Novel antimicrobial peptide from Bufo bufo gargarizans. *Biochemical and Biophysical Research Communications*, *218*, 408–413.
22. Oh, H., Hedberg, M., Wade, D., Edlund, C. (2000). Activities of synthetic hybrid peptides against anaerobic bacteria: Aspects of methodology and stability. *Antimicrobial Agents and Chemotherapy*, *44*, 68–72.
23. Holak, T. A., Engström, Â., Kraulis, P. J., Lindeberg, G., Bennich, H., Jones, T. A., Gronenborn, A. M., Clore, G. M. (1988). The solution conformation of the antibacterial peptide cecropin A: A nuclear magnetic resonance and dynamical simulated annealing study. *Biochemistry*, *27*, 7620–7629.
24. De Lucca, A. J., Bland, J. M., Grimm, C., Jacks, T. J., Cary, J. W., Jaynes, J. M., Cleveland, T. E., Walsh, T. J. (1998). Fungicidal properties, sterol binding, and proteolytic resistance of the synthetic peptide D4E1. *Canadian Journal of Botany*, *44*, 514–520.
25. Steven, M. D., Hotchkiss, J. H. (2008). Covalent immobilization of an antimicrobial peptide on poly(ethylene) film. *Journal of Applied Polymer Science*, *110*, 2665–2670.
26. Cherkasov, A., Hilpert, K., Jenssen, H., Fjell, C. D., Waldbrook, M., Mullaly, S. C., Volkmer, R., Hancock, R. E. W. (2009). Use of artificial intelligence in the design of small peptide antibiotics effective against a broad spectrum of highly antibiotic-resistant superbugs. *ACS Chemical Biology*, *4*, 65–74.
27. Xu, T., Levitz, S. M., Diamond, R. D., Oppenheim, F. G. (1991). Anticandidal activity of major human salivary histatins. *Infection and Immunity*, *59*, 2549–2554.
28. Rozek, A., Friedrich, C. L., Hancock, R. E. W. (2000). Structure of the bovine antimicrobial peptide indolicidin bound to dodecylphosphocholine and sodium dodecyl sulfate micelles. *Biochemistry*, *39*, 15765–15774.
29. Nguyen, L. T., Schibli, D. J., Vogel, H. J. (2005). Structural studies and model membrane interactions of two peptides derived from bovine lactoferricin. *Journal of Peptide Science*, *11*, 379–389.
30. Gudmundsson, G. H., Agerberth, B., Odeberg, J., Bergman, T., Olsson, B., Salcedo, R. (1996). The human gene FALL39 and processing of the cathelin precursor to the antibacterial peptide LL-37 in granulocytes. *European Journal of Biochemistry*, *238*, 325–332.
31. Gesell, J., Zasloff, M., Opella, S. J. (1997). Two-dimensional 1H NMR experiments show that the 23-residue magainin antibiotic peptide is an a-helix in dodecylphosphocholine micelles, sodium dodecylsulfate micelles, and trifluoroethanol/water solution. *Journal of Biomolecular NMR*, *9*, 127–135.
32. Raghuraman, H., Chattopadhyay, A. (2007). Melittin: A membrane-active peptide with diverse functions. *Bioscience Reports*, *27*, 189–223.

33. DeGray, G., Rajasekaran, K., Smith, F., Sanford, J., Daniell, H. (2001). Expression of an antimicrobial peptide via the chloroplast genome to control phytopathogenic bacteria and fungi. *Plant Physiology*, *127*, 852–862.

34. López-García, B., Pérez-Payá, E., Marcos, J. F. (2002). Identification of novel hexapeptides bioactive against phytopathogenic fungi through screening of a synthetic peptide combinatorial library. *Applied and Environmental Microbiology*, *68*, 2453–2460.

35. Jung, H. J., Park, Y., Hahm, K. S., Lee, D. G. (2006). Biological activity of Tat (47–58) peptide on human pathogenic fungi. *Biochemical and Biophysical Research Communications*, *345*, 222–228.

36. Moreland, J., Gramada, A., Buzko, O., Zhang, Q., Bourne, P. (2005). The Molecular Biology Toolkit (MBT): A modular platform for developing molecular visualization applications. *BMC Bioinformatics*, *6*, 1472–2105.

37. Batta, G., Barna, T., Gáspári, Z., Sándor, S., Kövér, K. E., Binder, U., Sarg, B., Kaiserer, L., Chhillar, A. K., Eigentler, A., Leiter, E., Hegedús, N., Pócsi, I., Lindner, H., Marx, F. (2009). Functional aspects of the solution structure and dynamics of PAF—A highly-stable antifungal protein from Penicillium chrysogenum. *FEBS Journal*, *276*, 2875–2890.

38. Haney, E. F., Hunter, H. N., Matsuzaki, K., Vogel, H. J. (2009). Solution NMR studies of amphibian antimicrobial peptides: Linking structure to function? *Biochimica et Biophysica Acta—Biomembranes*, *1788*, 1639–1655.

39. Zasloff, M., Martin, B., Chen, H. C. (1988). Antimicrobial activity of synthetic magainin peptides and several analogs. *Proceedings of the National Academy of Sciences of the USA*, *85*, 910–913.

40. Boman, H. G., Hultmark, D. (1987). Cell-free immunity in insects. *Annual Review of Microbiology*, *41*, 103–126.

41. Hultmark, D., Steiner, H., Rasmuson, T., Boman, H. G. (1980). Insect immunity. Purification and properties of three inducible bactericidal proteins from hemolymph of immunized pupae of Hyalophora cecropia. *European Journal of Biochemistry*, *106*, 7–16.

42. Boman, H. G., Wade, D., Boman, I. A., Wahlin, B., Merrifield, R. B. (1989). Antibacterial and antimalarial properties of peptides that are cecropin-melittin hybrids. *FEBS Letters*, *259*, 103–106.

43. Wade, D., Andreu, D., Mitchell, S. A., Silveira, A. M., Boman, A., Boman, H. G., Merrifield, R. B. (1992). Antibacterial peptides designed as analogs or hybrids of cecropins and melittin. *International Journal of Peptide and Protein Research*, *40*, 429–436.

44. Sato, H., Felix, J. B. (2006). Peptide-membrane interactions and mechanisms of membrane destruction by amphipathic α-helical antimicrobial peptides. *Biochimica et Biophysica Acta—Biomembranes*, *1758*, 1245–1256.

45. Saugar, J. M., Rodríguez-Hernández, M. J., de la Torre, B. G., Pachón-Ibañez, M. E., Fernández-Reyes, M., Andreu, D., Pachón, J., Rivas, L. (2006). Activity of cecropin A-melittin hybrid peptides against colistin-resistant clinical strains of Acinetobacter baumannii: Molecular basis for the differential mechanisms of action. *Antimicrobial Agents and Chemotherapy*, *50*, 1251–1256.

46. Bals, R., Wang, X., Zasloff, M., Wilson, J. M. (1998). The peptide antibiotic LL-37/hCAP-18 is expressed in epithelia of the human lung where it has broad

antimicrobial activity at the airway surface. *Proceedings of the National Academy of Sciences of the USA*, *95*, 9541–9546.

47. Durr, U. H. N., Sudheendra, U. S., Ramamoorthy, A. (2006). LL-37, the only human member of the cathelicidin family of antimicrobial peptides. *Biochimica et Biophysica Acta—Biomembranes*, *1758*, 1408–1425.
48. Murakami, M., López-García, B., Braff, M., Dorschner, R. A., Gallo, R. L. (2004). Postsecretory processing generates multiple cathelicidins for enhanced topical antimicrobial defense. *Journal of Immunology*, *172*, 3070–3077.
49. Selsted, M. E., Novotny, M. J., Morris, W. L., Tang, Y. Q., Smith, W., Cullor, J. S. (1992). Indolicidin, a novel bactericidal tridecapeptide amide from neutrophils. *Journal of Biological Chemistry*, *267*, 4292–4295.
50. Falla, T. J., Karunaratne, D. N., Hancock, R. E. W. (1996). Mode of action of the antimicrobial peptide indolicidin. *Journal of Biological Chemistry*, *271*, 19298–19303.
51. Schibli, D. J., Epand, R. F., Vogel, H. J., Epand, R. M. (2002). Tryptophan-rich antimicrobial peptides: Comparative properties and membrane interactions. *Biochemistry and Cell Biology*, *80*, 667–677.
52. Ganz, T. (2003). Defensins: Antimicrobial peptides of innate immunity. *Nature Reviews Immunology*, *3*, 710–720.
53. Carvalho, A., Gomes, V. M. (2009). Plant defensins—Prospects for the biological functions and biotechnological properties. *Peptides*, *30*, 1007–1020.
54. Broekaert, W. F., Cammue, B. P. A., De Bolle, M. F., Thevissen, K., De Samblanx, G. W., Osborn, R. W. (1997). Antimicrobial peptides from plants. *Critical Reviews in Plant Sciences*, *16*, 297–323.
55. Campos-Olivas, R., Bruix, M., Santoro, J., Lacadena, J., Martínez del Pozo, A., Gavilanes, J. G., Rico, M. (1995). NMR solution structure of the antifungal protein from Aspergillus giganteus: Evidence for cysteine pairing isomerism. *Biochemistry*, *34*, 3009–3021.
56. Mygind, P. H., Fischer, R. L., Schnorr, K. M., Hansen, M. T., Sonksen, C. P., Ludvigsen, S., Raventos, D., Buskov, S., Christensen, B., De Maria, L., Taboureau, O., Yaver, D., Elvig-Jorgensen, S. G., Sorensen, M. V., Christensen, B. E., Kjaerulff, S., Frimodt-Moller, N., Lehrer, R. I., Zasloff, M., Kristensen, H. H. (2005). Plectasin is a peptide antibiotic with therapeutic potential from a saprophytic fungus. *Nature*, *437*, 975–980.
57. Schneider, T., Kruse, T., Wimmer, R., Wiedemann, I., Sass, V., Pag, U., Jansen, A., Nielsen, A. K., Mygind, P. H., Raventos, D. S., Neve, S., Ravn, B., Bonvin, A. M. J. J., De Maria, L., Andersen, A. S., Gammelgaard, L. K., Sahl, H. G., Kristensen, H. H. (2010). Plectasin, a fungal defensin, targets the bacterial cell wall precursor Lipid II. *Science*, *328*, 1168–1172.
58. Gálvez, A., Abriouel, H., López, R. L., Omar, N. B. (2007). Bacteriocin-based strategies for food biopreservation. *International Journal of Food Microbiology*, *120*, 51–70.
59. Bierbaum, G., Sahl, H .G. (2009). Lantibiotics: Mode of action, biosynthesis and bioengineering. *Current Pharmaceutical Biotechnology*, *10*, 2–18.
60. Tiwari, B. K., Valdramidis, V. P., O'Donnell, C. P., Muthukumarappan, K., Bourke, P., Cullen, P. J. (2009). Application of natural antimicrobials for food preservation. *Journal of Agricultural and Food Chemistry*, *57*, 5987–6000.

61. Finking, R., Marahiel, M. A. (2004). Biosynthesis of nonribosomal peptides. *Annual Review of Microbiology, 58*, 453–488.
62. Prenner, E. J., Lewis, R. N. A. H., Mc Elhaney, R. N. (1999). The interaction of the antimicrobial peptide gramicidin S with lipid bilayer model and biological membranes. *Biochimica et Biophysica Acta—Biomembranes, 1462*, 201–221.
63. Gifford, J. L., Hunter, H. N., Vogel, H. J. (2005). Lactoferricin: A lactoferrin-derived peptide with antimicrobial, antiviral, antitumor and immunological properties. *Cellular and Molecular Life Sciences, 62*, 2588–2598.
64. Bellamy, W., Takase, M., Yamauchi, K., Wakabayashi, H., Kawase, K., Tomita, M. (1992). Identification of the bactericidal domain of lactoferrin. *Biochimica et Biophysica Acta—Protein Structure and Molecular Enzymology, 1121*, 130–136.
65. van der Kraan, M. I. A., Groenink, J., Nazmi, K., Veerman, E. C. I., Bolscher, J. G. M., Nieuw Amerongen, A. V. (2004). Lactoferrampin: A novel antimicrobial peptide in the N1-domain of bovine lactoferrin. *Peptides, 25*, 177–183.
66. Mine, Y., Ma, F. P., Lauriau, S. (2004). Antimicrobial peptides released by enzymatic hydrolysis of hen egg white lysozyme. *Journal of Agricultural and Food Chemistry, 52*, 1088–1094.
67. Froidevaux, R., Krier, F., Nedjar-Arroume, N., Vercaigne-Marko, D., Kosciarz, E., Ruckebusch, C., Dhulster, P., Guillochon, D. (2001). Antibacterial activity of a pepsin-derived bovine hemoglobin fragment. *FEBS Letters, 491*, 159–163.
68. Blondelle, S. E., Lohner, K. (2000). Combinatorial libraries: A tool to design antimicrobial and antifungal peptide analogues having lytic specificities for structure-activity relationship studies. *Biopolymers, 55*, 74–87.
69. Matsuzaki, K. (2009). Control of cell selectivity of antimicrobial peptides. *Biochimica et Biophysica Acta—Biomembranes, 1788*, 1687–1692.
70. Hallock, K. J., Lee, D. K., Ramamoorthy, A. (2003). MSI-78, an analogue of the magainin antimicrobial peptides, disrupts lipid bilayer structure via positive curvature strain. *Biophysical Journal, 84*, 3052–3060.
71. Blondelle, S. E., Houghten, R. A. (1991). Hemolytic and antimicrobial activities of the 24 individual omission analogs of melittin. *Biochemistry, 30*, 4671–4678.
72. Friedrich, C. L., Rozek, A., Patrzykat, A., Hancock, R. E. W. (2001). Structure and mechanism of action of an indolicidin peptide derivative with improved activity against gram-positive bacteria. *Journal of Biological Chemistry, 276*, 24015–24022.
73. Farnaud, S., Spiller, C., Moriarty, L., Patel, A., Gant, V., Odell, E. W., Evans, R. (2004). Interactions of lactoferricin-derived peptides with LPS and antimicrobial activity. *FEMS Microbiology Letters, 233*, 193–199.
74. Strom, M. B., Erik Haug, B., Rekdal, O., Skar, M. L., Stensen, W., Svendsen, J. S. (2002). Important structural features of 15-residue lactoferricin derivatives and methods for improvement of antimicrobial activity. *Biochemistry and Cell Biology, 80*, 65–74.
75. Andreu, D., Ubach, J., Boman, A., Wahlin, B., Wade, D., Merrifield, R. B., Boman, H. G. (1992). Shortened cecropin A-melittin hybrids. Significant size reduction retains potent antibiotic activity. *FEBS Letters, 296*, 190–194.
76. Ferre, R., Badosa, E., Feliu, L., Planas, M., Montesinos, E., Bardají, E. (2006). Inhibition of plant-pathogenic bacteria by short synthetic cecropin A-melittin hybrid peptides. *Applied and Environmental Microbiology, 72*, 3302–3308.

77. Chen, Y., Mant, C. T., Farmer, S. W., Hancock, R. E. W., Vasil, M. L., Hodges, R. S. (2005). Rational design of a-helical antimicrobial peptides with enhanced activities and specificity/therapeutic index. *Journal of Biological Chemistry, 280*, 12316–12329.

78. Zhang, L., Benz, R., Hancock, R. E. W. (1999). Influence of proline residues on the antibacterial and synergistic activities of α-helical peptides. *Biochemistry, 38*, 8102–8111.

79. Hwang, P. M., Vogel, H. J. (1998). Structure-function relationships of antimicrobial peptides. *Biochemistry and Cell Biology, 76*, 235–246.

80. Lejon, T., Stiberg, T., Strom, M. B., Svendsen, J. S. (2004). Prediction of antibiotic activity and synthesis of new pentadecapeptides based on lactoferricins. *Journal of Peptide Science, 10*, 329–335.

81. Frecer, V., Ho, B., Ding, J. L. (2004). De novo design of potent antimicrobial peptides. *Antimicrobial Agents and Chemotherapy, 48*, 3349–3357.

82. Houghten, R. A., Pinilla, C., Blondelle, S. E., Appel, J. R., Dooley, C. T., Cuervo, J. H. (1991). Generation and use of synthetic peptide combinatorial libraries for basic research and drug discovery. *Nature, 354*, 84–86.

83. Blondelle, S. E., Pérez-Payá, E., Houghten, R. A. (1996). Synthetic combinatorial libraries: Novel discovery strategy for identification of antimicrobial agents. *Antimicrobial Agents and Chemotherapy, 40*, 1067–1071.

84. Hilpert, K., Volkmer-Engert, R., Walter, T., Hancock, R. E. W. (2005). High-throughput generation of small antibacterial peptides with improved activity. *Nature Biotechnology, 23*, 1008–1012.

85. Wade, D., Boman, A., Wahlin, B., Drain, C. M., Andreu, D., Boman, H. G., Merrifield, R. B. (1990). All-D amino acid-containing channel-forming antibiotic peptides. *Proceedings of the National Academy of Sciences of the USA, 87*, 4761–4765.

86. Muñoz, A., López-García, B., Pérez-Payá, E., Marcos, J. F. (2007). Antimicrobial properties of derivatives of the cationic tryptophan-rich hexapeptide PAF26. *Biochemical and Biophysical Research Communications, 354*, 172–177.

87. Gonzalez, C. F., Provin, E. M., Zhu, L., Ebbole, D. J. (2002). Independent and synergistic activity of synthetic peptides against thiabendazole-resistant Fusarium sambucinum. *Phytopathology, 92*, 917–924.

88. Badosa, E., Ferre, R., Planas, M., Feliu, L., Besalu, E., Cabrefiga, J., Bardaji, E., Montesinos, E. (2007). A library of linear undecapeptides with bactericidal activity against phytopathogenic bacteria. *Peptides, 28*, 2276–2285.

89. Blondelle, S. E., Takahashi, E., Houghten, R. A., Pérez-Payá, E. (1996). Rapid identification of compounds with enhanced antimicrobial activity by using conformationally defined combinatorial libraries. *Biochemical Journal, 313*, 141–147.

90. Raventós, D., Taboureau, O., Mygind, P. H., Nielsen, J. D., Sonksen, C. P., Kristensen, H. H. (2005). Improving on nature's defenses: Optimization & high throughput screening of antimicrobial peptides. *Combinatorial Chemistry & High Throughput Screening, 8*, 219–233.

91. Soltani, S., Keymanesh, K., Sardari, S. (2007). In silico analysis of antifungal peptides: Determining the lead template sequence of potent antifungal peptides. *Expert Opinion on Drug Discovery, 2*, 837–847.

92. Fjell, C. D., Hancock, R. E. W., Jenssen, H. (2010). Computer-aided design of antimicrobial peptides. *Current Pharmaceutical Analysis, 6*, 66–75.

93. Oren, Z., Shai, Y. (2000). Cyclization of a cytolytic amphipathic alpha-helical peptide and its diastereomer: Effect on structure, interaction with model membranes, and biological function. *Biochemistry, 39*, 6103–6114.

94. Fernandez-Lopez, S., Kim, H. S., Choi, E. C., Delgado, M., Granja, J. R., Khasanov, A., Kraehenbuehl, K., Long, G., Weinberger, D. A., Wilcoxen, K. M., Ghadiri, M. R. (2001). Antibacterial agents based on the cyclic D,L-a-peptide architecture. *Nature, 414*, 329–329.

95. Makovitzki, A., Avrahami, D., Shai, Y. (2006). Ultrashort antibacterial and antifungal lipopeptides. *Proceedings of the National Academy of Sciences of the USA, 103*, 15997–16002.

96. Serrano, G. N., Zhanel, G. G., Schweizer, F. (2009). Antibacterial activity of ultrashort cationic lipo-b-peptides. *Antimicrobial Agents and Chemotherapy, 53*, 2215–2217.

97. Makovitzki, A., Baram, J., Shai, Y. (2008). Antimicrobial lipopolypeptides composed of palmitoyl di- and tricationic peptides: In vitro and in vivo activities, self-assembly to nanostructures, and a plausible mode of action. *Biochemistry, 47*, 10630–10636.

98. Bracci, L., Falciani, C., Lelli, B., Lozzi, L., Runci, Y., Pini, A., De Montis, M. G., Tagliamonte, A., Neri, P. (2003). Synthetic peptides in the form of dendrimers become resistant to protease activity. *Journal of Biological Chemistry, 278*, 46590–46595.

99. Tam, J. P., Lu, Y. A., Yang, J. L. (2002). Antimicrobial dendrimeric peptides. *European Journal of Biochemistry, 269*, 923–932.

100. Rotem, S., Mor, A. (2009). Antimicrobial peptide mimics for improved therapeutic properties. *Biochimica et Biophysica Acta—Biomembranes, 1788*, 1582–1592.

101. Scott, R. W., De Grado, W. F., Tew, G. N. (2008). De novo designed synthetic mimics of antimicrobial peptides. *Current Opinion in Biotechnology, 19*, 620–627.

102. Chongsiriwatana, N. P., Patch, J. A., Czyzewski, A. M., Dohm, M. T., Ivankin, A., Gidalevitz, D., Zuckermann, R. N., Barron, A. E. (2008). Peptoids that mimic the structure, function, and mechanism of helical antimicrobial peptides. *Proceedings of the National Academy of Sciences of the USA, 105*, 2794–2799.

103. Oren, Z., Shai, Y. (1998). Mode of action of linear amphipathic alpha-helical antimicrobial peptides. *Biopolymers, 47*, 451–463.

104. Yeaman, M. R., Yount, N. Y. (2003). Mechanisms of antimicrobial peptide action and resistance. *Pharmacological Reviews, 55*, 27–55.

105. Brogden, K. A. (2005). Antimicrobial peptides: Pore formers or metabolic inhibitors in bacteria? *Nature Reviews Microbiology, 3*, 238–250.

106. Jenssen, H., Hamill, P., Hancock, R. E. W. (2006). Peptide antimicrobial agents. *Clinical Microbiology Reviews, 19*, 491–511.

107. Hale, J. D., Hancock, R. E. W. (2007). Alternative mechanisms of action of cationic antimicrobial peptides on bacteria. *Expert Review of Anti-Infective Therapy, 5*, 951–959.

108. Marcos, J. F., Gandía, M. (2009). Antimicrobial peptides: To membranes and beyond. *Expert Opinion on Drug Discovery, 4*, 659–671.

109. Breukink, E., Wiedemann, I., van Kraaij, C., Kuipers, O. P., Sahl, H. G., de Kruijff, B. (1999). Use of the cell wall precursor lipid II by a pore-forming peptide antibiotic. *Science, 286*, 2361–2364.

110. Thevissen, K., Warnecke, D. C., François, I. E. J. A., Leipelt, M., Heinz, E., Ott, C., Zahringer, U., Thomma, B. P. H. J., Ferkel, K. K. A., Cammue, B. P. A. (2004). Defensins from insects and plants interact with fungal glucosylceramides. *Journal of Biological Chemistry*, *279*, 3900–3905.

111. Kragol, G., Lovas, S., Varadi, G., Condie, B. A., Hoffmann, R., Otvos, L., Jr. (2001). The antibacterial peptide pyrrhocoricin inhibits the ATPase actions of DnaK and prevents chaperone-assisted protein folding. *Biochemistry*, *40*, 3016–3026.

112. Sun, J. N. N., Li, W. S., Jang, W. S., Nayyar, N., Sutton, M. D., Edgerton, M. (2008). Uptake of the antifungal cationic peptide Histatin 5 by Candida albicans Ssa2p requires binding to non-conventional sites within the ATPase domain. *Molecular Microbiology*, *70*, 1246–1260.

113. Leiter, E., Szappanos, H., Oberparleiter, C., Kaiserer, L., Csernoch, L., Pusztahelyi, T., Emri, T., Posci, I., Salvenmoser, W., Marx, F. (2005). Antifungal protein PAF severely affects the integrity of the plasma membrane of Aspergillus nidulans and induces an apoptosis-like phenotype. *Antimicrobial Agents and Chemotherapy*, *49*, 2445–2453.

114. Morton, C. O., dos Santos, S. C., Coote, P. (2007). An amphibian-derived, cationic, a-helical antimicrobial peptide kills yeast by caspase-independent but AIF-dependent programmed cell death. *Molecular Microbiology*, *65*, 494–507.

115. Narasimhan, M. L., Damsz, B., Coca, M. A., Ibeas, J. I., Yun, D. J., Pardo, J. M., Hasegawa, P. M., Bressan, R. A. (2001). A plant defense response effector induces microbial apoptosis. *Molecular Cell*, *8*, 921–930.

116. Kramer, N. E., Van Hijum, S. A. F. T., Knol, J., Kok, J., Kuipers, O. P. (2006). Transcriptome analysis reveals mechanisms by which Lactococcus lactis acquires nisin resistance. *Antimicrobial Agents and Chemotherapy*, *50*, 1753–1761.

117. Li, M., Lai, Y. P., Villaruz, A. E., Cha, D. J., Sturdevant, D. E., Otto, M. (2007). Gram-positive three-component antimicrobial peptide-sensing system. *Proceedings of the National Academy of Sciences of the USA*, *104*, 9469–9474.

118. Hong, R. W., Shchepetov, M., Weiser, J. N., Axelsen, P. H. (2003). Transcriptional profile of the Escherichia coli response to the antimicrobial insect peptide Cecropin A. *Antimicrobial Agents and Chemotherapy*, *47*, 1–6.

119. Tomasinsig, L., Scocchi, M., Mettulio, R., Zanetti, M. (2004). Genome-wide transcriptional profiling of the *Escherichia coli* response to a proline-rich antimicrobial peptide. *Antimicrobial Agents and Chemotherapy*, *48*, 3260–3267.

120. Vylkova, S., Jang, W. S., Li, W.S., Nayyar, N., Edgerton, M. (2007). Histatin 5 initiates osmotic stress response in Candida albicans via activation of the Hog1 mitogen-activated protein kinase pathway. *Eukaryotic Cell*, *6*, 1876–1888.

121. Lis, M., Fuss, J. R., Bobek, L. A. (2009). Exploring the mode of action of antimicrobial peptide MUC7 12-mer by fitness profiling of Saccharomyces cerevisiae genomewide mutant collection. *Antimicrobial Agents and Chemotherapy*, *53*, 3762–3769.

122. Morton, C. O., Hayes, A., Wilson, M., Rash, B. M., Oliver, S. G., Coote, P. (2007). Global phenotype screening and transcript analysis outlines the inhibitory mode(s) of action of two amphibian-derived, a-helical, cationic peptides on Saccharomyces cerevisiae. *Antimicrobial Agents and Chemotherapy*, *51*, 3948–3959.

123. López-García, B., Gandía, M., Muñoz, A., Carmona, L., Marcos, J. F. (2010). A genomic approach highlights common and diverse effects and determinants of susceptibility on the yeast Saccharomyces cerevisiae exposed to distinct antimicrobial peptides. *BMC Microbiology*, *10*, 289–

124. Otvos, L., Jr. (2005). Antibacterial peptides and proteins with multiple cellular targets. *Journal of Peptide Science*, *11*, 697–706.

125. Peschel, A., Otto, M., Jack, R. W., Kalbacher, H., Jung, G., Gotz, F. (1999). Inactivation of the dlt operon in Staphylococcus aureus confers sensitivity to defensins, protegrins, and other antimicrobial peptides. *Journal of Biological Chemistry*, *274*, 8405–8410.

126. Guo, L., Lim, K. B., Poduje, C. M., Daniel, M., Gunn, J. S., Hackett, M., Miller, S. I. (1998). Lipid A acylation and bacterial resistance against vertebrate antimicrobial peptides. *Cell*, *95*, 189–198.

127. Gamberi, T., Cavalieri, D., Magherini, F., Mangoni, L., De Filippo, C., Borro, M., Gentile, G., Simmaco, M., Modesti, A. (2007). An integrated analysis of the effects of Esculentin 1–21 on Saccharomyces cerevisiae. *Biochimica et Biophysica Acta—Proteins and Proteomics*, *1774*, 688–700.

128. Rosenfeld, Y., Barra, D., Simmaco, M., Shai, Y., Mangoni, M. L. (2006). A synergism between temporins toward gram-negative bacteria overcomes resistance imposed by the lipopolysaccharide protective layer. *Journal of Biological Chemistry*, *281*, 28565–28574.

129. van der Weerden, N. L., Hancock, R. E. W., Anderson, M. A. (2010). Permeabilization of fungal hyphae by the plant defensin NaD1 occurs through a cell wall-dependent process. *Journal of Biological Chemistry*, *285*, 37513–37520.

130. Hsu, S. T. D., Breukink, E., Tischenko, E., Lutters, M. A. G., de Kruijff, B., Kaptein, R., Bonvin, A. M. J. J., van Nuland, N. A. J. (2004). The nisin-lipid II complex reveals a pyrophosphate cage that provides a blueprint for novel antibiotics. *Nature Structural & Molecular Biology*, *11*, 963–967.

131. Wiedemann, I., Breukink, E., van Kraaij, C., Kuipers, O. P., Bierbaum, G., de Kruijff, B., Sahl, H. A. (2001). Specific binding of nisin to the peptidoglycan precursor lipid II combines pore formation and inhibition of cell wall biosynthesis for potent antibiotic activity. *Journal of Biological Chemistry*, *276*, 1772–1779.

132. Dielbandhoesing, S. K., Zhang, H., Caro, L. H., van der Vaart, J. M., Klis, F. M., Verrips, C. T., Brul, S. (1998). Specific cell wall proteins confer resistance to Nisin upon yeast cells. *Applied and Environmental Microbiology*, *64*, 4047–4052.

133. Park, C. B., Yi, K. S., Matsuzaki, K., Kim, M. S., Kim, S. C. (2000). Structure-activity analysis of buforin II, a histone H2A-derived antimicrobial peptide: The proline hinge is responsible for the cell-penetrating ability of buforin II. *Proceedings of the National Academy of Sciences of the USA*, *97*, 8245–8250.

134. Muñoz, A., López-García, B., Marcos, J. F. (2006). Studies on the mode of action of the antifungal hexapeptide PAF26. *Antimicrobial Agents and Chemotherapy*, *50*, 3847–3855.

135. Palm, C., Netzerea, S., Hallbrink, M. (2006). Quantitatively determined uptake of cell-penetrating peptides in non-mammalian cells with an evaluation of degradation and antimicrobial effects. *Peptides*, *27*, 1710–1716.

136. Park, C. B., Kim, H. S., Kim, S. C. (1998). Mechanism of action of the antimicrobial peptide buforin II: Buforin II kills microorganisms by penetrating the cell membrane and inhibiting cellular functions. *Biochemical and Biophysical Research Communications*, *244*, 253–257.

137. Subbalakshmi, C., Sitaram, N. (1998). Mechanism of antimicrobial action of indolicidin. *FEMS Microbiology Letters*, *160*, 91–96.

138. Castle, M., Nazarian, A., Yi, S. S., Tempst, P. (1999). Lethal effects of apidaecin on Escherichia coli involve sequential molecular interactions with diverse targets. *Journal of Biological Chemistry*, *274*, 32555–32564.

139. Lehrer, R. I., Barton, A., Daher, K. A., Harwig, S. S. L., Ganz, T., Selsted, M. E. (1989). Interaction of human defensins with Escherichia coli. Mechanism of bactericidal activity. *Journal of Clinical Investigation*, *84*, 553–561.

140. Boman, H. G., Agerberth, B., Boman, A. (1993). Mechanisms of action on Escherichia coli of cecropin P1 and PR-39, two antibacterial peptides from pig intestine. *Infection and Immunity*, *61*, 2978–2984.

141. Mochon, A. B., Liu, H. P. (2008). The antimicrobial peptide Histatin-5 causes a spatially restricted disruption on the Candida albicans surface, allowing rapid entry of the peptide into the cytoplasm. *PLoS Pathogens*, *4*, e100 0190.

142. Lobo, D. S., Pereira, I. B., Fragel-Madeira, L., Medeiros, L. N., Cabral, L. M., Faria, J., Bellio, M., Campos, R. C., Linden, R., Kurtenbach, E. (2007). Antifungal Pisum sativum defensin 1 interacts with Neurospora crassa cyclin F related to the cell cycle. *Biochemistry*, *46*, 987–996.

143. Chadha, P., Das, R. H. (2006). A pathogenesis related protein, AhPR10 from peanut: An insight of its mode of antifungal activity. *Planta*, *225*, 213–222.

144. de Arauz, L. J., Jozala, A. F., Mazzola, P. G., Vessoni Penna, T. C. (2009). Nisin biotechnological production and application: A review. *Trends in Food Science and Technology*, *20*, 146–154.

145. Alexander, D. C., Rock, J., He, X., Brian, P., Miao, V., Baltz, R. H. (2010). Development of a genetic system for combinatorial biosynthesis of lipopeptides in Streptomyces fradiae and heterologous expression of the A54145 biosynthesis gene cluster. *Applied and Environmental Microbiology*, *76*, 6877–6887.

146. Coeffet-Le Gal, M. F., Thurston, L., Rich, P., Miao, V., Baltz, R. H. (2006). Complementation of daptomycin dptA and dptD deletion mutations in trans and production of hybrid lipopeptide antibiotics. *Microbiology-SGM*, *152*, 2993–3001.

147. Chiocchini, C., Linne, U., Stachelhaus, T. (2006). In vivo biocombinatorial synthesis of lipopeptides by COM domain-mediated reprogramming of the surfactin biosynthetic complex. *Chemistry and Biology*, *13*, 899–908.

148. Ingham, A. B., Moore, R. J. (2007). Recombinant production of antimicrobial peptides in heterologous microbial systems. *Biotechnology and Applied Biochemistry*, *047*, 1–9.

149. Zhang, L., Falla, T., Wu, M., Fidai, S., Burian, J., Kay, W., Hancock, R. E. W. (1998). Determinants of recombinant production of antimicrobial cationic peptides and creation of peptide variants in bacteria. *Biochemical and Biophysical Research Communications*, *247*, 674–680.

150. Zorko, M., Jerala, R. (2010). Production of recombinant antimicrobial peptides in bacteria. *Methods in Molecular Biology*, *618*, 61–76.

151. Bommarius, B., Jenssen, H., Elliott, M., Kindrachuk, J., Pasupuleti, M., Gieren, H., Jaeger, K. E., Hancock, R. E. W., Kalman, D. (2010). Cost-effective expression and purification of antimicrobial and host defense peptides in Escherichia coli. *Peptides*, *31*, 1957–1965.
152. Vidovic, V., Prongidi-Fix, L., Bechinger, B., Werten, S. (2009). Production and isotope labeling of antimicrobial peptides in Escherichia coli by means of a novel fusion partner that enables high-yield insoluble expression and fast purification. *Journal of Peptide Science*, *15*, 278–284.
153. Piers, K. L., Brown, M. H., Hancock, R. E. W. (1993). Recombinant DNA procedures for producing small antimicrobial cationic peptides in bacteria. *Gene*, *134*, 7–13.
154. Kim, H. K., Chun, D. S., Kim, J. S., Yun, C. H., Lee, J. H., Hong, S. K., Kang, D. K. (2006). Expression of the cationic antimicrobial peptide lactoferricin fused with the anionic peptide in Escherichia coli. *Applied Microbiology and Biotechnology*, *72*, 330–338.
155. Lee, J. H., Minn, I., Park, C. B., Kim, S. C. (1998). Acidic peptide-mediated expression of the antimicrobial peptide buforin II as tandem repeats in Escherichia coli. *Protein Expression and Purification*, *12*, 53–60.
156. Morin, K., Arcidiacono, S., Beckwitt, R., Mello, C. (2005). Recombinant expression of indolicidin concatamers in Escherichia coli. *Applied Microbiology and Biotechnology*, *70*, 698–704.
157. Metlitskaia, L., Cabralda, J. E., Suleman, D., Kerry, C., Brinkman, J., Bartfeld, D., Guarna, M. M. (2004). Recombinant antimicrobial peptides efficiently produced using novel cloning and purification processes. *Biotechnology and Applied Biochemistry*, *39*, 339–345.
158. Kim, S. J., Quan, R. S., Lee, S. J., Lee, H. K., Choi, J. K. (2009). Antibacterial activity of recombinant hCAP18/LL37 protein secreted from Pichia pastoris. *Journal of Microbiology*, *47*, 358–362.
159. Katsikogianni, M., Missirlis, Y. F. (2004). Concise review of mechanisms of bacterial adhesion to biomaterials and of techniques used in estimating bacteria-material interactions. *European Cells and Materials*, *8*, 37–57.
160. Costerton, J. W., Cheng, K. J., Geesey, G. G., Ladd, T. I., Nickel, J. C., Dasgupta, M., Marrie, T. J. (1987). Bacterial biofilms in nature and disease. *Annual Review of Microbiology*, *41*, 435–464.
161. Donlan, R. M., Costerton, J. W. (2002). Biofilms: Survival mechanisms of clinically relevant microorganisms. *Clinical Microbiology Reviews*, *15*, 167–193.
162. Humblot, V., Yala, J. F., Thebault, P., Boukerma, K., Héquet, A., Berjeaud, J. M., Pradier, C. M. (2009). The antibacterial activity of magainin I immobilized onto mixed thiols self-assembled monolayers. *Biomaterials*, *30*, 3503–3512.
163. Vreuls, C., Zocchi, G., Thierry, B., Garitte, G., Griesser, S. S., Archambeau, C., Van de Weerdt, C. V., Martial, J., Griesser, H. (2010). Prevention of bacterial biofilms by covalent immobilization of peptides onto plasma polymer functionalized substrates. *Journal of Materials Chemistry*, *20*, 8092–8098.
164. Etienne, O., Gasnier, C., Taddei, C., Voegel, J. C., Aunis, D., Schaaf, P., Metz-Boutigue, M. H., Bolcato-Bellemin, A. L., Egles, C. (2005). Antifungal coating by biofunctionalized polyelectrolyte multilayered films. *Biomaterials*, *26*, 6704–6712.

165. Etienne, O., Picart, C., Taddei, C., Haikel, Y., Dimarcq, J. L., Schaaf, P., Voegel, J. C., Ogier, J. A., Egles, C. (2004). Multilayer polyelectrolyte films functionalized by insertion of defensin: A new approach to protection of implants from bacterial colonization. *Antimicrobial Agents and Chemotherapy*, *48*, 3662–3669.

166. Guyomard, A. Dé E., Jouenne, T., Malandain, J. J., Muller, G., Glinel, K. (2008). Incorporation of a hydrophobic antibacterial peptide into amphiphilic polyelectrolyte multilayers: A bioinspired approach to prepare biocidal thin coatings. *Advanced Functional Materials*, *18*, 758–765.

167. Orivel, J., Redeker, V., Le Caer, J. P., Krier, F., Revol-Junelles, A. M., Longeon, A., Chaffotte, A., Dejean, A., Rossier, J. (2001). Ponericins, new antibacterial and insecticidal peptides from the venom of the ant Pachycondyla goeldii. *Journal of Biological Chemistry*, *276*, 17823–17829.

168. Shukla, A., Fleming, K. E., Chuang, H. F., Chau, T. M., Loose, C. R., Stephanopoulos, G. N., Hammond, P. T. (2010). Controlling the release of peptide antimicrobial agents from surfaces. *Biomaterials*, *31*, 2348–2357.

169. Yoshinari, M., Kato, T., Matsuzaka, K., Hayakawa, T., Inoue, T., Oda, Y., Okuda, K., Shimono, M. (2006). Adsorption behavior of antimicrobial peptide histatin 5 on PMMA. *Journal of Biomedical Materials Research*, *77*, 47–54.

170. De Prijck, K., De Smet, N., Rymarczyk-Machal, M., Van Driessche, G., Devreese, B., Coenye, T., Schacht, E., Nelis, H. J. (2010). Candida albicans biofilm formation on peptide functionalized polydimethylsiloxane. *Biofouling*, *26*, 269–275.

171. Bai, L., Zhu, L., Min, S., Liu, L., Cai, Y., Yao, J. (2008). Surface modification and properties of Bombyx mori silk fibroin films by antimicrobial peptide. *Applied Surface Science*, *254*, 2988–2995.

172. Izquierdo-Barba, I., Vallet-Regí, M., Kupferschmidt, N., Terasaki, O., Schmidtchen, A., Malmsten, M. (2009). Incorporation of antimicrobial compounds in mesoporous silica film monolith. *Biomaterials*, *30*, 5729–5736.

173. Zampa, M. F., Araújo, I. M. S., Costa, V., Nery Costa, C. H., Santos, J. R., Jr., Zucolotto, V., Eiras, C., Leite, J. R. S. A. (2009). Leishmanicidal Activity and Immobilization of dermaseptin 01 antimicrobial peptides in ultrathin films for nanomedicine applications. *Nanomedicine: Nanotechnology, Biology, and Medicine*, *5*, 352–358.

174. Bower, C. K., McGuire, J., Daeschel, M. A. (1995). Suppression of Listeria monocytogenes colonization following adsorption of nisin onto silica surfaces. *Applied and Environmental Microbiology*, *61*, 992–997.

175. Siragusa, G. R., Cutter, C. N., Willett, J. L. (1999). Incorporation of bacteriocin in plastic retains activity and inhibits surface growth of bacteria on meat. *Food Microbiology*, *16*, 229–235.

176. Scannell, A. G. M., Hill, C., Ross, R. P., Marx, S., Hartmeier, W., Arendt, E. K. (2000). Development of bioactive food packaging materials using immobilised bacteriocins Lacticin 3147 and Nisaplin®. *International Journal of Food Microbiology*, *60*, 241–249.

177. Neetoo, H., Ye, M., Chen, H. Q. (2007). Effectiveness and stability of plastic films coated with nisin for inhibition of Listeria monocytogenes. *Journal of Food Protection*, *70*, 1267–1271.

178. Limjaroen, P., Ryser, E., Lockhart, H., Harte, B. (2003). Development of a food packaging coating material with antimicrobial properties. *Journal of Plastic Film & Sheeting*, *19*, 95–109.
179. Miltz, J., Rsydlo, T., Mor, A., Polyakov, V. (2006). Potency evaluation of a dermaseptin S4 derivative for antimicrobial food packaging applications. *Packaging Technology and Science*, *19*, 345–354.
180. Tan, M. L., Choong, P. F. M., Dass, C. R. (2010). Recent developments in liposomes, microparticles and nanoparticles for protein and peptide drug delivery. *Peptides*, *31*, 184–193.
181. Taylor, T. M., Davidson, P. M., Bruce, B. D., Weiss, J. (2005). Liposomal nanocapsules in food science and agriculture. *Critical Reviews in Food Science and Nutrition*, *45*, 587–605.
182. Gouin, S. (2004). Microencapsulation: Industrial appraisal of existing technologies and trends. *Trends in Food Science and Technology*, *15*, 330–347.
183. Laridi, R., Kheadr, E. E., Benech, R. O., Vuillemard, J. C., Lacroix, C., Fliss, I. (2003). Liposome encapsulated nisin Z: Optimization, stability and release during milk fermentation. *International Dairy Journal*, *13*, 325–336.
184. Benech, R. O., Kheadr, E. E., Laridi, R., Lacroix, C., Fliss, I. (2002). Inhibition of Listeria innocua in cheddar cheese by addition of nisin Z in liposomes or by in situ production in mixed culture. *Applied and Environmental Microbiology*, *68*, 3683–3690.
185. Were, L. M., Bruce, B. D., Davidson, P. M., Weiss, J. (2003). Size, stability, and entrapment efficiency of phospholipid nanocapsules containing polypeptide antimicrobials. *Journal of Agricultural and Food Chemistry*, *51*, 8073–8079.
186. da Silva Malheiros, P., Micheletto, Y. M. S., Silveira, N. P., Brandelli, A. (2010). Development and characterization of phosphatidylcholine nanovesicles containing the antimicrobial peptide nisin. *Food Research International*, *43*, 1198–1203.
187. Taylor, T. M., Bruce, B. D., Weiss, J., Davidson, P. M. (2008). Listeria monocytogenes and Escherichia coli O157:H7 inhibition in vitro by liposome-encapsulated nisin and ethylene diaminetetraacetic acid. *Journal of Food Safety*, *28*, 183–197.
188. Al-Nabulsi, A. A., Ran, J. H., Liu, Z. Q., Rodrigues-Vieira, E. T., Holley, R. A. (2006). Temperature-sensitive microcapsules containing lactoferrin and their action against Carnobacterium viridans on Bologna. *Journal of Food Science*, *71*, M208-M214.
189. Fairman, R., Âkerfeldt, K. S. (2005). Peptides as novel smart materials. *Current Opinion in Structural Biology*, *15*, 453–463.
190. Rajagopal, K., Schneider, J. P. (2004). Self-assembling peptides and proteins for nanotechnological applications. *Current Opinion in Structural Biology*, *14*, 480–486.
191. Gazit, E. (2007). Self-assembled peptide nanostructures: The design of molecular building blocks and their technological utilization. *Chemical Society Reviews*, *36*, 1263–1269.
192. Hartmann, B. M., Kaar, W., Falconer, R. J., Zeng, B., Middelberg, A. P. J. (2008). Expression and purification of a nanostructure-forming peptide. *Journal of Biotechnology*, *135*, 85–91.
193. Moers, A. P. H. A., Wolbert, E. J. H., de Wolf, F. A., Werten, M. W. T. (2010). Secreted production of self-assembling peptides in Pichia pastoris by fusion to an artificial highly hydrophilic protein. *Journal of Biotechnology*, *146*, 66–73.

194. Riley, J. M., Aggeli, A., Koopmans, R. J., McPherson, M. J. (2009). Bioproduction and characterization of a pH responsive self-assembling peptide. *Biotechnology and Bioengineering*, *103*, 241–251.

195. Chen, C., Pan, F., Zhang, S., Hu, J., Cao, M., Wang, J., Xu, H., Zhao, X., Lu, J. R. (2010). Antibacterial activities of short designer peptides: A link between propensity for nanostructuring and capacity for membrane destabilization. *Biomacromolecules*, *11*, 402–411.

196. Eby, D. M., Farrington, K. E., Johnson, G. R. (2008). Synthesis of bioinorganic antimicrobial peptide nanoparticles with potential therapeutic properties. *Biomacromolecules*, *9*, 2487–2494.

197. Liu, L., Xu, K., Wang, H., Jeremy Tan, P. K., Fan, W., Venkatraman, S. S., Li, L., Yang, Y. Y. (2009). Self-assembled cationic peptide nanoparticles as an efficient antimicrobial agent. *Nature Nanotechnology*, *4*, 457–463.

198. Wang, H., Xu, K., Liu, L., Tan, J. P. K., Chen, Y., Li, Y., Fan, W., Wei, Z., Sheng, J., Yang, Y. Y., Li, L. (2010). The efficacy of self-assembled cationic antimicrobial peptide nanoparticles against Cryptococcus neoformans for the treatment of meningitis. *Biomaterials*, *31*, 2874–2881.

CHAPTER 9

RECOMBINANT ANTIMICROBIAL PEPTIDES

J. CARLOS RODRÍGUEZ-CABELLO, CARMEN GARCÍA-ARÉVALO, ALESSANDRA GIROTTI, LAURA MARTÍN, and MERCEDES SANTOS
GIR Bioforge, University of Valladolid, Valladolid, Spain

CONTENTS

9.1 Introduction . . . 227
9.2 Recombinant Routes for the Generation of Novel AMPs . . . 229
 9.2.1 AMP Biosynthesis in Prokaryotic Expression Systems . . . 229
 9.2.2 AMP Biosynthesis in Yeast Expression Systems . . . 231
 9.2.3 AMP Biosynthesis in Plant Expression Systems . . . 232
9.3 Compositional and Structural Requirements for AMP Activity . . . 233
 9.3.1 Hydrophilic Face: Charge . . . 235
 9.3.2 Hydrophobic Face . . . 238
 9.3.3 Structure–Activity Relationships . . . 239
9.4 Applications of AMPs . . . 244
 9.4.1 Therapeutic Applications . . . 244
 9.4.2 Agriculture, Textiles, Food Industry, and Other Applications . . . 247
 9.4.3 Limitations and Future Trends . . . 250
References . . . 251

9.1 INTRODUCTION

The innate immune response, which involves an interactive network of cellular and molecular systems responsible for recognizing and eradicating pathogens

Antimicrobial Polymers, First Edition.
Edited by José M. Lagarón, María J. Ocio, and Amparo López-Rubio.

by means of a variety of signaling pathways and biological responses, is the first line of defense against infectious agents. Neutrophils, monocytes, macrophages, complements, cytokines, chemokines, and host defense proteins and peptides with antimicrobial properties organize themselves in a complex and highly orchestrated response to infection to kill, block, or eliminate incoming pathogens and prevent an excessive or prolonged inflammation that could cause widespread damage and even death [1].

This rapid yet nonspecific innate immune response, which can be expressed either constitutive or inductively, is crucial because the activation and deployment of pathogen-specific immune responses occurs solely relative to the potential kinetics of microbial proliferation [2, 3].

Natural host-defense peptides (HDPs) or antimicrobial peptides (AMPs) belong to an evolutionary component of the innate immune system of most multicellular organisms [4]. In the last 30 years, hundreds of AMPs have been isolated from virtually all species, including bacteria, fungi, insects, tunicates, amphibians, crustaceans, birds, fish, mammals, and humans [5–8]. The roles of AMPs in innate immune system defense, their distribution, and their potency rely heavily on the specific needs of each organism. Thus, bacterial AMPs (or bacteriocins), which are extremely potent, do not usually protect against infections but contribute to the survival of individual bacterial cells that compete for nutrients in the same environment in a "competitive exclusion" manner. In contrast, plant AMPs play a fundamental role in the defense against bacteria and fungi, and AMPs from invertebrates, which lack an adaptive immune system, play a prominent role in counteracting invasive pathogens. In vertebrate species, in the presence of an adaptive immune response, AMPs are related with different degrees of first line of host defense [3]. In addition to being a successful form of chemical defense for eukaryotic cells with antimicrobial activities, some AMPs have been proposed to have more potent immunomodulatory activities than antimicrobial functions and may be involved in the orchestration of the innate immune and inflammatory responses or may play an active role in the transition to the active immune response. These actions include the ability to recruit the adaptive immune response by inducing or modulating chemokine and cytokine production, the alteration of gene expression in host cells, and the inhibition of proinflammatory responses of host cells to bacterial components such as lipopolysaccharides (LPSs), neutralizing their activity both *in vivo* and *in vitro* [9]. Recent studies have shown AMPs to be promising candidates for anticancer therapies in humans because of their unique features [10].

Natural, artificial and recombinant AMPs have been cataloged in several databases, including the AMSdb (http://www.bbcm.univ.trieste.it/~tossi/antimic.html), the Indian Collection of antimicrobial peptides database CAMP (http://www.bicnirrh.res.in/antimicrobial), the antimicrobial peptide database (ADP database) (http://aps.unmc.edu/AP/main.html), the recombinantly produced antimicrobial peptides database (RADP) (http://faculty.ist.unomaha.edu/chen/rapd/index.php), the antimicrobial plant peptides database (PhytAMP) (http://phytamp.pfba-lab.org.), the bacteriocin database (BACTIBASE) (http://bactibase.pfba-lab.org), the defensins knowledgebase (http://defensins.bii.a-star.edu.sg/), the shrimp antimicrobial peptide penaeidin database (PenBase) (http://penbase.immunaqua.com/), the database of sequences and structures of the peptaibols (Peptaibol database) (http://www.cryst.bbk.ac.uk/peptaibol), the synthetic antibiotic peptides database (SAPD) (http://oma.terkko.helsinki.fi:8080/~SAPD), and the bacteriocin genome web-base (BAGEL) (http://bioinformatics.biol.rug.nl/websoftware/bagel), with hundreds of synthetic variants or *de novo* synthesized AMPs continuing to increase the repertory. To date, more than 3500 AMPs of different origins have been reported [11].

9.2 RECOMBINANT ROUTES FOR THE GENERATION OF NOVEL AMPs

AMPs have the potential to be a viable alternative to other novel therapeutic approaches although because of their peptidic nature, they suffer from poor bioavailability and poor proteolitic stability, two features that have severely hampered clinical progress to date [12]. In addition, large quantities of AMPs are required for such applications. The extraction of AMPs from natural sources or their chemical synthesis is not efficient, thereby resulting in high manufacturing costs [7]. The relatively low cost and easy scale-up of the recombinant approach makes it the most attractive for the large-scale production of AMPs. The proteinaceous nature of AMPs allows them to be made using DNA recombinant techniques, which use the natural biosynthetic machinery of heterologous hosts to produce these compounds as recombinant peptides. Indeed, in the past few years, numerous efforts have been made to produce recombinant AMPs using bacteria and fungi as well as plant- and animal-based production systems.

9.2.1 AMP Biosynthesis in Prokaryotic Expression Systems

The production of recombinant AMPs in prokaryotic systems has successfully demonstrated the viability of this approach. Indeed, despite the difficulty of producing recombinant AMPs in bacterial cells, *Escherichia coli* is one of the most widespread expression systems used for the production of recombinant AMPs, accounting for more than 80% of all cases [13–15]. One of the most common strategies used to prevent the inherent toxicity of AMPs toward host bacterial cells limiting or impeding the expression of these compounds in *Escherichia coli* involves the inclusion of their sequences as part of fusion proteins. The sequence of these hybrid recombinant proteins possesses a specific proteolytic site between the modular components to separate them. The recombinant fusion proteins are, in general, highly expressed and do not acquire antimicrobial activity until the AMPs and fusion proteins are cleaved by means of chemical reagents or enzymatic activity [16–18]. The fusion protein approach has also been shown to result in increased AMP expression levels and to preserve the recombinant peptides from proteolytic degradation, with many carrier proteins having been used to both neutralize their innate toxicity and to preserve them.

Commercial expression vectors are available to subclone the genes codifying AMPs in such fusion systems. The plasmid pET32a(+) (Novagen; Merck KGaA, Darmstadt, Germany), for example, which contains the Thioredoxin (Trx) gene, has been employed as a fusion system for several AMP genes, including perinerin [19], CM4 (ABPCM4, which was first isolated from the Chinese silkmoth *Bombys mori* [14]), the hybrid peptide LFB15 [20], the bovine lactoferricin derivative LfcinB15-W4,10 [21], the recombinant dermcidin-1L [22], the hybrid peptide hinnavin II/α-melanocyte-stimulating hormone [23], and the antimicrobial human cathelicidin LL-37 [24]. The carrier protein Trx can alleviate the toxic effects of recombinant proteins in host cells and prevent the target peptide from proteolytic degradation. Trx can also accelerate the high-level expression of soluble fusion proteins, thus resulting in the production of up to 1 g of fusion protein per liter of culture [24]. The recombinant proteins are generally accumulated as soluble products and purified by Ni^{2+}-chelating affinity chromatography.

Another widely employed fusion system makes use of glutathione-S-transferase by cloning the AMP sequence into the pGEX-4T-1 plasmid (GE Healthcare, Waukesha, WI). This system facilitates purification of the resulting fusion proteins by affinity chromatography, as was demonstrated by the coexpression of AMPs PR-39 and protegrin-1 (PG-1) [25] and the bovine lactoferricin derivative peptide LfcinB-W10 [16]. Likewise, cloning in the pET-31b (+) vector (Novagen) allowed the high-level expression of human dermcidin, which is an anionic AMP fused with the 125-amino-acid ketosteroid isomerase protein [26].

The ability to clone AMPs in alternative carrier proteins opens up the possibility of improving AMP production; AMP biosynthesis as a soluble or insoluble recombinant product could improve the functionality or production yield, respectively. Lactococcin K, for example, has been formed as an inclusion body in recombinant *E. coli*, whereas a fusion protein containing lactococcin K and maltose-binding protein has been produced in a soluble form in pMAL (New England Biolabs, Ipswich, MA) [27].

The intein-mediated expression of some AMPs has been successfully employed for the production of cathelicidin-like peptide SMAP-29, the human antimicrobial anionic peptide dermcidin (DCD), CM4 (ABP-CM4), and some beta-defensins [28–30]. Likewise, AMPs cloned in pTYB11 vector (New England Biolabs) have been overexpressed as a fusion with a self-cleavable affinity tag (intein) to mediate their expression and purification. This system has the advantage of simplifying the cleavage and purification steps. Indeed, chemical treatments such as cyanogen bromide or dilute acid after enzyme treatment are employed in most systems to cleave the recombinant AMP from its fusion partners. However, these methods involve the use of expensive enzymes and run the risk of producing an inactive AMP because of incorrect treatment or cleavage. The main advantage of the intein protein purification strategy resides in the fact that the carrier protein can induce its own excision from the in-frame fusion of flanking protein sequences, thus meaning that the recombinant proteins can be recovered free from the affinity tag in a single cleavage step.

Recombinant synthesis provides additional strategies for the ready construction of multiple, combined genes from antimicrobial agents. Indeed, hybrid AMPs with different properties have been designed and expressed to develop more potent peptides with enhanced antimicrobial activities and a higher yield of heterologous products. The results of these studies indicate that hybrid compounds often have better properties than either parent compound in terms of higher antimicrobial activity and broader antimicrobial spectrum [23, 25, 31–34].

A similar improvement of recombinant AMPs was demonstrated using a peptide-polymerization approach. Similarly, the sequence of natural AMPs has been cloned in tandem repeats using the seamless cloning technique, which involves the use of type IIS restriction endonucleases to ensure unidirectional ligation and thereby avoid the insertion of extraneous DNA at the ligation joints, thus eliminating extraneous amino acid residues within the polymer sequence. A subsequent analysis of the properties of the serial AMPs obtained by this approach allowed the most appropriate version to be chosen [20, 21, 35].

Novel carrier molecules such as elastin-like proteins (ELPs) are used to facilitate the expression and purification of recombinant AMPs. These carriers solve one of the bottlenecks in the production and purification of AMPs on a larger scale, namely the high cost of purification (affinity chromatography), which limits the use of this technology. ELP constructs allow the purification of AMPs by making use of the reversible soluble–insoluble temperature-dependent phase transition. This property is based on a molecular transition of the polymer chains known as the inverse temperature transition [36], ELPs, or silk-elastin-like proteins, have been used as a fusion partner to express the antimicrobial polypeptides AMP MBI-28, CAM [37], halocidin18 (Hal18) [38], and moricin CM4. Similarly, human b-defensins 4(HbD4) have been synthesized by using an ELP purification tag in a self-cleaving intein system [39], and Shen *et al.* constructed a self-cleaving ELP carrier that results in a chromatography- and protease-free system. The use of ELPs as carriers therefore offers the possibility to simplify and reduce the cost of the production and purification steps of both the fusion protein and its free AMPs.

9.2.2 AMP Biosynthesis in Yeast Expression Systems

The methylotrophic yeast *Pichia pastoris* is one of the most highly successful systems for the production of a variety of heterologous proteins. Both this particular expression system and *E. coli* require the use of only simple molecular genetic manipulation techniques, which are commercially available as protein-expression kits and can produce high levels of recombinant proteins either intracellularly or extracellularly. Moreover, the *P. pastoris* system can be used to perform numerous eukaryotic posttranslational modifications, such as glycosylation and disulfide bond formation. This versatility makes *P. pastoris* an ideal heterologous expression host for cysteine-rich cationic AMPs such as perinerin, a linear peptide with 51 amino acids, which was first found and purified from the homogenate of an Asian marine clamworm. The four cysteine residues in this peptide can form two intramolecular disulfide bridges that drive its correct conformation [40]. The AMP defensin has also been produced in *P. pastori* [41, 42]. Furthermore, the arginine-rich cationic peptide HD5, which possesses six cysteines that form three intramolecular disulfide bonds, has been expressed in this eukaryotic system to prevent refolding of the peptide [43]. Other classes of AMPs that expressed successfully in *P. pastori* include lactoferricin-related peptide [44, 45], ABP-CM4, and the human AMP CAP18/LL37 (hCAP18/LL37), although the production yield is lower than that obtained with *E. coli* in some cases [46, 47].

9.2.3 AMP Biosynthesis in Plant Expression Systems

During the last 20 years, transgenic plants expressing AMPs have been produced to generate resistance to attack by microorganisms or as alternative eukaryotic "plant bioreactors." Several diseases caused by viral, bacterial or fungal infections affect plant crops, thereby decreasing the quality and safety of agricultural products and resulting in economic losses. Gene constructs including sequences which code for AMPs have been expressed in model or crop plants and have provided different degrees of protection against plant pathogens [8]. Animal defensins have been expressed in several plants to protect crops from infection. For example, cecropins A and B have been expressed in rice to confer protection against *Magneporthe grisea* [48], Alf-AFP alfalfa defensin has been expressed in potato to protect against *V. dahlia* [49], and Mj-AMP1 jalapa defensin has been expressed in tomato to protect against *Alternaria solani* [50, 51]. Although numerous transgenic plants expressing AMPs that confer different degrees of disease protection have been developed, these crops have not been marketed commercially because of current genetically modified organism regulations.

The increasing demand for antibiotic products has encouraged the search for alternative and commercially competitive systems for the production of recombinant AMPs. From this point of view, plants are an interesting source of recombinant proteins as it is possible to replace almost any eukaryotic posttranslational modification and their use presents several advantages, such as low production costs and lack of contamination from transmissible pathogenic agents [52]. Furthermore, plants allow the cost-effective production of recombinant proteins on an agricultural scale while eliminating the risk of product contamination with endotoxins or human pathogens [53]. However, although plants present numerous advantages, the current limitations of plant bioreactor technology include the greater complexity of the recombinant host produced, the low production yields achieved with many AMPs (often as a result of poor protein stability), and difficulties with downstream processing, especially the copurification of plant contaminants, which leads to inconsistent product quality [54]. Researchers have successfully employed both conventional *Agrobacterium tumefaciens* nuclear transformation methods and the regeneration of transgenic plant lines, or a simpler transient expression technique based on the engineering of plant virus vectors such as PVX, to produce AMPs in plants [51, 55]. As an example, transgenic tobacco plants expressing a modified vacuolar membrane ATPase (VMA) intein expression/purification system [56] have been used to obtain recombinant SMAP-29, a K-helical cathelicidin peptide with powerful antimicrobial activity [57].

9.3 COMPOSITIONAL AND STRUCTURAL REQUIREMENTS FOR AMP ACTIVITY

The molecular diversity of AMPs is so great that it is very difficult to categorize them definitively in terms of structure [58]. The most prominent characteristics shared by almost all AMPs include their small size (most of them are less than 10 kDa), ranging from 6 to more than 100 residues (usually containing L-amino acids), their overall positive net charge, their high content of hydrophobic residues (~50%), and their ability to fold into amphipathic structures, often after contact with membranes, which usually makes them membrane active [9, 59]. An appropriate balance among hydrophobicity, the net charge and volume of the molecule, its structure, and its oligomeric state in solution and in membranes determines the activity of AMPs and will be discussed briefly below [2].

Most AMPs are synthesized as a prepropeptide consisting of an N-terminal signal sequence, a pro-segment, and a C-terminal cationic peptide that demonstrates antimicrobial activity after it is cleaved from the rest of the protein [60].

As far the amino acid composition is concerned, there is apparently little preference for any residue other than the abundant lysines and arginines. Furthermore, no clear consensus patterns exist among the primary sequences of AMPs; therefore, it is not easy to rationalize on the basis of their length, their charge, or the presence/absence of S-S bridges. However, the prevalence of certain amino acid motifs in the preference of some AMPs for their targets has been reported. Thus, antiviral peptides often seem to be highly cationic and amphiphilic, the latter being in accordance with the presence of aromatic amino acids. In contrast, the peptides with primarily antifungal activity tend to be relatively rich in polar and neutral amino acids, and antiprotozoan activity may be dependent on peptide motifs fundamentally different from those required for bacterial, viral, and fungal activities [3, 61].

AMPs can be grouped into the following five classes, all of which have different amino acid compositions and 3D structures:

1. **Anionic AMPs.** These are small peptides (721.6–823.8 Da, around six amino acid residues) that are generally more active against both gram-positive and gram-negative bacteria when complexed with zinc or highly cationic peptides. Members of this family include maximin H5, small anionic glutamic and aspartic acid-rich peptides, and dermcidin. They are similar to the charge-neutralizing propeptides of larger zymogens, which also have antimicrobial activity when synthesized alone.
2. **Linear cationic α-helical peptides.** These are short cationic peptides (<40 amino acids) that lack cysteine residues and sometimes have a hinge or "kink" in the middle. Many of these peptides are disordered in aqueous solutions, but all or part of the molecule is converted into an α-helix in the presence of trifluoroethanol, sodium dodecyl sulfate micelles, phospholipid vesicles and liposomes, or lipid A. The degree of α-helicity correlates with the antibacterial activity against both gram-positive and gram-negative bacteria, with increasing α-helix content leading to stronger antimicrobial activity. Examples of this group include cecropins (A), andropin, moricin, ceratotoxin, melitin, cecropin P1, magainin-2, dermaseptin, bombinin, brevinin-1, esculentins, buforin II, pleurocidin, seminal plasmin, BMAP, SMAP (SMAP29, ovispirin), PMAP, CAP18, and LL-37.
3. **Cationic peptides enriched in specific amino acids.** This third subgroup includes cationic peptides that lack cysteine residues and are linear, although some can form extended coils. Thus, proline-containing peptides include abaecin; proline- and arginine-containing peptides include apidaecins, drosocin, pyrrhocorcin, bactenecins, and PR-39; glycine-containing peptides include hymenoptaccin; glycine- and proline-containing peptides include indolicin; and finally, small histidine-rich polypeptides include histatins.
4. **Anionic and cationic peptides that contain cysteine and form disulfide bonds.** The disulfide bounds formed by cysteine residues in this fourth subgroup lead to the formation of stable β-sheets. Peptides with one disulfide bond include brevinins; those with two disulfide bonds include protegrins and tachyplexins; those with three disulfide bonds include α-defensins, β-defensins, and θ-defensins; and those with more than three disulfide bonds include drosomycin and plant antifungal defensins.
5. **Anionic and cationic peptide fragments of larger proteins.** These fragments have antimicrobial activity and are similar in composition and structure to the microbial peptides described previously. Their role in innate immunity is, however, not yet clear. Examples include lactoferricin from lactoferrin, casodicin I from human casein, and the antimicrobial domains from bovine α-lactalbumin, human hemoglobin, lysocyme, and ovalbumin [62].

However, despite showing few preferences as far as amino acid sequence is concerned, small variations in the primary structure of peptides can lead to

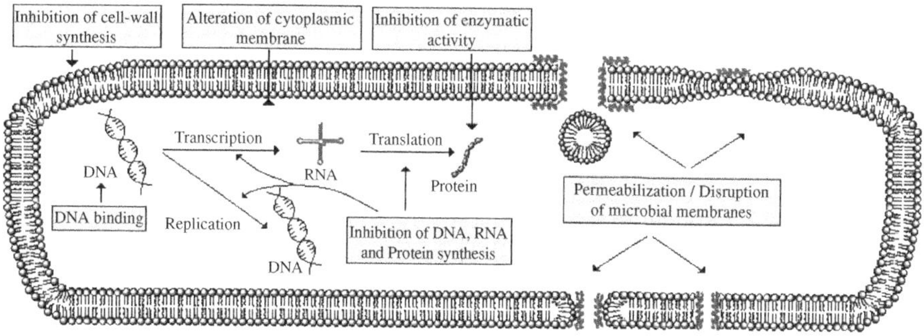

FIGURE 9.1 Cellular targets proposed as the mode of action of AMPs.

drastic changes in their specificity and activity, thereby rendering the peptides inactive or changing their minimum inhibitory concentrations (MICs) [63].

Numerous structure/activity studies on both natural and synthetic AMPs have identified several factors believed to be important for antimicrobial activity. However, one of the most common structural properties of AMPs by far is the presence of both hydrophobic and basic residues and an inducible preformed secondary structure [64].

Selective antimicrobial activity results from a delicate balance between the three-dimensional hydrophobic and electrostatic interactions, which occur between an AMP and its target. A moderate level of amphipathicity independent of, or concomitant with, charge polarization seems to favor cationic peptide transitions or peptide translocations to the cytoplasm, thereby increasing selective toxicity against microorganisms. It is well known that the amphipathicity of AMPs is necessary for their mechanism of action as the positively charged polar face helps the molecules reach their targets (biomembranes) by interacting electrostatically with the negatively charged head groups of phospholipids (Figure 9.1). The nonpolar face of the peptides allows their insertion into the target membrane as a result of hydrophobic interactions. This insertion results in increased permeability and loss of the barrier function of the target cells, thereby killing the microbes directly or allowing membrane translocation to target receptors inside the cells [65].

9.3.1 Hydrophilic Face: Charge

The net positive charge is the most preserved property of AMPs. Extensive studies on endogenous AMPs and their peptide derivatives have indicated that the presence of a positive charge is perhaps the most important determinant of activity as the mode of action of AMPs is central to their ability to interact with membranes [66]. The number of positive charges required for activity is, however, variable. The presence of basic residues means that most of the cationic AMPs characterized to date display a net positive charge ranging from +2 to +9 at pH 7, sometimes with highly defined cationic domains. As mentioned previously, this property is important for the initial electrostatic attraction. Indeed, a minimum threshold of charge as low as +2 seems to be necessary for AMP selectivity toward microorganisms. A strong correlation exists between peptide cationicity and antimicrobial activity [67, 68], although this relation is not entirely linear and there is a limit beyond which an additional increase in the positive charge no longer confers increased activity and may induce an excessively strong peptide interaction, thereby leading to a loss of antimicrobial activity and increased toxicity [67, 69]. In addition, the spatial positioning of the charged amino acids often seems to be more important than the actual net charge [3]. The preferential attraction for the binding of AMPs to microbial membranes is based on the difference between the compositional and architectural characteristics of prokaryotic and mammalian membranes. These are neither static nor symmetric and confer selective toxicity against bacterial targets on AMPs. The differences between the asymmetric distribution, compositional stoichiometry, and saturation of phospholipid bilayers significantly influence the membrane phase transition and fluidity. The LPSs in gram-negative bacteria and teichoic acids in gram-positive bacteria, as well as the negatively charged phospholipids in the inner membrane of gram-negative bacteria and the single membrane of gram-positive bacteria, such as phosphatidilglycerol (PG), cardiolipin (effectively a dimer of PG) and phosphatidylserine, mean that both the outermost and inner leaflets of the bilayer are heavily populated by lipids with negatively charged phospholipid headgroups [67, 70]. In contrast, the asymmetric distribution in eukaryotic cells and the distinct composition of mammalian membranes offer some protection. Thus, the outer leaflet of the membranes of plants and animals is mainly composed of zwitterionic lipids with no net charge, such as phosphatidylcholine (PC) and phosphatidylethanolamine. Moreover, sphingomyelin, a close analog of PC containing a palmitoyl residue, is also neutrally charged, and most of the lipids with negatively charged headgroups are segregated into the inner leaflet, facing the cytoplasm. In addition, sterols such as cholesterol and ergosterol, which are found in eukaryotic but rarely in prokaryotic membranes, are also neutral and, in general, the presence of cholesterol in the target membrane reduces the activity of AMPs as a result of either stabilization of the lipid bilayer or interactions between cholesterol and the peptide [4, 67].

The charge and amphipathicity of the inner and outer membrane leaflets also vary considerably and may also be compounded by differences in electrochemical gradients between prokaryotic and eukaryotic cells. Thus, the transmembrane potentials of bacterial pathogens, for example, range between −130 and −150 mV, whereas those of normal mammalian cells range from −90 to −110 mV. This difference, which has been termed "self-promoted uptake," has been proposed to be an additional factor affecting the selective toxicity of AMPs [4, 67]. The relatively nonspecific nature of the interactions between AMPs and the anionic lipids of microbial membranes means that many AMPs have a broad range of antimicrobial properties, targeting both gram-positive and gram-negative bacteria, fungi, protozoa, and some viruses [59].

Interestingly, despite the many similarities between AMPs, their spectrum of activity differs significantly and they can be classified into two main groups, which in turn can be subdivided into several subgroups.

1. Peptides that are toxic to microorganisms but not to normal mammalian cells. This group includes peptides which are active against gram-positive bacteria, such as cecropin, others that are active against gram-positive and gram-negative bacteria, such as magainins, dermaseptins, cathelicidins, and many others, still others which are active against both bacteria and fungi, or fungi only, and so on.
2. Peptides that are toxic to both microorganisms and mammalian cells, such as melittin, pardaxin, LL-37 and other cathelicidins, and defensins. Those peptides that can interact with mammalian membranes at high concentrations and with the outer surface of epithelial cells, as well as erythrocytes, are also negatively charged, which could partially explain their broad toxicity [2, 59].

Although, as mentioned previously, a positive net charge seems to be a common feature of almost all AMPs, a clear example that this property is not always essential is the recombinant AMP dermcidin-1L (rDCD-1L), a human skin AMP isolated from human sweat that has a broad spectrum of antimicrobial activity and has been found to play a role in tumorigenesis in breast carcinomas. The net negative charge and acidic pI of rDCD-1L have little effect on membrane binding; therefore, the affinity with which it binds to bacterial-mimetic membranes is a result primarily of its flexible amphipathic alpha-helical structure and its length (>30 residues), which allows it to adopt a helix-hinge-helix motif, a common molecular fold among AMPs. In addition, it has been suggested that the abundance of negatively charged sialic acids on erythrocytes may protect these cells against lysis as a result of electrostatic repulsion between negatively charged molecules as dermcidin-1L has a net negative charge and no hemolytic activity [22, 71].

The cationic fragment of some AMPs could be involved in processes other than the initial electrostatic interaction with the anionic phospholipids present on the outer leaflet of the plasma membranes of pathogens. One example of this is found with Bac7, where the cationic fragment could play a role in the internal targeting of the peptide. Bac-7 has been shown recently to inactivate bacteria via two different, concentration-dependent modes: by stereospecific-dependent uptake followed by binding to an unknown intracellular target, possibly DNA, at near-MIC concentrations, and by nonstereospecific membranolytic mechanisms at concentrations higher than the MIC [72]. Furthermore, it has been found that upon separation of the cationic and hydrophobic segments of the Bac-7 molecule, short fragments containing only a cationic or hydrophobic region can permeate the cell but have no microbicidal activity when measured in a fluorescence quantification assay and by confocal microscopy. In addition, the highly cationic fragments can cross the cell membrane and reside within the nucleus [73].

9.3.2 Hydrophobic Face

In addition to their charge, the conformation, amphipathicity, hydrophobic moment, and polar angle of AMPs are interdependent molecular determinants of peptide structure that can also affect antimicrobial activity and selective toxicity [67]. Amphipathicity reflects the relative abundance and polarization of hydrophobic and hydrophilic domains within a protein. A quantitative measure of amphipathicity is the hydrophobic moment (M_H), which is calculated as the vectorial sum of individual amino acid hydrophobicities normalized to an ideal helix. A moderate level of amphipathicity independent of, or in the context of, charge polarization seems to favor cationic peptide transitions or peptide translocations to the cytoplasm and increase selective toxicity against microorganisms. An increased hydrophobic moment has been linked to predominant host cell toxicity and has only a modest effect on peptide interactions with the typically highly negative pathogenic membranes [3, 67].

The proportion of hydrophobic residues in most AMPs is approximately 50%. This level of hydrophobicity is required for effective membrane permeabilization as higher levels of hydrophobicity are correlated strongly with mammalian cell toxicity and loss of antimicrobial specificity. In fact, antibacterial activity has been observed from amphiphilic helical structures in hydrophobic environments when only a small fraction of molecules populate the helical conformation.

The polar angle is a measurement of the relative proportion of polar versus nonpolar facets of a peptide with an amphipathic helix conformation that determines the segregation between hydrophobic and charged residues. Smaller polar angles, that is, greater hydrophobic surfaces, are associated with a greater ability to permeabilize membranes and a greater stability and half-life of the resulting peptide-induced membrane pores [67].

Conservation of the secondary structure (α-helix and β-sheet are the predominant conformations) might be key to the three-dimensional configuration facilitating the antimicrobial activity of various peptides. Such structures have been conserved throughout 2.5 billion years of evolution and are present in membrane-active toxins and chemokines, thus indicating possible evolutionary relationships and the potential role of such a motif in their effector functions [7]. Thus, the hydrophobic and hydrophilic stereogeometries in AMPs play a significant role in influencing the membrane-interaction and membrane-disruption process and its consequences. Extreme values of features such as charge, amphipathicity, hydrophobic moment, or polar angle tend to disfavor peptide antimicrobial activity and selective toxicity [67]. High amphipathicity, high hydrophobicity, and high helicity or β-sheet structure have all been correlated with increased toxicity, as measured by hemolytic activity. In contrast, antimicrobial activity has been found to be less dependent on these factors [74]. Thus, the modification of peptides possessing both antibacterial and hemolytic activity to decrease their propensity to form helical structures has tended to leave the antibacterial activity or spectrum largely unaffected but with considerable loss in hemolytic activity. However, antibacterial activity was also observed in the linearized peptides in the two disulfide bridge-containing peptides; hence, it is evident that the requirements with regard to charge and amphiphilic helical structure or β-structure for antimicrobial activity are not stringent [68].

9.3.3 Structure–Activity Relationships

It seems to be impossible to predict antimicrobial activity based primarily on the secondary structures of peptides, mostly because of the lack of sequence-specific interactions, as has been demonstrated with scrambled variants of some peptides such as Bac2a and other α-helical peptides [66, 75].

AMPs are usually grouped into five structural classes, namely (1) linear α-helical peptides, (2) peptides with several disulfide bonds stabilizing a β-sheet or α/β structure, (3) those with an unusual predominance of certain amino acid (s) (such as histidine, proline, tryptophan or arginine) or an unusual structure (loop and extended structures), (4) peptides generated by proteolysis of larger proteins in the host, and (5) those containing rare modified amino acids such as lanthionine, 3-methyllanthionine, dehydroalanine, and dehydrobutyrine, based on their secondary structures [58, 60, 61, 76]. Three distinct forms of amphipathicity can be identified within these structures. The first of these forms exhibits primary amphiphilicity with a core segment composed of a short central apolar sequence flanked at each end by cationic residues. Examples of this form include the extended Pro-rich indolicidin and the Trp-rich tritrpticin. Studies of both suggest that this core segment along these highly symmetrical sequences is essential for membrane–hydrophobic core interaction and peptide orientation during the antimicrobial action. In contrast to the previous form, cysteine-rich β-sheet AMPs, such as defensins, form a major group of peptides that invade microbial membranes by using tertiary amphiphilicity. In this case, the residues that are distal in the primary structure of the molecule, namely the N- and C-terminal residues of the molecule, are brought together to form its polar face, whereas the folding pattern of its middle region generates the apolar face. Disulfide bridges seem to play an important role in stabilizing their tertiary structures. Finally, the third group includes α-helical AMPs, such as magainin, which possess secondary amphiphilicity, in other words a spatial segregation of hydrophobic and hydrophilic residues about the long axis of the α-helix. This group of AMPs generally requires the presence of an amphiphilic interface, such as that of the membrane, to trigger helix formation [77].

Most experiments carried out to evaluate the mechanisms of action of AMPs after they reach the membrane have tended to focus on the interaction of cationic peptides with model membrane systems, consisting of single or mixed lipids prepared as membranes or vesicles, or on whole microbial cells, and have used mainly membrane potential-sensitive dyes and fluorescent-labeled peptides. In addition, the use of several microscopic techniques to visualize the effects of AMPs on microbial cells has helped to identify general target sites

[62, 78]. These models provide a simple, sensitive model for mimicking biological membranes and assessing the membrane insertion of peptides, but they have sometimes been criticized because of their use of very high peptide/lipid ratios, thereby hindering the effects that occur under physiological conditions, especially as there is no definitive evidence that such concentrations are present at the surface of the cytoplasmic membrane during bacterial killing [79]. It is worth noting that the antimicrobial activity of cationic host-defense peptides is antagonized to a variable extent by divalent cations such as Mg^{2+} and Ca^{2+} (at physiological concentrations of 1–2 mM), monovalent cations such as Na^+ and K^+ (100 mM), and polyanions such as glycosaminoglycans (heparin and others) and mucins. It has therefore been proposed that the term AMP should be reserved for those peptides that have been clearly shown to kill microbes directly under such physiological conditions [7].

Although AMP molecules are water soluble, their molecular surfaces are amphiphilic and have a high affinity for binding to lipid bilayers [80]. The pore dynamics in lipid bilayers are based on the interplay between a membrane tension, which tends to enlarge the pore, and a line tension, which tends to close the pore by the action of amphihilic peptides bound to lipid bilayers at the interface between the hydrophilic headgroups and the hydrophobic chains. The bound peptides insert an additional area into the headgroup side of the interface area, thereby causing a local deformation and creating an internal membrane tension. The local deformations produced by individual peptides overlap and because these peptides do not penetrate the hydrocarbon region, this region becomes thinner more or less uniformly because of the collective stretching induced by the peptides embedded in the headgroup region. As peptides preferentially bind to the edges of the pores, this results in a loosening of the internal membrane tension and subsequent pore formation [80].

Membrane-disruptive AMPs are reported generally to be of the α-helical class, although not all α-helical AMPs are membrane disruptive [78]. Unlike β-sheet peptides, α-helical and extended AMPs tend to be highly flexible in solution and tend to form linear, cationic, unstructured monomers in an aqueous environment (except LL-37, which is helical and aggregated). After interaction with target membranes or in structure-promoting solvents such as hexafluoroisopropanol and trifluoroethanol, however, AMPs adopt their characteristic secondary structures [10]. The lack of a secondary structure in solution results in an extremely durable peptide that can survive exposure to high temperatures, organic solvents, and incorporation into fibers and films while maintaining its antibacterial activity [81]. In addition, the fact that for many peptides (e.g., indolicidin), these amphipathic secondary structures are only observed when the peptides interact with membranes could prevent broad-spectrum membranolytic activity until the peptide identifies an appropriate target surface, especially when considering the potent antibiotic activity (in the low nanomolar concentration range) of many bacterially derived peptides. Thus, the lack of a bioactive structure at nontarget sites may be an important means by which AMPs minimize host-cell toxicity [7, 67].

Although all AMPs interact with membranes by either spontaneous lipid-assisted translocation, a non-receptor-mediated mechanism or, to a lesser extent, by stereospecific receptor-mediated membrane translocation (such as nisin Z, which uses the membrane-anchored cell-wall precursor lipid II as receptor, and mesenterin Y, which uses a receptor only found on the bacteria Listeria), they are heterogeneous in their impact on these membranes. Broadly speaking, AMPs can be divided into two mechanistic classes, namely membrane disruptive and non-membrane disruptive [70, 78]. Over a certain concentration threshold, peptides perturb membranes by forming transient pores via one of the various models proposed to account for this step, namely barrel-stave, carpet-like, toroidal (or "wormhole") pore formation, detergent-type micellization, and induction of nonlamellar phases, thus leading to membrane permeabilization and either leakage of the cell content and osmotic instability, and/or peptide diffusion to intracellular targets. A unifying model, the so-called Shai-Matsuzaki-Huang (SMH) model, has also been proposed, which envisages entry of the peptide into the microbial cell. It can also explain the action of short peptides that cannot span the membrane. Thus, presumably it cannot form a classic multimeric transmembrane pore. Briefly, the SMH model proposes that the peptide initially covers the outer membrane leaflet, integrating itself into the membrane and causing a thinning of the outer leaflet itself. A subsequent phase transition leads to the formation of transient "wormhole" pores, which allow the transport of lipids and peptides into the inner leaflet and the eventual diffusion of peptides onto intracellular fragments [4, 12].

Recent observations have shown that a significant correlation exists between the conformation adopted by the peptide in solution, in other words before interaction with the membrane, and its antimicrobial activity, at least for some peptides, such as temporin A, temporin L, and Vitr-p-13. Thus, a higher propensity to assume a folded conformation in aqueous solution may plausibly facilitate peptide folding in the microbial membrane (both thermodynamically and kinetically) and seems to be related to a greater antimicrobial and hemolytic activity. In other words, an interaction with lipid membranes is somehow precluded for those peptides that cannot adopt a partially folded structure in aqueous solution [12].

Although there is no doubt that virtually any cationic peptide AMP can act at high concentrations by penetrating/permeabilizing/disrupting the microbial membrane, growing evidence indicates that membrane permeabilization by AMPs is not always a lethal step. Indeed, recent studies have resulted in an ever-growing list of peptides that are presumed to affect microbial viability at low-to-moderate concentrations by interacting with one or more intracellular targets and interfering with their metabolic function. This list even includes peptides that have a total lack of pore-forming activity, such as apidaecin [70, 82]. Most AMPs that are thought to attack internal targets contain an excess of one or two amino acids, normally Trp and Pro, His, or Arg. Hence, these amino acids resemble cell-penetrating peptides. Mammalian cell-penetrating peptides, including TP-10, pVEC, Tat, ep-1, MAP, and penetratin, have the ability to act as either cell-penetrating peptides or AMPs; the threshold between these two properties relies on the composition of the membrane and the peptide concentration. However, their microbicidal activity is thought to be caused by their ability to inhibit key intracellular functions by crossing the microbial membranes rather than creating pores in the cell surface. Other examples include Pro/Arg-rich AMPs such as pyrrhocoricin, drosocin, apidaecin, bactenecin, PR-39, and prophenin, whose proposed mechanism of action involves initial stereospecific interaction with a receptor/docking molecule, which may be a permease-type transporter system on the inner membrane, followed by interaction with the intracellular target, which has been proposed to be the chaperone Dnak, or interference with DNA or protein synthesis on binding to nucleic acids [70]. Uptake of the His-rich AMP histatin is a dichotomous event. Thus, at concentrations below the MIC, the peptide binds to the cell wall heat shock protein 70 (Ssa1/2) and is subsequently transferred to a membrane permease, which transports the peptide across the cytoplasm without harmful effects on the parasite leishmania, whereas at physiological concentrations, it produces a bioenergetic collapse of the parasite as a result of a decrease of mitochondrial ATP synthesis by inhibition of F1F0-ATPase, which results in subsequent fast ATP exhaustion. However, use of the Hst5 enantiomer showed that the key steps of its lethal mechanism involve no chiral recognition [83]. Likewise, at concentrations close to the MIC, the Trp-rich AMP indolicidin adopts an extended wedge-type conformation when bound to biological membranes because of the presence of Trp and binds strongly to DNA, thereby inhibiting DNA synthesis rather than RNA and protein synthesis and inducing bacterial filamentation in *E. coli* [84].

As a general rule, a strong link exists between cell death and membrane-disruptive events for AMPs with MICs in the micromolar range. Indeed, most members of this group can be subjected to pronounced sequence alterations that do not drastically affect their function [2]. At low peptide/lipid ratios, usually in the nanomolar concentration range, AMPs can translocate across the plasma membrane, thereby perturbing its structure in a transient, nonlethal manner, and the AMPs reach the cell interior [70].

Studies with all-D AMPs have been widely considered as proof that many peptides act directly on the membrane lipid bilayer rather than on a receptor protein. Peptides that lack a specific bacterial target and instead bind to membranes are expected to be just as potent as the all-D-amino acid enantiomer, as was shown in original studies with α-helical peptides such as magainin, cecropin, and mellitin. These studies showed that the all-D peptides were equipotent to the naturally occurring all-L peptides [85]. In contrast, the all-D enantiomer of apidaecin is completely devoid of antibacterial activities, thus indicating a significant degree of stereospecificity [61]. However, the example of histatin-5, whose mechanism of action involves no chiral recognition (see previous section), disproves the previous statement. Membrane translocation, therefore, seems to be a corollary of transient membrane permeabilization, and peptides probably use more than one method to translocate across the bilayers.

As already anticipated previously, membrane interactions and antimicrobial properties are strongly dependent on the composition of the media and are inhibited by divalent cations and serum. Therefore, their immunomodulatory activities are more relevant in those physiological settings in which their microbicidal activity is strongly inhibited by the presence of divalent cations, serum, and high salt concentrations (such as sites of acute inflammation) [59]. It has been shown that some AMPs activate the host's innate immunity and counteract the harmful inflammatory/septic responses induced by bacterial endotoxins, which is extremely important because the rapid killing of bacteria and subsequent liberation of bacterial components such as LPSs or peptidoglycans can induce potentially fatal immune deregulation [66]. Indeed, it has been reported recently that some AMPs, such as the linear innate defense regulator peptide (IDR-1), selectively modulate the immune response and have no direct antimicrobial activity. Such AMPs are thought to influence the host's innate immune response by modulating several signaling pathways, including the global gene expression pattern of human macrophages and monocytes, thereby protecting against a broad range of *in vivo* infection models by local and systemic administration. Such modulation leads to the induction of key chemokines, such as RANTES and MCP-1, which are likely responsible for monocyte and/or macrophage recruitment to the site of the infection. IDR-1 can also stimulate the activation of several transcription factors, including C/EBP, STAT1, several zinc fingers, and Hox transcription factors; adhesion molecules such as ICAM, NCAM, and integrin-α; and genes involved in actin polymerization and cytoskeletal remodeling. Finally, IDR-1 can sustain or enhance the levels of pathogen infection-clearing chemokines while suppressing the levels of pathogen-associated proinflammatory cytokines, such as tumor necrosis factor without displaying undesirable toxicity (mast cell degranulation, hypersensitivity, complement activation, and cytotoxicity/induction of apoptosis), and it can induce the anti-inflammatory cytokine IL-10, thereby helping to balance inflammation rather than suppress it [1].

Additionally, an increasing number of studies have shown that some cationic AMPs, such as the α-helical BMAP-27 and BMAP-28, cecropins A and B, LL-37/hCAP-18, magainins and melittin, the β-sheet defensins and lactoferricin, the linear Pro-Arg rich PR-39 and other hybrid anticancer peptides such as cecropin A-based hybrids and other synthetic versions, which are toxic to bacteria but not to normal mammalian cells, exhibit a broad spectrum of cytotoxic activity against cancer cells [10]. Three mechanisms to explain this activity, namely lysis of the cell membrane, activation of extrinsic apoptotic pathways, and inhibition of angiogenesis, have been proposed. AMPs have also been shown to play an important role in wound repair and angiogenesis [11].

Our growing understanding of the structure of AMPs and their diverse biological effects can be used to complement the therapeutic arsenal with attractive candidates for novel and specific applications. The main drawback to this application of the potential and desirable properties of AMPs, and improved synthetic variants of naturally occurring AMPs, is still, however, the large-scale production of highly purified peptides at a competitive price as many of the natural host defense peptides (HDPs or AMPs) currently being studied can be obtained only using recombinant synthesis technology.

9.4 APPLICATIONS OF AMPs

9.4.1 Therapeutic Applications

Since their discovery in the early 1980s, AMPs have attracted interest as prospective antibiotic agents because of the rapid emergence of antibacterial drug resistance in the treatment of infectious diseases and the decrease in the number of new classes of antibiotics approved by the Food and Drug Administration in recent years. In addition, recent novel biological effects of AMPs, such as endotoxin neutralization, chemotactic and immunomodulating activities, induction of angiogenesis, and wound repair, have been documented, making these molecules outstanding candidates for novel therapeutic approaches.

9.4.1.1 AMPs and Antimicrobial Activity. Whereas occurrences of multidrug-resistant infections continue to increase, the discovery and development of drugs to combat these bacterial strains have slowed. The driving force for the development of newer anti-infective agents is the inevitable emergence of bacterial resistance to antibiotics following widespread clinical, veterinary and agricultural usage [86]. Some examples include the growing emergence of methicillin-resistant *Staphylococcus aureus* (MRSA), which has become a serious problem in chemotherapy [87], and vancomycin-resistant enterococci (VRE).

Despite coevolving with microbes over millions of years, AMPs have retained their activity and microbes have yet to develop resistance to their potent effects. The fact that AMPs display antimicrobial activity against both gram-positive and gram-negative bacteria, as well as against fungi and different viruses, has made these peptides attractive targets for therapeutic use. Table 9.1 lists some peptides that are being tested currently in clinical and preclinical trials [88]. However, their clinical application has not been encouraging. In addition,

TABLE 9.1 Peptides that are currently being tested in clinical and preclinical trials

Peptide	Company	Medical use
MBI-226	Migenix	Prevention of catheter-related infections
MX-594AN	Migenix	Dermatology-related infections
Plectasin	Novozymes	Treatment for anti–gram-positive antibiotic-resistant bacterial infections
P113	Dermegen	Treatment for oral candidiasis
Daptomycin (Cubicin)	Cubist Pharmaceuticals	Treatment for gram-positive skin infections
Polymixin B-Colistin-Colomycin	RX Generic Drugs	Treatment for gram-negative infections

the use of such peptides as innate immunity modulators in the treatment of infectious disease and inflammation has added another dimension to the field. Nonetheless, the use of AMPs as antimicrobial therapeutic agents, either alone or in combination with classic chemotherapeutic agents, continues to provide new possible alternative approaches.

The antibacterial potency of recombinant mature human α-defensin 5 (rmHD5) against clinical pathogenic strains was evaluated recently, with abnormal morphological changes and increased permeabilization of the cytomembrane being observed in both gram-positive bacteria and gram-negative bacteria treated with rmHD5. These results confirmed the bactericidal activity of rmHD5 and constitute a substantial basis for the potential clinical application of rmHD5 as an antibacterial agent for digestive/urogenital tract infections [89]. Another example from a recent study involves the AMP Arminin 1a from *Hydra*, which was found to exhibit significant and widespread activity against bacteria, including MRSA and enterococci, a common cause of nosocomial infections that is also drug resistant [90].

The broad antimicrobial spectrum of AMPs has resulted also in promising advances against viruses and parasites. Thus, AMPs have been found to inhibit the replication of enveloped viruses such as influenza A virus, vesicular stomatitis virus (VSV), and HIV-1 [91–93]. In addition, the entry of human papillomavirus, human cytomegalovirus, and herpes simplex virus into human cells is inhibited by lactoferricin [94–96]. A good example of the use of AMPs to combat parasite infections can be observed from the activity shown by dermaseptin peptides against human erythrocytes infected by the malaria parasite *Plasmodium falciparum* [97, 98]. Other AMPs have been isolated recently from aquatic animals to test their trypanocidal and leishmanicidal activities [99].

9.4.1.2 AMPs and Immunomodulatory Properties. The fundamental role of AMPs during the control of infectious and inflammatory diseases arises as a result of two complementary functions, namely direct antimicrobial activity and immunomodulatory properties (Figure 9.2). Indeed, it has been shown that an imbalance in the activity of AMPs can lead to the development of such diseases [11].

The immunomodulatory properties of AMPs have been widely studied in two skin diseases, namely atopic dermatitis and psoriasis. Beta defensins and cathelicidins are the two main groups of AMPs in human skin. However, although only cathelicidins have been conclusively demonstrated to protect against skin infections [100, 101] at least a dozen more peptides in the skin have been described to have antimicrobial properties [102]. These peptides, together with other molecules such as enzymes, free fatty acids, and reactive oxygen species, make up the innate chemical cutaneous antimicrobial barrier. Patients with psoriasis and rosacea display high levels of LL-37 and hBD-2, both of which seem to be responsible for promoting undesirable inflammation [103]. Indeed, a mechanism whereby LL-37 combines with self-DNA released by damaged cells to form a complex that activates TLR-9 in dendritic cells to

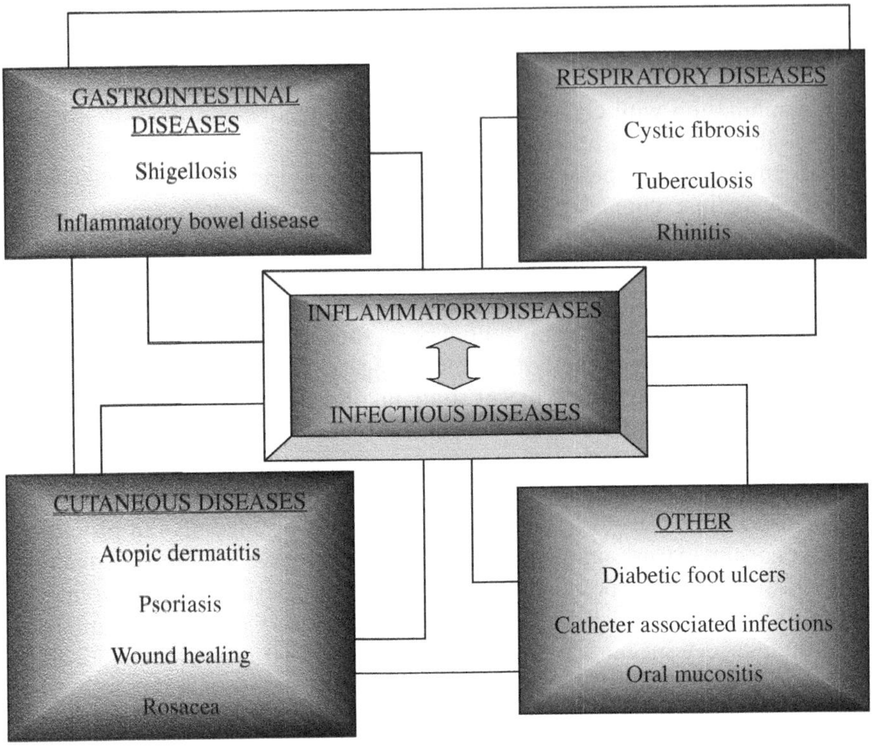

FIGURE 9.2 An imbalance in the activity of AMPs has been related to the development of infectious/inflammatory diseases.

produce type I interferons, which in turn sustain the expression of LL-37 in a feedback loop that maintains the inflammatory process in psoriasis, has been proposed [104].

The fact that vitamin D induces the expression of cathelicidin has promoted its use in dermatology. However, it has been suggested that continuous supplementation may lead to a worsening of the disease as constant levels of AMPs could induce chronic inflammation [105]. Another example of this dual effect occurs in the case of Crohn's colitis versus ulcerative colitis. Thus, although both are inflammatory bowel diseases, the former is associated with an impaired induction of the beta defensins hBD-2 and hBD-3, whereas high levels of these AMPs (approximately 1000-fold and 300-fold, respectively) are present in the latter [106, 107].

Abnormal levels of beta defensins have been reported in several respiratory diseases such as asthma, chronic obstructive pulmonary disease, sepsis, or pulmonary fibrosis, among others [108, 109].

9.4.1.3 AMPs and Wound Healing. The standard therapy to prevent and treat wound infection is still antibiotic based; therefore, a deeper understanding of wound healing is required to improve the search for potential antimicrobial substances. The use of AMPs as alternative treatments to combat wound infections, which has shown its potential in both preclinical and clinical trials, is steadily increasing [110, 111]. Thus, the proline-novispirin G10 (P-novispirin G10) peptide, which is a variant of ovispirin-1, has been shown to be promising in topical applications because of the speed with which it kills bacteria. In these studies, P-novispirin G10 was less active against gram-positive bacteria, but showed a similar activity against the gram-negative strains tested as protegrin-1, a naturally occurring host-defense peptide with superior antimicrobial properties but greater cytotoxicity [112].

9.4.1.4 AMPs and Cancer. In addition to the antibacterial and immunomodulatory activities of AMPs, a significant number of these bactericidal peptides has shown anticancer activity [113]. However, it is still unclear why some host-defense peptides can kill cancer cells whereas others do not. In this sense, AMPs can be placed into two broad categories: those that are highly potent against bacteria and cancer cells but *not* against normal mammalian cells, and those that are cytotoxic for bacteria, cancer cells, and normal mammalian cells. Changes in the membrane of a cell have important implications in terms of the progression of a tumor as they often influence its ability to grow and to attach and respond to neighboring cells differently. The cell membrane may also affect the motility of cancer cells, thereby aiding in invasion and metastasis. It is important to determine the differences between normal cells, nonmalignant tumor cells, and malignant tumor cells, as well as how these differences arise during the progression from a normal cell to a tumor cell. These differences are likely to provide an insight into possible treatments as it is not yet clear whether the molecular mechanism(s) underlying the antibacterial and anticancer activities of AMPs are the same or different [114–117]. Solid-state nuclear magnetic resonance approaches and model membranes can be used to solve the high-resolution structure and orientation of AMPs [118]. Cecropins and magainins, which are currently the most widely studied AMPs in this field, have shown cytotoxic effects against cells from melanoma, lung cancer, lymphomas, and leukemias, although other AMPs, such as defensins, cathelicidin, and lactoferricin, are being studied also [10]. The clinical use of AMPs in cancer treatment requires certain characteristics, including very high specificity, low cytolytic activity to normal mammalian and red blood cells, and an ability to target and kill cancer cells, as well as stability in serum.

9.4.2 Agriculture, Textiles, Food Industry, and Other Applications

Plants and other organisms protect themselves from bacterial, fungal, and viral attack by producing AMPs. The long use of pesticides and antibiotics in agriculture and veterinary has led to the development of resistant pathogens and therefore increasing environmental and health risks [119, 120]. Many plants and animals have been genetically manipulated to include AMP genes, and several pesticides and drugs have been produced based on these peptides. Recombinant DNA techniques have provided a new tool to obtain transgenic plants with AMP-expressing genes [121]. Tobacco, for example, is widely used as a model crop for the improvement of genetic-engineering techniques for the commercial and experimental expression of proteins, and the genetic modification of plants such as tobacco, potato, and tomato is being used to improve their resistance against different diseases. MSI-99, which is a synthetic analog of magainin II, has provided enhanced resistance to bacterial and fungal diseases in transgenic tomato lines [122], and its expression via genetic manipulation of the chloroplast genome in tobacco has led to improved antifungal and antibacterial properties. This expression via the chloroplast genome instead of via the nuclear genome is desirable to achieve high expression levels and to prevent the pollen-mediated escape of transgenes [123]. Numerous reports have been published concerning the genetic modification of potato. Thus, the expression of alfalfa antifungal peptide (alfAFP) defensin, isolated from the seeds of *Medicago sativa*, in transgenic potato plants was found to result in strong activity against the agronomically important fungal pathogen *Verticillium dahlia* [49]. The application of AMPs to control post-harvest diseases might be a suitable alternative to conventional methods involving chemical fungicides. For example, expression of a recombinant version of the peach defensin gene rDFN1 in the yeast *P. pastoris* was found to inhibit spore germination of the fungal pathogens *Penicillium expansum* and *Botrytis cinerea* in apple [124].

Different antibiotic-free strategies to improve animal health, productivity, and microbial food safety have been explored over the years. Thus, in an attempt to prevent bovine mastitis, a plasmid-mediated gene-transfer technique has been used to enable mammary cells to synthesize and secrete bovine lactoferricin and bovine tracheal antibacterial peptides. The results of this study demonstrated the secretion of a bioactive form of antibacterial peptide in the milk, which suppressed different bacterial pathogens [125]. More than a dozen different porcine AMPs have been identified, with most belonging to the cathelicidin family [126]. These antimicrobial agents could be added directly to the pigs' diet, or it may be possible to engineer genetically the animals to produce greater amounts of these compounds. Alternative cost-effective strategies may include the transfer of the genes responsible for synthesising these peptides into a microorganism, thereby enabling large quantities of these antimicrobial compounds to be produced at a lower cost.

Consumer demand for natural preservative-free or minimally processed foods requires the development of new types of antimicrobial agents. In this respect, another class of AMPs known as lantibiotics has attracted attention in recent years because of the success of nisin as a food preservative and in poultry products. Nisin is an antimicrobial protein produced naturally by a bacterium called *Lactococcus lactis*, and its use has been approved for dairy products (especially cheeses) and canned goods such as vegetables and soups [127]. Although much information exists regarding its ability to inhibit pathogenic and spoilage bacteria in many food products, nisin nevertheless has several drawbacks, such as low solubility at physiological pH or inactivity against yeast and gram-negative bacteria. The incubation conditions can also affect the peptides' performance and limit their activity, with secondary structure, solubility, aggregation, binding properties, and cytotoxicity, among others, all known to be affected by the temperature, salt concentration, and variations in pH [128]. Thus, new combined processing technologies and more promising AMPs might need to be developed.

The development of antimicrobial packaging, which involves the incorporation of volatile and nonvolatile antimicrobial agents into packages and polymers, the coating of polymer surfaces by adsorption or immobilization via ionic and covalent linkages, and the use of inherently antimicrobial polymers, is one of the most challenging topics in the field of active packaging [129]. Many antimicrobial agents, including AMPs such as cecropins, defensins, and magainins; enzymes such as lactoperoxidase and lysozyme; bacteriocins; antibiotics; polysaccharides and some organic acids; and combinations of such agents, have the potential to be incorporated into polymers [130]. Indeed, different antimicrobial food coatings containing the AMP dermaseptin K4K20-S4, which displays cytolytic activity *in vitro* against a broad spectrum of pathogenic microorganisms, including bacteria, protozoa, yeast, and filamentous fungi, have been investigated and found to show significant growth inhibition against moulds and aerobic bacteria [131].

Immobilization requires the presence of functional groups on both the antimicrobial agent and the polymer, and a spacer is desirable to link the polymer surface to the antimicrobial agent, thereby allowing the active agent to move freely when interacting with microorganisms. One example of this is the surface modification of polystyrene containing poly(ethylene glycol) grafted as a spacer by the covalent attachment of a synthetic AMP composed of eight lysine and seven leucine (6K8L) residues [132]. The effect of incorporating the antimicrobial into the polymer matrix must also be considered as it can change the mechanical, optical, and barrier properties. Plant extracts, for example, usually provide color and opacity to polymers [133]. Biopolymers are promising candidates because of the wide variety of proteins, carbohydrates and lipids (which act as plasticizers) that have been used in fields as diverse as composites, food safety, wood products, inhibition of biofouling, filters, wound dressing and textile preservation (tentage, parachutes). Likewise, some polymers are inherently antimicrobial and have been used in films and coatings. For example, many attempts have been made to use chitosan in several fields related to the food, medical, cosmetic, and textile industries [134, 135]. Many biomimetic peptides are stable at high temperatures, in organic solvents, after autoclaving or lyophilization, and within polymeric matrices. Recombinant DNA technology is a promising approach for the production of functional protein polymers using the tremendous potential diversity of amino acid combinations. Thus, repeat sequence protein polymer technology has been used to produce protein-based materials combining three diverse examples with antimicrobial, textile targeting, and ultraviolet-protective properties, thus suggesting that the regular placement of functional peptides along the polymer chain could be the key to achieving binding and targeting efficacy [37]. Recently, a relatively new class of biocidal polymers, known as cyclic *N*-halamines, has demonstrated properties such as biocidal efficacy and long-term stability, and the activity of these polymers could be regenerated once it had decreased with use. Several such materials have been prepared and tested for antimicrobial behavior [136, 137]. Likewise, cationic amphiphilic polymers prepared from different functionalized poly-ethylene imines have been studied to determine their biocidal properties and hemolytic potential. A structure–property relationship for these polymers was established by analyzing their antimicrobial activity as a function of the ratio of alkyl to cationic groups, length of the alkyl chains, and molecular weight of the polymers [138]. Polyelectrolyte multilayers are versatile material platforms that can be tailored to provide different properties such as bacterial adhesion resistance, contact killing, and biocide leaching [139].

The development of antimicrobial polymers represents a major challenge for both the academic world and industry. The use of such polymers might allow the efficacy of some existing antimicrobial agents to be improved and their environmental problems to be minimized by reducing their residual toxicity, thereby increasing their efficiency and selectivity and prolonging their lifetimes [140].

9.4.3 Limitations and Future Trends

The pharmaceutical exploitation of AMPs for clinical use has been limited mainly by the high production cost of these peptides. Indeed, the current unit cost of synthetic peptides is several times higher than that for antibiotics. One of the cheapest methods for producing such peptides involves the use of recombinant DNA technologies in bacterial, fungal, plant, and animal production systems, although none of these expression systems has proven commercially feasible to date. Nevertheless, the emergence of multiresistant pathogenic bacteria and the subsequent demand for antibiotics with novel modes of action has promoted renewed interest in these fascinating compounds. Although a few AMPs such as bacitracin, polymixins, and gramicidins have been used as efficient topical antibiotics, our current limited understanding of their complex mechanisms of action, along with their potential toxicity and protease lability, has limited their systemic and oral clinical applications [141]. Many strategies have been proposed to increase peptide half-life, including the incorporation of D- and non-natural amino acids, chemical modification at the N-terminus, peptide cyclization, or the use of non-peptidic backbones (peptidomimetics) or combinations of natural peptides [142–144]. For instance, the peptaibols, which contain high levels of nonstandard amino acids as well as peptidomimetic compounds, tend to be protease resistant [145, 146]. Another interesting approach to produce proteolysis resistance involves the development of dendrimeric peptides, whose improved *in vivo* and *in vitro* efficiency is a result of their multimeric nature, which allows polyvalent interactions. Thus, lysine dendrimers have been used as carriers of two to eight copies of tetrapeptide or octapeptide fragments of the AMP tachyplesin. The resulting dendrimers expressed high potency against a broad range of microorganisms while being less toxic than their natural precursors at the same concentration [147]. Likewise, new polyfunctional low-molecular-weight amphiphilic peptide dendrimers have also been synthesized recently [148]. This approach therefore seems to be highly promising for the design of AMPs to obtain multifunctional molecules.

REFERENCES

1. Scott, M. G., Dullaghan, E., Mookherjee, N., Glavas, N., Waldbrook, M., Thompson, A., Wang, A., Lee, K., Doria, S., Hamill, P., Yu, J. J., Li, Y., Donini, O., Guarna, M. M., Finlay, B. B., North, J. R., Hancock, R. E. (2007). An anti-infective peptide that selectively modulates the innate immune response. *Nature Biotechnology*, *25*, 465–472.
2. Shai, Y. (2002). Mode of action of membrane active antimicrobial peptides. *Biopolymers*, *66*, 236–248.
3. Jenssen, H., Hamill, P., Hancock, R. E. (2006). Peptide antimicrobial agents. *Clinical Microbiology Reviews*, *19*, 491–511.
4. Zasloff, M. (2002). Antimicrobial peptides of multicellular organisms. *Nature*, *415*, 389–395.
5. Lehrer, R. I., Ganz, T. (2002). Defensins of vertebrate animals. *Current Opinion in Immunology*, *14*, 96–102.
6. Hancock, R. E., Lehrer, R. (1998). Cationic peptides: A new source of antibiotics. *Trends in Biotechnology*, *16*, 82–88.
7. Hancock, R. E., Sahl, H. G. (2006). Antimicrobial and host-defense peptides as new anti-infective therapeutic strategies. *Nature Biotechnology*, *24*, 1551–1557.
8. Montesinos, E. (2007). Antimicrobial peptides and plant disease control. *FEMS Microbiology Letters*, *270*, 1–11.
9. Bowdish, D. M., Davidson, D. J., Scott, M. G., Hancock, R. E. (2005). Immunomodulatory activities of small host defense peptides. *Antimicrobial Agents and Chemotherapy*, *49*, 1727–1732.
10. Hoskin, D. W., Ramamoorthy, A. (2008). Studies on anticancer activities of antimicrobial peptides. *Biochimica et Biophysica Acta*, *1778*, 357–375.
11. Guani-Guerra, E., Santos-Mendoza, T., Lugo-Reyes, S. O., Teran, L. M. (2010). Antimicrobial peptides: General overview and clinical implications in human health and disease. *Clinical Immunology*, *135*, 1–11.
12. Giuliani, A., Pirri, G., Bozzi, A., Di Giulio, A., Aschi, M., Rinaldi, A. C. (2008). Antimicrobial peptides: Natural templates for synthetic membrane-active compounds. *Cellular and Molecular Life Sciences*, *65*, 2450–2460.
13. Sun, C., Du, X. J., Xu, W. T., Zhang, H. W., Zhao, X. F., Wang, J. X. (2010). Molecular cloning and characterization of three crustins from the Chinese white shrimp, Fenneropenaeus chinensis. *Fish and Shellfish Immunology*, *28*, 517–524.
14. Li, B. C., Zhang, S. Q., Dan, W. B., Chen, Y. Q., Cao, P. (2007). Expression in *Escherichia coli* and purification of bioactive antibacterial peptide ABP-CM4 from the Chinese silk worm, Bombyx mori. *Biotechnology Letters*, *29*, 1031–1036.

15. Kovalskaya, N., Hammond, R. W. (2009). Expression and functional characterization of the plant antimicrobial snakin-1 and defensin recombinant proteins. *Protein Expression and Purification*, *63*, 12–17.
16. Feng, X., Liu, C., Guo, J., Bi, C., Cheng, B., Li, Z., Shan, A. (2010). Expression and purification of an antimicrobial peptide, bovine lactoferricin derivative LfcinB-W10 in Escherichia coli. *Current Microbiology*, *60*, 179–184.
17. Zorko, M., Jerala, R. (2010). Production of recombinant antimicrobial peptides in bacteria. *Methods in Molecular Biology*, *618*, 61–76.
18. Kim, J. M., Jang, S. A., Yu, B. J., Sung, B. H., Cho, J. H., Kim, S. C. (2008). High-level expression of an antimicrobial peptide histonin as a natural form by multimerization and furin-mediated cleavage. *Applied Microbiology and Biotechnology*, *78*, 123–130.
19. Zhou, Q. F., Luo, X. G., Ye, L., Xi, T. (2007). High-level production of a novel antimicrobial peptide perinerin in Escherichia coli by fusion expression. *Current Microbiology*, *54*, 366–370.
20. Tian, Z. G., Dong, T. T., Yang, Y. L., Teng, D., Wang, J. H. (2009). Expression of antimicrobial peptide LH multimers in Escherichia coli C43 (DE3). *Applied Microbiology and Biotechnology*, *83*, 143–149.
21. Tian, Z. G., Teng, D., Yang, Y. L., Luo, J., Feng, X. J., Fan, Y., Zhang, F., Wang, J. H. (2007). Multimerization and fusion expression of bovine lactoferricin derivative LfcinB15-W4, 10 in Escherichia coli. *Applied Microbiology and Biotechnology*, *75*, 117–124.
22. Lai, Y. P., Peng, Y. F., Zuo, Y., Li, J., Huang, J., Wang, L. F., Wu, Z. R. (2005). Functional and structural characterization of recombinant dermcidin-1L, a human antimicrobial peptide. *Biochemical and Biophysical Research Communications*, *328*, 243–250.
23. Bang, S. K., Kang, C. S., Han, M. D., Bang, I. S. (2010). Expression of recombinant hybrid peptide hinnavin II/alpha-melanocyte-stimulating hormone in Escherichia coli: Purification and characterization. *Journal of Microbiology*, *48*, 24–29.
24. Krahulec, J., Hyrsova, M., Pepeliaev, S., Jilkova, J., Cerny, Z., Machalkova, J. (2010). High level expression and purification of antimicrobial human cathelicidin LL-37 in Escherichia coli. *Applied Microbiology and Biotechnology*, *88*, 167–175.
25. Fan, F., Wu, Y., Liu, J. (2010). Expression and purification of two different antimicrobial peptides, PR-39 and Protegrin-1 in Escherichia coli. *Protein Expression and Purification*, *73*, 147–151.
26. Cipakova, I., Gasperik, J., Hostinova, E. (2006). Expression and purification of human antimicrobial peptide, dermcidin, in Escherichia coli. *Protein Expression and Purification*, *45*, 269–274.
27. Kim, Y. S., Kim, M. J., Kim, P., Kim, J. H. (2006). Cloning and production of a novel bacteriocin, lactococcin K, from Lactococcus lactis subsp. lactis MY23. *Biotechnology Letters*, *28*, 357–362.
28. Morassutti, C., De Amicis, F., Bandiera, A., Marchetti, S. (2005). Expression of SMAP-29 cathelicidin-like peptide in bacterial cells by intein-mediated system. *Protein Expression and Purification*, *39*, 160–168.
29. Chen, Y. Q., Zhang, S. Q., Li, B. C., Qiu, W., Jiao, B., Zhang, J., Diao, Z. Y. (2008). Expression of a cytotoxic cationic antibacterial peptide in Escherichia coli using two fusion partners. *Protein Expression and Purification*, *57*, 303–311.

30. Diao, H., Guo, C., Lin, D., Zhang, Y. (2007). Intein-mediated expression is an effective approach in the study of beta-defensins. *Biochemical and Biophysical Research Communications*, *357*, 840–846.

31. Saugar, J. M., Rodríguez-Hernández, M. J., de la Torre, B. G., Pachón-Ibañez, M. E., Fernández-Reyes, M., Andreu, D., Pachón, J., Rivas, L. (2006). Activity of cecropin A-melittin hybrid peptides against colistin-resistant clinical strains of Acinetobacter baumannii: Molecular basis for the differential mechanisms of action. *Antimicrobial Agents and Chemotherapy*, *50*, 1251–1256.

32. Plunkett, R. M., Murray, S. I., Lowenberger, C. A. (2009). Generation and characterization of the antibacterial activity of a novel hybrid antimicrobial peptide comprising functional domains from different insect cecropins. *Canadian Journal of Microbiology*, *55*, 520–528.

33. Deslouches, B., Islam, K., Craigo, J. K., Paranjape, S. M., Montelaro, R. C., Mietzner, T. A. (2005). Activity of the de novo engineered antimicrobial peptide WLBU2 against Pseudomonas aeruginosa in human serum and whole blood: Implications for systemic applications. *Antimicrobial Agents and Chemotherapy*, *49*, 3208–3216.

34. Rossi, L. M., Rangasamy, P., Zhang, J., Qiu, X. Q., Wu, G. Y. (2008). Research advances in the development of peptide antibiotics. *Journal of Pharmaceutical Sciences*, *97*, 1060–1070.

35. Feng, X. J., Wang, J. H., Shan, A. S., Teng, D., Yang, Y. L., Yao, Y., Yang, G. P., Shao, Y. C., Liu, S., Zhang, F. (2006). Fusion expression of bovine lactoferricin in Escherichia coli. *Protein Expression and Purification*, *47*, 110–117.

36. Rodríguez-Cabello, J. C., Reguera, J., Girotti, A., Alonso, M., Testera, A. M. (2005). Developing functionality in elastin-like polymers by increasing their molecular complexity: The power of the genetic engineering approach. *Progress in Polymer Science*, *30*, 1119–1145.

37. Kumar, M., Sanford, K. J., Cuevas, W. A., Du, M., Collier, K. D., Chow, N. (2006). Designer protein-based performance materials. *Biomacromolecules*, *7*, 2543–2551.

38. Hu, F., Ke, T., Li, X., Mao, P. H., Jin, X., Hui, F. L., Ma, X. D., Ma, L. X. (2010). Expression and purification of an antimicrobial peptide by fusion with elastin-like polypeptides in Escherichia coli. *Applied Biochemistry and Biotechnology*, *160*, 2377–2387.

39. Shen, Y., Ai, H. X., Song, R., Liang, Z. N., Li, J. F., Zhang, S. Q. (2010). Expression and purification of moricin CM4 and human beta-defensins 4 in Escherichia coli using a new technology. *Microbiological Research*, *165*(8),713–718.

40. Zhou, Q., Li, M., Xi, T. (2009). Cloning and expression of a clamworm antimicrobial peptide perinerin in Pichia pastoris. *Current Microbiology*, *58*, 384–388.

41. Chen, H., Xu, Z., Peng, L., Fang, X., Yin, X., Xu, N., Cen, P. (2006). Recent advances in the research and development of human defensins. *Peptides*, *27*, 931–940.

42. Burrowes, O. J., Diamond, G., Lee, T. C. (2005). Recombinant expression of pleurocidin cDNA using the Pichia pastoris expression system. *Journal of Biomedicine and Biotechnology*, 374–384.

43. Hsu, K. H., Pei, C., Yeh, J. Y., Shih, C. H., Chung, Y. C., Hung, L. T., Ou, B. R. (2009). Production of bioactive human alpha-defensin 5 in Pichia pastoris. *The Journal of General and Applied Microbiology*, *55*, 395–401.

44. Chen, G. H., Chen, W. M., Huang, G. T., Chen, Y. W., Jiang, S. T. (2009). Expression of recombinant antibacterial lactoferricin-related peptides from Pichia

pastoris expression system. *Journal of Agricultural and Food Chemistry*, *57*, 9509–9515.

45. Chen, G. H., Yin, L. J., Chiang, I. H., Jiang, S. T. (2007). Expression and purification of goat lactoferrin from Pichia pastoris expression system. *Journal of Food Science*, *72*, M67–71.
46. Zhang, J., quan Zhang, S., Wu, X., qing Chen, Y., yu Diao, Z. (2006). Expression and characterization of antimicrobial peptide ABP-CM4 in methylotrophic yeast Pichia pastoris. *Process Biochemistry*, *41*, 251–256.
47. Kim, S. J., Quan, R., Lee, S. J., Lee, H. K., Choi, J. K. (2009). Antibacterial activity of recombinant hCAP18/LL37 protein secreted from Pichia pastoris. *The Journal of Microbiology*, *47*, 358–362.
48. Coca, M., Bortolotti, C., Rufat, M., Penas, G., Eritja, R., Tharreau, D., del Pozo, A. M., Messeguer, J., San Segundo, B. (2004). Transgenic rice plants expressing the antifungal AFP protein from Aspergillus giganteus show enhanced resistance to the rice blast fungus Magnaporthe grisea. *Plant Molecular Biology*, *54*, 245–259.
49. Gao, A. G., Hakimi, S. M., Mittanck, C. A., Wu, Y., Woerner, B. M., Stark, D. M., Shah, D. M., Liang, J., Rommens, C. M. (2000). Fungal pathogen protection in potato by expression of a plant defensin peptide. *Nature Biotechnology*, *18*, 1307–1310.
50. Schaefer, S. C., Gasic, K., Cammue, B., Broekaert, W., van Damme, E. J., Peumans, W. J., Korban, S. S. (2005). Enhanced resistance to early blight in transgenic tomato lines expressing heterologous plant defense genes. *Planta*, *222*, 858–866.
51. Saitoh, H., Kiba, A., Nishihara, M., Yamamura, S., Suzuki, K., Terauchi, R. (2001). Production of antimicrobial defensin in Nicotiana benthamiana with a potato virus X vector. *Molecular Plant-Microbe Interactions*, *14*, 111–115.
52. Giddings, G., Allison, G., Brooks, D., Carter, A. (2000). Transgenic plants as factories for biopharmaceuticals. *Nature Biotechnology*, *18*, 1151–1155.
53. Ma, J. K., Drake, P. M., Christou, P. (2003). The production of recombinant pharmaceutical proteins in plants. *Nature Reviews Genetics*, *4*, 794–805.
54. Fischer, R., Stoger, E., Schillberg, S., Christou, P., Twyman, R. M. (2004). Plant-based production of biopharmaceuticals. *Current Opinion in Plant Biology*, *7*, 152–158.
55. Baulcombe, D. C., Chapman, S., Santa Cruz, S. (1995). Jellyfish green fluorescent protein as a reporter for virus infections. *The Plant Journal*, *7*, 1045–1053.
56. Morassutti, C., De Amicis, F., Skerlavaj, B., Zanetti, M., Marchetti, S. (2002). Production of a recombinant antimicrobial peptide in transgenic plants using a modified VMA intein expression system. *FEBS Letters*, *519*, 141–146.
57. Saiman, L., Tabibi, S., Starner, T. D., San Gabriel, P., Winokur, P. L., Jia, H. P., McCray, P. B., Jr., Tack, B. F. (2001). Cathelicidin peptides inhibit multiply antibiotic-resistant pathogens from patients with cystic fibrosis. *Antimicrobial Agents and Chemotherapy*, *45*, 2838–2844.
58. Won, H. S., Kang, S. J., Lee, B. J. (2009). Action mechanism and structural requirements of the antimicrobial peptides, gaegurins. *Biochimica et Biophysica Acta*, *1788*, 1620–1629.
59. Nijnik, A, Hancock, R. (2009). Host defence peptides: antimicrobial and immunomodulatory activity and potential applications for tackling antibiotic-resistant infections. *Emerging Health Threats Journal*, 2:e1. doi: 10.3134/ehtj.09.001

60. Reddy, K. V., Yedery, R. D., Aranha, C. (2004). Antimicrobial peptides: Premises and promises. *International Journal of Antimicrobial Agents*, *24*, 536–547.

61. Cudic, M., Otvos, L., Jr. (2002). Intracellular targets of antibacterial peptides. *Current Drug Targets*, *3*, 101–106.

62. Brogden, K. A. (2005). Antimicrobial peptides: Pore formers or metabolic inhibitors in bacteria? *Nature Reviews Microbiology*, *3*, 238–250.

63. Thomas, S., Karnik, S., Barai, R. S., Jayaraman, V. K., Idicula-Thomas, S. (2010). CAMP: A useful resource for research on antimicrobial peptides. *Nucleic Acids Research*, *38*, D774–780.

64. Chen, Y., Mant, C. T., Farmer, S. W., Hancock, R. E., Vasil, M. L., Hodges, R. S. (2005). Rational design of alpha-helical antimicrobial peptides with enhanced activities and specificity/therapeutic index. *The Journal of Biological Chemistry*, *280*, 12316–12329.

65. Ludtke, S. J., He, K., Wu, Y., Huang, H. W. (1994). Cooperative membrane insertion of magainin correlated with its cytolytic activity. *Biochimica et Biophysica Acta*, *1190*, 181–184.

66. Hilpert, K., Elliott, M. R., Volkmer-Engert, R., Henklein, P., Donini, O., Zhou, Q., Winkler, D. F., Hancock, R. E. (2006). Sequence requirements and an optimization strategy for short antimicrobial peptides. *Chemistry and Biology*, *13*, 1101–1107.

67. Yeaman, M. R., Yount, N. Y. (2003). Mechanisms of antimicrobial peptide action and resistance. *Pharmacological Reviews*, *55*, 27–55.

68. Sitaram, N., Nagaraj, R. (1999). Interaction of antimicrobial peptides with biological and model membranes: Structural and charge requirements for activity. *Biochimica et Biophysica Acta*, *1462*, 29–54.

69. Dathe, M., Nikolenko, H., Meyer, J., Beyermann, M., Bienert, M. (2001). Optimization of the antimicrobial activity of magainin peptides by modification of charge. *FEBS Letters*, *501*, 146–150.

70. Nicolas, P. (2009). Multifunctional host defense peptides: Intracellular-targeting antimicrobial peptides. *FEBS Journal*, *276*, 6483–6496.

71. Jung, H. H., Yang, S. T., Sim, J. Y., Lee, S., Lee, J. Y., Kim, H. H., Shin, S. Y., Kim, J. I. (2010). Analysis of the solution structure of the human antibiotic peptide dermcidin and its interaction with phospholipid vesicles. *Biochemistry and Molecular Biology Reports*, *43*, 362–368.

72. Podda, E., Benincasa, M., Pacor, S., Micali, F., Mattiuzzo, M., Gennaro, R., Scocchi, M. (2006). Dual mode of action of Bac7, a proline-rich antibacterial peptide. *Biochimica et Biophysica Acta*, *1760*, 1732–1740.

73. Sadler, K., Eom, K. D., Yang, J. L., Dimitrova, Y., Tam, J. P. (2002). Translocating proline-rich peptides from the antimicrobial peptide bactenecin 7. *Biochemistry*, *41*, 14150–14157.

74. Kitamura, A., Kiyota, T., Tomohiro, M., Umeda, A., Lee, S., Inoue, T., Sugihara, G. (1999). Morphological behavior of acidic and neutral liposomes induced by basic amphiphilic alpha-helical peptides with systematically varied hydrophobic-hydrophilic balance. *Biophysical Journal*, *76*, 1457–1468.

75. Papo, N., Oren, Z., Pag, U., Sahl, H. G., Shai, Y. (2002). The consequence of sequence alteration of an amphipathic alpha-helical antimicrobial peptide and its diastereomers. *The Journal of Biological Chemistry*, *277*, 33913–33921.

76. Yan, H., Hancock, R. E. (2001). Synergistic interactions between mammalian antimicrobial defense peptides. *Antimicrobial Agents and Chemotherapy*, *45*, 1558–1560.
77. Dennison, S. R., Wallace, J., Harris, F., Phoenix, D. A. (2005). Amphiphilic alpha-helical antimicrobial peptides and their structure/function relationships. *Protein and Peptide Letters*, *12*, 31–39.
78. Powers, J. P. S., Hancock, R. E. W. (2003). The relationship between peptide structure and antibacterial activity. *Peptides*, *24*, 1681–1691.
79. Zhang, L., Rozek, A., Hancock, R. E. (2001). Interaction of cationic antimicrobial peptides with model membranes. *The Journal of Biological Chemistry*, *276*, 35714–35722.
80. Huang, H. W., Chen, F. Y., Lee, M. T. (2004). Molecular mechanism of Peptide-induced pores in membranes. *Physics Review Letters*, *92*, 198–304.
81. Soares, J. W., Mello, C. M. (2004). Antimicrobial peptides: A review of how peptide structure impacts antimicrobial activity. *Proceedings of the Society of Photo-Optical Instrumentation Engineers (SPIE)*, *5271*, 20–27.
82. Li, W. F., Ma, G. X., Zhou, X. X. (2006). Apidaecin-type peptides: Biodiversity, structure-function relationships and mode of action. *Peptides*, *27*, 2350–2359.
83. Luque-Ortega, J. R., van't Hof, W., Veerman, E. C. I., Saugar, J. M., Rivas, L. (2008). Human antimicrobial peptide histatin 5 is a cell-penetrating peptide targeting mitochondrial ATP synthesis in Leishmania. *Journal of the Federation of American Societies for Experimental Biology*, *22*, 1817–1828.
84. Hsu, C. H., Chen, C., Jou, M. L., Lee, A. Y., Lin, Y. C., Yu, Y. P., Huang, W. T., Wu, S. H. (2005). Structural and DNA-binding studies on the bovine antimicrobial peptide, indolicidin: Evidence for multiple conformations involved in binding to membranes and DNA. *Nucleic Acids Research*, *33*, 4053–4064.
85. Epand, R. M., Vogel, H. J. (1999). Diversity of antimicrobial peptides and their mechanisms of action. *Biochimica et Biophysica Acta Biomembranes*, *1462*, 11–28.
86. Gordon, Y. J., Romanowski, E. G., McDermott, A. M. (2005). A review of antimicrobial peptides and their therapeutic potential as anti-infective drugs. *Current Eye Research*, *30*, 505–515.
87. Maple, P. A. C., Hamilton-Miller, J. M. T., Brumfitt, W. (1989). World-wide antibiotic resistance in methicillin-resistant staphylococcus aureus. *The Lancet*, *333*, 537–540.
88. Marr, A. K., Gooderham, W. J., Hancock, R. E. W. (2006). Antibacterial peptides for therapeutic use: obstacles and realistic outlook. *Current Opinion in Pharmacology*, *6*, 468–472.
89. Wang, A. P., Su, Y. P., Wang, S., Shen, M. Q., Chen, F., Chen, M., Ran, X. Z., Chen, T. M., Wang, J. P. (2010). Antibacterial activity and mechanism of recombinant human alpha-defensin 5 against clinical antibiotic-resistant strains. *African Journal of Microbiology Research*, *4*, 626–633.
90. Augustin, R., Anton-Erxleben, F., Jungnickel, S., Hemmrich, G., Spudy, B., Podschun, R., Bosch, T. C. G. (2009). Activity of the novel peptide Arminin against multiresistant human pathogens shows the considerable potential of phylogenetically ancient organisms as drug sources. *Antimicrobial Agents and Chemotherapy*, *53*, 5245–5250.
91. Murakami, T., Niwa, M., Tokunaga, F., Miyata, T., Iwanaga, S. (1991). Direct virus inactivation of tachyplesin I and its isopeptides from horseshoe crab hemocytes. *Chemotherapy*, *37*, 327–334.

92. Morimoto, M., Mori, H., Otake, T., Ueba, N., Kunita, N., Niwa, M., Murakami, T., Iwanaga, S. (1991). Inhibitory effect of tachyplesin I on the proliferation of human immunodeficiency virus in vitro. *Chemotherapy*, *37*, 206–211.

93. Masuda, M., Nakashima, H., Ueda, T., Naba, H., Ikoma, R., Otaka, A., Terakawa, Y., Tamamura, H., Ibuka, T., Murakami, T. (1992). A novel anti-HIV synthetic peptide, T-22 (Tyr5,12,Lys7]-polyphemusin II). *Biochemical and Biophysical Research Communications*, *189*, 845–850.

94. Andersen, J. H., Osbakk, S. A., Vorland, L. H., Traavik, T., Gutteberg, T. J. (2001). Lactoferrin and cyclic lactoferricin inhibit the entry of human cytomegalovirus into human fibroblasts. *Antiviral Research*, *51*, 141–149.

95. Andersen, J. H., Jenssen, H., Sandvik, K., Gutteberg, T. J. (2004). Anti-HSV activity of lactoferrin and lactoferricin is dependent on the presence of heparan sulphate at the cell surface. *Journal of Medical Virology*, *74*, 262–271.

96. Mistry, N., Drobni, P., Näslund, J., Sunkari, V. G., Jenssen, H., Evander, M. (2007). The anti-papillomavirus activity of human and bovine lactoferricin. *Antiviral Research*, *75*, 258–265.

97. Krugliak, M., Feder, R., Zolotarev, V. Y., Gaidukov, L., Dagan, A., Ginsburg, H., Mor, A. (2000). Antimalarial activities of dermaseptin S4 derivatives. *Antimicrobial Agents and Chemotherapy*, *44*, 2442–2451.

98. Ghosh, J. K., Shaool, D., Guillaud, P., Ciceron, L., Mazier, D., Kustanovich, I., Shai, Y., Mor, A. (1997). Selective cytotoxicity of dermaseptin S3 toward intraerythrocytic Plasmodium falciparium and the underlying molecular basis. *Journal of Biological Chemistry*, *272*, 31609–31616.

99. Lofgren, S. E., Miletti, L. C., Steindel, M., Bachere, E., Barracco, M. A. (2008). Trypanocidal and leishmanicidal activities of different antimicrobial peptides (AMPS) isolated from aquatic animals. *Experimental Parasitology*, *118*, 197–202.

100. Howell, M. D., Gallo, R. L., Boguniewicz, M., Jones, J. F., Wong, C., Streib, J. E., Leung, D. Y. (2006). Cytokine milieu of atopic dermatitis skin subverts the innate immune response to vaccinia virus. *Immunity*, *24*, 341–348.

101. Nizet, V., Ohtake, T., Lauth, X., Trowbridge, J., Rudisill, J., Dorschner, R. A., Pestonjamasp, V., Piraino, J., Huttner, K., Gallo, R. L. (2001). Innate antimicrobial peptide protects the skin from invasive bacterial infection. *Nature*, *414*, 454–457.

102. Braff, M. H., Gallo, R. L. (2006). Antimicrobial peptides: an essential component of the skin defensive barrier. *Current Topics in Microbiology and Immunology*, *306*, 91–110.

103. Ong, P. Y., Ohtake, T., Brandt, C., Strickland, I., Boguniewicz, M., Ganz, T., Gallo, R. L., Leung, D. Y. (2002). Endogenous antimicrobial peptides and skin infections in atopic dermatitis. *The New England Journal of Medicine*, *347*, 1151–1160.

104. Lai, Y. P., Gallo, R. L. (2009). AMPed up immunity: How antimicrobial peptides have multiple roles in immune defense. *Trends in Immunology*, *30*, 131–141.

105. Hata, T. R., Kotol, P., Jackson, M., Nguyen, M., Paik, A., Udall, D., Kanada, K., Yamasaki, K., Alexandrescu, D., Gallo, R. L. (2008). Administration of oral vitamin D induces cathelicidin production in atopic individuals. *Journal of Allergy and Clinical Immunology*, *122*, 829–831.

106. Fellermann, K., Stange, D. E., Schaeffeler, E., Schmalzl, H., Wehkamp, J., Bevins, C. L., Reinisch, W., Teml, A., Schwab, M., Lichter, P., Radlwimmer, B., Stange, E. F. (2006). A chromosome 8 gene-cluster polymorphism with low human beta-defensin 2 gene copy number predisposes to Crohn disease of the colon. *The American Journal of Human Genetics*, *79*, 439–448.

107. Wehkamp, J., Schmid, M., Stange, E. F. (2007). Defensins and other antimicrobial peptides in inflammatory bowel disease. *Current Opinion in Gastroenterology*, *23*, 370–378.

108. Herr, C., Shaykhiev, R., Bals, R. (2007). The role of cathelicidin and defensins in pulmonary inflammatory diseases. *Expert Opinion on Biological Therapy*, *7*, 1449–1461.

109. Beisswenger, C., Bals, R. (2005). Antimicrobial peptides in lung inflammation. *Chemical Immunology and Allergy*, *86*, 55–71.

110. Ulvatne, H. (2003). Antimicrobial peptides–Potential use in skin infections. *American Journal of Clinical Dermatology*, *4*, 591–595.

111. Koczulla, A. R., Bals, R. (2003). Antimicrobial peptides: Current status and therapeutic potential. *Drugs*, *63*, 389–406.

112. Jacobsen, F., Mohammadi-Tabrisi, A., Hirsch, T., Mittler, D., Mygind, P. H., Sonksen, C. P., Raventos, D., Kristensen, H. H., Gatermann, S., Lehnhardt, M., Daigeler, A., Steinau, H. U., Steinstraesser, L. (2007). Antimicrobial activity of the recombinant designer host defence peptide P-novispirin G10 in infected full-thickness wounds of porcine skin. *Journal of Antimicrobial Chemotherapy*, *59*, 493–498.

113. Mader, J. S., Hoskin, D. W. (2006). Cationic antimicrobial peptides as novel cytotoxic agents for cancer treatment. *Expert Opinion on Investigational Drugs*, *15*, 933–946.

114. Utsugi, T., Schroit, A. J., Connor, J., Bucana, C. D., Fidler, I. J. (1991). Elevated expression of phosphatidylserine in the outer membrane leaflet of human tumor cells and recognition by activated human blood monocytes. *Cancer Research*, *51*, 3062–3066.

115. Dobrzynska, I., Szachowicz-Petelska, B., Sulkowski, S., Figaszewski, Z. (2005). Changes in electric charge and phospholipids composition in human colorectal cancer cells. *Molecular and Cellular Biochemistry*, *276*, 113–119.

116. Leuschner, C., Hansel, W. (2004). Membrane disrupting lytic peptides for cancer treatments. *Current Pharmaceutical Design*, *10*, 2299–2310.

117. Mader, J. S., Salsman, J., Conrad, D. M., Hoskin, D. W. (2005). Bovine lactoferricin selectively induces apoptosis in human leukemia and carcinoma cell lines. *Molecular Cancer Therapeutics*, *4*, 612–624.

118. Durr, U. H., Sudheendra, U. S., Ramamoorthy, A. (2006). LL-37, the only human member of the cathelicidin family of antimicrobial peptides. *Biochimica et Biophysica Acta*, *1758*, 1408–1425.

119. Rekha, Naik, S. N., Prasad, R. Pesticide residue in organic and conventional food-risk analysis. *Journal of Chemical Health and Safety*, *13*, 12–19.

120. Molina, A., Molina, M. P., Althaus, R. L., Gallego, L. (2003). Residue persistence in sheep milk following antibiotic therapy. *The Veterinary Journal*, *165*, 84–89.

121. van der Biezen, E. A. (2001). Quest for antimicrobial genes to engineer disease-resistant crops. *Trends in Plant Science*, *6*, 89–91.

122. Alan, A. R., Earle, E. D. (2002). Sensitivity of bacterial and fungal plant pathogens to the lytic peptides, MSI-99, magainin II, and cecropin B. *Molecular Plant-Microbe Interactions*, *15*, 701–708.

123. DeGray, G., Rajasekaran, K., Smith, F., Sanford, J., Daniell, H. (2001). Expression of an antimicrobial peptide via the chloroplast genome to control phytopathogenic bacteria and fungi. *Plant Physiology*, *127*, 852–862.

124. Wisniewski, M., Bassett, C., Artlip, T., Janisiewicz, W., Norelli, J., Goldway, M., Droby, S. (2005). Overexpression of a peach defensin gene can enhance the activity of postharvest biocontrol agents. *Acta Horticulturae*, *682*, 1999–2006.

125. Zhang, J. X., Zhang, S. F., Wang, T. D., Guo, X. J., Hu, R. L. (2007). Mammary gland expression of antibacterial peptide genes to inhibit bacterial pathogens causing mastitis. *Journal of Dairy Science*, *90*, 5218–5225.

126. Zhang, G., Ross, C. R., Blecha, F. (2000). Porcine antimicrobial peptides: New prospects for ancient molecules of host defense. *Veterinary Research*, *31*, 277–296.

127. 21 CFR Part 184. Fed Reg. (1988). FDA (U.S. Food and Drug Administration)/ Federal Register, 53, 11247–11251.

128. Rydlo, T., Miltz, J., Mor, A. (2006). Eukaryotic antimicrobial peptides: Promises and premises in food safety. *Journal of Food Science*, *71*, R125-R135.

129. Appendini, P., Hotchkiss, J. H. (2002). Review of antimicrobial food packaging. *Innovative Food Science & Emerging Technologies*, *3*, 113–126.

130. Suppakul, P., Miltz, J., Sonneveld, K., Bigger, S. W. (2003). Active packaging technologies with an emphasis on antimicrobial packaging and its applications. *Journal of Food Science*, *68*, 408–420.

131. Miltz, J., Rydlo, T., Mor, A., Polyakov, V. (2006). Potency evaluation of a dermaseptin S4 derivative for antimicrobial food packaging applications. *Packaging Technology and Science*, *19*, 345–354.

132. Appendini, P., Hotchkiss, J. H. (2001). Surface modification of poly(styrene) by the attachment of an antimicrobial peptide. *Journal of Applied Polymer Science*, *81*, 609–616.

133. Hong, S., Park, J., Kim, D. (2000). Antimicrobial and physical properties of food packaging films incorporated with some natural compounds. *Food Science and Biotechnology*, *9*, 38–42.

134. Lim, S. H., Hudson, S. M. (2004). Synthesis and antimicrobial activity of a water-soluble chitosan derivative with a fiber-reactive group. *Carbohydrate Research*, *339*, 313–319.

135. Rabea, E. I., Badawy, M. E., Stevens, C. V., Smagghe, G., Steurbaut, W. (2003). Chitosan as antimicrobial agent: Applications and mode of action. *Biomacromolecules*, *4*, 1457–1465.

136. Liang, J., Wu, R., Wang, J. W., Barnes, K., Worley, S. D., Cho, U., Lee, J., Broughton, R. M., Huang, T. S. (2007). N-halamine biocidal coatings. *Journal of Industrial Microbiology and Biotechnology*, *34*, 157–163.

137. Qian, L., Sun, G. (2004). Durable and regenerable antimicrobial textiles: Improving efficacy and durability of biocidal functions. *Journal of Applied Polymer Science*, *91*, 2588–2593.

138. Pasquier, N., Keul, H., Heine, E., Moeller, M., Angelov, B., Linser, S., Willumeit, R. (2008). Amphiphilic branched polymers as antimicrobial agents. *Macromolecular Bioscience*, *8*, 903–915.
139. Lichter, J. A., Van Vliet, K. J., Rubner, M. F. (2009). Design of antibacterial surfaces and interfaces: Polyelectrolyte multilayers as a multifunctional platform. *Macromolecules*, *42*, 8573–8586.
140. Kenawy el, R., Worley, S. D., Broughton, R. (2007). The chemistry and applications of antimicrobial polymers: A state-of-the-art review. *Biomacromolecules*, *8*, 1359–1384.
141. van't Hof, W., Veerman, E. C., Helmerhorst, E. J., Amerongen, A. V. (2005). Antimicrobial peptides: Properties and applicability. *Biological Chemistry*, *382*, 597–619.
142. Li, Q., Lawrence, C. B., Maelor Davies, H., Everett, N. P. (2002). A tridecapeptide possesses both antimicrobial and protease-inhibitory activities. *Peptides*, *23*, 1–6.
143. Osusky, M., Zhou, G., Osuska, L., Hancock, R. E., Kay, W. W., Misra, S. (2000). Transgenic plants expressing cationic peptide chimeras exhibit broad-spectrum resistance to phytopathogens. *Nature Biotechnology*, *18*, 1162–1166.
144. McPhee, J. B., Scott, M. G., Hancock, R. E. W. (2005). Design of host defence peptides for antimicrobial and immunity enhancing activities. *Combinatorial Chemistry & High Throughput Screening*, *8*, 257–272.
145. Wiest, A., Grzegorski, D., Xu, B. W., Goulard, C., Rebuffat, S., Ebbole, D. J., Bodo, B., Kenerley, C. (2002). Identification of peptaibols from Trichoderma virens and cloning of a peptaibol synthetase. *The Journal of Biological Chemistry*, *277*, 20862–20868.
146. Freidinger, R. M. (2003). Design and synthesis of novel bioactive peptides and peptidomimetics. *Journal of Medicinal Chemistry*, *46*, 5553–5566.
147. Tam, J. P., Lu, Y. A., Yang, J. L. (2002). Antimicrobial dendrimeric peptides. *European Journal of Biochemistry*, *269*, 923–932.
148. Polcyn, P., Jurczak, M., Rajnisz, A., Solecka, J., Urbanczyk-Lipkowska, Z. (2009). Design of antimicrobially active small amphiphilic peptide dendrimers. *Molecules*, *14*, 3881–3905.

CHAPTER 10

NOVEL ANTIMICROBIALS OBTAINED BY ELECTROSPINNING METHODS

SERGIO TORRES-GINER
Instituto de Agroquímica y Tecnología de Alimentos (IATA), CSIC, Paterna (Valencia), Spain

CONTENTS

10.1 Fundamentals of Electrospinning 261
10.2 Physical and Morphological Properties of Electrospun Polymers 264
10.3 Development of Biocide Polymeric Fiber Mats 267
10.3.1 Antimicrobial Polymers 267
10.3.2 Biocide-Loaded Nanofibers 270
10.4 Electrospinning as an Emerging Technology to Develop Antimicrobial Functionalities 276
10.4.1 Antimicrobial Bone Interfaces 276
10.4.2 Protective Wound Dressings 279
10.4.3 Other Antimicrobial Applications 280
References 282

10.1 FUNDAMENTALS OF ELECTROSPINNING

Electrospinning is a process that was originally developed in the early 1930s, but it did not receive much attention until recent decades because more research groups shifted their focus toward nanotechnology. This is a broadly used and efficient technology for fiber formation that uses electrical forces to produce ultrathin fibers using solutions of both natural and synthetic polymers. Overall,

Antimicrobial Polymers, First Edition.
Edited by José M. Lagarón, María J. Ocio, and Amparo López-Rubio.

this is a relatively robust and simple technique to produce fibers from a wide range of polymers. Electrospinning, however, has recently gained much attention not only because of its versatility in processing a wide variety of polymer materials but also because of its ability to produce fibers consistently within the submicron range that is otherwise not viable to achieve by using standard fiber-generating technologies. Over recent years, nanofibers from more than 200 polymers have been reported to be electrospun and characterized with respect to their applications [1]. These include various synthetic polymers, natural polymers, or a blend of both including proteins and polysaccharides. The electrospun fibers have ultrathin diameters, ranging from a few nanometers to several micrometers, which imply the formation of membranes with a larger surface area than those obtained from conventional spinning processes. In particular, fibers with a diameter of 100 nm have a ratio of geometrical surface area to mass of approximately 100 m^2/g [2]. This high specific surface area and the small pore size of the electrospun membranes make them interesting candidates for a wide variety of applications.

In electrospinning, a DC voltage in the range of 10–30 kV is necessary to generate the fine fibers from polymer solutions or melts. Currently, there are two standard electrospinning setups, vertical and horizontal. The typical setup of electrospinning apparatus is shown in Figure 10.1. Basically, the apparatus consists of three major components: a high-voltage power supply, a digitally controlled system pump ending at a tip of a metallic needle, and a grounded collecting plate that is usually a metal screen. The process generally involves dissolving the polymer of interest in an appropriate solvent. This solution is then placed in a syringe and a high voltage is applied. A small amount of the polymer solution is drawn out of the syringe, forming the so-called Taylor cone. Increasing the applied voltage further results in the initiation of a charged fluid jet that becomes stable when the charge is increased above a critical voltage. At this value, the repulsive electrical forces overcome the surface tension forces of the liquid. The process typically results in the drawing of a virtually endless fiber with a nanometer- to micrometer-sized diameter.

With the expansion of this technology, several research groups have developed sophisticated systems that can fabricate more complex nanofibrous structures in a more controlled and efficient manner. One of these aspects is the type of collector used. Generally, an aluminium foil is used as a collector but because of difficulty in transferring the collected fibers, and with the need for aligned fibers for various applications, other collectors such as conductive paper and cloth, wire mesh, rotating wheel, or bath of nonsolvent liquid can also be used. Other innovations include modifications of the solution spinneret such as coaxial electrospinning, mixing and multiple electrospinning, blow-assisted electrospinning, and others. Coaxial electrospinning includes fabrication of nanofibers from two polymers that uses a coaxial capillary spinneret, and as a result, a core of one polymer and shell of the other are formed [3]. With this technology, some polymers that are difficult to process can be co-electrospun and form a core inside the shell of the other polymer. This method gains

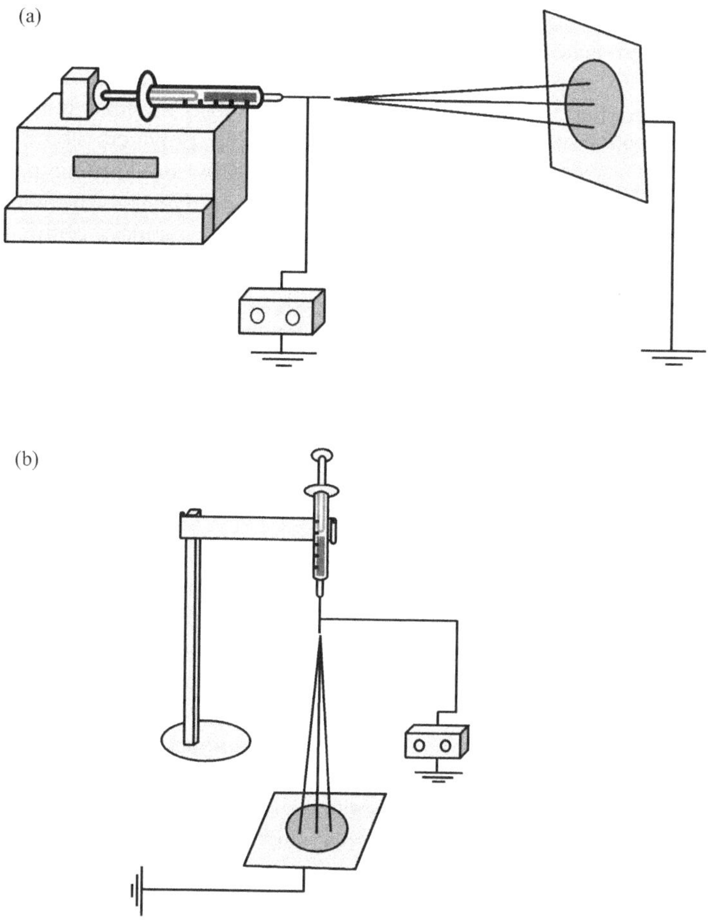

FIGURE 10.1 Schematic diagram of a typical electrospinning apparatus setup: (a) in horizontal and (b) in vertical.

attention as it provides novel properties and functionalities of nanoscale devices through the combination of polymeric materials in the axial and radial direction.

Other novel innovations can also offer various advantages. With the advent of copolymerization and polymer mixtures, attainment of the desired physical and biological properties of nanofibrous mesh has become possible now. The use of copolymers or blends of different polymers is then a feasible way to increase the electrospinnability and to generate new materials with desirable or improved properties. Blending different polymers is a straightforward pathway to create unique physical and mechanical properties that neat polymers do not possess. Therefore, when properly implemented, the performance of electrospun materials based on copolymers or polymer blends can be significantly implemented when compared with that of homopolymers. For instance, because of their generally good flexibility and toughness, synthetic polymers can be used to improve the physical properties of other polymers with bioactive properties but poor physical behavior, such as most biopolymers. Therefore, blends of synthetic and nature-derived polymers combine two desired characteristics, i.e., the strength and durability of synthetic polymers, and natural bioactivity and biodegradability of biopolymers. Another option consists of combining two very dissimilar polymers to achieve optimal properties from both. Melt electrospinning or blowing-assisted electrospinning can further help in processing high-viscosity polymer solutions that were otherwise difficult to process by solution electrospinning. In general, there is ongoing research for the improvement of nanofiber properties and the scale-up of this process.

10.2 PHYSICAL AND MORPHOLOGICAL PROPERTIES OF ELECTROSPUN POLYMERS

The electrospinning process is governed by many parameters, classified broadly into solution parameters, process parameters, and ambient parameters. Each one of these significantly affects the fibers morphology obtained, and through their proper manipulation, the desired morphology can be attained. Solution parameters include viscosity, conductivity, molecular weight (MW), and surface tension. Process parameters are referred to the applied electric field, tip-to-collector distance, and feeding or flow rate. As an example, Figure 10.2 shows different fiber morphologies obtained for chitosan when changing the MW of this carbohydrate [4]. This work further shows that by the correct selection of the polymer/solution pair or by modifying the processing conditions, a three-dimensional, nonwoven mat of entangled chitosan nanofibers can be fully feasible.

Among all factors, solution parameters play the main role in the fiber production, and among them, solution concentration produces the highest impact [5]. First of all, for fiber formation to occur, a minimum solution concentration is required. Then, at low concentrations, beads or capsule-like structures are usually formed instead of fibers, whereas at high concentrations, very thick fibers can be observed. The MW of the polymer also has a significant effect on the rheological and electrical properties such as viscosity, surface tension, conductivity, and dielectric strength. As shown in Figure 10.2, a low MW polymer solution can tend to form beads rather than fibers, and conversely, the high MW solution would give fibers with larger average diameters [4]. The improvement of the process parameters, especially the use of short tip-to-collector distances, may favor the right performance of the electrospinning process [4, 5]. However, electrospinning is habitually conducted at room temperature in normal atmosphere conditions. However, the ambient parameters may play a significant role in determining the morphology and diameter

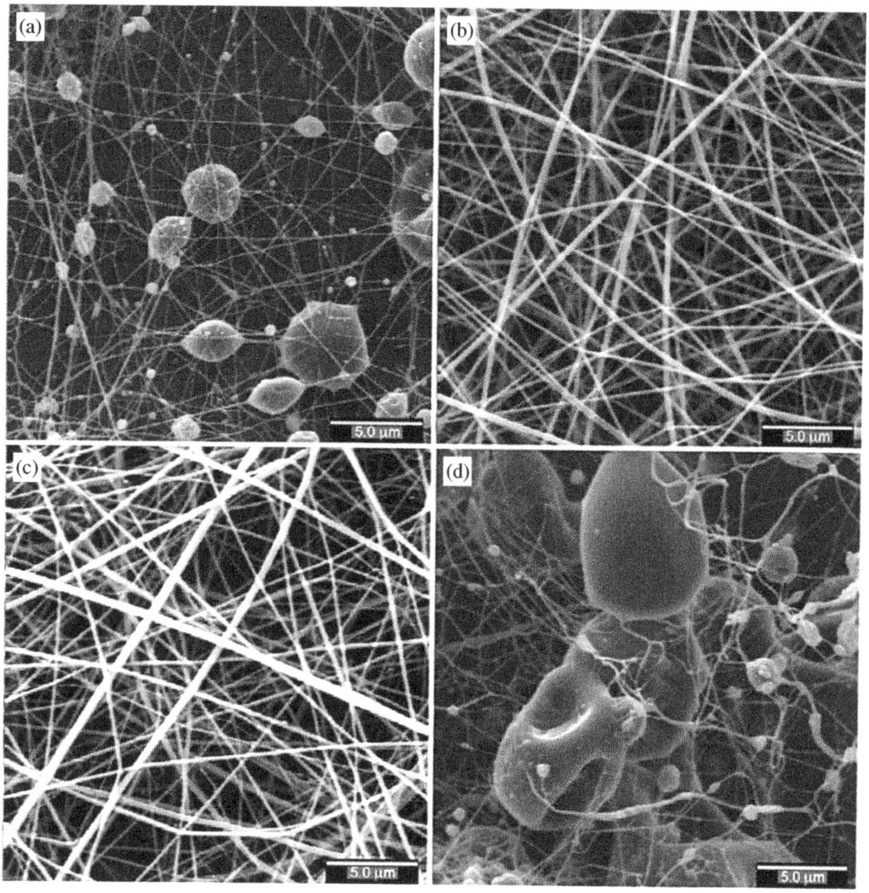

FIGURE 10.2 Different structures observed in the electrospinning of chitosan depending on the polymer source: (a) low MW, (b) medium MW, (c) high MW, and (d) chitosan from crab shells. Scale markers of 5 μm in all cases [4].

of electrospun nanofibers. For instance, high humidity can result in circular pores on the fibers and an increase of temperature can result in a decrease in the fiber diameter [6].

For some applications, it is desirable to control not only the diameter of the fibers but also the electrospun surface structure. Fibers with hollow interiors can result in interest for applications such as filtration or the preparation of functional nanotubes by fiber templates. This porous formation can be achieved by the partial dissolution of one component in the electrospinning of a polymer blend. In Figure 10.3, chitosan-based nanoporous structures are presented, in which the fiber pores are relatively large in size and fully interconnected [4]. Other works have previously showed that the use of volatile solvents such as dichloromethane can also yield nanofibers with porous structures [7].

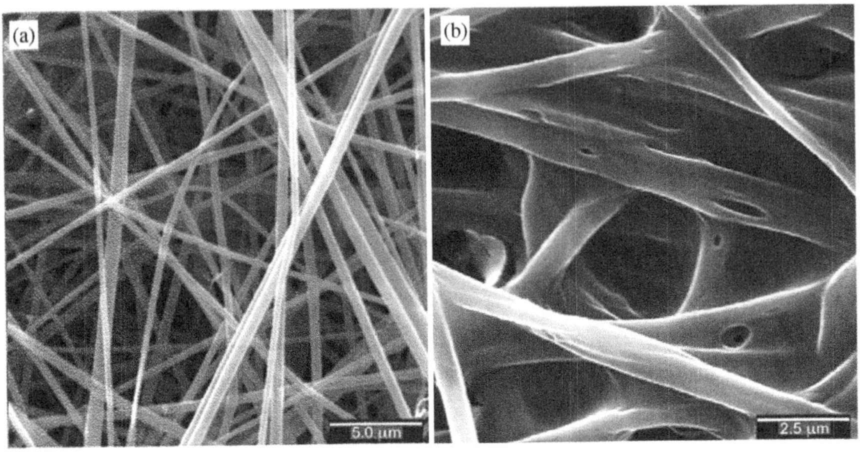

FIGURE 10.3 Electrospun chitosan nanofibers from a 50/50 chitosan and poly(lactic) acid blend: (a) as spun with a scale marker of 5 μm and (b) porous structure resulting from dissolving previous fibers with trichloromethane and with a scale marker of 2.5 μm [4].

This particular morphology is considered to be generated by a rapid-phase separation during the electrospinning process in which the solvent-rich regions were apparently transformed into pores. When the structure of the nanofibers is modified from a solid form to a porous one, its surface area is greatly increased.

Compared with conventional electrospinning, coaxial electrospinning is essentially a modification or extension of the former with a major difference in the configuration of the spinneret. This consists of a setup with two solutions independently fed through two concentric needles to facilitate the formation of a stable jet that results in core-shell fibers made of different polymers (see Figure 10.4 [8]). As drug carriers, the bilayer structure from coaxial electrospinning can potentially provide a better therapeutic effect, reduced toxicity, and sustained drug release [9]. In addition, the shell layer can avoid the contact between biological components encapsulated in the core and the organic solvents used in the electrospinning process. This fact can be extremely important, for instance, for the encapsulation of sensitive substances. In other cases, the round-shape formation can be interesting. As mentioned, ultrathin beads or fibers with beads, along its length, can be formed when the solution viscosity for electrospinning goes below a threshold value. This is produced when the molecular entanglements of the polymer solution are not strong enough to prevent the jet from breaking into droplets. This particular variation of the electrospinning process is habitually called electrospraying because it acts like an atomization procedure to form polymer droplets using high electricity [10]. This method can be interesting for encapsulating bioactives in ultrathin round-like structures that can then be easier dispersed into other materials such as food and beverages than fibers and without inducing major changes in their organoleptic properties.

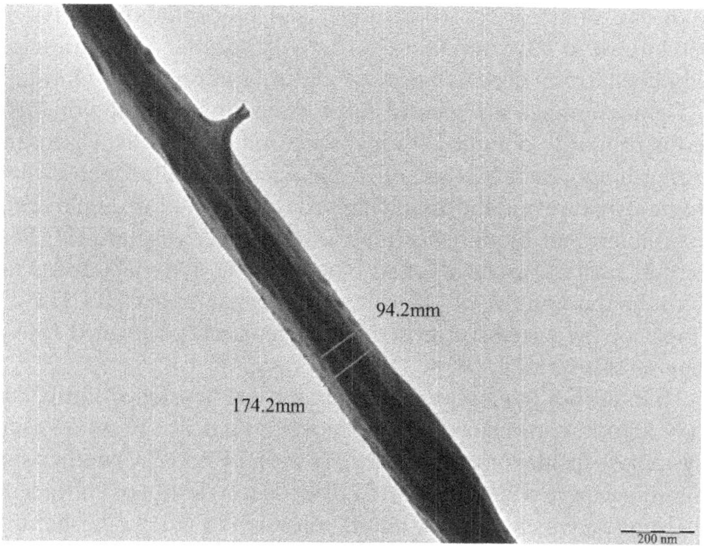

FIGURE 10.4 Electrospun bilayer fiber made of PLA (shell) and collagen (core) used for encapsulating antibiotic particles. Scale marker of 200 nm [8].

10.3 DEVELOPMENT OF BIOCIDE POLYMERIC FIBER MATS

In recent years, there has been an increasing interest in exploiting the electrospinning technology to produce nanoscale fibers, especially for the fabrication of biocide materials. Two basic approaches can be currently considered in the generation of biocide fiber-based systems using electrospinning: The use of originally antimicrobial polymers and the incorporation of exogenous biocides in the fibers.

10.3.1 Antimicrobial Polymers

The first method for the development of biocide fibers by electrospinning includes the use of antimicrobial polymers. These, which are also referred as polymeric biocides, are a class of polymers with the inherent ability to inhibit the growth of microorganisms such as bacteria, fungi, or protozoans. They can be divided into two main groups: natural polymers, also known as biopolymers, and engineered polymers. On the one hand, biopolymers are macromolecules produced in nature by living organisms and plants, and because of their participation in the natural biocycles, they can display inherently bioactive properties. The most widespread biopolymers are polysaccharides and proteins. On the other hand, engineered antimicrobial polymers are artificial repetitive arrangements generally produced by attaching or inserting an active antimicrobial agent onto a polymer backbone via an alkyl or acetyl linker. These polymers are then structurally modified to mimic antimicrobial peptides that are used by the immune systems of the living matter to kill bacteria.

Chitosan is a nontoxic, biocompatible, and biodegradable polysaccharide that is well known to have antimicrobial activity because of its cationic nature. This is obtained from deacetylation of chitin, composed of (1,4)-linked N-acetyl-β-D-glucosamine, the second most abundant natural polymer in the world that is primarily obtained from shrimp and crab shells. Chitosan is also an interesting biopolymer because of its physicochemical properties, including its solid-state structure and the dissolving state conformation. So far, chitosan is the only antimicrobial biopolymer that has been electrospun [11]. The possibility of using such chitosan fibers in biocomposite materials opens up novel applications in the control of the microorganisms. Table 10.1 [12–22] summarizes most of the successful experimental research performed for the electrospinning of chitosan.

Electrospun fabrics from pure chitosan were not developed until 2004, from 7 to 8 wt% solutions of chitosan in trifluoroacetic acid (TFA) or using a co-solvent system of dichloromethane (DCM) with TFA [12]. Another successful electrospinning experience was reported from deacetylation of chitin nanofibers obtained from electrospinning of chitin solutions in 1,1,1,3,3,3-hexafluoro-2-propanol (HFP) [14] and from chitosan in 90 wt% aqueous acetic acid solutions [15]. Although chitosan is soluble in most acids, the protonation of chitosan changes it into a polyelectrolyte in acidic solutions as a result of the many amino groups in its backbone. This makes the repulsive forces between ionic groups that arise as a result of the application of a high electric field during electrospinning increase the surface tension and restrict the formation of continuous fibers. In addition, and similarly to other polysaccharides, the polymer solution still presents other limitations for attaining stable electrospinning such as a limited solubility in other solvents and a high viscosity inherently caused by the MWs and electrical charges [23]. These issues have been particularly overcome for chitosan by varying the concentration, MW, and the solvent composition. Successful work on the direct electrospinning of pure chitosan was performed by some researchers using TFA as a main solvent [4, 13]. Chitosan initially forms stable salts in TFA, which may destroy strong interactions between the

TABLE 10.1 Electrospun chitosan-based fibers

Polymer(s)	Solvent	References
Chitosan	TFA, TFA/DCM	[4, 12, 13]
Chitosan	HFP	[14]
Chitosan	Acetic acid	[15]
Chitosan/PEO	Acetic acid	[16, 17]
Chitosan/silk	Formic acid	[18]
Chitosan/PVA	Acetic acid	[19]
Chitosan/collagen	TFA/HFP	[20]
Chitosan/PET	TFA/HFP	[21]
Chitosan/PLA	TFA/TCM	[4]
Chitosan/zein	TFA/ethanol	[22]

chitosan chains, making them more adequate for electrospinning. Additionally, its high volatility is advantageous for the rapid solidification of the electrified jet. Furthermore, the electrospinning conditions were restrictive to the use of very specific concentration, MW, and degree of deacetylation (DD).

Because pure chitosan proved to be difficult to electrospin into a continuous fibrous structure, other research has been focused on blending chitosan with synthetic or natural polymers. Some authors have reported the electrospinning of blends of chitosan and poly(ethylene oxide) (PEO) [16, 17]. The results showed that PEO aided in the electrospinning of chitosan, and as the proportion of chitosan increased, the fiber diameter decreased. It was also found that a chitosan/PEO ratio of 9/1 was an appropriate condition for the preparation of nanofibers for bone tissue engineering because these fibers retained good structural integrity in water and promoted good cellular adhesion. Novel membranes of chitosan for medical applications have also been prepared using silk fibroin [18]. Up to a 30/70 chitosan/silk in formic acid could be electrospun. Some research has also been aimed at preparing nanofibers of poly(vinyl alcohol) (PVA) with chitosan by electrospinning because of the good fiber-forming characteristics of PVA [19]. The electrospun chitosan/PVA composites were proposed in different ratios for biomedical applications. Blends with collagen to develop a better matrix for tissue engineering of functional biomaterials have also been reported using the co-solvent TFA/HFP [20]. Specifically, electrospun chitosan/poly(ethylene terephthalate) (PET) matrices were fabricated using a co-solvent of TFA/HFP, which inhibited the growth of *Staphylococcus aureus* and *Klebsiella pneumoniae* much more effectively than pure PET matrices [21]. Chitosan porous matrices from chitosan/poly(lactic) acid (PLA) blends were also prepared by electrospinning [4]. The hybrid matrices had antibacterial activity against *S. aureus*. Finally, novel antimicrobial ultrathin structures of zein/chitosan blend by electrospinning for biomedical and bioactive packaging applications were also recently developed [22].

Antibacterial tests carried out in the latter studies indicated that low amounts of chitosan nanofibers, particularly from 25 mg into 10-mL tubes containing 10^5 CFU/mL, were able to inhibit the growth of bacteria [4]. Antimicrobial activity was associated with the presence of protonated glucosamine fractions of chitosan in the solution, which are thought to inhibit bacterial growth and inactivate microorganisms. The exact biocide mechanism of chitosan is, however, still not properly understood, and several hypotheses are contemplated. It is often considered that the polycationic nature of chitosan would interfere, even at these low amounts, with the negatively charged residues of macromolecules at the cell membrane surface that results in the death of microorganisms. The chitosan presence in the polymer blends proved to be a very influencing factor controlling bacterial growth, and chitosan-containing nanofibers were much more effective than control nanofibers without chitosan. Additionally, the little amount of remnant acid, which is naturally carried in the biopolymer electrospun structure, could help to increase the biocide capacity of the fibers [4, 22]. Although during electrospinning the solidification of fibers is extremely fast and more than 99% of the solvents evaporate during or shortly after the process [24], they could still provide a certain extent of solvent release. For instance, it was shown that when zein/chitosan 95/5 (wt/wt) blends carrying free TFA were put in contact with the bacterial solution, they released acid groups resulting in a pH drop and in high antimicrobial performance [22]. This work further suggested that nanofibers made of a polymer such as zein that itself is not a natural antimicrobial material can exert a biocide activity as a result of the release of trapped TFA. Nevertheless, at mild acidic conditions, provided by using lower amounts of fibers in the tests, the antibacterial action of electrospun fibers of neat chitosan or of zein containing chitosan were found to be much stronger and efficient as biocides than the neat zein fibers.

Furthermore, there are some antimicrobials suitable for immobilization on polymer surfaces. Cationic polymers with quaternary ammonium salts can be prepared by polymerization of the base polymers with chain extenders having a tertiary amino group and subsequent quaternization of the tertiary nitrogen atoms with various acids or alkyl halides. Such covalent attachments seem attractive because their target is primarily the microbial membrane and may provide a powerful tool to introduce antimicrobial functionality to electrospun polymer fiber surfaces at reasonably high efficiency. For instance, polyurethanes cationomers (PUCs) containing different amounts of quaternary ammonium groups have been synthesized and successfully electrospun into nonwoven nanofiber mats for use in antimicrobial applications [25]. Specifically, 18 wt% of PUC solutions were electrospun in a 7/3 (wt/wt) mixed solvent consisting of N,N-dimethylacetamide (DMA) and tetrahydrofuran (THF). The resultant PUC nanofibers showed adhesion between nanofibers with various bonding sites, thus yielding mats with a film-like character and structural integrity. These morphologies were attributed to the lower evaporation rate of the electrospinning solvent used. The average fiber diameters decreased with increasing quaternary ammonium group content. In other work [26], PCU nanofibers were produced from various electrospinning solutions with concentrations ranging from 5% to 11% (wt/v) of polyurethane dissolved in a mixed solvent of THF and N,N-dimethyl formamide (DMF) at 1/1 (v/v). The resultant material surface was modified by a process that involved plasma pretreatment, ultraviolet (UV)-induced graft copolymerization of 4-vinylpyridine and quaternization of the grafted pyridine groups with hexylbromide. PCU nanofibers have showed very strong antimicrobial activities against *S. aureus* and *Escherichia coli*. Antibacterial fibers of self-quaternized block copolymers via atom transfer radical polymerization have as well been synthesized using electrospinning [27].

10.3.2 Biocide-Loaded Nanofibers

As a second route, the electrospun polymer nanofiber surfaces can display specific antimicrobial characteristics by encapsulation of exogenous biocide substances. This is a straightforward strategy to get nanofibers with improved protection against bacteria. The method to incorporate biocide substances into nanofibers habitually involves solubilizing the substances into the polymer solution prior to electrospinning. After the fiber formation, the biocide is embedded within the electrospun nanofibers in droplets of liquid or solid particles. Later, depending on the biocide nature and according to the desired application, the biocide could remain in the interior or on the surface or be delivered from the electrospun matrices. Antibiotics and silver are the most preferred biocides to generate nanofibers with induced antimicrobial properties. Table 10.2 [28–44] presents some interesting research on the encapsulation of biocides into electrospun nanofibers.

The high-surface-area/volume ratio of electrospun mats allows the efficient delivery of previously loaded antibiotics. Thus, electrospun nanofibers can potentially be used as antibiotic delivery vehicles. Various biocide drugs have been successfully introduced into electrospun nanofiber mats, without altering the fiber morphology and maintaining their antimicrobial activity. This includes powerful antibiotics with a common therapeutic use such as gentamicin sulphate [32] or tetracycline hydrochloride (TCH) [29] and, more recently, also naturally occurring antimicrobial compounds such as allyl isothiocyanate in soy protein and PEO blends and in PLA [34] and the essential oil eugenol in PVA [35]. Additionally, as Table 10.2 shows, the electrospun matrices to incorporate antibiotics are usually based on synthetic polymers such as PLA, poly(lactide-co-glycolide) acid (PLAGA), and poly(ε-caprolactone) (PCL). The reason for using synthetic polymeric matrices instead of biopolymers is that natural polymers add an extra level of complexity because such release devices habitually increase the drug delivery rates during biodegradation. Generally, when selecting a material as a drug platform, it is desirable to be able to tailor the active agent release kinetics; however, this may be difficult if the material degrades as the drug is being released. In such rapid biodegradable nanofibers,

TABLE 10.2 Electrospun biocide-loaded nanofibers

Polymer(s)	Biocide	References
PLA, PEVA, PLA/PEVA, PCL/PLA	Tetracycline hydrochloride	[28, 29]
PLA	Rifampin	[30]
PLAGA and PLAGA/PEG-b-PLA/PLA	Mefoxin	[31]
PCL, PLA, and PLA/collagen	Gentamicin sulphate	[8, 32]
PCL	Biteral	[33]
Soy protein/PEO, PLA	Allyl isothiocyanate	[34]
PVA	Eugenol	[35]
Cellulose acetate	nAg	[36]
PLA	nAg	[37]
PCL	nAg/Z	[38]
PVP and PVA	nAg	[39–41]
Silk fibroin	nAg	[42]
PVP	nAgCl	[43]
Gelatin	nAg	[44]

the antibiotic would be released by diffusion through the material, as well as through the voids left as a consequence of the matrix degradation, which in some cases can lead to dose dumping, thus resulting in local drug concentrations that can reach toxic levels. In contrast, nondegradable and low degradable materials would tend to release the antibiotic primarily by diffusion. Therefore, special care must be taken to tailor both the release rate and the degradation rate if a relative fast biodegradable material is to be used in, for instance, a biomedical application.

Different release rates may be obtained by simply varying the antibiotic loading dosage. For instance, the incorporation of 9–15 wt% of TCH in electrospun PCL nanofibers showed that the delivery rates enlarged with the increase of the antibiotic amounts [29]. A similar electrospinning approach was used for the controlled release of gentamicin in PCL [32]. It was also observed that PCL released the antibiotic fairly fast and nearly completely, whereas PLA could hold it much longer. This proved that the chemical affinity between the antibiotic and the polymer plays a crucial role in the antibiotic release design, and depending on the chemical affinity between the carrier and the antibiotic, the release profile changes and, thus, can be shaped. Therefore, the electrospinning technology can further contribute to the drug-releasing tunability by selecting the right carrier for the loaded drug. The polymer choice may be dependent on the desired drug to be released that supposes that hydrophobic drugs should be dispersed within lipophilic fibers and vice versa. For instance, controlled release of a 5 wt% hydrophilic commercial antibiotic Mefoxin, also known as Cefoxitin (Merck & Co., US), was very dissimilar from electrospun PLAGA and PLAGA/PEG-b-PLA/PLA 80:15:5 (wt%) mats [31]. Both mats demonstrated greater than 90% inhibition of *S. aureus* growth. However, the *in vitro* release study demonstrated that the addition of the amphiphilic PEG-b-PLA block copolymer allowed the hydrophilic antibiotic particles to be more embedded in the fiber interiors. The blend structure produced a more sustained release than did the neat PLGA mat in which the antibiotic was primarily located at the fiber surfaces.

The addition of anionic, cationic, nonionic surfactants can be used to improve the release profile of antibiotics. Electrospun fibers containing various percents of rifampin were prepared, and along with degradation of the fibers, the antibiotic was released constantly and no burst release was observed [30]. Also, blends of biodegradable polymers with different degradation rates can be used to control the antibiotic release. The release kinetics of tetracycline hydrochloride changed from burst to a more sustained release when 50/50 blends of PLA and poly(ethylene-co-vinyl acetate) (PEVA) were used compared with either PLA or PEVA alone [28]. It was found that the initial release rate of all formulations was high during the first 10–12 hours, and the electrospun mats gave relatively smooth release of drug over approximately 5 days.

As anticipated, it is nontrivial to achieve a homogeneous distribution for the encapsulated antibiotics within each nanofiber. In addition to the amount of antibiotic and the polymer/antibiotic nature, the polymer morphology can affect antibiotic location and, consequently, the release profile. Figure 10.5 shows typical antibiotic release curves from electrospun fibers of different composition and structure. The curves follow a typical dual-stage profile: an initial fast diffusion of the particles on or near the fiber surface and then a quick stabilization to a slow diffusion from subsequent diffusion mechanisms such as pores and cracks formed in the fibers. The relatively fast initial release of the antibiotic can represent a common problem, and a tighter control of release kinetics may be needed depending on the application. In this sense, the use of the coaxial electrospinning method in the nanofibers' production with a core-shell structure can enhance the control of the antibiotic release rate. For instance, recent encapsulation of gentamicin antibiotic in PLA/collagen coaxial fibers proved that the protective shell layer could provide an initial sustained release from the particles incorporated into the core [8]. This is habitually followed by a relatively steady and smoother release stage than the one observed in uniaxial fibers of same composition, as shown in Figure 10.5. In another example, the core-shell structured fibers fabricated by coaxial electrospinning of PCL have proved to avoid the burst release of gentamicin [32]. As compared with uniaxial fibers, coaxial electrospun fibers offered better drug stability by means of a more complete drug encapsulation that may imply a better therapeutic effect and reduced toxicity. The sustained released of the antibiotic from the nanofibers could last several days and present a short burst release over the first 24 hours. Remaining amounts of antibiotic would be delivered later during fiber degradation. In addition, the bilayer structure can offer the added advantage of largely reducing the contact between the

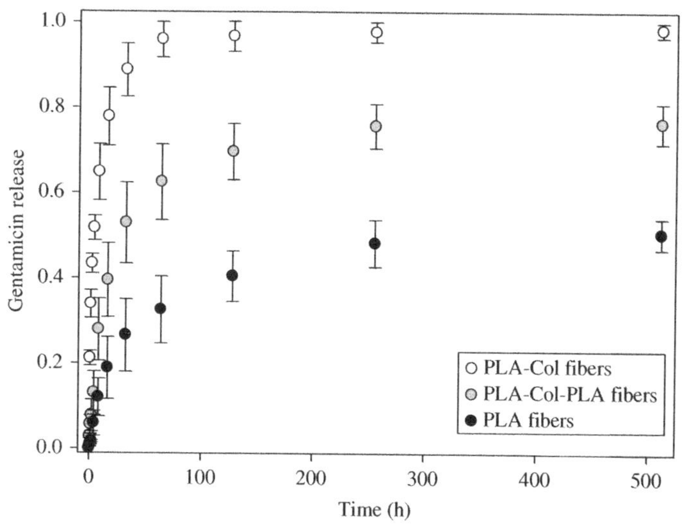

FIGURE 10.5 Release profiles of gentamicin electrospun fibers from pure PLA, uniaxial (PLA-Col) and coaxial (PLA-Col-PLA) blends of PLA and collagen in water-based medium [8].

encapsulated antibiotics and the organic solvents used in the electrospinning process, a fact that is extremely important, for instance, when these are sensitive substances.

In other cases, the antibiotic loading has been achieved by absorption of the antibiotic solution into the fiber mats after electrospinning. As a result of this process, the antibiotic is habitually located at the electrospun fiber surfaces. For instance, the potential of PCL electrospun mats loaded with the commercial antibiotic Biteral (Roche, France) was examined [33]. PCL 13% (wt/v) in chloroform/DMF (30/70) was initially electrospun to form a nonwoven fibrous mat, which was then covered with a solution of the antibiotic. *In vitro* release studies demonstrated that nearly 80% of the antibiotic was released after the first 3 hours, with complete release occurring after almost 18 hours. The authors suggested that this rapid burst release of drug is advantageous as most infections occur within a few hours after surgery.

As newer antimicrobial agents, electrospun antibacterial and antifungal nanofibers based on silver have attracted a great interest. The release of silver ions Ag^+ to aqueous and moistened media associates a broad-spectrum antimicrobial activity, and microorganisms with resistance to this antimicrobial agent are rare. However, it is difficult to disperse silver homogeneously into a polymer matrix because of the easy agglomeration of silver nanoparticles. Electrospinning represents a convenient and effective way of preparing silver-based antimicrobial polymeric materials. Binding of silver nanoparticles on the electrospun nanofibers can be confirmed with field emission scanning and transmission electron microscopy. For instance, Figure 10.6 shows the presence of silver nanoparticles in the copolymer of ethylene and vinyl alcohol, EVOH.

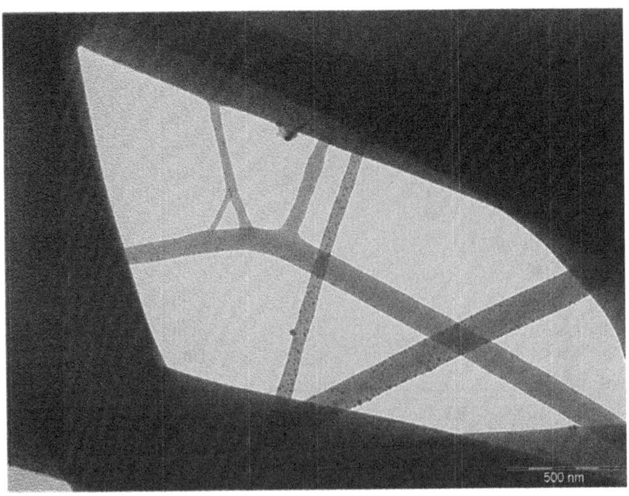

FIGURE 10.6 TEM image of electrospun EVOH fibers containing silver nanoparticles. Scale marker of 500 nm. From our laboratory work.

Although the precise antimicrobial mechanism remains unclear, it is generally accepted that silver in the reduced state Ag^0 may be able to kill bacteria by surface contact. This relies on the fact that when silver is presented in the nanoscale range, it can penetrate and interact with the bacterial membrane. Specifically, it was reported that silver inactivates microbes through the following mechanisms [41]: (1) Ag interacts with enzymes and proteins important for the bacterial respiration and the transport of important substances across the cell membrane and within the cell; (2) Ag interacts with DNA, thereby inhibiting cell division; and (3) Ag ions are bound to the cell wall and outer cell, thereby altering the functionality of the cell membrane.

In contrast to previously mentioned antibiotics, which are incorporated into the nanofibers before or after the electrospinning process and then released, the introduction of silver nanoparticles into polymer nanofibers has been accomplished by adding silver salts as precursors to the solution for electrospinning followed by postspinning treatment. Thus, silver ions Ag^+ must be first reduced into silver nanoparticles, also referred as nAg, to be activated. For instance, silver nitrate ($AgNO_3$) was added to a PLA solution to produce nanofibers with antibacterial properties [37]. Annealing of the electrospun silver-loaded PLA nanofibers in a tubular furnace at 80 °C with an atmosphere of hydrogen of 0.8 MPa for 48 hours led to the formation of silver nanoparticles within each fiber. Photoreduction with UV irradiation was used to prepare electrospun cellulose acetate fibers of 610–680 nm in diameter containing nAg of 3–21 nm on their surfaces from a 10 wt% polymer solution in 80:20 (v/v) acetone/water containing $AgNO_3$ at 0.05–0.5 wt% [36]. Heat treatment and UV irradiation was also used to prepare electrospun PVA fibers containing nAg of 5.9–107 nm from 10 wt% PVA aqueous solution containing $AgNO_3$ at a concentration of 0.1 wt% of PVA [41]. The activation process has been then normally performed by heat or irradiation, although the same polymers or solvents used for electrospinning may also serve as reducing agents. Then, electrospun nanocomposite fibers of poly(vinyl pyrrolidone) (PVP) were prepared by a one-step reduction process of $AgNO_3$ using ethanol [40]. PVP was inherently able to immobilize silver nanoparticles because of the strong affinity of the pyridyl group to metals and its ability to undergo hydrogen bonding with polar species. In other work [39], DMF was used to reduce silver ions Ag^+ into nAg of 3.4–6.0 nm in the preparation of electrospun PVP and PVA fibers from 47 wt% PVP solution in DMF containing $AgNO_3$ at 0.5 wt% of PVP or from 12 wt% PVA aqueous solution containing 5 wt% solution of PVP containing 15 wt% $AgNO_3$, respectively. This certainly makes silver a complex antimicrobial system because, depending on the chemical environment, silver cations can be rapidly inactivated by biomolecules or remains insoluble, a fact that could extremely limit the antimicrobial applications.

Some of these nAg-loaded electrospun fiber mats have been tested for their antibacterial activity against *E. coli*, *K. pneumoniae*, *Pseudomonas aeruginosa*, and *S. aureus*. Some electrospun systems based on nAg, apart from presenting a strong antibacterial activity, also have the advantage of using biodegradable and biocompatible polymers as the support of the particles that makes them interesting candidates for some biomedical applications. In a recent work [38], PCL nanofibers containing silver-loaded nanoparticles were prepared by electrospinning from polymer solutions containing silver-zirconium phosphate nanoparticles (nAg/Z). The same nanofibers also showed good cell attachment and proliferation in the absence of cytotoxicity. Thus, the nAg/Z-containing PCL nanofibers prepared by electrospinning are believed to have great potential in the application of wound dressings. Ultrafine gelatin fiber mats from a 22% (wt/v) gelatin solution in 70 vol% acetic acid containing 2.5 wt% $AgNO_3$ were also proposed as antibacterial dressings [44]. Silver nanoparticles have also been incorporated onto the surface of silk electrospun matrices to develop antimicrobial wound dressings [42]. Silver nanoparticles were incorporated by dipping the silk matrices into an aqueous $AgNO_3$ solution and followed by photoreduction. PVP/nAgCl nanocomposite fibers were also successfully prepared from hydrochloric acid through a combined method of electrospinning and the reaction of solid and gas by means of the acid vapor [43].

10.4 ELECTROSPINNING AS AN EMERGING TECHNOLOGY TO DEVELOP ANTIMICROBIAL FUNCTIONALITIES

With smaller pores and higher surface areas than regular fibers, electrospun fiber-based membranes offer the desired results from the viewpoint of properties and functionality. In addition, the electrospun fibers are generally required in a small amount and the electrospinning process itself is a versatile process as fibers can be produced into any shape using a wide range of polymers. With such versatility, electrospun fibers can be successfully applied in various fields. One of them is the bioactive segment, which is generally related to functionalized materials intended to produce a beneficial impact on a human's health. Such newly employed technologies include antimicrobial materials and systems that are becoming increasingly important not only in hospital and health-care environments but also for laboratory, home, and many industrial applications. In general, these are being developed to prevent the buildup, spread, and transfer of harmful bacteria and viruses.

Electrospinning currently represents a facile and effective approach to produce membrane or multicomponent materials with comparable or even higher antibacterial activity than other methods generally based on powder-form materials with a more difficult or impractical application. In Figure 10.7, some applications related to the development of novel electrospun materials with antimicrobial properties are presented.

10.4.1 Antimicrobial Bone Interfaces

Tissue engineering is a multidisciplinary field that combines both the principles of engineering and life sciences for the development of biological substitutes

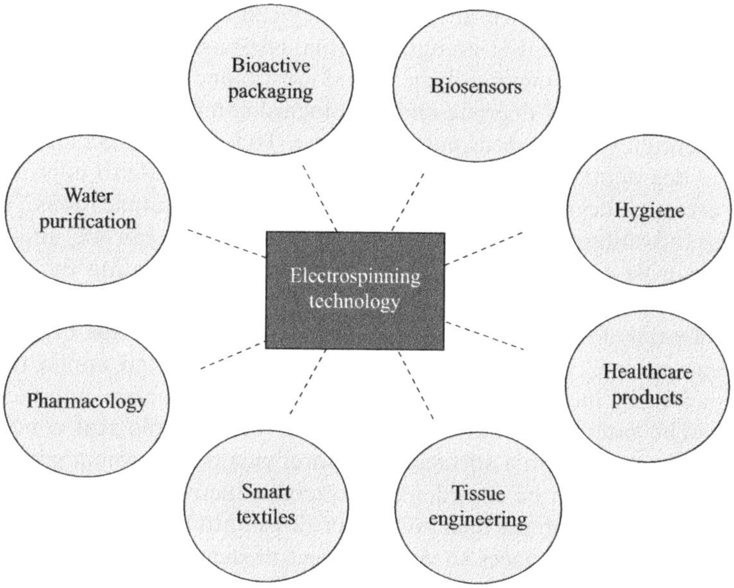

FIGURE 10.7 Some examples of potential applications of the bioactive antimicrobial materials that are developed by the electrospinning technology.

and for the restoration, maintenance, or improvement of tissue function. For engineering living tissues, the biodegradable scaffold is generally considered an indispensable element as these are used as temporary templates for cell seeding prior to the regeneration of biologically functional tissue or natural extracellular matrix (ECM). The current biomedical scientific community has mainly looked at the exploit of the electrospinning because the electrospun mats can approximate the nanostructural morphology of ordinary human tissues displaying appropriate porosity levels and pore sizes to allow cell migration, sufficient surface area, a variety of surface chemistries to encourage cell adhesion, growth, migration, differentiation, and adjustable degradation rates to match regeneration of the natural tissue [45].

Many medical reports have demonstrated the effectiveness of electrospinning techniques to create nanofibrous membranes to support cell proliferation in, for instance, uses as bone scaffolds for osteoblasts [46]. This is suggested because the diameter of electrospun fibers is of similar magnitude as that of ECM fibrils; thus, such fibers mimic the natural tissue environment, a fact that is known as biomimetics. The high surface porosity presented in nanofiber mats is yet another important characteristic of the electrospun scaffolds that contributes to the creation of a better milieu for the cells to express a correct ECM formation. Additionally, some polymers widely used in electrospinning such as PLA, PCL, and poly(glycolic acid) (PGA) already are extensively presented as the base materials for U.S. Food and Drug Administration (FDA)-approved devices, including sutures, drug delivery systems, as well as orthopedic screws and plates.

In addition to the mentioned advantages, they can offer regulatory approved resorbability and high tensile strength, thermal resistance, and hydrophobicity. Furthermore, these synthetic polymers, and most biopolymers, are biodegradable; that is to say, they degrade under biological conditions into by-products that are eventually reabsorbed/reused in nature. To fulfill the specific requisites, the *in vivo* degradation must take place at a well-defined rate to nontoxic and readily excreted degradation products. In biomedical applications, this is an added advantage because biodegradable polymers can be resorbable and thus can be completely eliminated from the body avoiding the use of a secondary surgery.

The potential of the electrospun resorbable nanofibers in tissue/organ repair and regeneration is very promising. Possibly most potential applications of electrospun mats include the development of osteconductive interfaces for implants. The osteocondution term is referred to the biological concept by which the bone grows on a surface. Therefore, membranes composed of electrospun nanofibers can be considered as osteoconductive because they permit osteoblasts proliferation on their surface or down into their pores. The use of these mats as bone interfaces to cover current medical implants, as shown in Figure 10.8, is expected to play an important role in the bone growth. First, it can increase the implant fixation and reduce the risks for loosening and rejection. Second, this would act as a scaffold providing a framework to foster bone growth and implant attachment, a concept also known as osseo-integration.

However, it is desirable that the biomedical material apart from providing improved biocompatibility would also prevent postsurgical infections. This is an emerging problem, which frequently develops after surgical procedures and may even requïre secondary intervention. The use of antimicrobial electrospun compositions can result, not only in nanofiber membranes that display enhanced cell biocompatibility but also in antibacterial performance. Therefore, antimicrobials can be conveniently introduced into the implantable tissue engineering scaffolds to prevent infection while repair and regeneration occur. The biocide release from the implant area can be designed as rapid, immediate, delayed, or modified dissolution. Because of the large surface area and high porosity of the nanofibers, a significant burst release is generally observed. However, this large initial burst must be actually ideal because it is important to eliminate the intruding bacteria before they begin to proliferate. For the few organisms that may survive the initial burst, a continued release of antibiotic

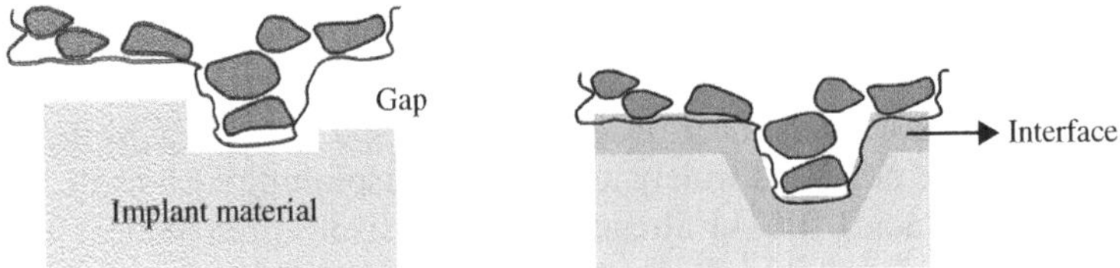

FIGURE 10.8 The importance of the use of bone-implant interfaces.

should be necessary to prevent their further proliferation This can be achieved by using the core-shell configuration offered in the coaxial electrospinning, as the shell of the polymer acts as a diffusion barrier for the incorporated biocides and smoother releases of them over longer periods of time can be obtained. Another advantage is that it offers site-specific delivery of the biocide substances from the scaffold into the body that can reduce the risk of infection, inflammation, and further aid in the healing process. The combination of therapeutic effect and the road map for self-repair makes the antimicrobial electrospun mats as an excellent candidate for bone-tissue interfaces.

10.4.2 Protective Wound Dressings

Several research groups are also actively exploring other biomedical uses for electrospun nanofibers. Some of them are applied to the development of wound dressings. These are adjuncts used for direct contact with a wound to promote healing and prevent further harm. A wound is considered to be any injury or tear on the skin surface produced by physical, chemical, mechanical, and/or thermal damages. The main functions of the wound dressings are to facilitate wound healing and minimize scarring. The decontamination of any exogenous organisms is habitually considered a critical factor for a wound healing material.

Electrospun porous membranes made of nanofibers are good candidates as wound dressings because they can protect the wound from bacterial penetration while providing a good path for the vapor transport. The wound dressing materials produced by electrospinning have special properties as compared with the dressings produced by other conventional methods [47]. First, they provide a physical barrier to protect the wound from further physical damages and from contamination with exogenous organisms. Moreover, they are also permeable to moisture and air and allow the extraction of extra body fluid from the wound area to maintain a partially immobilized moist environment. In addition, nanofibers can conform to the contour of wound resulting in a better coverage and protection. Finally, they could also heal wounds without leaving scars because of their good cellular promoting properties.

It has been recently established that the antibacterial property may play a small role in the electrospun-based wound dressing membranes and no direct relationship between the antibacterial ability and the wound healing performance was found [48]. However, the interconnected nanofibers do create perfect blocks and pores in the electrospun membrane, which can be able to prevent any bacteria from penetrating, therefore avoiding exogenous infections effectively. For instance, an electrospun mat of poly (vinylidene fluoride-co-hexafluoropropene) (PVdF-HFP) cannot inhibit the growth of bacteria, but it stopped the bacteria from spreading in the areas covered probably because of its high hydrophobicity [48].

From a bioactive point of view, the main advantage of using electrospinning is that, unlike other commercial dressings that use multilayer configuration to attain desired objectives and functionalities, this can be achieved by processing various functional materials into one blended layer to obtain an all-in-one wound dressing. This brings an extra benefit of reduced frequency in changing dressings that may disturb the regeneration of neotissue. For instance, although the small size of the pores presented in the electrospun mats can effectively protect the wound from bacterial infection, the addition of antimicrobials to the electrospun membrane composition is a significant parameter to be considered. To produce an active wound dressing material, one of the most applicable additives is the nAg that will probably cause the wound environment to have an antibacterial property preventing it from infections [47].

Such electrospun wound dressings would be as well indicated for medical treatments after surgery. Postsurgical adhesion is a common problem that frequently develops after surgical procedures. This may require secondary surgical intervention, which is not only time-consuming and costly, but also it increases the potential for further adhesions, scar tissue formation, and bacterial infection. Prevention of these adhesions can thus decrease the overall hospital costs and the severity of complications, and it can ease the technical complexity and/or the need for subsequent surgery. The use of a systemic antibiotic therapy is less effective than direct physical intervention at the injury site, and excess dosage may cause adverse side effects in the body. Therefore, the best strategy can be to design and construct a new barrier system that also possesses local delivery capability. Electrospun nanofibers mats based on antimicrobial biodegradable polymers can also be used for this aim.

10.4.3 Other Antimicrobial Applications

With the growing public health awareness of the pathogenic effects, malodors, and stain formations caused by microorganisms, there is an increasing need for antibacterial materials in many other application areas like, health care, hygienic application, water purification, biofilters, smart textiles, bioactive packaging, biosensors, and so on.

Electrospun nanofibers have been also presented to generate bioactive packaging materials incorporated as a reinforcement to functionalize plastic matrices [49]. The concept of functional or, more precisely, bioactive packaging is referred to a matrix that would be capable of withholding desired health-enhancing principles in optimum conditions until their eventual release into the food product [50]. This can be made either through controlled or fast release during storage, or just before consumption, taking into account the specific product/functional substance characteristics or requirements. The conventionally adopted approach in the preparation of antimicrobial polymeric materials is to impregnate antimicrobial agents as additives into the target polymeric plastics. However, the maximal potential of antimicrobial functionality can hardly be achieved in most cases because the antimicrobial agents often have low solubility/compatibility with the polymers, leading to the aggregation/segregation of antimicrobial agents, and the prepared antimicrobial polymeric materials present low surface-to-mass ratios. The layer deposition of antimicrobial nanofibers during electrospinning on packaging plastics would avoid the technical difficulties encountered by industries in the development of such composites. The antimicrobial layer produced by this mean would be exposed on the matrix surface to the environmental conditions, introducing biocide functionalities in the plastic packaging material that would protect the contained material, such as food, from microbial growth. In other cases, the biocide nanofibers can be intercalated into the plastic matrix, or even on the surface exposed to the food, but these may require the existence of high humidity to obtain a suitable release and the use of FDA-approved biocides. This can result in composites with high antimicrobial efficacy for the purpose of improving the preservation of foods because of the high surface contact offered by the nanofibers between the biocides and the targeted microorganisms. Similar systems may be also used for the design and fabrication of chemical biosensors for detection of microbial contamination.

Other uses of antimicrobial electrospun nanofibers are referred to smart textiles. These are clothing materials that sense and react to environmental conditions or *stimuli*. Electrospun fibers with biocides can be incorporated into textile materials with a view to protect the wearer and the textile substrate itself. Textile fibers with built-in antimicrobial properties can then act as a barrier for bacteria and fungi. This becomes especially interesting in the use of nonleaching biocides that can add durability and may not cause any health problems by migration. For instance, the feasibility of imparting antimicrobial functions to cotton substrates by applying zinc oxide nanoparticles using electrospinning has been examined [51]. The biocide nanoparticles were first electrospun with polyurethane fibers, and then, to form a layered fabric system, a 100% cotton, lightweight, plain-weave fabric was used as a substrate. Various concentrations of zinc oxide and web area densities were examined to optimize the antimicrobial functions. Textiles exhibited high rates of reduction in the number of colonies of both *S. aureus* and *K. pneumoniae* when tested under dynamic contact conditions. However, there was no distinct inhibition area surrounding the samples. This implied that the antimicrobial agent functions by direct contact of microbes with the treated textiles, rather than by the diffusion of the agent into its environment. Smart textiles based on electrospun fibers can be applied to avoid infestation by microbes that may cause cross infection by pathogens and, thus, the development of odor where the fabric is worn next to the skin. In addition, they can reduce the risk of staining and loss of the performance properties of textile substrates by microbial attack.

Another possible application of the antimicrobial nanofibers is water microfiltration. In contrast with other uses such as bioactive packaging or smart textiles, for this application, the nanofiber membrane itself can be directly used as a flat sheet membrane without the need for coupling other matrix material. Nanofibers, because of their higher porosities and interconnected pore structures, offer a higher permeability to water filtration over conventional materials being used. The low size porosity of the electrospun membranes can allow water to pass retaining suspended solids and even microorganisms such as bacteria, yeast, and fungi. The use of antimicrobial polymers or the incorporation of biocides in the fibers would further increase their performance for pathogen removal in the water-producing facility. For instance, polyamide nanofibers containing silver nanoparticles were developed for water filtration applications [52]. It was concluded that neat nanofibers without silver were not efficient in the removal of pathogens from wastewater. However, functionalization with nAg produced a log 4 removal.

To conclude, a huge amount of research is currently being focused on the electrospinning of nanofibers with biocide performance. Such ultrathin materials can be useful in the manufacture of what are variously described as "anti-infective" or "antimicrobial" devices. Most promising results have been currently presented in the biomedical field, for instance, in the development of bone interfaces and wound dressings. However, increasingly, scientists are turning their attention to the use of electrospun materials to combat infection problems in other fields such as food packaging, textiles, health care, and so on. The manufacture of such ultrathin fiber-based antimicrobials will be probably applied to advanced biocide platforms enhancing their performance, introducing control release capacity, or simply making possible their use in novel devices.

REFERENCES

1. Jiang, H. L., Fang, D. F., Hsiao, B. S., Chu, B., Chen, W. L. (2004). Optimization and characterization of dextran membranes prepared by electrospinning. *Biomacromolecules*, *5*, 326–333.
2. Frenot, A., Chronakis, S. (2003). Polymer nanofibers assembled by electrospinning. *Current Opinion in Colloid & Interface Science*, *8*, 64–75.
3. Sun, Z. C., Zussman, E., Yarin, A. L., Wendorff, J. H., Greiner, A. (2003). Compound core-shell polymer nanofibers by co-electrospinning. *Advanced Materials*, *15*, 1929–1932.
4. Torres-Giner, S., Ocio, M. J., Lagaron, J. M. (2008). Development of active antimicrobial fiber based chitosan polysaccharide nanostructures using electrospinning. *Engineering of Life Science*, *8*, 303–314.
5. Torres-Giner, S., Gimenez, E., Lagaron, J. M. (2008). Characterization of the morphology and thermal properties of zein prolamine nanostructures obtained by electrospinning. *Food Hydrocolloids*, *22*, 601–614.
6. Li, D., Xia, Y. (2004). Electrospinning of nanofibers: Reinventing the wheel. *Advanced Materials*, *16*, 1151–1170.
7. Bognitzki, M., Czado, W., Frese, T., Schaper, A., Hellwig, M., Steinhart, M., Greiner, A., Wendorff, J. H. (2001). Nanostructured fibers via electrospinning. *Advanced Materials*, *13*, 70–72.
8. Torres-Giner, S., Martinez-Abad, A., Ocio, M. J., Lagaron, J. M. (In press). Controlled delivery of gentamicin antibiotic from bioactive electrospun polylactide-based ultrathin fibers. *Advanced Biomaterials*.
9. Huang, H. -H., He, C. -L., Wang, H. -S., Mo, X. -M. (2009). Preparation of core-shell biodegradable microfibers for long-term drug delivery. *Journal of Biomedical Materials Research Part A*, *90*, 1243–1251.

10. Torres-Giner, S., Martinez-Abad, A., Ocio, M. J., Lagaron, J. M. (2010). Stabilization of a nutraceutical omega-3 fatty acid by encapsulation in ultrathin electrosprayed zein prolamine. *Food Science*, *75*, N69–N79.

11. Jayakumar, R., Prabaharan, M., Nair, S. V., Tamura, H. (2010). Novel chitin and chitosan nanofibers in biomedical applications. *Biotechnology Advances*, *28*, 142–150.

12. Ohkawa, K., Cha, D., Kim, H., Nishida, A., Yamamoto, H. (2004). Electrospinning of chitosan. *Macromolecular Rapid Communications*, *25*, 1600–1605.

13. Sangsanoh, P., Supaphol, P. (2006). Stability improvement of electrospun chitosan nanofibrous membranes in neutral or weak basic aqueous solutions. *Biomacromolecules*, *7*, 2710–2714.

14. Min, B. M., Lee, S. W., Lim, J. N., You, Y., Lee, T. S., Kang, P. H., et al. (2004). Chitin and chitosan nanofibers: Electrospinning of chitin and deacetylation of chitin nanofibers. *Polymer*, *45*, 7137–7142.

15. Geng, X., Kwon, O., Jang, J. (2005). Electrospinning of chitosan dissolved in concentrated acetic acid solution. *Biomaterials*, *26*, 5427–5432.

16. Spasova, M., Manolova, N., Paneva, D., Rashkov, I. (2004). Preparation of chitosan-containing nanofibres by electrospinning of chitosan/poly(ethylene oxide) blend solutions. *e-Polymers*, *56*, 1–12.

17. Duan, B., Dong, C., Yuan, X., Yao, K. (2004). Electrospinning of chitosan solutions in acetic acid with poly(ethylene oxide). *Journal of Biomaterials Science, Polymer Edition*, *15*, 797–811.

18. Park, W. H., Jeong, L., Yoo, D. I., Hudson, S. (2004). Effect of chitosan on morphology and conformation of electrospun silk fibroin nanofibers. *Polymer*, *45*, 7151–7157.

19. Li, L., Hsieh, Y. -L. (2006). Chitosan biocomponent nanofibers and nanoporous fibers. *Carbohydrate Research*, *341*, 374–381.

20. Chen, Z., Mo, X., Qing, F. (2006). Electrospinning of collagen-chitosan complex. *Materials Letters*, *61*, 3490–3494.

21. Jung, K., Huh, M., Meng, W., Yuan, J., Hyun, S. H., Bae, J., Hudson, S. M., Kang, I. (2007). Preparation and antibacterial activity of PET/chitosan nanofibrous mats using an electrospinning technique. *Journal of Applied Polymer Science*, *105*, 2816–2823.

22. Torres-Giner, S., Ocio, M. J., Lagaron, J. M. (2009). Novel antimicrobial ultrathin structures of zein-chitosan blends obtained by electrospinning. *Carbohydrate Polymers*, *77*, 261–266.

23. Lee, K. Y., Jeong, L., Kang, Y. O., Lee, S. J., Park, W. H. (2009). Electrospinning of polysaccharides for regenerative medicine.*Advanced Drug Delivery Reviews*, *61*, 1020–1032.

24. Fong, H. (2005). *Polymeric Nanostructures and Their Applications*. American Scientific Publishers, Los Angeles, CA.

25. Jeong, E. H., Yang, J., Youk, J. H. (2007). Preparation of polyurethane cationomer nanofiber mats for use in antimicrobial nanofilter applications. *Materials Letters*, *61*, 3991–3994.

26. Yao, C., Li, X., Neoh, K. G., Shi, Z., Kang, E. T. (2008). Surface modification and antibacterial activity of electrospun polyurethane fibrous membranes with quaternary ammonium moieties. *Journal of Membrane Science*, *320*, 259–267.

27. Guo-Dong, F., Fang, Y., Zhigang, L., Xinsong, L. (2008). Solvent-resistant antibacterial microfibers of self-quaternized block copolymers from atom transfer radical polymerization and electrospinning. *Journal of Materials Chemistry*, *18*, 859–867.
28. Kenawy, E. -R., Bowlin, G. L., Mansfield, K., Layman, J., Simpson, D. G., Sanders, E. H., Wnek, G. E. (2002). Release of tetracycline hydrochloride from electrospun poly(ethylene-co-vinylacetate), poly(lactic acid), and a blend. *Journal of Controlled Release*, *81*, 57–64.
29. Buschle-Diller, G., Cooper, J., Xie, Z., Wu, Y., Waldrup, J., Ren, X. (2007). Release of antibiotics from electrospun bicomponent fibers. *Cellulose*, *14*, 553–562.
30. Zeng, J., Xu, X., Chen, X., Liang, Q., Bian, X., Yang, L., Jing, X. (2003). Biodegradable electrospun fibers for drug delivery. *Journal of Controlled Release*, *92*, 227–231.
31. Kim, K., Luu, Y. K., Chang, C., Fang, D., Hsiao, B. S., Chu, B., et al. (2004). Incorporation and controlled release of a hydrophilic antibiotic using poly(lactide-coglycolide)-based electrospun nanofibrous scaffolds. *Journal of Controlled Release*, *23*, 47–56.
32. Huang, Z. -M., He, C. -L., Yang, A., Zhang, Y., Han, X. -J., Yin, J., Wu, Q. (2006). Encapsulating drugs in biodegradable ultrafine fibers through co-axial electrospinning. *Journal of Biomedical Materials Research Part A*, *77*, 169–179.
33. Bolgen, N., Vargel, I., Korkusuz, P., Menceloglu, Y. Z., Piskin, E. (2007). In vivo performance of antibiotic embedded electrospun PCL membranes for prevention of abdominal adhesions. *Journal of Biomedical Materials Research Part B: Applied Biomaterials*, *81*, 530–543.
34. Vega-Lugo, A. -C., Lim, L. -T. (2009). Controlled release of allyl isothiocyanate using soy protein and poly(lactic acid) electrospun fibers. *Food Research International*, *42*, 933–940.
35. Kriegel, C., Kit, K. M., McClements, D. J., Weiss, J. (2009). Nanofibers as carrier systems for antimicrobial microemulsions. Part I: Fabrication and characterization. *Langmuir*, *25*, 1154–1161.
36. Son, W. K., Youk, J. H., Lee, T. S., Park, W. H. (2004). Preparation of antimicrobial ultrafine cellulose acetate fibers with silver nanoparticles. *Macromolecular Rapid Communications*, *25*, 1632–1637.
37. Xu, X., Yang, Q., Wang, Y., Yu, H., Chen, X., Jing, X. (2006). Biodegradable electrospun poly(l-lactide) fibers containing antibacterial silver nanoparticles. *European Polymer Journal*, *42*, 2081–2087.
38. Duan, Y. Y., Jia, J., Wang, S. H., Yan, W., Jin, L., Wang, Z. Y. (2007). Preparation of antimicrobial poly(e-caprolactone) electrospun nanofibers containing silver-loaded zirconium phosphate nanoparticles. *Journal of Applied Polymer Science*, *106*, 1208–1214.
39. Jin, W. J., Lee, H. K., Jeong, E. H., Park, W. H., Youk, J. H. (2005). Preparation of polymer nanofibers containing silver nanoparticles by using poly(N-vinylpyrrolidone). *Macromolecular Rapid Communications*, *26*, 1903–1907.
40. Wang, Y., Li, Y., Yang, S., Zhang, G., An, D., Wang, C., Yang, Q., Chen, X., Jing, X., Wei, Y. (2006). A convenient route to polyvinyl pyrrolidone/silver nanocomposite by electrospinning. *Nanotechnology*, *17*, 3304–3307.

41. Hong, K. H. (2007). Preparation and properties of electrospun poly (vinyl alcohol)/silver fiber web as wound dressings. *Polymer Engineering & Science*, *47*, 43–49.
42. Kang, M., Jung, R., Kim, H. S., Youk, J. H., Jin, H. J. (2007). Silver nanoparticles incorporated electrospun silk fibers. *Journal of Nanoscience and Nanotechnology*, *7*, 3888–3891.
43. Bai, J., Li, Y., Li, M., Wang, S., Zhang, C., Yang, Q. (2008). Electrospinning method for the preparation of silver chloride nanoparticles in PVP nanofiber. *Applied Surface Science*, *254*, 4520–4523.
44. Rujitanaroj, P., Pimpha, N., Supaphol, P. (2008). Wound-dressing materials with antibacterial activity from electrospun gelatin fiber mats containing silver nanoparticles. *Polymer*, *49*, 4723–4732.
45. Lannutti, J., Reneker, D., Ma, T., Tomasko, D., Farson, D. (2007). Electrospinning for tissue engineering scaffolds. *Materials Science and Engineering: C*, *27*, 504–509.
46. Torres-Giner, S., Gimeno-Alcaniz, J. V., Ocio, M. J., Lagaron, J. M. (2009). Comparative performance of electrospun collagen cross-linked by means of different methods. *Applied Materials and Interfases*, *1*, 218–223.
47. Zahedi, P., Rezaeian, I., Ranaei-Siadat, S. -O., Jafari, S. -H., Supaphol, P. (2010). A review on wound dressings with an emphasis on electrospun nanofibrous polymeric bandages. *Polymers for Advanced Technologies*, *21*, 77–95.
48. Liu, X., Lin, T., Fang, J., Yao, G., Zhao, H., Dodson, M., Wang, X. (2010). *In vivo* wound healing and antibacterial performances of electrospun nanofibre membranes. *Journal of Biomedical Materials Research Part A*, *94*, 499–508.
49. Huang, Z., Zhang, Y., Kotaki, M., Ramakrishna, S. (2003). A review on polymer nanofibers by electrospinning and their applications in nanocomposites. *Composite Science and Technology*, *63*, 2223–2253.
50. Lopez-Rubio, A., Gavara, R., Lagaron, J. M. (2006). Bioactive packaging: Turning foods into healthier foods through biomaterials. *Trends in Food Science & Technology*, *17*, 567–575.
51. Lee, S. (2009). Multifunctionality of layered fabric systems based on electrospun polyurethane/zinc oxide nanocomposite fibers. *Journal of Applied Polymer Science*, *114*, 3652–3658.
52. Bjorge, D., Daels, N., De Vrieze, S., Dejans, P., Van Camp, T., Audenaert, W., Hogie, J., Westbroek, P., De Clerck, K., Van Hulle, S. (2009). Performance assessment of electrospun nanofibers for filter applications. *Desalination*, *249*, 942–948.

CHAPTER 11

SILVER- AND NANOSILVER-BASED PLASTIC TECHNOLOGIES

ANTONIO MARTÍNEZ-ABAD
Instituto de Agroquímica y Tecnología de Alimentos (IATA), CSIC Paterna, Valencia, Spain

CONTENTS

11.1 Introduction . . . 288
11.1.1 Historical Use of Silver as an Antimicrobial Agent . . . 288
11.1.2 Renewed Interest in Silver: Antimicrobial Silver for Infinite Applications . . . 288
11.2 Antimicrobial Silver: How Does It Work? . . . 289
11.2.1 Molecular Understanding of the Mechanism of Action . . . 290
11.2.2 Silver Nanoparticles: Optimizing Antimicrobial Silver . . . 292
11.2.3 Advantages and Limitations of the Use of Silver . . . 293
11.3 Obtention and Incorporation of Silver into Coatings and Polymer Matrices . . . 297
11.3.1 Introduction . . . 297
11.3.2 Preparation of Silver Nanoparticles: Mastering the Size and Shape of Nanosilver . . . 299
11.3.3 Incorporating Nanosilver: Techniques and Materials . . . 301
11.4 Properties of Silver-Based Polymers . . . 302
11.4.1 Characterization . . . 302
11.4.2 Release Issues . . . 303

Antimicrobial Polymers, First Edition.
Edited by José M. Lagarón, María J. Ocio, and Amparo López-Rubio.

11.5 Applications and Future Trends . . . 305
11.5.1 Regulatory Issues . . . 305
11.5.2 Current Applications . . . 305
11.5.3 Future Trends . . . 307
References . . . 308
Websites . . . 316

11.1 INTRODUCTION

11.1.1 Historical Use of Silver as an Antimicrobial Agent

Silver has been attributed antimicrobial properties since ancient times. Even before the Neolithic era, it was known that cooking or storing water or wine in silver pots would keep it safe [1, 2]. Alexander the Great, for example, used to drink only from silver vessels. Silver nitrate, known in antiquity as "lapis infernalis," was first recorded to be used for therapeutic purposes in the 8th century. In 980, Avicenna reports the use of silver as a blood purifier, for bad breath, and for heart palpitations. In the 17th and 18th centuries, the use of silver nitrate was quickly generalized in the treatment of venereal diseases, fistulas, and abscesses. By the 19th century, physicians realized that silver not only prevented wounds from being infected, but it also allowed for epithelization and promoted crust formation [3, 4]. Consequently, pencils charged with a 0.5% silver nitrate solution were introduced in the basic surgical equipment, and silver nitrate became the most useful tool in the treatment of burns or wounds. However, in 1928, penicillin was discovered, and with the advent of antibiotics after World War II, a new system was developed that could fight infections systemically as well as superficially. As a result, the use of silver against microbes was pushed to the background, being limited to the occasional treatment of burns and ulcers until present time.

11.1.2 Renewed Interest in Silver: Antimicrobial Silver for Infinite Applications

A combination of multiple factors has promoted the rediscovery of silver as an antimicrobial agent and its rapid rise to the forefront in the development of antimicrobial systems. First, the emergence of antibiotic-resistant microbes manifests ever more rapidly every time a new version is introduced in the healthcare system. Therefore, the faith so long deposited on antibiotics is waning and the health-care sector is in need of finding new antimicrobial systems to fight the increasing number of nosocomial infections [5] and to ensure the proper functioning of the public health system. Second, our society is increasingly more aware of the ubiquitous presence of microbes in all aspects of daily life. A public opinion ever more concerned with sterility and safety is demanding antimicrobials that can be safely and cost-effectively applied to any materials in our environment. Both the medical and consumer demands represent an immense potential market that promotes the fast development of new technologies based on antibacterial materials. Although other natural substances with antimicrobial properties are either volatile, do not withstand thermal processing, or are not cost-effectively synthesized or purified, the excellent thermal and chemical stability of silver and its relatively low cost make it an ideal candidate for its incorporation in a wide variety of materials. In addition to these advantages, the discovery of the antibacterial properties of silver nanoparticles has further increased the interest in silver as a key component in antimicrobial materials. Because of the extreme increase in their surface-area-to-volume ratio, metal nanoparticles have unique mechanical, optical and chemical properties as compared with the bulk material [6]. The combination of nanotechnology with its antimicrobial effect has caused silver to become one of the most promising fields of scientific as well as of commercial interest. Of all nanomaterials, and even though regulatory issues, are still unclear, nanosilver has the highest degree of commercialization [2, 7]. Not only restricted to medical applications, more than 100 consumer products with nanosilver could already be found in the market in 2007 [8] and the number might still be probably rising. In the scientific field, research trends focus on incorporating silver to new materials and on enhancing its efficacy by changing the size and shape of the silver particles, and improving their stability and release [9, 10]. As an example, the number of scientific publications dealing with the formation of silver nanoparticles for antimicrobial purposes has increased about ten times from 2004 to 2009 (search on scopus with the keywords "silver antimicrobial nanoparticles").

The extraordinary impact of silver-based antimicrobial systems is best reflected in its widespread use. In the medical field, orthopedic implants, prostheses, vascular grafts, and wound dressings are the most common application, but it can also be found in creams, gels, contraceptive devices, recipient sites, and all kinds of surgical instruments [2, 11, 12]. The success and outstanding cost-effectiveness of these materials in the medical field has been the driving force for their implantation in consumer product markets, being already present in many aspects of our daily life. Antimicrobial silver can be found in home appliances like inner liners in refrigerators or washing machines, in cosmetics or hygiene products (creams, lotions, soaps, deodorants, toothbrushes, or toothpaste), in cutlery or food contact surfaces (cutting boards, counter tops, containers, conveyors), in textiles (clothing, underwear, socks, upholstery), in toys, in air or water purification systems, and so on [2, 11, 12]. Silver has become so fashionable that even its ingestion in the form of a colloidal solution is marketed as a nutrition supplement (www.mesosilver.com).

11.2 ANTIMICROBIAL SILVER: HOW DOES IT WORK?

In this section, the molecular basis for the antimicrobial efficacy of silver will be discussed, in either the ionic or the elemental nanoparticulated form. The advantages and drawbacks of using silver technology will be considered, taking a look at the effects on the environment and on human health.

11.2.1 Molecular Understanding of the Mechanism of Action

In dealing with its antimicrobial efficacy, a difference must be made between silver at the macroscale or microscale and silver nanoparticles. The biocidal properties of the bulk material, which itself has no antibacterial effect, rely on the sustained oxidation and release of very small quantities of silver cations (Ag^+) to an aqueous or moistured environment. It is the presence of a tiny fraction of these oxidized cations that is responsible for the empirical appraisal of antimicrobial efficacy of silver cups, vessels, pots or cutlery, and their historical use.

Silver cations interact with atoms with a high electronic density, in particular having an extreme chemical affinity for sulphur groups, like thiol groups (-SH) in biomolecules [13, 14]. The interaction of silver with L-Cysteine residues causes denaturation and loss of enzymatic functions [13, 15]. This unspecific mechanism affects bacterial viability at different levels. First, with inactivation of enzymes in the outer membrane, permeability and transmembranous energy metabolism are disrupted leading to a loss in the proton motive force [16–18]. This electrolyte imbalance is reflected in a massive loss of potassium out [15, 19]. When examined by electron microscopy, a light region with dense granules is formed from enzyme aggregation and the membrane is noted to shrink and detach from the cell wall [16–18, 20] (Figure 11.1). Second, once the ions have entered the cell, they inhibit dehydrogenases of the respiratory chain, which depletes intracellular ATP levels [17, 18]. Furthermore, the ions can bind DNA components, which stabilize DNA closed conformation, evidenced by a condensed region in the center of the cell, preventing replication (Figure 11.1). In this respect, it remains disputed whether the silver ions are bond to the phosphate residues [14, 16] or whether they intercalate between N-H bonds in purines and pyrimidines bases [17, 21]. Additionally, it has been postulated that the depletion of protective enzymes produces an increase in reactive oxygen species (ROS), which further contributes to damage vital functions of the cell in aerobic conditions [22].

Bacterial defense mechanisms against the damage consist mainly of the overproduction of the targeted proteins. Lok et al., for example, detected the overexpression of envelope protein precursors in silver-treated bacteria [18]. Feng et al. describe the dense granules around the DNA as a protective proteinic envelope to prevent Ag^+ from getting to the DNA molecule [16]. Several genes have been isolated and identified that promote a higher resistance to silver ions [23]. These genes imply the production of silver binding proteins and efflux Ag^+/H^+ exchange pumps. Accordingly, when silver ions are present at low concentrations, damaged bacteria can reestablish homeostasis and survive. Above a certain concentration, however, bacterial damage would be irreversible, leading to cell death.

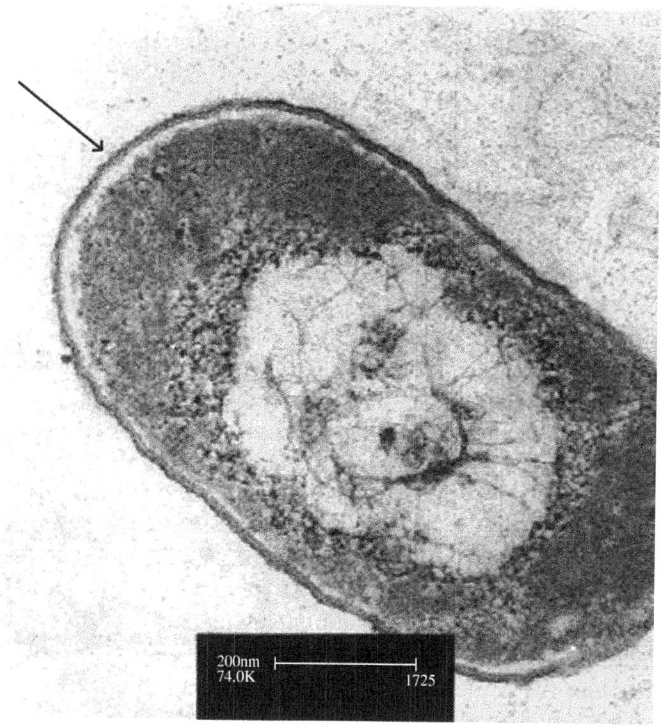

FIGURE 11.1 TEM image of *E. coli* treated with silver nanoparticles. Arrow indicates detachment of the membrane from the cell wall. Condensation of DNA is noted in the center of the cytoplasm. (*Source*: Reprinted from Reference 16 with permission from John Wiley & Sons, Inc.)

Establishing breakpoints for the inactivation of bacteria with silver is a particularly challenging task, as silver ions will not only interact with bacterial constituents but also with any chemical substance drawing silver ions, like sulphur groups or halides [24]. Moreover, bacterial susceptibility has been found to differ highly among different strains of a single species [25]. Generally speaking, it seems that gram-negative bacteria are more sensitive to the presence of silver than gram-positive bacteria [16]. This has been attributed to the presence of the negatively charged lipopolysaccharide (LPS), which would attract the positively charged silver ions. However, gram positives possess a thick peptidoglycan in their cell wall that would act as a protective shield against the ions, and high content in teichoic acids, known to be good chelators. Hwang et al. found concentrations below 10 ppb to show a bactericidal effect against the gram-negative *Legionella pneumophila*, *Pseudomonas aeruginosa*, and *Escherichia coli* [26]. They further estimated only about 0.5–2.5pg · h/cell Ag^+ was the real amount of silver needed to inactivate these bacteria. This was done by subtracting the silver content before and after sterile filtration. These findings put forth the extraordinary potential of silver as an antimicrobial, as well as the need for understanding the complexity of its antibacterial activity.

11.2.2 Silver Nanoparticles: Optimizing Antimicrobial Silver

11.2.2.1 Mechanism of Action. Nanotechnology is emerging as a rapidly growing field with its application in science and technology for the purpose of manufacturing new materials at the nanoscale level. In the case of silver, the reduction to the nanoscale confers the material better conductivity and improves the enhancement for surface-enhanced Raman scattering (SERS) or metal-enhanced fluorescence techniques [27]. But notably, the most interesting feature of silver nanoparticles is their antibacterial activity. Thermodynamic reactivity is dependent on the surface-to-volume ratio or active surface of the particles. When reduced to the nanoscale ($<$100 nm) and most remarkably less than 25 nm, silver nanoparticles are a metastable, high-energy form of elemental silver, which leads to similar effects on bacteria as compared with silver ions [28]. These imply, as mentioned, interaction with bacterial membrane constituents, disturbing permeability and forming pits; penetration inside the cell, unbalancing respiratory functions, leading to an increase of ROS and depletion of ATP levels; and intercalating between DNA bases, interfering replication [27, 29, 30]. Although the explanation remains disputed, it is somewhat accepted that these effects could be related to either an increased reactivity of the particles resulting from the high active surface or the increase in released free silver cations or radicals when exposed to water [27, 28]. The enhanced antibacterial effect of nanosilver as compared with silver ions, in the nano- and micromolar levels, respectively, could be a result of the higher chemical stability of the nanoparticles, which would enable them to penetrate more efficiently to the inside of the cells.

11.2.2.2 Advanced Spectrum: Effects on Viruses and Fungi. The nature of the mechanism of action, combining many unspecific action levels, makes silver active not only against bacteria but also against other harmful microorganisms, like fungi or viruses. Although not many studies have focused on the inactivation of fungi, it has been found that the antimicrobial activity of silver nanoparticles of about 10 nm was similar to that of itraconazole, exerting a biocidal effect with concentrations of 6.6–13.2 μM [31].

More interesting is, however, the inhibitory effect that silver nanoparticles exert on viruses. Among the various studies, it has been found that silver inhibits respiratory syncticial virus [32], murine norovirus, and bacteriophages like MS-2 [33] and UZ-1 [34]. Several studies have focused on the inactivation of the HIV virus. In this respect, silver nanoparticles exclusively in the range of 1–10 nm were found to complex gp120 glycoprotein knobs, preventing CD4-dependent virion binding and postentry stages [35, 36]. This could pose further opening perspectives for the use of nanosilver in alternative therapies against AIDS and the HIV virus or application for transmission-proof devices.

11.2.2.3 Size and Shape Dependence. The size is a very important aspect considering that thermodynamic properties are directly proportional to the diameter of the nanoparticles. Therefore, it is generally accepted that the antimicrobial efficacy of elemental silver increases with decreasing nanoparticle size. Although antibacterial efficacy has been proven with particles of more than 100 nm [9], reviews set the optimum range between 3 and 10 nm [11]. Apart from the size, particles may vary in surface charge or agglomeration state. Stabilizers or "capping agents" used to prevent coalescence of the nanoparticles usually have a certain surface charge, which will modify the physicochemical properties of the particles. Concerning the shape, particle morphology as investigated by transmission electron microscopy (TEM), high-resolution TEM (HRTEM), or X-ray diffraction (XRD) is predominantly spherical, but other shapes like nanocubes or nanoprisms cannot be overlooked [29, 37, 38]. Pal et al. produced truncated triangular nanoplates and nanorods by the solution-phase method and nanospheres by the seeded growth method. Comparing their efficacy on *E. coli*, they found bactericidal concentrations of about 1, 12.5, and 50–100 ppm for triangular nanoplates, nanospheres, and nanorods, respectively [39]. These results demonstrated that bacterial adhesion of silver is favored by high-atom-density facets such as {111} as proposed previously [6, 40] (Figure 11.2).

11.2.3 Advantages and Limitations of the Use of Silver

11.2.3.1 Stability vs. Efficiency. It is inevitable to realize that if the antibacterial efficacy of silver relies on a mechanism of action so unspecific, it will be subject to many variables. Indeed, silver ions not only bind L-Cys residues of bacterial enzymes but any atom with relatively high electronegativity and high atomic radius. This is reflected in the solubility constants for various compounds shown in Table 11.1. Considering the extreme affinity of silver for sulphides and halides, it is clear antibacterial activity will be strongly influenced by the presence of these ligands in the environment of action.

Liau et al. reported on the interaction of silver with different groups containing sulphur. They found silver antibacterial activity was lost when compounds with free-SH (thiol) groups were present, like in cysteine, glutathione, or thioglicolate, whereas others like sulfates, thiosulfates, taurine, or methionine with S-O or S-C bonds did not substantially alter its efficacy [13]. Any environment with natural organic matter (NOM) can be an important source of thiol-containing ligands, which can fully void or significantly reduce antibacterial efficacy. Efforts to increase the chemical stability of silver ions in this respect must take into account that the mechanism of bactericidal activity relies on this affinity to thiol groups. Therefore, it is crucial to find a compromise between stability and efficiency.

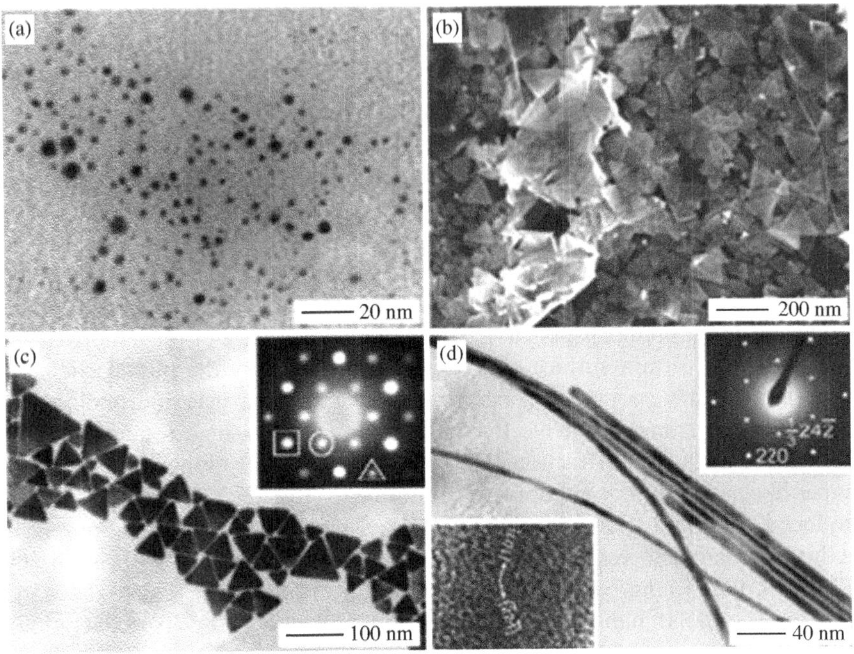

FIGURE 11.2 (a) TEM image of the initial silver nanoparticles with a diameter of ~3.5 nm. (b) Scanning electron microscope (SEM) image of the product that was obtained by refluxing these small silver nanospheres in the presence of PVP and sodium citrate, showing the formation of a mixture of triangular nanoplates (with sharp corners) and nanobelts of silver. Some nanoplates were oriented with their triangular faces perpendicular to the supporting substrate. (c) TEM image of triangular nanoplates self-assembled into a stripe of monolayer, clearly indicating the bimodal sizes of nanoplates. The inset gives a typical electron microdiffraction pattern taken from an individual nanoplate supported on the TEM grid against one of its triangular faces. (d) TEM image of silver nanobelts whose flat faces were oriented parallel to the surface of the supporting substrate. The inset in the lower left demonstrates the nanobelts were single crystals and grew along the [101] direction. The other inset shows a typical SAED pattern taken from an individual nanobelt. (*Source*: Reprinted from Reference 38 with permission from John Wiley & Sons, Inc.)

TABLE 11.1 Silver ion (Ag^+) solubility constants for different selected anions [14, 23]

Anion selected as counterion for silver	Solubility constant (mol/L)
NO_3^-	51.60 (soluble)
SO_4^{2-}	$1.58 \cdot 10^{-5}$
PO_4^{3-}	$2.51 \cdot 10^{-18}$
Cl^-	$1.58 \cdot 10^{-10}$
Br^-	$7.70 \cdot 10^{-13}$
I^-	$1.50 \cdot 10^{-16}$
S^{2-}	$7.94 \cdot 10^{-51}$

Among the halides, which also form very stable complexes with silver, attention must be paid to chloride, as it will be present in practically any substrate susceptible to bacterial contamination. According to the solubility constant, the soluble concentration of silver chloride would be of about 6.3 ppb, above which saturation is reached and the equilibrium is gradually shifted toward silver chloride complexes, a very low percentage of silver ions remaining active in solution [41]. These quantities, although very low, are very similar to bactericidal concentrations (10 ppb) observed for some gram-negative bacteria in synthetic water [26]. Hence, in the presence of chloride, a very small fraction of free silver ions will remain soluble, which would be enough to exert an antibacterial effect. This could explain the reported antimicrobial activity of silver chloride [42, 43]. The existence of an equilibrium between silver chloride complexes and free silver ions further increases the complexity of the system, as new other factors like the ionic strength and presence of common or non-common ions come into play.

In addition, silver has a relative low redox potential, being easily reduced to metal particles in the presence of weak reducing gents, ultraviolet (UV) light, or increased temperature. According to Kim et al., UV reduction could take place even after the silver ions have been inactivated by -SH groups, disrupting the Ag-S bond to form nanoparticles. Exposure to light should therefore be taken into consideration, as it would increase antibacterial efficacy [33]. In the form of elemental silver, the antibacterial efficacy as well as the stability of silver will also be governed by the size, shape, agglomeration state, or other electrostatic forces from the chemical environment affecting the particles.

All these speciation issues make it very difficult to standardize biocidal tests and establish breakpoints for antimicrobial silver or nanosilver, and moreover, they limit the range of possible applications and point out the complexity of the aspects to face when designing a silver-based antimicrobial system.

11.2.3.2 Effects on Human Health: Benefits and Toxicity. When studying the potential toxicity of silver, it is compelling to differentiate among silver, silver ions, or silver nanoparticles. Silver and the ions that naturally leach from it in minute concentrations are not a cause of concern to humans, as the ions are rapidly inactivated by biomolecules. Historical evidence throughout the centuries in contact with it has proven this form of silver to be innocuous. However, if big quantities of silver ions are ingested, like in the form of silver nitrate, deposits of silver sulfides can result in a brownish discoloration of the skin called argyria. This discoloration resolves with cessation of the therapy, remaining a cosmetic concern [15].

However, the possible side effects of the recently discovered silver nanoparticles on human health are mostly unknown and constitute a topic of dispute and concern [2]. In the gastrointestinal tract, silver nanoparticles may encounter radical pH changes and a complex mixture of food, enzymes, electrolytes, and bacterial flora. In addition, constant renewal of the epithelium may prevent accumulation and possible toxicity as well as systemic access of the particles.

Therefore, systemic toxicity of ingested nanoparticles is not to be expected and scarcely seen. In the form of aerosol, however, silver nanoparticles easily reach alveoli and increase the oxidative stress of lung epithelial cells, increasing the recruitment of macrophages. Takenaka et al. found 9–21% of silver inhaled nanoparticles by rats accumulated in the liver, demonstrating systemic distribution and pointing at the liver as a target for circulatory clearance [44]. In the skin, the results on nanosilver toxicity remain disputed. Although many studies confirm good dermal biocompatibility, some other studies have found morphology and proliferation of the keratinocytes is affected by nanoparticles. The molecular basis underlying most reported tissue affectations consists of a decrease in the mitochondrial function and of an increase in oxidative stress of the cells, leading in the worst case to apoptosis and activation of the inflammatory response. This is explained by the interaction of silver with thiol-containing antioxidant compounds and enzymes, like glutathione, superoxide dismutase, or rheodoxine, in charge of neutralizing the ROS produced during mitochondrial activity [2]. These hints on possible toxicity do not threaten the use of nanosilver in the medical field at any rate, where a highly positive benefit–risk balance has enabled its approval and widespread use, but they will question whether its use on other fashion-oriented applications, like in textiles, cosmetics, or as nutrition supplements, are sufficiently justified.

The same remarkable mechanism of action to which some authors attribute toxicity may allow promising benefits to human health to be gathered from the study of silver nanoparticles. As mentioned in Section 11.1.1, since the 19th century, it has been empirically observed that silver application on wounds promotes fast epithelization and scarring. Recently, it has been discovered that silver nanoparticles also exhibit excellent anti-inflammatory properties. The nanoparticles seem to promote selectively apoptosis of damaged and pro-inflammatory cells, by interfering with the electron transport chain and the intrinsic signaling pathway of downstream pro-caspases, without targeting keratinocytes [1]. A consequent decrease in IL-12, TNF-α, and matrix metalloproteinase has been reported [27] that reduces erythema and edema and promotes scarring and epithelization [45]. The overall citokyne regulation and anti-inflammatory effectiveness was shown to be comparable and more rapid than that of steroids [3, 27].

Apart from anti-inflammatory properties, the use of silver nanoparticles poses great expectations in other fields of interest, like in cancer therapy. In the induction of apoptosis, it has been discovered that the nanosilver predominantly targets cancer cells rather than normal functioning cells. Although the mechanism is still unclear, it seems that the action is receptor dependent, as only nanoparticles in the range of 5–20 nm exhibit this activity, and because surfactants reduce the action, probably preventing the union to the receptor [1]. Even without taking into consideration its intrinsical properties, nanosilver could be used a carrier in drug delivery. The release could be triggered by internal (e.g., glutathione) or external (e.g., UV light) stimulus. The nanocarrier would recognize the target cell and be internalized before the drug is released inside the cell.

11.2.3.3 Effects on the Environment. The wide range of action of silver can be seen as a threat to the environment if we consider nonpathogenic bacteria or other susceptible organisms. Environmentally acute toxicity of silver is known to be derived exclusively from the free Ag^+ ion concentration in water [46, 47]. These ions are shown to inhibite Na^+/K^+ ATPase and carbonic anhydrase and, thus, the respiratory functions of many organisms, like bacteria, algae, and other small organisms [48]. Different crustacean, algae, and fish species have been used as markers for environmental toxicity studies, trying to determine how silver complexation to chlorides or to NOM affect the accumulation of silver at the gill of these species or their mortality rates, either *in vivo* or with the help of the biotic ligand model. In general, it has been found that NOM decreases the toxicity of waterborne Ag depending on the rate of uptake, most probably because of complexation [49–51]. The contribution of chlorides to toxicity is disputed probably because bioavailability might be more influenced by the uptake mechanism and the nature of the studied organism in each case than by silver speciation itself in solution [47, 52]. The emergence of silver nanoparticles and their application for antimicrobial purposes increases concern about their toxicity, as they seem to be more toxic than the silver ions themselves [43]. This could be explained in terms of stability, considering silver ions are more reactive and could therefore be inactivated in greater extent before reaching the internal environment.

11.3 OBTENTION AND INCORPORATION OF SILVER INTO COATINGS AND POLYMER MATRICES

11.3.1 Introduction

As the antibacterial activity of silver only occurs when it is in free ionic form (Ag^+) or in the form of elemental silver nanoparticles (nano-Ag^0), the design of a silver-based antimicrobial polymer must be oriented to one of two ways, each posing different challenges.

In the first case, the matrix incorporating the silver ions must chemically hold them while enabling a slow migration to a moistured environment. This has been successfully accomplished with the use of silver zeolites, silver glasses, or silver zirconium phosphate resins. Zeolites are zinc sodium ammonium aluminosilicate minerals with a three-dimensional microporous structure commonly used as commercial adsorbents. In silver zeolites, naturally occurring sodium ions are partially replaced for silver ions by immersing the zeolite in a silver nitrate solution [9]. The substituted zeolites can be then incorporated or coated into a wide variety of polymers and other surfaces. In contact with moisture, silver ions are again substituted by sodium ions present in the release

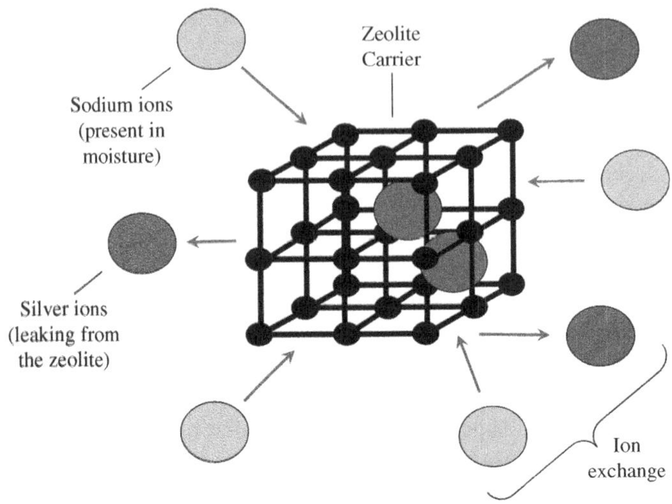

FIGURE 11.3 Schematic representation of the ion-exchange-induced release mechanism in silver zeolites.

environment and sustainably leach from the surface (Figure 11.3). This is practical, as release of silver ions will be dependent on the amount of saline moisture, which is a crucial risk factor for the development of microbes on surfaces. Silver zeolites are used in a large variety of materials in many sectors, being the most widely used silver-based antimicrobial system for consumer products [8]. This same principle of sodium–silver ion exchange applies for the manufacture of silver glasses (silver–magnesium–aluminium–sodium phosphosilicates) and silver zirconium phosphates, also available under many forms in the market.

In the case of incorporating reduced silver in a material, the challenge resides in the production of silver particles at the nanoscale. This step is most critical, as it will govern the size, shape, size distribution, and stability of the nanoparticles, which in turn play a determinant role in the rate of release and antibacterial efficacy. As mentioned, silver has a great advantage against other antimicrobial substances, like peptides, bacteriocins, chitosan, essential oils (thymol, carvacrol, cynnamaldehyde), enzymes, and all other organic substances. Its outstanding thermal and chemical stability allows it to withstand any kind of processing. This stability, of course, is more feasible in the form of elemental silver, as silver ions readily reduce with thermal treatment, exposure to light, or very weak reducing agents. Moreover, the sustained release of nanoparticles can be achieved with matrices less complex than when dealing with silver ions. Therefore, the recent discovery of antibacterial nanosilver has raised many expectations concerning new possible applications and research predominantly focuses on the production of stable silver nanoparticles with controlled size and shape and narrow size distribution. The different approach to this challenging task by researchers deserves special attention and is therefore dealt with in the next subsection.

11.3.2 Preparation of Silver Nanoparticles: Mastering the Size and Shape of Nanosilver

The production of elemental silver particles (nano-Ag^0) can be approached in two different ways. The first one, the so-called top-down technique, consists of mechanically reducing the size of silver metal in its bulk form to the nanoscale [11]. This can be done with the help of specialized methodologies such as nanolithography or laser ablation. These techniques have lower running costs, are easy to automate, and no solvents are used in the process. However, until August 2008, less than 5% of all scientific articles related to silver nanoparticles focused on these methods [11]. This is mainly attributed to the surface imperfection of the formed particles as well as to the difficulties in adapting experimental conditions to shift the shape and the size of the particles, which in turn leads to a deficient release and thus the loss of antimicrobial efficacy. The second and predominantly used approach is the bottom-up or self-assembly technique, which involves the dissolution of a silver salt into a solvent and the subsequent reduction of it. As most studies deal with the latter approach, the rest of the subsection will refer to the particularities of chemical, physical, and biological reduction of silver salts to form nanoparticles; their stabilization; and the influence of this process on the size and shape of the nanoparticles.

Nitrate, although considered a drinking water contaminant, is the most commonly used counteranion for silver in the preparation of nanoparticles from silver salts. This is because of its low cost, chemical stability, and high solubility in water. It seems from studies with other counteranions like acetate, perchlorate, or sulphate that its role is not crucial in the process [11]. Water is the most commonly used solvent, followed by ethanol, dimethylformamide (DMF), or ethyleneglycol (EG), among others, depending on the application [11]. The reduction of the salt is the critical step and can be approached by physical, chemical, or biological means.

11.3.2.1 Physical Reduction. The yield of silver nanoparticles by physical irradiation of a silver salt solution has been demonstrated with UV, gamma, microwave, and ultrasonic irradiation [27, 29]. The size of the particles is dose-dependent so that higher intensities yield smaller particles [27]. On the one hand, the main advantage over chemical or biological methods is easily controlled reduction rates without excess from the reducing agent, oxidation by-products, or other unwished interfering compounds [27]. On the other hand, these techniques are usually associated with low production rates and high expense. Among the most interesting studies, photo-induced conversion of silver nanospheres to nanoprisms with fluorescent light has proven the feasibility of tailoring the size and shape of nanoparticles with irradiation methods that in turn determines the scattering properties, reactivity, or antibacterial effectiveness of the nanoparticles.

11.3.2.2 Chemical Reduction. The chemical reduction is one of the most prevalent methods to synthesize colloidal metal particles, because of its convenient operation and simple equipments needed. Among all chemical reducing agents, sodium borohydride, as a strong reductant, and citrate, as a weak reductant, are most commonly used [11]. Strong reducing agents yield smaller nanoparticles compared with weak reducing agents. However, because of the excess surface energy and high thermodynamic instability, smaller particles rapidly undergo alternating nucleation and Ostwald ripening leading to agglomeration into oligomers, clusters, and eventually precipitates [27]. Therefore, different strategies have been developed to increase the stability of the silver colloids. One approach is a two-step process where a strong reductant, mostly $NaBH_4$, is used to initiate the reaction forming small particles. These particles are in a second step enlarged with a weaker reductant [29]. Other approaches are Tollens or modified Tollens methods. These imply the stabilization of the silver cations with ammonia. According to the basic Tollens method, the $Ag(NH_3)_2$ complex is afterward reduced with an aldehyde. However, research has focused more deeply on so-called "green" synthesis, where the silver ammoniacal solution is reduced by environmentally friendlier reductants like saccharides, ascorbic acids, or polyols [29]. When using weak reductants, the reaction rate can be intensified if the temperature is increased. Whatever the strategy, the stability of the nanoparticles is usually further supported with a stabilizer or "capping agent." The capping agent introduces electrostatic or steric impediments preventing the coalescence of the particles, but it can also play an important role in the size, size distribution, and biocompatibility of the nanomaterials [27]. Among the different compounds used as capping agents, we find mostly surfactants like cetyl trimethyl ammonium bromide (CTAB), sodium bis(2-ethylhexyl)sulfossuccinate (AOT), and triton X-100, which allow for the formation of micelles, vesicles, or microemulsions; polymers like poly (vinyl pirrolidone) (PVP), polyacrylates, polyacrylamides, poly(ethylene glycol) (PEG), or poly(vinyl alcohol) (PVA); and biomolecules like heparin or starch [11, 27, 53]. Many of these not only act as capping agents but also take an active part in the reduction mechanism of silver, like PVP. Refluxing PVP and sodium citrate, Wiley et al. were able to convert spherical nanoparticles into triangular nanoplates and nanobelts, evidencing the importance of capping agents in the tuning of size and shape of the nanoparticles (Figure 11.2) [38]. In combination with heating, some capping agents can reduce the silver cations even in the absence of reducing agents (with a lower yield), like some microemulsions, starch, or heparin.

11.3.2.3 Biological Reduction. The interest in biological synthesis of silver nanoparticles versus physical or chemical methods has been growing in the last few years. Since early studies on biosorption of metals on bacteria inferred that bacterial cells could bind metals at the nanoscale [1], researchers have put forth the capability of a great number of biological substrates to reduce silver. Most studies focus on bacteria, predominantly bacilli, and fungi, but there are also studies with plant extracts from alfalfa, lemongrass, geranium leaves, or cinnamon [1, 29]. The particles are either deposited onto microbial constituents or secreted to the culture supernatant, their size varying from tenths to hundreds of nanometers depending on the substrate. The mechanism suggested for the process implies the enzymatic reduction of silver nitrate by nitrate reductase together with other electron shuttling compounds [1]. Biological methods present an interesting alternative over physical or chemical methods. They usually involve less amount of capital, avoid the use of hazardous chemicals, and provide a biomolecularly rich environment that naturally acts a capping agent. Furthermore, the usually lower kinetics offers the possibility of better manipulation and control over crystal growth [1, 29]. In this sense, genetic engineering techniques could constitute a very useful tool in turning the production of nanosilver into a profitable cost-effective market.

11.3.3 Incorporating Nanosilver: Techniques and Materials

Once the elemental silver nanoparticles are obtained, their high thermal stability enables them to be introduced into polymers by many well-established techniques which assure homogeneous distribution of the nanoparticles on the surface. These include plasma immersion ion implantation (PIII), pulse filtered cathodic vacuum arc deposition, autocatalytic electroless chemical plating (AECP), magnetron sputtering, or physical vapor deposition (PVD). The latter method, used for example in the fabrication of Acticoat (Smith & Nephew, London, UK), allows deposition of a 15-nm-thick sheet of silver nanoparticles homogeneously distributed [54]. AECP has been used to coat silver nanoparticles on silica spheres with adsorbed tin. When exposed to silver ions, tin oxidizes to form and adsorb simultaneously silver nanoparticles [27]. Magnetron sputtering enables tuning of the dissolution kinetics of the particles adjusting the pressure and distance during the treatment [55], but this technique usually lacks adhesion to the substrate [27].

However, taking into account the rising demand for day-to-day antibacterial consumer products, other techniques must be investigated that can provide cost-effective production of nanosilver polymers for applications with less economical impact. Being a simple and relatively efficient process, the reduction of silver once inside the material, so-called self-assembly, could be a more cost-effective solution in the fabrication of nanosilver products.

Although casting techniques experience aggregation and thus inhomogeneous size distribution [29], physical techniques like UV, microwave irradiation, sonochemical processes, or simply applying high temperatures, so-called thermal annealing, are easy to apply with good results in materials like polycarbonate [27], PVP, or PVA [29]. Thermal annealing of silver nanoparticles has been demonstrated in polyacrylonitriles (PAN), polyacrylates, and PVA.

In PAN, the reduction conveniently occurs during *in situ* polymerization [29]. In the sonochemical process, the reduction seems to rely on the transient high temperatures and fast cooling rates during ultrasonic radiation under argon atmosphere. Polyelectrolyte multilayers allow the immobilization of silver ions, silver reduction, and alternative deposition of oppositely charged polymers where particle concentration sand size can be varied with the processing conditions [56]. Chemical *in situ* reduction of silver-based polymers has also been studied in a Tollens process (explained in Section 11.3.2), mostly on PVP. In these cases, size and size distribution can be tuned by varying the stechiometry of silver and ammonia [37]. These processes can be as well combined with UV irradiation or thermal annealing, enabling the chemical treatment to be softened. Typically, the immersion of the polymer in a silver colloidal solution is followed by enlargement of these silver seeds with a weak reductant under UV light [57]. In addition to all these strategies, the creation of functional groups in the polymer can promote or enhance adhesion of the nanoparticles. Polycarbonate was UV-etched and silanized to produce amino groups that allowed the deposition of nanoparticles on the surface [58]. Chromia surfaces have been modified with 3-mercaptopropyl trimethoxy silane. The polyols formed were able to reduce surface-adsorbed silver ions into nanoparticles [59]. Hydroxyl groups in hydroxyapatite are able to bind silver by electron transfer to the ions.

One promising field of interest is the combination of silver with TiO_2. TiO_2 in itself causes photocatalytic inactivation of bacteria (see Chapter 19 for detailed information). Recently, it was discovered that the antimicrobial activity of TiO_2 is greatly enhanced when the material is reduced to the nanoscale, being even able to kill viruses [30]. Furthermore, it has been found that doping TiO_2 surfaces with silver has a synergistic effect on their antimicrobial efficacies. Silver absorption of UV light induces electron transfer to TiO_2 resulting in charge separation and thus activation by visible light, improving the antibacterial effect of TiO_2 [30]. In view of these findings, much interest has developed in formulating silver-titania doped materials or coatings either in the micro- or nanoscale.

11.4 PROPERTIES OF SILVER-BASED POLYMERS

11.4.1 Characterization

Once the particles have been embedded in the polymer, indication of their presence and information about their size, shape, size distribution, and dispersion in the matrix can be obtained from the different characterization techniques available, like images from scanning and transmission electron microscopy (SEM, TEM, and HRTEM), X-ray diffraction patterns (XRD), or UV-absorbtion spectra. XRD patterns of silver nanoparticles, as an example, show characteristic peaks at 38, 44, 64, and 77 a.u. for the crystalline facets {111}, {200}, {220}, and {311}, respectively. Although the intensities of the bands increase with aggregation of the particles, XRD is a useful tool to determine the shape of the particles and to monitor the possible changes in it after a given treatment. This technique revealed that bacterial adhesion of silver is favored by high-atom-density facets such as {111} [6, 55, 60]. Most interesting is the possibility of roughly characterizing the nanoparticles with the help of their UV-visible absorption spectra. When the colloidal particles are smaller than the wavelength of visible light, the solutions acquire a yellow to brownish color with an intense band at the 380-420-nm range. This unique property, known as surface plasmon resonance, is attributed to collective excitation of the electron gas in the particles with a periodic change in electron density at the surface. The band shifts to longer wavelengths with increased particle size [29], accompanied with a change in color. Accordingly, monitoring the appearance, position, and thickness of this band constitutes a very useful tool to predict roughly the yield, size, and size distribution. Pal et al. furthermore observed different plasmon absorbance depending on the shape of the particles (418 nm for triangular, 420 for spherical, and 514 for rod shaped particles), whereas Wiley et al. found 430 nm and 690 nm for spherical and truncated triangular nanoplates, respectively, demonstrating the usefulness of this relatively simple, fast, and easily affordable approach [38, 39].

Thermal properties as investigated by differential scanning calorimetry will, of course, be dependent on the material where the silver is incorporated and the amount of it. In general, alterations resulting from the presence of silver observed in this sense consist mostly of a slight rigidization of the amorphous phase, as noted by the increase in the glass transition temperature (Tg), for example in polyamines, polyolefins, or PVA [61–63]. In the case of PVA, a decrease in crystallinity is also observed. Thermal gravimetrical analysis (TGA) is only affected when with several materials, like cellulose, decrease in stability below 250°C [64], while others remain unaffected throughout the heating process [62]. In any case, with silver having such thermal stability, TGA is mainly used to calculate the yield of incorporation in the material.

Although silver cannot be detected by means of Fourier transformed infrared analysis (FTIR), its presence produces alterations in the normal patterns of the material. Generally speaking, coordination of silver with elements in the polymers stretches the bondlength, usually shifting the frequency to longer wavelengths. This is for example noticeable with imidazole or N-H groups in polyamines, rhodamines, or chitosan [60, 65, 66]. The secondary structure of proteins seems nonetheless unaffected [67].

11.4.2 Release Issues

Apart from the fabrication of the material and its characterization, there are other issues that have to be faced by manufacturers, legislators, researchers, and consumers. As explained in Section 11.2, the antimicrobial efficacy of the material relies on the leaching of silver nanoparticles or silver ions to the surface and surrounding environment. Clinical studies on collagen cuffs [68] and silicon discs [5, 9] incorporating nanosilver have demonstrated that the mere presence of nanoparticles does not assure antimicrobial efficacy. Thus, it is crucial to elucidate the release kinetics of the material in question and to evaluate the equilibrium between reactivity of the particles on bacteria and their stability in a specific environment. These are governed by factors depending on the material and by factors depending on the environment in contact with it.

Concerning the polymer, the release will be mainly dependent on its water-uptake capacity, as release is sorption induced. Kumar and Münstedt published several studies with silver-polyamide nanocomposites measuring silver release by anodic stripping voltammetry for up to three months. They found release kinetics could be modified based on the crystallinity of the polymer. Furthermore, introducing hygroscopic fillers into the matrix or into a multilayer coating greatly enhanced the release kinetics of the nanocomposite and its biocidal efficacy [69–72]. Dowling et al. approached the challenge of tuning release kinetics by incorporating platinum into a coating on polyurethane. As platinum has a higher redox potential, silver oxidation is enhanced when the two metals are in contact, increasing the release of free silver ions from the polymer [73].

The release from the material can also depend on the characteristics of the aqueous or moistured environment in contact with the surface of the material. As an example, tests on human plasma have revealed that a greater amount of ions is released in these conditions than in deionized water [9]. But the role of the release environment in release kinetics is not so important compared with the enormous influence it can exert on the stability of silver ions or nanoparticles. If the material is to show antimicrobial efficacy, the released silver species must be reactive enough as to be able to bind bacterial constituents. At the same time, however, they have to remain stable enough not to be inactivated by biomolecules or organic matter probably present in the release environment. Despite the widespread use of nanosilver-based materials, there is still much to learn about factors affecting silver stability. Although studies focus on the characterization of the silver particles and the release rates from the different materials, bactericidal concentrations prove to be more strongly related to the growth conditions of the bacterial culture. This is easily evidenced when looking at released silver bactericidal concentrations among the different publications. These go from the ppb range when water or salt buffers are used [26, 74, 75] to hundreds of ppm when complex growth media, like LB, TSB, or MHB come into play [20, 25, 60]. These huge differences in antibacterial response (up to four orders of magnitude) put forth the need for standardizing biocidal tests if the potential of different materials throughout the literature is to be compared [24]. This pending task is difficult because of the complexity of silver bioavailability and speciation issues. It is well known that silver ions form very stable complexes with sulphur groups and chlorides. However, silver chloride colloids retain antimicrobial activity [43] and interactions between naturally organic matter (NOM) and human plasma remain unclear [76, 77].

11.5 APPLICATIONS AND FUTURE TRENDS

11.5.1 Regulatory Issues

When looking at regulation of silver, the fact must be taken into consideration that the rediscovery of the use of silver for antimicrobial purposes is relatively new. On the one hand, historical use of silver has proved its benefits and safety for society. On the other hand, the new applications imply new ways of contact with the silver ion. In the case of nanosized silver, there is no previous experience or evidence of its safety or risks for the environment or for human health. Therefore, the topic is differently confronted by the various legislative bodies in Europe, the United States, Japan, or Australia, becoming even contradictory in some cases.

In the United States, and especially in Japan, the use of silver zeolites and silver zirconium phosphate resins is well established with several commercial brands incorporating silver in textiles or as coatings in different consumer products [7, 78] (see the list of websites at the end of this chapter). The silver content in these materials can be up to a maximum of 3% of their formulation. In the food area, only silver nitrate is regulated with a maximum limit of 0.017 mg/kg in foodstuffs and 0.1 mg/kg for drinking waters [79]. As far as nanosilver is concerned, colloidal solutions are accepted in the United States and commercialized as nutrition supplements (e.g., Mesosilver), claiming to have important benefits on human health (www.mesosilver.com). In the medical field, different dressings containing nanocrystalline silver or silver ion releasing systems are widely spread as well as indwelling devices, like prostheses or catheters.

The confrontation between past and present circumstances is best noticed in European law. Medical devices and dressings containing nanosilver are not widespread, and the council of Europe does not recommend silver for medicinal use, because of a lack of sufficient information about health risk assessment. The European Food Safety Authority (EFSA) provisionally accepts the use of silver in food-contact materials with a maximum of 5% silver in the form of silver zeolites or silver zirconium phosphate glasses until an official evaluation of their safety is stated. Migration is restricted to a maximum 0.05 mg/kg in food for the whole group [80]. Silver hydrosols are not included in the list of food additives or supplements "because of lack of appropriate information about silver bioavailability" from them [81]. Paradoxically, under Directive 94/36/EC, still in force, silver is considered a coloring agent used in confectionery without any restriction limits [82]. Other countries, like Australia, have totally banned the use of silver in foodstuffs or food-contact materials.

11.5.2 Current Applications

11.5.2.1 Medical Devices. As discussed in Section 11.1.2, although nanosilver is being marketed for countless applications, the driving force in the innovative search for new silver-based antimicrobial nanocomposites was and still relies mainly on the medical and health-care sector. Half of all nosocomial infections in hospitals are caused as a consequence of the implantation of an indwelling device. Typical antibiotic therapies are ineffective against biofilm formation and promote fast development of resistances [5]. The high demand for medical devices with antimicrobial properties and the high economical impact of these products promote research attention in this sense. Silver products used to treat infection can be classified into two categories depending on the presence of either silver ions (Ag^{+}) or silver nanoparticles (Ag^{0}). Initially, catheters and fixation pins were coated with metallic silver with little or no success, as silver ions were not sufficiently released from the materials [5]. The incorporation of silver sulfadiazine (a silver nitrate and sulfamide combination historically used for the treatment of burns) in hydrogels, alginates, or foam formulations has enabled the rise of different wound dressings (Silvercell, Urgotul, Silverex, etc.) or catheters (Bardex Ic) releasing silver ions. On the second category, different dressings have been embedded or coated with nanocrystalline silver, e.g., Actisorb (charcoal), Aquacel-Ag (carboxymethyl cellulose), SilverIon, Contreet (polyurethane foam), or Acticoat (Polyethylene) [3, 83, 84]. Nanocrystalline silver dressings prove to be generally more effective than silver ion release dressings [83]. Furthermore, bone cement (PMMA) and hydroxyapatite doped with nanosilver have shown good antibacterial efficacy without compromising ostheoblasts and epithelial cells [85].

11.5.2.2 Consumer Products. The outstanding success of nanosilver in the medical field has been the driving force for its implantation in other consumer product markets, where silver ion releasing technologies, mainly in the form of silver zeolites, have already colonized many sectors. Commercial examples of silver-substituted zeolites are Zeomic, Apacider (Sangi Co), Croslite Ag+, Bactekiller, RepelaCOAT, D2P, Novaron, SPE, Biomaster (Addmaster), Irgaguard (Ciba Specialty Chemicals), AgIon, Biocote, or Zeargol [78, 86] (see also the list of websites at the end of this chapter). The combination of low migration rates with a high melting point makes them able to withstand any kind of plastic processing or operating temperature as well as indoor and outdoor conditions, in contrast to all other antimicrobial natural or synthetic substances, including triclosan [86]. The zeolites are usually manufactured as a 3–6-μm-thick layer with 2–5% silver content. This layer is then coated on polymeric or stainless steel surfaces, preventing any biofilm formation on the treated surfaces. One of the most typical and worldwide applications is the coating of inner liners in household appliances, mostly refrigerators, but also washing machines, dishwashers, microwave ovens, tiles, knobs and handles, and so on. Food contact surfaces are another usual application, with silver being the most commonly used antimicrobial additive in the food industry. It can be applied on practically any food processing equipment: cutting boards, counter tops, conveyors, cutlery, containers, or other machines (see the list of websites at the end of this chapter). Silver zeolites can also be incorporated inside the polymer in the manufacture of, for example, antibacterial textiles. In this respect, socks and T-shirts with silver zeolites have been much hyped by AgIon technologies, claiming to eliminate body odor from sweating (www.stinkatnothing.com). Finally, we find silver zeolites in hygiene products, like in deodorants (Nivea Silver Protect), toothbrushes, or toothpaste [87, 88]. Apart from silver zeolites, silver zirconium phosphates, under commercial brands like Alphasan (Milliken Chemical), Forfargol, among others obey the same principle as the moisture-induced, silver-sodium exchange mechanism and are manufactured likewise and for the same applications (www.alphasan.com).

11.5.3 Future Trends

As discussed in the previous section, silver zeolites and resins containing and releasing silver ions have newly hit the market and experience an upward trend, diversifying into more and more applications. For researchers, the challenge now lies in the mastering of nanosilver, which seems to offer better efficacy, stability, and cost-effectiveness and thus opens new expectations for the range of applicability. When reviewing scientific literature concerning silver nanoparticles, two clear goals seem to predominate. First, there is the effort to discover the method to synthesize silver nanoparticles that best matches in size, stability, simplicity, and cost-effectiveness. In this respect, chemical reduction methods still predominate in number, but green synthesis has newly emerged and attracted much attention. Second, there is the search for new materials with nanosilver technology. Table 11.2 [89–120] presents a classification of scientific articles according to the material chosen for incorporation of the nanoparticles. The table is based on an electronic search conducted on the scopus database with the keywords "silver nanoparticles" and including "polymer, plastic or nanocomposite" in the abstract. According to the table, the most commonly used polymers are polyacrylamides and polyacrylates. The reason why these are mostly selected is probably the simplicity in the preparation of these silver hydrogels and the control over the reduction process. Cellulose and cotton are materials in which a silver ion solution can be easily absorbed to be subsequently reduced to silver nanoparticles, being third and fourth in predominance. The group of polyamides is second in importance probably because it includes polymers that could be used for prostheses or implants as well as for textile applications. Polymers typically used for packaging applications, like polyolefins, do not have great representation. Among these publications, we also find combinations of silver with other additives like carbon nanotubes, genipin, antibiotics, activated carbon, and others, but the most used combination is silver-TiO_2 nanocomposites.

The table illustrates that the process of developing new nanosilver-based materials is still in its early stage. The search for new materials not only implies the mastering of the nanoparticle procuration (size, size distribution, and stability), but it should also take into account all issues affecting the release and stability of silver nanoparticles. These include the sizes, shapes, and distribution of the particles, the specific material in which they are incorporated, the chemical environment where the material has to exert its effect, and even the conditions to which the material will be exposed (humidity, light, redox

TABLE 11.2 Relative frequency of scientific articles dealing with silver-based nanocomposites according to the material chosen for incorporation (only the three most recent references are cited)

Materials	Prevalence	References
Polyacrylates/polyacrylamides	11.8%	[89–91]
Polyamides	8.8%	[64, 92–93]
Cellulose	8.1%	[94–96]
Cotton	5.9%	[97–99]
Chitosan	5.1%	[100–102]
PE/PEO	5.1%	[55, 103, 104]
Gelatines/agars/alginates	4.4%	[105–107]
Glass	4.4%	[108–110]
Ceramics/polyphosphates	2.9%	[111–113]
Clays	2.9%	[114–116]
PU	2.9%	[91, 117–118]
PS	2.9%	[62, 66, 119]
PVA	2.9%	[63, 116, 120]
Others	31.9%	

reactants, etc.). The combination of all these factors will govern the antibacterial efficacy on the surface of the material as well as the release to the environment of particles or free cations and its reactivity.

Nanosilver has already established itself in the health-care markets and seems to be expanding to the consumer products markets. It is difficult to estimate to what extent the population is already exposed to silver. Some authors state that there are more than 100 consumer products with nanosilver [8]; therefore, the high demand for these products is evident. Considering health and environmental risks are still uncertain, it is difficult to predict how legislators and the public opinion will confront the introduction of nanosilver in their lives. Future findings will help ascertain whether this problem will derive in an expansion or a limitation of silver applications.

REFERENCES

1. Vaidyanathan, R., Kalishwaralala, K., Gopalrama, S., Gurunathan, S. (2009). Nanosilver: The burgeoning therapeutic molecule and its green synthesis. *Biotechnology Advances*, *27*(6), 924–937.
2. Chen, X., Schluesener, H. J. (2008). Nanosilver: A nanoproduct in medical application. *Toxicology Letters*, *176*(1), 1–12.
3. Bhattacharya, R., Mukherjee P. (2008). Biological properties of "naked" metal nanoparticles. *Advanced Drug Delivery Reviews*, *60*(11), 1289–1306.
4. Klasen, H. J. (2000). Historical review of the use of silver in the treatment of burns. Early uses. *Burns*, *26*(2), 117–130.

5. Hetrick, E. M., Schoenfisch, M. H. (2006). Reducing implant-related infections: Active release strategies. *Chemical Society Reviews*, *35*(9), 780–789.
6. Morones, J. R., Elechiguerra, J. L., Camacho, A., Holt, K., Kouri, J. B., Ramírez, J. T., Yacaman, M. J. (2005). The bactericidal effect of silver nanoparticles. *Nanotechnology*, *16*(10), 2346–2353.
7. Appendini, P. Hotchkiss, J. H. (2002). Review of antimicrobial food packaging. *Innovative Food Science and Emerging Technologies*, *3*(2), 113–126.
8. Maynard, A. D. (2007). Nanoparticle safety: A perspective from the United States. *Issues in Environmental Science and Technology 24*, pp. 118–131.
9. Rai, M., Yadav, A., Gade, A. (2009). Silver nanoparticles as a new generation of antimicrobials. *Biotechnology Advances*, *27*(1), 76–83.
10. Damm, C., Münstedt, H., Rösch, A. (2007). Long-term antimicrobial polyamide 6/silver-nanocomposites. *Journal of Materials Science*, *42*(15), 6067–6073.
11. Tolaymat, T. M., El Badawy, A. M., Genaidy, A., Scheckel, K. G., Luxton, T. P., Suidan, M. (2010). An evidence-based environmental perspective of manufactured silver nanoparticle in syntheses and applications: A systematic review and critical appraisal of peer-reviewed scientific papers. *Science of the Total Environment*, *408*(5), 999–1006.
12. Gupta, A., Silver, S. (1998). Silver as a biocide: Will resistance become a problem? *Nature Biotechnology*, *16*(10), 888.
13. Liau, S. Y., Read, D. C., Pugh, W. J., Furr, J. R., Russell, A. D. (1997). Interaction of silver nitrate with readily identifiable groups: Relationship to the antibacterial action of silver ions. *Letters in Applied Microbiology*, *25*(4), 279–283.
14. Choi, O., Clevenger, T. E., Deng, B., Surampalli, R. Y., Ross, Jr. L., Hua, Z. (2009). Role of sulfide and ligand strength in controlling nanosilver toxicity. *Water Research*, *43*(7), 1879–1886.
15. Russell, A. D., Hugo, W. B. (1994). Antimicrobial activity and action of silver. *Progress in Medicinal Chemistry*, *31*, 351–370.
16. Feng, Q. L., Wu, J., Chen, G. Q., Cui, F. Z., Kim, T. N., Kim, J. O. (2000). A mechanistic study of the antibacterial effect of silver ions on Escherichia coli and Staphylococcus aureus. *Journal of Biomedical Materials Research*, *52*(4), 662–668.
17. Guggenbichler, J. P., Böswald, M., Lugauer, S., Krall, T. (1999). A new technology of microdispersed silver in polyurethane induces antimicrobial activity in central venous catheters. *Infection*, *27*(Suppl 1).
18. Lok, C. N., Ho, C. M., Chen, R., He, Q. Y., Yu, W. Y., Sun, H. Z., Tam, P. K. H., Chiu, J. F. (2006). Proteomic analysis of the mode of antibacterial action of silver nanoparticles. *Journal of Proteome Research*, *5*(4), 916–924.
19. Semeykina, A. L., Skulachev, V. P. (1990). *Submicromolar* Ag^+ increases passive Na^+ permeability and inhibits the respiration-supported formation of Na^+ gradient in Bacillus FTU vesicles. *FEBS Letters*, *269*(1), 69–72.
20. Sondi, I., Salopek-Sondi, B. (2004). Silver nanoparticles as antimicrobial agent: A case study on E. coli as a model for Gram-negative bacteria. *Journal of Colloid and Interface Science*, *275*(1), 177–182.
21. Jung, R., Kim, Y., Kim, H.-S., Jin, H.-J. (2009). Antimicrobial properties of hydrated cellulose membranes with silver nanoparticles. *Journal of Biomaterials Science, Polymer Edition*, *20*(3), 311–324.

22. Park, H. J., Kim, J. Y., Kim, J., Lee, J.-H., Hahn, J.-S., Gu, M. B., Yoon, J. (2009). Silver-ion-mediated reactive oxygen species generation affecting bactericidal activity. *Water Research*, *43*(4), 1027–1032.
23. Silver, S. (2003). Bacterial silver resistance: Molecular biology and uses and misuses of silver compounds. *FEMS Microbiology Reviews*, *27*(2–3), 341–353.
24. Chopra, I. (2007). The increasing use of silver-based products as antimicrobial agents: A useful development or a cause for concern? *Journal of Antimicrobial Chemotherapy*, *59*(4), 587–590.
25. Ruparelia, J. P., Chatterjee, A. K., Duttagupta, S. P., Mukherji, S. (2008). Strain specificity in antimicrobial activity of silver and copper nanoparticles. *Acta Biomaterialia*, *4*(3), 707–716.
26. Hwang, M. G., Katayama, H., Ohgaki, S. (2007). Inactivation of Legionella pneumophila and Pseudomonas aeruginosa: Evaluation of the bactericidal ability of silver cations. *Water Research*, *41*(18), 4097–4104.
27. Nair, L. S., Laurencin, C. T. (2007). Silver nanoparticles: Synthesis and therapeutic applications. *Journal of Biomedical Nanotechnology*, *3*(4), 301–316.
28. Li, W. R., Xie, X. B., Shi, Q. S., Zeng, H. Y., Ou-Yang, Y. S., Chen, Y. B. (2010). Antibacterial activity and mechanism of silver nanoparticles on Escherichia coli. *Applied Microbiology and Biotechnology*, *85*(4), 1115–1122.
29. Sharma, V. K., Yngard, R. A., Lin, Y. (2009). Silver nanoparticles: Green synthesis and their antimicrobial activities. *Advances in Colloid and Interface Science*, *145* (1–2), 83–96.
30. Li, Q., Mahendra, S., Lyon, D. Y., Brunet, L., Liga, M. V., Li, D., Alvarez, P. J. J. (2008). Antimicrobial nanomaterials for water disinfection and microbial control: Potential applications and implications. *Water Research*, *42*(18), 4591–4602.
31. Kim, J. S., Kuk, E., Yu, K. N., Kim, J. H., Park, S. J., Lee, H. J., Kim, S. H., Park, Y. K., Park, Y. H., Hwang, C. Y., Kim, Y. K., Lee, Y. S., Jeong, D. H., Cho, M. H. (2007). Antimicrobial effects of silver nanoparticles. *Nanomedicine: Nanotechnology, Biology, and Medicine*, *3*(1), 95–101.
32. Sun, L., Singh, A. K., Vig, K., Pillai, S. R., Singh, S. R. (2008). Silver nanoparticles inhibit replication of respiratory syncytial virus. *Journal of Biomedical Nanotechnology*, *4*(2), 149–158.
33. Kim, J. Y., Lee, C., Cho, M., Yoon, J. (2008). Enhanced inactivation of E. coli and MS-2 phage by silver ions combined with UV-A and visible light irradiation. *Water Research*, *42*(1–2), 356–362.
34. De Gusseme, B., Sintubin, L., Baert, L., Thibo, E., Hennebel, T., Vermeulen, G., Uyttendaele, M., Verstraete, W., Boon, N. (2010). Biogenic silver for disinfection of water contaminated with viruses. *Applied and Environmental Microbiology*, *76*(4), 1082–1087.
35. Lara, H. H., Ayala-Nuñez, N. V., Ixtepan-Turrent, L., Rodriguez-Padilla, C. (2010). Mode of antiviral action of silver nanoparticles against HIV-1. *Journal of Nanobiotechnology*, *8*, 1.
36. Elechiguerra, J. L., Burt, J. L., Morones, J. R., Camacho-Bragado, A., Gao, X., Lara, H. H., Yacaman, M. J. (2005). Interaction of silver nanoparticles with HIV-1. *Journal of Nanobiotechnology*, *3*, 6.

37. Wang, H., Qiao, X., Chen, J., Wang, X., Ding, S. (2005). Mechanisms of PVP in the preparation of silver nanoparticles. *Materials Chemistry and Physics, 94*(2–3), 449–453.
38. Wiley, B., Sun, Y., Mayers, B., Xia, Y. (2005). Shape-controlled synthesis of metal nanostructures: The case of silver. *Chemistry—A European Journal, 11*(2), 454–463.
39. Pal, S., Tak, Y. K., Song, J. M. (2007). Does the antibacterial activity of silver nanoparticles depend on the shape of the nanoparticle? A study of the gram-negative bacterium Escherichia coli. *Applied and Environmental Microbiology, 73*(6), 1712–1720.
40. Chen, Q., Yue, L., Xie, F., Zhou, M., Fu, Y., Zhang, Y., Weng, J. (2008). Preferential facet of nanocrystalline silver embedded in polyethylene oxide nanocomposite and its antibiotic behaviors. *Journal of Physical Chemistry C, 112*(27), 10004–10007.
41. Ward, T. J., Kramer, J. R. (2002). Silver speciation during chronic toxicity tests with the mysid, Americamysis bahia. *Comparative Biochemistry and Physiology—C Toxicology and Pharmacology, 133*(1–2), 75–86.
42. Hu, W., Chen, S., Li, X., Shi, S., Shen, W., Zhang, X., Wang, H. (2009). In situ synthesis of silver chloride nanoparticles into bacterial cellulose membranes. *Materials Science and Engineering C, 29*(4), 1216–1219.
43. Choi, O., Deng, K. K., Kim, N. J., Ross, L. Jr., Surampalli, R. Y., Hu, Z. (2008). The inhibitory effects of silver nanoparticles, silver ions, and silver chloride colloids on microbial growth. *Water Research, 42*(12), 3066–3074.
44. Takenaka, S., Karg, E., Roth, C., Schulz, H., Ziesenis, A., Heinzmann, U., Schramel, P., Heyder, J. (2001). Pulmonary and systemic distribution of inhaled ultrafine silver particles in rats. *Environmental Health Perspectives, 109*(Suppl 4), 547–551.
45. Tian, J., Wong, K. K. Y., Ho, C.-M., Lok, C.-N., Yu, W.-Y., Che, C.-M., Chiu, J.-F., Tam, P. K. H. (2007). Topical delivery of silver nanoparticles promotes wound healing. *ChemMedChem, 2*(1), 129–136.
46. Peng, H., Brooks, B. W., Chan, R., Chyan, O., La Point, T. W. (2002). Response of Ceriodaphnia dubia to ionic silver: Discrepancies among model predictions, measured concentrations and mortality. *Chemosphere, 46*(7), 1141–1146.
47. Reinfelder, J. R., Chang, S. I. (1999). Speciation and microalgal bioavailability of inorganic silver. *Environmental Science and Technology, 33*(11), 1860–1863.
48. Bielmyer, G. K., Brix, K. V., Grosell, M. (2008). Is Cl^- protection against silver toxicity due to chemical speciation? *Aquatic Toxicology, 87*(2), 81–87.
49. Glover, C. N., Sharma, S. K., Wood, C. M. (2005). Heterogeneity in physicochemical properties explains differences in silver toxicity amelioration by natural organic matter to Daphnia magna. *Environmental Toxicology and Chemistry, 24*(11), 2941–2947.
50. Kolts, J. M., Boese, C. J., Meyer, J. S. (2006). Acute toxicity of copper and silver to Ceriodaphnia dubia in the presence of food. *Environmental Toxicology and Chemistry, 25*(7), 1831–1835.
51. Nichols, J. W., Brown, S., Wood, C. M., Walsh, P. J., Playle, R. C. (2006). Influence of salinity and organic matter on silver accumulation in Gulf toadfish (Opsanus beta). *Aquatic Toxicology, 78*(3), 253–261.

52. Lee, D. Y., Fortin, C., Campbell, P. G. C. (2005). Contrasting effects of chloride on the toxicity of silver to two green algae, Pseudokirchneriella subcapitata and Chlamydomonas reinhardtii. *Aquatic Toxicology*, *75*(2), 127–135.

53. Wang, B., Tian, C., Zheng, C., Wang, L., Fu, H. (2009). A simple and large-scale strategy for the preparation of Ag nanoparticles supported on resin-derived carbon and their antibacterial properties. *Nanotechnology*, *20*(2), art. no. 025603.

54. Dunn, K., Edwards-Jones, V. (2004). The role of Acticoat™ with nanocrystalline silver in the management of burns. *Burns*, *30*(Suppl. 1), pp. S1–S9.

55. Chen, Q., Zhou, M., Fu, Y., Weng, J., Zhang, Y., Yue, L., Xie, F., Huo, C. (2008). Magnetron sputtering synthesis silver and organic PEO nanocomposite. *Surface and Coatings Technology*, *202*(22–23), 5576–5578.

56. Shi, X., Shen, M., Möhwald, H. (2004). Polyelectrolyte multilayer nanoreactors toward the synthesis of diverse nanostructured materials. *Progress in Polymer Science (Oxford)*, *29*(10), 987–1019.

57. Jia, H., Zeng, J., Song, W., An, J., Zhao, B. (2006). Preparation of silver nanoparticles by photo-reduction for surface-enhanced Raman scattering. *Thin Solid Films*, *496*(2), 281–287.

58. Aslan, K., Holley, P., Geddes, C. D. (2006). Metal-enhanced fluorescence from silver nanoparticle-deposited polycarbonate substrates. *Journal of Materials Chemistry*, *16*(27), 2846–2852.

59. Kim, D.W. , Lee, J. -M., Lee, J. -J., Kang, P. -Y., Kim, Y. C., Oh, S. -G. (2007). Formation and immobilization of silver nanoparticles onto chromia surface by novel preparation route involving polyol process. *Surface and Coatings Technology*, *201*(18), 7663–7667.

60. Thomas, V., Yallapu, M. M., Sreedhar, B., Bajpai, S. K. (2007). Versatile strategy to fabricate hydrogel-silver nanocomposites and investigation of their antimicrobial activity. *Journal of Colloid and Interface Science*, *315*(1), 389–395.

61. Mahapatra, S. S., Karak, N. (2008). Silver nanoparticle in hyperbranched polyamine: Synthesis, characterization and antibacterial activity. *Materials Chemistry and Physics*, *112*(3), 1114–1119.

62. Dutra, R. L., Segala, K., De Oliveira, E. M. N., De Souza, E. P., Rossi, L. M., Matos, J. R., Noda, L. K., Paula, M. M. S., Franco, C. V. (2008). Preparation and characterization of the novels terpolymers of poly-{trans-[$RuCl_2(vpy)_4$]-styrene-divinylbenzene} and styrene-divinylbenzene-vinylpiridine impregnated with silver nanoparticles. *Polymer Bulletin*, *60*(6), 809–819.

63. Galya, T., Sedlařík, V., Kuřitka, I., Novotný, R., Sedlaříková, J., Sáha, P. (2008). Antibacterial poly(vinyl alcohol) film containing silver nanoparticles: preparation and characterization. *Journal of Applied Polymer Science*, *110*(5), 3178–3185.

64. Park, S. W., Bae, H. -S., Xing, Z. -C., Kwon, O. H., Huh, M. -W., Kang, I. -K. (2009). Preparation and properties of silver-containing nylon 6 nanofibers formed by electrospinning. *Journal of Applied Polymer Science*, *112*(4), 2320–2326.

65. Kong, H., Song, J., Jang, J. (2009). One-step preparation of antimicrobial polyrhodanine nanotubes with silver nanoparticles. *Macromolecular Rapid Communications*, *30*(15), 1350–1355.

66. An, J., Wang, D., Luo, Q., Yuan, X. (2009). Antimicrobial active silver nanoparticles and silver/polystyrene core-shell nanoparticles prepared in room-temperature ionic liquid. *Materials Science and Engineering C*, *29*(6), 1984–1989.

67. Fayaz, A. M., Balaji, K., Girilal, M., Kalaichelvan, P. T., Venkatesan, R. (2009). Mycobased synthesis of silver nanoparticles and their incorporation into sodium alginate films for vegetable and fruit preservation. *Journal of Agricultural and Food Chemistry*, *57*(14), 6246–6252.

68. Pascual, A. (2002). Pathogenesis of cathether-related infections: Lessons for new designs. *Clinical Microbiology and Infection*, *8*(5), 256–264.

69. Kumar, R., Münstedt, H. (2005). Silver ion release from antimicrobial polyamide/silver composites. *Biomaterials*, *26*(14), 2081–2088.

70. Kumar, R., Münstedt, H. (2005). Polyamide/silver animicrobials: Effect of crystallinity on the silver ion release. *Polymer International*, *54*(8), 1180–1186.

71. Kumar, R., Howdle, S., Münstedt, H. (2005). Polyamide/silver antimicrobials: Effect of filler types on the silver ion release. *Journal of Biomedical Materials Research—Part B Applied Biomaterials*, *75*(2), 311–319.

72. Damm, C., Münstedt, H. (2008). Kinetic aspects of the silver ion release from antimicrobial polyamide/silver nanocomposites. *Applied Physics A: Materials Science and Processing*, *91*(3), 479–486.

73. Dowling, D. P., Betts, A. J., Pope, C., McConnell, M. L., Eloy R., Arnaud, M. N. (2003). Anti-bacterial silver coatings exhibiting enhanced activity through the addition of platinum. *Surface and Coatings Technology*, *163–164*, 637–640.

74. Kim, T. N., Feng, Q. L., Kim, J. O., Wu, J., Wang, H., Chen, G. C., Cui, F. Z. (1998). Antimicrobial effects of metal ions (Ag^+, Cu^{2+}, Zn^{2+}) in hydroxyapatite. *Journal of Materials Science: Materials in Medicine*, *9*(3), 129–134.

75. Bjarnsholt, T., Kirketerp-Møller, K., Kristiansen, S., Phipps, R., Nielsen, A. K., Jensen, P. Ø., Høiby, N., Givskov, M. (2007). Silver against Pseudomonas aeruginosa biofilms. *APMIS*, *115*(8), 921–928.

76. Jain, J., Arora, S., Rajwade, J. M., Omray, P., Khandelwal, S., Paknikar, K. M. (2009). Silver nanoparticles in therapeutics: Development of an antimicrobial gel formulation for topical use. *Molecular Pharmaceutics*, *6*(5), 1388–1401.

77. Schierholz, J. M., Lucas, L. J., Rump, A., Pulverer, G. (1998). Efficacy of silver-coated medical devices. *Journal of Hospital Infection*, *40*(4), 257–262.

78. Quintavalla, S., Vicini, L. (2002). Antimicrobial food packaging in meat industry. *Meat Science*, *62*(3), 373–380.

79. FDA/CFSAN, U. (2010). Listing of food additive status: Silver nitrate-172.167, U. FDA/CFSAN, Editor.

80. EFSA. (2006). 1–21 Opinion of the Scientific Panel on food additives, flavourings, processing aids and materials in contact with food (AFC) on a request related to a 12th list of substances for food contact materials. *The EFSA Journal*, 395–401.

81. EFSA. (2008). Scientific Statement of the Panel on Food Additives and Nutrient Sources added to Food (ANS): Question n EFSA-Q-2005–169. *The EFSA Journal*, (12).

82. European Parliament and Council Directive 94/36/EC of 30 June 1994 on colours for use in foodstuffs.

83. Ip, M., Lui, S. L., Poon, V. K. M., Lung, I., Burd, A. (2006). Antimicrobial activities of silver dressings: An in vitro comparison. *Journal of Medical Microbiology*, *55*(1), 59–63.

84. Lo, S. F., Hayter, M., Chang, C.-J., Hu, W.-Y., Lee, L.-L. (2008). A systematic review of silver-releasing dressings in the management of infected chronic wounds. *Journal of Clinical Nursing*, *17*(15), 1973–1985.

85. Zhao, L., Chu, P. K., Zhang, Y., Wu, Z. (2009). Antibacterial coatings on titanium implants. *Journal of Biomedical Materials Research—Part B Applied Biomaterials*, *91*(1), 470–480.
86. Simpson, K. (2003). Using silver to fight microbial attack. *Plastics, Additives and Compounding*, *5*(5), 32–35.
87. Nivea Spray Silver Protect. (2010). [cited 2010 10/04/2010]; Available from: http://www.niveaformen.es/productos/silver_protect_range.html.
88. Silver Care Plus Self Sanitizing Toothbrush. (2010). Available from: www.dentist.net/silver-care-plus.asp; accessed June 15, 2011.
89. Häntzschel, N., Hund, R.-D., Hund, H., Schrinner, M., Lück, C., Pich, A. (2009). Hybrid microgels with antibacterial properties. *Macromolecular Bioscience*, *9*(5), 444–449.
90. Mohan, Y. M., Vimala, K., Varsha, T., Varaprasad, K., Sreedhar, B., Bajpai, S. K., Raju, K. M. (2010). Controlling of silver nanoparticles structure by hydrogel networks. *Journal of Colloid and Interface Science*, *342*(1), 73–82.
91. Thimma, R. T., Arihiro, K., Atsushi, M., Atsushi, T. (2009). Novel silver-loaded semi-interpenetrating polymer network gel films with antibacterial activity. *Journal of Polymer Science, Part A: Polymer Chemistry*, *47*(19), 4950–4962.
92. Dong, H., Wang, D., Sun, G., Hinestroza, J. P. (2008). Assembly of metal nanoparticles on electrospun nylon 6 nanofibers by control of interfacial hydrogen-bonding interactions. *Chemistry of Materials*, *20*(21), 6627–6632.
93. Rangari, V. K., Mohammad, G. M., Jeelani, S. (2008). Fabrication of nylon nanocomposite fibers for antimicrobial textile applications. *Nanotechnology*, *21*, 745–749.
94. Tankhiwale, R., Bajpai, S. K. (2009). Graft copolymerization onto cellulose-based filter paper and its further development as silver nanoparticles loaded antibacterial food-packaging material. *Colloids and Surfaces B: Biointerfaces*, *69*(2), 164–168.
95. Pinto, R. J. B., Marques, P. A. A. P., Neto, C. P., Trindade, T., Daina, S., Sadoccoc, P. (2009). Antibacterial activity of nanocomposites of silver and bacterial or vegetable cellulosic fibers. *Acta Biomaterialia*, *5*(6), 2279–2289.
96. Zhu, C., Xue, J., He, J. (2009). Controlled in-situ synthesis of silver nanoparticles in natural cellulose fibers toward highly efficient antimicrobial materials. *Journal of Nanoscience and Nanotechnology*, *9*(5), 3067–3074.
97. Ilić, V., Šaponjić, Z., Vodnik, V., Potkonjak, B., Jovančić, P., Nedeljković, J., Radetić, M. (2009). The influence of silver content on antimicrobial activity and color of cotton fabrics functionalized with Ag nanoparticles. *Carbohydrate Polymers*, *78*(3), 564–569.
98. Shateri Khalil-Abad, M., Yazdanshenas, M. E., Nateghi, M. R. (2009). Effect of cationization on adsorption of silver nanoparticles on cotton surfaces and its antibacterial activity. *Cellulose*, *16*(6), 1147–1157.
99. Yazdanshenas, M. E., Shateri Khalilabad, M., Nateghi, M. R. (2009). A new method based on exhaustion for immobilization of silver nanoparticles on the surface of cotton fabrics. *Asian Journal of Chemistry*, *21*(7), 5741–5748.
100. Yoksan, R., Chirachanchai, S. (2009). Silver nanoparticles dispersing in chitosan solution: Preparation by γ-ray irradiation and their antimicrobial activities. *Materials Chemistry and Physics*, *115*(1), 296–302.

101. Wei, D., Sun, W., Qian, W., Ye, Y., Ma, X. (2009). The synthesis of chitosan-based silver nanoparticles and their antibacterial activity. *Carbohydrate Research, 344*(17), 2375–2382.

102. Vimala, K., Mohan, M., Sivudu, K., Samba, Varaprasad, K., Ravindra, S., Reddy, N., Narayana, Padma, Y., Sreedhar, B., MohanaRaju, K. (2010). Fabrication of porous chitosan films impregnated with silver nanoparticles: A facile approach for superior antibacterial application. *Colloids and Surfaces B: Biointerfaces, 76*(1), 248–258.

103. An, J., Zhang, H., Zhang, J., Zhao, Y., Yuan, X. (2009). Preparation and antibacterial activity of electrospun chitosan/poly(ethylene oxide) membranes containing silver nanoparticles. *Colloid and Polymer Science*, 1–10.

104. Sánchez-Valdes, S., Ortega-Ortiz, H., Ramos-de Valle, L. F., Medellín-Rodríguez, F. J., Guedea-Miranda, R. (2009). Mechanical and antimicrobial properties of multilayer films with a polyethylene/silver nanocomposite layer. *Journal of Applied Polymer Science, 111*(2), 953–962.

105. Ayyad, O., Muñoz-Rojas, D., Oró-Solé, J., Gómez-Romero, P. (2010). From silver nanoparticles to nanostructures through matrix chemistry. *Journal of Nanoparticle Research, 12*(1), 337–345.

106. Rattanaruengsrikul, V., Pimpha, N., Supaphol, P. (2009). Development of gelatin hydrogel pads as antibacterial wound dressings. *Macromolecular Bioscience, 9*(10), 1004–1015.

107. Tofoleanu, F., Balau, M. T., Brinza, F., Sulitanu, N., Sandu, I. G., Raileanu, D., Floristean, V., Hagiu, B. A., Ionescu, C., Sandu, I., Tura, V. (2008). Electrospun gelatin nanofibers functionalized with silver nanoparticles. *Journal of Optoelectronics and Advanced Materials, 10*(12), 3512–3516.

108. Lv, Y., Liu, H., Wang, Z., Hao, L., Liu, J., Wang, Y., Du, G., Liu, D., Zhan, J., Wang, J. (2008). Antibiotic glass slide coated with silver nanoparticles and its antimicrobial capabilities. *Polymers for Advanced Technologies, 19*(11), 1455–1460.

109. Perkas, N., Amirian, G., Applerot, G., Efendiev, E., Kaganovskii, Y., Ghule, A. V., Chen, B.-J., Ling, Y.-C., Gedanken, A. (2008). Depositing silver nanoparticles on/in a glass slide by the sonochemical method. *Nanotechnology, 19*(43), 5604.

110. Esteban-Tejeda, L., Malpartida, F., Esteban-Cubillo, A., Pecharromán, C., Moya, J. S. (2009). The antibacterial and antifungal activity of a soda-lime glass containing silver nanoparticles. *Nanotechnology, 20*(8), 5103.

111. Loher, S., Schneider, O. D., Maienfisch, T., Bokorny, S., Stark, W. J. (2008). Micro-organism-triggered release of silver nanoparticles from biodegradable oxide carriers allows preparation of self-sterilizing polymer surfaces. *Small, 4*(6), 824–832.

112. Lv, Y., Liu, H., Wang, Z., Liu, S., Hao, L., Sang, Y., Liu, D., Wang, J., Boughton, R. I. (2009). Silver nanoparticle-decorated porous ceramic composite for water treatment. *Journal of Membrane Science, 331*(1–2), 50–56.

113. Du, W. L., Niu, S. -S., Xu, Y. -L., Xu, Z. -R., Fan, C. -L. (2009). Antibacterial activity of chitosan tripolyphosphate nanoparticles loaded with various metal ions. *Carbohydrate Polymers, 75*(3), 385–389.

114. Magaña, S. M., Quintana, P., Aguilar, D. H., Toledo, J. A., Ángeles-Chávez, C., Cortés, M. A., Léon, L., Freile-Pelegŕn, Y., Ĺopez, T., Torres Śanchez, R. M. (2008). Antibacterial activity of montmorillonites modified with silver. *Journal of Molecular Catalysis A: Chemical, 281*(1–2), 192–199.

115. Carja, G., Kameshima, Y., Nakajima, A., Dranca, C., Okada, K. (2009). Nanosized silver-anionic clay matrix as nanostructured ensembles with antimicrobial activity. *International Journal of Antimicrobial Agents*, *34*(6), 534–539.

116. Park, J. H., Karim, M. R., Kim, I. K., Cheong, I. W., Kim, J. W., Bae, D. G., Cho, J. W., Yeum, J. H. (2010). Electrospinning fabrication and characterization of poly (vinyl alcohol)/montmorillonite/silver hybrid nanofibers for antibacterial applications. *Colloid and Polymer Science*, *288*(1), 115–121.

117. Jeon, H. J., Kim, J. S., Kim, T. G., Kim, J. H., Yu, W.-R., Youk, J. H. (2008). Preparation of poly(ε-caprolactone)-based polyurethane nanofibers containing silver nanoparticles. *Applied Surface Science*, *254*(18), 5886–5890.

118. Sheikh, F. A., Barakat, N. A. M., Kanjwal, M. A., Chaudhari, A. A., Jung, I. -H., Lee, J. -H., Kim, H. -Y. (2009). Electrospun antimicrobial polyurethane nanofibers containing silver nanoparticles for biotechnological applications. *Macromolecular Research*, *17*(9), 688–696.

119. da Silva Paula, M. M., Franco, C. V., Baldin, M. C., Rodrigues, L., Barichello, T., Savi, G. D., Bellato, L. F., Fiori, M. A., da Silva, L. (2009). Synthesis, characterization and antibacterial activity studies of poly-{styrene-acrylic acid} with silver nanoparticles. *Materials Science and Engineering C*, *29*(2), 647–650.

120. Liu, S., He, J., Xue, J., Ding, W. (2009). Efficient fabrication of transparent antimicrobial poly(vinyl alcohol) thin films. *Journal of Nanoparticle Research*, *11*(3), 553–560.

WEBSITES

www.irgaguard.com
www.agion-tech.com
www.biocote.com
www.alphasan.com
www.mesosilver.com
www.sangi-co.com
www.crocsrx.eu
www.astp.com
www.ecopolynae.com
www.silverphase.fi
www.laboratorios-argenol.com

CHAPTER 12

ANTIMICROBIAL PLASTICS BASED ON METAL-CONTAINING NANOLAYERED CLAYS

JOSÉ M. LAGARÓN[1] and MARIA A. BUSOLO[1,2]

[1] Instituto de Agroquímica y Tecnología de Alimentos (IATA), CSIC, Valencia, Spain
[2] Instituto de Agroquímica y Tecnología de Alimentos (IATA), NanoBioMatters R&D S.L., Paterna, (Valencia), Spain

CONTENTS

12.1 Introduction 317
12.2 Silver-Containing Nanolayered Clays 320
12.3 Conclusions 324
References 325

12.1 INTRODUCTION

This chapter presents an antimicrobial technology supported on nanolayered clays also called nanoclays, which provides in addition to the well-known benefits associated with the reinforcing effect of nanolayered clays, the capacity to deliver active new functionalities to plastic materials. More specifically, it is shown how active metals, typically silver, or their compounds can be nanoscaled and stabilized on the surface of nanoclays to provide efficient antimicrobial (AM) capacity while being able to disperse nicely within polymers.

Antimicrobial activity can be realized by adding AM agents to a polymeric system during manufacturing or by using AM polymeric materials [1]. These technologies remove the essential factors of microbial growth from the product

Antimicrobial Polymers, First Edition.
Edited by José M. Lagarón, María J. Ocio, and Amparo López-Rubio.

and inhibit the growth of microorganisms. The immobilization systems are not intended to release AM agents and hence limit the biocide action to microorganisms existing at the contact surface. The release systems allow the migration of the AM agent (to the liquid or gas phase) into the product or surface to inhibit the growth of microorganisms. Whereas a gaseous AM agent can penetrate through any space, a solute AM agent cannot migrate through the air space between the product and the contaminated part. The release kinetics of an antimicrobial polymeric system are typically studied by measuring the release rate of the AM agent into the product or product simulant or by measuring the effectiveness in inhibiting microbial growth and extending the shelf life of goods.

Control release plastic is thus a new generation of materials that can release active compounds at different controlled rates suitable for enhancing the quality and safety of products and surfaces. The antimicrobial substances in the releasing plastic permit the gradual migration of the biocide agent to the product during storage and use. These technologies are very effective in minimizing the superficial contamination of products.

This chapter deals with the introduction of a new nanotech toolbox based on the natural dispersability and good properties of nanoclays to impart new active functionalities to plastics and bioplastics of interest in food packaging applications.

Nanotechnology in the form of nanocomposites can be designed to control the release of, for instance, antimicrobial natural components from polymeric materials. One recent example is the release of natural antimicrobial agents such as thymol and linalool. Thymol is a phenolic monoterpene that has received considerable attention as an antimicrobial agent with very high antifungal activity, very low minimum inhibitory concentration (MIC) values [2], and as a possible food antioxidant [3]. Linalool is another essential oil that has been previously reported to have effective antibacterial [4] and antifungal [5] properties that would make it suitable for the development of antimicrobial films. The combination of active technologies such as antimicrobials and of nanotechnologies such as clay-based nanocomposites can synergistically lead to bioplastic formulations with balanced properties and functionalities for their implementation in plastics containing applications. As an example of antimicrobial plastic packaging, the formulation of novel antimicrobial nanocomposites of polycaprolactone (PCL) was presented as a way to control solubility and diffusion of natural biocides such as thymol [6]. The antimicrobial nanocomposites of biodegradable PCL were processed by a solution casting method. The diffusion kinetics of the released biocide was determined by attenuated total reflection Fourier transformed infrared (ATR-FTIR) spectroscopy.

The enhancement in the antimicrobial solubility resulting from the presence of the nanoplatelets of mica possibly could be attributed to retention of the apolar biocide agent over the engineered nanofiller surface (see Figure 12.1). However, the thymol diffusion coefficient was shown to decrease (from ca. $2.8\ 10^{-15}$ to $1.1\ 10^{-15}\ m^2/s$) with the addition of the nanoadditive in the biocomposite. This is likely the result of the larger tortuosity effect imposed on the diffusion of the

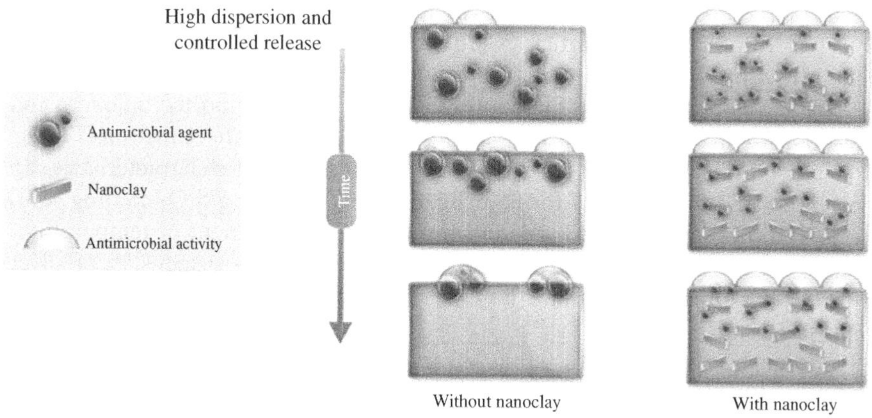

FIGURE 12.1 Schematics of the functioning of antimicrobial nanolayered clays.

biocide by the dispersed nanoclay. As a result, the incorporation of nanoclays led not only to enhance the solubility of natural biocides into polymeric matrices but also to control the release of natural antimicrobials with interest in the design of novel active antimicrobial film and coating polymeric systems.

With the exposure of the first commercial antimicrobial packaging materials, certain concerns were raised by authorities, legislators, and consumers with respect to the release of chemical antimicrobial agents such as triclosan or other organic molecules from packaging to, for instance, foods. For this reason, there has been a strong push toward the development of natural antimicrobial technologies derived from mineral, plant, or animal sources [7-9].

Besides the use of natural extracts, silver is a mineral with very efficient biocide properties known since ancient times. The use of silver-based antimicrobial additives for plastics used in food production and medical equipment is today permitted and regulated [10].

Thus, silver nanoparticles as well as silver compounds are widely used as efficient biocides. In fact, many commercial antimicrobial products include silver in their formulations as the active ingredient. In this context, many products have been developed for specific applications in several different areas, such as medical devices, liquid disinfectants for large surfaces, personal care products, electronics, as well as food and water storage materials to extend shelf life.

Recent technical innovations and findings facilitate the availability and incorporation of silver products in a wide range of materials at the manufacturing stage, providing novel antimicrobial formulations. Nevertheless, a specific form of efficient silver does not exist for every application, procedure, or matrix. In this sense, nanotechnology is a key factor because of the capability of modulating metals, compounds, and materials into the nanosize, which often changes their chemical, physical, and optical properties, as well of those of the matrixes into which they are incorporated. Stable silver nanoparticles can be obtained by way of using soluble starch as both the reducing and the stabilizing agent [11], or being synthesized via the regular borohydride reduction of Ag+ ions [12, 13]. Silver nanoparticles were synthesized in the interlamellar space of kaolin by ultraviolet (UV) radiation or chemical-induced reduction [14, 15], in layered laponite suspensions via photoreduction [16], or supported on micro and mesoporous structures after ion exchange followed by *in situ* reduction [17, 18].

Silver (I) nitrate adducts with diverse electronic and steric characteristics can be synthesized with N- and P-donor ligands [19]. Thus, Ag/SiO_2 coating solutions have been prepared to serve as antimicrobial refinement of temperature-sensitive materials like fabrics or wood [20]. Moreover, a suspension of silver nitrate in an ammonium salt medium has been reported as a precursor of stable nanoscale AgBr particles [21].

On another line of work, many efforts have also been carried out for the development of inorganic materials, such as zeolites, for supporting Ag^+ ions as a result of their capability to incorporate and release ionic species. Coleman et al. [22] prepared Ag^+- and Zn^+-exchanged tobermorites and demonstrated that they have a marked bacteriostatic effect and that they can be potentially used as antimicrobial materials for *in situ* bone tissue regeneration. The thermal stability of Ag^+-supported clinoptilolite and possible applications were tested by Akdeniz and Ülkü [23].

Some reports based on silver-modified clays by a cation exchange method have been published before. Oya et al. [24] reported the antimicrobial properties of Ag^+-exchanged montmorillonite on 1991. Keller-Besrest et al. [25] prepared a silver-loaded montmorillonite for possible topic uses in the treatment of burns. They obtained the coexistence of both silver metal particles and Ag^+ ions, and they observed significant differences on the final silver content in clays using a exchange resin procedure (of up to 10 wt%) with regard to the standard cation exchange capacity (CEC) methodology performed in solution (of up to 1 wt%). Quintana et al. [26] studied the effects of calcination and mechanical grinding on silver-exchanged Na-MMT and its antimicrobial performance. They reported metallic silver nanoparticles on the clay aggregates, and they pointed out that the antibacterial performance is affected by the availability of the ionic silver to be in contact with the bacteria. Praus et al. [27] compared the antimicrobial activities of some chemical compounds, silver ions, and elemental silver immobilized on montmorillonite. They demonstrated that antibacterial compounds are effective just when they are released from the inorganic carrier, and they concluded that intercalated silver ions are the most effective antibacterial elements, whereas elemental silver does not show any antibacterial effects. In any case, silver species provide color when incorporated into inorganic carriers and is unstable against temperature.

12.2 SILVER-CONTAINING NANOLAYERED CLAYS

A recent new patented technology [28] that makes use of silver strongly stabilized on nanolayered clays either in the elemental nanoform or in the ionic form (see Figures 12.2 and 12.3) and aimed at dispersing in polymers has been developed, scaled up to an industrial production level, and is commercially available under the registered trademark of Bactiblock (NanoBioMatters Ltd., Paterna, Spain). This is a white powder material, heat stable, and readily dispersible in all kinds of plastics with strong biocide capacity at low dosages (see Table 12.1).

The particle size of the nanomaterial before implementation into plastics was analyzed by a Mastersizer 2000, Particle Size Analyzer (Worcestershire, U.K.). The D_{50} (i.e., the average particle size) of the micropowder material was typically found to be of around 5 microns, and therefore, the handling is similar to that of a micropowder. This offers significant advantages for a nanostructure material in industrial applications.

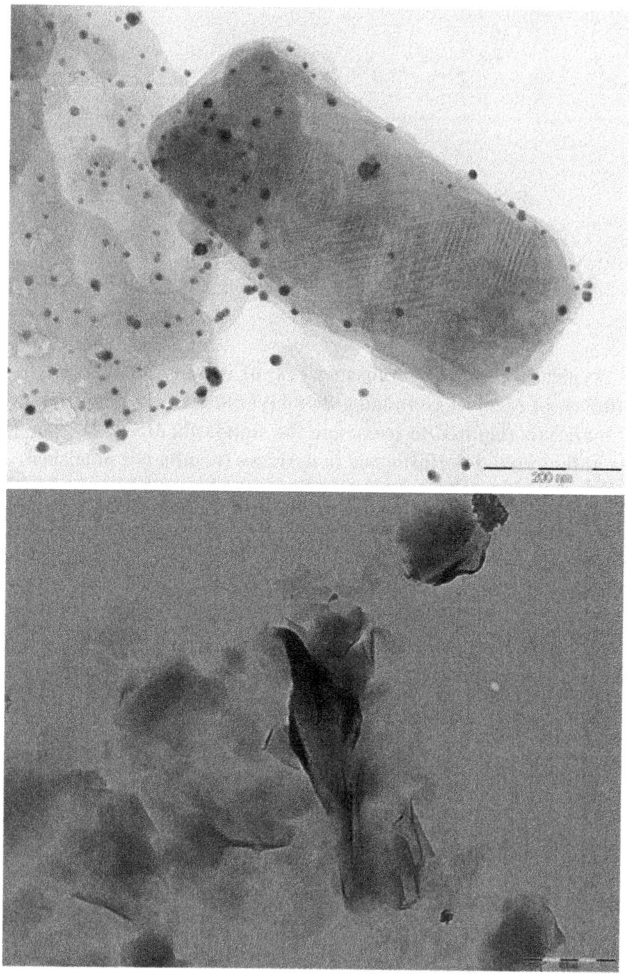

FIGURE 12.2 Typical TEM pictures of silver containing nanolayered montmorillonite clays containing elemental silver nanoparticles (top image) and with ionically exchanged silver (bottom picture). Scale markers are 200 nm.

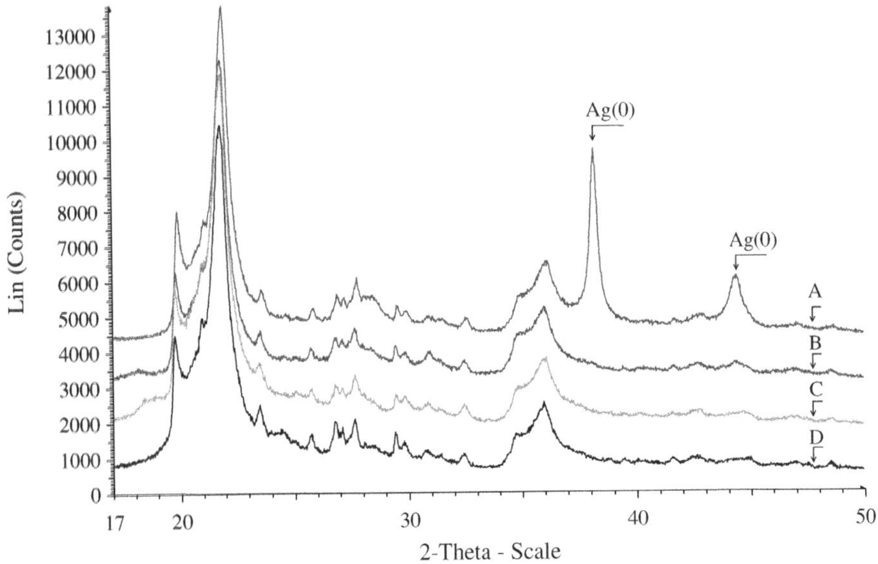

FIGURE 12.3 Typical WAXS difractograms of silver containing organomodified nanolayared montmorillonite clays (samples A–C) and of the reference organomodified layered clay (sample D).

TABLE 12.1 Typical recommended dosages (%) of various commercial grades (so-called bactiblock®) of silver containing nanolayared clays in nanocomposites of various plastic materials required to overcome the standards JIS Z 2801 or ISO 22196:2007 for antimicrobial performance in surfaces (results not published)

	101 R1.43	101 R1.47	101 S1.19	101 V1.47
THERMOSETS				
Epoxy based	1.5-3%	1.5-3%	0.5-1.5%	0.5-1.5%
Polyester based	1.5-3%	1.5-3%	0.5-1.5%	0.5-1.5%
THERMOPLASTICS				
Polypropylene	1-3%	1-3%	0.75-2.5%	0.75-2.5%
Polyethylene	1-3%	1-3%	0.75-2%	0.75-2%
Polystyrene	1-3%	1-3%	0.5-2%	0.5-2%
PA	1-3%	1-3%		0.75-2%
PVC			0.5-3%	
PHB	1-2%	1-2%	1-2%	1-2%
ELASTOMERS				
EVA	1-2.5%			0.5-1.5%
COATINGS/PAINTS				
Solvent Based	0.25-1%**	0.25-1%**	0.25-1%	
Water Based	0.25-1%**	0.25-1%**	0.25-1%	
Powder Coatings	1-3%	1-3%	0.5-1.5%	0.5-1.5%

**Available in the form of a predispersed gel.

Some of these nanolayered clays were analyzed by transmission electron microscopy (TEM) (see Figure 12.2) and by wide-angle X-ray scattering (WAXS) (see Figure 12.3). The WAXS experiments were carried out directly on the micropowder of the samples (Bruker Advance D8 diffractometer, Bruker GmbH, Karlsruhe, Germany), whereas TEM (JEOL1010 Transmission electron microscope, equipped with a digital Bioscan image acquisition system, Jeol Ltd., Tokio, Japan) was carried out by dispersing 0.001 wt% of the micropowders in water followed by direct casting of the copper grids used for TEM observation.

Figure 2 shows two typical TEM pictures of a silver containing organomodified nanolayered clay with nanoparticles of elemental silver and of a sample showing no silver nanoparticles because silver is in its ionic form. In the top picture, nanosilver particles are distributed homogeneously over the surface of the clay.

In Figure 3, the WAXS difractograms of some of these material show that the two clear strong elemental silver reflections can only be observed in sample A. Sample A contains silver nanoparticles on the clay surface (see also Figure 2). The rest of the samples (B and C) show no indication of the presence of elemental or metallic silver because silver is designed to stay in its ionic form.

Regarding nanocomposites, the value of this technology was additionally demonstrated in polylactide (PLA) films [29]. From the results, the silver-based nanoclay showed a strong antimicrobial effectiveness against gram-negative *Salmonella* spp. with minimum inhibitory concentration and minimum bactericide concentration less than 1 mg $10\,mL^{-1}$.

PLA nanobiocomposites with different antimicrobial nanofiller loadings were trial by casting or by melt compounding showing excellent optical properties. An improved barrier to water was measured for the nanobiocomposites because of the presence of the nanoclay, which also exhibited strong antimicrobial performance (see Table 12.2). To carry out the antimicrobial tests, a pathogenic microorganism of food origin, *Salmonella* spp. (CECT 554, from Spanish Type Culture Collection, Valencia, Spain) was used. The strain was stored in tryptone soy broth (TSB) with 20% glycerol at −80 °C until needed. For experimental use, the stock cultures were maintained by regular subculture on agar tryptone soy

TABLE 12.2 Viable cells counts before and after 24 h incubation in antimicrobial activity tests and water permeability of solvent cast PLA-1% bactiblock nanocomposite (0.75 g of Film)

Sample	InitialCFU/mL	CFU/mL after 24 h incubation	WVTR ($g\ m/m^2s\ Pa$)
Control without film	2.0×10^5	2.44×10^7	—
PLA control film	2.0×10^5	4.64×10^7	$1.6\ 10^{-11}$
PLA-1% Bactiblock® nanocomposite film	2.0×10^5	5.24×10^4	$9.7\ 10^{-12}$
PLA-5% Bactiblock® nanocomposite film	2.0×10^5	0	$1.3\ 10^{-11}$

agar (TSA) slants at 4 °C and transferred monthly. A loopful of bacteria was removed to 10 ml of TSB and incubated at 37 °C overnight. A 100-μL aliquote from the overnight culture was again transferred to TSB and grown at 37 °C to the mid-exponential phase of growth. This culture served as the inoculum in the susceptibility study, starting with approximately 10^5 CFU/mL in the test tubes. These colony-forming unit (CFU) counts were accurately and reproducibly obtained by inoculation of 0.1 mL of the culture having an absorbance value of 0.2, as determined by optical density at 600 nm by UV visible spectroscopy. To perform the antimicrobial tests, 0.75 g of the composite was added to the 10-mL Mueller Hinton broth (MHB). After 5 h, the samples were inoculated with 0.1 mL of *Salmonella* spp. culture. Control experiments were performed under the same conditions. Samples and control were incubated to 37 °C for 24 h. Afterward, 0.1 mL of each one was sowed in agar-triptone-soybean (TSA) plates and incubated for 24 h at 37 °C. Finally, viable cells were counted.

The water vapor transmission rate (WVTR) of the films was determined according to the ASTM method E96, using aluminium cells with internal and external diameters of 3.5 and 6 cm, respectively. Films samples were previously dried in a vacuum oven at 50 °C during 4 h in order to remove solvent residues. Each sample film was sealed to a permeation cell containing liquid water, and then the permeation cells were placed under controlled environmental conditions (40 °C and 0% relative humidity) and weighed regularly until steady state was reached. The water vapor transmission rate was easily determined from the slope of the cell weight loss versus the time plot in the following equation:

$$\mathrm{WVP} = (G \times L)/(A \times \Delta p)$$

where G is the slope of weight loss versus the time straight line, L is the film thickness, and Δp is the vapor partial pressure differential across the film. All tests were conducted in duplicate, and aluminium films were used as blanks to evaluate water vapor loss through the sealing.

Regarding the use of silver as an antimicrobial, the European Food Safety Authority (EFSA) has recently evaluated the use of several silver-based biocides intended to come into contact with foods, and it has defined a general specific migration limit of 0.05 mg of silver per kg of food (see the EFSA Journals). The study of Busolo et al. [29], making use of the cited silver containing nanolayered clays, proved that those levels of permitted migration can be sufficient to exert strong biocide performance.

12.3 CONCLUSIONS

In summary, the addition of engineered antimicrobial phyllosilicates (i.e., layered clays) to polymers through innovative nanotechnology is now available as a formidable tool for improving the properties of plastics and, therefore, to enhance product quality and safety aspects. Interestingly, these antimicrobial additives provide plastics not only with novel antimicrobial functionalities but also with enhanced physical performance such as impermeability to gases and vapors. The fact that these technologies have become commercially available makes them even more interesting for its widespread implementation.

REFERENCES

1. Hotchkiss, J. H. (1997). Food packaging interactions influencing quality and safety. *Food Additives and Contaminants, 14*, 601–607.
2. Thompson, D. P. (1989). Fungitoxic activity of essential oil components on food storage. *Mycologia, 81*, 151–153.
3. Youdim, K. A., Deanes, S. G. (2000). Effect of thyme oil and thymol dietary supplementation on the antioxident status and fatty acid composition of the ageing rat brain. *Journal of Nutrition, 83*, 87–93.
4. Onawunmi, G. O., Yisak, W. A., Ogunlana, E. O. (1984). Antibacterial constituent in essential oil of cymbopogon citratus. *Journal of Ethnapharmacology, 12*, 279–286.
5. Reuveni, R., Fleischer, A., Putievsk, E. (1984). Fungistatic activity of essential oilsfrom ocimum basilicum. *Journal of Essential Oil, 110*, 20–22.
6. Sanchez-Garcia, M. D., Ocio, M. J., Gimenez, E., Lagaron, J. M. (2008). Novel polycaprolactone nanocomposites containing thymol of interest in antimicrobial film and coating applications. *Journal of Plastic Film and Sheeting, 24* (3–4), 239–250.
7. Sun Lee, D. (2005). *Packaging containing natural antimicrobial or antioxidative agents*. In: *Innovations in Food Packaging, Part 2*. Ed.: Han, Jung H. pp. 108–122. Elsevier, New York.
8. Fernandez-Saiz, P., Ocio, M. J., Lagaron, J. M. (2006). Film-forming process and biocide assessment of high-molecular-weight chitosan as determined by combined ATR-FTIR spectroscopy and antimicrobial assays. *Biopolymers, 83*, 577–583.
9. Sanchez-Garcia, M. D., Gimenez, E., Ocio, M. J., Lagaron, J. M. (2008). *Technical Papers, Regional Technical Conference Society of Plastics Engineers, Vol. 4*, pp. 2084–2088.
10. Simpson, K. (2003). Plastics. *Additives and Compounding, 5*, 32.
11. Vigneshwaran, N., Nachane, R. P., Balasubramanya, R. H., Varadarajan, P. V. (2006). A novel one-pot "green" synthesis of stable silver nanoparticles using soluble starch. *Carbohydrate Research, 34*, 2012–2018.
12. Lok, C., Ho, C., Chen, R., He, Q., Yu, W., Sun, H., Kwong-Hang Tam, P., Chiu, J., Che, C. (2007). Silver nanoparticles: partial oxidation and antibacterial activities. *Journal of Biological Inorganic Chemistry, 12*, 527–534.
13. Oh, S. G., Lee, G. J., Shin, S. I., Kim, I. C. (2004). Preparation of silver nanorods through the control of temperature and pH of reaction medium. *Materials Chemistry and Physics, 84*, 197–204.
14. Dékány, I., Patakfalvi, R., Oszkó, A. (2003). Synthesis and characterization of silver nanoparticle/kaolinite composites. *Colloids and Surfaces A: Physicochemical and Engineering Aspects, 220*, 45–54.
15. Dékány, I., Patakfalvi, R. (2004). Synthesis and intercalation of silver nanoparticles in kaolinite/DMSO complexes. *Applied Clay Science, 25*, 149–159.

16. Huang, H., Yang, Y. (2007). Preparation of silver nanoparticles in inorganic clay suspensions. *Composite Science and Technology*, *68*(14), 2948–2953.
17. Yang, X., Yang, L., Wang, X., Yang, F. (2008). Excellent antimicrobial properties of mesoporous anatase TiO2 and Ag/TiO2 composite films. *Microporous and Mesoporous Materials*, *114*, 431–439.
18. Lv, L., Luo, Y., Ng, W. J., Zhao, X. S. (2009). Bactericidal activity of silver nanoparticles supported on microporous titanosilicate ETS-10. *Microporous and Mesoporous Materials*, *120*, 304–309.
19. Pettinari, C., Di Nicola, C., Effendy, Marchetti, F., Skelton, B. W., White, A. H. (2007). Synthesis and structural characterization of adducts of silver (I) nitrate with ER3 (E = P, As, Sb; R = Ph, cy, o-tolyl, mes)and oligodentate aromatic bases derivative of 2,20-bipyridyl, L, AgNO3:ER3:L (1:1:1). *Inorganica Chimica Acta*, *360*, 1433–1450.
20. Mahltig, B., Gutmann, E., Meyer, D. C., Reibold, M., Bund, A., Böttcher, H. (2009). Thermal preparation and stabilization of crystalline silver particles in SiO2-based coating solutions. *Journal of Sol-Gel Science and Technology*, *49*, 202–208.
21. Zhang, J., Liu, X., Luo, X., Lu, S., Cao, W. (2007). A novel cetyltrimethyl ammonium silver bromide complex and silver bromide nanoparticles obtained by the surfactant counterion. *Journal of Colloid and Interface Science*, *307*, 94–100.
22. Coleman, N. J., Bishop, A. J., Booth, S. E., Nicholson, J. W. (2009). Ag+– and Zn2+-exchange kinetics and antimicrobial properties of 11Å tobermorites. *Journal of the European Ceramic Society*, *29*, 1109–1117.
23. Akdeniz, Y., Ülkü, S. (2008). Thermal stability of Ag-exchanged clinoptilolite rich mineral. *Journal of Thermal Analysis and Calorimetry*, *3*, 703–710.
24. Oya, A., Banse, T., Ohashi, F., Otani, S. (1991). An antimicrobial agent derived from montmorillonite. *Applied Clay Science*, *6*, 135–142.
25. Keller-Besrest, F., Bénazeth, S., Souleau, Ch. (1995). EXAFS structural investigation of a silver-added montmorillonite clay. *Materials Letters*, *24*, 17–21.
26. Quintana, P., Magaña, S. M., Aguilar, D. H., Toledo, J. A., Angeles-Chavez, C., Cortés, M. A., León, L., Freile-Pelegrín. Y., López, T., Sánchez, R. M. (2008). Antibacterial activity of montmorillonites modified with silver. *Journal of Molecular Catalysis A: Chemical*, *281*, 192–199.
27. Praus, P., Malachová, K., Pavlíčková, Z., Turicová, M. (2009). Activity of antibacterial compounds immobilised on montmorillonite. *Applied Clay Science*, *43*, 364–368.
28. Lagaron J. M., Busolo M., Fernandez-Saiz, P. (2010). Patent Application ES2331640.
29. Busolo, M. A., Fernandez, P., Ocio, M. J., Lagaron, J. M. (2010). Novel silver-based nanoclay as an antimicrobial in polylactic acid food packaging coatings. *Food Additives & Contaminants*, *27*, 1617–1626.

CHAPTER 13

NANOMETALS AS ANTIMICROBIALS

R. P. ALLAKER,[1] M. A. VARGAS-REUS,[1] and G. G. REN[2]
[1] Queen Mary University of London, Barts and The London School of Medicine and Dentistry, Institute of Dentistry, London, U.K.
[2] School of Engineering and Technology, University of Hertfordshire, Hatfield, U.K.

CONTENTS

13.1 Introduction 328
13.2 Nanometals as Antimicrobial Agents 328
13.2.1 Silver (Ag) 331
13.2.2 Copper (Cu) 332
13.2.3 Gold (Au) 333
13.3 Nanoparticulate Metal Oxides as Antimicrobial Agents 335
13.3.1 Copper Oxide (CuO) 335
13.3.2 Zinc Oxide (ZnO) 335
13.3.3 Titanium Dioxide (TiO_2) 336
13.4 Nanometals and the Control of Biofilms 336
13.5 Properties of Polymeric Matrix-Nanocomposites 337
13.5.1 Methods of Combining Nanometals with Polymers 339
13.5.2 Polymeric Films Incorporating Nanometals 340
13.6 Toxicity Issues and Nanometals 341
13.7 Concluding Remarks 345
Acknowledgments 345
References 345

Antimicrobial Polymers, First Edition.
Edited by José M. Lagarón, María J. Ocio, and Amparo López-Rubio.

13.1 INTRODUCTION

Metals have been used for centuries as antimicrobial agents. Silver, copper, gold, and zinc have attracted particular attention, each with different properties and spectra of activity. Some of the most fundamental breakthroughs in medicinal history can be attributed to the antimicrobial properties of metals. The use of mercury as a medicinal agent can be traced back to the 10th century in Europe and the 2nd century BC in China. Skin diseases and syphilis were treated with inorganic mercury compounds, and more recently, organomercurial compounds have been used as antiseptics and disinfectants [1]. Copper and zinc salts have also been investigated with respect to their use as antiseptics and as antifungal agents in the treatment of tinea pedis (athlete's foot).

The precious metals platinum and silver have been reported as being able to exert an antimicrobial effect [2, 3]. The antimicrobial properties of silver have been used for centuries; for example, silver-lined containers were used for water storage in the past, and silver nitrate was introduced in the late 19th century as a topical treatment for blindness in infants. The idea of using gold as an antibacterial treatment dates back to the 8th century [4]. Historically, gold compounds have been tested against *Mycobacterium tuberculosis* and have been used in the treatment of rheumatoid arthritis where it was assumed that bacteria were responsible. More recently, gold complexes have been reported to possess a wide range of antimicrobial activities [5].

The use of metals in a nanoparticulate form as antimicrobials is gaining considerable attention. Technologies are now able to fabricate spherical, cubic, and needle-like nanoscaled particles (approximately 5 to 100 nm) [6]. The whole field of nanotechnology is currently experiencing rapid growth, with many and diverse potential applications being explored in the biomedical field, including the control of infectious diseases. Nanotechnology has the potential to offer not only improvements to current approaches for immunization, antimicrobial drug design and delivery, diagnostics, and cross-infection control, but also it is unexpectedly delivering many new tools and capabilities [7]. Polymer nanocomposites with an antimicrobial function are now finding applications in the biomedical device, food, and personal hygiene sectors.

This chapter will consider the use of nanoparticulate metals and metal oxides as antimicrobial agents that have the potential to provide an additional functionality to polymeric materials.

13.2 NANOMETALS AS ANTIMICROBIAL AGENTS

Functionalized metal nanoparticles have been the subject of intense research interest for the past 15 years. As particles are reduced from a micrometer to a nanometer size, the resultant properties can change dramatically. For example, electrical conductivity, hardness, active surface area, chemical reactivity, and biological activity are all known to be altered. The biocidal effectiveness of metal nanoparticles has been suggested to be a result of both their size and their high surface-to-volume ratio. Such characteristics should allow them to interact closely with bacterial membranes, rather than the effect being solely a result of the release of metal ions [8]. Metal nanoparticles are now being combined with polymers or coated onto surfaces, generating materials that may have a variety of potential antimicrobial applications.

Exploitation of the toxic properties of nanoparticulate metals and metal oxides, in particular those that produce reactive oxygen species under ultraviolet (UV) light, such as titanium dioxide (TiO_2; Figure 13.1) and zinc oxide (ZnO; Figure 13.1), are finding increasing use in antimicrobial formulations, with silver metal nanoparticles (5–40 nm) having been reported to inactivate most microorganisms, including HIV-1 [9]. The high reactivity of nano titanium dioxide and nano silicon dioxide (SiO_2) is exploited extensively for their

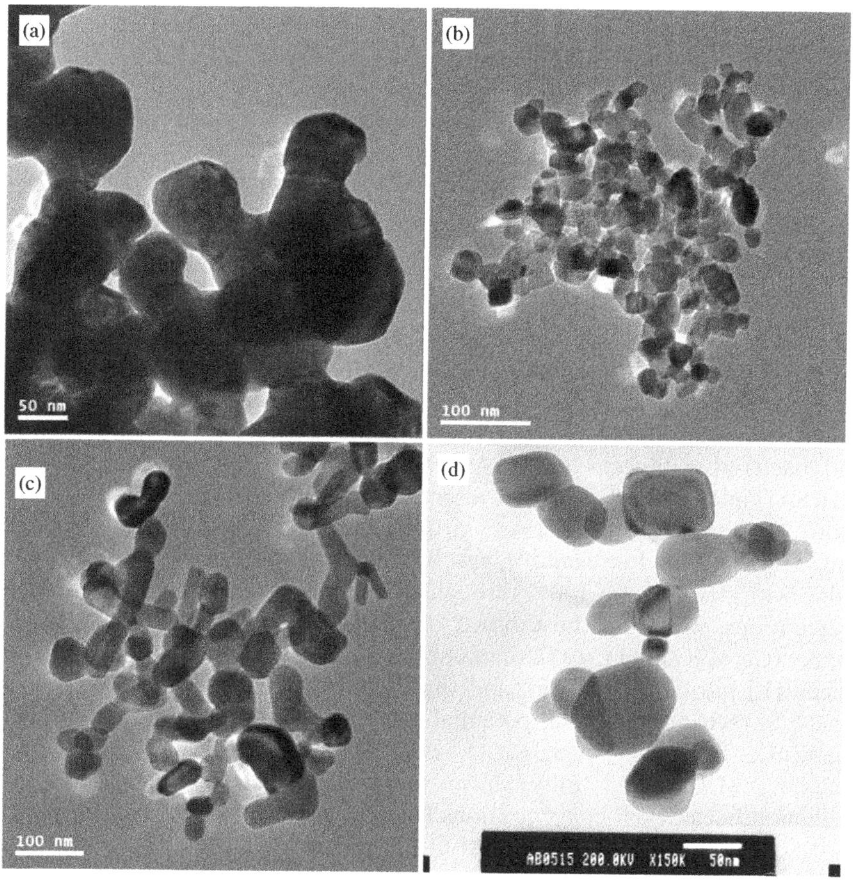

FIGURE 13.1 Transmission electron microscopy images of agglomerated silver (a), titanium dioxide (b), zinc oxide (c), and copper oxide (d) nanoparticles.

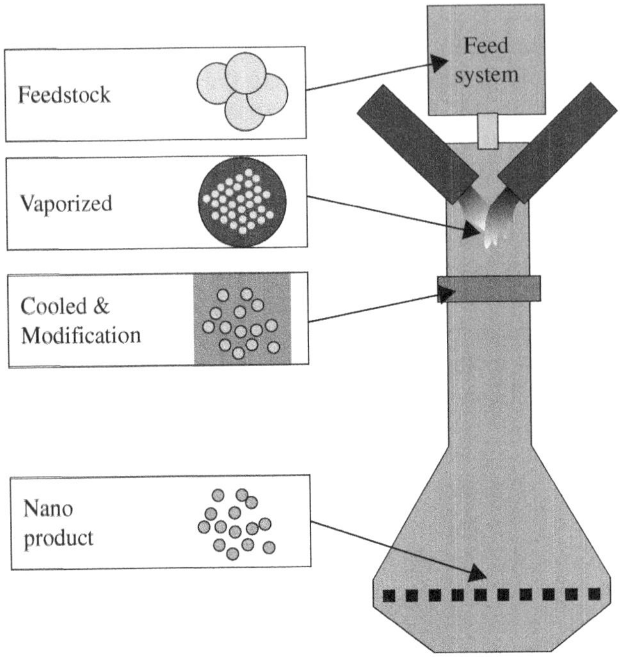

FIGURE 13.2 Tesima (QinetiQ Nanomaterials Ltd, Farnborough, Hampshire, U.K.) plasma process to generate nanoparticles.

bactericidal properties in filters and coatings on substrates such as polymers, ceramics, glasses, and alumina [10]. Significant activity using the latest generation of nanoparticles, such as nano Cu, Cu_2O, CuO, and their compound clusters produced by thermal plasma technology (Figure 13.2), against fungal and bacterial pathogens such as meticillin-resistant *Staphylococcus aureus* (MRSA) and *Escherichia coli*, has recently been demonstrated. These have also shown the capability to inactivate viruses, including SARS, H1N1 influenza, and H5N1 bird flu. For example, new broad-spectrum materials (5–60 nm) can reduce virus levels by 80–100% through direct or indirect contact. Nanoparticle preparations, including those based on nickel (Ni, NiO), zirconium (ZrO), copper (Cu, CuO, and Cu_2O), titanium (TiO_2), zinc (ZnO), aluminum (Al_2O_3), silicon(IV) nitride, silver (Ag), and tungsten carbide have been compared in regard to their antimicrobial potential (Figure 13.1). Significant activity with Ag, ZnO, TiO_2, SiO_2, Cu, Cu_2O, and CuO against bacterial pathogens, including MRSA and *Pseudomonas aeruginosa*, has been demonstrated. Minimum bactericidal concentrations (MBCs) were found to be in the range of 0.1 to 5 mg mL^{-1}. NiO, Ni, Al_2O_3, TiO_2 (in the absence of ultraviolet light), silicon (IV) nitride, tungsten carbide, and ZrO were found to lack antimicrobial activity at the concentrations tested (Allaker, Vargas-Reus, and Ren, 2009).

The antimicrobial properties of both silver [11] and copper nanoparticles [12] have received the greatest attention, and both of these have been coated onto or incorporated into various materials [13]. A diverse range of materials is now being considered as potential substrates for nanoparticulate metals, including polymethyl methacrylate (PMMA) [14] and hydrogels (water-insoluble networks of polymer chains) [15]. The incorporation of nanoparticles into materials that cannot withstand high sterilization temperatures, for example, textiles, wood, and some plastics, offers an obvious benefit. Although some antimicrobial nanoparticles currently under investigation could possess a toxic effect, the development of stable nanoparticle-containing polymer composites may help to address this issue [11].

Nanoparticles in the size range of 5–10 nm have been shown to have the greatest interaction with bacteria with an inverse relationship between nanoparticle size and antimicrobial activity demonstrated [8, 16]. As a result of their small size, nanoparticles may well offer other advantages to the biomedical field through improved biocompatibility [17]. Also, it seems that bacteria are far less likely to acquire resistance against metal nanoparticles compared with other conventional and narrow-target antibiotics [18]. This is thought to occur as metals may act against a broad range of microbial targets, and many mutations would have to occur in order to resist their antimicrobial activity. It is thus proposed that the scope for materials incorporating or coated with nanoparticles is vast. Current applications within the biomedical field include wound dressings [19], coatings on medical implants [20], and ultrafiltration membranes [21].

13.2.1 Silver (Ag)

The antibacterial and antiviral actions of silver, silver ions, and silver compounds have been extensively investigated [22]. In comparison with other metals, silver is relatively less toxic to human cells, albeit at very low concentrations. Silver ions have been considered for a range of biomedical applications, including their use within the dental field as an antibacterial component in dental resin composites [23]. Silver also exhibits a strong affinity for zeolite, a porous crystalline material of hydrated aluminosilicate that can bind up to 40% silver ions within its structure. Silver zeolite has been incorporated into tissue conditioners, acrylic resins, and mouth rinses within the dental field [24–27]. Silver nanoparticles (Figure 13.1), either alone or together with other antimicrobial agents, have shown particularly encouraging results [11, 28, 29]. The use of silver salt nanoparticles instead of elemental silver or complex silver compounds to prevent biofilm formation on surfaces for both biomedical and more general use has been investigated. Using silver bromide precipitation to synthesize polymer-nanocomposites, surfaces were shown to resist biofilm formation. It was also shown to be possible, through controlling the size of the embedded AgBr, to modify the release of biocidal silver ions [30].

Surprisingly little is known about how nanoparticles behave in relation to microorganisms, particularly at the cellular level. The mechanism of the antimicrobial activity of silver is not completely understood. Studies have shown that the positive charge on the silver ion is critical for antimicrobial activity, allowing the electrostatic attraction between the negative charge of the bacterial cell membrane and the positively charged nanoparticles [17]. In terms of the molecular mechanisms of inhibitory action of silver ions on microorganisms, it has been shown that DNA loses its ability to replicate [31], and the expression of ribosomal subunit proteins, and other cellular proteins and enzymes necessary for ATP production, become inactivated [32]. It has also been hypothesized that silver ions affect membrane-bound respiratory enzymes [33]. However, the precise mechanism(s) of biocidal activity of silver nanoparticles against bacteria remains to be fully elucidated. The work of Sondi and Salopek-Sondi [11] demonstrated structural changes and damage to bacterial membranes resulting in cell death. These particular studies suggest that sulfur-containing proteins in the membrane or inside the cells and phosphorus-containing elements, such as DNA, are likely to be the preferential binding sites for silver nanoparticles. It remains unclear as to the contribution of silver ion release from nanoparticles to the overall antimicrobial activity. It is suggested that a bacterial cell in contact with silver nanoparticles will take up silver ions, which possibly in turn will inhibit respiratory enzymes, helping to generate reactive oxygen species/free radicals with a subsequent free-radical–induced damage to the cell membrane [17]. To determine the relationship between free-radical formation and antimicrobial activity, the use of antioxidants does suggest that free radicals may be derived from the surface of silver nanoparticles [17].

Shape is also thought to be important as regards the activity of nanoparticles in general. Indeed, it has been demonstrated that the shape of silver nanoparticles can influence antimicrobial activity, as shown with *E. coli* [18]. Truncated triangular silver nanoplates with a {111} lattice plane as the basal plane showed the greatest biocidal activity compared with spherical and rod-shaped nanoparticles. The differences seem to be explained by the proportion of active facets present in nanoparticles of different shapes.

13.2.2 Copper (Cu)

Alongside silver, copper is a traditionally well-known antimicrobial material. In comparison with silver, relatively few studies have reported the antimicrobial properties of copper. It is speculated that copper may well have a similar mode of action to that of silver. However, it remains unclear as to the precise mechanism by which copper nanoparticles exert their antimicrobial activity. As with silver, it is thought that copper partly elicits its antimicrobial activity by combining with the –SH groups of key enzymes. Yoon et al. [34] demonstrated superior antimicrobial activity with copper nanoparticles against *E. coli* and *Bacillus subtilis* when compared with silver nanoparticles. However, in the author's laboratory, silver consistently demonstrated superior activity to copper with a wide range of different species/strains [35].

The antimicrobial properties of both silver and copper nanoparticles were also investigated by Ruparelia et al. [36] using *E. coli* (4 strains), *B. subtilis*, and *Staphylococcus aureus* (three strains). The bactericidal effect of the nanoparticles was compared using disc diffusion tests as well as the minimum inhibitory concentration (MIC) and MBC determinations in batch cultures. Bacterial sensitivity was found to differ according to the species tested and the test system employed. For all cultures of *S. aureus* and *E. coli*, the action of silver nanoparticles was found to be superior. Strain-specific variation for *S. aureus* was negligible, whereas some strain-specific variation was observed for *E. coli*. A higher sensitivity as shown with *B. subtilis* [36] may be attributed to more amine and carboxyl groups, in comparison with other species, on the cell surface, which have a greater affinity for copper [37]. Released copper ions within the cell may then disrupt nucleic acid and key enzymes [38]. In theory, a combination of silver and copper nanoparticles may give rise to a more complete bactericidal effect, especially against a mixed population of bacteria. Indeed, the studies of Ren et al. [35] demonstrated that populations of gram-positive (4 strains) and gram-negative (3 strains) bacteria could be reduced by 68% and 65%, respectively, in the presence of 1000 $\mu g\,mL^{-1}$ nano copper oxide within 2 h. This was significantly increased to 88% and 100%, respectively, with the addition of a relatively low concentration (50 $\mu g\,mL^{-1}$) of nano silver.

13.2.3 Gold (Au)

In comparison with silver and copper, gold generally shows a weak antimicrobial effect. However, gold nanoparticles are employed in multiple applications involving biological systems. The binding properties of gold are exceptional, and this makes it particularly suitable for attaching ligands to enhance biomolecular interactions. Gold nanoparticles also exhibit an intense color in the visible range and contrast strongly for imaging by electron microscopy [39]. Despite all the current and potential applications for gold nanoparticles, there remains little information as to how these particles affect microorganisms. Growth inhibition studies, to measure the effect of gold nanoparticles (polyethylene glycol [PEG]-coated to allow dispersion) on *E. coli* at various concentrations, demonstrated no significant activity [40]. Studies in the author's laboratory with PEG-coated gold nanoparticles also showed no activity against *E. coli*. However, the growth of the gram-negative species *Proteus* and *Pseudomonas aeruginosa* was inhibited at a concentration of 1 $mg\,mL^{-1}$.

13.2.3.1 Use of Gold Nanoparticles to Deliver Antimicrobials. Nanoparticles can provide additional stability, selectivity, and functionality to a drug molecule. Gold nanoparticles are able to provide stable and relatively nontoxic surfaces for the attachment and delivery of drug molecules [41]. Such loaded nanoparticles could possibly be more targeted than a soluble, systemically administered agent. In theory, gold nanoparticle conjugates should have improved binding and cell wall penetration properties, and so deliver a higher concentration of antibiotic molecules. It remains to be fully established if such conjugates will show an increased antibacterial activity when compared with more conventional treatments [42].

13.2.3.2 Photodynamic Therapy and the Use of Nanoparticulate Gold.

Photodynamic therapy (PDT) for microbial control involves the application of a photosensitizing agent and light source to those areas requiring treatment. This approach is now being actively used within the clinical setting to help control periodontal infection in some countries. The killing of microorganisms with light depends on cytotoxic singlet oxygen and free-radical generation, which are formed by the excitation of a photactivateable agent or sensitizer. The result of excitation is that the sensitizer moves from an electronic ground state to a triplet state that then interacts with microbial components to generate cytotoxic species [43]. One of the advantages of light-activated killing is that resistance to the action of singlet oxygen is unlikely to become widespread, in comparison with that experienced with more traditional chemical antimicrobial agents. A sensitizer ideally should absorb light at red to near-infrared wavelengths because these wavelengths are able to penetrate more. The most commonly tested sensitizers on bacteria have been tricyclic dyes (for example, methylene blue and erythrosine), tetrapyroles (for example, porphyrins), and furocoumarins (for example, psoralen) [43]. The use of nanoparticles within this area is now under investigation. For example, biodegradable and biocompatible poly(lactic-co-glycolic acid) (PLGA) nanoparticles complexed with colloidal gold particles, loaded with methylene blue and exposed to red light at 665 nm, have been tested against planktonic *E. faecalis* and in experimentally infected root canals of extracted teeth [44].

Considerations in relation to the therapeutic use of light-activated killing of biofilms on host surfaces include (1) direct toxicity of the sensitizer, (2) indirect toxicity of the sensitizer in terms of "by-stander" damage to adjacent host cells, (3) penetration into the biofilm, (4) light exposure time required to kill bacteria within *in vivo* biofilms, and (5) widespread relatively nonspecific bacterial killing.

Certain polymers incorporating nanoparticles have been investigated with regard to the prevention of catheter-related infections as these are a major cause of morbidity and mortality [20]. Perni et al. [45] investigated a catheter polymer (silicone elastomer) containing the light-activated antimicrobial agent methylene blue together with gold nanoparticles. These workers demonstrated that these novel polymers had excellent killing properties against MRSA and *E. coli* when exposed to light, enhanced by the presence of gold nanoparticles. The study indicated that the gold nanoparticles synergistically enhanced the antimicrobial activity of the methylene blue-containing polymer while having no antimicrobial activity on their own. It is thought that the bacterial killing observed was as a result of the production of reactive oxygen species (ROS), which are produced close to the surface of the polymer where any bacteria present are killed. The nanogold itself is thought not to exert a bactericidal effect but to enhance the light-activated antimicrobial activity of methylene blue, possibly by increasing the overall yield of ROS.

13.3 NANOPARTICULATE METAL OXIDES AS ANTIMICROBIAL AGENTS

Highly ionic nanoparticulate metal oxides have been of particular interest as antimicrobial agents as they can be prepared with extremely high surface areas and unusual crystal morphologies that have a high number of edges/corners and other potentially reactive sites [46]. However, metal oxides are now coming under close scrutiny because of their potential toxic effects [47]. Oxides under consideration as antimicrobial agents include oxides of copper, zinc oxide, titanium dioxide (titania), and tungsten oxide (WO_3).

13.3.1 Copper Oxide (CuO)

Copper oxide (CuO) is a semiconducting compound with a monoclinic structure. CuO has attracted particular attention because it is the simplest member of the family of copper compounds and exhibits a range of potentially useful physical properties, such as high temperature superconductivity, electron correlation effects, and spin dynamics [48, 49]. Limited information on the possible antimicrobial activity of nano CuO is available. Copper oxide is much cheaper than silver, easily mixed with polar liquids (i.e., water) and polymers, and relatively stable in terms of both chemical and physical properties.

Copper oxide nanoparticles have been physically and chemically characterized and investigated with respect to potential antimicrobial applications [35]. It was found that nanoscaled CuO, as generated by thermal plasma technology, demonstrated particle sizes in the range 20 to 95 nm with a mean surface area of $15.7\,m^2\,g^{-1}$ (Figure 13.1). CuO nanoparticles in suspension showed activity against a range of bacterial pathogens, including MRSA and *E. coli*, with minimum bactericidal concentrations ranging from 100 to 5 000 $\mu g\,mL^{-1}$. The ability of CuO nanoparticles to reduce bacterial populations to zero was enhanced in the presence of sub-MBCs of silver nanoparticles. As with silver, studies of CuO nanoparticles incorporated into polymers suggest that release of ions may be required for optimum killing [35]. Incorporation of nano CuO into porous elastomeric polyurethane films has demonstrated the potential for a number of applications. Recent studies have shown this approach to be effective against MRSA within 4 hours of contact [50].

13.3.2 Zinc Oxide (ZnO)

The antimicrobial mechanisms of zinc are not completely understood. In recent years, nano zinc oxide (ZnO; Figure 13.1) has received increasing attention, partly because it is stable under harsh processing conditions, but also because it is generally regarded as safe to humans [46]. Studies have shown that some nanoparticulate metal oxides, such as ZnO, have a degree of selective toxicity to bacteria with a minimal effect on human cells [51–53]. The proposed mechanisms of antibacterial activity include induction of reactive oxygen species [54, 55] and damage to the cell membrane with subsequent interaction of the nanoparticle with the intracellular contents [51].

In a study by Liu et al. [56], the antimicrobial properties of ZnO nanoparticles were investigated against *E. coli* O157:H7. This strain was significantly inhibited as shown using scanning electron microscopy (SEM) and transmission electron microscopy (TEM) analyses to assess the morphological changes of bacterial cells. Leakage of intracellular contents and a degree of membrane disorganization were observed. Using Raman spectroscopy, the intensities of lipid and protein bands were shown to increase after exposure to ZnO nanoparticles, whereas no significant change to nucleic acid was indicated.

13.3.3 Titanium Dioxide (TiO_2)

Titanium dioxide (TiO_2) is the most common titanium compound, and its ability to act as a photocatalytic antimicrobial compound is well established [57]. TiO_2 is widely used in several applications, as a powder and increasingly in a nanoparticulate form, and it is considered to be nontoxic at the concentrations normally employed. The anatase form of nano TiO_2 and UV light excitation are required to ensure maximum antimicrobial activity. It is known that TiO_2 photocatalysis can promote the peroxidation of the polyunsaturated phospholipid component of the microbial lipid membrane, induce the loss of respiratory activity, and elicit cell death [58]. In comparison with silver and copper nanoparticles, there have been relatively few studies on nano titanium and nano titanium dioxide. The study of Tsuang et al. [59] demonstrated TiO_2-mediated photocatalytic (in the presence of ultraviolet light) and bactericidal activities against obligate aerobes (*Ps. aeruginosa*), facultative anaerobes (*S. aureus*, *E. coli*, and *Enterococcus hirae*) and obligate anaerobes (*Bacteroides fragilis*).

13.4 NANOMETALS AND THE CONTROL OF BIOFILMS

Modification of materials to prevent biofilm growth or reduce adhesion of microorganisms to surfaces is of particular importance. Within this context, a biofilm can be classified as an aggregate of microorganisms in which cells adhere to each other and to a surface. Silver-containing materials are of particular interest in regard to biofilm control and have been widely used where biofilm formation is of concern, in catheters, dental materials, medical devices, implants, and wound and burn dressings.

To reduce bacterial adhesion and prevent biofilm formation, silver-based antimicrobial agents are increasingly being considered as alternatives to more conventional antimicrobials. It has been clearly demonstrated that silver is effective against a broad range of bacterial species and mature biofilms [22]. Elementary silver, silver zeolite, and silver nanoparticles, incorporated into polymers, have all been considered effective biomaterials with an antimicrobial function for a variety of biofilm reduction applications, the release of silver ions from these materials being dependent on the nature and concentration of the silver component as well as on the polymer matrix. Silver nanoparticles, with a large surface area-to-volume ratio, can thus be used at a relatively low concentration.

The superiority of silver nanoparticles has been clearly demonstrated by comparing the antibacterial properties of poly (methyl methacrylate) (PMMA) nanofiber containing silver nanoparticles with silver sulfadiazine and silver nitrate, all incorporating the same concentration of silver [60]. PMMA nanofiber was shown to be more effective in the killing of both gram-positive and gram-negative species of bacteria under planktonic growth conditions than either silver sulfadiazine or silver nitrate, which is more likely to be active through the release of silver ions. It is thought that such silver ions, on contact with bacteria, encourage the formation of black precipitates as a result of ion reduction or salt formation; in turn, these precipitates may then reduce the overall antimicrobial activity of silver sulfadiazine and silver nitrate. PMMA nanofiber was found to release silver nanoparticles with a small diameter (approximately 7 nm) and did not form precipitates when in contact with bacteria. Therefore, in theory the concentration of silver nanoparticles to be effective in the control of biofilm growth could be much lower than that required for silver ions.

13.5 PROPERTIES OF POLYMERIC MATRIX-NANOCOMPOSITES

Nanocomposites are usually solid combinations of a bulk matrix and particles with at least one of their dimensions in the nanorange. The addition of these nanoparticles confers different structural and chemical properties upon the bulk phase material. The physical properties of the nanocomposite will thus differ markedly from that of the component materials. In mechanical terms, this is attributed to the high surface-to-volume ratio of the nanoconstituents. With polymer-nanocomposites, properties related to local chemistry, thermoset cure, polymer chain mobility, conformation, and ordering can all vary markedly and continuously from the interface with the nanophase into the bulk of the matrix.

Polymer-matrix nanocomposites (nanofilled polymer composites) are, in their simplest case, made by appropriately adding nanoparticulates to a polymer matrix to enhance its functionality [61]. This can be particularly effective in producing high-performance composites when optimum dispersion of the nanofiller is achieved, and the properties of such filler are markedly superior to that of the matrix, for example, the reinforcement of a polymer matrix with more rigid nanoparticles of ceramics or carbon nanotubes. The high aspect ratio and/or the high surface area-to-volume ratio of nanoparticulates provide such enhanced properties.

Silver nanoparticles have been investigated to improve both the physical and antimicrobial properties of dental polymeric materials in, for instance, denture materials (Figure 13.3) [22] and orthodontic adhesives [62]. Lackovic et al. (2008) investigated the use of silver nanoparticles in an attempt to improve the physical and antimicrobial properties of orthodontic bracket-bonding cement.

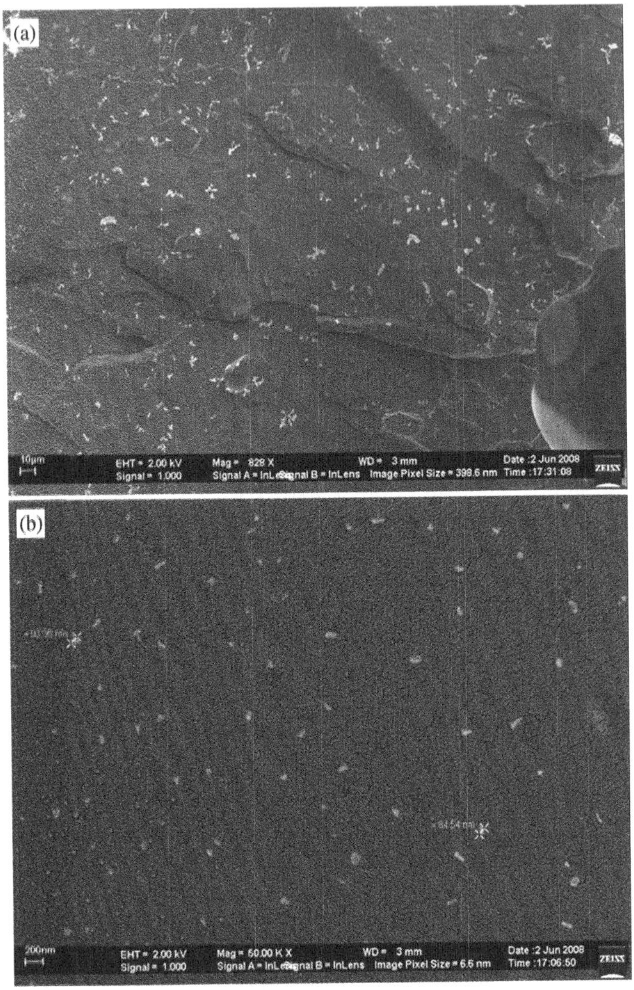

FIGURE 13.3 Scanning electron micrograph of a fractured polymethyl methacrylate PMMA/Ag nanocomposite containing approximately 0.04% wt/wt silver. Distribution of silver particles in the PMMA acrylic resin shown. (a) White areas are agglomerated silver nanoparticles distributed in the PMMA (× 828 magnification). (b) Silver nanoparticles (white dots) with approximate mean size 88 nm distributed in the PMMA matrix (× 50,000 magnification).

Source: Reprinted from Reference 22 with permission.

Incorporation of silver nanoparticles at a concentration of less than 1% wt/v was found not to decrease the modulus of the cement tested. However, no significant effect on either the attachment or the growth of the cariogenic bacterium *Streptococcus mutans* was observed. Thus, an optimum amount of silver nanoparticles used within polymer materials may well be of critical importance to avoid an adverse effect upon the physical properties. The study by Ahn et al. [62] clearly demonstrated that experimental composite adhesives (ECAs) had rougher surfaces than conventional adhesives as a result of the addition of silver nanoparticles. Bacterial adhesion to ECAs was shown to be less than that to conventional adhesives and was not influenced by saliva coating. No significant difference between ECAs and conventional adhesives was shown as regards to bond shear strength.

13.5.1 Methods of Combining Nanometals with Polymers

Combining nanoparticulate metals within a polymer matrix, so that they remain adequately dispersed, is a process that has presented some problems. Three general methods to formulate polymer-matrix nanocomposites have been used:

1. *In situ* synthesis of nanoparticles in the polymer matrix by the reduction of a metal salt in the matrix or the evaporation of the metal at the heated surface of the matrix.
2. The polymerization of the matrix around the nanoparticles.
3. The incorporation of presynthesized nanoparticles into a presynthesized polymer matrix with the aid of a blending solvent [63].

However, the first and second methods have a tendency to produce undesirable nanospecies and polydisperse (range of particle sizes) polymer matrices. An approach to impregnate silicone (as used in a wide variety of devices) with nanoparticulate silver, using supercritical carbon dioxide, has been investigated [20]. This may be particularly applicable in implantable devices, as their use is a major risk factor for hospital-acquired infection. The initiation of infection involving biomaterials requires an initial adhesion event to the device or more often to the patient-derived glycoprotein coating (conditioning film) that is deposited from the time of implantation [64]. After adhesion, microbial proliferation leads to the development of a biofilm that is more resistant to the effect of antimicrobials. Although approaches to prevent this type of infection have been proposed, including the use of coatings with nanoparticulate silver, unsatisfactory clinical results and, on occasion, further complications have occurred [65]. Possible reasons for the lack of activity include the inactivation of the antimicrobial coating by plasma components and a lack of inherent durability of the coatings used. Impregnation of polymers may well have benefits over the use of coatings or addition into the mixture. A significant level of surface/near-surface deposited silver nanoparticles should provide an initial "burst" effect. However, Furno et al. [20] showed that the majority of the antimicrobial activity, against both biofilms and planktonic cells, was simply removed by washing the polymer discs impregnated with silver nanoparticles. Further sustained release with significant antimicrobial activity would need to be carefully evaluated. The protection of both inner and outer surfaces against bacterial colonization by impregnation of an antimicrobial agent is advantageous [66] and has been demonstrated in the clinical setting [67].

The continued release of silver ions at antimicrobial concentrations even in the presence of a plasma protein conditioning film, and the ability to offer protection to both the inner and outer surfaces of a catheter, are two distinct potential advantages of polymer impregnation [20].

13.5.2 Polymeric Films Incorporating Nanometals

Nanotechnology is now beginning to be able to provide the tools required to synthesize films/layers with embedded metal nanoparticles. For copper nanoparticles embedded in an inert, teflon (polytetrafluoroethylene)-like matrix, Cioffi et al. [68] demonstrated significant antimicrobial activity *in vitro* as a result of ion release. It is suggested that in such an experimental situation a bacterial growth medium facilitates the release of metallic ions, possibly as a consequence of a reaction with the nutrient media constituents. A greater release of copper ions into a liquid medium could well be from the presence of an oxide layer on the copper nanoparticles and reaction with chloride ions in the medium. The copper-fluoropolymer (Cu-CFx) nanocomposite films in this study were deposited by dual ion-beam sputtering, a technique to deposit different polymeric and inorganic materials in a very controlled manner. Analysis of the layers revealed that the inorganic Cu (II) nanoparticles were evenly dispersed in the branched fluoropolymer matrix. Through the use of electrochemical atomic absorption spectroscopy, copper release kinetics in solutions was measured. A correlation was shown among the chemical composition (copper loading) of the material surface, the concentration of copper ions released into the microbial culture broths, and the bioactivity of the Cu-CFx coating. The Cu-CFx layers, used as coatings, were shown to inhibit the yeast *Saccharomyces cerevisiae* and the bacterial species *E. coli*, *S. aureus*, and *Listeria* spp. Such use of Cu-CFx coatings, which allow variable metal loading, demonstrate good stability upon storage, and the microbial inhibitory activity is being explored in a variety of applied fields, including food chemistry and biomedicine.

Sacrificial-anode electrochemical synthesis of metal nanoparticles in the presence of tetraoctylammonium (TAO) salts have allowed Cioffi et al. [69] to prepare copper- and silver-containing coatings by combining the metal colloid with a polymer-dispersing matrix. The surfactants used were capable of providing physical and chemical stabilization of the metal nanoparticles through the formation of a protective organic shell. These materials showed a significant inhibitory effect on the growth of *E. coli* and *S. cerevisiae*. The marked effects of such nanocomposites can be attributed to the synergistic effect of the antimicrobial metal and the TAO salts. This study has provided further support for the use of metal nanoparticles in disinfecting/antifouling paint and other coating formulations.

To determine whether the localization of controlled loadings of silver nanoparticles within nanometer-thick polymeric films could kill bacteria while supporting the growth of mammalian cells, poly (allylamine hydrochloride) and poly (acrylic acid) were prepared using layer-by-layer deposition and were then loaded with silver nanoparticles in the range 0.4–23.6 μg cm^{-2} [70]. Suspensions of *Staphylococcus epidermidis* in contact with the film were reduced 6 log fold with only 0.4 μg nanoparticles cm^{-2}. Polymeric films containing this concentration of silver nanoparticles were also shown to be nontoxic to mammalian fibroblast cells, allowing both growth and attachment.

13.6 TOXICITY ISSUES AND NANOMETALS

Although the development and applications of nanotechnology are of major importance in both industrial and consumer areas, the knowledge on human exposure and possible toxicity of nanotechnology products is limited. The greatest level of risk is posed by free nanoparticles as present in a powder or liquid. When fixed into a structure, for example, in a solid polymer nanocomposite or in a coating, it is far less likely that nanoparticles could be absorbed into the body or enter into the environment. However, upon disposal, it is not clear whether fixed nanoparticles will remain so. There is thus a possibility that they could leach into the environment, and so uncertainties remain as to the degree of risk they pose. Research into the behavior of free and fixed nanoparticles in relation to a possible toxicity risk is required.

To understand the mechanism of toxicity, a thorough knowledge of the toxico-kinetic properties of nanoparticles is required. This includes information on the absorption, distribution, metabolism, and excretion (ADME) of nanoparticles [71]. In theory, certain nanoparticles may be retained within the body for longer than is desirable, and thus, the safety profile becomes a matter of overriding significance. Nanomaterials can cross biological membranes and access cells, tissues, and organs that larger sized particles normally cannot [72]. Nanomaterials can enter the blood stream after inhalation or ingestion, and some can even penetrate the skin. Studies with lung epithelial cells, enterocytes, and skin keratinocytes indicate marked differences in susceptibility to metallic nanoparticles according to cell type tested [73]. However, a particle's surface chemistry, which in some cases can be modified, can govern whether it should be considered further for biomedical applications.

The safe use of nanotechnology and the design of nanomaterials, as used in polymer nanocomposites, particularly for biomedical applications, should involve a complete understanding of the interface between these materials and the biological systems. The interface comprises three interacting components: (1) the surface of the nanoparticle, (2) the solid–liquid interface and the effects of the surrounding medium, and (3) the contact zone with biological substrates (Figure 13.4; Table 13.1). The nanoparticle characteristics of most importance regarding the interaction with biological systems, whether mammalian or microbial, are chemical composition, surface function, shape and number of sides, porosity and surface crystallinity, heterogeneity, roughness, and hydrophobicity or hydrophilicity [74]. For example, it has been shown that titanium dioxide nanoparticles [75] act to resist the formation

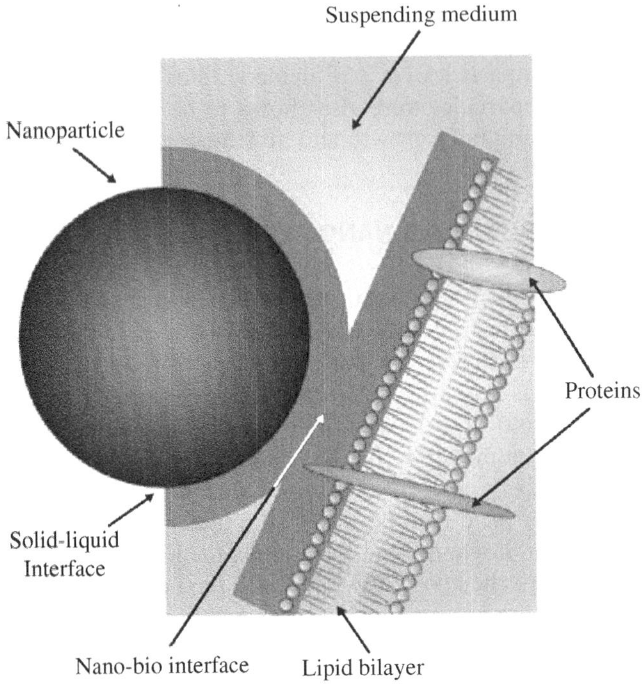

FIGURE 13.4 Representation of the interface between a nanopartcle and a lipid bilayer. *Source:* Reprinted from Reference 77 with permission.

of surface biofilms through increased hydrophilicity in comparison with an unmodified surface.

The characteristics of the surface layer, such as zeta potential (surface charge), nanoparticle aggregation, dispersion state, stability, and hydration, as influenced by the characteristics of the surrounding medium (including the ionic strength, pH, temperature, and presence of organic molecules or detergents), are also of critical importance. The importance of surface charge to both mammalian and microbial interactions has been clearly illustrated using surfactant-coated nanoparticles [76]. Both anti-adherent and antifungal effects using buccal epithelial cells in culture were demonstrated with non–drug-loaded poly (ethylcyanoacrylate) nanoparticle-treated cells. Nanoparticles were prepared using emulsion polymerization and stabilized with cationic, anionic, or nonionic surfactants. Cationic surfactants, for example cetrimide, which are known antimicrobial agents, were the most effective in reducing *Candida albicans* blastospore adhesion and showed a growth inhibitory and biocidal effect against the yeast. Production of nanoparticles with an anionic surfactant provided a lower yield and wide particle size distribution. No evidence of killing against *C. albicans* was shown. Nonionic surfactant-coated nanoparticles produced intermediate killing rates. These studies clearly demonstrate the importance of surface charge on the nanoparticle surface.

TABLE 13.1 Main biophysicochemical influences on the interface between nanomaterials and biological systems in humans

Nanoparticle
Size, shape, and surface area
Surface charge, energy, roughness, and porosity
Valence and conductance states
Functional groups
Ligands
Crystallinity and defects
Hydrophobicity and hydrophilicity
Suspending Media
Water molecules
Acids and bases
Salt and multivalent ions
Natural organic matter (proteins, lipids)
Surfactants
Polymers
Polyelectrolytes
Solid–Liquid Interface
Surface hydration and dehydration
Surface reconstruction and release of free surface energy
Ion adsorption and charge neutralization
Electrical double-layer formation, zeta potential, isoelectric point
Sorption of steric molecules and toxins
Electrostatic, steric, and electrosteric interactions
Aggregation, dispersion, and dissolution
Hydrpophilc and hydrophobic interactions
Nano-Bio Interface
Membrane interactions: specific and nonspecific forces
Receptor-ligand binding interactions
Membrane wrapping: resistive and promotive forces
Biomolecule interactions (lipids, proteins, DNA) leading to structural and functional effects
Free energy transfer to biomolecules
Conformational change in biomolecules
Oxidant injury to biomolecules
Mitochondrial and lysosomal damage, decrease in ATP

Source: Adapted from Reference 77 with permission.

The *in vivo* screening of around 130 nanoparticles intended for therapeutic use has allowed detailed assessments in regard to biocompatibility [77]. It was shown that the main independent particle variables that determine compatibility are size, surface charge, and dispersibility (particularly the effect of hydrophobicity). Cationic particles or particles with a high surface reactivity are more likely to be toxic (both to eukaryotes and to prokaryotes). Larger, more hydrophobic, or poorly dispersed particles, which are rapidly removed by the reticuloendothelial system, were shown to be relatively less toxic. Karlsson et al. [47] have recently shown that metal oxide nanoparticles are more toxic than at first envisaged, at concentrations down to 40 μg mL^{-1}, and show a high variation, depending on the nanoparticle species, in cytotoxicity, DNA damage, and oxidative DNA lesions. Copper oxide was found to be the most toxic and, therefore, may pose the greatest health risk. Nanoparticulate ZnO and TiO_2, both ingredients in sunscreens and cosmetics, showed cytotoxic and DNA-damaging effects. Iron oxides (Fe_3O_4 and Fe_2O_3) were shown to demonstrate no or relatively low toxicity. The mechanisms of toxicity for selected nanoparticles are listed in Table 13.2.

To help prevent nanoparticles from aggregation, stabilizing (capping) agents that bind to the entire nanoparticle surface can be used; these include water-soluble polymers, oligo- and poly-saccharides, sodium dodecyl sulfate, polyethylene glycols, and glycolipids. The specific roles of surface capping, size scale, and aspect ratio of ZnO particles toward antimicrobial activity and cytotoxicity have been investigated [78]. Polyethylene glycol-capped ZnO nanoparticles demonstrated an increase in antimicrobial efficacy with a reduction in particle size. Gram-negative bacteria were more affected than gram-positive, which suggests that a membrane damage mechanism of action rather than a mechanism through the production of ROS is of overriding significance. Polyethylene glycol-capped nanoparticles were highly toxic to cancer cells with a very low concentration (100 μM) threshold for cytotoxic action, whereas the

TABLE 13.2 Mechanisms of nanoparticulate cytotoxicity to mammalian cells

Nanomaterial	Cytotoxicity mechanism
TiO_2	ROS (reactive oxygen species) production
	Glutathione depletion and toxic oxidative stress
	Cell membrane disruption
ZnO	ROS production
	Dissolution and release of toxic cations
	Lysosomal damage
	Inflammation
Ag	Dissolution and silver ion release inhibits respiratory enzymes and ATP production
	ROS production
	Disruption of membrane integrity and transport processes
Gold	Disruption of protein conformation
SiO_2	ROS production
	Protein unfolding
	Membrane disruption
Cu/CuO	DNA damage and oxidative stress

Source: Adapted from Reference 77 with permission.

concentration for antibacterial activity was 50 times greater (5 mM). It is hypothesized that the toxicity to eukaryotic cells is related to nanoparticle-enhanced apoptosis by upregulation of the Fas ligand on the cell membrane.

An understanding of the interface between biological systems and nanomaterials should enable design features to be used to control the exposure, bioavailabilty, and biocatalytic activities. Several possible approaches are starting to be identified [77], including changing ability to aggregate, application of surface coatings, and altering charge density and oxidative state. However, this may well compromise the intended selective toxicity of antimicrobial metal nanoparticles.

13.7 CONCLUDING REMARKS

The use of antimicrobial nanoparticulate metals and metal oxides in polymer-nanocomposites is an area of considerable interest. These agents have received particular attention as a result of their durability and the lack of significant microbial resistance. However, if released from a polymer matrix/coating certain nanoparticles may pose environmental and direct toxicity risks to mankind. It remains to be determined how these issues will fully impact the use of polymer-nanocomposites.

ACKNOWLEDGMENTS

The author is also grateful to the *International Journal of Antimicrobial Agents* and *Nature Materials* for permission to use material from *34*, 103-110 and *8*, 543-557, respectively.

REFERENCES

1. O'Shea, J. G. (1990). "Two minutes with venus, two years with mercury"—mercury as an antisyphilitic chemotherapeutic agent. *Journal of the Royal Society of Medicine*, *83*, 392–395.
2. Ferguson, C. A., Murray, R. G. E., Lancy, P. (1979). Effects of some platinum (IV) complexes on cell division of Escherichia coli. *Canadian Journal of Microbiology*, *25*, 545–559.
3. Thurman, R. B., Gerba, C. P. (1989). The molecular mechanisms of copper and silver ion disinfection of bacteria and viruses. *CRC Critical Reviews of Environment Control*, *18*, 205–315.
4. Sadler, P. J. (1976). The biological chemistry of gold: A metallo-drug and heavy-atom label with variable valency. *Structure and Bonding*, *29*, 171–214.
5. Elsome, A. M., Hamilton-Miller, J. M. T., Brumfitt, W., Noble, W. C. (1996). Antimicrobial activities in vitro and in vivo of transition element complexes containing gold (I) and osmium (VI). *Journal of Antimicrobial Chemotherapy*, *37*, 911–918.

6. Cushing, B. L., Kolesnichenko, V. L., O'Connor, C. J. (2004). Recent advances in the liquid-phase syntheses of inorganic nanoparticles. *Chemical Reviews*, *104*, 3893–3946.
7. Allaker, R. P., Ren, G. (2008). Potential impact of nanotechnology on the control of infectious diseases. *Transactions of the Royal Society of Tropical Medicine and Hygiene*, *102*, 1–2.
8. Morones, J. R., Elechiguerra, J. L., Camacho, A., Holt, K., Kouri, J. B., Ramirez, J. T. (2005). The bactericidal effect of silver nanoparticles. *Nanotechnology*, *16*, 2346–2353.
9. Elechiguerra, J. L., Burt, J. L., Morones, J. R., Camacho-Bragado, A., Gao X., Lara, H. H., Yacaman, M. J. (2005). Interaction of silver nanoparticles with HIV-1. *Journal of Nanobiotechnology*, *3*, 6.
10. Han, J., Chen, L., Duan, S., Yang, Q. X., Yang, M., Gao, C., Zhang, B. Y., He, H., Dong, X. P. (2005). Efficient and quick inactivation of SARS coronavirus and other microbes exposed to the surfaces of some metal catalysts. *Biomedical and Environmental Sciences*, *18*, 176–180.
11. Sondi, I., Salopek-Sondi, B. (2004). Silver nanoparticles as an antimicrobial agent: A case study on E. coli as a model for Gram-negative bacteria. *Journal of Colloid and Interface Science*, *275*, 177–182.
12. Cioffi, N., Torsi, L., Ditaranto, N., Tantillo, G., Ghibelli, L., Sabbatini, L., Bleve-Zacheo, T., D'Alessio, M., Zambonin, P. G., Traversa, E. (2005). Copper nanoparticle/polymer composites with antifungal and bacteriostatic properties. *Chemistry of Materials*, *17*, 5255–5262.
13. Li, Z., Lee, D., Sheng, X., Cohen, R. E., Rubner, M. F. (2006). Two-level antibacterial coating with both release-killing and contact-killing capabilities. *Langmuir*, *22*, 9820–9823.
14. Boldyryeva, H., Umeda, N., Plaskin, O. A., Takeda, Y., Kishimoto, N. (2005). High-fluence implantation of negative metal ions into polymers for surface modification and nanoparticle formation. *Surface and Coatings Technology*, *196*, 373–377.
15. Lee, W. F., Tsao, K. T. (2006). Preparation and properties of nanocomposite hydrogels containing silver nanoparticles by ex situ polymerization. *Journal of Applied Polymer Science*, *100*, 3653–3661.
16. Verran, J., Sandoval, G., Allen, N. S., Edge, M., Stratton, J. (2007). Variables affecting the antibacterial properties of nano and pigmentary titania particles in suspension. *Dyes and Pigments*, *73*, 298–304.
17. Kim, J. S., Kuk, E., Yu, K. N., Kim, J. H., Park, S. J., Lee, H. J., Kim, S. H., Park, Y. K., Park, Y. H., Hwan,g C. Y., Kim, Y. K., Lee, Y. S., Jeong, D. H., Cho, M. H. (2007). Antimicrobial effects of silver nanoparticles. *Nanomedicine*, *3*, 95–101.
18. Pal, S., Tak, Y. K., Song, J. M. (2007). Does the antibacterial activity of silver nanoparticles depend on the shape of the nanoparticle? A study of the Gram-negative bacterium Escherichia coli. *Applied and Environmental Microbiology*, *73*, 1712–1720.
19. Tarimala, S., Kothari, N., Abidi, N., Hequet, E., Fralick, J., Dai, L. L. (2006). New approaches to antibacterial treatment of cotton fabric with silver nanoparticle-doped silica using sol-gel process. *Journal of Applied Polymer Science*, *101*, 2938–2943.
20. Furno, F., Morley, K. S, Wong, B., Sharp, B. L., Arnold, P. L., Howdle, S. M., Bayston, R., Brown, P. D., Winship, P. D., Reid, H. J. (2004). Silver nanoparticles

and polymeric medical devices: a new approach to prevention of infection? *Journal of Antimicrobial Chemotherapy*, *54*, 1019–1024.

21. Bae, T. H., Tak, T. M. (2005). Effect of TiO_2 nanoparticles on fouling mitigation of ultrafiltration membranes for activated sludge filtration. *Journal of Membrane Science*, *249*, 1–8.
22. Monteiro, D. R., Gorup, L. F., Takamiya, A. S., Ruvollo-Filho, A. C., de Camargo, E. R., Barbosa, D. B. (2009). The growing importance of materials that prevent microbial adhesion: Antimicrobial effect of medical devices containing silver. *International Journal of Antimicrobial Agents*, *34*, 103–110.
23. Herrera, M., Carrion, P., Baca, P., Liebana, J., Castillo, A. (2001). In vitro antibacterial activity of glass-ionomer cements. *Microbios*, *104*, 141–148.
24. Casemiro, L. A., Gomes-Martins, C. H., Pires-de-Souza, F. deC., Panzeri, H. (2008). Antimicrobial and mechanical properties of acrylic resins with incorporated silver-zinc zeolite - part 1 *Gerodontology*, *25*, 187–194.
25. Kawahara, K., Tsuruda, K., Morishita, M., Uchida, M. (2000). Antibacterial effect of silver-zeolite on oral bacteria under anaerobic conditions. *Dental Materials*, *16*, 452–455.
26. Matsuura, T., Abe, Y., Sato, Y., Okamoto, K., Ueshige, M., Akagawa, Y. (1997). Prolonged antimicrobial effect of tissue conditioners containing silver-zeolite. *Journal of Dentistry*, *25*, 373–377.
27. Morishita, M., Miyagi, M., Yamasaki, Y., Tsuruda, K., Kawahara, K., Iwamoto, Y. (1998). Pilot study on the effect of a mouthrinse containing silver zeolite on plaque formation. *Journal of Clinical Dentistry*, *9*, 94–96.
28. Li, P., Li, J., Wu, C., Wu, Q., Li, J. (2005). Synergistic antibacterial effects of β-lactam antibiotic combined with silver nanoparticles. *Nanotechnology*, *16*, 1912–1917.
29. Rai, M., Yadav, A., Gade, A. (2009). Silver nanoparticles as a new generation of antimicrobials. *Biotechnology Advances*, *27*, 76–83.
30. Sambhy, V., MacBride, M. M., Peterson, B. R., Sen, A. (2006). Silver bromide nanoparticle / polymer composites: Dual action tunable antimicrobial materials. *Journal of the American Chemical Society*, *128*, 9798–9808.
31. Feng, Q. L., Wu, J., Chen, G. Q., Cui, F. Z., Kim, T. M., Kim, J. O. (2000). A mechanistic study of the antibacterial effect of silver ions on Escherichia coli and Staphylococcus aureus. *J Biomedical Materials Research*, *52*, 662–668.
32. Yamanaka, M., Hara, K., Kudo, J. (2005). Bactericidal actions of a silver ion solution on Escherichia coli, studied by energy-filtering transmission electron microscopy and proteomic analysis. *Applied and Environmental Microbiology*, *71*, 7589–7593.
33. Bragg, P. D., Rainnie, D. J. (1974). The effect of silver ions on the respiratory chain of Escherichia coli. *Canadian Journal of Microbiology*, *20*, 883–889.
34. Yoon, K. Y., Byeon, J. H., Park, J. H., Hwang, J. (2007). Susceptibility constants of Escherichia coli and Bacillus subtilis to silver and copper nanoparticles. *Science of the Total Environment*, *373*, 572–575.
35. Ren, G., Hu, D., Cheng, E. W. C., Vargas-Reus, M. A., Reip, P., Allaker, R. P. (2009). Characterisation of copper oxide nanoparticles for antimicrobial applications. *International Journal of Antimicrobial Agents*, *33*, 587–590.

36. Ruparelia, J. P., Chatterje, A. K., Duttagupta, S. P., Mukherji, S. (2008). Strain specificity in antimicrobial activity of silver and copper nanoparticles. *Acta Biomaterialia, 4,* 707–716.
37. Beveridge, T. J., Murray, R. G. E. (1980). Sites of metal deposition in the cell wall of Bacillus subtilis. *Journal of Bacteriology, 141,* 876–878.
38. Stohs, S. J., Bagchi, D. (1995). Oxidative mechanisms in the toxicity of metal ions. *Free Radical Biology and Medicine, 18,* 321–336.
39. Lin, C., Yeh, Y., Yang, C., Chen, C., Chen, G., Chen, C. C., Wu, Y. C. (2002). Selective binding of mannose-encapsulated gold nanoparticles to type I pili in Escherichia coli. *Journal of the American Chemical Society, 13,* 155–168.
40. Williams, D. N., Ehrman, S. H., Pulliman Holoman, T. R. (2006). Evaluation of the microbial growth response to inorganic nanoparticles. *Journal of Nanobiotechnology, 4,* 3.
41. Murakami, T., Tsuchida, K. (2008). Recent advances in inorganic nanoparticle-based drug delivery systems. *Mini-Reviews in Medicinal Chemistry, 8,* 175–183.
42. Pissuwan, D., Cortie, C. H., Valenzuela, S. M., Cortie, M. B. (2010) Functionalised gold nanoparticles for controlling pathogenic bacteria. *Trends in Biotechnology, 28,* 207–213.
43. Allaker, R. P., Douglas, C. W. I. (2009). Novel anti-microbial therapies for dental plaque-related diseases. *International Journal of Antimicrobial Agents, 33,* 8–13.
44. Pagonis, T. C., Chen, J., Fontana, C. R., Devalapally, H., Ruggiero, K., Song, X., Foschi, F., Dunham, J., Skobe, Z., Yamazaki, H., Kent, R., Tanner, A. C. R., Amiji, M. M., Soukos, N. S. (2010). Nanoparticle-based endodontic antimicrobial photodynamic therapy. *Journal of Endodontics, 36,* 322–328.
45. Perni, S., Piccirillo, C., Pratten, J., Prokopovich, P., Chrzanowski, W., Parkin, I., Wilson, M. (2009). The antimicrobial properties of light-activated polymers containing methylene blue and gold nanoparticles. *Biomaterials, 30,* 89–93.
46. Stoimenov, P. K., Klinger, R. L., Marchin, G. L., Klabunde, K. J. (2002). Metal oxide nanoparticles as bactericidal agents. *Langmuir,18,* 6679–6686.
47. Karlsson, H. L., Cronholm, P., Gustafsson, J., Moller, L. (2008). Copper oxide nanoparticles are highly toxic: A comparison between metal oxide nanoparticles and carbon nanotubes. *Chemical Research in Toxicology, 21,* 1726–1732.
48. Cava, R. J. (1990). Structural chemistry and the local charge picture of copper oxide superconductors. *Science, 247,* 656–662.
49. Tranquada, J. M., Sternlieb, B. J., Axe, J. D., Nakamura, Y., Uchida, S. (1995). Evidence for stripe correlations of spins and holes in copper oxide superconductors. *Nature, 375,* 561.
50. Ahmad, Z., Vargas-Reus, M. A., Bakhshi, R., Ryan, F., Ren, G., Oktar, F., Allaker, R. P. (2010). Antimicrobial properties of electrically formed elastomeric polyurethane-copper oxide nanocomposites. Unpublished Observations.
51. Brayner, R., Ferrari-Iliou, R., Brivois, N., Djediat, S., Benedetti, M. F., Fievet, F. (2006). Toxicological impact studies based on Escherichia coli bacteria in ultrafine ZnO nanoparticles colloidal medium. *Nano Letters 6,* 866–870.
52. Reddy, K. M., Feris, K., Bell, J., Wingett, D. G., Hanley, C., Punnoose, A. (2007). Selective toxicity of zinc oxide nanoparticles to prokaryotic and eukaryotic systems. *Applied Physics Letters, 90,* 213902.

53. Zhang, L. L., Jiang, Y. H., Ding, Y. L., Povey, M., York, D. (2007). Investigation into the antibacterial behaviour of suspensions of ZnO nanoparticles (ZnO nanofluids). *Journal of Nanoparticle Research*, *9*, 479–489.

54. Sawai, J. (2003). Quantitative evaluation of antibacterial activities of metallic oxide powders (ZnO, MgO and CaO) by conductimetric assay. *Journal of Microbiological Methods*, *54*, 177–182.

55. Jones, N., Ray, B., Ranjit, K. T., Manna, A. C. (2008). Antibacterial activity of ZnO nanoparticle suspensions on a broad spectrum of microorganisms. *FEMS Microbiology Letters*, *279*, 71–76.

56. Liu, Y., He, L., Mustapha, A., Li, H., Hu, Z. Q., Lin, M. (2009). Antibacterial activities of zinc oxide nanoparticles against Escherichia coli O157:H7. *Journal of Applied Microbiology*, *107*, 1193–1201.

57. Blake, D. M., Maness, P. -C., Huang, Z., Wolfrum, E. J., Jacoby, W. A., Huang, J. (1999). Application of the photocatalytic chemistry of titanium dioxide to disinfection and the killing of cancer cells. *Separation and Purification Methods*, *28*, 1–50.

58. Maness, P. C., Smolinski, S., Blake, D. M., Huang, Z., Wolfrum, E. J., Jacoby, W. A. (1999). Bactericidal activity of photocatalytic TiO_2 reaction: Toward an understanding of its killing mechanism. *Applied and Environmental Microbiology*, *65*, 4094–4098.

59. Tsuang, Y. H., Sun, J. S., Huang, Y. C., Lu, C. H., Chang, W. H. S., Wang, C. C. (2008). Studies of photokilling of bacteria using titanium dioxide nanoparticles. *Artificial Organs*, *32*, 167–174.

60. Kong, H., Jang, J. (2008). Antibacterial properties of novel poly(methyl methacrylate) nanofiber containing silver nanoparticles. *Langmuir*, *24*, 2051–2056.

61. Manias, E. (2007). Nanocomposites: Stiffer by design. *Nature Materials*, *6*, 9–11.

62. Ahn, S. J., Lee, S. J., Kook, J. K., Lim, B. S. (2009). Experimental antimicrobial orthodontic adhesives using nanofillers and silver nanoparticles. *Dental Materials*, *25*, 206–213.

63. Corbierre, M. K., Cameron, N. S., Sutton, M., Laaziri, K., Lennox, R. B. (2005). Gold nanoparticle / polymer nanocomposites: Dispersion of nanoparticles as a function of capping agent molecular weight and grafting density. *Langmuir*, *21*, 6063–6072.

64. Green, R. J., Davies, M. C., Roberts, C. J., Tendler, S. J. (1999). Competitive protein adsorption as observed by surface plasmon resonance. *Biomaterials*, *20*, 385–391.

65. Riley, D. K., Classen, D. C., Stevens, L. E., Burke, J. P. (2002). A large randomised clinical trial of a silver-impregnated urinary catheter: Lack of efficacy and staphylococcal superinfection. *American Journal of Medicine*, *98*, 349–356.

66. Wilcox, M., Kite, P., Dobbins, B. (1988). Antimicrobial intravascular catheters—which surface to coat? *Journal of Hospital Infection*, *40*, 322–323.

67. Darouiche, R. O., Raad, I. I., Heard, S. O., Thornby, J. I., Wenker, O. C., Gabrielli, A., Berg, J., Khardori, N., Hanna, H., Hachem, R., Harris, R. L., Mayhall, G. (1999). A comparison of two antimicrobial-impregnated central venous catheters. Catheter Study Group. *New England Journal of Medicine*, *340*, 1–8.

68. Cioffi, N., Ditaranto, N., Torsi, L., Picca, R. A., Sabbatini, L., Valentini, A, Novello, L., Tantillo, G., Bleve-Zacheo, T., Zambonin, P. G. (2005). Analytical

characterization of bioactive fluoropolymer ultra-thin coatings modified by copper nanoparticles. *Analytical and Bioanalytical Chemistry*, *381*, 607–616.

69. Cioffi, N., Ditaranto, N., Torsi, L., Picca, R. A., De Giglio, E., Sabbatini, L., Novello, L., Tantillo, G., Bleve-Zacheo, T., Zambonin, P. G. (2005). Synthesis, analytical characterization and bioactivity of Ag and Cu nanoparticles embedded in poly-vinyl-methyl-ketone films. *Analytical and Bioanalytical Chemistry*, *382*, 1912–1918.
70. Agarwal, A., Weis, T. L., Schurr, M. J., Faith, N. G., Czuprynski, C. J., Mc Anulty, J. F., Murphy, C. J., Abbott, N. L. (2010). Surfaces modified with nanometer-thick silver-impregnated polymeric films that kill bacteria but support growth of mammalian cells. *Biomaterials*, *31*, 680–690.
71. Hagens, W. I., Oomen, A. G., de Jong, W. H., Cassee, F. R., Sips, A. J. (2007). What do we (need to) know about the kinetic properties of nanoparticles in the body? *Regulatory Toxicology and Pharmacology*, *49*, 217–229.
72. Yang, Z., Liu, Z. W., Allaker, R. P., Reip, P., Oxford, J., Ahmad, Z., Ren, G. (2010). A review of nanoparticle functionality and toxicity on the central nervous system. *Journal of Royal Society Interface*, *7*, 411–422.
73. Allaker, R. P. (2010). The use of nanoparticles to control oral biofilm formation. *Journal of Dental Research*, *89*, 1175–1186.
74. Nel, A., Xia, T., Madler, I., Li, N. (2006). Toxic potential of materials at the nanolevel. *Science*, *311*, 622–627.
75. Luo, M. L., Zhao, J. Q., Tang, W., Pu, S. (2005). Hydrophilic modification of poly (ether sulfone) ultrafiltration membrane surface by self-assembly of TiO_2 nanoparticles. *Applied Surface Science*, *49*, 76–84.
76. McCarron, P. A., Donnelly, R. F., Marouf, W., Calvert. D. E. (2007). Anti-adherent and antifungal activities of surfactant-coated poly (ethylcyanoacrylate) nanoparticles. *International Journal of Pharmaceutics*, *340*, 182–190.
77. Nel, A. E., Madler, L., Velegol, D., Xia, T., Hoek, E. M. V., Somasundaran, P., Klaessig, F., Castranova, V., Thompson, M. (2009). Understanding biophysicochemical interactions at the nano-bio interface. *Nature Materials*, *8*, 543–557.
78. Nair, S., Sasidharan, A., Rani, V. V. D., Menon, D., Nair, S., Manzoor, K., Raina, S. (2008). Role of size scale of ZnO nanoparticles and microparticles on toxicity toward bacteria and osteoblast cancer cells. *Journal of Materials Science: Materials in Medicine*, *20*, 235–241.

CHAPTER 14

TITANIUM DIOXIDE-BASED PLASTIC TECHNOLOGIES

MARTA FERNÁNDEZ-GARCÍA,[1] MARÍA L. CERRADA,[1]
ANNA KUBACKA,[2] and MARCOS FERNÁNDEZ-GARCÍA[2]
[1] Instituto de Ciencia y Tecnología de Polímeros (ICTP-CSIC), Madrid, Spain
[2] Instituto de Catálisis y Petroleoquímica, (ICP-CSIC), Madrid, Spain

CONTENTS

14.1 Titanium Dioxide: General Introduction . . . 352
14.2 Titanium Dioxide: Disinfection . . . 354
14.3 Titanium Dioxide-Polymeric Nanocomposites . . . 356
14.4 TiO_2-Nanocomposites with Biocidal Properties . . . 366
References . . . 369

The idea of this chapter is to impart some notions about the preparation and physical properties of titania-based polymeric nanocomposites. To do so, this contribution highlights selected works among an extensive literature. Before getting into the heart of the discussion, we first review titanium dioxide characteristics, paying attention to its photocatalytic properties. Afterward, the different methodologies that can be used to obtain titanium dioxide-polymeric nanocomposites are described. Representative examples of materials with different properties, i.e., high transparency, auto-degradability, antimicrobial, among others, obtained by optimization of the preparation protocol in both components, inorganic nanoparticles and polymeric matrix, are presented. Finally, we widely discuss the antimicrobial properties of nanocomposites containing titania.

Antimicrobial Polymers, First Edition.
Edited by José M. Lagarón, María J. Ocio, and Amparo López-Rubio.

14.1 TITANIUM DIOXIDE: GENERAL INTRODUCTION

Photoinduced processes are studied in several industrial-oriented applications, which have been developed since their first descriptions in the scientific literature. Despite the difference in character and utilization, all photoinduced processes have the same origin. A semiconductor can be excited by light energy higher than the band gap inducing the formation of energy-rich electron-hole pairs. It is commonly understood that the term *photocatalysis* is referred to any chemical process catalyzed by a solid where the external energy source is an electromagnetic field with wavenumbers in the ultraviolet (UV)-visible range [1–6]. Customarily, photocatalysts are solid semiconductors that are (1) able to absorb visible and/or UV light, (2) chemically and biologically inert and photostable, (3) inexpensive, and (4) nontoxic. Titania (TiO_2), zinc oxide (ZnO), strontium titanate ($SrTiO_3$), ceria (CeO_2), tungstena (WO_3), iron oxide (Fe_2O_3), cadmium sulphide (CdS), and zinc sulphide (ZnS) can act as photoactive materials for redox processes as a result of their electronic structure, which is characterized by a filled valence band and an empty conduction band. Among these possible semiconductors, titanium dioxide, TiO_2, is the most used photocatalytic material as it fulfils all of these requirements and exhibits adequate conversion values [7].

TiO_2 occurs in nature in three crystallographic phases: rutile, anatase, and brookite, with anatase being the most commonly employed in photocatalytic applications because of its inherent superior photocatalytic properties [6–8]. Anatase is the least thermodynamically stable polymorph of TiO_2 as a bulk phase, although from energy calculations, it appears as the more likely phase when the grain size is smaller than 15–20 nm [9]. High surface TiO_2 materials would thus present the anatase polymorph as a general rule. The crystalline structure of the TiO_2 oxides can be described in terms of TiO_6 octahedral chains differing by the distortion of each octahedron and the assembly pattern of such octahedra chains. The Ti-Ti distances in the anatase structure are greater than in rutile, whereas Ti-O distances are shorter [10]. These structural differences cause different mass densities and lead to a different electronic structure of the bands. The anatase phase is 9% less dense than rutile, presenting more pronounced localization of the Ti $3d$ states and, therefore, a narrower $3d$ band. Also the O $2p$-Ti $3d$ hybridization is different in the two structures (more covalent mixing than in rutile), with anatase exhibiting a valence and conduction band with, respectively, more pronounced O $2p$-Ti $3d$ characters and less nonbonding self-interaction between similar ions (e.g., anion–anion and cation–cation interactions) [11]. The importance of the covalent versus ionic contributions to the metal-oxygen bond is discussed in a more general context for several photocatalytic oxides by Wiswanathan [12]. In any case, these differential structural features, which distinguish anatase and rutile, are presumably responsible for the difference in the mobility of charge carriers during light excitation.

Nanostructured anatase-TiO_2 has been synthesized by an enormous number of techniques including liquid-phase sol–gel, microemulsion, hydro and solvothermal, aerogel, sonochemical, or surfactant-templated methods. Also solid-state chemistry methods related to the oxidation or chemical vapor deposition of (solid) precursors have been used. Review articles summarized the details of such methods and the physicochemical properties of the resulting titania solids [13, 14], In brief, TiO_2 nanoparticles would appear essentially amorphous smaller than 3–4 nm and will display the anatase structure in the stability region mentioned previously, e.g., smaller than ca. 15–20 nm. Truncated bi-pyramidal shapes mainly exposing (101), (001), and (100) surfaces are essentially expected for anatase nanoparticles, although very recent studies show unexpected shapes for "molecular" (e.g., smaller than 1.5 nm) TiO_2 entities, which may be potentially present near the low 3–4 nm limit [15]. Figure 14.1 shows that anatase nanoparticles can have two main or "extreme" morphologies: elongated (along the *c* crystallographic axis) ones dominated by (100) and (101) faces, and more isotropically ones with certain additional main contribution from (001) faces. Typical anatase nanoparticle morphologies can be thus represented by an aspect parameter (B/A in Figure 14.1) that usually gets values of 0.3–0.4 and may display a maximum value of 0.57 [16]. Modulation/control of the *c*-axis elongation of the particle by presence of alien surface species such as fluorine ions during preparation conditions [17] as well as morphology control with formation of nanostructures (other than nanoparticles as, for example, nanosheets) having a dominant presence of (001) faces [18–21] were also recently reported. Recent work indicates that nanoparticle morphology (primary particle size/shape) is governed, on the one hand, by thermodynamic (specifically the initial local order, around 3–6 Angstroms, in the solid precursor state of the oxide at nucleation states) and not only kinetic parameters of the anatase genesis process [22] and, on the other hand, that surface OH density and net charge (e.g., hydration of surface layers) [16, 23] together with the number of kinks/edges [24] are key factors to control surface energy and, thus, morphology and can be envisaged to be key parameters in

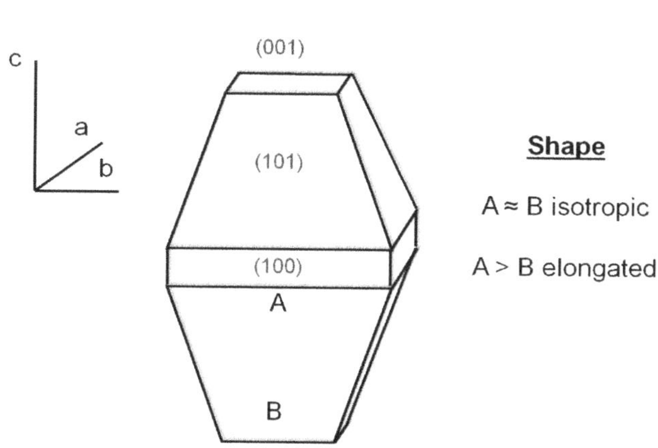

FIGURE 14.1 Morphologies of anatase nanoparticles.

controlling the elongation along the *c* axis (e.g., particle shape) in the case of OH species.

In the quest to optimize the photoactivity of anatase-TiO_2 systems, several paths have been pursued. Almost from the conception of the use of photocatalysis, the photoactivity of anatase-TiO_2 systems was typically improved with the addition of surface noble metals such as Pt, Pd, and Ag, which act as electron trapping centers, and/or oxide–oxide contact using SnO, ZrO_2 and others with appropriate electronic structure to influence the electron-hole charge separation process positively [1, 5, 6, 25]. Another methodology, essential in the current research, considers the extension of the solid light-absorption spectrum to the visible region. This would facilitate the use of sunlight, an inexpensive, renewable energy source, as the excitation energy of the photocatalytic processes. Several alternatives, mainly based on the use of inorganic solids like metals, sulphides, nitrides, and other non-Ti-based oxides with adequate optical properties, other (polymeric or molecular species) sensitizers, or hypothetical cubic-type TiO_2 polymorphs have been tested in pursuing this goal [1, 26]. Additional studies involving TiO_2 and aimed at improving its optical absorption and photocatalytic performance are focused on anatase modification by cation and/or anion doping [1, 4–6, 27]. Doping may also turn out in surface modification as well as changes in the electron-hole charge handling properties of the anatase-TiO_2 systems.

14.2 TITANIUM DIOXIDE: DISINFECTION

Photocatalytic disinfection has undergone a boost in the last decades, and promising results have been obtained with a significant number of microorganisms including gram-negative bacteria (*Escherichia coli, Enterobacter cloacae, Erwinia caratovora, Salmonella typhimurimm, enterica, and faecalis, Pseudonomas aeruginosa and fluorescens, Listeria monocytogenes, Klebsiella pneumonae*, and *Microbacterium sp.*), gram-positive bacteria (*Staphylococuss aureus, Streptococuss sobrinus, Bacillus stearopthermophilus, pumilus, subtilis*, and sp., *Lactobacillus helveticus*, and *plantarum*), yeasts (*Zygosaccharomices rouxii* and *Pichia jadini*), fungus (*Aspergillus niger, Candida albicans*, and *Fusarium solani*), protozoa (*Giardia lambia* and *Acanthamoeba polyphaga*), and viruses (*Lactobacillus case phage PL-1, Bacteriophage MS2, polvirus 1, Avian A/H5N2*, and *hepatitis B*) [5, 6, 28–36]. Titania is used as a powder [5, 6, 28–32, 34, 36] but also supported [5, 6, 30, 33, 35] on plastics, metals, ceramics, and other materials, which facilitate its use and recovery. These studies showed that photokilling performance is sensitive to several experimental factors; among others of importance are included:

- Excitation energies, fluences (power × time), and irradiation conditions (pulsed, continuous) used in experiments with titania, as they all significantly influence the photokilling performance, possibly in a more pronounced way than in the traditional use of anatase in the photodegradation of organic waste. For example, although *E. coli* and *B. fragilis* require prolonged illumination for effective killing, *B. pumilis* inactivation is better obtained by using intermittent illumination [37–39]. The photon energy is also important as TiO_2 materials are mostly effective upon ultraviolet light excitation, but their interaction with certain polymeric matrices [33, 35] and/or their modification by adding additional components (like Ag, PdO, etc.) or doping allows the use of visible and sunlight as excitation sources [40, 41].

- State and nature of the biocidal oxide. TiO_2 works by surface/near-surface contact and displays significant variability in efficiency when used as powders or immobilized on a support. As it is well known, that it may present typically a 1-fold order of magnitude greater activity as a powder (for example, in the case of *E. coli*) because nanoparticles in suspension/powders can be ingested by microorganisms by phagocytosis, causing rapid cellular damage in addition to that triggered by photoactivity [39].
- Temperature also has an effect as typically the inactivation rate increases/decreases with temperature for gram-positive/gram-negative bacteria. Coliforms are the exception to this rule. All microorganisms display, in any case, very narrow temperature ranges where photocatalytic disinfection activity reaches maximum values [5, 6, 39].
- Titania photokilling performance seems to decrease in the order virus $>$ gram negative $>$ gram positive $>$ bacterial spores $\approx$ yeasts $>$ fungus [5, 6, 42]. This appears to be connected with the increasing cell wall complexity, which goes from the thin peptidoglycan layer of gram-negative bacteria to the thicker and more compact walls of gram-positive bacteria and cocci, and ends with the thick eukaryotic cell membrane containing sugar polymers for yeasts and/or complex peptidoglycan chains for fungus.
- The last minor point is specific for the measurements of cell inactivation reaction rates as they show a linear relationship with the initial bacteria concentration, a fact that should be considered when comparing results.

Despite these aspects, good efficiencies close to those considered useful for bacteria disinfection (4–5 log reduction in the temporal range of minutes) are customarily reported in studies reviewed here. Note, however, that wastewater or other real targets for disinfection having significant turbidity (water, for example, contains significant amounts of solids in suspension) as well as dissolved organic matter may complicate the functionality of titania photocatalysts. Tests of titania performance with real municipal wastewater have been proven nevertheless successful [6]. Also, the presence of microorganism aggregation forming biofilms is another point of relevance because of the limited uses of current technologies. In this aspect, only few titania-based photocatalysts show adequate performance for effectively reducing the microorganism population [33].

The mechanism leading to cell death appears as key information in order to optimize the photokilling process but is not yet fully understood. Adherence of the microorganism seems the first important stage in the process when using TiO_2 powders [43] but may not be as dramatic in the case of polymer-based composites as the surface contact area is here orders of magnitude superior than the one corresponding to the powders. The photocatalytic attack seems to be carried out by light-induced, hole-related species upon UV light [44–46], whereas oxide (e.g., superoxide and singlet oxygen) species come out as the key intermediate upon visible light [46]. Early proposals suggest that photokilling mechanisms imply the use of such charge carrier species in the oxidation of the intracellular coenzyme A (CoA), inhibiting cell respiration. Subsequent direct TiO_2–microorganism contact would result in cell death [37]. However, the photokilling sensitivity that seems to be related to the microorganism structural surface properties, particularly to the chemical complexity and thickness of the cell wall, suggests that it is initiated by a cell wall and/or cytoplasmatic membrane attack. Kinetic studies with *E. coli* spheroplats, which do not have cell walls, strongly support such a mechanism [47]. Recent microscopy studies indicate that cell walls undergo a continuous collapse by which the photoprocess leads initially to lysed and round-shaped cells with a restricted number of breaks in their walls. This fact decreases the cell volume by a factor of two thirds, but the microorganism may subsequently regrow under dark conditions. In a subsequent step, the viability of cells is, however, fully lost by further attack of TiO_2-derived radicals [33]. The viability loss would be a function of the cell wall chemical complexity and thickness and of the efficiency of the microorganism repair/protection mechanisms using the superoxide dismutase (an enzyme that dismutates superoxide radicals to H_2O_2 and O_2) and catalase (an enzyme that reduces intracellular concentration of H_2O_2 and converts it into H_2O and O_2) enzymes.

14.3 TITANIUM DIOXIDE-POLYMERIC NANOCOMPOSITES

In the last few years, hybrid or nanocomposite organic–inorganic materials that combine attractive qualities of dissimilar components have received great interest for a wide range of mechanical, electronic, magnetic, biological, and optical properties. These novel materials are not simply physical mixtures but can be broadly defined as complex materials having both organic and inorganic components intimately connected at nanometric length scale. The methodologies used for their preparation are mainly: sol–gel, *in situ* polymerization, solution, and melt processes. They can be classified as Novak [48] established as follows: type I for soluble, preformed organic polymers embedded in an inorganic network; type II for embedded, preformed organic polymers possessing covalent bonds to the inorganic network; type III for mutually interpenetrating organic-inorganic networks; type IV for mutually interpenetrating networks with covalent bonds between the organic and inorganic phases; and type V for "nonshrinking" sol–gel composite materials. Obviously, there are materials falling between several categories, which would make difficult their classification.

Titania-containing polymer-based nanocomposites find application in a significant number of fields related to the exploitation of several physical and chemical properties of the oxide. Particularly important features are those related to its optical properties, but here a short review will be made of all the systems presented in literature leaving for the end, in a small section, the analysis of those contributions based on the use of the photocatalytic properties of the oxide in biocidal processes.

Focusing on titania-nanocomposites, Lantelme et al. [49] in 1996 obtained a hybrid organic–inorganic material by *in situ* polymerization of titanium isopropoxide in polyvinylacetate (PVAc). The use of a titanium alkoxide as the inorganic precursor enables a chemical cross-linking of PVAc resulting from transesterification reactions. The strong chemical interactions prevent a macroscopic phase separation and allow manufacturing composites at nanometer scale (size < 10 nm) where the clusters are randomly dispersed in the PVAc matrix and are bound to the matrix through an interphase made of a titanium dioxide network interpenetrated by polymer chains. However, the main disadvantage of this nanocomposite is that Ti-based clusters contain some residual isopropoxy groups and the TiO_2 phase could display limited crystallinity and thus high electron-hole recombination rates, limiting the conversion of light into chemical energy.

Hu and Marand [50] prepared poly(amide–imide)/TiO_2 (PAI/TiO_2) composite films by the *in situ* sol–gel process using tetraethyltitanate. In this case, the hydrogen bonding interactions between the amide group in the PAI polymer and the hydroxyl groups on the inorganic oxide allow obtaining a high degree of mixing of the organic–inorganic system and the formation of the nanosized TiO_2 domains. The size of these domains increases from 5 to 50 nm when the TiO_2 content rises from 3.7% to 17.9% by weight. These composite films exhibit higher optical transparency, higher glass transition temperature, an increase and flattening of the rubbery plateau modulus, and a decrease in polymer crystallinity in comparison with the pure PAI. This is a constant while using preparation methods that do not allow calcination of titania higher than 673 K in order to obtain reasonable crystalline oxide phases, and will not be mentioned again for subsequent works.

Transparent nanohybrid thin films (250 to 350 nm in thickness) consisting of nanocrystalline TiO_2 particles in PMMA were also successfully synthesized by *in situ* sol–gel and spin coating assisted polymerization, using titanium isoproproxide as precursors and methyl methacrylate and 3-(trimethoxysilyl)propyl methacrylate as monomers [51]. All nanohybrid thin film samples exhibit a nonlinear optical behavior with a very fast characteristic relaxation time of ~ 1.5 ps, which has been measured by a pump probe technique. Similar nanomaterials using the same silane coupling agent have been developed by *in situ* polymerization in supercritical carbon dioxide [52].

Ahmad et al. [53] demonstrated the effect of a nanosized TiO_2 addition in different concentrations to PMMA-based gel polymeric electrolytes on the conductivity as well as on the thermal and rheological properties. The addition of nanoparticles enhances the ionic conductivity with negligible effect on other electrochemical properties. Furthermore, no changes appear in the X-ray diffraction pattern, but an increase in the glass transition temperature (T_g) of the resulting electrolytes is noticeable. These nanocomposite polymeric electrolytes are prepared by PMMA addition into a homogenous dispersion of commercial anatase TiO_2 nanoparticles ($d = 15$ nm), in different weight percentages with respect to liquid electrolyte weight. Free radical suspension polymerization in water medium with *in situ* sol–gel transformation is also used by these authors to prepare PMMA/TiO_2 polymeric nanocomposites [54]. Electrolytes based on these polymeric materials show superior properties compared with those found in conventional polymer electrolytes. Acrylamide has been also polymerized in aqueous solution in the presence of titania nanoparticles (average diameter of 30 nm and a specific surface area of 50 m^2/g) and a cross-linking agent to obtain photocatalytic-degradable nanocomposites [55].

Surface-modified TiO_2 nanoparticles, obtained with 6-palmitate ascorbic acid, were encapsulated in poly(methyl methacrylate), PMMA, by *in situ* radical polymerization thermally initiated by 2,2′-azobisisobutyronitrile in toluene solutions [56]. These nanoparticles were previously synthesized from controlled hydrolysis of titanium tetrachloride, having an average diameter of 4.5 nm. The resulting nanocomposites have similar T_g, which could be unexpected, since the presence of inorganic nanoparticles is reported in literature that leads to an increase of T_g [50]. This fact is explained as a result of two opposing effects. Specifically, the existing palmitoyl chains in the modified nanoparticles could act as a plasticizer, reducing the T_g of PMMA, whereas the presence of TiO_2 particles might trigger an increase of T_g. However, the thermal stability analyzed either in inert or in air atmospheres exhibits a significant increase. The nanocomposite degradation curves under nitrogen atmosphere do not show any peak at 180 °C indicating the lack of head-to-head bonds (polymer chain double bond at the end) in PMMA chains, which is responsible for the first degradation stage. The improvement of the degradation under thermo-oxidative atmosphere is attributed to the presence of ascorbid acid chains.

Acrylic acid and allyl acetylacetone have been also chosen as coupling agents to modify the titania surface, and then, methyl methacrylate is a conventionally free radical polymerized using benzoyl peroxide, BPO, as the initiator [57]. When acrylic acid is used as compatibilizer, the starting metal alkoxide is titanium butoxide, whereas this is titanium isopropoxide when allyl acetylacetone is selected, because it is difficult to obtain transparent hybrid materials using titanium butoxide. The hybrid materials achieved with the latest one show a reversible thermochromic effect. They are red and transparent at room temperature and change to yellow and opaque when they are cooled. A similar approach is used to manufacture hybrid materials with different morphologies by self-assembly of trifluoroacetic acid modified titania with poly(ethylene oxide)-*b*-poly(propylene oxide)-*b*-poly(ethylene oxide), PEO-PPO-PEO, amphiphilic triblock copolymers [58]. Information about molecular proximities and interfaces, which is unavailable from transmission electron microscopy (TEM) and small angle X-ray scattering (SAXS) experiments, has been attained by solid-state nuclear magnetic resonance (NMR) techniques, such as cross-polarization and two-dimensional (2D) experiments.

Beads milling with centrifugal bead separation has been used to prepare PMMA/TiO_2 nanocomposite materials from stable nanoparticle-monomer dispersions with the addition of (3-acryloxypropyl)trimethoxysilane as a coupling agent, followed by monomer polymerization [59]. These well-dispersed titania nanoparticles have little effect on the transmittance of visible light through MMA but enhance the UV absorbing properties of MMA. Improvements on achieving monodisperse nanocomposite particles are made by coupling of an electrospray method following the beads mill protocol [60]. In this work, authors found that an increase in concentration of titania nanoparticles ($\sim$15 nm, rutile phase) leads to a decrease in secondary particle size (aggregation size), their distribution being made uniform.

Commercial titania nanoparticles (diameter approximately 30 nm) have been also encapsulated by using styrene microemulsion polymerization [61]. This synthesis method requires two conditions to attain stable microemulsions: (1) the TiO_2 nanoparticles must be successfully dispersed in the monomer phase and (2) this phase must be dispersed in an aqueous surfactant solution to form stable submicron droplets. In this work, the effect of nanoparticle concentration on the miniemulsification step is proved by increasing the droplet size as the amount of titania is added to miniemulsion.

In situ suspension polymerization has also been used to obtain TiO_2/PMMA composite cross-linked microspheres [62] with a size range of 1–10 μm using 10–50 wt% of titania. The interfacial compatibility is studied using dimethyl silicone treated (hydrophobically treated) or untreated TiO_2 nanoparticles. The sedimentation stability of the nanoparticles in the monomer mixture proves that the untreated nanoparticles show a tendency to separate in solvents easily because of their high density and low compatibility with organic materials. The sun protection factor indexes of microspheres in water-in-oil emulsions show a negligible decrease, their value being only slightly lower than that for the pure titania. Therefore, these materials present the ability to protect against UV rays, and consequently, they could be used in cosmetic formulations. However, the effect of titania nanoparticle surface modification on the physical properties of the polypropylene-based, PP/TiO_2, nanocomposites after processing by injection molding has been studied by Zhang et al. [63]. They describe that the incorporation of a small amount of TiO_2 nanoparticles, lower than 1% in weight, does not promote any toughening effect in the injection-molded nanomaterials although their ductility is enhanced.

Moreover, PMMA/TiO_2 nanocomposites have been prepared by the polymerization with photoexcited TiO_2 nanoparticles as initiator [64]. The titania, which is an UV semiconductor (e.g., band gap in the UV region), catalyzes reactions on the surface [65]. Polymerization initiated by the excited nanoparticles is called photocatalytic polymerization. X-ray photoelectron spectroscopy (XPS) results show that the Ti 2p doublet of the TiO_2 in the composites shifts to a lower binding energy by 1.0 eV. This downshift could reveal that the Ti^{4+} cation interacts with PMMA.

Afterward, the photopolymerization kinetics of PMMA initiated by TiO_2 nanoparticles in aqueous suspensions has been studied [66], demonstrating the dependence of the polymerization rate on pH while the tacticity of the polymer remains unchanged.

Polyaniline PANI/TiO_2 nanocomposites [67] have been prepared by the chemical oxidative polymerization in a reverse micelle of aniline/TiO_2/hydrochloride acid precursor solution with ammonium peroxydisulfate used as an oxidant. The nanodimensional TiO_2 colloidal dispersion is previously prepared by the hydrolysis of titanium chloride. After 60 days, the nanocomposites create spherical nanoparticles self-organized in different morphologies depending on the surfactant system used. Then, a sea urchin-like morphology is developed when this nonionic system is Triton-x100; a triangle shape is promoted in a sodium bis (2-ethylhexyl) sulfosuccinate anionic system; and nanoclusters are observed in a cetyltrimethyl ammonium cationic system. These nanoclusters are built up by needle-like nanowires with the same aspect ratio than the nanowires in the Triton-x100 surfactant system, and these needle-like nanowires are composed of spherical particles of PANI/TiO_2.

Li et al. [68] have also prepared PANI/TiO_2 nanocomposites using *in situ* chemical oxidative polymerization of aniline in the TiO_2 suspension. The nanoparticles, prepared by sol–gel using tetrabutyl titanate are mostly anatase with an average particle size of ca. 15 nm and BET specific surface area of ca. 70 m^2/g. PANI exhibits high absorption in the UV range and in the visible light region. As a consequence, the absorption of PANI/TiO_2 nanocomposite increases over the whole range of visible light compared with the one presented by neat TiO_2, whereas it decreases in the UV range. Therefore, this method can be effective to extend the absorption of TiO_2 to visible light range. In addition, these nanocomposites show good stability under irradiation conditions and they keep their perfect photocatalytic activity after several cycles with only a small decrease after each cycle resulting from slight aggregation of nanoparticles during the catalytic process.

Tang's group [69] also used oxidative reaction to chemically graft PANI on the surface of the self-assembled monolayer (SAM)-coated TiO_2 nanoparticles (commercial anatase with an average particle size of 15 nm). The γ-aminopropyltriethoxysilane is used as a coupling agent to make a dense aminopropylsilane monolayer with active sites for grafting polymerization of aniline. These approaches to produce PANI/TiO_2 nanocomposites improve the thermal stability at high temperatures ($>400\,°C$), resulting from the covalent bonds formed between PANI chains and titania nanoparticles. As happens with the previous system, photocatalitic activity is observed upon visible light excitation.

This group [70] also used *in situ* photocatalytic polymerization to achieve molecular imprinting polymer coated photocatalysts. In this work, *o*-phenylenediamine (OPDA) is selected as the functional monomer, since it has two NH_2

groups in its molecule capable to form hydrogen bonds with the target molecules or templates, 4-chlorophenol (4CP) or 2-chlorophenol (2CP). The photodegradation of the mixture of 2 mg/L 4CP and 500 mg/L phenol is confirmed in the nanocomposites, where a rapid degradation with a half-time of approximately 8 min occurs for 4CP, whereas the relative concentration of phenol apparently remains almost unchanged. Therefore, these formed nanocomposites can promote the selectivity of TiO_2 photocatalysis (see Figure 14.2).

Nanocomposite particles can be also prepared through *in situ* emulsion polymerization in the presence of nano-TiO_2 colloids obtained by the hydrolysis of titanium tetrachloride [71]. These particles are based on PMMA or poly(*n*-butyl acrylate) as core and TiO_2 as shell. The diameter of nanocomposite particles is about 150 nm, and the thickness of the TiO_2-shell is approximately 4–10 nm. However, the pH value has to be strictly controlled to obtain a tight TiO_2 coating. These materials can be used in the field of photocatalytic coating, keeping in mind, the coexistence of rutile and anatase structures in the nanocomposite.

In a recent article [72], TiO_2 nanoparticles were modified by adding a small amount of polythiophene to improve the dispersion of TiO_2 nanoparticles and to enhance the photocatalytic activity of the resulting nanocomposites. These nanomaterials have been synthesized via oxidative polymerization of thiophene using $FeCl_3$ in the presence of anionic, cationic, and nonionic surfactants. Here, the nanocrystal titania is the core and the polymer is the shell. The best semiconductor property is observed when an anionic surfactant is used, which is attributed to differences in the stability of the nanocomposites. Consequently, the decomposition temperature when the anionic system is used is higher than that obtained in the hybrid materials prepared using the nonionic, cationic, or doped ones.

Surface modification by brushes allows dispersion and stability of the nanoparticles in various solvents or polymeric matrices while they maintain their physical characteristics. The modification of TiO_2 nanoparticles (primary particle diameter between 70 and 100 nm) by grafting polystyrene (PS) on its

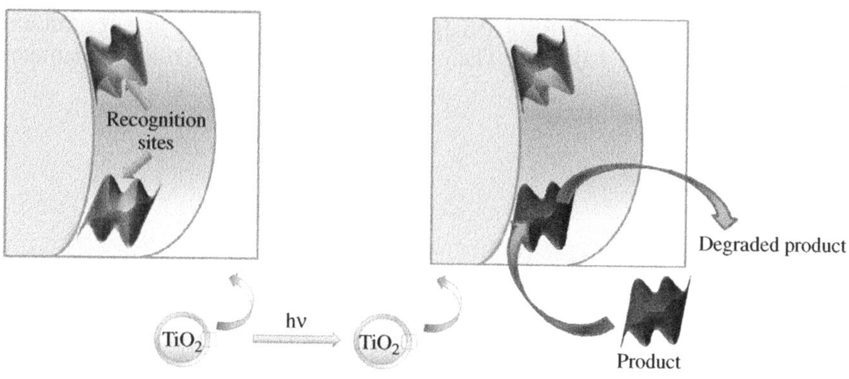

FIGURE 14.2 Molecular imprinting polymer-coated titania and its use in photocatalytic degradation.

surface, and further polymerization of styrene allows preparing photodegradable PS/TiO_2 nanocomposite films [73]. The presence of TiO_2 nanoparticles promotes the photocatalytic oxidation of polystyrene films [74]. Accordingly, the mass loss rate of the nanocomposite films is much higher than that for the neat PS film.

PS/TiO_2 nanocomposites can also be built up through a direct polymer grafting reaction from the surfaces of titanium oxide nanoparticles (d $\approx$ 15 nm) [75]. In this case, a controlled radical polymerization method, nitroxide mediated radical polymerization (NMP), is used to obtain the polystyrene and poly (3-vinylpyridine) (P3VP) brushes. The initiator for NMP with a phosphoric acid group is chemisorbed onto the nanoparticles, and then, a monomer is polymerized giving controlled polymer graft layers on the surface (Figure 14.3 illustrates the general procedure). The PS- and P3VP-modified nanoparticles are stably dispersed in organic solvents, such as tetrahydrofurane, toluene, and ethyl acetate. The dispersion of grafted and nongrafted nanoparticles in the PS matrix is later studied. The nanoparticles dispersed in chloroform are mixed with a PS matrix chloroform solution. It is found that the nanocomposite film becomes clouded and that the transparency is drastically decreased along with an increase in the weight ratio of the nanoparticle, when nongrafted TiO_2 particles are added to the PS matrix. On the contrary, the PS-grafted TiO_2 nanoparticles are finely dispersed in the PS matrix.

Transparent PMMA/TiO_2 nanocomposite particles have been prepared by graft polymerization of MMA from the surface of the modified TiO_2 particles with [γ-(methacryloxy)-propyl]trimethoxysilane by *in situ* emulsion polymerization [76]. These nanocomposite particles present either a glass transition temperature or thermal stability higher than the pure PMMA. This coupling agent is also used to obtain PS/TiO_2 nanocomposite by *in situ* polymerization in toluene solution [77, 78].

Anatase TiO_2 nanoparticles have been also modified with 3-(trimethoxysilyl) propyl methacrylate to obtain PMMA/TiO_2 nanocomposites with high transparency and UV-absorption characteristics [79]. The anatase TiO_2 nanocrystals are prepared by a nonaqueous sol–gel approach involving the mixing of titanium isopropoxide and benzyl alcohol. The nanoparticles obtained exhibit a diameter

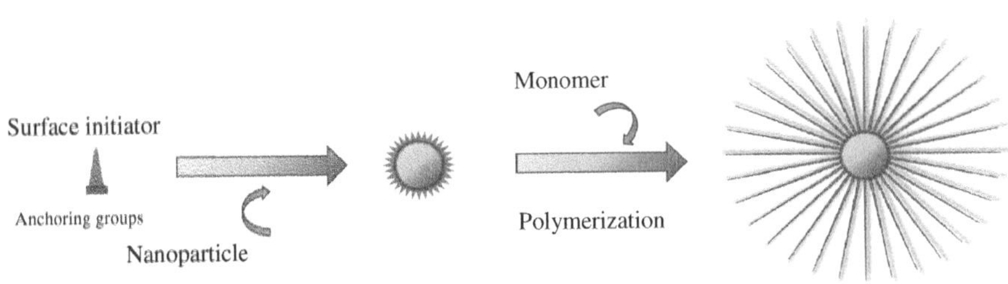

FIGURE 14.3 Surface-initiated radical polymerization on the surface of TiO_2 nanoparticles.

ranging between 10 and 20 nm. As commented in the previous materials, the surfaces are modified to make the nanoparticles easily dispersible in nonpolar media like xylene and dichloromethane and compatible with the PMMA matrix. The synthesis of PMMA/TiO_2 functional nanocomposite particles has been performed by atom transfer radical polymerization (ATRP) using commercial titanium dioxide nanoparticles (80% anatase and 20% rutile, with an average particle size of 34 nm and a BET specific surface area of 45 m^2/g) where the catechol-terminated initiator is anchored by chemisorption [80]. This specific biomimetic initiator is synthesized because of mussels secreting adhesive proteins that are mainly formed by catecholic amino acid, L-3,4-dihydroxyphenylalanine, which is believed to interact strongly with a variety of metals, metal oxides, and polymers. Therefore, it has a robust anchoring to the titania surface. However, the controlled polymerization is not achieved when higher initiator concentrations are used. This group also studies the effect on the final properties of these nanocomposite particles when they are incorporated into a PMMA matrix [81]. Their glass transition temperature and elastic modulus increase with respect to those presented in the pure PMMA. On the contrary, nanocomposites obtained using unmodified nanoparticles lead to a decrease in these properties. This behavior demonstrates the influence that interfacial modification of nanoparticles bears on the nanocomposite homogeneity.

Another way to produce hybrid materials is using functional polymers that can interact with the inorganic component. This is the case of the poly(styrene-*alt*-maleic anhydride) (PSMA) alternating copolymer [82] with highly regular anhydride groups on the backbone chains that may provide regular sites for combination. This copolymer is dissolved along with the tetrabutyl titanate precursor, and both are hydrolyzed at the same time. Acrylate groups are created from the reaction between uncondensed Ti-OH and maleic acid during the sol–gel process. Consequently, TiO_2 is covalent bonded with PSMA. Similar processes are described for nanocomposites based on the poly(methyl methacrylate-*co*-butyl methacrylate-*co*-methacrylic acid) statistical copolymer [83] and other polymers [84] using the same precursor. In addition, functional poly(butyl acrylate-co-methyl methacrylate-*co*-(3-methacryloxypropyl)trimethoxysilane) terpolymer [85] and titanium *n*-butoxide are used for the synthesis of hybrid materials with higher hardness, elastic modulus, thermal stability, and refractive index than those presented by the neat acrylic resin.

Whang and Chiang [86] prepared hybrid nanocomposite films of titania in a polyimide (PI) matrix. The synthesis was performed by imidazion of the homogeneous mixture of the tetraethyl titanate precursor and the poly(amic acid) functional polymer. The titania particle size on these nanocomposites increases from 10 to 40 nm as the TiO_2 content does from 5 to 30 wt%. All the films exhibit good optical transparency within this range. Other dianhydrides and/or diamines, which produce poly(amic acid), are used for the synthesis of PI/TiO_2 nanocomposites [87, 88]. Polyurethane/titania hybrid materials [89] with high refractive indexes are also attained by modification of polyurethane with (3-isocyanatopropyl) triethoxysilane, which allows the formation of covalent bonds between titania moiety and polymer segments restricting the polymer chain motion and preventing the phase separation.

The solution process has been used to synthesize hybrid films of poly(methyl methacrylate) (PMMA) containing titanium dioxide nanoparticles of around 20 nm in diameter [90]. These nanoparticles are produced by sol–gel from titanium tetraisopropoxide, within the hydrophilic core of micelles, formed using poly(acrylic acid-*block*-methyl methacrylate) in toluene solution. The thermal stabilities of the obtained nanocomposites increase in comparison with the one shown by the pure PMMA. In addition, they exhibit high transparency in the visible light region even at 30 wt% titania, over 87% at 500 nm (see Figure 14.4). The refractive index of the hybrid films at 633 nm linearly increases with TiO_2 content up to attain a value of 1.579, which is 0.1 higher than that observed in PMMA [91]. Other nanocomposites based on poly(vinyl alcohol), polyvinylpyrrolidone, and poly(4-vinylpyridine) and titania also exhibit an enhancement of their refractive indexes by the addition of nanoparticles [92].

Membranes of polyethersulfone/TiO_2 with application for direct methanol fuel cells [93] are obtained by dissolving sulfonated polyethersulfone with a sulfonation degree of 70% and titania obtained from hydrolysis of titanium isopropoxide. It is well known that barrier properties in membranes, in terms of mass transport, improve by incorporation of inorganic fillers because of their higher tortuosity. However, the addition of nanoparticles produces in this case the contrary effect, i.e., a diminishment of proton conductivity [93, 94].

PS/TiO_2 nanocomposite membranes for ultrafiltration can be prepared by blending titania sol (initial state of the sol–gel preparation procedure) in a polymer solution and posterior cast-evaporation [95]. The resulting membranes

FIGURE 14.4 Photograph of 30 wt% TiO_2-contained hybrid film. (Reproduced from Reference 90 with permission from the Chemical Society of Japan.)

develop a network of pores that avoids the formation of macrovoids. They show higher hydrophilicity, porosity, and permeability in comparison with the unfilled polymeric membranes. Polysulfone has been used as well for ultrafiltration technology [96].

Nanocomposite films of poly(vinyl chloride) (PVC) and neat and modified TiO_2 have been casted from tetrahydrofuran solutions [97, 98]. Hyperbranched poly(3-caprolactone) with carboxylic acid groups are used to increase the compatibility of titania with the PVC matrix. The resulting well-dispersed nanocomposites show photocatalytic degradation where chloride acid is not formed. Therefore, they present a potential application as an eco-friendly alternative for the disposal of PVC wastes to the current waste landfill and dioxin-generating incineration.

Photoactive nanomaterials are also obtained by spin-coating of homogeneous solutions of low-molecular-weight poly[2-methoxy-5-(2′-ethyl-hexyloxy) phenylenevinylene] (MEH-PPV) and oleic acid surface-coating TiO_2 anatase nanocrystals onto low resistivity indium tin oxide (ITO) substrates [99]. Authors report photoinduced transfer of electrons and holes at the interface between anatase TiO_2 nanorods or nanoparticles and MEH-PPV conjugated polymer within these solution blends, being higher for spherical TiO_2 nanoparticles.

There is in the literature much less articles dealing with the preparation of TiO_2 nanocomposites by the melting process, although it is the most attractive methodology from an industrial standpoint. The effects of the content and nature of TiO_2 nanoparticle on the crystalline structure of high-density polyethylene and isotactic polypropylene have been studied [33, 35, 41, 45, 100, 101]. The degree of crystallinity, the unit cell dimensions, the average lamellar thickness, or the average spherulite size are not usually altered in the polymer by the presence of nanoparticles. Moreover, the location of the glass transition as well as melting temperatures in these nanocomposites is rather independent of the nanoparticle content. However, their photodegradation is enhanced by the nanoparticle presence.

The effect of nanoparticle coating and the nanoparticle amount on the rheological properties have been studied in high impact polystyrene nanocomposites performed by this melting method [102]. These coated nanoparticles are obtained by *in situ* emulsion polymerization of styrene. The nanocomposite melts with coated-nanoparticles present higher shear viscosity because of their stronger interfacial interaction than in the uncoated ones.

Very recently, polypropylene-titania nanocomposites, PP/TiO_2, have been prepared [103] by this process using titanium *n*-butoxide premixed with PP in the molten state. Under these conditions, the nanocomposites present TiO_2 nanoparticles with primary particle sizes with a mean diameter as low as 5 nm. The authors found that a fractal structure of these particles is observed at the highest concentrations, with a characteristic aggregates size $d_{aggr} \approx 130$ nm. Furthermore, the rheological properties demonstrate that, in this case, the inorganic filler concentration at the percolation threshold is lower than 6 wt% (volume fraction ≈ 0.014). The appearance of a second plateau modulus at low frequency is mainly correlated to the formation of an aggregate-particle network.

14.4 TiO_2-NANOCOMPOSITES WITH BIOCIDAL PROPERTIES

There is a significant interest in the development of antimicrobial materials for application in the health and biomedical areas, as well as in the food packaging, and personal hygiene industries [104–116]. Among several possibilities currently explored, titania can be spot out as a potential candidate for polymer modification with a significant number of advantages, particularly, the absence of releasing of dangerous materials to the media. However, there are only a few articles exploring the use of TiO_2-nanocomposites for biocide purposes.

Kubacka et al. [33] describe the preparation by melting process of an ethylene-vinyl alcohol copolymer, EVOH/TiO_2 nanocomposites, with different amounts of the inorganic TiO_2 component with anatase structure, primary particle size of ~9 nm, and BET area of 104 m^2/g. These nanoparticles are synthesized by microemulsion method, and the EVOH copolymer is selected because of its extensive commercial applications in food packaging, health, and biomedical industrial areas. The use of compatibilizer or coupling agent is not needed in these nanocomposites because of the partial amphiphilic nature of EVOH copolymers, which is able to include titania nanoparticles in the final material with a good adhesion at interfaces between the two components. The good homogeneity of these nanocomposites can be visualized by TEM for the nanocomposite with a 2% titania content, as depicted in Figure 14.5. These nanocomposites present extraordinary antimicrobial properties against gram-negative *P. aeruginosa* and gram-positive *E. faecalis*. Figure 14.6 illustrates how the incorporation of TiO_2 nanoparticles influences the biokilling potential of the

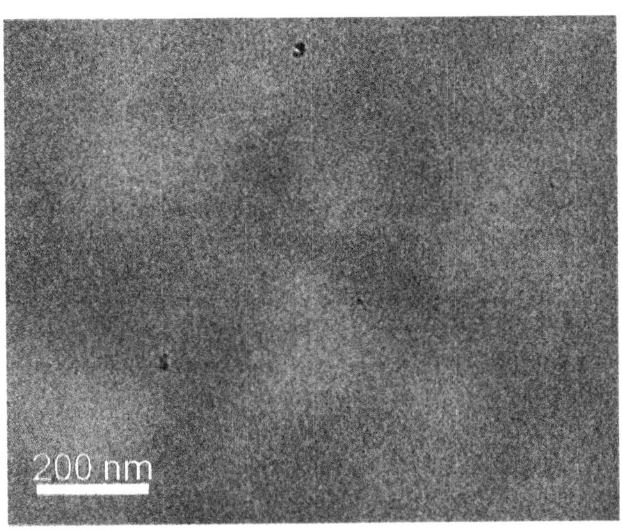

FIGURE 14.5 TEM image of the TiO_2-EVOH nanocomposite with a 2 wt% nanoparticle content.

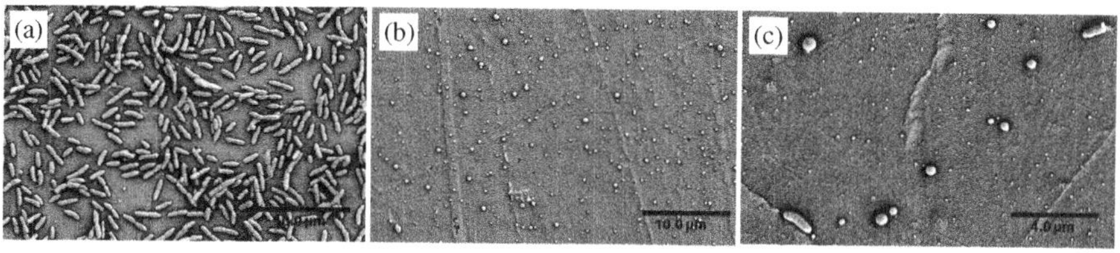

FIGURE 14.6 SEM images of the *P. aeruginosa* cells sited at the surface of the TiO_2-EVOH nanocomposite with a 2 wt% of nanoparticle content in the absence (a) and presence of UV light (b, c).

nanocomposites, not only affecting the cell viability (round shapes indicating the partial loss of microbe membrane integrity) but also bacterial aggregation and biofilm formation. In addition, these nanocomposites are easily photodegradaded by exposure to sunlight. Therefore, they could be considered as environmentally friendly polymeric nanomaterials with potential applicability in the industrial sector.

Moreover, Cerrada et al. [35] have later studied the biocidal characteristics of these EVOH/TiO_2 nanocomposites against gram-positive and gram-negative bacteria (*E. coli*, *E. caratovora*, and *P. fluorescens*), gram-positive cocci (*S. aureus*), and yeasts (*Z. rouxii* and *P. jadini*). They find that the maximum biocidal performance is obtained for nanoparticle content ranging between 2 and 5 wt%, which is attributed to differences among samples in terms of the available oxide surface area by effect of changes in the oxide secondary particle size (e.g., the morphological observable describing the oxide agglomeration) within the polymeric matrix and subsequent optimization of charge carrier transfer from the oxide to the polymer upon light excitation. The study showed that a key advantage of these polymer-based materials with biocidal properties (with respect to oxide-alone ones or even to polymer-based systems incorporating biocidal agents like Ag) relies in the fact that the whole nanocomposite surface becomes biocidal and thus relaxes the requirement for direct biocidal agent to microorganism contact. In such a way, these nanocomposites (in this composition interval scanned) exhibit an astonishing biocidal behavior in comparison with that presented by well-known biocide agents such as Ag-based systems, AgBr particles coated with poly(vinylpyridine), or simple chemicals (glutaraldehyde, formaldehyde, H_2O_2, phenol, cupric ascorbate, or sodium hypochlorite) [31, 110, 113, 117, 118]. The outstanding biocidal capability comes from a close contact at the nanometric scale between polymer and oxide nanoparticles, this contact being proved by the presence of new electronic states in the nanocomposite systems, which are nonexisting in the individual components [119]. The physico-chemical characterization of these nanocomposites has also shown that the EVOH crystallinity does not practically vary throughout the titania nanoparticles incorporation, although the crystal size slightly increases with nanoparticle content at low additions. The values of glass transition and melting temperatures do not differ significantly with the increase of titania content, indicating a slight effect of TiO_2 nanoparticles on the transitions related to the amorphous and crystalline phases within the polymeric component. However, a considerable microhardness improvement is observed as the TiO_2 amount rises in the nanocomposites [35].

In additional works, these authors have evaluated isotactic polypropylene, iPP, as a matrix and, once more, anatase titania nanoparticles to verify their straightforward and cost-effective approach as well as the biocidal possibilities of these other nanocomposites [45, 120], which also have applications in the packaging sector. In this case, the addition of an interfacial agent (a low-molecular-weight polypropylene with maleic anhydride grafted, PP-*g*-MAH) is required to improve the interfacial contact between matrix and nanoparticles because of the hydrophobic nature of iPP [121]. The antimicrobial activity is substantially enhanced in the resulting nanocomposites by efficiently managing charge carrier handling through the organic–inorganic interface, making biocidal the whole system. As occurred in EVOH/TiO_2 nanocomposites, the crystallinity degree of the iPP matrix remains practically unchanged by the presence of nanoparticles in the polymer.

The antimicrobial performance exhibited by these TiO_2-polymer nanocomposites can be even improved by extending the oxide absorption power into the visible region through an oxide-surface doping process [41]. Consequently, TiO_2-doped nanoparticles have been prepared by photodeposition of silver in a content of 1 wt%. Afterward, different amounts of these doped nanoparticles have been incorporated into an EVOH copolymer through, once more, a melting process without using any compatibilizer. These nanocomposites demonstrate very effective germicide activity toward gram-negative and gram-positive bacteria/cocci (*E. coli*, *P. putida*, *S. aureus*) and yeasts (*P. jadini*). Figure 14.7 shows the film performance of TiO_2-EVOH and doped-TiO_2-EVOH nanocomposites both with a 2 wt% nanoparticle content. As expected, they appeared effective with respect to the TiO_2 oxide alone (comparison of killing performance per mass unit) upon all illumination conditions but particularly while submitted to visible light irradiation. In addition, they showed exceptional resistance to biofilm formation, which is responsible for the microorganisms' antibiotic resistance. In summary, the presence of minuscule amounts of Ag greatly boosts the antimicrobial power through a plasmonic effect in comparison with TiO_2-EVOH systems under UV light and introduces the successful use of visible-light sources.

Very recently, Kong et al. [122] reported on the preparation of titania nanocomposites exhibiting antimicrobial performances even under dark conditions, through the incorporation of a biocidal polymer. These materials are obtained by photoinduced copolymerization of 2-(*tert*-butylamino)ethyl methacrylate and ethylene glycol dimethacrylate to form core/shell poly(tBAM-*co*-EGDMA)/commercial anatase needle-like TiO_2 nanoparticles. They show enhanced photocatalytic antibacterial properties rather than neat TiO_2 nanoparticles against both *E. coli* and *S. aureus* as a result of the synergic antibacterial

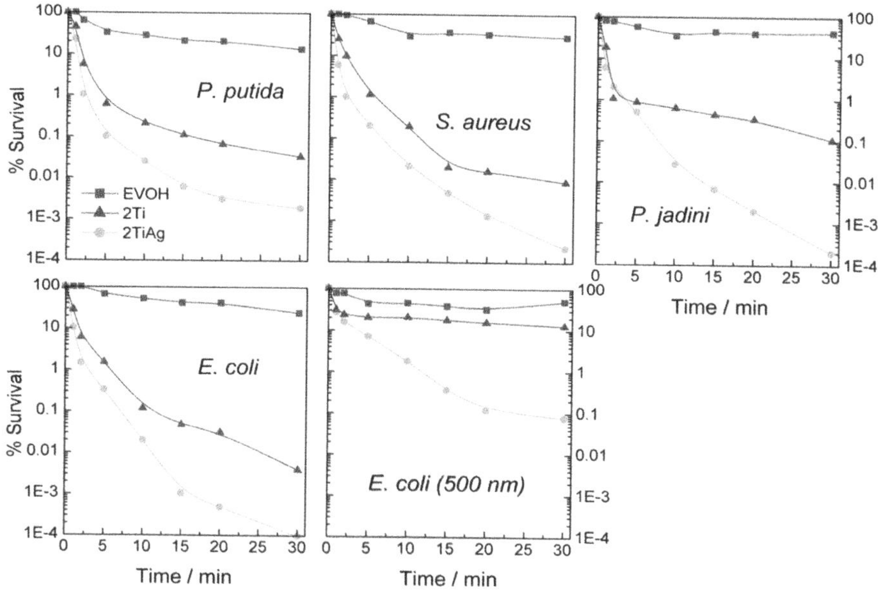

FIGURE 14.7 Survival of microorganism population according to the microorganisms tested. Survival curves for four microorganisms suspended in the appropriate liquid media as a function of the irradiation time for pure EVOH, TiO_2-EVOH, and doped-TiO_2-EVOH nanocomposites both with a 2 wt% nanoparticle content.

performances of the biocidal polymer shell and light-activated disinfective TiO_2 core. Furthermore, nanocomposites display excellent antibacterial properties in dark conditions because the secondary amine-containing antifouling copolymer shell provides additional antimicrobial activity to the one provided by TiO_2 nanoparticles.

REFERENCES

1. Hoffmann, M. R., Martin, B. T., Choi, W., Bahnemann, D. W. (1995). Environmental applications of semiconductor photocatalysis. *Chemical Reviews*, *95*, 69.
2. Linsebigler, A. L., Lu, G., Yates, J. T. (1995). Photocatalysis on TiO_2 surfaces—principles, mechanisms, and selected results. *Chemical Reviews*, *95*, 735.
3. Bahnemann, D. W. (2004). Photocatalytic water treatment: Solar energy applications. *Solar Energy*, *77*, 445.
4. Serpone, N. (2006). Is the band gap of pristine TiO_2 narrowed by anion- and cation-doping of titanium dioxide in second-generation photocatalysts? *Journal of Physical Chemistry B*, *110*, 24287.
5. Colón, G., Belver, C., Fernández-García, M. (2007). *Nanostructured Oxides in Photocatalysis in Synthesis, Properties and Applications of Solid Oxides*. Eds.: Rodríguez, J. A., Fernández-García, M. Wiley, NewYork, NY.

6. Carp, O., Huisman, C. L., Reller, A. (2004). Photoinduced reactivity of titanium dioxide. *Progress in Solid State Chemistry*, *32*, 33.
7. Fox, M. A., Dulay, M. T. (1993). Heterogeneous photocatalysis. *Chemical Reviews*, *93*, 341.
8. Fujishima, A., Rao, T. N., Tryk, D. A. (2000). Titanium dioxide photocatalysis. *Journal of Photochemistry and Photobiology C*, *1*, 1.
9. Barnard, A. S., Xu, H. (2008). An environmentally sensitive phase map of titania nanocrystals. *American Chemical Society Nano*, *2*, 2237.
10. Burdett, J. K., Hughbands, T., Gordon, J. M., Richarson, J. W., Smith, J. (1987). Structural electronic relationships in inorganic solids—powder neutron-diffraction studies of the rutile and anatase polymorphs of titanium-dioxide at 15 and 295-K. *Journal of the American Chemical Society*, *109*, 3639.
11. Asahi, R., Taga, Y., Mannstadt, W., Freem, A. J. (2000). Electronic and optical properties of anatase TiO_2. *Physical Review B*, *61*, 7459.
12. Wiswanathan, B. (2003). Photocatalytic processes—selection criteria for the choice of materials. *Bulletin of the Catalysis Society of India*, *2*, 71.
13. Chen, X., Mao, S. S. (2007). Titanium dioxide nanomaterials: Synthesis, properties, modifications, and applications. *Chemical Reviews*, *107*, 2891.
14. Luan, J., Ma, K., Zhang, L., Li, M., Li, Y., Pan, B. (2010). Research on different preparation methods of new photocatalysts. *Current Organic Chemistry*, *14*, 683.
15. Catlow, C. R. A., Bromley, S. T., Hamad, S., Mora-Fonz, M., Sokol, A. A., Woodley, S. M. (2010). Modelling nano-clusters and nucleation. *Physical Chemistry Chemical Physics*, *12*, 786.
16. Barnard, A. S. Curtiss, L. A. (2005). Prediction of TiO_2 nanoparticle phase and shape transitions controlled by surface chemistry. *Nano Letters*, *5*, 1261.
17. Yang, H. G., Sun, C. H., Qiao, S. Z., Zou, J., Liu, G., Smith, S. C., Cheng, M. H., Lu, G. Q. (2008). Anatase TiO_2 single crystals with a large percentage of reactive facets. *Nature*, *453*, 638.
18. Sakai, N., Ebina, K., Takada, K., Sasaki, T. (2004). Electronic band structure of titania semiconductor nanosheets revealed by electrochemical and photoelectrochemical studies. *Journal of the American Chemical Society*, *126*, 5851.
19. Marshall, M. S. J., Castell, M. R. (2009). Shape transitions of epitaxial islands during strained layer growth: Anatase $TiO_2(001)$ on $SrTiO_3(001)$. *Physical Review Letters*, *102*, 146102.
20. Dai, Y. Q., Cobley, C. M., Zeng, J., Sun, Y. M., Xia, Y. N. (2009). Synthesis of anatase TiO_2 nanocrystals with exposed {001} facets. *Nano Letters*, *9*, 2455.
21. Zhang, D. Q., Li, G. S., Wang, H. B., Chang, K. M., Yu, J. C. (2010). Biocompatible anatase single-crystal photocatalysts with tunable percentage of reactive facets. *Crystal Growth Design*, *10*, 1130.
22. Fernández-García, M., Belver, C., Hanson, J. C., Wang, X., Rodríguez, J. A. (2007). Anatase-TiO_2 nanomaterials: Analysis of key parameters controlling crystallization. *Journal of the American Chemical Society*, *129*, 13604.
23. Li, G., Li, L., Boerio-Goates, J., Woodfield, B. F. (2005). High purity anatase TiO_2 nanocrystals: Near room-temperature synthesis, grain growth kinetics, and surface hydration chemistry. *Journal of the American Chemical Society*, *127*, 8659.

24. Hummer, D. R., Kubicki, J. D., Kent, P. R. C., Post, J. E., Heaney, P. J. (2009). Origin of nanoscale phase stability reversals in titanium oxide polymorphs. *Journal of Chemical Physics C*, *113*, 4240.

25. Lee, S.-K., Mills, A. (2003). Platinum and palladium in semiconductor photocatalytic system. *Platinum Metals Review*, *47*, 61.

26. Mattesini, M., de Almeida, J. S., Dubrovinsky, L., Dubrovinskaia, N., Johansson, B., Ahuja, R. (2004). Cubic TiO_2 as a potential light absorber in solar-energy conversion. *Physical Review B*, *70*, 115101.

27. Kubacka, A., Fernández-García, M., Colón, G. (2008). Nanostructured Ti-M mixed-metal oxides: Toward a visible light-driven photocatalyst. *Journal of Catalysis*, *254*, 272.

28. Robertson, J. M. C., Robertson, P. K. J., Lawton, L. A. (2005). A comparison of the effectiveness of TiO_2 photocatalysis and UVA photolysis for the destruction of three pathogenic micro-organisms. *Journal of Photochemistry and Photobiology A*, *175*, 51.

29. Lonnen, J., Kilvington, S., Kehoe, S. C., Al-Touati, F., McGuigan, K. G. (2005). Solar and photocatalytic disinfection of protozoan, fungal and bacterial microbes in drinking water. *Water Research*, *39*, 877.

30. Vohra, A., Goswami, D. Y., Desphande, D. A., Block, S. S. (2006). Enhanced photocatalytic disinfection of indoor air. *Applied Catalysis B*, *65*, 57.

31. Pal, A., Pehkonen, S. O., Yu, L. E., Ray, M. B. (2007). Photocatalytic inactivation of Gram-positive and Gram-negative bacteria using fluorescent light. *Journal of Photochemistry and Photobiology A*, *186*, 335.

32. Mitoraj, D., Janczyk, A., Strus, M., Kisch, H., Stochel, G., Heeczko, P. B. (2007). Visible light inactivation of bacteria and fungi by modified titanium dioxide. *Photochemical & Photobiological Sciences*, *6*, 642.

33. Kubacka, A., Serrano, C., Ferrer, M., Lundsford, H., Bielecki, P., Cerrada, M. L., Fernández-García, M., Fernández-Garcia, M. (2007). High-performance dual-action polymer-TiO_2 nanocomposite films via melting processing. *Nano Letters*, *7*, 2529.

34. Guillard, C., Bui, T. H., Felix, C., Moules, V., Lina, B., Lejeune, P. (2008). Microbiological disinfection of water and air by photocatalysis. *Comptes Rendus Chimie*, *11*, 107.

35. Cerrada, M. L., Serrano, C., Fernández-García, M., Fernández-Martín, F., Jiménez-Rioboo, R. J., de Andrés, A., Kubacka, A., Ferrer, M., Fernández-Garcia, M. (2008). Self-sterilized EVOH-TiO_2 nanocomposites: Interface effects on biocidal properties. *Advanced Functional Materials*, *18*, 1949.

36. Li, Q. L., Mahendra, S., Lyon, D. Y., Brunet, L., Liga, M. V., Li, D., Alvarez, P. J. J. (2008). Antimicrobial nanomaterials for water disinfection and microbial control: Potential applications and implications. *Water Research*, *42*, 4591.

37. Matsunaga, T., Tomoda, R., Nakajima, T., Wake, H. (1985). Photoelectrochemical sterilization of microbial cells by semiconductor powders. *FEMS Microbiology Letters*, *29*, 211.

38. Phan, H. N., McDowell, T., Wilkins, E. (1995). Photocatalytically-mediated disinfection of water using TiO_2 as a catalyst and spore-forming bacillus-pumilus as a model. *Journal of Environmental Science and Health A*, *30*, 627.

39. Rincón, A. G., Pulgarin, C. (2003). Photocatalytical inactivation of E. coli: Effect of (continuous-intermittent) light intensity and of (suspended-fixed) TiO_2 concentration. *Applied Catalysis B*, *44*, 263.
40. Wu, P. G., Xie, R. C., Shang J. K. (2008). Enhanced visible-light photocatalytic disinfection of bacterial spores by palladium-modified nitrogen-doped titanium oxide. *Journal of the American Ceramic Society*, *91*, 2957.
41. Kubacka, A., Cerrada, M. L., Serrano, C., Fernández-García, M., Ferrer, M., Fernández-García, M. (2009). Plasmonic nanoparticle/polymer nanocomposites with enhanced photocatalytic antimicrobial properties. *Journal of Physical Chemistry C*, *113*, 9182.
42. Zhang, D., Li, G., Yu, J. C. (2010). Inorganic materials for photocatalytic water disinfection. *Journal of Materials Chemistry*, *20*, 4529.
43. Elahifard, M. R., Rahimnejad, S., Haghighi, S., Gholami, M. R. (2007). Apatite-coated $Ag/AgBr/TiO_2$ visible-light photocatalyst for destruction of bacteria. *Journal of the American Chemical Society*, *129*, 9552.
44. Yan, G., Chen, J., Hua, Z. (2009). Roles of H_2O_2 and OH radical in bactericidal action of immobilized TiO_2 thin-film reactor: An ESR study. *Journal of Photochemistry and Photobiology A*, *207*, 153.
45. Kubacka, A., Ferrer, M., Cerrada, M. L., Serrano, C., Sánchez-Chaves, M., Fernández-García, M., de Andrés, A., Jiménez-Riobóo, R. J., Fernández-Martín, F., Fernández-García, M. (2009). Boosting TiO_2-anatase antimicrobial activity: Polymer-oxide thin films. *Applied Catalysis B: Environmental*, *89*, 441.
46. Rengifo-Herrera, J. A., Pierzchala, K., Sienkiewicz, A., Forró, L., Kiwi, J., Pulgarin, C. (2009). Abatement of organics and Escherichia coli by N, S co-doped TiO_2 under UV and visible light. Implications of the formation of singlet oxygen (O-1(2)) under visible light. *Applied Catalysis B*, *88*, 398.
47. Sunada, K., Watanabe, T., Hasimoto, K. (2003). Studies on photokilling of bacteria on TiO_2 thin film. *Journal of Photochemistry and Photobiology A*, *156*, 227.
48. Novak, B. M. (1993). Hybrid nanocomposite materials—between inorganic glasses and organic polymers. *Advanced Materials*, *5*, 422.
49. Lantelme, B., Dumon, M., Mai, C., Pascault, J. P. (1996). In situ polymerization of titanium alkoxides in polyvinylacetate. *Journal of Non-Crystalline Solids 194*, 63.
50. Hu, Q., Marand, E. (1999). In situ formation of nanosized TiO_2 domains within poly(amide-imide) by a sol–gel process. *Polymer*, *40*, 4833.
51. Yuwono, A. H., Xue, J., Wang, J., Elim, H. I., Ji, W., Li, Y., White, T. J. (2003). Transparent nanohybrids of nanocrystalline TiO_2 in PMMA with unique nonlinear optical behavior. *Journal of Materials Chemistry*, *13*, 1475.
52. Matsuyama, K., Mishima, K. (2009). Preparation of poly(methyl methacrylate)–TiO_2 nanoparticle composites by pseudo-dispersion polymerization of methyl methacrylate in supercritical CO_2. *Journal of Supercritical Fluids*, *49*, 256.
53. Ahmad, S., Ahmad, S., Agnihotry, S. A. (2006). The effect of nanosized TiO_2 addition on poly(methylmethacrylate) based polymer electrolytes. *Journal of Power Sources*, *159*, 205.
54. Ahmad, S., Agnihotry, S. A., Ahmad, S. (2008). Nanocomposite polymer electrolytes by in situ polymerization of methyl methacrylate: For electrochemical applications. *Journal of Applied Polymer Science*, *107*, 3042.

55. Tang, Q., Lin, J., Wu, Z., Wu, J., Huang, M., Yang, Y. (2007). Preparation and photocatalytic degradability of TiO_2/polyacrylamide composite. *European Polymer Journal*, *43*, 2214.

56. Džunuzović, E., Jeremić, K., Nedeljković, J. M. (2007). In situ radical polymerization of methyl methacrylate in a solution of surface modified TiO_2 and nanoparticles. *European Polymer Journal*, *43*, 3719.

57. Zhang, J, Luo, S., Gui, L. (1997). Poly(methyl methacrylate)–titania hybrid materials by sol–gel processing. *Journal of Materials Science*, *32*, 1469.

58. Boettcher, S. W., Bartl, M. H., Hu, J. G., Stucky, G. D. (2005). Structural Analysis of Hybrid Titania-Based Mesostructured Composites, *Journal of The American Chemical Society*, *127*, 9721.

59. Inkyo, M., Tokunaga, Y., Tahara, T., Iwaki, T. Iskandar, F., Hogan, C. J. Jr., Okuyama, K. (2008). Beads mill-assisted synthesis of poly methyl methacrylate (PMMA)-TiO_2 nanoparticle composites. *Industrial & Engineering Chemistry Research*, *47*, 2597.

60. Yu, K. M., Suryamas, A. B., Hirakawa, C., Iskandar, F., Okuyama, K. (2009). A new physical route to produce monodispersed microsphere nanoparticle-polymer composites. *Langmuir*, *25*, 11038.

61. Erdem, B., Sudol, E. D., Dimonie, V. L., El-Aasser, M. S. (2000). Encapsulation of inorganic particles via miniemulsion polymerization. II.Preparation and characterization of styrene miniemulsion droplets containing TiO_2 particles. *Journal of Polymer Science A: Polymer Chemistry*, *38*, 4431.

62. Kim, J. -W., Shim, J. -W., Bae, J. -H., Han, S. -H., Kim, H. -K., Chang, I. -S., Kang, H. -H., Suh, K. -D. (2002). Titanium dioxide/poly(methyl methacrylate) composite microspheres prepared by in situ suspension polymerization and their ability to protect against UV rays. *Colloid & Polymer Science*, *280*, 584.

63. Zhang, H., Zhang, Z., Park, H.-W., Zhu, X. (2008). Influence of surface-modified TiO_2 nanoparticles on fracture behavior of injection molded polypropylene. *Frontiers of Material Science in China*, *2*, 9.

64. Wang, J., Ni, X. (2008). Interfacial structure of poly(methyl methacrylate)/TiO_2 nanocomposites prepared through photocatalytic polymerization. *Journal of Applied Polymer Science*, *108*, 3552.

65. Hoffman, A. J., Yee, H., Mills, G., Hoffman, M. R. (1992). Photoinitiated polymerization of methyl-methacrylate using q-sized ZnO colloids. *Journal of Physical Chemistry*, *96*, 5540.

66. Ni, X., Ye, J., Dong, C. (2006). Kinetics studies of methyl methacrylate photopolymerization initiated by titanium dioxide semiconductor nanoparticles. *Journal of Photochemistry and Photobiology A: Chemistry*, *181*, 19.

67. Sui, X., Chu, Y., Xing, S., Yu, M., Liu, C. (2004). Self-organization of spherical PANI/TiO_2 nanocomposites in reverse micelles. *Colloids and Surfaces A: Physicochemical and Engineering Aspects*, *251*, 103.

68. Li, X., Wang, D., Cheng, G., Luo, Q., An, J., Wang, Y. (2008). Preparation of polyaniline-modified TiO_2 nanoparticles and their photocatalytic activity under visible light illumination. *Applied Catalysis B: Environmental*, *81*, 267.

69. Li, J., Zhu, L., Wu, Y., Harima, Y., Zhang, A., Tang, H. (2006). Hybrid composites of conductive polyaniline and nanocrystalline titanium oxide prepared via self-assembling and graft polymerization. *Polymer*, *47*, 7361.

70. Shen, X., Zhu, L., Li, J., Tang, H. (2007). Synthesis of molecular imprinted polymer coated photocatalysts with high selectivity. *Chemical Communications*, 1163.
71. Ai, Z., Sun, G., Zhou, Q., Xie, C. (2006). Polyacrylate-core/TiO_2-shell nanocomposite particles prepared by in situ emulsion polymerization. *Journal of Applied Polymer Science*, *102*, 1466.
72. Uygun, A., Turkoglu, O., Sen, S., Ersoy, E., Yavuz, A. G., Batir, G. G. (2009). The electrical conductivity properties of polythiophene/TiO_2 nanocomposites prepared in the presence of surfactants. *Current Applied Physics*, *9*, 866.
73. Zan, L., Tian, L., Liu, Z., Peng, Z. (2004). A new polystyrene–TiO_2 nanocomposite film and its photocatalytic degradation. *Applied Catalysis A: General*, *264*, 237.
74. Zan, L., Wang, S., Fa, W., Hu, Y., Tian, L., Deng, K. (2006). Solid-phase photocatalytic degradation of polystyrene with modified nano-TiO_2 catalyst. *Polymer*, *47*, 8155.
75. Kobayashi, M., Matsuno, R., Otsuka, H., Takahara, A. (2006). Precise surface structure control of inorganic solid and metal oxide nanoparticles through surface-initiated radical polymerization. *Science and Technology of Advanced Materials*, *7*, 617.
76. Yang, M., Dan, Y. (2005). Preparation and characterization of poly(methyl methacrylate)/titanium oxide composite particles. *Colloid & Polymer Science*, *284*, 243.
77. Zan, L., Liu, Z., Zhong, J., Peng, Z. (2004). Organic modification on TiO_2 nanoparticles by grafting polymer. *Journal of Materials Science*, *39*, 3261.
78. Rong, Y., Chen, H. -Z., Wu, G., Wang, M. (2005). Preparation and characterization of titanium dioxide nanoparticle/polystyrene composites via radical polymerization. *Materials Chemisty and Physics*, *91*, 370.
79. Koziej, D., Fischer, F., Kränzlin, N., Caseri, W. R., Niederberger, M. (2009). Nonaqueous TiO_2 nanoparticle synthesis: A versatile basis for the fabrication of self-supporting, transparent, and UV-absorbing composite films. *Applied Materials and Interfaces*, *1*, 1097.
80. Fan, X., Lin, L., Messersmith, P. B. (2006). Surface-initiated polymerization from TiO_2 nanoparticle surfaces through a biomimetic initiator: A new route toward polymer–matrix nanocomposites. *Composites Science and Technology*, *66*, 1198.
81. Hamming, L. M., Qiao, R., Messersmith, P. B., Brinson, L. C. (2009). Effects of dispersion and interfacial modification on the macroscale properties of TiO_2 polymer–matrix nanocomposites. *Composites Science and Technology*, *69*, 1880.
82. Wang, S., Wang, M., Lei, Y., Zhang, L. (1999). "Anchor effect" in poly(styrene maleic anhydride)/TiO_2 nanocomposites. *Journal of Materials Science Letters*, *18*, 2009.
83. Perrin, F. X., Nguyen, V., Vernet, J. L. (2002). Mechanical properties of polyacrylic–titania hybrids—microhardness Studies. *Polymer*, *43*, 6159.
84. Su, B., Liu, X., Peng, X., Xiao, T., Su, Z. (2003). Preparation and characterization of the TiO_2/polymer complex nanomaterial. *Materials Science and Engineering* A, *349*, 59.
85. Xiong, M., Zhou, S., Wu, L., Wang, B., Yang, L. (2004). Sol–gel derived organic–inorganic hybrid from trialkoxysilane-capped acrylic resin and titania: Effects of preparation conditions on the structure and properties. *Polymer*, *45*, 8127.

86. Chiang, P. -C., Whang, W. -T. (2003). The synthesis and morphology characteristic study of BAO-ODPA polyimide/TiO_2 nano hybrid films. *Polymer*, *44*, 2249.

87. Chiang, P. C., Whang, W. T., Tsai, M. H., Wu, S. C. (2004). Physical and mechanical properties of polyimide/titania hybrid films. *Thin Solid Films*, 447–448, 359.

88. Liou, G. -S., Lin, P. -H., Yen, H. -J., Yu, Y. -Y., Chen, W. -C. (2010). Flexible nanocrystalline-titania/polyimide hybrids with high refractive index and excellent thermal dimensional stability. *Journal of Polymer Science Part A: Polymer Chemistry*, *48*, 1433.

89. Lü, C., Cui, Z., Guan, C., Guan, J., Yang, B., Shen, J. (2003). Research on preparation, structure and properties of TiO_2/Polythiourethane hybrid optical films with high refractive index. *Macromolecular Materials and Engineering*, *288*, 717.

90. Yamada, S., Wang, Z., Yoshinaga, K. (2009). Incorporation of TiO_2 nanoparticles, formed via sol–gel process in micelle of block copolymer, into poly(methyl methacrylate) to fabricate high refractive and transparent hybrid materials. *Chemical Letters*, *38*, 828.

91. Aspnes, D. E. (1982). Local-field effects and effective-medium theory—a microscopic perspective. *American Journal of Physics*, *50*, 704.

92. Nussbaumer, R. J., Caseri, W. R., Smith, P., Tervoort, T. (2003). Polymer-TiO_2 nanocomposites: A route towards visually transparent broadband UV filters and high refractive index materials. *Macromolecular Materials and Engineering*, *288*, 44.

93. Prashantha, K., Park, S. G. (2005). Nanosized TiO_2-filled sulfonated polyethersulfone proton conducting membranes for direct methanol fuel cells. *Journal of Applied Polymer Science*, *98*, 1875.

94. Wu, G., Gan, S., Cui, L., Xu, Y. (2008). Preparation and characterization of PES/TiO_2 composite membranes. *Applied Surface Science*, *254*, 7080.

95. Yang, Y., Wang, P. (2006). Preparation and characterizations of a new PS/TiO_2 hybrid membranes by sol–gel process. *Polymer*, *47*, 2683.

96. Yang, Y., Zhang, H., Wang, P., Zheng, Q., Li, J. (2007). The influence of nanosized TiO_2 fillers on the morphologies and properties of PSF UF membrane. *Journal of Membrane Science*, *288*, 231.

97. Cho, S., Choi, W. (2001). Solid-phase photocatalytic degradation of PVC–TiO_2 polymer composites. *Journal of Photochemisrty and Photobiology A: Chemistry*, *143*, 221.

98. Kim, S. H., Kwak, S.-Y., Suzuki, T. (2006). Photocatalytic degradation of flexible PVC/TiO_2 nanohybrid as an eco-friendly alternative to the current waste landfill and dioxin-emitting incineration of post-use PVC. *Polymer*, *47*, 3005.

99. Petrella, A., Tamborra, M., Cozzoli, P. D., Curri, M. L., Striccoli, M., Cosma, P., Farinola, G. M., Babudri, F., Naso, F., Agostiano, A. (2004). TiO_2 nanocrystals–MEH-PPV composite thin films as photoactive material. *Thin Solid Films*, 451–452, 64.

100. Ma, D., Akpalu, Y. A., Li, Y., Siegel, R. W., Schadler, L. S. (2005). Effect of titania nanoparticles on the morphology of low density polyethylene. *Journal of Polymer Science Part B: Polymer Physics*, *43*, 488.

101. Zan, L., Fa, W., Wan, S. (2006). Novel photodegradable low-density polyethylene-TiO_2 nanocomposite film. *Environmental Science & Technology*, *40*, 681.

102. Tang, C. Y., Yue, T. M., Chen, D. Z., Tsui, C. P. (2007). Effect of surface coating on the rheological properties of a highly opaque nano-TiO_2/HIPS composite. *Materials Letters*, *61*, 4618.

103. Bahloul, W., Bounor-Legaré, V., David, L., Cassagnau, P. (2010). Morphology and viscoelasticity of PP/TiO_2 nanocomposites prepared by In Situ Sol–Gel Method. *Journal of Polymer Science Part B: Polymer Physics*, *48*, 1213.

104. Appendini, P., Hotchkiss, J. H. (2002). Review of antimicrobial food packaging. *Innovative Food Science & Emerging Technologies*, *3*, 113.

105. Devlieghere, F., Vermeiren, L., Debevere, J. (2004). New preservation technologies: Possibilities and limitations. *International Dairy Journal*, *14*, 273.

106. Brody, A. L. (2003). "Nano, nano" food packaging technology. *Food Technology*, *57*, 52.

107. Tang, H. Z., Doerksen, R. J., Tew, G. N. (2005). Synthesis of urea oligomers and their antibacterial activity. *Chemical Communications*, *12*, 1537.

108. Ignatova, M., Voccia, S., Gilbert, B., Markova, N., Cossement, D., Gouttebaron, R., Jerome, R., Jerome, C. (2006). Combination of electrografting and atom-transfer radical polymerization for making the stainless steel surface antibacterial and protein antiadhesive. *Langmuir*, *22*, 255.

109. Brayner, R., Ferrari-Iliou, R., Brivois, N., Djediat, S., Benedetti, M. F., Fievet, F. (2006). Toxicological impact studies based on Escherichia coli bacteria in ultrafine ZnO nanoparticles colloidal medium. *Nano Letters*, *6*, 866.

110. Sambhy, V., MacBride, M., Peterson, B. R., Sen, A. (2006). Silver bromide nanoparticle/polymer composites: Dual action tunable antimicrobial materials. *Journal of the American Chemical Society*, *128*, 9798.

111. Dizman, B., Elasri, M. O., Mathias, L. J. (2005). Synthesis, characterization, and antibacterial activities of novel methacrylate polymers containing norfloxacin. *Biomacromolecules*, *6*, 514.

112. Iconomopoulou, S. M., Voyiatzis, G. A. (2005). The effect of the molecular orientation on the release of antimicrobial substances from uniaxially drawn polymer matrixes. *Journal of Controlled Release*, *103*, 451.

113. Sagripanti, J. C., Bonifacio, A. (2000). Resistance of pseudomonas aeruginosa to liquid disinfectants on contaminated surfaces before formation of biofilms. *Journal of the Association of Official Analytical Chemists*, *83*, 1415.

114. Chen, Y., Worley, S. D., Kim, J., Wei, T. Y., Santiago, J. I., Williams, J. F., Sun, G. (2003). Biocidal poly(styrenehydantoin) beads for disinfection of water. *Industrial & Engineering Chemistry Research*, *42*, 280.

115. Robbins, M. E., Hopper, E. D., Schoenfisch, M. H. (2004). Synthesis and characterization of nitric oxide-releasing sol–gel microarrays. *Langmuir*, *20*, 10296.

116. Klibanov, A. M. (2007). Permanently microbicidal materials coatings. *Journal of Materials Chemistry*, *17*, 2479.

117. Cowan, M. M., Abshire, K. Z., Houk, S. L., Evans, S. M. (2003). Antimicrobial efficacy of a silver-zeolite matrix coating on stainless steel. *Journal of Industrial Microbiology and Biotechnology*, *30*, 102.

118. Mazzola, P. G., Martins, A. M. S., Penna, T. C. V. (2006). Chemical resistance of the gram-negative bacteria to different sanitizers in a water purification system. *BMC Infectious Disease*, *6*, 131.

119. Jiménez Riobóo, R. J., De Andrés, A., Kubacka, A., Fernández-García, M., Cerrada, M. L., Serrano, C., Fernández-García, M. (2010). Influence of nanoparticles on elastic and optical properties of a polymeric matrix: Hypersonic studies on ethylene–vinyl alcohol copolymer–titania nanocomposites. *European Polymer Journal*, *46*, 397.

120. Kubacka, A., Cerrada, M. L., Serrano, C., Fernández-García, M., Ferrer, M., Fernández-García, M. (2008). Light-driven novel properties of TiO_2-modified polypropylene-based nanocomposite films. *Journal of Nanoscience and Nanotechnology*, *8*, 3241.

121. Cerrada, M. L., Fernández-García, M., Serrano, C., Sánchez-Chaves, M., Fernández-García, M., De Andrés, A., Jiménez Riobóo, R. J., Fernández-Martín, F., Kubacka, A., Ferrer, M., Fernández-García, M. (2009). Biocidal capability optimization in organic-inorganic nanocomposites based on titania. *Environmental Science & Technology*, *43*, 1630.

122. Kong, H., Song, J., Jang, J. (2010). Photocatalytic antibacterial capabilities of TiO_2−biocidal polymer nanocomposites synthesized by a surface-initiated photopolymerization. *Environmental Science & Technology*, *44*, 5672.

CHAPTER 15

TISSUE-IMPLANT ANTIMICROBIAL INTERFACES

E. MARSICH, A. TRAVAN, I. DONATI, G. TURCO, F. BELLOMO, and S. PAOLETTI
Department of Life Sciences, University of Trieste, Trieste, Italy

CONTENTS

15.1 Implant-Related Infections: Background Concepts 380
15.2 Peri-Implant Infections 382
 15.2.1 Bacterial Adhesion to Surfaces and Processes Governing Biofilm Formation 382
 15.2.2 Pathogenesis of Implant-Associated Infections 385
 15.2.3 Biofilm Resistance Mechanisms on Implants 388
15.3 Biofilm-Controlling Materials 388
 15.3.1 Designing Efficient Antimicrobial Implant/Tissue Interfaces 388
15.4 Polymer-Based Antimicrobial Surfaces 396
 15.4.1 Polyelectrolyte Multilayers (PEMS) 397
 15.4.2 Release of Antibiotic Agents from Neutral Polymer Surfaces . . . 401
 15.4.3 Sterile-Surface Polymers 402
15.5 Novel Engineering Approaches for Treatment of Implant-Related Infections 405
15.6 Future Trends 412
 References 412

Antimicrobial Polymers, First Edition.
Edited by José M. Lagarón, María J. Ocio, and Amparo López-Rubio.

15.1 IMPLANT-RELATED INFECTIONS: BACKGROUND CONCEPTS

The increase of the average population age leads the way of the researchers to the development of new and more efficient biomedical implants. Medical prostheses constitute an indispensable component of modern health care, and the extensive use of such medical devices has affected mostly critically ill patients, resulting in a greater survival rate.

Like other approaches of medical intervention, the insertion of medical prostheses can be associated with serious complications; infections remain the most common ones. One of the principal complications of the implants is represented by the associated infections that affect a large percentage (5% to 10%) of the implants inserted into humans [1]. Approximately half of the 2 million cases of nosocomial infections that occur each year in the United States are associated with indwelling devices [2, 3]. Infections associated with surgical implants have clinical and economical consequences, often result in major morbidity, and can be life threatening. For instance, the mortality associated with prosthetic valve endocarditis ranges from 30% to 80% in patients with early onset infection and from 20% to 40% in patients with late-onset endocarditis. Eradication of infections is expensive and often requires removal of the prosthesis [4].

Mortality attributable to such an infection is the highest among patients with cardiovascular implants, particularly prosthetic heart valves and aortic grafts. Infections associated with orthopedic devices and ventricular shunts often result in serious disabilities. Although infections of mammary and penile implants are rarely life threatening, they can cause major disfigurement and psychological trauma. The nature and the number of stages of surgical intervention vary according to the type of implant. Except for combination pacemaker-defibrillator systems, the cost of an implant itself usually constitutes a small fraction of the cost of treatment implant-associated infection.

Infection associated with devices involves the formation of biofilms by microorganisms on the surface of the implants. The most commonly cultured microorganisms are coagulase-negative Staphylococci (30% to 43% of cases), mainly *Staphylococcus aureus* (12% to 23%), followed by mixed flora (10% to 11%), Streptococci (9% to 10%), gram-negative bacilli (3% to 6%) enterococci (3% to 7%), and anerobes (2% to 4%) [5]. An increasing proportion of device-related infections, particularly those that involve the blood stream and the urinary tract are caused by *Candida* spp. There are several risk factors also for particular non-*Albicans* species: *C. glabrata* is related to azole prophylaxis, surgery and urinary or vascular catheters; *C. krusei* is related to azole prophylaxis and, along with *C. tropicalis*, to neutropenia and bone marrow transplantation; and *C. parapsilosis* infections are related to foreign-body insertion in neonates [6, 7].

Bacterial colonization of surgical implants does not necessarily indicate infection. Microorganisms are not detected in approximately 11% of apparent infections [6, 8, 9]. The growth of a virulent microorganism usually indicates infection, whereas low-virulence microorganisms that are typically part of normal skin flora may be either contaminants or pathogens.

The ultimate proof of implant-associated infection requires the presence of clinical manifestations, intraoperative signs of infection adjacent to the implant, and the growth of pathogens in cultures of surgical specimens. To obviate the need to operate on a patient who may not have an implant-associated infection, implant-specific microbiology and imaging studies are required. Although bacteremia is the hallmark of prosthetic-valve endocarditis and some infections associated with vascular graft, blood cultures are negative in most cases of infection associated with pacemaker-defibrillator systems (unless endocarditis is present), ventricular-assist devices, orthopedic devices, ventricular shunts, mammary implants, and penile implants.

Infection is confirmed if at least one of the following criteria is present, according to a previously described classification system: (1) growth of the same microorganism on two or more cultures of either a preoperative aspirate or intraoperative tissue specimens; (2) purulence of preoperative aspirate or intraoperative tissue, as determined by surgeon; (3) acute inflammation on histopathological examination of intraoperative tissue sections.

According to the route, the infection can be classified as perioperative, hematogenous, or contiguous. Perioperative infections were divided according to the presence of clinical symptoms after implantation, into those with early (within 3 months after surgery) or late onset (more than 3 months after surgery). Hematogenous infection is diagnosed if a documented or suspected bacteremia preceded the clinical onset of infection and the original site of infection was identified. Contiguous infections referred to spread from an adjacent focus of infection.

The essential factor in the evolution and persistence of infection is the formation of biofilm around implanted devices.

The most important clinical objectives in treating infections associated with surgical implants are to cure infections, prevent their recurrence, preserve body function, and reduce the risk of death.

In many cases, the objective is achieved with both antibiotic therapy and surgical intervention.

Approximately two thirds of infections are caused by either *S. aureus* or coagulase-negative staphylococci. Methicillin-resistant staphylococci are susceptible to older antibiotics (doxycicline, quinolones, and clindamycin), and they are almost universally sensitive to the newer agent linozolid, but the clinical efficacy of these drugs for the treatment of infection associated with surgical devices has not been compared prospectively with the efficacy of vancomycin [10, 11].

Most implants that are infected by *S. aureus* or *Candida* require surgical removal. Patients with an established response to medical therapy for an implant infection caused by less virulent coagulase-negative staphylococci may not require surgery to remove the implant. If an infected implant must be removed, the complete extraction of all components is essential if surgically feasible, regardless of the type of infecting organism [1]. Although surgical removal of infected implant is generally associated with a better outcome than

TABLE 15.1 Antibiotic therapies related to device-centered infections

Microorganism	Initial treatment	Duration	Follow-up treatment
Staphylococcus spp. Methicillin resistant	Cloxacillin or nafcillin+rifampicin	2 weeks	Ciprofloxacin+rifampicin
Streptococcus spp. (except *Streptococcus agalactiae*)	Penicillin G or ceftriaxone	4 weeks	Amoxicillin
Staphylococcus spp.	Vancomicin+rifampicin	4 weeks	Fusidic acid or minocucline + rifampicin
Enterococcus spp. (and *Streptococcus agalactiae*)	Penicillin G + gentamicin	4 weeks	Amoxicillin
Enterobactenaceae (ciprofloxacin susceptible)	Ceftriaxone	3 to 5 days	Ciprofloxacin
Pseudomonas aeruginosa	Ceftazidime or cefepime + gentamicin	4 weeks	Ciprofloxacin
Anerobes	Clindamycin	4 weeks	According to pathogen

retention of the infected implant, medical treatment alone may be warranted in patients who are at high risk for intraoperative or postoperative complications. Table 15.1 summarizes the antibiotic therapy related to the pathogens.

The difficulty and the costs associated to the treatment of prosthesis have led to heightened interests in prevention. The use of perioperative antimicrobial prophylaxis and a laminar airflow surgical environment has reduced the risk of intraoperative infection. In the case of bone surgery, antibiotics should be administrated intravenously 30 min before incision of skin and for no longer than 24 h after the operation. Because colonization of the prosthesis is a prelude to clinical infection, one important approach has been to coat the surface of the prosthesis devices with an antimicrobial substance.

15.2 PERI-IMPLANT INFECTIONS

15.2.1 Bacterial Adhesion to Surfaces and Processes Governing Biofilm Formation

Bacteria can switch between two distinct "forms of life": a planktonic, free-living lifestyle (not attached to a surface) and a sessile state (forming communities). When a bacterium "decides" to colonize a surface, pattern gene expression undergoes a deep and radical change to produce a unique biofilm phenotype that differs from that of a single planktonic bacterial cell for more than 70% of expressed proteins [12–16]. A biofilm is commonly defined as a structured complex community of one or more species of microorganisms attached to a substrate and to one another and embedded in a highly hydrated extracellular matrix (up to 97% of water content) consisting mainly of exopolysaccharide(s) bound together by proteins and DNA [17–21].

Both gram-positive and gram-negative bacteria can produce biofilm structures on every environmental surface: from inorganic substrates as minerals and carapaces of dead animals, to living organisms as plants and animals. Potentially in the human body every niche, either external (skin) and internal (oropharyngeal, mouth, nose, intestine, and artificial devices), is colonized by bacteria. This bacterial flora includes both commensal nonpathogenic microorganisms and pathogenic and opportunistic bacteria. Biofilm represents an evolved mechanism for microbial survival that promotes the transmission of pathogens by providing a stable protective environment and by acting as a reservoir for cells dissemination [22].

There are as many different types of biofilms as bacteria, and even one bacterium can produce various types of biofilms. The activation of genes involved in formation and maintenance of biofilm, as well as biofilm architecture and composition, are influenced strictly by environmental and surface conditions.

When a bacterium approaches a surface that can be potentially colonized, a transient attachment takes place, a process that implies weak attractive forces. Most bacteria break the bond with the surface and only a few, according to a process that is apparently stochastic, remain attached for a longer period. This event is defined as a transition from transient to permanent attachment. Although it is hypothesized that this transition is modulated by environmental signals, most of such signals are still unidentified. A recent work by Van Dellen et al. [23] reports that dissipation of the membrane potential ($\Delta\Psi$) completely blocks the transition from transient to permanent attachment. Transient attachment is influenced by electrical charges carried on the bacteria, by Van der Waals forces, and by electrostatic attraction, although the precise nature of the interaction is still a matter of intense debate. The classic Derjaguin, Landau, Verwey, Overbeek (DLVO) theory [24, 25], which is used extensively to describe the deposition of bacteria and to calculate qualitatively and quantitatively adhesion free energy changes involved in microbial adhesion, is now considered as limited [26–28]. This classic model visualizes the planktonic bacterial cells as smooth colloidal particles, whose adhesion on natural or artificial surfaces is guided by the interaction potential taking place between charged surfaces. Bacteria deposition mainly occurs in correspondence to a secondary energy minimum, when the charge repulsions are overcame (see Surface Characteristics Influencing Bacterial Adhesion). This theory assumes that microbial cells behave as inert particles, whereas experimental data clearly suggest the involvement of interactions between macromolecular structures localized on the bacterial cell membrane and on the material surface. Flagella and pili [17, 29–34] and conditionally synthesized adhesins (proteins or polysaccharides) [35–38] play a key role as surface-sensing structures to mediate the initial transient interaction and to stabilize the bacterium attachment onto the surface, respectively. Motility induced by flagella can be necessary to reach the surface by allowing the cell to overcome the repulsive forces between the cell and surface. This mechanism is possibly more important under stagnant conditions rather than under flow conditions. In addition, motility can be required to move along the surface, thereby facilitating growth and spread of a developing biofilm. Finally, flagella and pili themselves (as surface appendages) can directly mediate the attachment to surfaces.

Transition from isolated cells to biofilm is regulated by a variety of environmental and physiological cues, such as bacterial cell density, nutrient availability, and cellular stress. Some of these signals have been identified, and many others are still unknown. Moreover, this process requires the activation and repression of a wide number of genes. An extended and detailed review of cellular and transcriptional programs that guide biofilm formation is beyond the scope of this chapter; an exhaustive description can be found in many reviews devoted to this subject [39–41].

The identified signals inducing biofilm formation include mechanical signals [17, 29, 30, 33, 34], nutritional and metabolic cues [42, 43], presence of inorganic compounds [44–47], host-derived signals [48, 49], and a specific mechanism called "quorum sensing."

Flagellar arrest represents one of the many checkpoints that drive along the pathway for biofilm formation [17, 29, 30, 33, 34]. One hypothesis on this mechanism [40] is that the microorganism perceives an increased drag on the flagellar motor, which is caused by the interaction with substrate, and it induces a specific signal transduction that stimulates biofilm matrix production.

A biofilm promotion signal that is considered critical for many bacteria is the quorum sensing. Quorum sensing bacteria produce and release chemical signal molecules called autoinducers, which are secreted into the extracellular environment and increase in concentration as a function of cell density. The detection of a minimal threshold stimulatory concentration of an autoinducer triggers the alteration and coordination of the gene expression of bacterial population [50]. The two known different quorum systems are the LuxR-LuxI system in gram-negative bacteria, which uses an *N*-acyl-homoserine lactone as signal, and the *agr* system in gram-positive bacteria, which uses a peptide-tiolactone as signal and the RNAIII as effector molecules [51, 52]. Quorum sensing modulates both intraspecies and interspecies cell–cell communication, and it plays a major role in enabling bacteria to architect complex community structures. Gram-positive and gram-negative bacteria use these communication circuits to regulate a wide array of physiological functions (virulence, competence, conjugation, antibiotic and bacteriocin production, motility, and sporulation), including biofilm formation [53–55].

The final stage in the irreversible adhesion of bacterial cells to a surface is associated with the production of exopolysaccharides (EPS), which leads to the formation of a mature biofilm architecture. Some EPS are neutral or polycationic polymers, but most are polyanionic because of the presence of either uronic acids (*D*-glucuronic acid being the most common one, although *D*-galacturonic and *D*-mannuronic acids are also found) or ketal-linked pyruvate. Inorganic residues,

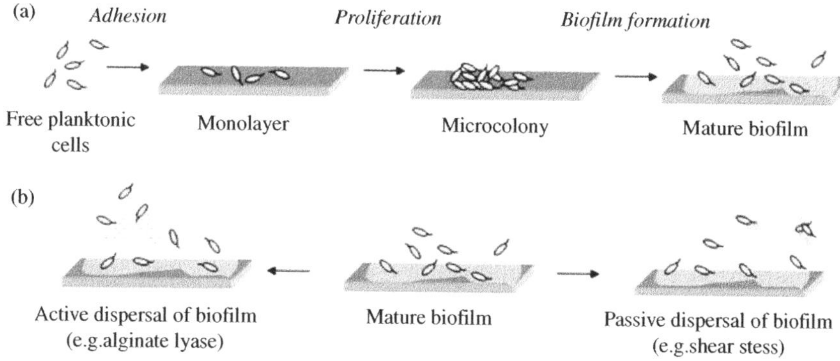

FIGURE 15.1 (a) Models of the development of a mature *P. aeruginosa* biofilm from planktonic cells. (b) Dispersal of bacteria from a biofilm. Two mechanisms for the detachment and dispersal of cells from a biofilm are presented. Active dispersal takes place through a highly regulated process involving many sensory circuits. Dispersion mechanisms include the synthesis of enzymes that degrade biofilm matrix, return to motility, surfactant production, and cell lysis. The passive dispersal of biofilm is the result of hydrodynamic parameters such as shear stress that trigger biofilm fragmentation and bacteria carrying by the bulk fluid until they lodge in a new site and initiate a new sessile population.
Source: Adapted from Costerton et al. [37].

such as phosphate or, rarely sulfate, may also confer polyanionic status. Although the composition of the polysaccharides differ among different bacteria, one of the most common exopolysaccharidic component of the biofilm matrix is the polymer β-1,6-*N*-acetyl-D-glucosamine (indicated as poly(L-glutamic acid) [PGA] or PNAG). Clinically relevant bacteria as *Staphylococcus epidermidis*, *S. aureus*, *Bordetella* spp., and *Escherichia coli* express this polysaccharide [56]. Other exopolysaccharides commonly found are cellulose-like ones in *E. coli* and alginate-like in *Pseudomonas* spp. Matrices [19].

The complex architectural structure of a biofilm consists of EPS-embedded microcolonies of bacteria scattered in cellular less dense regions of matrix, which include pores and channels for nutrients and waste exchange. Each biofilm bacterium lives in a customized microniche, in a complex microbial community that has a primitive homeostasis, a primitive circulatory system, and metabolic cooperativity. Each of these sessile cells reacts to this special environment so that it differs fundamentally from a planktonic cell of the same species [39]. Sessile biofilm communities can give rise to nonsessile free-floating bacteria that can rapidly disperse in the environment (Figure 15.1).

15.2.2 Pathogenesis of Implant-Associated Infections

The major barrier to the extended use of implant devices is bacterial adhesion to biomaterials, which causes biomaterial-centered infection and the lack of successful tissue integration or compatibility with biomaterial surfaces.

Adhesion-mediated infections are extremely resistant to antibiotics and host defenses, and frequently they persist until the device is removed.

Early implant-related infections (approximately less than 1–2 months postoperatively) are mostly caused by contaminations contracted during surgery procedures. In these cases, the most common sources of bacteria are the skin of patients and physicians, air, nonsterile surgical instruments, or postoperative complications such as infections of wounds. In contrast, late biomaterial-centered infections (those that occur months to years after deep tissue implantation) are primarily initiated by bacteria spread hematogenously from endogenous foci. Most hematogenous infections are believed to arise from infected skin lesions producing relapsing bacteremia. Dental or other surgical interventions, bacteriuria, intestinal surgery, and pneumonia have also been proposed as possible causes of the hematogenous spreading of bacteria. A low percentage of infections, which are defined as delayed infections, involve bacteria inserted during implantation that stay clinically unnoticed for a long time.

The principal implants that can be compromised by biofilm-associated infections are central venous catheters, heart valves, ventricular assist devices, coronary stents, neurosurgical ventricular shunts, implantable neurological stimulators, arthroprostheses, fracture-fixation devices, inflatable penile implants, breast implants, cochlear implants, intraocular lenses, and dental implants. Different species, both gram-positive and gram-negative, have been isolated from colonized devices. Microbiological profile of bacteria involved in contamination mainly include nonpathogen or opportunistic pathogens whose species vary according to type and localization of device [39, 57]. Staphylococci (*S. aureus* and *S. epidermidis*) and other gram-positive cocci commonly infect sutures, arteriovenous shunts, vascular grafts, central venous catheters, scleral buckles, and orthopedic devices. Enterobacteriaceae and enterococci (commonly *E. coli*) colonize urinary catheters. *Pseudomonas* spp. contaminations are spread on contact lenses and others intraocular devices, as well as vascular and urinary catheters. Endotracheal tubes colonization in patients who undergo mechanical ventilation involves a variety of bacteria and fungi, which are the first cause of nosocomial pneumonia.

The transformation of nonpathogen or opportunistic pathogen bacteria to highly virulent subpopulations is triggered both by phenotypic phase variation of microorganism from planktonic to sessile forms producing defensive barrier structures (biofilms) and by local altered immune host response. The two events are closely associated. Intact host immune systems usually eliminate transient bacterial contamination unless the microorganism inoculum exceeds a defined threshold. Biomaterials, despite being inert and nontoxic, often draw in the sites of insertion an adverse physiological phenomenon known as a "foreign body reaction," which induces local immunosuppression [58, 59]. In this condition, even low microorganism counts are not eliminated by a proper host immune response and the threshold of bacteria that can generate an infection decreases dramatically [60–63].

When a device is inserted into a tissue or organ, the induced body injury initiates a cascade of molecular and cellular events similar to, but ultimately diverging from, a normal wound-healing process. This physiological response sequentially comprises the following interdependent and overlapping phases: blood-material interaction, acute and chronic inflammation, formation of a granulation tissue, foreign body reaction, and finally fibrous encapsulation of the implant. The foreign-body reaction to biomaterials is composed primarily of giant cells and components of granulation tissue (macrophages, fibroblasts, and capillaries). This fibroinflammatory structure may persist at the tissue-implant interface for the lifetime of the implant in case the tissue repair is not solved by replacement of injured fibrotic structures with regenerated native tissues.

As reported by Gristina [64, 65], granulation and fibroinflammatory tissues surrounding implants represent an immunoincompetent zone. Macrophages and foreign body giant cells as well as neutrophils have been shown to exhibit a reduced antibacterial activity when adhered to biomaterial surfaces because of a locally acquired phagocytic-bactericidal defect [66–68]. Zimmerli et al. [68] demonstrated that local polymorphonucleated (PMN) cells display a low ability to ingest particles, low levels of granular enzymes, and a decreased ability to mount respiratory burst when in contact with poly(tetrafluoroethylene) tubes implanted subcutaneously into guinea pigs. This functional damage could be secondary to partial exhaustion of PMN and macrophages functions by "frustrated" phagocytosis and prolonged stimulation of cells along the foreign and nonphagocytable implant surface. The massive release of mediators of degradation such as reactive oxygen intermediates (oxygen free radicals), degradative enzymes, and cytokines induced by repeated macrophage priming by biomaterial particulates [69, 70], result into mediated tissue damage, macrophage exhaustion, inhibition of device integration with the body tissues, and therefore, increased susceptibility to infection or aseptic loosening [64].

Another mechanism that may be involved in the weakening of the immune response is macrophage and giant-cell apoptosis induced by interaction with biomaterials. An extensive apoptotic process reduces the capability of phagocytic cells to react properly to the attack of foreign organisms. Shive et al. [71] demonstrated that the shear stress-induced apoptosis of adherent neutrophils is critical for the persistence of infections in cardiovascular devices. Moreover, programmed cellular death has been shown to be influenced strongly by biomaterial surface chemistry [71, 72].

The first crucial step that starts and actuates the complex pathway of peri-implant colonization is always the contact and the adherence of bacteria on the device surface. Stable interactions of bacteria with biomaterials are directed by specific receptors and outer membrane molecules on bacterial and by blood factors absorbed onto implant surface. The latter ones include mainly plasma proteins such as fibrinogen, fibronectin, and adherent platelets forming the conditioning film [73–75]. In *S. epidermidis*, factors binding to host extracellular matrix components like fibrinogen-binding protein Fbe [76], vitronectin-binding autolysins AtlE [77] and Aae [78], and extracellular matrix-binding protein (Embp) [79] are thought to determine primary attachment critically.

15.2.3 Biofilm Resistance Mechanisms on Implants

Recalcitrant infections caused by antibiotic-resistant bacterial biofilms are one preeminent and fundamental cause of implant failure. The molecular strategies of biofilm resistance are manifold [39].

Exopolysaccharides and other molecules constituting the biofilm matrix may entrap antimicrobial agents by physical-chemical interactions. As a result, antibiotics, disinfectants, or germicides are either inactivated or delayed and blocked in their diffusion and penetration into the biofilm interior. Several studies have examined the penetration and interaction of antimicrobial agents with the extracellular polymeric components of biofilm [80–83]. The uptake of antimicrobial agents varies with drug size, charge, hydrophilicity, and chemical composition. As an example, aminoglycosides as gentamicin and tobramycin bound avidly to alginate matrix of *Pseudomonas* blocking almost completely their diffusion [84], whereas beta-lactams exhibit a higher diffusion coefficient [85].

Phenotype changes in growth and metabolism and genetic mutations also contribute to generate resistance. Cells that are biofilm-associated grow significantly more slowly than planktonic cells and have a reduced metabolic activity. As a result, they display increased insensitivity to antimicrobial compounds [86], 87]. Moreover, high rates of genetic material exchange and hypermutations within biofilms, promote the development of genetic survival mechanism like expression of specialized efflux pumps and upregulation of porin proteins, which can rapidly propagate throughout the population [88].

An important and fundamental characteristic of implant-centered infection is their high rate of reoccurring, which flows into more aggressive diseases. It is now believed that it is attributable to the existence of *persister cells* [89]. These cells might have a disabled programmed-death mechanism that renders them resistant to the actions of all antibiotics, thus conferring them multidrug tolerance (MDT). At the same time, persister cells can originate an MDT progeny that repopulates the community after initial apparently successful therapies, being responsible for the infection persistence. The mechanism of MDT and the nature of persisters are still elusive. As studied and reported by Lewis [89], MDT may be associated to the expression of toxin–antitoxin modules and other genes that can block important cellular functions, such as translation, which are the main targets of antibiotic compounds.

Because of these various and still unknown resistant properties displayed by bacterial biofilm, it is clear that is extremely difficult to find a single substance or strategy that can overcome the problems and difficulties faced by the medical community in the treatment of bacterial-based implant infections.

15.3 BIOFILM-CONTROLLING MATERIALS

15.3.1 Designing Efficient Antimicrobial Implant/Tissue Interfaces

The adhesion of bacterial cells to material surfaces and interfaces is the first step in bacterial colonization and can be followed by the maturation of a bacterial

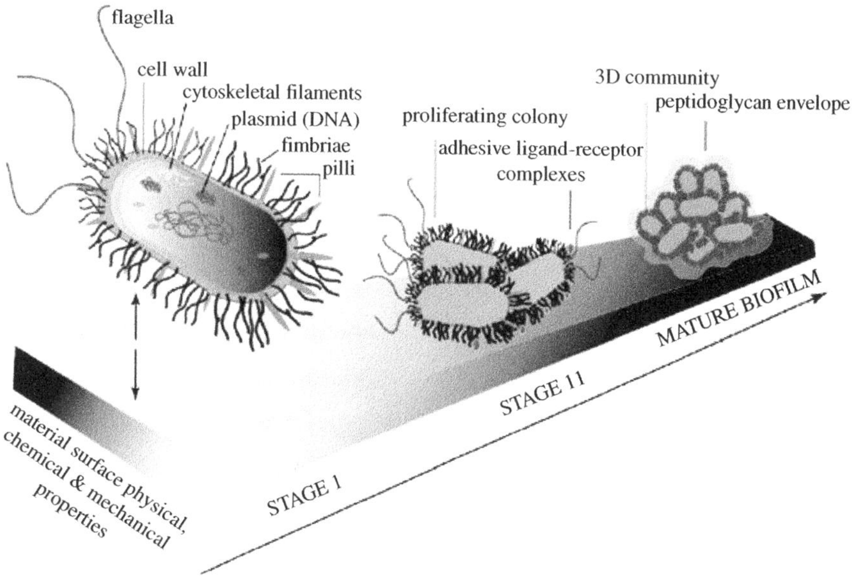

FIGURE 15.2 Schematic of prokaryotic bacterial cell structure and the two-stage bacterial adhesion model that precedes the organization of a mature biofilm.
Source: Reprinted from Lichter et al. [90] with permission.

biofilm. Figure 15.2 shows the main steps of bacterial adhesion which can be generally summarized in a two-stage model: an initial reversible interaction between the bacterial membrane and the material surface, followed by a second stage that includes specific and nonspecific interactions between adhesin proteins expressed on bacterial surface structures and binding molecules on the material surface. The second step is slowly reversible and often termed as irreversible [90].

Bacterial adhesion is a complex interplay of different mechanisms such as Brownian motion, gravitation, diffusion, convection, and intrinsic motility of microorganisms. When a biomaterial is inserted into the body, a so-called "conditioning film" is deposited on the biomaterials surface that comes from organic matter present in the surrounding fluid. It mostly consists of adsorbed proteins and blood platelets; the composition of this organic matter is influenced by the physical-chemical and morphological properties of the surface. This represents an intermediate layer connecting the substratum and the adhering microorganisms, and it mediates the direct interaction among the latter two ones [91]. Microbial adhesion to inert surfaces is described in terms of both specific and nonspecific interactions (Figure 15.3). Although the former are highly directional and operate over small distances (about 5 nm), the latter arise from interaction forces between all molecules of the entire cell and substratum, and they have a long-range character. Once the general macroscopic interaction forces (long range and nonspecific) allow the bacteria to come in contact with the surface, interaction among the stereochemical

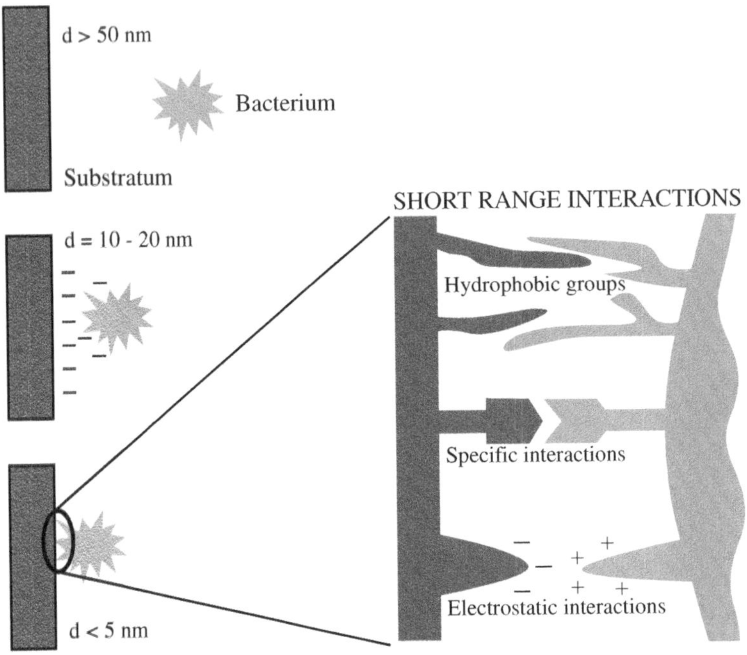

FIGURE 15.3 Distance dependency of bacteria–material surface interactions. At separation distances > 50 nm, only attractive Van der Waals forces occur while at 10–20 nm; Van der Waals and electrostatic interactions influence adhesion. At a distance <5 nm, short-range interactions can occur, irreversibly binding of a bacterium to a surface.
Source: Adapted from Busscher et al. [91].

groups (short range and specific) take place with a much higher attractive interaction energy [92].

Several characteristics of the material surface have been proposed to affect the first stage of bacterial adhesion. These factors include surface roughness, stiffness, charge, degree of hydrophobicity, Lewis acid-base character, hydrogen bonding capacity [93], van der Waals forces, and specific receptor-adhesion interactions [94]; the influence of surface characteristics on bacterial adhesion is addressed more specifically in the section Surface Characteristics Influencing Bacterial Adhesion.

Because the colonization of implant/bacteria interfaces is influenced by so many factors, it is not easy to highlight and weight their relative contribution to bacterial adhesion. Moreover, it is difficult to reach general conclusions of bacterial adhesion mechanisms because the environmental conditions over which bacteria strains proliferate differ considerably, including variables such as temperature, flow conditions, presence of adsorbed proteins, and concentrations of glucose and oxygen. Consequently, even for a single strain and material surface, environmental stimuli can affect the relative importance of both adhesion mechanisms and material surface characteristics.

Given these premises, the presence of contradictory reports in the literature, which evaluate the most significant factors in bacterial adhesion, proves that no singular material feature or bacterial characteristic completely describes or controls bacterial adhesion [90]. However, it is possible to analyze common strategies and draw general conclusions in the study of polymer-based biomaterial interfaces, as it will be discussed in the following paragraph of this chapter.

15.3.1.1 Main Strategies for Antibacterial Surface Design. Three main strategies (Figure 15.4) have been pursued to limit or prevent colonization of material surfaces, by selecting (1) their anti-adhesiveness, (2) their activity by contact, and (3) their biocide-release activity [95].

1. This strategy is focused on preventing bacterial adhesion or at least reducing the capacity of bacteria to achieve the two adhesion stages. To this end, both hydrophilic and ultrahydrophobic polymer surfaces are being explored [94, 96–98].
2. This approach aims at killing bacteria by contact with the biomaterial surface, thus inducing death of bacteria that have stably adhered to a surface. In general, it occurs via conjugation of a polymeric surface with antibiotic functional molecules like antimicrobial peptides [99], quaternary ammonium compounds [100], guanidine polymers [101], phosphonium salts [102], or the use of the polycations [100].
3. This design is based on the leaching from the material surface of biocide compounds which diffuse over time inducing bacterial death. Polymeric surfaces can be tuned to allow a controlled release of compounds like antibiotics [60, 103], silver ions [97, 104] and nanoparticles, or peptides [105].

15.3.1.2 Surface Characteristics Influencing Bacterial Adhesion. A physical-chemical approach for modeling the adhesive interactions between microbes and surfaces has been tackled by several authors, and a correlation has been attempted between the theoretical predictions and the knowledge of the adhesion mechanism. The different physical-chemical approaches proposed in the literature are based on a thorough analysis of the interaction energy between

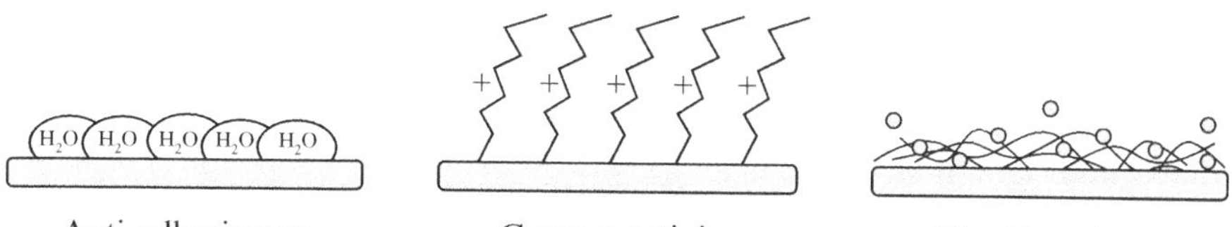

FIGURE 15.4 Three main strategies for antibacterial surface design.
Source: Adapted from Lichter et al. [90].

the microbial system and the surface characteristics. However, they are all affected by a flaw of the intrinsic characteristics of the systems under investigation. In fact, microbial cell surfaces are heterogeneous in composition and with a high level of complexity, which is normally not taken into account when considering a physical-chemical modeling at the molecular level. Moreover, simple concepts like, e.g., distance partially lose their meaning considering that appendages on microbial cell surfaces can become as long as 1 μm [106]. Microbial adhesion can be envisaged as the process of transferring cells from an unbound state (bulk phase) to a more or less firmly attached state at the interface [107].

Nevertheless, some physical-chemical approaches showed some ability to predict and explain the behavior of bacteria on different surfaces. Two main physical-chemical analyses have been proposed, namely the thermodynamic approach [108] and the DLVO (Derjaguin, Landau, Verwey, Overbeek) theory [24, 25, 109] (for an extended discussion see Reference 106).

The thermodynamic approach is based on the calculation of the Gibbs energy associated with the bacterial adhesion (ΔG^{adh}) using the Dupré equation:

$$\Delta G^{\mathrm{adh}} = \gamma_{BS} - \gamma_{BL} - \gamma_{SL}$$

where γ_{BS}, γ_{BL}, and γ_{SL} are the bacterium-surface (substratum), bacterium-liquid, and substratum-liquid interfacial energies, respectively, as evaluated by contact angle measurements [108]. A negative value of the free energy ΔG^{adh} indicates the adhesion of the bacterium onto the substratum. In this approach, a purely thermodynamic equilibrium is assumed to take place. In other words, the adsorption is reversible, which is often not the case [110]. Several limitations have been underlined in the thermodynamic model, including the lack of a distance dependence of the adhesion strength, its nonapplicability when a new cell-substratum interface is not formed (secondary minimum) [111], and the complete lack of kinetics considerations. This model was applied on several cases of bacterial adhesion providing for agreement with experimental determinations in some cases [112], but evidence was found for its strong limitations in some other cases [113] (for a more detailed discussion see [107]).

The DLVO approach extends the thermodynamic treatment by including, in the description of the free energy of microbial adhesion (ΔG^{adh}), a balance between Lifshitz-van der Waals forces (ΔG^{LW}, generally attractive) and electrostatic forces generated from the overlap of the electrical double layer of the microbial cell and the substratum (ΔG^{EL}, generally repulsive). The strength and range of the latter contribution is markedly influenced by the presence of surrounding ions. An increase of the ionic strength would act compressing the double layer and, thus, inducing an increase in the bacterial adhesion [114, 115]. Both these free energy terms are distance dependent (Figure 15.5a) and their determination requires the measurement of zeta-potential and contact angles.

$$\Delta G^{\mathrm{adh}}(d) = \Delta G^{\mathrm{LW}}(d) + \Delta G^{\mathrm{EL}}(d)$$

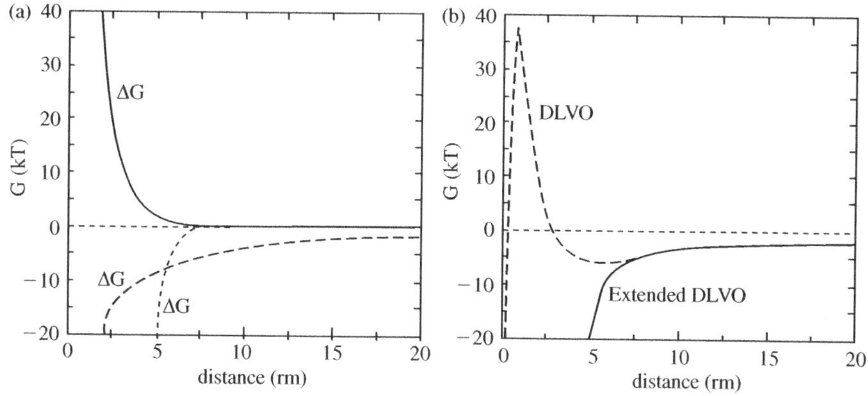

FIGURE 15.5 (a) Contributions of Lifshitz-van der Waals (GLW), electrostatic (G^{EL}) and Lewis acid-base (G^{AB}) interaction energies. (b) Comparison of the interaction energies, as calculated by means of the DLVO (. . .) and extended DLVO (?) theories, between a spherical particle (microorganism) and a solid substratum as a function of distance.

Source: Adapted from Bos et al. [106].

The dependence of both ΔG^{LW} and ΔG^{EL} on distance is reported in Figure 15.5a.

The DLVO theory has been extended by van Oss et al. [92, 116] to include also the Lewis acid-base interaction ΔG^{AB}.

$$\Delta G^{adh}(d) = \Delta G^{LW}(d) + \Delta G^{EL}(d) + \Delta G^{AB}(d)$$

The acid-base interactions are based on electron-donating and electron-accepting interactions between polar moieties in aqueous solutions, and they show an exponential decay function with the distance between the cell and the substratum (Figure 15.5a). The term ΔG^{AB} should account for the attractive hydrophobic interactions[117, 118] and repulsive hydration effects, which have been evaluated to be 10 to 100 times stronger than the Lifshitz-van der Waals ones (Figure 15.5a) [119]. An example of the comparison between the theoretical predictions by means of the DLVO and of the extended DLVO theory is reported in Figure 15.5b. Depending on the ionic strength, a so-called secondary minimum of a few kT is found in the adsorption of the bacteria on the substratum. This represents a reversible capturing of the microorganisms and precedes the primary minimum of adhesion that requires a closer approach. It should be noted that the energetic term for Brownian motion is approximately 1 kT.

It is obvious that the determination of the free energy contribution for the electron-donor/electron-acceptor and hydrophobic interactions is of paramount importance.

Microbial cell surface hydrophobicity is without any doubt the most studied property of the cell surface with regard to their adhesion to surfaces [120, 121].

As the cell-surface hydrophobicity is related with its wettability, it can be estimated by contact angle measurements [122], although criticism on the dependence of the latter from the dehydration of the bacterial lawn has been raised [123]. The surface hydrophobicity has also been studied by means of the microbial adhesion to hydrocarbons (MATH) experiments [124]. This is based on a partition of the microorganisms between water and hydrocarbons; hydrophilic organisms would remain in the aqueous phase, whereas the hydrophobic ones would adhere to the hydrophobic phase. Therefore, the basic rationale for the adhesion of organisms to hydrocarbon or any other hydrophobic ligand is driven by their dislike for water. One of the key aspects of the MATH approach is that electrostatic interactions in hydrocarbons are neglected, although this assumption does not always hold [125]. Thus, the affinity of the microorganism for the hydrocarbons is solely caused by the Lifshitz-van der Waals forces.

The work by van Oss [119] led researchers to revise the hydrophobic interactions underlying the importance of the acid-base features of microbial cell surface. As a consequence, the surface tension γ of the microbial layer is determined by two contributions, namely the Lifshitz-van der Waals (γ^{LW}) and the acid-base ones (γ^{AB}).

$$\gamma = \gamma^{LW} + \gamma^{AB}$$

in which

$$\gamma^{AB} = 2\sqrt{\gamma^{+}\gamma^{-}}$$

where γ^{+} and γ^{-} are the electron-donating and electron-accepting surface free energy parameters. γ^{LW}, γ^{+}, and γ^{-} can be determined by measuring the contact angles using three different probe liquids with known surface tension parameters [126, 127]. The values of γ^{+} and γ^{-} can be directly determined by the microbial adhesion to solvents (MATS) approach, which is based on the fact that microbial cell surfaces can exert acid-base interactions, thus proving their electron-donating feature [128]. In this case, the measurement of microbial adhesion to hexadecane and hexane (unable to exert acid-base interaction), to chloroform (electron-accepting solvent), and to diethyl ether (electron-donating organic solvent) demonstrated the fundamental importance of acid-base interactions and their involvement in the cell surface hydrophobicity. For these reasons, MATS seems a promising method to determine the electron-donor and electron-acceptor features of microbial cells.

The knowledge of these basic principles has led to design specific surfaces that impede or limit bacterial adhesion based on their physical-chemical characteristics. The overall energy of cohesion is the interaction free energy (per unit area) that is often used to describe the hydrophobicity or hydrophilicity of a substrate and bacteria. Positive values of the cohesive energy account for hydrophilic surfaces, whereas negative values indicate hydrophobic surfaces.

The interpretation of the bacterial adhesion in terms of the DLVO theories has been attempted in several cases. In general, the theoretical force values do not always match with experimental values reported in the literature [129]. Adhesion of hydrophobic cells is favored on low-energy substrates and was found to increase after increasing the hydrophobicity of the substrate [130, 131]. Conversely, the cell wall hydrophilicity induces a repulsive interaction with the surface of the material. These repulsive interactions become weaker when the cell exhibits an hydrophobic surface. This effect is largely determined by the acid-base interactions. The inclusion of the latter interactions in the computation affects mainly the primary interaction energy given its short-range character [127]. Along the same line, it was found that the increase of hydrophilicity of the substratum correlated with a decrease of microbial adhesion with the same trend predicted by the extended DLVO theory [98]. However, it has been reported that a marked variation of the hydrophobicity of the substratum is required to spot variations in the bacterial adhesion and that impurities might also affect the final outcome, especially if these are represented by proteins or lipopolysaccharides [132]. The modification of the surface with poly(ethylene-oxide) was reported to reduce the bacterial adhesion on polyurethane [94]. The increase of hydrophilicity of poly(vinyl chloride) by oxygen plasma was found to reduce the adhesion of *P. aeruginosa* strains, although not to impede the bacterial biofilm formation completely [133]. It is interesting to note that after oxygen plasma treatment, and the subsequent increase of hydrophilicity of the material, the profile of the calculated free energy of adhesion is reversed from a negative primary minimum to a net positive maximum of ΔG^{adh}, which reveals a lack of adhesion of the bacterium. Once more, the lack of a primary minimum, which guides the irreversible adhesion of bacteria, is impaired by the Lewis acid-base repulsion at small separation distances. However, a secondary energy minimum of approximately −3 kT was found, determined primarily by the acid-base interactions. This value is slightly lower than forces exerted by the Brownian motion of the bacteria that could determine their detachment [133].

A correlation between the substratum hydrophobicity and adhesion of *S. epidermidis* was found in phosphate-buffered solution, confirming once more that short-range (acid-base) interactions seem to dictate the nonspecific interaction. This overall picture varies if the adhesion is considered in the presence of plasma proteins [134]. This is basically because of the adsorption of the latter on the surface of the substratum varying its physical-chemical properties [134].

As a general consideration, it can be stated that for the same substratum, the higher the hydrophobicity of the bacteria, the higher their adhesion and the biofilm formed. This was proved, as an example, comparing the more hydrophilic *S. aureus* with the more hydrophobic *E. coli* [132]. It is to be stressed that different strains of the same bacterium show marked differences in the hydrophobicity/hydrophilicity properties [135, 136]. In general, the hydrophobicity of the bacterial cell wall, which is determined basically by its chemical composition in terms of the biomolecules synthesized, has a lower effect when compared to substratum hydrophobicity [136]. The importance of the Lewis acid-base interactions in driving the bacterial biofilm formation was also evident in the comparison of the adhesion of *E. coli* toward materials characterized by different hydrophobicity, namely poly(vinyl chloride), low-density polyethylene, and poly(methyl methacrylate) [137].

The prediction of the bacterial adhesion based on the chemical characteristics of the substratum based on the (extended) DLVO approach shows several limitations. First, the use of medium containing proteins, because of the adsorption of the latter on the surface of the materials, complicates the analysis. Moreover, bacterial cells show peculiar features that could affect their adhesion to surfaces. This finding implies that the prediction ability of the extended DLVO theory was questioned in several occasions because of the structural and chemical cell surface heterogeneities of the microorganisms. It is well known that most bacteria can excrete polymeric substances associated with the cell wall that could inhibit [138] or enhance [139] the bacterial adhesion to substrates. The steric effect exerted by the lipopolysaccharides present in the bacterial cell wall was reported to affect and contribute to the adhesion of the cells over the surface of a material [131]. In general, it has been considered that the presence and packing of surface polymers on bacterial cells causes deviation from the predictions of the extended DLVO approach [135].

The importance of surface roughness on the adhesion of bacteria was recently reviewed [140, 141]. It is generally agreed that irregularities on the polymeric surface promote bacterial adhesion, whereas ultrasmooth surfaces prevent biofilm deposition [137, 142, 143]. This is because the initial bacterial adhesion probably locates where the cells are sheltered from the shear forces and because an increase of the roughness leads to an increase of the total surface available for adhesion [132]. It has been reported that roughening of the surface increased bacterial adhesion and biofilm formation [144], although a direct correlation did not hold for all the materials tested [145].

The influence of the stiffness of the surface of the material was also investigated. It has been reported that this physical parameter modulates the mechano-selective adhesion of *S. epidermidis* and *E. coli*. In particular, the increase of the Young's modulus of the polyelectrolyte layers on the substrate correlates positively with the adhesion of the bacterial strains in the range of 1 MPa to 100 MPa [146].

15.4 POLYMER-BASED ANTIMICROBIAL SURFACES

Polymers are widely used for the construction of antimicrobial surfaces of implant biomaterials. They can be synthetic, naturally derived, or blended forms, and they can be used alone or in association with other materials or molecules to obtain composite materials. Surface modifications of interface polymers offer many possibilities to face the problem of biomaterials infections; in fact, polymeric materials engineered at the nanostructure and microstructure levels play a fundamental role in the design and creation of surfaces that can influence the attachment of both prokaryotic and eukaryotic cells [147, 148]. Many efforts are devoted to develop polymer-based surfaces lethal to pathogens, especially to prevent biofilm formation on indwelling devices [100].

Although it is not always considered in antimicrobial studies, a fundamental aspect that must be taken into account when designing antimicrobial implantable materials active against prokaryotic cells (bacteria) is their biocompatibility; implantable biomaterials should not cause cytotoxic effects toward the eukaryotic cells of the surrounding tissues. The choice of the antimicrobial material is also related to the technique used for the modification of the biomaterial surface; different methods are attempted to match the different requirements of antimicrobial interfaces, as reported in the following sections.

15.4.1 Polyelectrolyte Multilayers (PEMS)

In the study of bioactive coatings of implant materials, polymeric systems that mimic natural highly hydrated environments are appealing [149, 150], in particular both synthetic and natural hydrophilic polymers that can be modified with respect to chemical, physical, and mechanical properties. Among these materials, charged polymers or oligomers are being employed in the development of antibacterial surfaces; in fact, the presence of cationic groups has been generally correlated with potent antimicrobial effects in various studies [100].

A convenient route to bind charged polymers like polyelectrolytes to biomaterials is represented by the surface adsorption driven by electrostatic interactions. This technique is at the basis of PEMs; in the field of biomaterials, this type of surface modification offers numerous opportunities for studying and developing materials whose surfaces can inhibit or promote cell adhesion and proliferation. The preparation technique of PEMs is typically based on the assembly of one "molecular layer" at a time by an adsorption process from aqueous solution [90]; this "layer-by-layer" (LbL) assembly is typically accomplished by dipping a substrate material alternately into solutions containing positively or negatively charged species to obtain a surface coating with nanoscale control over surface chemistry [151], thickness [152], and molecular architecture [153, 154].

The LbL assembly process is particularly suited for coating biomaterials with complex shapes such as scaffolds, stents, and implantable electrode arrays. Moreover, with this technique, it is possible to incorporate a broad range of polymers, nanoparticles [97, 97, 155, 156], proteins [157], drugs [154, 158], and even cells [159]. Typical examples of polyelectrolytes used for the coating of biomaterials include poly(acrylic acid) (PAA), poly(allylamine hydrochloride) (PAH), poly-(methacrylic acid), sulfonated poly(styrene) [160], poly(ethyleneimine) (PEI) [154], chitosan [161], and hyaluronic acid [162].

In the preparation of PEMs, the main processing parameters used to manipulate the molecular organization of layers are ionic strength and solution pH. Varying the preparation conditions it is possible to prepare surfaces with different layer thickness and compositions [152], densities of free ionic functional groups [160], and elastic moduli [147], which affects bacterial response to the surface. For antimicrobial applications, PEMs usually consist of weak polyelectrolytes in the form of antimicrobial synthetic polymers, polysaccharides, or polypeptides, or they contain free functional groups that bind releasable antimicrobial agents [90, 99].

In general, the challenge is to obtain a surface coating presenting its relevant functional groups in a biologically or chemically active form. The three dominant antibacterial strategies mentioned in the section Main Strategies for Antibacterial Surface Design have been followed in the preparation of antimicrobial PEMs as discussed in the following subsections.

15.4.1.1 Antiadhesive PEMS. These surfaces are intended to avoid stable bacterial adhesion on the surface of materials by the use of polyelectrolytes that prevent the early attachment of cells.

One approach for creating antiadhesive PEMs is based on the exploitation of hydrophilic polymers that can establish strong interactions with water molecules *via* hydrogen bonding; such hydrophilic tissue-material interface is supposed to reduce protein and platelet adsorption and curtail bacterial adhesion. For instance, poly(ethylene glycol) (PEG) immobilized or grafted onto surfaces was shown to form a highly hydrated layer that significantly decreased adsorption of *E. coli* cells [94] because of PEG's strong affinity for water molecules. PEG has also been used in association with PGA and poly(L-lysine) (PLL) to create PEMs, which efficiently reduced *E. coli* adhesion [163]. However, the susceptibility of PEG to oxidation damage has limited its long-term application in complex media [164]. The negatively charged, antiadhesive polymer heparin was assembled into multilayers with the positively charged, antibacterial biopolymer chitosan [161] to build surfaces effective against *E.coli* cells *in vitro*. Hyaluronan has been incorporated into many multilayers [165, 166] and has been shown to decrease adhesion of *S. epidermidis* because of its hydrophilicity [162]. Lichter et al. [146] recently demonstrated that manipulating the assembly conditions of PEMs to create compliant hydrated surfaces can significantly inhibit bacterial attachment on poly(acrylic acid)/poly(allylamine) hydrochloride multilayers.

15.4.1.2 Contact-active PEMS. Contact-active PEMs are suggested to induce damage or death of bacteria that have stably adhered to their surface. In the construction of such multilayers, antibacterial strategies based on surface immobilized polycations have been widely explored [100, 167].

Cationic contact-killing surfaces often contain polymers with hydrophobic alkyl side chains that enhance antimicrobial activity [100, 168], although emerging evidence suggests that the bactericidal effect can occur also with polymers of sufficient charge density even without hydrophobic alkyl chains [169]. Various mechanisms have been proposed to explain how cationic molecules ultimately lead to bacteria cell death; however, it is still partially understood how these antimicrobial agents function [100, 167, 169]. The most accepted hypothesis on the mechanism invokes the destabilization and permeation of bacteria caused by charge displacement on cell membrane [100].

A key factor for bacteria-membrane disruption is considered to be polymer chain mobility; recent studies have pointed out that cationic charges along rigid polymer backbones have lower activity than cationic charges along flexible backbones because the rigidity makes it more difficult for multiple charges to interact simultaneously with the cell membrane [170]. Charge mobility has also been related to antibacterial efficacy [171]. The use of PEMs offers the possibility to modulate these factors; in particular, by varying PEM assembly and/or post-assembly pH conditions, PEMs can be engineered to present a sufficient density of cationic charge associated with highly mobile chain segments [90].

Recently, various contact-active PEMs have been prepared with encouraging results. Lichter et al. [146] prepared films composed of poly-(allylamine hydrochloride) effective against *S. epidermidis* and *E. coli*, whereas Thompson et al. [172] used sulfonated poly(styrene) and polyacrylamide assembled with poly(acrylic acid) to study cell adhesion. Many researchers focused on incorporating into contact-active multilayers the polycation chitosan; chitosan is a naturally occurring biocompatible antibacterial polysaccharide that causes the disruption of bacterial cell membrane integrity. Chua et al. [166] functionalized titanium by means of PEMs based on hyaluronic acid and chitosan with high antimicrobial efficacy against *S. aureus*. Bratskaya et al. [173] reported the use of chitosan coatings covalently grafted or applied as chitosan/κ-carrageenan multilayers against strains of *E. faecalis*. More details about the use of chitosan as antimicrobial agent are given in Chapter 4 of the book.

An emerging strategy to create nonleaching biocidal surfaces exploits the positive charge of quaternary ammonium compounds (QACs) [174]. QACs have been grafted on polymers used in the PEM assembly process [175]; these multilayers displayed killing efficacies higher than 99% with respect to *E. coli* and *S. epidermidis*. Shi et al. [176] developed PEMs that confer antibacterial properties to stainless steel *via* the alternate deposition of quaternized PEI or quaternized polyethylenimine-silver complex and PAA; these coatings were shown to be effective toward both gram-positive and gram-negative strains and to display minimal toxic effects toward mammalian cells. Wong et al. [177] incorporated polymeric hydrophobic quarternary ammonium salts into LbL films to design contact-killing thin films; in this PEM, the polycation *N*,N-dodecyl,methyl-polyethylenimine was layered with the polyanion poly(acrylic acid) to create LbL films active against both airborne and waterborne *E. coli* and *S. aureus*.

Additional methods have been pursued to design contact-active multilayers. Rudra et al. [157] developed PEMs with poly(L-glutamic acid) (PLGA) and the positively charged hen egg white lysozyme exploiting its antimicrobial properties, whereas Pan et al. [178] made use of polyguanidines as biocidal polycations for the construction of multilayers. Etienne et al. [179] inserted the antimicrobial peptide defensin into PEMs, which inactivated both gram-positive and gram-negative bacteria. More details about the use of antibacterial peptides are given in Chapter 8. Various other contact-killing species are being incorporated into PEMs, such as titania particles [180] and polyoxometalates (clusters of transition metals) [181], proving the versatility of the LbL technique for this class of materials.

15.4.1.3 Biocide Leaching PEMs. PEMs have also been endowed with biocidal properties by the use of leachable antimicrobial agents. Releasable antibiotics have been often incorporated into PEMs. Chuang et al. [182] created degradable PEMs based on poly(β-amino ester)s with tunable release of gentamicin, which is an aminoglycoside antibiotic. Nguyen et al. [183] incorporated into LbL films micelles encapsulating a hydrophobic bactericide (triclosan) that allowed the release of the drug over a period of several weeks, accounting for long-term inhibition of *S. aureus* growth.

PEMs are being exploited for the release of metallic antimicrobial compounds; in fact, with a suitable choice of polymer and assembly conditions, it is possible to create multilayers that contain free (non–polymer-bound) functional groups that can be used to bind metallic ions or particles. Among metals, the most widely used biocidal leaching agent in PEMs is silver both in the form of ions and nanoparticles [156, 165, 175, 176, 184]; when incorporated into PEMs, the biocidal effect can be exerted by release of silver ions or it can also involve a direct contact in the nanoparticle form [185]. Recently, Travan et al. [186] developed an antimicrobial coating based on silver nanoparticles embedded within a modified-chitosan layer deposited on methacrylic thermosets by the layer-by-layer technique; such coatings were proved to be effective against both g-positive (*S. aureus*) and gram-negative (*P. aeruginosa*) bacteria without displaying toxicity toward mesenchimal stem cells that adhered and proliferated on the nanocomposite layer. Silver-related antimicrobial materials are discussed more in detail in Chapter 11.

Quaternary ammonium compounds have also been incorporated into PEMs in the form of release agents. Grunlan et al. [154] prepared a coating based on the alternate deposition of poly(acrylic acid) and poly(ethylenimine) with cetrimide, which is an organic quaternary ammonium molecule that displays potent antimicrobial effects. The elution of cetrimide and silver ions accounted for inactivation on *S. aureus* and *E. coli.*

The incorporation of leachable antimicrobial peptides into PEMs has also been explored. Shukla et al. [105] took advantage of the LbL assembly of polymer thin films with ponericin G1, which is an antimicrobial peptide active against *S. aureus*. In this study, PEMs based on poly(β-amino esters) and different polyanions (chondroitin sulfate, dextran sulfate, and alginic acid) were tuned to allow a controlled release of the peptide through hydrolytical degradation. Such coatings displayed also a good biocompatibility with respect to NIH 3T3 fibroblasts and human umbilical vein endothelial cells. Boulmedais et al. [163] developed polyelectrolyte multilayer films with PEGylated polypeptides as antimicrobial protection for biomaterials. In this study, PEMs based on poly(l-glutamic acid) grafted by poly(ethylene glycol) were shown to reduce both protein adsorption and bacterial adhesion drastically.

15.4.2 Release of Antibiotic Agents from Neutral Polymer Surfaces

The adhesion of bacteria to the implant surfaces and its overwhelming resistance to conventional antibiotic therapy is now one of the principal issues of the biomedical research field to develop a new generation of synthetic polymer surfaces and coatings that resist bacterial colonization [1]. Two main groups of substrates are now available: passive and active. The former category, which includes, for example, surfaces modified with poly(ethylene glycol) [187], poly(ethylene oxide) brushes [188], and hydrophilic polyurethanes [189], refers to surfaces that reduce bacterial adhesion by altering the physicochemical properties of the substrate. In this way, biofilms and bacteria–substrate interactions are not favorable. Bacterial adhesion on passive coatings is connected with the specie of the bacteria limiting the efficacy of the coating.

This branch of substrates is exceeded by a more recent "active" coating that can provide both short- and long-term release of antibacterial agents. The short-term postimplantation period, which occurs normally several hours from the surgery, is the most critical period in which infections could be originated. Active strategies reduced surface colonization and prevented the formation of biofilms. Moreover, the continued release of drugs from the surface over a longer lapse of time (long term release) avoids the formation of protective fibrous capsule [190].

The controlled delivery of antibiotics from the implant surface is the most efficient way of preventing implant-associated biofilms. Administrating antibiotics directly at the site of implantation allows maintaining the concentration of the drug under the systemic toxicity level. Wu and Grainger [191], in their review, reported the advantages of the local drug release strategies over the systemic drug therapy. Wu and Grainger [191] presented combination devices. According to the U.S. Food and Drug Administration's definition, "a combination device comprises two or more regulated components, i.e. drug/device, biologic/device, or drug/device/biologic, that are physically, chemically, or otherwise combined or mixed and produced as a single entity; or two or more separate products packaged together in a single package or as a unit and comprised of drug and device products, device and biological products, or biological and drug products" [191].

Combination devices match the purpose of the structural replacement of host tissue with the principle of local controlled drug delivery from the implant. In this fashion, enhanced efficacy can be achieved at the implant site. The main target of an ideal drug delivery system consists of a continuous release of therapeutic drug over prolonged periods. The rates and the duration of the drug release may vary in function of the clinical context, thus requiring a careful assessment of the pharmacokinetics at the implant site.

Active localized administration avoids fostering resistance by the selection of specific antibiotics (i.e., vancomycin, tobramycin, cefamandole, cephalothin, carbenicillin, amoxicillin, and gentamicin) related to the pathogens associated with implant infections, therefore preventing dangerous side reactions in other parts of the body [192, 193]. The rate and the fashion in which the drug is released can be tuned by means of the physicochemical properties of the matrix into which the drug is loaded.

The chemical similarity between the drug and the biocompatible polymer matrix influences the effectiveness of the release. Similar hydrophobicity between polymer and drug resulted in homogenous drug distribution within the matrix. The use of hydrophilic drugs mixed with hydrophobic polymers originated an initial "burst" of drugs followed by a significant decrease in the release rates at prolonged periods. In contrast, less substantial burst followed by a sustained release for extended periods characterized the release profile of more similar (e.g., both hydrophobic or hydrophilic) drug/polymer coatings [193–195]. Prolonged antibiotic release can be achieved by adding an additional polymer layer on the surface of the drug/polymer matrix. This outer layer acts as a barrier to drug diffusion, reducing the initial burst and extending the overall release duration. The release profile of the antibiotics can be tuned as a function of the extent of polymeric cross-linking of the outer barrier and of the polymer matrix [196].

An improvement in the field of drug release from polymer matrix is represented by the usage of biodegradable polymers such as poly(propylenefumarate/methyl)methacrylate, collagen, polyanhydrides, polyorthoesters, and PLGA. Biodegradable coatings can actively release higher doses of antibiotics during prolonged periods, and their degradation products can be easily metabolized by the body [192].

Immunotherapy (i.e., the clinical delivery of externally derived antibodies) exhibited general therapeutic benefits by the controlled release of bioactive antibodies from a polymer matrix. High efficacy in the bacterial killing was achieved by means of antibodies released from polyurethane coatings. Rojas et al. [197] found that the release of antibodies from a polymer coating represents a solution that is more successful at mitigating implant-associated infection than exogenously administered antibody treatments. A more detailed discussion is given in the section Novel Engineering Approaches for Treatment of Implant-Related Infections.

15.4.3 Sterile-Surface Polymers

As discussed in the previous section, there are some possible limitations for materials impregnated with leaching antibacterial agents, which include the risk of toxicity for eukaryotic cells, the issue of bacteria resistance, and the problem of time-dependent antimicrobial efficacy. In fact, materials that kill bacteria by releasing bactericidal agents are eventually depleted of the active substance after its complete elution into the surrounding environment.

An alternative approach aims at creating a permanently sterile, nonleaching material by covalently functionalizing its surface with an antimicrobial compound [100, 198, 199]. In the design of such a material, it must be considered that when the antimicrobial molecule is attached to the surface, it might lose its mobility and, thus, the capability to penetrate into bacteria, hence becoming biologically inactive. A possible solution is to bind the antimicrobial agent to a flexible polymeric chain of appropriate length anchored covalently to the material surface (Figure 15.6). Among various antimicrobials that can be immobilized to a surface, as mentioned previously, QACs represent an interesting class of molecules because they can interact with the microbial membrane and accumulate in the cell driven by the membrane potential [200]; this mechanism was referred to as the "hole-poking" mechanism [100]. In the case of gram-negative bacteria, antimicrobial cationic compounds are supposed to displace the divalent cations that hold together the negatively charged surface of the lipopolysaccharide network, thus disrupting the outer membrane. It is also possible that, having destroyed the outer membrane permeability barrier, the cationic groups linked to the polymers penetrate even more into the inner membrane, producing leakage. The mechanism of action of immobilized polycations against the gram-positive bacteria probably

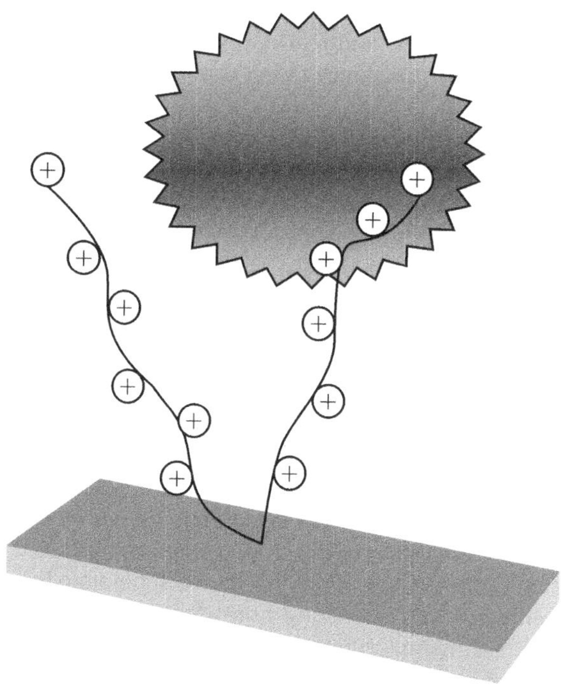

FIGURE 15.6 Design of a sterile-surface material: An antimicrobial moiety is attached to a long flexible polymeric chain that, in turn, is attached to the surface covalently. The antimicrobial monomers are hydrophobic cations that can be actively electrophoresed into the negatively charged cellular membranes of a microbial pathogen.

Source: Adapted from Lewis et al. [14].

requires penetration of the cationic groups across the cell wall to reach the cytoplasmic membrane [198].

Lewis and Kibanov [100] reported the use of *N*-alkyl-pyridinium linked to poly(4-vinylpyridine) or poly(ethylene-imine) for the preparation of contact active antibacterial surfaces. The main points of their approach are summed as follows:

- The immobilized polymeric chains must exist as individual entities (not entangled agglomerates) to ensure antimicrobial activity.
- A minimum density of positively charged groups must be reached to boost the bactericidal effect.
- Hydrophobicity and charge of the polymer affect antibacterial properties.
- The immobilized polycation must have sufficient long and flexible chains to penetrate bacterial membrane.
- The effect is bactericidal and bacteria are killed by membrane damage.
- This strategy is effective also toward persister cells of biofilms invulnerable to conventional antibiotics.
- QACs are effective disinfectants when they are applied at doses high enough to overwhelm the multi–drug-resistance pumps, which is the only known mechanism of resistance to hydrophobic cations.

Tiller et al. [198] developed surfaces functionalized with poly(4-vinyl-N-alkylpyridinium bromide), which can kill airborne bacteria like *S. aureus*, *S. epidermidis*, *P. aeruginosa*, and *E. coli* by direct contact. Although the toxicity of QACs toward eukaryotic cells can represent an issue for implantable biomaterials, the preparation of noncytotoxic QACs-based surfaces was recently reported [104].

Quaternary phosphonium salts can also be immobilized on biomaterials to create sterile-surface polymers [102]. Kenawy et al. [201] functionalized methacrylate-based substrates with quaternary phosphonium compounds that exert biocidal activity against both gram-positive and gram-negative strains and the fungus *Trichophyton rubrum*. In contrast, an issue related to the use of quaternary ammonium- and phosphonium-based surfaces can be the attachment of dead microorganisms remaining on antimicrobial coatings, which can trigger an immune response and inflammation, and can block its antimicrobial functional groups. To overcome such possible limitation, Cheng et al. [164] developed a self-sterilizing non-fouling polymer coating; in this system, poly(*N*,N-dimethyl-*N*-(ethoxycarbonylmethyl)-*N*-[2'-(methacryloyloxy)ethyl]-ammonium bromide) was used to create a surface that will effectively kill bacteria and release dead bacterial cells when the cationic derivatives are hydrolyzed to nonfouling zwitterionic polymers. The resulting nonfouling zwitterionic surface can prevent the attachment of proteins and microorganisms, and it can reduce the formation of a biofilm on the biomaterial surface.

15.5 NOVEL ENGINEERING APPROACHES FOR TREATMENT OF IMPLANT-RELATED INFECTIONS

Responsive polyelectrolyte multilayers (responsive PEMs) are a broad area that is attracting scientists' interest. The challenge is the preparation of films showing a predictable response of their functional properties such as wettability, adhesivity, or porosity when an external condition is varied. A common approach to produce such systems consists on using the propensity of some polyelectrolytes to change their macromolecular characteristics after a variation of external stimuli like pH, temperature, ionic strength, magnetic field, light, or specific biological moieties [202]. The construction of responsive PEMs coatings offers the advantage of delaying their functionality (i.e., biocidal activity) until a specific stimulus activates the multilayer. Lichter and Rubner [203] described PEMs comprising PAH and poly(sodium 4-styrene sulfonate) (SPS), which undergo a reversible pH-dependent swelling transition that allows turning on and off the antimicrobial functionality with suitable pH treatment (Figure 15.7). In particular, the reversible change in swelling behavior is associated with the acid-activated opening of hydrophobic PAH clusters containing free amino groups. The viability of two strains of gram-positive and gram-negative strains were effectively reduced in SPS/PAH multilayers displaying accessible cationic charges on PAH chains.

Other interesting examples of antibacterial activity tuned by responsive-PEM cosean be found in the work performed by Pan et al. [204], who associated antimicrobial poly(guanidine) with temperature-responsive polyanions to create films that undergo morphological modification after heating. In the work by Kim et al. [205], it is described the development of copolymer micelles assembled under acid conditions by weak hydrogen bonding between poly(acrylic acid) and poly(ethylene oxide)-block-poly(caprolactone). These micelles enable the incorporation and pH-sensitive release of drugs.

A subsequent innovative upgrade of responsive-PEM technology is the so defined "lift-off" PEMs that can improve the functionality of antibacterial PEMs. In the traditional form, lift-off methods are microstructuring techniques used to etch the surface of target materials (mainly metals) using a sacrificial material (i.e., photoresist) that can be dissolved to create the desired patterning (hence the term lift-off) [206]. Taking this conceptual strategy, the lift-off PEMs are chemically designed to release the layered biofilm upon application of a stimulus. Lichter et al. [90] described the design of a lift-off PEM with a hydrogen-bonded region consisting of poly(methacrylic acid) and temperature-sensitive poly(*N*-isopropylacrylamide) (PNIPAAm), topped with a supportive electrostatic PEM composed of PAH and iron oxide nanoparticles. At low-temperature and neutral pH conditions, the hydrogen-bonded film degraded, lifting off the PAH/iron oxide layer in addition to anything adhered to the top of that layer. A biofilm of *S. epidermidis* grown on these PEMs under high-nutrient conditions, exposing a clean, unfouled underlying surface was efficiently released (Figure 15.8).

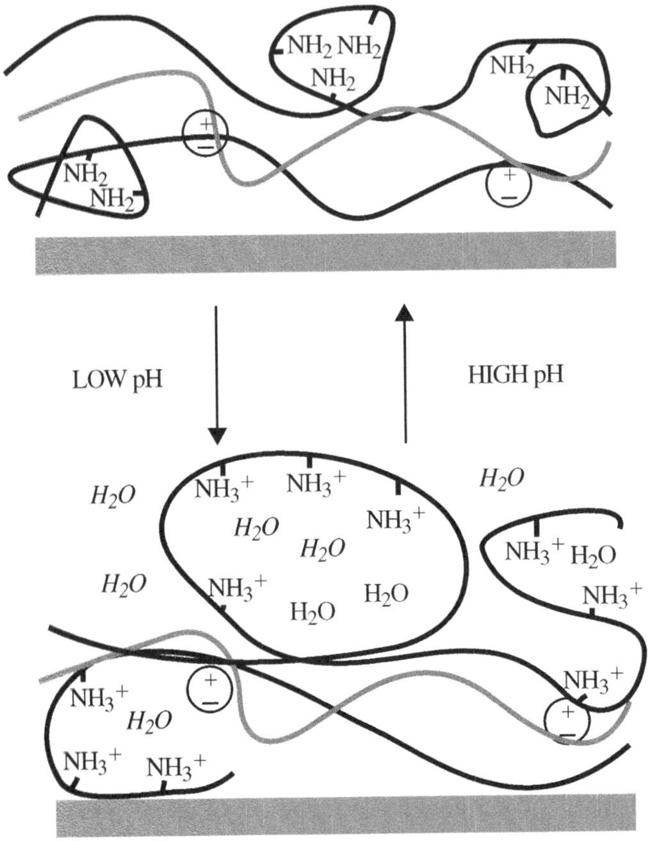

FIGURE 15.7 Schematic of acid-activated opening of hydrophobic clusters in SPS/PAH PEMs. The hydrophobic clusters can be reformed with subsequent treatment at high pH.

Source: Reprinted from Lichter et al. [204] with permission.

Several authors have described the construction of hydrogen-bonded multilayer films based on a combination of (1) poly(vinylpyrrolidone) and poly(methylacrylic acid), (2) poly(ethylene oxide) and poly(methylacrylic acid), (3) poly-(ethylene oxide) and poly(acrylic acid) [207, 208], and (4) on poly(acrylic acid) with poly(ethyleneglycol) [209]. These systems are disintegrated by pH changes. Dubas and Schlenoff [210] reported that polyelectrolyte multilayer films were released from a substrate whose surface was covered with multilayers formed via electrostatic interaction between poly(acrylic acid)/poly(diallyldimethylammonium), which decompose at high salt concentrations or at a high pH. Nevertheless, these systems have not been used or tested for biofilm removal.

One obvious disadvantage of lift-off design is that once the PEM is removed, the underlying surface can become fouled and therefore colonized by bacteria.

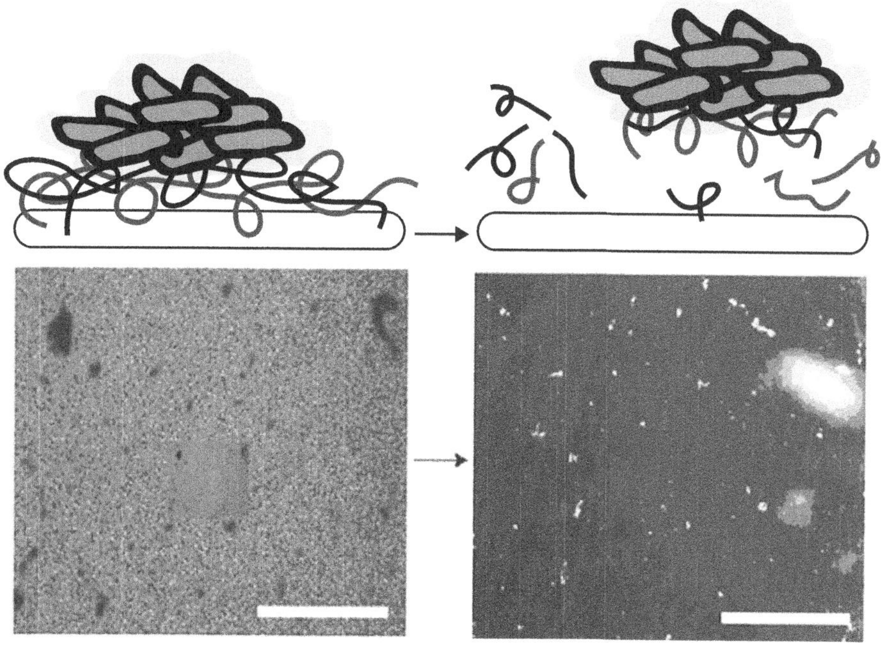

FIGURE 15.8 A contact-resistant PEM (comprising PMAA/PNIPAAm/PAH and iron oxide nanoparticles) is eventually biofouled by *E. coli* in nutrient-rich environments (left), but PEM lift-off removes the intact biofilm from the underlying surface (right). Scale bar = 10 mm. The white spots observed in the right-hand image are attributed to residual salt deposits from the buffer solution (PBS).

Source: Reprinted from Lichter et al. [204] with permission.

This may be acceptable for some short-term applications but not for long-term methods.

In this perspective, the large variety of antibacterial PEMs supports the idea that they are suitable for developing multifaceted and multifunctional antibacterial layers designed so that once a bactericidal factor is exhausted, another one can exert antimicrobial activity. As clearly presented schematically in Figure 15.9 by Lichter et al. [90], a PEM strategy can be based on the integrated use of antibacterial and lift-off PEMs. The potential methods of lift-off include tunable cleavage via pH changes or reduction-oxidation reversible cross-linking, electrical bias, hydrolytic erosion, or enzymatic degradation of polyelectrolytes or crosslinkers. Both distinct release lift-off mechanisms and single-stage mechanisms can be employed as reported in Figure 15.9.

A noteworthy and innovative strategy was proposed by Amitai et al. [211]. They described the preparation of polyurethane-based leukocyte-inspired biocidal materials which release biocidal halogens in the presence of renewable resources such as water, sugar, and salt. These enzyme-polymer matrices are prepared by electrospinning of polyurethane in the presence of the enzymes

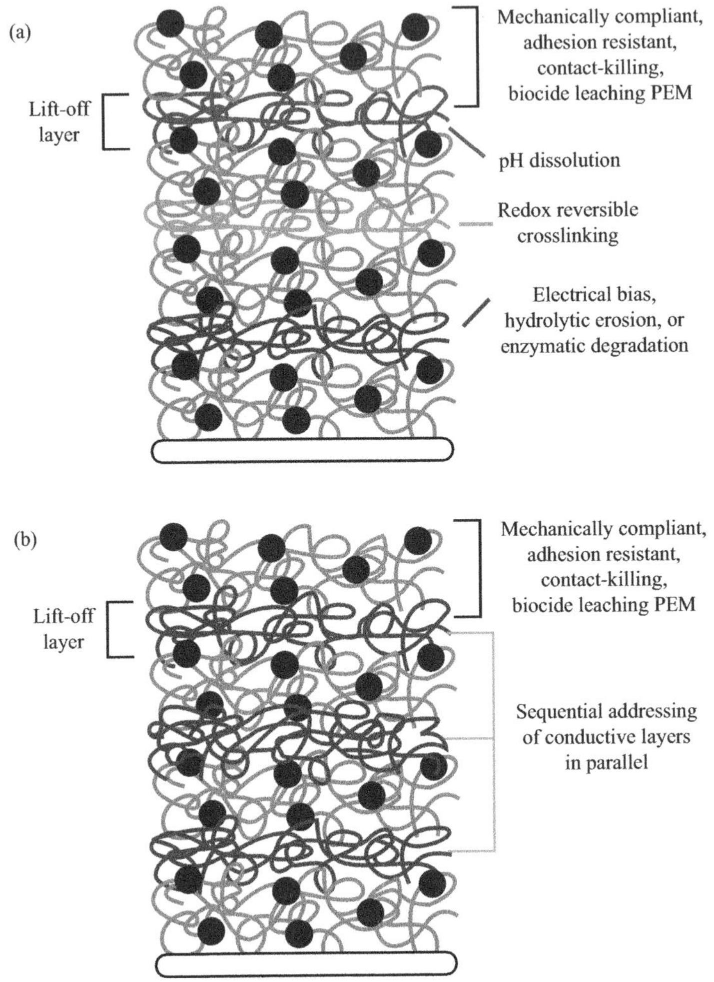

FIGURE 15.9 PEM lift-off strategy to integrate the main strategies for antibacterial surface design, even for conditions where eventual biofilm formation occurs. This can be achieved via (a) a series of distinct release mechanisms or (b) a staged single mechanism. *Source:* Reprinted from Lichter et al. [90] with permission.

glucose oxidase and horseradish peroxidase. These enzymes catalyze two reactions in tandem using glucose, hydrogen peroxide, and sodium halide (iodide or bromide). The final products of these two consecutive enzymatic reactions are free halogens (iodine or bromine), which are highly active and rapid disinfectants.

As widely reported by the scientific literature and as described in the previous sections of this chapter, during the past 20 years, researchers have attempted to develop bacteria-resistant biomaterials based almost exclusively on the use of compounds that kill target bacteria or inhibit their growth. These drugs include mostly antibiotics but also nontraditional compounds like antimicrobial peptides [105, 212] and heavy metals [213] administered by employing several strategies and technologies. In the latest years, the field of biomaterial design has begun to focus on other approaches to control medical biofilms. Interest in using alternatives to treat device-related infections is being fuelled by the wide dissemination of drug-resistant microorganisms, the emergence of new microorganisms, the relative inefficacy of biocide molecules, and the low responsiveness to antimicrobial drugs of biofilm-associated bacteria. Part of the research is now directed toward the employment of substances or other biological weapons that inhibit virulence factors and mechanisms, thus preventing the onset of infections from the first steps of bacterial invasion. Such drugs are called "antipathogenic" because in contrast to biocidal drugs, they do not kill bacteria and are believed to reduce antibiotic resistance.

Some approaches employ antiadhesive strategies aimed at baring bacterial attachment through the modification of material surfaces (see the section Antiadhesive PEMS). To reduce bacterial adhesion and colonization, one method is to coat surfaces with nonfouling materials as PEG derivatives [214] or zwitterionic polymers [164]. A recent study demonstrates that in a rabbit model, methoxy-polyethylene-glycol-dihydroxyphenylalanine-coated ureteral stents resist bacterial adherence, infection development, and encrustation of uropathogenic *E. coli* cystitis [215]. Likewise, it was reported that surfaces coated with 2-methacryloyloxyethyl phosphorylcholine reduced the attachment of bacteria by 90% [216]. Zwitterionic poly(sulfobetaine methacrylate) (pSBMA) [217, 218] and poly(2-carboxy-N,N-dimethyl-*N*-(20-(methacryloyloxy) ethyl)-ethanaminium) (pCBMA-2) [219] efficiently reduce the colonization of *P. aeruginosa* and *S. epidermidis*. Both materials are proven to be highly resistant to protein adsorption from 100% serum and blood plasma.

Although not belonging to the field of polymers, carbonated hydroxyapatite coatings form strong chemical bonds with bone and can be modified with surface-absorbed antibiotics by the immersion of the coating in the antibiotic solution or by means of coprecipitation processes. The drug release rate is influenced by the chemical structure of the antibiotics in relation with the hydroxyapatite. Acid antibiotics can be easily incorporated in the carbonated hydroxyapatite coatings because of the calcium-chelating properties of the carboxylate groups [193].

Promising investigations have been conducted on the role of nitric oxide (NO) released from polymer coatings. NO is a diatomic free radical that is produced by several sources in the body and serves as a bioregulator for cardiovascular, respiratory, gastrointestinal, and nervous functions [220]. NO is an endogenous product that has a short life in the biological contest and acts as an antithrombogenic substance that regulates vasodilation and platelet activation [221]. Intravascular sensors and extracorporeal tubes have been integrated with NO-releasing polymers coatings thromboresistant matrices that improve the biocompatibility of the devices via NO release [222–224]. Recently, antimicrobial properties have been ascribed to NO, which suggests this substance as a potential antimicrobial agent [225].

Antibody-based therapies to treat chronic infectious diseases or cancer are currently undergoing a renaissance, and many antibody preparations are now in clinical use [226]. These therapies are based on the administration of chimeric and humanized monoclonal antibodies (mAb) that specifically bind to target cells stimulating in this way the patient's immune system to attack them. However, currently only few studies have evaluated in the field of biomaterial-associated infections the applicability of antibody therapy against biofilm formation [227]. In most bacterial infections, the interaction between host and pathogen initiates the virulence process. This initial attachment is therefore the most attractive target for immunoprophylactic interventions. For instance, two clinically applied antistaphylococcal antibody therapies target microbial surface components recognizing adhesive matrix molecules that mediates adherence of *S. aureus* to the extracellular matrix of the host [228]. Proteins involved in biofilm formation are also promising alternative targets for passive antibody therapies. For instance, Sun et al. [229] described the inhibition of biofilm formation by mAb against the accumulation-associated protein of *S. epidermidis*. Similarly, polyclonal rabbit antibodies raised against the extracellular protein SesC of *S. epidermidis* reduce the numbers of bacteria in biofilms on subcutaneously implanted polyurethane catheters [230].

Elegant immunotherapeutic approaches aimed at inhibiting bacterial virulence are based on the use of artificial opsonins to promote and enhance the phagocytosis of bacteria. These molecules, as reported in Figure 15.10a, might be used to decorate biomaterial surfaces and, thus, to enhance long- and short-term local immune response. Opsonization is the process by which opsonin molecules, as antibodies and C3b complement fragment, mark pathogens for phagocytosis by simultaneously binding to receptors on the pathogen's cell membrane and to receptors on phagocytic cells. Bacteria developed several strategies to escape from natural opsonins, thus circumventing immune system reaction. Artificial opsonins are usually bifunctional molecules (fusion proteins [231] or synthetically derived opsonins [232, 233]) that are designed to bind both target bacteria surface proteins and phagocyte receptors.

Krishnamurthy *et al.* [232] described a bifunctional polyacrylamide presenting both vancomycin and fluorescein groups. The vancomycin group allows the specific recognition of peptides terminated in D–Ala–D–Ala on the bacterial cell wall whereas fluorescein allows the imaging of polymer and serves as a hapten that is recognized by antibodies immunoglobulin G (antifluor). This compound can carry out the recognition of the surface of some gram-positive bacteria to modify their surface to generate specific binding sites recognized by an antibody and to promote phagocytosis of the opsonized bacteria.

Among the various papers by Taylor and colleagues about the study of bi-specific fusion antibodies to target different bacteria, a recent one by Gyimesi [234] reports the characterization of a bispecific monoclonal antibody complex specific for the primate erythrocyte complement C3b receptor (CR1) and type 5

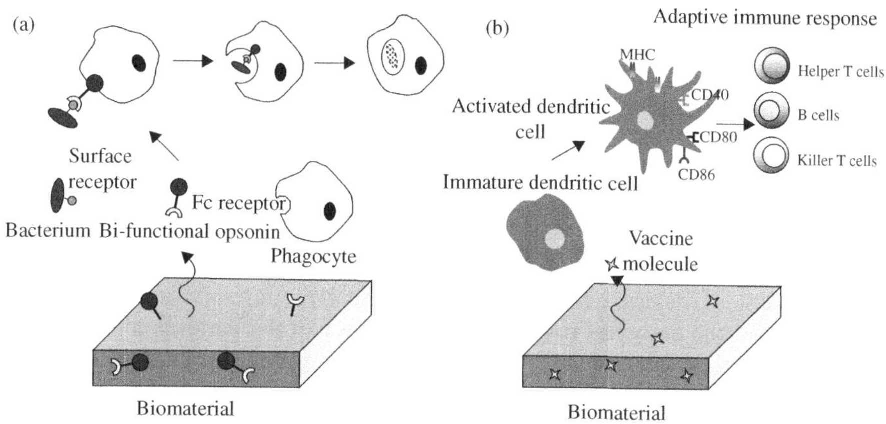

FIGURE 15.10 Hypothetical biomaterial engineered to enhance short and long-term infection immune response. (a) Immunotherapy based on the use of artificial opsonins to promote and enhance phagocytosis of bacteria. (b) Self-vaccinating biomaterials. *Source:* Adapted from Bryers [235].

capsular polysaccharide of *S. aureus*. These fusion proteins have been shown to promote binding of the pathogen first to erythrocytes that enhance macrophage phagocytosis.

The research performed by Bryers [235] is noteworthy. It is based on the fundamental concept that innate and adaptive immunity represents the most efficient life-long defense barrier against bacterial colonization. In this perspective, the innovative research is addressed to develop polymeric biomaterial platforms that engineer infection immunity through the release of antigen bacterial targets as vaccines (Figure 15.10b) [236]. The ideal antigens to elicit promotion of both innate and adaptive systems are bacterial surface proteins ("adhesins"), which are involved in mediation of bacterial adhesion to host tissues or device surfaces. It must be underlined that for most microorganisms, initial stable attachment leads to the activation of molecular signal cascade that up-regulates virulence factors and induces biofilm formation. Immunity responses against an adhesin may prevent bacterial adhesion and, therefore, can impede colonization and block infection [237]. The concept of biomaterials designed to activate dendritic cells, targeting specific adhesins, could be applied to numerous situations: direct antigen presentation (e.g., attenuated pathogen, isolated or recombinant antigen), DNA vaccine (i.e., pDNA encoding for the antigen protein), messenger RNA (mRNA) vaccine (i.e., RNA encoding for recombinant antigenic protein) [238]. Brayer's group is currently developing a platform of constructs that would release condensed RNA vaccines meant to transfect dendritic cells arriving after implantation of a biomedical device. Such polymer constructs will release nanoparticles of condensed DNA or mRNA vaccine that will target some selected specific adhesion protein employed by the microorganism to initiate colonization.

Other potential virulence targets that can be employed in antipathogenic systems comprise quorum-sensing and cell-cell signaling [239], bacterial iron metabolism [240], and biofilm detachment [241]. However, the success and the applicability of these strategies still require much development and optimization.

15.6 FUTURE TRENDS

As a result of more than two decades of research, the mechanism involved in the development of device infections has been clearly addressed and identified. Nevertheless, the design of truly biocidal surfaces is still a major goal for public health and scientific research. Although the interest in this field is sharply increasing, the main limitation in most of the proposed systems is the lack of long-term durability of the antibacterial activity. Surfaces that leach antibacterial agents have the disadvantage of releasing the active components only for a limited period of time. Surfaces that contain nonleaching biocides that kill on contact are challenged by the build-up of dead cells which diminish their efficacy over time. A surface with broad spectrum of biocidal activity is needed that can sustain bactericidal properties for an extended time.

Therefore, the problem of biofilm prevention requires new ideas and strategies before one or more successful solutions arise. Novel potential technical approaches, some of which have been described in this chapter, are continuously sought and proposed by scientists to overcome the previously cited limits. The most promising research seems now oriented toward different immunotherapeutic approaches that aim to exploit the physiological processes of immune defense [235, 238] to guarantee a long-term barrier against bacterial colonization.

REFERENCES

1. Hetrick, E. M., Schoenfisch, M. H. (2006). Reducing implant-related infections: active release strategies. *Chemical Society Reviews*, *35*(9), 780–789.
2. Trebse, R., Pisot, V., Trampuz, A. (2005). Treatment of infected retained implants. *Journal of Bone and Joint Surgery British Volume*, *87*(2), 249–256.
3. Fridkin, S. K., Jarvis, W. R. (1996). Epidemiology of nosocomial fungal infections. *Clinical Microbiology Reviews*, *9*(4), 499–511.
4. Darouiche, R. O. (2004). Treatment of infections associated with surgical implants. *New England Journal of Medicine*, *350*(14), 1422–1429.
5. Segawa, H., Tsukayama, D. T., Kyle, R. F., Becker, D. A., Gustilo, R. B. (1999). Infection after total knee arthroplasty. A retrospective study of the treatment of eighty-one infections. *Journal of Bone and Joint Surgery American Volume*, *81*(10), 1434–1445.
6. Kojic, E. M., Darouiche, R. O. (2004). Candida infections of medical devices. *Clinical Microbiology Reviews*, *17*(2), 255–267.

7. Richards, M. J., Edwards, J. R., Culver, D. H., Gaynes, R. P. (2000). Nosocomial infections in combined medical-surgical intensive care units in the United States. *Infection Control & Hospital Epidemiology*, *21*(8), 510–515.
8. Darouiche, R. O. (2001). Device-associated infections: a macroproblem that starts with microadherence. *Clinical Infectious Disease*, *33*(9), 1567–1572.
9. Trampuz, A., Zimmerli, W. (2005). Prosthetic joint infections: Update in diagnosis and treatment. *Swiss Medical Weekly*, *135*(17–18), 243–251.
10. Perlroth, J., Kuo, M., Tan, J., Bayer, A. S., Miller, L. G. (2008). Adjunctive use of rifampin for the treatment of Staphylococcus aureus infections: A systematic review of the literature. *Archives of Internal Medicine*, *168*(8), 805–819.
11. Tverdek, F. P., Crank, C. W., Segreti, J. (2008). Antibiotic therapy of methicillin-resistant Staphylococcus aureus in critical care. *Crititcal Care Clinics*, *24*(2), 249–260.
12. Sauer, K., Camper, A. K. (2001). Characterization of phenotypic changes in Pseudomonas putida in response to surface-associated growth. *Journal of Bacteriology*, *183*(22), 6579–6589.
13. Sauer, K., Camper, A. K., Ehrlich,. G. D., Costerton, J. W., Davies, D. G. (2002). Pseudomonas aeruginosa displays multiple phenotypes during development as a biofilm. *Journal of Bacteriology*, *184*(4), 1140–1154.
14. Tremoulet, F., Duche, O., Namane, A., Martinie, B., Labadie, J. C. (2002). A proteomic study of Escherichia coli O157:H7 NCTC 12900 cultivated in biofilm or in planktonic growth mode. *FEMS Microbiology Letters*, *215*(1), 7–14.
15. Tremoulet, F., Duche, O., Namane, A., Martinie, B., Labadie, J. C. (2002). Comparison of protein patterns of Listeria monocytogenes grown in biofilm or in planktonic mode by proteomic analysis. *FEMS Microbiology Letters*, *210*(1), 25–31.
16. Vilain, S., Cosette, P., Hubert, M., Lange, C., Junter, G. A., Jouenne, T. (2004). Comparative proteomic analysis of planktonic and immobilized Pseudomonas aeruginosa cells: A multivariate statistical approach. *Analytical Biochemistry*, *329* (1), 120–130.
17. Blair, K. M., Turner, L., Winkelman, J. T., Berg, H. C., Kearns, D. B., (2008). A molecular clutch disables flagella in the Bacillus subtilis biofilm. *Science*, *320*(5883), 1636–1638.
18. Jabbouri, S., Sadovskaya, I. (2010). Characteristics of the biofilm matrix and its role as a possible target for the detection and eradication of Staphylococcus epidermidis associated with medical implant infections. *FEMS Immunology and Medical Microbiology*, *59*(3), 280–291.
19. Ryder, C., Byrd, M., Wozniak, D. J. (2007). Role of polysaccharides in Pseudomonas aeruginosa biofilm development. *Current Opinion In Microbiology*, *10*(6), 644–648.
20. Marrie, T. J., Nelligan, J., Costerton, J. W., (1982). A scanning and transmission electron microscopic study of an infected endocardial pacemaker lead. *Circulation*, *66*(6), 1339–1341.
21. Peters, G., Locci, R., Pulverer, G. (1982). Adherence and growth of coagulase-negative staphylococci on surfaces of intravenous catheters. *Journal of Infectious Diseases*, *146*(4), 479–482.
22. Hall-Stoodley, L., Stoodley, P. (2005). Biofilm formation and dispersal and the transmission of human pathogens. *Trends in Microbiology*, *13*(1), 7–10.

23. Van Dellen, K. L., Houot, L., Watnick, P. I. (2008). Genetic analysis of Vibrio cholerae monolayer formation reveals a key role for DeltaPsi in the transition to permanent attachment. *Journal of Bacteriology*, *190*(24), 8185–8196.
24. Derjaguin, B., Landau, L. (1941). Theory of the stability of strongly charged lyophobic sols and of the adhesion of strongly charged particles in solution of electrolytes. *Acta Physicochimica URSS*, *14*, 633–662.
25. Verwey, E. J. W., Overbeek, J. T. G. (1948). *Theory of the Stability of Lyophobic Colloids*. Elsevier, Amsterdam, the Netherlands.
26. de Kerchove, A. J., Elimelech, M. (2007). Impact of alginate conditioning film on deposition kinetics of motile and nonmotile Pseudomonas aeruginosa strains. *Applied and Environmental Microbiology*, *73*(16), 5227–5234.
27. de Kerchove, A. J., Weronski, P., Elimelech, M., (2007). Adhesion of nonmotile Pseudomonas aeruginosa on "soft" polyelectrolyte layer in a radial stagnation point flow system: Measurements and model predictions. *Langmuir*, *23*(24), 12301–12308.
28. Hermansson, M. (1999). The DLVO theory in microbial adhesion. *Colloids and Surfaces B: Biointerfaces*, *14*(1–4), 105–119.
29. Garrett, E. S., Perlegas, D., Wozniak, D. J. (1999). Negative control of flagellum synthesis in Pseudomonas aeruginosa is modulated by the alternative sigma factor AlgT (AlgU). *Journal of Bacteriology*, *181*(23), 7401–7404.
30. Soriani, M., Telford, J. L. (2010). Relevance of pili in pathogenic streptococci pathogenesis and vaccine development. *Future Microbiology*, *5*(5), 735–747.
31. Pratt, L. A., Kolter, R. (1998). Genetic analysis of Escherichia coli biofilm formation: Roles of flagella, motility, chemotaxis and type I pili. *Molecular Microbiology*, *30*(2), 285–293.
32. Vatanyoopaisarn, S., Nazli, A., Dodd, C. E., Rees, C. E., Waites, W. M. (2000). Effect of flagella on initial attachment of Listeria monocytogenes to stainless steel. *Applied and Environmental Microbiology*, *66*(2), 860–863.
33. Houry, A., Briandet, R., Aymerich, S., Gohar, M. (2010). Involvement of motility and flagella in Bacillus cereus biofilm formation. *Microbiology*, *156*(4), 1009–1018.
34. Lemon, K. P., Higgins, D. E., Kolter, R. (2007). Flagellar motility is critical for Listeria monocytogenes biofilm formation. *Journal of Bacteriology*, *189*(12), 4418–4424.
35. Agladze, K., Wang, X., Romeo, T. (2005). Spatial periodicity of Escherichia coli K-12 biofilm microstructure initiates during a reversible, polar attachment phase of development and requires the polysaccharide adhesin PGA. *Journal of Bacteriology*, *187*(24), 8237–8246.
36. Caiazza, N. C., Merritt, J. H., Brothers, K. M., O'Toole, G. A. (2007). Inverse regulation of biofilm formation and swarming motility by Pseudomonas aeruginosa PA14. *Journal of Bacteriology*, *189*(9), 3603–3612.
37. Hinsa, S. M., Espinosa-Urgel, M., Ramos, J. L., O'Toole, G. A. (2003). Transition from reversible to irreversible attachment during biofilm formation by Pseudomonas fluorescens WCS365 requires an ABC transporter and a large secreted protein. *Molecular Microbiology*, *49*(4), 905–918.
38. Li, G., Smith, C. S., Brun, Y. V., Tang, J. X. (2005). The elastic properties of the caulobacter crescentus adhesive holdfast are dependent on oligomers of N-acetylglucosamine. *Journal of Bacteriology*, *187*(1), 257–265.

39. Costerton, J. W., Stewart, P. S., Greenberg, E. P. (1999). Bacterial biofilms: A common cause of persistent infections. *Science*, *284*(5418), 1318–1322.
40. Karatan, E., Watnick, P. (2009). Signals, regulatory networks, and materials that build and break bacterial biofilms. *Microbiology and Molecular Biology Reviews*, *73*(2), 310–347.
41. Moscoso, M., Garcia, E., Lopez, R. (2009). Pneumococcal biofilms. *International Microbiology*, *12*(2), 77–85.
42. Lim, Y., Jana, M., Luong, T. T., Lee, C. Y. (2004). Control of glucose- and NaCl-induced biofilm formation by rbf in Staphylococcus aureus. *Journal of Bacteriology*, *186*(3), 722–729.
43. Shemesh, M., Tam, A., Steinberg, D. (2007). Expression of biofilm-associated genes of Streptococcus mutans in response to glucose and sucrose. *Journal of Medical Microbiology*, *56*(11), 1528–1535.
44. Bollinger, N., Hassett, D. J., Iglewski, B. H., Costerton, J. W., McDermott, T. R. (2001). Gene expression in Pseudomonas aeruginosa: Evidence of iron override effects on quorum sensing and biofilm-specific gene regulation. *Journal of Bacteriology*, *183*(6), 1990–1996.
45. Patriquin, G. M., Banin, E., Gilmour, C., Tuchman, R., Greenberg, E. P., Poole, K. (2008). Influence of quorum sensing and iron on twitching motility and biofilm formation in Pseudomonas aeruginosa. *Journal of Bacteriology*, *190*(2), 662–671.
46. Wu, Y., Outten, F. W. (2009). IscR controls iron-dependent biofilm formation in Escherichia coli by regulating type I fimbria expression. *Journal of Bacteriology*, *191*(4), 1248–1257.
47. Monds, R. D., Newell, P. D., Gross, R. H., O'Toole, G. A. (2007). Phosphate-dependent modulation of c-di-GMP levels regulates Pseudomonas fluorescens Pf0-1 biofilm formation by controlling secretion of the adhesin LapA. *Molecular Microbiology*, *63*(3), 656–679.
48. Hung, D. T., Zhu, J., Sturtevant, D., Mekalanos, J. J. (2006). Bile acids stimulate biofilm formation in Vibrio cholera. *Molecular Microbiology*, *59*(1), 193–201.
49. Mathee, K., Ciofu, O., Sternberg, C., Lindum, P. W., Campbell, J. I., Jensen, P., Johnsen, A. H., Givskov, M., Ohman, D. E., Molin, S., Hoiby, N., Kharazmi, A. (1999). Mucoid conversion of Pseudomonas aeruginosa by hydrogen peroxide: A mechanism for virulence activation in the cystic fibrosis lung. *Microbiology*, *145*(6), 1349–1357.
50. Bassler, B. L. (1999). How bacteria talk to each other: Regulation of gene expression by quorum sensing. *Current Opinion in Microbiology*, *2*(6), pp. 582–587.
51. Boyer, M., Wisniewski-Dye, F. (2009). Cell-cell signalling in bacteria: Not simply a matter of quorum. *FEMS Microbiology Ecology*, *70*(1), 1–19.
52. Novick, R. P., Geisinger, E. (2008). Quorum Sensing in Staphylococci. *Annual Review of Genetics*, *42*(1), 541–564.
53. Novak, E. A., Shao, H., Daep, C. A., Demuth, D. R. (2010). Autoinducer-2 and QseC control biofilm formation and in vivo virulence of Aggregatibacter actinomycetemcomitans. *Infection and Immunity*, *78*(7), 2919–2926.
54. Xiao-Gang, W., Hui-Mei, D., Tao, T., Nan, Y., Hong-You, Z., Li-Qun, Z. (2010). Effect of the hfq gene on 2,4-diacetylphloroglucinol production and the PcoI/PcoR

quorum-sensing system in Pseudomonas fluorescens 2P24. *FEMS Microbiology Letters*, *309*(1), 16–24.

55. Bjarnsholt, T., Tolker-Nielsen, T., Hoiby, N., Givskov, M. (2010). Interference of Pseudomonas aeruginosa signalling and biofilm formation for infection control. *Expert Reviews in Molecular Medicine*, *12*, e11.
56. Cerca, N., Jefferson, K. K., (2008). Effect of growth conditions on poly-N-acetylglucosamine expression and biofilm formation in Escherichia coli. *FEMS Microbiology Letters*, *283*(1), 36–41.
57. Schierholz, J. M., Beuth, J. (2001). Implant infections: A haven for opportunistic bacteria. *Journal of Hospital Infection*, *49*(2), 87–93.
58. Hu, W. J., Eaton, J. W., Ugarova, T. P., Tang, L. (2001). Molecular basis of biomaterial-mediated foreign body reactions. *Blood*, *98*(4), 1231–1238.
59. Wang, J. Y., Wicklund, B. H., Gustilo, R. B., Tsukayama, D. T. (1997). Prosthetic metals impair murine immune response and cytokine release in vivo and in vitro. *Journal of Orthopaedic Research*, *15*(5), 688–699.
60. Aviv, M., Berdicevsky, I., Zilberman, M. (2007). Gentamicin-loaded bioresorbable films for prevention of bacterial infections associated with orthopedic implants. *Journal of Biomedical Materials Research*, *83*(1), 10–19.
61. Del Pozo, J. L., Rouse, M. S., Euba, G., Kang, C. I., Mandrekar, J. N., Steckelberg, J. M., Patel, R. (2009). The electricidal effect is active in an experimental model of Staphylococcus epidermidis chronic foreign body osteomyelitis. *Antimicrobial Agents and Chemotherapy*, *53*(10), 4064–4068.
62. Lemperle, G., Gauthier-Hazan, N. (2009). Foreign body granulomas after all injectable dermal fillers: Part 2. Treatment Options. *Plastic and Reconstructive Surgery*, *123*(6), 1864–1873.
63. Uçkay, I., Pittet, D., Vaudaux, P., Sax, H., Lew, D., Waldvogel, F. (2009). Foreign body infections due to Staphylococcus epidermidis. *Annals of Medicine*, *41*(2), 109–119.
64. Gristina, A. G. (1994). Implant failure and the immuno-incompetent fibro-inflammatory zone. *Clinical Orthopaedics and Related Research*, Jan (298), 106–118.
65. Gristina, A. G. (1987). Biomaterial-centered infection: Microbial adhesion versus tissue integration. *Science*, *237*(4822), 1588–1595.
66. Kaplan, S. S., Basford, R. E., Mora, E., Jeong, M. H., Simmons, R. L. (1992). Biomaterial-induced alterations of neutrophil superoxide production. *Journal of Biomedical Materials Research*, *26*(8), 1039–1051.
67. Kaplan, S. S., Basford, R. E., Jeong, M. H., Simmons, R. L. (1996). Biomaterial-neutrophil interactions: Dysregulation of oxidative functions of fresh neutrophils induced by prior neutrophil-biomaterial interaction. *Journal of Biomedical Materials Research*, *30*(1), 67–75.
68. Zimmerli, W., Lew, P. D., Waldvogel, F. A. (1984). Pathogenesis of foreign body infection. Evidence for a local granulocyte defect. *Journal of Clinical Investigation*, *73*(4), 1191–1200.
69. Wettero, J., Bengtsson, T., Tengvall, P. (2001). C1q-independent activation of neutrophils by immunoglobulin M-coated surfaces. *Journal of Biomedical Materials Research*, *57*(4), 550–558.
70. Wettero, J., Bengtsson, T., Tengvall, P. (2001). C1q-independent activation of neutrophils by immunoglobulin M-coated surfaces. *Journal of Biomedical Materials Research*, *57*(4), 550–558.

71. Shive, M. S., Brodbeck, W. G., Colton, E., Anderson, J. M. (2002). Shear stress and material surface effects on adherent human monocyte apoptosis. *Journal of Biomedical Materials Research*, *60*(1), 148–158.

72. Brodbeck, W. G., Shive, M. S., Colton, E., Nakayama, Y., Matsuda, T., Anderson, J. M. (2001). Influence of biomaterial surface chemistry on the apoptosis of adherent cells. *Journal of Biomedical Materials Research*, *55*(4), 661–668.

73. Herrmann, M., Suchard, S. J., Boxer, L. A., Waldvogel, F. A., Lew, P. D. (1991). Thrombospondin binds to Staphylococcus aureus and promotes staphylococcal adherence to surfaces. *Infection Immunity*, *59*(1), 279–288.

74. Herrmann, M., Lai, Q. J., Albrecht, R. M., Mosher, D. F., Proctor, R. A. (1993). Adhesion of Staphylococcus aureus to surface-bound platelets: Role of fibrinogen/fibrin and platelet integrins. *Journal of Infectious Diseases*, *167*(2), 312–322.

75. Lima, E. M., Koo, H., Vacca Smith, A. M., Rosalen, P. L., Del Bel Cury, A. A. (2008). Adsorption of salivary and serum proteins, and bacterial adherence on titanium and zirconia ceramic surfaces. *Clinical Oral Implants Research*, *19*(8), 780–785.

76. Nilsson, M., Frykberg, L., Flock, J. I., Pei, L., Lindberg, M., Guss, B. (1998). A fibrinogen-binding protein of Staphylococcus epidermidis. *Infection Immunity*, *66*(6), 2666–2673.

77. Heilmann, C., Hussain, M., Peters, G., Gotz, F. (1997). Evidence for autolysin-mediated primary attachment of Staphylococcus epidermidis to a polystyrene surface. *Molecular Microbiology*, *24*(5), 1013–1024.

78. Heilmann, C., Thumm, G., Chhatwal, G. S., Hartleib, J., Uekotter, A., Peters, G. (2003). Identification and characterization of a novel autolysin (Aae) with adhesive properties from Staphylococcus epidermidis. *Microbiology*, *149*(10), 2769–2778.

79. Christner, M., Franke, G. C., Schommer, N. N., Wendt, U., Wegert, K., Pehle, P., Kroll, G., Schulze, C., Buck, F., Mack, D., Aepfelbacher, M., Rohde, H. (2010). The giant extracellular matrix-binding protein of Staphylococcus epidermidis mediates biofilm accumulation and attachment to fibronectin. *Molecular Microbiology*, *75*(1), 187–207.

80. Suci, P. A., Mittelman, M. W., Yu, F. P., Geesey, G. G. (1994). Investigation of ciprofloxacin penetration into Pseudomonas aeruginosa biofilms. *Antimicrobial Agents and Chemotherapy*, *38*(9), 2125–2133.

81. Duguid, I. G., Evans, E., Brown, M. R., Gilbert, P., (1992). Effect of biofilm culture upon the susceptibility of Staphylococcus epidermidis to tobramycin. *Journal of Antimicrobial Chemotherapy*, *30*(6), 803–810.

82. Cerca, N., Martins, S., Cerca, F., Jefferson, K. K., Pier, G. B., Oliveira, R., Azeredo, J. (2005). Comparative assessment of antibiotic susceptibility of coagulase-negative staphylococci in biofilm versus planktonic culture as assessed by bacterial enumeration or rapid XTT colorimetry. *Journal of Antimicrobial Chemotherapy*, *56*(2), 331–336.

83. Sadovskaya, I., Vinogradov, E., Li, J., Hachani, A., Kowalska, K., Filloux, A. (2010). High-level antibiotic resistance in Pseudomonas aeruginosa biofilm: The ndvB gene is involved in the production of highly glycerol-phosphorylated beta-(1- > 3)-glucans, which bind aminoglycosides. *Glycobiology*, *20*(7), 895–904.

84. Hatch, R. A., Schiller, N. L. (1998). Alginate lyase promotes diffusion of aminoglycosides through the extracellular polysaccharide of mucoid Pseudomonas aeruginosa. *Antimicrobial Agents and Chemotherapy*, *42*(4), 974–977.

85. Gordon, C. A., Hodges, N. A., Marriott, C. (1988). Antibiotic interaction and diffusion through alginate and exopolysaccharide of cystic fibrosis-derived Pseudomonas aeruginosa. *Journal of Antimicrobial Chemotherapy, 22*(5), 667–674.
86. Duguid, I. G., Evans, E., Brown, M. R., Gilbert, P. (1992). Growth-rate-independent killing by ciprofloxacin of biofilm-derived Staphylococcus epidermidis: Evidence for cell-cycle dependency. *Journal of Antimicrobial Chemotherapy, 30*(6), 791–802.
87. Evans, D. J., Allison, D. G., Brown, M. R., Gilbert, P. (1990). Effect of growth-rate on resistance of gram-negative biofilms to cetrimide. *Journal of Antimicrobial Chemotherapy, 26*(4), 473–478.
88. Zhang, L., Mah, T. F., (2008). Involvement of a novel efflux system in biofilm-specific resistance to antibiotics. *Journal of Bacteriology, 190*(13), 4447–4452.
89. Lewis, K. (2005). Persister cells and the riddle of biofilm survival. *Biochemistry, 70*(2), 267–274.
90. Lichter, J. A., Van Vliet, K. J., Rubner, M. F. (2009). Design of antibacterial surfaces and interfaces: Polyelectrolyte multilayers as a multifunctional platform. *Macromolecules, 42*(22), 8573–8586.
91. Busscher, H. J., Doornbusch, G. I., Van der Mei, H. C. (1992). Adhesion of mutans streptococci to glass with and without a salivary coatings as studied in a parallel plate flow chamber. *Journal of Dental Research, 71*, 491–500.
92. van Oss, C. J. (1995). Hydrophobicity of biosurfaces - origin, quantitative determination and interaction energies. *Colloids and Surface B, 5*, 91–110.
93. bu-Lail, N. I., Camesano, T. A. (2006). Specific and nonspecific interaction forces between escherichia coli and silicon nitride, determined by poisson statistical analysis. *Langmuir, 22*(17), 7296–7301.
94. Ki, D. P., Young, S. K., Dong, K. H., Young, H. K., Eun, H. B. L., Hwal, S., Kyu, S. C. (1998). Bacterial adhesion on PEG modified polyurethane surfaces. *Biomaterials, 19*(7–9), 851–859.
95. Travan, A., Marsich, E., Donati, I., Paoletti, S. (2010). *Silver Nanocomposites and their Biomedical Applications.* Wiley, New York, NY.
96. Genzer, J., Efimenko, K. (2006). Recent developments in superhydrophobic surfaces and their relevance to marine fouling: A review. *Biofouling, 22*, 339–360.
97. Ho, C. H., Tobis, J., Sprich, C., Thomann, R., Tiller, J. C. (2004). Nanoseparated polymeric network with multiple antimicrobial properties. *Advanced Materials, 16*(12), 957–961.
98. Kang, S., Choi, H., (2005). Effect of surface hydrophobicity on the adhesion of S. cerevisiae onto modified surfaces by poly(styrene-ran-sulfonic acid) random copolymers. *Colloids and Surfaces B: Biointerfaces, 46*(2), 70–77.
99. Toppazzini, M., Coslovi, A., Boschelle, M., Marsich, E., Benincasa, M., Gennaro, R., Paoletti, S., (2010). Can the interaction between LL-37 and alginate be exploited for the formulation of new biomaterials with antimicrobial properties? *Carbohydrate Polymers, 298*, 106–118.
100. Lewis, K., Klibanov, A. M., (2005). Surpassing nature: Rational design of sterile-surface materials. *Trends in Biotechnology, 23*(7), 343–348.
101. Bromberg, L., Hatton, T. A. (2007). Poly(N-vinylguanidine): Characterization, and catalytic and bactericidal properties. *Polymer, 48*(26), 7490–7498.

102. Kenawy, E. R., Worley, S. D., Broughton, R., (2007). The Chemistry and applications of antimicrobial polymers: A state-of-the-art review. *Biomacromolecules, 8*(5), 1359–1384.

103. Darouiche, R. O. (2003). Antimicrobial approaches for preventing infections associated with surgical implants. *Clinical Infectious Diseases, 36*(10), 1284–1289.

104. Milovic, N. M., Wang, J., Lewis, K., Klibanov, A. M. (2005). Immobilized N-alkylated polyethylenimine avidly kills bacteria by rupturing cell membranes with no resistance developed. *Biotechnology and Bioengineering, 90*(6), 715–722.

105. Shukla, A., Fleming, K. E., Chuang, H. F., Chau, T. M., Loose, C. R., Stephanopoulos, G. N., Hammond, P. T. (2010). Controlling the release of peptide antimicrobial agents from surfaces. *Biomaterials, 31*(8), 2348–2357.

106. Bos, R., van der Mei, H. V., Busscher, H. J. (1999). Physico-chemistry of initial microbial adhesive interactions - its mechanisms and methods for study. *FEMS Microbiology Reviews, 23*, 179–230.

107. Hermansson, M. (1999). The DLVO theory in microbial adhesion. *Colloid and Surface B, 14*, 105–119.

108. Absolom, D. R., Lamberti, F. V., Policova, Z., Zingg, W., van Oss, C. J., Neumann, A. W. (1983). Surface thermodynamics of bacterial adhesion. *Applied and Environmental Microbiology, 46*(1), 90–97.

109. Rutter, P. R., Vincent, B. (1980). *The adhesion of microorganisms to surfaces, physico-chemical aspects.* In: *Microbial Adhesion to Surfaces.* Eds.: Berkeley, R. C. W., Lynch, J. M., Melling, J., Rutter, P. R., Vincent, B. pp. 79–91. Ellis Horwood, London, U.K.

110. Rijnaarts, H. H. M., Norde, W., Bouwer, E. J., Lyklema, J., Zehnder, A. J. B. (1995). Reversibility and mechanism of bacterial adhesion. *Colloid and Surface B, 4*, 5–22.

111. van Loosdrecht, M. C., Lyklema, J., Norde, W., Schraa, G., Zehnder, A. J. (1987). The role of bacterial cell wall hydrophobicity in adhesion. *Applied and Environmental Microbiology, 53*(8), 1893–1897.

112. Uyen, M., Busscher, H. J., Weerkamp, A. H., Arends, J., (1985). Surface free energies of oral streptococci and their adhesion to solids. *FEMS Microbiology Letters, 30*, 103–106.

113. Pratt-Terpstra, I. H., Weerkamp, A. H., Busscher, H. J., (1988). On the relation between interfacial free energy-dependent and noninterfacial free energy-dependent adherence of oral streptococci to solid substrata. *Current Microbiology, 16*, 311–313.

114. Marshall, K. C., Stout, R., Mitchell, R., (1971). Mechanism of the initial events in the sorption of marine bacteria to surfaces. *Journal of General Microbiology, 68*, 337–348.

115. Simoni, S. F., Harms, H., Bosma, T. N. P., Zehnder, A. J. B., (1998). Population heterogeneity affects transport of bacteria through sand columns at low flow rates. *Environmental Science & Technology, 32*(14), 2100–2105.

116. van Oss, C. J., Good, R. J., Chaudhury, M. K. (1986). The role of van der Waals forces and hydrogen bonds in "hydrophobic interactions" between biopolymers and low energy surfaces. *Journal of Colloid and Interface Science, 111*(2), 378–390.

117. Wood, J., Sharma, R. (1995). How long is the long-range hydrophobic attraction. *Langmuir, 11*(12), 4797–4802.

118. Ducker, W. A., Xu, Z., Israelachvili, J. N. (1994). Measurements of hydrophobic and dlvo forces in bubble-surface interactions in aqueous solutions. *Langmuir, 10*(9), 3279–3289.

119. van Oss, C. J. (1989). Energetics of cell-cell and cell-biopolymer interactions. *Cell Biochemistry and Biophysics, 14*, 1–16.

120. Jones, D. S., McGovern, J. G., Woolfson, A. D., Gorman, S. P., (1997). Role of physiological conditions in the oropharynx on the adherence of respiratory bacterial isolates to endotracheal tube poly(vinyl chloride). *Biomaterials, 18*(6), 503–510.

121. Marshall, K. C., Cruickshank, R. H., (1973). Cell surface hydrophobicity and the orientation of certain bacteria at interfaces. *Archives of Microbiology, 91*, 29–40.

122. Van der Mei, H. C., Bos, R., Busscher, H. J. (1998). A reference guide to microbial cell surface hydrophobicity based on contact angles. *Colloid and Surface B, 11*, 213–221.

123. Bellon-Fontaine, M. -N., Mozes, N., Van der Mei, H. C., Sjollema, J., Cerf, O., Rouxhet, P. G., Busscher, H. J. (1990). A comparison of thermodynamic approaches to predict the adheison of dairy microorganisms to solid substrata. *Cell Biochemistry Biophysics, 17*, 93–106.

124. Rosenberg, M., Gutnick, D., Rosenberg, E. (1980). Adherence of bacteria to hydrocarbons: A simple method for measuring cell-surface hydrophobicity. *FEMS Microbiology Letters, 9*, 29–33.

125. Busscher, H. J., van de Brlt-Gritter, B., van der Mai, H. C. (1995). Implications of microbial adhesion to hydrocarbons for evaluating cell surface hydrophobicity 1. Zeta potentials of hydrocarbon droplets. *Colloid and Surface B, 5*, 111–116.

126. Bayoudh, S., Othmane, A., Bettaieb, F., Bakhrouf, A., Ouada, H. B., Ponsonnet, L. (2006). Quantification of the adhesion free energy between bacteria and hydrophobic and hydrophilic substrata. *Materials Science and Engineering: C, 26*(2–3), 300–305.

127. Meinders, J. M., Van der Mei, H. C., Busscher, H. J. (1995). Deposition efficiency and reversibility of bacterial adhesion under flow. *Journal of Colloid Interface Science, 176*, 329–341.

128. Bellon-Fontaine, M. -N., Rault, J., van Oss, C. J. (1996). Microbial adhesion to solvents: A novel method to determine the electron-donor/electron-acceptor or Lewis acid-base properties of microbial cells. *Colloids and Surface B, 7*, 47–53.

129. Boks, N. P., Norde, W., van der Mei, H. C., Busscher, H. J. (2008). Forces involved in bacterial adhesion to hydrophilic and hydrophobic surfaces. *Microbiology, 154*(10), 3122–3133.

130. Bayoudh, S., Othmane, A., Mora, L., Ouada, H. B. (2009). Assessing bacterial adhesion using DLVO and XDLVO theories and the jet impingement technique. *Colloids and Surfaces B: Biointerfaces, 73*(1), 1–9.

131. Ong, Y. L., Razatos, A., Georgiou, G., Sharma, M. M. (1999). Adhesion forces between E. coli bacteria and biomaterial surfaces. *Langmuir, 15*(8), 2719–2725.

132. Mauclaire, L., Brombacher, E., Bünger, J. D., Zinn, M. (2010). Factors controlling bacterial attachment and biofilm formation on medium-chain-length polyhydroxyalkanoates (mcl-PHAs). *Colloids and Surfaces B: Biointerfaces, 76*(1), 104–111.

133. Triandafillu, K., Balazs, D. J., Aronsson, B. -O., Descouts, P., Tu Quoc, P., van Delden, C., Mathieu, H. J., Harms, H. (2003). Adhesion of Pseudomonas

aeruginosa strains to untreated and oxygen-plasma treated poly(vinyl chloride) (PVC) from endotracheal intubation devices. *Biomaterials*, *24*(8), 1507–1518.

134. Mackintosh, E. E. Patel, J. D., Marchant, R. E., Anderson, J. M. (2006). Effects of biomaterial surface chemistry on the adhesion and biofilm formation of staphylococcus epidermidis in vitro. *Journal of Biomedical Materials Research Part A*, *78*, 836–842.

135. Jucker, B. A., Zehnder, A. J. B., Harms, H. (1998). Quantification of polymer interactions in bacterial adhesion. *Environmental Science & Technology*, *32*(19), 2909–2915.

136. Vacheethasanee, K., Temenoff, J. S., Higashi, J. M., Gary, A., Anderson, J. M., Bayston, R., Marchant, R. E. (1998). Bacterial surface properties of clinically isolated Staphylococcus epidermidis strains determine adhesion on polyethylene. *Journal of Biomedical Materials Research*, *42*(425), 432.

137. Speranza, G., Gottardi, G., Pederzolli, C., Lunelli, L., Canteri, R., Pasquardini, L., Carli, E., Lui, A., Maniglio, D., Brugnara, M., Anderle M., (2004). Role of chemical interactions in bacterial adhesion to polymer surfaces. *Biomaterials*, *25*(11), 2029–2037.

138. Velraeds, M. M., Van der Mei, H. C., Reid, G., Busscher, H. J. (1996). Inhibition of initial adhesion of uropathogenic *Enterococcus faecalis* by biosurfactants from *Lactobacillus* isolates. *Applied and Environmental Microbiology*, *62*, 1958–1963.

139. Williams, V., Fletcher, M. (1996). Pseudomonas fluorescens adhesion and transport through porous media are affected by lipopolysaccharide composition. *Applied and Environmental Microbiology*, *62*, 100–104.

140. An Y.H., Friedman, R. J. (1998). Concise review of mechanism of bacterial adhesion to biomaterial surfaces. *Journal of Biomedical Materials Research*, *43*, 338–348.

141. Katsikogianni, M., Missirlis, Y. F. (2004). Concise review of mechanism of bacterial adhesion to biomaterials and of techniques used in estimating bacteria-material interactions. *European Cells and Materials*, *8*, 37–57.

142. Scheuerman, T. R., Camper, A. K., Hamilton, M. A. (1998). Effects of substratum topography on bacterial adhesion. *Journal of Colloid and Interface Science*, *208*, 23–33.

143. McAllister, E. W., Carey, L. C., Brady, P. G., Heller, R., Kovacs, S. G., (1993). The role of polymeric surface smoothness of biliary stents in bacterial adherence, biofilm deposition, and stent occlusion. *Gastrointestinal Endoscopy*, *39*(3), 422–425.

144. Baker, A. S., Greenham, L. W., (1988). Release of gentamicin from acrylic bone cement. Elution and diffusion studies. *The Journal of Bone and Joint Surgery*, *70*(10), 1551–1557.

145. Teixeira, P., Trindade, A. C., Godinho, M. H., Azeredo, J., Oliveira, R., Fonseca, J. G. (2006). Staphylococcus epidermidis adhesion on modified urea/urethane elastomers. *Journal of Biomaterials Science, Polymer Edition*, *17*(1–2), 239–246.

146. Lichter, J. A., Thompson, M. T., Delgadillo, M., Nishikawa, T., Rubner, M. F., Van Vliet, K. J. (2008). Substrata mechanical stiffness can regulate adhesion of viable bacteria. *Biomacromolecules*, *9*(6), 1571–1578.

147. Thompson, M. T., Berg, M. C., Tobias, I. S., Rubner, M. F., Van Vliet, K. J. (2005). Tuning compliance of nanoscale polyelectrolyte multilayers to modulate cell adhesion. *Biomaterials*, *26*(34), 6836–6845.

148. Picart, C. (2008). Polyelectrolyte multilayer films: from physico-chemical properties to the control of cellular processes. *Current Medicinal Chemistry*, *15*(7), 685–697.

149. Donati, I., Borgogna, M., Turello, E., Cesaro, A., Paoletti, S. (2007). Tuning supramolecular structuring at the nanoscale level: Nonstoichiometric soluble complexes in dilute mixed solutions of alginate and lactose-modified chitosan (Chitlac). *Biomacromolecules*, *8*(5), 1471–1479.

150. Travan, A., Donati, I., Marsich, E., Bellomo, F., Achanta, S., Toppazzini, M., Semeraro, S., Scarpa, T., Spreafico, V., Paoletti, S. (2010). Surface modification and polysaccharide deposition on BisGMA/TEGDMA thermoset. *Biomacromolecules*, *11*(3), 583–592.

151. Dotzauer, D. M., Dai, J., Sun, L., Bruening, M. L. (2006). Catalytic membranes prepared using layer-by-layer adsorption of polyelectrolyte/metal nanoparticle films in porous supports. *Nano Letters*, *6*(10), 2268–2272.

152. Shiratori, S. S., Rubner, M. F. (2000). pH-Dependent thickness behavior of sequentially adsorbed layers of weak polyelectrolytes. *Macromolecules*, *33*(11), 4213–4219.

153. Yoo, D., Shiratori, S. S., Rubner, M. F. (1998). Controlling bilayer composition and surface wettability of sequentially adsorbed multilayers of weak polyelectrolytes. *Macromolecules*, *31*(13), 4309–4318.

154. Grunlan, J. C., Choi, J. K., Lin, A. (2005). Antimicrobial behavior of polyelectrolyte multilayer films containing cetrimide and silver. *Biomacromolecules*, *6*(2), 1149–1153.

155. Dai, J. H., Bruening, M. L. (2002). Catalytic nanoparticles formed by reduction of metal ions in multilayered polyelectrolyte films. *Nano Letters*, *2*(5), 497–501.

156. Lee, D., Cohen, R. E., Rubner, M. F. (2005). Antibacterial properties of Ag nanoparticle loaded multilayers and formation of magnetically directed antibacterial microparticles. *Langmuir*, *21*(21), 9651–9659.

157. Rudra, J. S., Dave, K., Haynie, D. T. (2006). Antimicrobial polypeptide multilayer nanocoatings. *Journal of Biomaterials Science, Polymer Edition*, *17*(11), 1301–1315.

158. Kim, B. S., Park, S. W., Hammond, P. T. (2008). Hydrogen-bonding layer-by-layer-assembled biodegradable polymeric micelles as drug delivery vehicles from surfaces. *ACS Nano*, *2*(2), 386–392.

159. Rajagopalan, P., Shen, C. J., Berthiaume, F., Tilles, A. W., Toner, M., Yarmush, M. L. (2006). Polyelectrolyte nano-scaffolds for the design of layered cellular architectures. *Tissue Engineering*, *12*(6), 1553–1563.

160. Mendelsohn, J. D., Yang, S. Y., Hiller, J., Hochbaum, A. I., Rubner, M. F. (2002). Rational design of cytophilic and cytophobic polyelectrolyte multilayer thin films. *Biomacromolecules*, *4*(1), 96–106.

161. Fu, J., Ji, J., Yuan, W., Shen, J. (2005). Construction of anti-adhesive and antibacterial multilayer films via layer-by-layer assembly of heparin and chitosan. *Biomaterials*, *26*(33), 6684–6692.

162. Cassinelli, C., Morra, M., Pavesio, A., Renier, D. (2000). Evaluation of interfacial properties of hyaluronan coated poly(methylmethacrylate) intraocular lenses. *Journal of Biomaterials Science, Polymer Edition*, *11*, 961–977.

163. Boulmedais, F., Frisch, B., Etienne, O., Lavalle, P., Picart, C., Ogier, J., Voegel, J. -C., Schaaf, P., Egles, C. (2004). Polyelectrolyte multilayer films with pegylated

polypeptides as a new type of anti-microbial protection for biomaterials. *Biomaterials*, *25*(11), 2003–2011.

164. Cheng, G., Xue, H., Zhang, Z., Chen, S., Jiang, S. (2008). A switchable biocompatible polymer surface with self-sterilizing and nonfouling capabilities. *Angewandte Chemie International Edition*, *47*(46), 8831–8834.
165. Malcher, M., Volodkin, D., ü. Heurtault, B., Andre¦ü, P., Schaaf, P., Mo¦êhwald, H., Voegel, J. C., Sokolowski, A., Ball, V., Boulmedais, F., Frisch, B. (2008). Embedded silver ions-containing liposomes in polyelectrolyte multilayers: Cargos films for antibacterial agents. *Langmuir*, *24*(18), 10209–10215.
166. Chua, P. H., Neoh, K. G., Kang, E. T., Wang, W. (2008). Surface functionalization of titanium with hyaluronic acid/chitosan polyelectrolyte multilayers and RGD for promoting osteoblast functions and inhibiting bacterial adhesion. *Biomaterials*, *29*(10), 1412–1421.
167. Murata, H., Koepsel, R. R., Matyjaszewski, K., Russell, A. J. (2007). Permanent, non-leaching antibacterial surfaces–2: How high density cationic surfaces kill bacterial cells. *Biomaterials*, *28*(32), 4870–4879.
168. Kurt, P., Wood, L., Ohman, D. E., Wynne, K. J. (2007). Highly effective contact antimicrobial surfaces via polymer surface modifiers. *Langmuir*, *23*(9), 4719–4723.
169. Kugler, R., Bouloussa, O., Rondelez, F. (2005). Evidence of a charge-density threshold for optimum efficiency of biocidal cationic surfaces. *Microbiology*, *151* (5), 1341–1348.
170. Wilson, J. T., Cui, W., Chaikof, E. L. (2008). Layer-by-layer assembly of a conformal nanothin peg coating for intraportal islet transplantation. *Nano Letters*, *8*(7), 1940–1948.
171. Bombelli, C., Bordi, F., Ferro, S., Giansanti, L., Jori, G., Mancini, G., Mazzuca, C., Monti, D., Ricchelli, F., Sennato, S., Venanzi, M. (2008). New cationic liposomes as vehicles of m-tetrahydroxyphenylchlorin in photodynamic therapy of infectious diseases. *Molecular Pharmaceutics*, *5*(4), 672–679.
172. Thompson, M. T., Berg, M. C., Tobias, I. S., Lichter, J. A., Rubner, M. F., Van Vliet, K. J. (2006). Biochemical functionalization of polymeric cell substrata can alter mechanical compliance. *Biomacromolecules*, *7*(6), 1990–1995.
173. Bratskaya, S., Marinin, D., Simon, F., Synytska, A., Zschoche, S., Busscher, H. J., Jager, D., Van der Mei, H. C. (2007). Adhesion and viability of two enterococcal strains on covalently grafted chitosan and chitosan/+¦-carrageenan multilayers. *Biomacromolecules*, *8*(9), 2960–2968.
174. Gottenbos, B., Van der Mei, H. C., Klatter, F., Nieuwenhuis, P., Busscher, H. J. (2002.) In vitro and in vivo antimicrobial activity of covalently coupled quaternary ammonium silane coatings on silicone rubber. *Biomaterials*, *23*(6), 1417–1423.
175. Li, Z., Lee, D., Sheng, X., Cohen, R. E., Rubner, M. F. (2006). Two-level antibacterial coating with both release-killing and contact-killing capabilities. *Langmuir*, *22*(24), 9820–9823.
176. Shi, Z., Neoh, K. G., Zhong, S. P., Yung, L. Y., Kang, E. T., Wang, W. (2006). In vitro antibacterial and cytotoxicity assay of multilayered polyelectrolyte-functionalized stainless steel. *Journal of Biomedical Materials Research Part A*, *76*(4), 826–834.
177. Wong, S. Y., Li, Q., Veselinovic, J., Kim, B. S., Klibanov, A. M., Hammond, P. T. (2010). Bactericidal and virucidal ultrathin films assembled layer by layer from

polycationic N-alkylated polyethylenimines and polyanions. *Biomaterials*, *31*(14), 4079–4087.

178. Pan, Y., Xiao, H., Zhao, G., He, B. (2008). Antimicrobial and thermal-responsive layer-by-layer assembly based on ionic-modified guanidine polymer and PVA. *Polymer Bulletin*, *61*(5), 541–551.
179. Etienne, O., Picart, C., Taddei, C., Haikel, Y., Dimarcq, J. L., Schaaf, P., Voegel, J. C., Ogier, J. A., Egles, C. (2004). Multilayer polyelectrolyte films functionalized by insertion of defensin: a new approach to protection of implants from bacterial colonization. *Antimicrobial Agents and Chemotherapy*, *48*(10), 3662–3669.
180. Weiyong, Y., Jian, J., Jinhong, F., Jiacong, S. (2008). A facile method to construct hybrid multilayered films as a strong and multifunctional antibacterial coating. *Journal of Biomedical Materials Research Part B: Applied Biomaterials*, *85B*(2), 556–563.
181. Feng, Y., Han, Z., Peng, J., Lu, J., Xue, B., Li, L., Ma, H., Wang, E. (2006). Fabrication and characterization of multilayer films based on Keggin-type polyoxometalate and chitosan. *Materials Letters*, *60*(13–14), 1588–1593.
182. Chuang, H. F., Smith, R. C., Hammond, P. T. (2008). Polyelectrolyte multilayers for tunable release of antibiotics. *Biomacromolecules*, *9*(6), 1660–1668.
183. Nguyen, P. M., Zacharia, N. S., Verploegen, E., Hammond, P. T. (2007). Extended release antibacterial layer-by-layer films incorporating linear-dendritic block copolymer micelles. *Chemistry of Materials*, *19*(23), 5524–5530.
184. Fu, J., Ji, J., Fan, D., Shen, J. (2006). Construction of antibacterial multilayer films containing nanosilver via layer-by-layer assembly of heparin and chitosan-silver ions complex. *Journal of Biomedical Materials Research Part A*, *79*(3), 665–674.
185. Stevens, K. N. J., Crespo-Biel, O., van den Bosch, E. E. M., Dias, A. A., Knetsch, M. L. W., Aldenhoff, Y. B. J., van der Veen, F. H., Maessen, J. G., Stobberingh, E. E., Koole, L. H. (2009). The relationship between the antimicrobial effect of catheter coatings containing silver nanoparticles and the coagulation of contacting blood. *Biomaterials*, *30*(22), 3682–3690.
186. Travan, A., Marsich, E., Donati, I., Benincasa, M., Giazzon, M., Felisari, L., Paoletti, S. (2011). Silver-polysaccharide nanocomposite antimicrobial coatings for methacrylic thermosets. *Acta Biomaterialia*, *7*(1), 337–46.
187. Kingshott, P., Wei, J., Bagge-Ravn, D., Gadegaard, N., Gram, L. (2003). Covalent attachment of poly(ethylene glycol) to surfaces, critical for reducing bacterial adhesion. *Langmuir*, *19*, 6912–6921.
188. Kaper, H. J., Busscher, H. J., Norde, W. (2003). Characterization of poly(ethylene oxide) brushes on glass surfaces and adhesion of Staphylococcus epidermidis. *Journal of Biomaterials Science, Polymer Edition*, *14*, 313–324.
189. Nagel, J. A., Dickinson, R. B., Cooper, S. L. (1996). Bacterial adhesion to polyurethane surfaces in the presence of pre-adsorbed high molecular weight kininogen. *Journal of Biomaterials Science, Polymer Edition*, *7*, 769–780.
190. Anderson, J. M. (2001). Biological responses to materials. *Annual Review of Materials Research*, *31*, 81–110.
191. Wu, P., Grainger, D. W. (2006). Drug/device combinations for local drug therapies and infection prophylaxis. *Biomaterials*, *27*(11), 2450–2467.

192. Price, J. S., Tencer, A. F., Arm, D. M., Bohach, G. A., (1996). Controlled release of antibiotics from coated orthopedic implants. *Journal of Biomedical Materials, 30*(3), 281–286.

193. Stigter, M., Bezemer, J., de Groot, K., Layrolle, P. (2004). Incorporation of different antibiotics into carbonated hydroxyapatite coatings on titanium implants, release and antibiotic efficacy. *Journal of Controled Release, 99*(1), 127–137.

194. Rossi, S., Azghani, A. O., Omri, A. (2004). Antimicrobial efficacy of a new antibiotic-loaded poly(hydroxybutyric-co-hydroxyvaleric acid) controlled release system. *Journal of Antimicrobial Chemotherapy, 54*(6), 1013–1018.

195. Schierholz, J. M., Steinhauser, H., Rump, A. F., Berkels, R., Pulverer, G. (1997). Controlled release of antibiotics from biomedical polyurethanes: Morphological and structural features. *Biomaterials, 18*(12), 839–844.

196. Kwok, C. S., Horbett, T. A., Ratner, B. D. (1999). Design of infection-resistant antibiotic-releasing polymers. II. Controlled release of antibiotics through a plasma-deposited thin film barrier. *Journal of Controled Release, 62*(3), 301–311.

197. Rojas, I. A., Slunt, J. B., Grainger, D. W. (2000). Polyurethane coatings release bioactive antibodies to reduce bacterial adhesion. *Journal of Controled Release, 63*(1–2), 175–189.

198. Tiller, J. C., Liao, C. J., Lewis, K., Klibanov, A. M. (2001). Designing surfaces that kill bacteria on contact. *The Proceedings of the National Academy of Sciences of the United States of America, 98*(11), 5981–5985.

199. Lee, S. B., Koepsel, R. R., Morley, S. W., Matyjaszewski, K., Sun, Y., Russell, A. J. (2004). Permanent, nonleaching antibacterial surfaces. 1. Synthesis by atom transfer radical polymerization. *Biomacromolecules, 5*(3), 877–882.

200. Severina, I. I., Muntyan, M. S., Lewis, K., Skulachev, V. P. (2001). Transfer of cationic antibacterial agents berberine, palmatine, and benzalkonium through bimolecular planar phospholipid film and Staphylococcus aureus membrane. *IUBMB Life, 52*(6), 321–324.

201. El-Refaie, K., Fouad, I. A., bd El-Raheem, R. E., Mohamed, H. E. (2002). Biologically active polymers. V. Synthesis and antimicrobial activity of modified poly(glycidyl methacrylate-co-2-hydroxyethyl methacrylate) derivatives with quaternary ammonium and phosphonium salts. *Journal of Polymer Science Part A: Polymer Chemistry, 40*(14), 2384–2393.

202. Sukhishvili, S. A. (2005). Responsive polymer films and capsules via layer-by-layer assembly. *Current Opinion in Colloid and Interface Science, 10*(1–2), 37–44.

203. Lichter, J. A., Rubner, M. F. (2009). Polyelectrolyte multilayers with intrinsic antimicrobial functionality: The importance of mobile polycations. *Langmuir, 25*(13), 7686–7694.

204. Pan, Y. F., Xiao, H. N., Zhao, G. L., He, B. H. (2008). Antimicrobial and thermal-responsive layer-by-layer assembly based on ionic-modified guanidine polymer and PVA. *Polymer Bulletin, 61*(5), 541–551.

205. Kim. B. S., Park. S. W., Hammond, P. T. (2008). Hydrogen-bonding layer-by-layer assembled biodegradable polymeric micelles as drug delivery vehicles from surfaces. *ACS Nano, 2*(2), 386–392.

206. Hatzakis, M., Canavello, B. J., Shaw, J. M. (1980). Single-step optical lift-off process. *IBM Journal of Research and Development, 24*(4), 452–460.

207. Kharlampieva, E., Sukhishvili, S. A., (2004). Competition of hydrogen-bonding and electrostatic interactions within hybrid polymer multilayers. *Langmuir*, *20*(24), 10712–10717.

208. Kharlampieva, E., Sukhishvili, S. A. (2004). Release of a dye from hydrogen-bonded and electrostatically assembled polymer films triggered by adsorption of a polyelectrolyte. *Langmuir*, *20*(22), 9677–9685.

209. Ono, S. S., Decher, G. (2006). Preparation of ultrathin self-standing polyelectrolyte multilayer membranes at physiological conditions using pH-responsive film segments as sacrificial layers. *Nano Letters.*, *6*(4), 592–598.

210. Dubas, S. T., Schlenoff, J. B. (2001). Polyelectrolyte multilayers containing a weak polyacid: Construction and deconstruction. *Macromolecules*, *34*(11), 3736–3740.

211. Amitai, G., Andersen, J., Wargo, S., Asche, G., Chir, J., Koepsel, R., Russell, A. J. (2009). Polyurethane-based leukocyte-inspired biocidal materials. *Biomaterials*, *30*(33), 6522–6529.

212. Glinel, K., Jonas, A. M., Jouenne, T., Leprince, J., Galas, L., Huck, W. T. (2009). Antibacterial and antifouling polymer brushes incorporating antimicrobial peptide. *Bioconjugate Chemistry*, *20*(1), 71–77.

213. Furno, F., Morley, K. S., Wong, B., Sharp, B. L., Arnold, P. L., Howdle, S. M., Bayston, R., Brown, P. D., Winship, P. D., Reid, H. J. (2004). Silver nanoparticles and polymeric medical devices: A new approach to prevention of infection? *Journal of Antimicrobial Chemotherapy*, *54*(6), 1019–1024.

214. Balakrishnan, B., Kumar, D. S., Yoshida, Y., Jayakrishnan, A. (2005). Chemical modification of poly(vinyl chloride) resin using poly(ethylene glycol) to improve blood compatibility. *Biomaterials*, *26*(17), 3495–3502.

215. Pechey, A., Elwood, C. N., Wignall, G. R., Dalsin, J. L., Lee, B. P., Vanjecek, M., Welch, I., Ko, R., Razvi, H., Cadieux, P. A. (2009). Anti-adhesive coating and clearance of device associated uropathogenic Escherichia coli cystitis. *Journal of Urology*, *182*(4), 1628–1636.

216. Fujii, K., Matsumoto, H. N., Koyama, Y., Iwasaki, Y., Ishihara, K., Takakuda, K. (2008). Prevention of biofilm formation with a coating of 2-methacryloyloxyethyl phosphorylcholine polymer. *Journal of Veterinary Medical Science*, *70*(2), 167–173.

217. Cheng, G., Zhang, Z., Chen, S., Bryers, J. D., Jiang, S. (2007). Inhibition of bacterial adhesion and biofilm formation on zwitterionic surfaces. *Biomaterials*, *28*(29), 4192–4199.

218. Cheng, G., Li, G., Xue, H., Chen, S., Bryers, J. D., Jiang, S. (2009). Zwitterionic carboxybetaine polymer surfaces and their resistance to long-term biofilm formation. *Biomaterials*, *30*(28), 5234–5240.

219. Cheng, G., Xue, H., Li, G., Jiang, S. (2010). Integrated Antimicrobial and nonfouling hydrogels to inhibit the growth of planktonic bacterial cells and keep the surface clean. *Langmuir*, *26*(13), 10425–10428.

220. Marxer, S. M., Rothrock, A. R., Nablo, B. J., Robbins, M. E., Schoenfisch, M. H. (2003). Preparation of nitric oxide (NO)-releasing sol gels for biomaterial applications. *Chemical Materials*, *15*(22), 4193–4199.

221. Oh, B. K., Meyerhoff, M. E. (2003). Spontaneous catalytic generation of nitric oxide from S-nitrosothiols at the surface of polymer films doped with lipophilic copperII complex. *Journal of American Chemical Society*, *125*(32), 9552–9553.

222. Frost, M. C., Reynolds, M. M., Meyerhoff, M. E. (2005). Polymers incorporating nitric oxide releasing/generating substances for improved biocompatibility of blood-contacting medical devices. *Biomaterials*, *26*(14), 1685–93.

223. Schoenfisch, M. H., Mowery, K. A., Rader, M. V., Baliga, N., Wahr, J. A., Meyerhoff, M. E. (2000). Improving the thromboresistivity of chemical sensors via nitric oxide release: Fabrication and in vivo evaluation of NO-releasing oxygen-sensing catheters. *Analytical Chemistry*, *72*(6), 1119–26.

224. Zhang, H., Annich, G. M., Miskulin, J., Osterholzer, K., Merz, S. I., Bartlett, R. H., Meyerhoff, M. E. (2002). Nitric oxide releasing silicone rubbers with improved blood compatibility: Preparation, characterization, and in vivo evaluation. *Biomaterials*, *23*(6), 1485–1494.

225. Fang, F. C. (1997). Perspectives series: Host/pathogen interactions. Mechanisms of nitric oxide-related antimicrobial activity. *Journal of Clinical Investigation* *99*(12), 2818–2825.

226. Waldmann, T. A. (2003). Immunotherapy: Past, present and future *Nature Medicine*, *9*(3), 269–277.

227. Saylor, C., Dadachova, E., Casadevall, A. (2009). Monoclonal antibody-based therapies for microbial diseases. *Vaccine*, *27*(Suppl 6), G38-G46.

228. Reynolds, J., Norgan, V. A., Bhambra, U., Smith, J., Cook, H. T., Pusey, C. D. (2002). Anti-CD8 monoclonal antibody therapy is effective in the prevention and treatment of experimental autoimmune glomerulonephritis. *Journal of the American Society of Nephrology*, *13*(2), 359–369.

229. Sun, D., Accavitti, M. A., Bryers, J. D. (2005). Inhibition of biofilm formation by monoclonal antibodies against Staphylococcus epidermidis RP62A accumulation-associated protein. *Clinical and Diagnostic Laboratory Immunology*, *12*(1), 93–100.

230. Shahrooei, M., Hira, V., Stijlemans, B., Merckx, R., Hermans, P. W., Van, E. J. (2009). Inhibition of Staphylococcus epidermidis biofilm formation by rabbit polyclonal antibodies against the SesC protein. *Infection Immunity*, *77*(9), 3670–3678.

231. Kobayashi, T., Takauchi, A., van Spriel, A. B., Vile, H. A., Hayakawa, M., Shibata, Y., Abiko, Y., van de Winkel, J. G., Yoshie, H. (2004). Targeting of Porphyromonas gingivalis with a bispecific antibody directed to FcalphaRI (CD89) improves in vitro clearance by gingival crevicular neutrophils. *Vaccine*, *23*(5), 585–594.

232. Krishnamurthy, V. M., Quinton, L. J., Estroff, L. A., Metallo, S. J., Isaacs, J. M., Mizgerd, J. P., (2006). Whitesides, G. M. Promotion of opsonization by antibodies and phagocytosis of Gram-positive bacteria by a bifunctional polyacrylamide. *Biomaterials*, *27*(19), 3663–3674.

233. Bruno, J. G., Carrillo, M. P., Crowell, R. (2009). Preliminary development of DNA aptamer-Fc conjugate opsonins. *Journal of Biomedical Materials Research Part A*, *90*(4), 1152–1161.

234. Gyimesi, E., Bankovich, A. J., Schuman, T. A., Goldberg, J. B., Lindorfer, M. A., Taylor, R. P. (2004). Staphylococcus aureus bound to complement receptor 1 on human erythrocytes by bispecific monoclonal antibodies is phagocytosed by acceptor macrophages. *Immunology Letters*, *95*(2), 185–192.

235. Bryers, J. D. (2008). Medical biofilms. *Biotechnology and Bioengineering*, *100*(1), 1–18.

236. Bennewitz, N. L., Babensee, J. E. (2005). The effect of the physical form of poly(lactic-co-glycolic acid) carriers on the humoral immune response to co-delivered antigen. *Biomaterials*, *26*(16), 2991–2999.

237. Wizemann, T. M., Adamou, J. E., Langermann, S. (1999). Adhesins as targets for vaccine development. *Emerging Infectious Diseases*, *5*(3), 395–403.

238. Xu, L., Anchordoquy, T. (2010). Drug delivery trends in clinical trials and translational medicine: Challenges and opportunities in the delivery of nucleic acid-based therapeutics. *Journal of Pharmaceutical Sciences*, *100*(1), 38–52.

239. Otto, M. (2004). Quorum-sensing control in Staphylococci–A target for antimicrobial drug therapy? *FEMS Microbiology Letters*, *241*(2), 135–141.

240. Kaneko, Y., Thoendel, M., Olakanmi, O., Britigan, B. E., Singh, P. K. (2007). The transition metal gallium disrupts pseudomonas aeruginosa iron metabolism and has antimicrobial and antibiofilm activity. *Journal of Clinical Investigation*, *117*(4), 877–888.

241. Boles, B. R., Thoendel, M., Singh, P. K. (2005). Rhamnolipids mediate detachment of Pseudomonas aeruginosa from biofilms. *Molecular Microbiology*, *57*(5), 1210–1223.

CHAPTER 16

CHARACTERIZING THE INTERACTIONS BETWEEN CELL MEMBRANES AND ANTIMICROBIALS VIA SUM-FREQUENCY GENERATION VIBRATIONAL SPECTROSCOPY

CHRISTOPHER W. AVERY and ZHAN CHEN
University of Michigan, Ann Arbor, Michigan, USA

CONTENTS

16.1 Introduction 430
16.2 Sum-Frequency Generation 430
16.3 Solid-Supported Lipid Bilayers 433
16.3.1 Methods and Materials 435
16.4 Magainin 2 436
16.5 Melittin 439
16.6 Tachyplesin I 440
16.7 Synthetic Antimicrobial Compound—Oligomer 1 443
16.8 Synthetic Antimicrobial Compounds—Oligomers 2, 3, and 4 446
16.9 Summary and Conclusions 452
Acknowledgments 453
References 454

Antimicrobial Polymers, First Edition.
Edited by José M. Lagarón, María J. Ocio, and Amparo López-Rubio.

16.1 INTRODUCTION

Interactions between a membrane bilayer and biological molecules such as peptides and proteins are omnipresent throughout a cell [1]. Each one of those proteins and peptides is designed to interact with lipid bilayers in a precise manner. Some proteins interact very weakly or even repel bilayers, whereas others are integral to the membrane itself and bind very tightly [2]. Antimicrobial peptides (AMPs) are specifically designed to interact very strongly with certain types of membrane bilayers so that they can disrupt the membrane and disable their natural function as a barrier to the external environment [3]. Here we will present our studies on molecular interactions between AMPs and model cell membranes to elucidate molecular mechanisms of AMP activity and selectivity. In addition, molecular interactions between several synthetic oligomers (which possess certain structural characters of AMPs) and cell membranes will also be discussed.

In the recent years, sum-frequency generation (SFG) vibrational spectroscopy has been developed into a powerful technique to investigate natural and synthetic membrane-active compounds (MACs) and their interactions with model cellular membranes [4–12]. Not only does SFG allow for a clear delineation of a model system, it is ideally suited to obtain detailed information regarding the actions of individual moieties on larger molecules, providing structural, mechanistic, and dynamic information [7, 11–14].

16.2 SUM-FREQUENCY GENERATION

Optical SFG describes the process in which two input laser beams of frequencies ω_1 and ω_2 overlap in a medium, which then generates an output beam with the sum of the two input frequencies ($\omega_{SF} = \omega_1 + \omega_2$) [15–16]. This process is diagrammed in Figure 16.1. For infrared-visible SFG, ω_1 describes the input visible

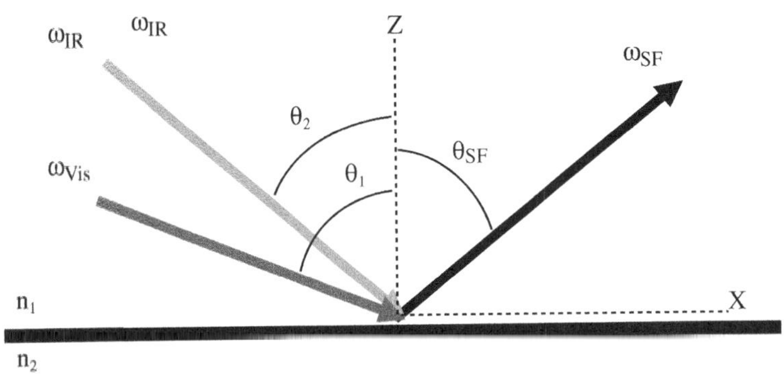

FIGURE 16.1 The SFG geometry. The reflected and transmitted infrared, visible, and transmitted SFG beams have been omitted for clarity.

beam, which is usually maintained at a fixed frequency, and ω_2 describes the frequency tunable input infrared (IR) beam. When ω_2 is scanned over the vibrational resonances of the molecules of the material under study, the SFG signal can be resonantly enhanced, thus generating a vibrational spectrum of the sample if the specific vibrational mode is SFG active.

Sum-frequency generation is a second-order nonlinear optical process, and under the electric-dipole approximation, SFG is forbidden in media that possess inversion symmetry. Most bulk materials, such as polymers or pure liquid samples, are centro-symmetric and therefore do not exhibit an SFG signal. However, at interfaces between sample layers, or at sample surfaces, the inversion symmetry of the bulk is broken and an SFG signal can usually be detected. Substrate supported lipid bilayers containing two leaflets are commonly used as models for cell membranes [17]. Normal lipid bilayers more or less have an inversion symmetry, therefore only generating weak SFG signals. However, lipid bilayers with two leaflets labeled with different isotopes have been used as model cell membranes in SFG studies [5]. Such lipid bilayers are ideal for SFG studies, as they are highly organized along a single surface with no inversion symmetry, resulting in significant SFG signal (more details will be presented in the subsequent discussion). SFG is also known to be submonolayer surface-sensitive, a determination that was made based on both theoretical calculations and experimental evidence [11, 15, 18]. In addition, the orientation and orientational distribution of functional groups on the sample surface can be deduced by collecting SFG spectra using different polarization combinations of the three input and output beams [19–23].

Mathematically speaking, SFG experiments are measuring the effective second-order nonlinear susceptibility, $\chi_{\mathrm{eff}}^{(2)}$, of a system, which is directly related to the SFG signal intensity by

$$I(\omega_{\mathrm{SF}}) \propto |\chi_{\mathrm{eff}}^{(2)}|^2 I(\omega_1) I(\omega_2) \tag{1}$$

where $\omega_{\mathrm{SF}} = \omega_1 + \omega_2$, and $I(\omega_{\mathrm{SF}})$, $I(\omega_1)$, and $I(\omega_2)$ are the intensities of each beam. For IR-visible SFG when the IR frequency (ω_2) is near a vibrational resonance, $\chi^{(2)}$ can be written as

$$\chi^{(2)} = \chi_{\mathrm{NR}}^{(2)} + \sum_{\mathrm{q}} \frac{\chi_{\mathrm{q}}}{\omega_2 - \omega_{\mathrm{q}} + i\Gamma_{\mathrm{q}}} \tag{2}$$

where the subscript NR designates the nonresonant contribution and χ_{q}, ω_{q}, and Γ_{q} denote the strength, resonant frequency, and damping constant of the qth vibrational mode, respectively. By obtaining various components $(\chi_{\mathrm{q}})_{\mathrm{ijk}}$ from the vibration mode associated with any particular molecule or chemical group on the surface in the SFG spectra collected using different polarization combinations of the input and output laser beams, the average orientation of that moiety can be determined if certain hyperpolarizability components of the vibrational mode $(\alpha)_{\mathrm{abc}}$ are known.

As an example, a common application of this method in this laboratory is to use the above information to determine the orientation of terminal methyl groups on a surface, or on a helical segment at an interface. In most cases, it is reasonable to assume that a surface or an interface is azimuthally isotropic; therefore the angle θ between the surface normal and the principal axis of a methyl group or helical structure can be used to describe its orientation. In addition, sometimes not all the methyl groups or helical structures adopt the same orientation, so the molecular orientation can be characterized by the distribution function $f(\theta)$ of the orientation angles [18].

The SFG hyperpolarizability is defined in the molecular coordinate system, which is directly proportional to the product of the IR dipole derivative and the Raman polarizability tensor of the vibrational mode q. This relationship is described as follows:

$$\alpha_{\mathrm{R}}^{(2)} \propto \frac{\partial \alpha_{\mathrm{Raman}}^{*}}{\partial Q} \frac{\partial \mu_{\mathrm{IR}}}{\partial Q} \tag{3}$$

As a result of what is shown in Eq. (3), only those vibrational modes that are both IR-active and Raman active will be SFG-active. Figure 16.2a diagrams a schematic energy level diagram for the IR-visible SFG process.

In summary, SFG can provide vibrational spectra of surfaces and interfaces (including each leaflet in the lipid bilayer) with an excellent surface specificity. It

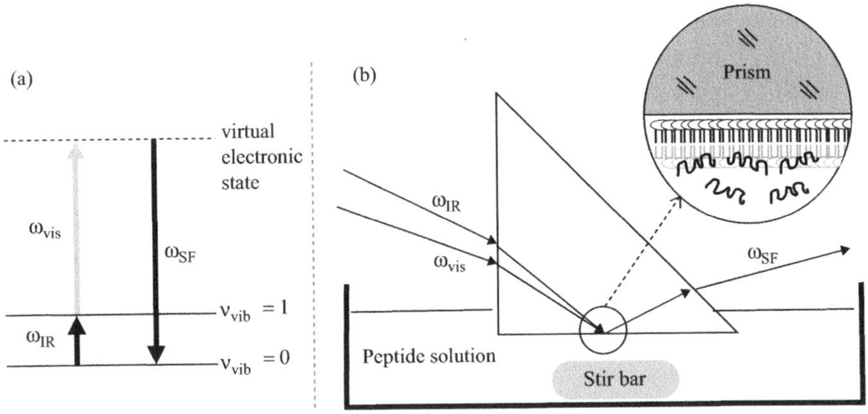

FIGURE 16.2 (a) SFG energy diagram. (b) SFG experimental setup using total reflection geometry. A bilayer is immersed in a small reservoir with an approximate volume of 1.8 mL. To achieve the desired concentration of a certain peptide, the appropriate volume of a stock peptide solution of appropriate concentration was injected into the water in the reservoir. Stirring was applied as needed to ensure a homogenous distribution of peptides and to eliminate the possible diffusion limit that may affect any interaction kinetics studies. Inset: peptides can interact with the bilayer in a carpet-like mechanism.

Source: Reproduced with permission from Reference 5.

can probe any surfaces/interfaces *in situ* where laser beams can reach, including the solid/liquid interfaces, which will be discussed in this chapter. SFG can measure the molecular orientation of molecules and functional groups on surfaces or at interfaces.

The SFG laser system used in this laboratory is a custom-designed EKSPLA SFG spectrometer. The 20-picosecond mode-locked Nd:YAG laser has a fundamental output of 1064 nm and a repetition rate of 20 Hz. The fundamental beam from the laser is directed to the harmonics unit, where two K*DP nonlinear crystals produce the second and third harmonics, which are at 532 nm and 355 nm, respectively. The 532-nm beam serves as the visible beam for the SFG experiment. The 355-nm beam, along with the fundamental 1064-nm beam, pump the optical parametric/amplification and difference frequency generation system (OPG/OPA/DFG), which is based on LBO and $AgGaS_2$ crystals. The IR beam generated from the OPG/OPA/DFG is tunable from 2.3 to 10 μm (4300 to 1000 cm^{-1}). The visible and the IR beams overlap on a sample surface/interface, and the SFG signal is collected using a monochromator and a photomultiplier tube (PMT).

16.3 SOLID-SUPPORTED LIPID BILAYERS

A common method of determining the activity of MACs is the dye-leakage test. Such a test is done by first synthesizing a series of lipid model systems, typically vesicles or liposomes containing a fluorescent dye. The liposomes are then exposed to known concentrations of MACs, causing penetration of the MACs into the liposomes, subsequently releasing the dye into the bulk whereupon it fluoresces, allowing for an accurate measurement of membrane activity of MACs. Although such tests are common and useful, two problems arise when translating results from dye-leakage tests to biologically relevant systems. First, liposomes are much smaller than cells, resulting in a very different level of curvature strain on the membrane. Second, and more importantly, dye-leakage tests only show whether the MAC interacts with the membrane, and at what concentrations. It gives limited mechanistic information, which is arguably more useful when trying to model an interaction scheme.

SFG is uniquely suited to address both of these deficiencies in dye-leakage tests using solid-supported lipid bilayers. The use of solid-supported lipid bilayers to model cellular membranes has long been established as both feasible and accurate to biological reality [17]. A typical SFG experimental setup, illustrated in Figure 16.2b, shows that by using a prism substrate, curvature strain on the lipid bilayer can be eliminated. This allows for a true planar bilayer, which much more closely models the cell membrane bilayer "seen" by individual MAC molecules. Second, and again, more importantly, SFG provides molecular level information of the surface/interface, and it allows us to strategically spectrally separate signal from individual bilayer leaflets using isotope labeled lipid bilayers.

A series of commonly used SFG bilayer structures and their corresponding SFG signals are shown in Figure 16.3. In the cases of both A and B, bilayers were built using lipids with hydrogenated acyl chains on the one side, and deuterated acyl chains on the other. As a result, the symmetry broken and significant signal from the terminal CH_3 (shown) and CD_3 (not shown) can be observed. This in turn allows the experimenter to monitor the effect of an MAC on individual leaflets, not just a bilayer as a whole by examining the CH_3 and CD_3 signals separately. This is a significant advantage over dye-leakage tests, as changes in the intensity of these individual leaflet signals during interaction with MACs can yield significantly more mechanistic information. Additionally, because SFG signal requires asymmetry at an interface, use of a symmetric bilayer without isotope labeling, as shown in Figure 16.3c, can be useful. Doing so eliminates signal from the bilayer itself initially when the bilayer is ordered, allowing time-dependent signal from the bilayer interacting with MACs to be

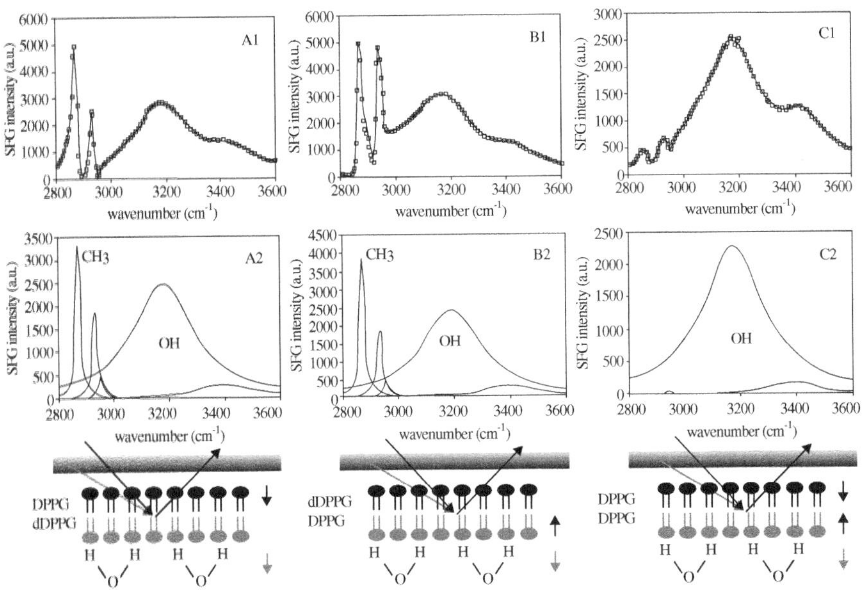

FIGURE 16.3 Deuterating a leaflet can lead to dramatically different SFG spectra of a SiO2-supported DPPG/DPPG bilayer. Spectra A1, B1, and C1 are the SFG spectra and the fitted results (solid line) from DPPG/d-DPPG and the d-DPPG/d-DPPG bilayer, respectively. The corresponding component peaks used in fitting are shown in A2, B2, and C2. In A2, peaks at ~2875 cm^{-1}, ~2940 cm^{-1}, ~3200 cm^{-1}, and ~3400 cm^{-1} have the same phase; the peak at ~2960 cm^{-1} has a different phase. In B2, the peaks at ~2875 cm^{-1} and ~2940 cm^{-1} have the same phase, which is different from that of the peaks at ~2960 cm^{-1}, ~3200 cm^{-1}, and ~3400 cm^{-1}. The phases of the peaks determine the interference (constructive/destructive) patterns that provide such dramatically different results. The last row depicts the setup and relative orientation of SFG active molecules involved.

Source: Reproduced with permission from Reference 5.

followed when different interactions between each leaflet in the bilayer and the MACs induce the "asymmetry" in the two leaflets. SFG can also be applied to study signals of MACs in other frequency regions (e.g., C=O stretching or amide I) than C-H and C-D stretches. Therefore, SFG can be used to study each leaflet in the bilayer and the MACs simultaneously during their interactions in real time *in situ*. A series of different peptides and artificially synthesized MACs have been studied using SFG and will be discussed in detail later in this chapter. Each study showcases useful investigatory properties that are unique to SFG [4, 7–9, 12, 24, 25].

16.3.1 Methods and Materials

Bacterial membranes and erythrocyte membranes have a fundamental difference that allows AMPs to differentiate between them: bacterial membranes are negatively charged, whereas erythrocyte membranes are zwitterionic, appearing neutral overall [1, 3]. To model these structural differences between membrane types, many different lipids are available for use. All lipids used in the experimentation performed here were purchased from Avanti Polar Lipids (Alabaster, AL). Because of long chemical names, lipids are often referred to by abbreviations, and a summary of the abbreviations used here are shown in Table 16.1.

Solid-supported lipid bilayers were built using the Langmuir–Blodgett–Schaeffer method on CaF_2 right-angle prisms [7–9, 25, 26]. Neutral CaF_2 prisms were used to minimize interaction between charged lipid headgroups and the substrate, which would have been a concern with fused silica or mica prisms. Use of this technique to build the bilayer system allowed for control over the identity of the lipids in each leaflet. This was a key benefit in that it allowed us to take advantage of the ability of SFG to monitor multiple spectral regions at once. More specifically, we frequently use DPPG bilayers as models for bacterial cell membranes. By using an asymmetric bilayer such as DPPG (proximal, or inner leaflet)/d-DPPG (distal, or outer leaflet), SFG allows us to monitor the effect any interacting compound has on each leaflet. Additionally, using symmetric leaflets (such as d-DPPG/d-DPPG) allows for the cancellation of all bilayer signals at the outset of experimentation, and changes to that signal as a peptide-bilayer

TABLE 16.1 Names and abbreviations for lipid molecules

Abbreviation	Lipid name
DPPG	1.2-dipalmitoyl-*sn*-glycero-3-phosphoglycerol
d-DPPG	1.2-dipalmitoyl-D62-*sn*-glycero-3-phosphoglycerol (deuterated acyl chains)
DPPE	1,2-dipalmitoyl-sn-glycero-3-phosphoethanolamine
DPPC	1,2-dipalmitoyl-sn-glycero-3-phosphocholine
POPG	1-palmitoyl-2-oleoyl-*sn*-glycero-3-[phospho-*rac*-(1-glycerol)]
POPC	1-palmitoyl-2-oleoyl-*sn*-glycero-3-phosphocholine

interaction evolves can yield further information about the reaction mechanism, which will be discussed in more detail later in the chapter. SFG spectra were collected directly from the lipid bilayer (deposited on the prism substrate) in contact with an AMP or a synthetic antimicrobial compound solution interface using different polarization combinations of the input and output beams. As discussed, SFG spectra were collected in C-H stretching, C-D stretching, C=O stretching, or amide I frequency regions. Sometimes SFG signals at a particular frequency were detected as a function of time to follow the kinetics.

16.4 MAGAININ 2

Magainin 2 is a widely studied antimicrobial peptide found in the African clawed frog *Zenopuslaevis* [25]. Like many other α-helical peptides, it converts from a random structure in solution to an α-helical structure upon interaction with a cell membrane [5]. Magainin 2 and a large variety of analogs have been under study for many years for pharmaceutical purposes [27–29], and the resulting conclusions from a multitude of studies have resulted in several different proposed interaction mechanisms to attempt to account for its antimicrobial activity [3, 5, 25, 30–36].

As discussed, isotopically asymmetric bilayers allow us to use SFG to probe each bilayer leaflet independently and simultaneously, providing crucial information regarding the interaction mechanisms of AMPs [9]. However, in real cell membranes, lipid bilayers are in the liquid phase. For substrate-supported lipid bilayers in the liquid phase, flip-flop will mix the lipids in two leaflets, creating a symmetric bilayer and becoming invisible to SFG. Flip-flop is the movement of lipids from one side of the bilayer to the other, whether by a natural mixing rate between the leaflets or induced by an MAC–bilayer interaction [12]. One simple way to avoid this problem is to use lipids in the "gel" phase, instead of in the "liquid" phase. The gel-phase lipids are more tightly packed, thus inhibiting transmembrane movement and reducing the likelihood of flip-flop [37]. In contrast, biological systems are typically in the liquid phase; this is often a result of membrane constituents, such as proteins or cholesterols, and not the identity of the lipids themselves [37]. Thus, although these gel-phase bilayers are somewhat more ordered and predictable than a true cellular membrane would be, this is as much a result of the simplification of the model as it is of the phase of the system. Despite this shortcoming, gel-phase models, especially DPPG-based bilayers (a negatively charged lipid used to model bacterial membranes), often provide us with extremely useful insights into these mechanistic models that we would be unable to observe if flip-flop were introduced into the system.

Experiments with a symmetric d-DPPG/d-DPPG bilayer were carried out with D-magainin 2, the results of which are shown in Figure 16.4. Instead of collecting SFG spectra, SFG signals at certain frequencies were monitored as a function of time to follow the kinetics of magainin 2–lipid bilayer interactions. A time-dependent SFG signal from the symmetric deuterated bilayer was

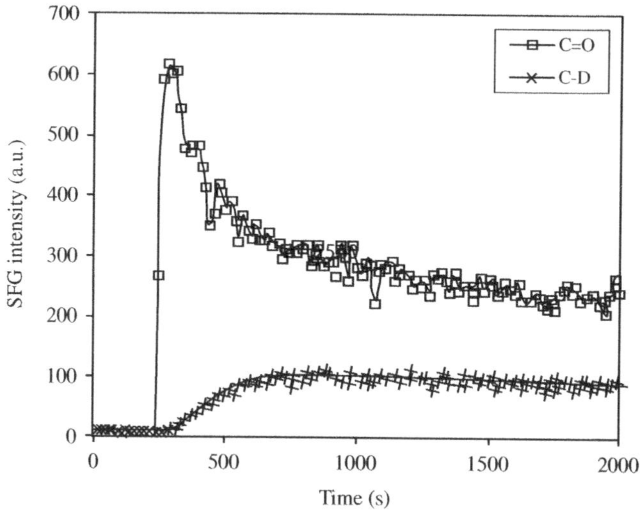

FIGURE 16.4 Peak intensities at 1655 cm^{-1} from magainin 2 amide bands and at 2070 cm^{-1} from deuterated methyl C-D symmetric stretching mode were monitored. A d-DPPG/d-DPPG bilayer was used, and the solution concentration of magainin 2 was 22.2 μg/mL.

Source: Reproduced with permission from Reference 5.

monitored at 2070 cm^{-1}, which is characteristic of the symmetric stretch of the terminal CD_3 of the acyl lipid chain [5]. Magainin 2 is known to generate a very strong amide I signal attributed to the α-helical structure of the peptide, and this was monitored at 1655 cm^{-1} as a function of time. This was particularly useful, as it allowed for direct observation of the signal from the peptide, rather than from the secondary indicators, such as a change to the bilayer signal. The intensity of the amide I signal reaches a peak within 20 seconds of peptide injection to the subphase, and subsequently, it decreases to a stable level. At the same time, the SFG signal from the deuterated bilayer increases until it too reaches stability.

It is important to note that the intensity of the SFG signal, unlike linear spectroscopy techniques such as infrared, is a complicated combination because of both the orientation of the SFG-active functional groups and the abundance of those groups [5]. In light of this, we concluded that magainin 2 associated very quickly with the bilayer, and then subsequently underwent a systemic change, either orientational or conformational in nature, causing the amide I signal to decrease until equilibrium was reached. This conclusion was supported by monitoring the O-H stretching mode of the water layer at the bilayer surface, which is an extremely sensitive probe for the interfacial electric field [5]. Experiments with multiple AMPs, magainin 2 included, showed a nearly instantaneous drop in O-H signal upon addition of peptide to the system, further supporting our conclusion that these peptides partition to the bilayer surface very quickly and then undergo a structural change that takes place over a far longer time scale. It is worth noting that these experiments were done over biologically relevant concentrations, making these mechanistic and kinetic studies highly relevant to understanding how these AMPs interact with living systems [5, 25, 38–39].

Signal from the amide I band of magainin 2 can be easily collected via SFG in multiple polarization combinations. This is particularly important for orientation analysis, which is one of the major goals of any SFG-based peptide study. For an α-helical peptide, SFG can be used to perform such an analysis by using different polarization combinations to calculate the angle from the bilayer surface normal, θ_0. To do this, we must assume that the distribution around that angle has some fixed width. By plotting multiple ratios using a variety of widths, we may determine which distribution width is most appropriate [26]. Figure 16.5 shows the χ_{zzz}/χ_{yyz} ratio assuming either a δ-distribution or a Gaussian distribution function of the alpha-helix orientation angle. Symmetric POPG and POPC bilayers, which model the bacterial and mammalian cell membranes, respectively, were used in the studies [25]. SFG results indicated that magainin 2 molecules insert into the POPG bilayer and adopt a transmembrane orientation with an angle of about 20 degrees from the POPG bilayer normal when the concentration is at 800 nM. For the POPC bilayer, even at a much higher peptide concentration of 2.0 μM, magainin 2 molecules adopt an orientation nearly parallel to the bilayer surface, with an orientation angle of 75 degrees from the surface normal. This shows that magainin 2 interact with different lipid bilayers quite differently, leading to the insertion into the bacterial cell membrane but not into the mammalian cell membrane [25]. This research demonstrates the power of using SFG to study the antimicrobial

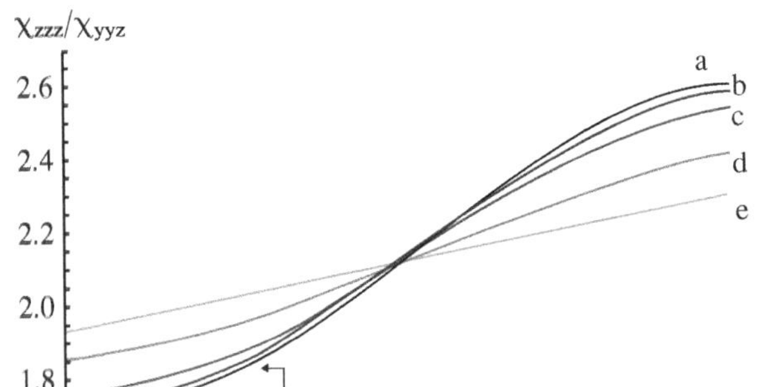

FIGURE 16.5 Relationships between the χ_{zzz}/χ_{yyz} ratio and θ for an α-helix in terms of different Gaussian distribution widths σ. (a) $\sigma = 0$, (b) $\sigma = 5°$, (c) $\sigma = 10°$, (d) $\sigma = 20°$, and (e) $\sigma = 30°$. When σ is zero, the distribution is a delta-distribution.
Source: Reproduced with permission from *J. Phys. Chem B.* 2009, 113, 12169-12180.

activities of AMPs. In this study, it was also shown that SFG is much more sensitive than the traditional ATR-FTIR technique.

16.5 MELITTIN

Melittin, a principle component in bee venom, is one of the most widely studied peptides [8,9]. Along with its analogs, it has long been used as a common model for α-helical peptides in studies of the structure and activity of membrane active compounds [8, 31, 40]. A multitude of studies have focused on understanding the interaction mechanism by which melittin operates, and only recently has a consensus begun to emerge. In general, melittin is considered to create toroidal pores within the bilayer via a destabilization effect on the bilayer system [9, 41, 42]. However, little experimental evidence exists to back up these claims. Studies carried out using SFG and solid-supported lipid bilayers have added strong evidence to these mechanisms [5, 8, 9].

Using a d-DPPG/DPPG (deuterated proximal leaflet, hydrogenated distal leaflet) asymmetric bilayer, the kinetics of the melittin–bilayer interaction can be studied. As shown in Figure 16.6, two leaflets had varied time-dependent behavior and the melittin concentration affects such behavior. SFG peak intensities were monitored at 2070 cm^{-1}, which corresponds to the symmetric stretch of the terminal CD_3 group of the proximal leaflet, and at 2875 cm^{-1}, corresponding to the terminal CH_3 group of the distal leaflet. Changes in the intensities of these peaks are indicative of either perturbation of the bilayer, or of transmembrane movement of lipids, which effectively destroys the asymmetry required for generation of the SFG signal. The addition of melittin to the

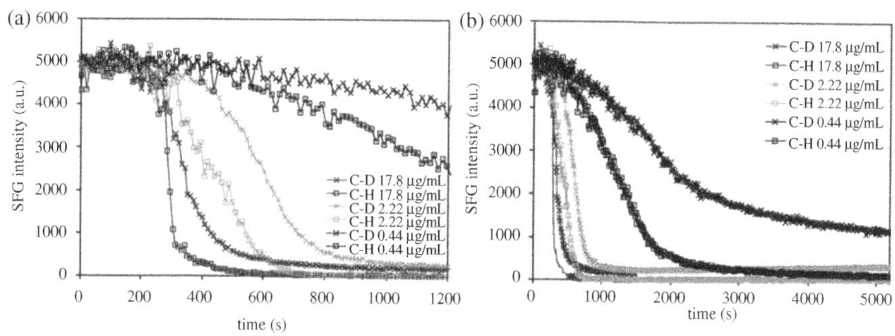

FIGURE 16.6 Real-time observation of melittin interacting with a d-DPPG/DPPG bilayer at three solution concentrations. Time-dependent spectra monitored the peak intensities at 2070 cm^{-1} for proximal methyl C-D symmetric stretching mode and at 2875 cm^{-1} for distal methyl C-H stretching mode. A 100-μL melittin solution of appropriate concentration was injected at 200 s to achieve the three solution concentrations specified in the figure. Panel a is a blowup of the initial 1000 seconds of panel b. *Source:* Reproduced with permission from Reference 5.

subphase that results in a concentration of 0.44 μg/mL leads to a near immediate, yet slow, decrease in signal from both leaflets, as shown in Figure 16.6. Different, larger concentrations of melittin in the subphase induce changes in the speed of membrane disruption. However, in every case, the distal, hydrogenated leaflet decreases at a faster pace than does the proximal, deuterated leaflet. This is indicative of a mechanism whereby the melittin AMP first associates with the bilayer in a carpet-like mechanism, disrupting the symmetry of the distal leaflet before penetrating to the proximal leaflet. This is especially clear at a lower concentration such as 0.44 μg/mL.

Different from the magainin 2 case, for melittin, the measured χ_{zzz}/χ_{yyz} ratio is not in the possible range of a delta or Gaussian orientation distribution (as shown in Figure 16.5). Initial concerns that the bent structure of melittin might be causing the discrepancy were found to be unable to explain this discrepancy [8]. In light of these results, we concluded that a simple delta or Gaussian orientation was insufficient to describe the orientation of melittin within a bilayer. The real orientation distribution is more complex, and it requires more measured parameters to determine. In addition to the χ_{zzz}/χ_{yyz} ratio, we also measured the absolute SFG signal intensity and carried out a complementary ATR-FTIR study [8]. Then, assuming a dual δ-distribution function, we were able to describe our experimental results. We also deduce the orientation distribution using a maximum entropy distribution function, without using the dual δ-distribution assumption. The orientation distribution deduced matches the dual δ-distribution quite well. Figures 16.7a and 16.7b show both the orientation distribution of melittin resulting from our dual δ-distribution calculations and the calculations using maximum entropy theory. This indicates that melittin in fact is adopting two orientational conformations, one at approximately 100° from the surface normal and one at approximately 6°. One group is therefore perpendicular to the surface normal, largely interacting with the lipid headgroups via a carpet-like mechanism, and one is parallel to it, spanning the hydrophobic core the lipid bilayer. A schematic of this interaction mechanism is diagrammed in Figure 16.7c. It is our belief that these results provide a much clearer view of the interaction mechanism of AMPs like melittin, including the fact that there should be some degree of equilibrium between these two orientational populations of melittin. This research also indicates the importance of combining different analytical techniques into the study.

16.6 TACHYPLESIN I

Another interesting application of SFG to study AMPs is tachyplesin I. It is a widely used model with an antiparallel β-sheet structure held together rigidly by two disulfide bonds [43–46]. Research has shown conflicting results as to the importance of those disulfide bonds; some researchers have witnessed a decrease in antimicrobial activity when they are broken, whereas others have used linear

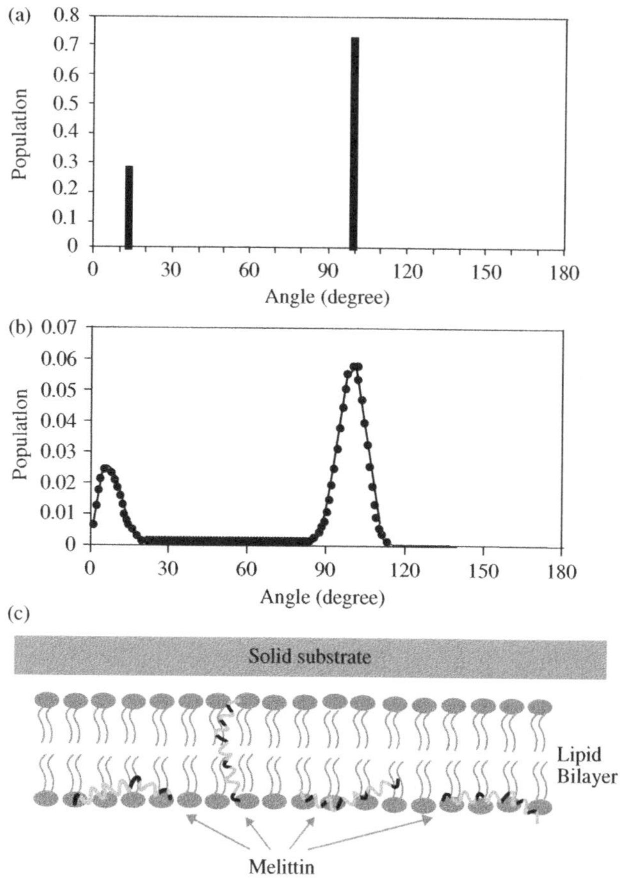

FIGURE 16.7 Orientation distribution of melittin derived based on (a) a dual δ-distribution and (b) the maximum entropy theory. (c) Schematic of the two orientations of melittin inside a lipid bilayer.

Source: Reproduced with permission from Reference 8.

analogs to show that the rigid β-sheet structure may be unimportant to the antimicrobial activity of the peptide as a whole [43, 46]. Because of this controversy, many research articles have focused on the role of those disulfide bonds in the structure and activity of tachyplesin I.

SFG experimentation with tachyplesin I found two interesting results that contribute to this discussion. First, we found that tachyplesin I induced a change to the bilayer that was peptide-concentration dependent. Increasing the peptide concentration corresponded to a decrease in signal from both the proximal and distal leaflets of the lipid bilayer. It should be noted that this is a common effect from AMP–bilayer interactions, and these results serve only to confirm that tachyplesin I was indeed bilayer-active. It is worth mentioning that we were questioned on whether SFG can be used to study the β-sheet structure because it has a more or less symmetric structure. Previous research showed that using dithiothreitol (DTT) to reduce the disulfide bonds allowed us to confirm that the SFG signal can be observed from antiparallel β-sheet structures using SFG with tachyplesisn I with and without treated by DTT [10].

The disulfide bonds can be reduced by preparing a solution of tachyplesin I with an excess of DTT. Comparison of the interaction patterns of this DTT-treated tachyplesin I with wild-type tachyplesin I with disulfide bonds still intact yields very different reaction patterns, as shown in Figure 16.8. The wild-type tachyplesin I caused a decrease in signal from both leaflets of a d- DPPG/DPPG bilayer simultaneously, roughly 200 seconds after injection of the peptide to the system subphase. By contrast, the DTT-treated tachyplesin I disrupted both leaflets, but not simultaneously, and not to the same extent. The distal leaflet saw a greater amount of disruption, indicative of association with the outer portion of the bilayer before penetration to the interior. In addition, the DTT-treated tachyplesin I acted faster than the wild type, inducing a nearly instantaneous disruption effect on the bilayer during the initial interaction period. Surprisingly, it seems that the absence of the two disulfide bonds does not prevent significant perturbation of the bilayer, even increasing the speed at which that perturbation occurs. However, at the tested solution concentrations, the extent of perturbation is not as complete for DTT-treated tachyplesin I as for the wild type, and this difference may be critical in the peptide's ability to act

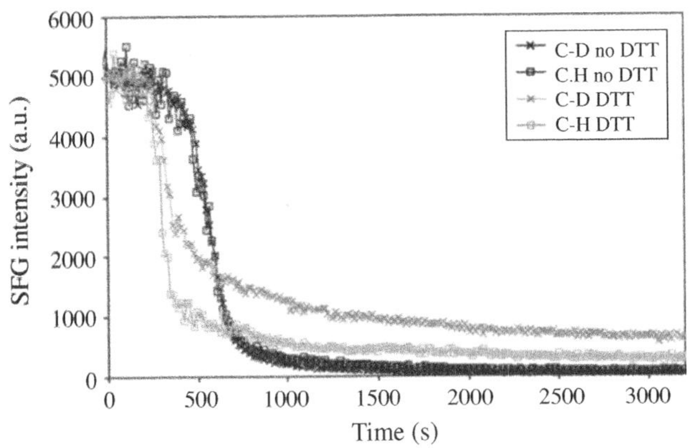

FIGURE 16.8 Tachyplesin I induced d-DPPG/DPPG bilayer structural perturbation is modulated by the disulfide linkage. Peak intensities at 2070 cm^{-1} for proximal methyl C-D symmetric stretching mode and at 2875 cm^{-1} for distal methyl C-H symmetric stretching mode were monitored simultaneously. Tachyplesin I stock solution was injected at 200 s, and the resultant solution concentration was 5.6 μg/mL. DTT treatment was carried out at 10 mg/mL for DTT and 0.1 mg/mL for tachyplesin I (DTT is about 1500-fold molar excess compared with tachyplesin I), for 30 minutes prior to injection. *Source:* Reproduced with permission from Reference 5.

as a biocide. Further experimentation with DTT and α-helical peptides showed no effect, indicating that these DTT-related interaction changes are unique to β-sheet peptides. Clearly, tachyplesin I is still capable of membrane interaction without its disulfide bonds; however, those bonds are necessary for the peptide to achieve its full potency.

Studies in our lab have shown that the amide I bands of peptides with an antiparallel β-sheet structure have a characteristic SFG signal, with a dominant band around 1685 cm^{-1} (because of the B_1/B_3 mode of an antiparallel β-sheet) [10]. We have observed that the addition of DTT to a tachyplesin I solution on a polystyrene surface results in loss of that characteristic band. However, the addition of DTT to tachyplesin I already adsorbed onto a bilayer surface causes no such signal loss [10]. This difference could be from one of two possibilities: either the membrane offers some level of protection to the disulfide bonds, shielding them from the effect of the reducing agent, or the membrane itself induces an antiparallel β-sheet structure, making the presence of the disulfide bonds superfluous after membrane interaction begins. This is an interesting question that SFG was uniquely able to observe and which further SFG experiments may be able to solve. Similar to the SFG study on helical peptides, the orientation of anti-parallel β-sheet has also been deduced using SFG spectra collected with different polarization combinations of the input and output laser beams [47]. It was found that the orientation angles of tachyplesin I on the DPPG/d-DPPG lipid bilayer can be deduced: the tilt angle (θ) has a range of 75–90 degrees, and the twist angle (Ψ) has a range of 75–90 degrees [47].

16.7 SYNTHETIC ANTIMICROBIAL COMPOUND—OLIGOMER 1

A larger goal of work with AMPs is to develop a way to bypass the growing problem of antibacterial resistance [7, 12]. Since antibiotics became a workhorse drug in the standard medical arsenal, the efficacy of those drugs has declined as bacteria most susceptible to them have been killed, leaving only those resistant strains to flourish [1]. As a result, bacterial resistance has become considered to be among the world's most pressing health problems. AMPs have the potential to dramatically affect this problem because of their unique mode of action. Instead of targeting specific membrane receptors, they target the lipid bilayer of the membrane itself, thus severely limiting the ability of any bacteria organism to develop resistance [1]. However, the cost for synthesizing AMPs can be quite high. Also, AMPs can have complex structures, making detailed mechanistic studies challenging. Further complicating interaction mechanism studies is the fact that many AMPs have a significantly different structure in a bulk solution than they do when actually interacting with a membrane. Often, AMPs do not assume a membrane-active structure until actually interacting with a membrane, making nonsurface-sensitive studies less useful [1]. However, as previously discussed, SFG is uniquely capable of investigating just such a system, where the bilayer is just as important as the AMP itself in determining the interaction mechanism.

A more fundamental goal of this branch of work is the realization of the same antimicrobial action in far simpler structures [7]. It has been demonstrated that simple compounds can have potent, broad-spectrum activity, and the ability to differentiate between a charged bacterial membrane and a zwitterionic erythrocyte membrane [48–53]. Not only is SFG well suited to these types of investigations, using it to study simpler structures allows us to make some initial determinations about those structural properties of antimicrobial compounds as well as AMPs that make them more active and/or less toxic than others.

One example of such a small, membrane-active compound is oligomer **1**, as shown in Figure 16.9. Oligomer **1** was intentionally, rationally designed to structurally mimic the amphiphilicstructure of natural AMPs—cationic segments are spatially separated from the nonpolar segments. This creates a facially amphiphilic structure, mimicking in a smaller scale the induced amphiphilicity of a natural AMP [7].

A DPPG/d-DPPG bilayer was used as a bacterial membrane model for these experiments. The spectral separation of the two leaflet layers allowed for observation of independent effects of **1** on the lipid bilayer. The bilayer was then exposed to concentrations of **1** below, at, and above the minimum inhibitory concentration (MIC) of 1, as shown in Figure 16.10. Spectral signals between 2800 cm^{-1} and 3000 cm^{-1} are primarily a result of symmetric stretching of the terminal methyl groups of the acyl chains in DPPG (the inner leaflet), whereas signals between 2000 cm^{-1} and 2300 cm^{-1} are contributed from the CD_3 groups on d-DPPG (the outer leaflet). SFG signals of CD_3decrease for all concentrations, even at 0.1 μg/mL, indicating that **1** can disrupt the outer leaflet in all cases. The decrease in SFG signals from CH_3 groups is substantial for high concentrations, but the decrease is minimal at 0.1 μg/mL, indicating that the inner leaflet was not affected at 0.1 μg/mL.

Because of the regular, predictable structure of **1**, all the carbonyl groups and t-butyl groups along the backbone must orient in a single direction. Therefore, if **1** has adopted a specific orientation relative to the bilayer surface normal, the SFG signal from those specific moieties should be theoretically observable. As shown in Figure 16.11, the signal from each of these moieties was observed, both at multiple concentrations of compound **1** and in spectra collected using different polarization combinations. To determine the orientation of **1** in the bilayer, signal from the methyl symmetric stretch (2907 cm^{-1}) of the t-butyl groups was used. The ratio of $\chi_{yyz,as}/\chi_{yyz,s}$ was used to determine a possible orientation distribution. The relative values of $\chi_{yzy,as}$ and $\chi_{yzy,s}$ were used to cross-check. The resulting value along with the possible distributions is shown in Figure 16.12. Experimentally calculated to be –0.72, the measured ratio corresponds to a restricted range of both the orientation angle and the angle distribution around the orientation angle. Oligomer **1** orients at a slight tile away from the surface normal, between two extremes of 35° of the tilt angle with a delta distribution and 0° tilt angle with an orientation distribution of 30°.

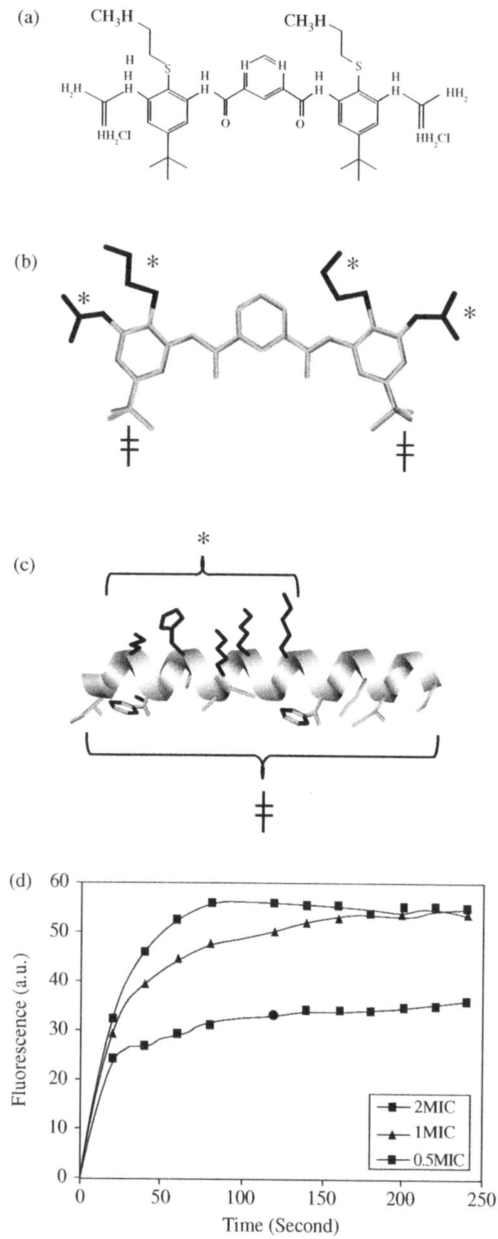

FIGURE 16.9 Structure of oligomer **1** (a) has a facial amphiphilicity (b), which mimics natural antimicrobial peptides such as magainin (c): * and ‡ represent hydrophobic and positively charged functional groups, respectively. (d) when S. aureus containing the osmotic potential sensitive dye, diSC3-5, was exposed to **1**, leakage of the dye was observed at concentrations very similar to the minimal inhibitory concentration (MIC) of 0.8 μg/mL. *Source:* Reproduced with permission from Reference 7.

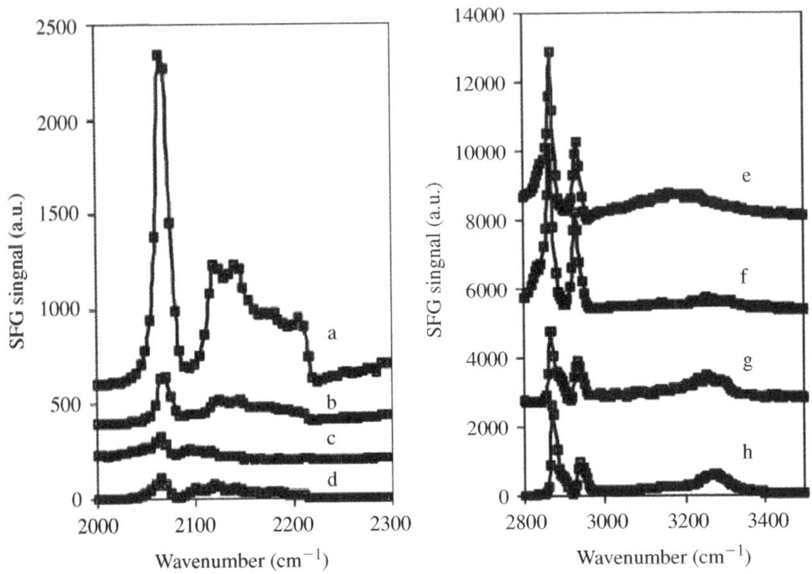

FIGURE 16.10 SFG spectra collected from a DPPG/d-DPPG bilayer in the C-D stretching region (left panel) and the C-H/O-H stretching region (right panel) before (a, e), and after contacting **1** at three solution concentrations: (b, f) 0.1 μg/mL; (c, g) 0.8 μg/mL; and (d, h) 4.0 μg/mL.

Source: Reproduced with permission from Reference 7.

A similar analysis of the C=O stretching signals confirmed these results. This more or less perpendicular orientation of **1** indicates a very specific interaction mechanism, one where the oligomer inserts into the bilayer in a similar way to a knife cutting into food. We have therefore termed this a "molecular knife" interaction mechanism [7].

16.8 SYNTHETIC ANTIMICROBIAL COMPOUNDS—OLIGOMERS 2, 3, AND 4

In light of the useful application of SFG to understanding the interaction mechanism of **1**, three more oligomers were designed and synthesized to determine the effect structural properties could have on the interaction mechanism of these small molecules [12]. These oligomers, shown in Figure 16.13, were named **2**, **3**, and **4**. Although these oligomers share a similar delocalized planar core, the surrounding side chains are very different from **1**. Instead of t-butyl groups, fluoro and trifluoromethyl groups were used as lipophilic groups. Amino, guanidine, and carboxylic groups were used as hydrophilic groups in all four oligomers; however, in **1** these groups are largely restricted to a specific orientation. In **2**, **3**, and **4**, these groups are far more flexible, attached via longer acyl chains. As the side chains contain the functional groups

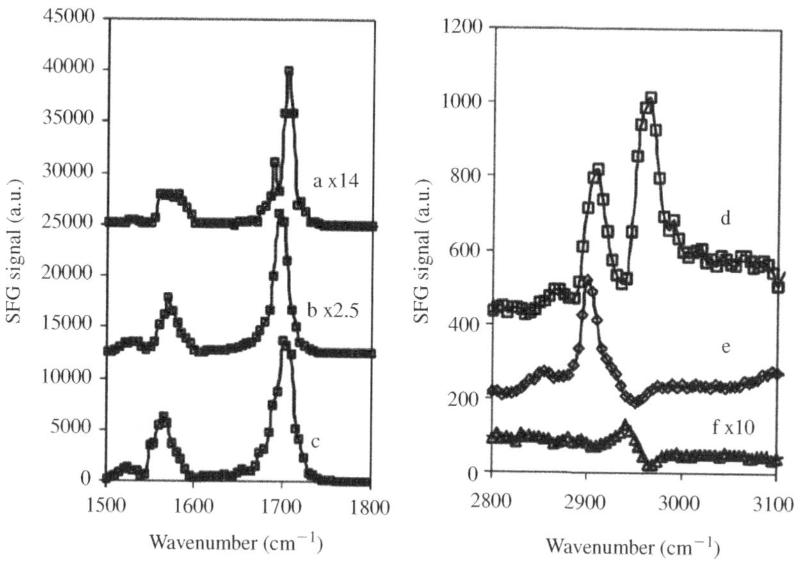

FIGURE 16.11 SFG spectra collected from **1** adsorbed onto a bilayer. (Left panel) C=O stretching signals collected at three solution concentrations of **1**: (a) 0.1 μg/mL; (b) 0.8 μg/mL; and (c) 4.0 μg/mL. (Right panel) C-H stretching signals from **1** at a solution concentration of 0.8 μg/mL using (d) ppp, (e) ssp, and (f) sps polarization combinations. A d-DPPG/d-DPPG symmetric bilayer was used to avoid spectral confusion. The SFG signals are multiplied by 14, 2.5, and 10 respectively for (a), (b), and (f) for comparison purposes.

Source: Reproduced with permission from Reference 7.

responsible for the interaction of the oligomer with the lipid bilayer, such changes to the structure of the side chains could reasonably be assumed to affect activity.

To determine the effect that oligomer concentration has on a lipid bilayer, DPPG/d-DPPG bilayers were constructed and exposed to concentrations of oligomer above, at, and below the MIC for each, just as previously described for **1**. These concentration progressions are shown in Figures 16.14, 16.15, and 16.16 for **2**, **3**, and **4**, respectively. Both **2** and **3** display clear activity toward the charged DPPG membrane. However, two significant differences were observed. First, the range of concentrations required to see the membrane fully disrupted was dramatically different for **1**, **2**, and **3**. As previously shown, **1** was membrane active between 0.1 and 4 μg/mL. Oligomer **2** showed membrane disruption at 0.1 μg/mL, and full disruption of the inner leaflet by 0.44 μg/mL. This is a far narrower range of membrane-active concentrations than seen in the case of **1**. Oligomer **3** also showed membrane activity, but again, over a very different concentration range, between 0.21 and 1.2 μg/mL. This is spectral evidence of different activity levels between the synthetic oligomers, with **2** being the most active, followed by **3** and **1**, in that order. The concentration progression of **4**, shown in Figure 16.16, is dramatically different than the other three oligomers.

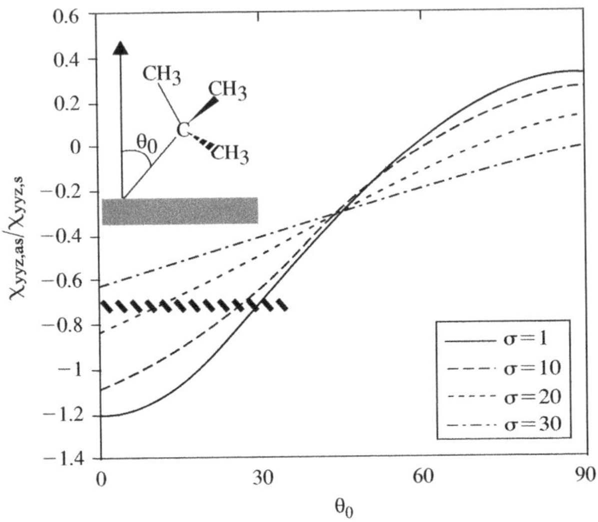

FIGURE 16.12 Relation between the $X_{yyz,as}/X_{yyz,s}$ and the orientation distribution of the *tert*-butyl group. $X_{yyz,as}/X_{yyz,s}$ was experimentally found to be −0.72, which is marked in the figure by the shaded area.

Source: Reproduced with permission from Reference 5.

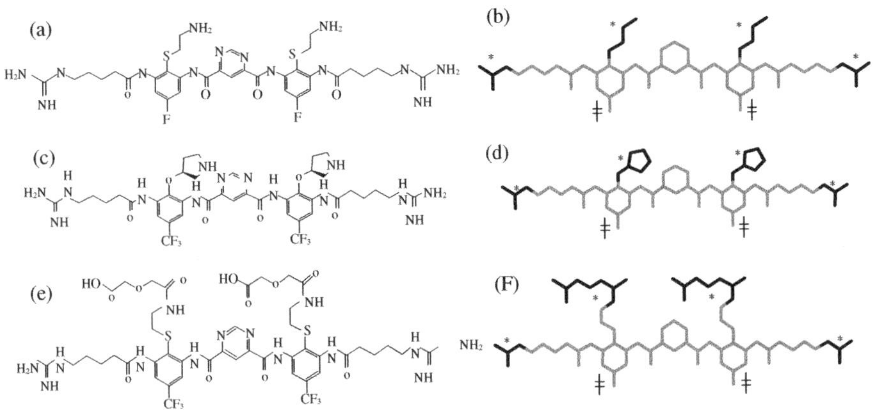

FIGURE 16.13 Oligomers **2** (a, b), **3** (c, d), and **4** (e, f). Sections denoted with * represent positively charged functional groups, and sections denoted with ‡ represent hydrophobic portions of the oligomers. All oligomers share a planar backbone structure.

Source: Reproduced with permission from Reference 12.

Despite exposure to concentrations more than 10 times higher than those used for **2**, lipid bilayers exposed to **4** showed no decrease in SFG signal from either leaflet, indicating a lack of activity. Dye-leakage tests were also performed on each oligomer, all of which indicated leakage at concentrations similar to those

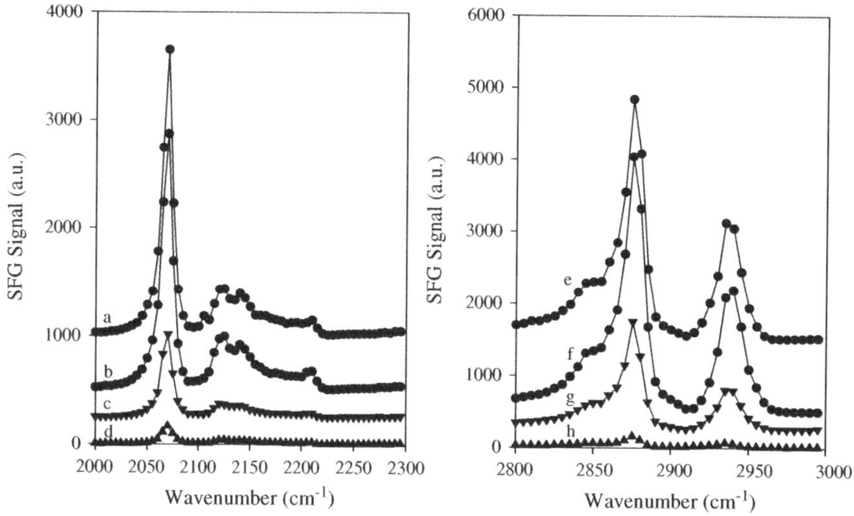

FIGURE 16.14 SFG spectra from a DPPG/d-DPPG bilayer in the C-D (left panel) and C-H (right panel) stretching regions before (a, e) and after being exposed to oligomer **2** at three solution concentrations: (b, f) 0.10 μg/mL; (c, g) 0.25 μg/mL; (d, h) 0.44 μg/mL.
Source: Reproduced with permission from Reference 12.

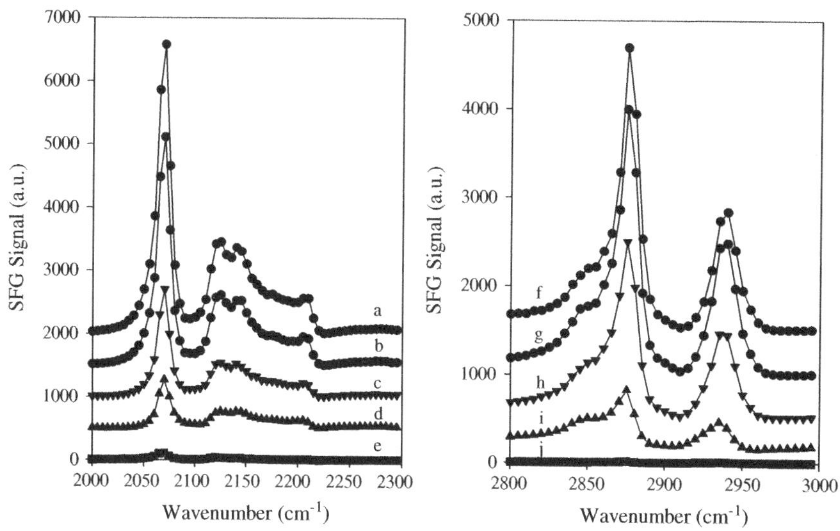

FIGURE 16.15 SFG spectra from a DPPG/d-DPPG bilayer in the C-D (left panel) and C-H (right panel) stretching regions before (a, f) and after being exposed to oligomer **3** at four solution concentrations: (b, g) 0.21 μg/mL; (c, h) 0.41 μg/mL; (d, i) 0.82 μg/mL; and (e, j) 1.2 μg/mL.
Source: Reproduced with permission from Reference 12.

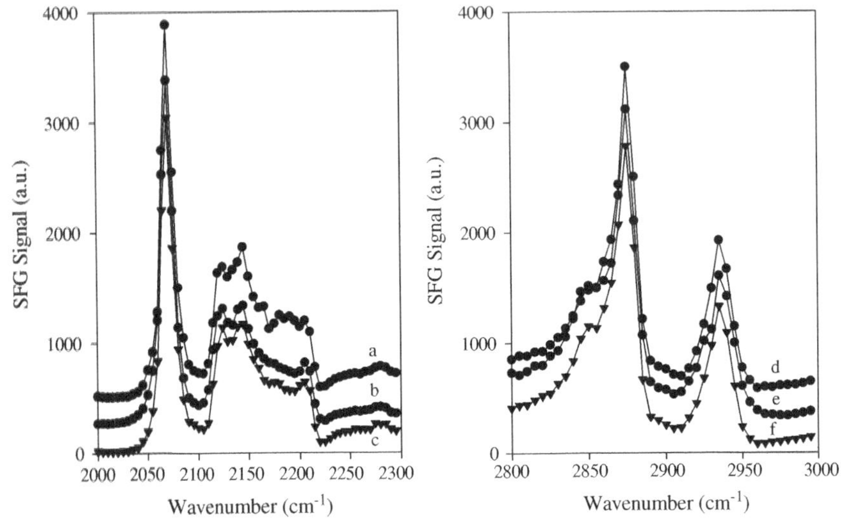

FIGURE 16.16 SFG spectra from a DPPG/d-DPPG bilayer in the C-D (left panel) and C-H (right panel) stretching regions before (a, d) and after being exposed to oligomer **4** at two solution concentrations: (b, e) 2.0 μg/mL; (c, f) 5.8 μg/mL.

Source: Reproduced with permission from Reference 12.

at which the inner leaflet was disrupted in our SFG studies. This makes sense, as the oligomers must penetrate the inner leaflet for the fluorescent dye to be released. However, this spectral confirmation of the MIC of these compounds, although useful as a secondary confirmation of activity, is not the only information gained from these studies.

What dye-leakage tests are unable to observe are the mechanistic differences among **1**, **2**, and **3** interacting with bilayers. Unlike **1**, which displayed independent action between the inner and outer leaflets, both **2** and **3** showed a simultaneous decrease in signal from both leaflets. This difference in interaction from one oligomer to another is a crucial detail regarding the bilayer–oligomer interaction mechanism that dye-leakage tests are incapable of observing.

There are several possible reasons for this loss of signal. One is the transmembrane movement of the lipids induced by the oligomers. As **2** and **3** saw a simultaneous signal decrease from both leaflets, the transmembrane movement of lipids could still technically be a feasible possibility. However, under the experimental conditions of these studies, the DPPG/d-DPPG bilayer is in a gel phase, making the intrinsic flip-flop of lipids across the bilayer negligibly slow. This was confirmed by experimental tests [9]. Furthermore, it has been confirmed that toroidal pores tend to form in bilayers when interacting with antimicrobial peptides [5, 9]. Formation of such pores significantly increases transmembrane flip-flop by creating a channel by which the hydrophilic lipid headgroup can move from one side of the membrane to the other without having to pass through the hydrophobic core of the bilayer. We contend that, given the small size and reasonably rigid structure of these oligomers, they are unlikely to effectively form a toroidal pore structure, making flip-flop of lipids a negligible concern and an unrealistic explanation for such a difference in mechanistic behavior. Therefore we believe that the similar behavior of the two leaflets is because the compounds **2** and **3** disrupted the two leaflets at the same time.

Orientation analysis of the C=O groups along the planar core could provide some additional information about the insertion mechanism observed here. As was previously done for **1**, both **2** and **3** were analyzed for orientation relative to the bilayer surface normal. The measured SFG signal strength ratios χ_{zzz}/χ_{yyz} of the C=O stretching modes are very similar for **2** and **3**, and they correspond to a roughly "stand-up" orientation (<35° from the bilayer surface normal) with a limited distribution width (<30°). As in the case of **1**, this orientation clearly shows **2** and **3** operate via the molecular knife mechanism. Despite this similarity, **2** and **3** affect the bilayer in a different way than does **1**, as shown by the concentration study shown earlier. A possible reason for this difference in leaflet behavior from oligomer to oligomer is the concentration range of activity. Because of the smaller range of concentrations needed to dramatically affect the bilayer are narrower for **2** and **3** than for **1**, it is likely that **2** and **3** are inserting into the bilayer faster than **1**, allowing for deeper penetration at lower concentrations, preventing us from observing independent action on the leaflets. Further testing of this hypothesis, with a focus on a set of oligomers with a wide range of activities and similar functional groups, would be ideal to address this important mechanistic question.

Of crucial importance when developing synthetic antimicrobial compounds is the need for such compounds to be selective for bacterial cells. As previously discussed, AMPs use an amphiphilic structure to preferentially interact with charged bacterial membranes over zwitterionic erythrocyte membranes. For any synthetic compound to effectively serve as an AMP mimic, it too must have the same functionality.

To test this, a study of mixed DPPG/DPPE/DPPC bilayers was done using **2** and **3**. In all cases, increasing the percentage of PC decreases the SFG C=O signal observed from the oligomer, despite the concentration being held constant. For **2**, a significant decrease in signal is observed once the %PE drops below 40%. A parallel decrease occurs for **3** when the %PE drops below 60%. It is theoretically possible that such a decrease in signal could be a result of a reorientation of the oligomer during interaction; essentially, the change in lipid identity could alter the interaction mechanism itself. Therefore, to determine whether that was indeed the case, orientational analysis of the carbonyl groups was repeated for each oligomer at each mixed lipid ratio. Again, in all cases, the oligomer orientation was found to match the previously discussed orientation of the oligomers in a DPPG/d-DPPG bilayer. Thus, it is highly unlikely that this signal decrease could be attributed to a reorientation effect, rather than to an interaction effect. Furthermore, this pattern is identical to the decrease in signal observed from dye-leakage tests using similar mixed bilayer ratios for each oligomer. This implies a significant correlation exists between the dye-leakage

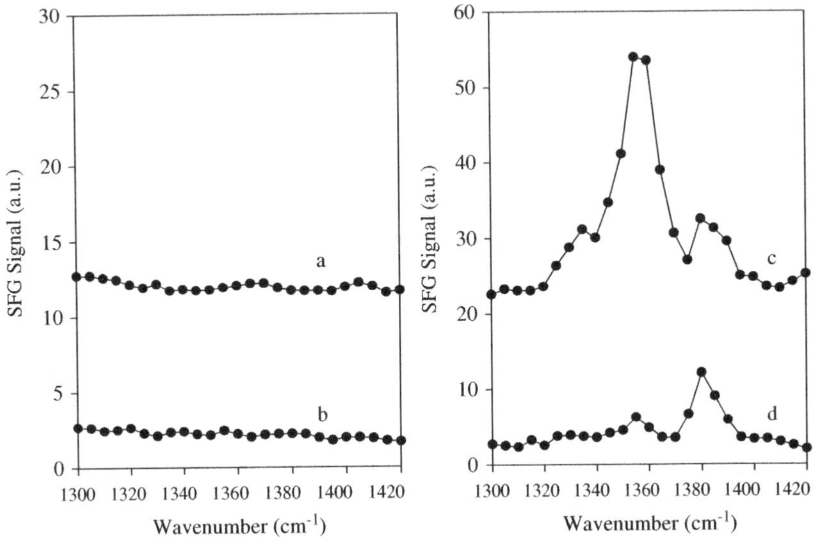

FIGURE 16.17 SFG spectra of C-F stretching region for oligomers **2** (left panel) and **3** (right panel). Spectra (a) and (c) were taken in the ssp polarization combination, whereas (b) and (d) were taken in ppp.

Source: Reproduced with permission from Reference 12.

results and our SFG studies, pointing yet again to the value of SFG as another tool for these characterization studies.

A valuable aspect of these oligomers is the side chain functional groups surrounding the planar core. Not only does this delocalized region maintain the side chains in a discrete pattern and known location, it also allows unique side chains to be arranged in such a way to maximize their usefulness for the oligomer to be membrane active. In the case of **2** and **3**, fluoro and trifluoromethyl groups were used as hydrophobic side chains. Previous studies in our lab have demonstrated that a trifluoromethyl group attached to a phenyl ring is SFG active [54]. The SFG spectra of **2** and **3** in the C-F stretching region are shown in Figure 16.17. Although the signal was not observed from the fluoro groups in **2**, the signal characteristics of trifluoromethyl groups attached to a benzyl ring were observed from **3**.

16.9 SUMMARY AND CONCLUSIONS

Sum-frequency generation vibrational spectroscopy has been shown to be a highly useful, powerful, and versatile technique for surface and interface studies. This is especially true for the studies of AMPs and their mimics described here. The inherently surface-specific aspect of the SFG technique makes it ideal for investigating properties of compounds at biological surfaces and interfaces.

Because SFG detects vibrational modes, which are ascribed to specific groups of atoms within the molecules, the relative orientation of different groups within the same molecule can be obtained and molecular formations might be deduced, as well as information about the absolute orientation of molecules with respect to the normal surface. In addition, SFG can be used to monitor the dynamics of both surface groups themselves and the interaction of surface groups with bulk compounds such as proteins or peptides. SFG also overcomes many of the operational shortcomings of other surface-sensitive techniques in that it can be applied to any interface that is accessible by light and does not require high vacuum or other environmental considerations to be used.

Particularly in the papers describing this work, we showed that SFG can be used to elucidate molecular interactions between various AMPs or AMP mimics and model cell membranes, as well as the molecular mechanisms of their antimicrobial activity and their selectivity. SFG results indicated different orientations of magainin 2 in model bacterial and model mammalian cell membranes, which can be used to interpret the selectivity of the peptide. The concentration-dependent magainin 2 orientation in the bacterial model membrane and the structural change of the model cell membrane led to understanding of the antimicrobial activity of magainin 2. The combined SFG and ATR-FTIR study quantitatively deduced the multiple orientations of melittin in the model cell membrane, showing that the combined vibrational spectroscopic study can provide a more complete picture on the melittin–cell membrane interactions. In addition to the α-helical AMPs like magainin 2 and melittin, we also demonstrated that SFG can be used to study molecular interactions among β-sheet AMP, tachyplesin I, and model cell membranes. The membrane orientation of tachyplesin I was deduced. SFG results indicated that tachyplesin I is still capable of membrane interaction without its disulfide bonds; however, those bonds are necessary for the peptide to achieve its full potency. Different from magainin 2 (which disrupts the outer leaflet of the lipid bilayer first, and then the inner leaflet), tachyplesin I disrupts both leaflets simultaneously. In addition, SFG has been used to study detailed interactions between AMP mimics and model cell membranes. Slight changes in the AMP mimic structures lead to substantial differences in antimicrobial selectivity and activity. SFG studies clearly showed different interactions between various AMP mimics and model cell membranes, providing molecular level understanding on the properties of AMP mimics. Clearly SFG is a powerful technique to elucidate molecular interactions among AMPs, AMP mimics, as well as other biocides and model cell membranes to provide molecular understanding on biocide–cell interactions, complimentary to the investigations on such interactions using biological approaches or other analytical techniques.

ACKNOWLEDGMENTS

This research is supported by National Institute of Health (1R01GM081655-01A2) and Office of Naval Research (N00014-08-1-1211).

REFERENCES

1. Zasloff, M. (2002). Antimicrobial peptides of multicellular organisms. *Nature, 415*, 389–395.
2. Rothman, J. E., Lenard, J. (1977). Membrane asymmetry. *Science, 195*, 743–753.
3. Matsuzaki, K. (1999). Why and how are peptide-lipid interactions utilized for self-defense? Magainins and tacyplesins as archetypes. *Biochemica et Biophysical Acta, 1462*, 1–10.
4. Chen, X., Boughton, A. P., Tesmer, J. J. G., Chen, Z. (2007). In situ investigation of heterotrimeric G protein $\beta\gamma$ subunit binding and orientation of membrane bilayers. *Journal of the American Chemical Society, 129*, 12658–12659.
5. Chen, X., Chen, Z. (2006). SFG studies on interactions between antimicrobial peptides and supported lipid bilayers. *Biochmica et Biophysica Acta, 1758*, 1257–1273.
6. Chen, X., Clarke, M. L., Wang, J., Chen, Z. (2005). Sum frequency generation vibrational spectroscopy studies on molecular conformation and orientation of biological membranes at interfaces. *International Journal of Modern Physics B, 19*, 691–713.
7. Chen, X., Tang, H., Even, M. A., Wang, J., Tew, G. N., Chen, Z. (2006). Observing a molecular knife at work. *Journal of the American Chemical Society, 128*, 2711–2714.
8. Chen, X., Wang, J., Boughton, A. P., Kristalyn, C. B., Chen, Z. (2007). Multiple orientation of melittin inside a single lipid bilayer determined by combined vibrational spectroscopic studies. *Journal of the American Chemical Society, 129*, 1420–1427.
9. Chen, X., Wang, J., Kristalyn, C. B., Chen, Z. (2007). Real-time sum frequency generation structural investigation of a lipid bilayer during its interaction with melittin. *Biophysical Journal, 93*, 1–10.
10. Chen, X., Wang, J., Sniadecki, J. J., Even, M. A., Chen, Z. (2005). Probing a-helical and b-sheet structures of peptides at solid/liquid interfaces with SFG. *Langmuir, 21*, 2662–2664.
11. Chen, Z., Shen, Y. R., Somorjai, G. A. (2002). Studies of polymer surfaces by sum frequency generation vibrational spectroscopy. *Annual Reviews of Physical Chemistry, 53*, 437–465.
12. Avery, C. W., Som, A., Xu, Y., Tew, G. N., Chen, Z. (2009). Dependence of antimicrobial selectivity and potency on oligomer structure investigated using substrate supported lipid bilayers and sum frequency generation vibrational spectroscopy. *Analytical Chemistry, 81*, 8365–8372.
13. Wang, J., Woodcock, S. E., Buck, S. M., Chen, C., Chen, Z. (2001). Difference surface-restructuring behaviors of poly(methacrylate)s detected by SFG in water. *Journal of the American Chemical Society, 123*, 9470–9471.
14. Clarke, M. L., Wang, J., Chen, Z. (2003). Sum frequency generation studies on the surface structures of plasticized and unplasticized polyurethane in air and in water. *Analytical Chemistry, 75*, 3275–3280.
15. Shen, Y. R. (1984). *The Principles of Nonlinear Optics*, Wiley, New York, NY.
16. Bain, C. D. (1995). Sum-frequency vibrational spectroscopy of the solid/liquid interface. *Journal of the Chemical Society, Dalton Transactions, 91*, 1281–1296.

17. Tamm, L. K., McConnell, H. M. (1985). Supported phospholipid bilayers. *Biophysical Journal*, *47*, 105–113.
18. Wang, J., Chen, C., Buck, S. M., Chen, Z. (2001). Molecular chemical structure on poly(methyl methacrylate) (PMMA) surface studied by sum frequency generation (SFG) vibrational spectroscopy. *Journal of Physical Chemistry B*, *105*, 12118–12125.
19. Hirose, C., Akamatsu, N., Domen, K. (1992). Formulas for the analysis of surface sum-frequency generation spectrum by CH stretching modes of methyl and methylene groups. *Journal of Chemical Physics*, *96*, 997–1004.
20. Hirose, C., Akamatsu, N., Domen, K. (1992). Formulas for the analysis of the surface SFG spectrum and transformation coefficients of Cartesian SFG tensor components. *Applied Spectroscopy*, *46*, 1051–1072.
21. Hirose, C., Yamamoto, H., Akamatsu, N., Domen, K. (1993). Orientation analysis by simulation of vibrational sum-frequency generation spectrum: CH stretching bands of the methyl group. *Journal of Physical Chemistry*, *97*, 10064–10069.
22. Zhuang, X., Miranda, P. B., Kim, D., Shen, Y. R. (1990). Mapping molecular orientation and conformation at interfaces by surface nonlinear optics. *Physical Review B*, *172*, 303–306.
23. Moad, A. J., Simpson, G. J. (2004). A unified treatment of selection rules and symmetry relations for sum-frequency and second harmonic spectroscopies. *Journal of Physical Chemistry B*, *108*, 3548–3562.
24. Wang, J., Chen, X., Clarke, M. L., Chen, Z. (2005). Detection of chiral sum frequency generation vibrational spectra of proteins and peptides at interfaces in situ. *Proceedings of the National Academy of the Sciences*, *102*, 4978–4983.
25. Nguyen, K. T., Le Clair, S. V., Ye, S., Chen, Z. (2009). Molecular interactions between magainin 2 and model membranes in situ. *Journal of Physical Chemistry B*, *113*, 12358–12363.
26. Nguyen, K., Le Clair, S. V., Ye, S., Chen, Z. (2009). Orientation determination of protein helical secondary structure using linear and nonlinear vibrational spectroscopy, Feature Article. *Journal of Physical Chemistry B*, *113*, 12358–12363.
27. Genco, C. A., Maloy, W. L., Kari, U. P., Motley, M. (2003). Antimicrobial activity of magainin analogues againstanaerobic oral pathogens. *International Journal of Antimicrobial Agents*, *21*, 75–78.
28. Giacometti, A., Ghiselli, R., Cirioni, O., Mocchegiani, F., D'Amato, G., Orlando, F., Sisti, V., Kamyscz, W., Silvestri, C., Naldoski, P., Lukasiak, J., Saba, V., Scalise, G. (2004). Therapeutic efficacy of the magainin analogue MSI-78 in different intra-abdominal sepsis rat models. *Journal of Antimicrobial Chemotherapy*, *54*, 654–660.
29. Kwak, H. B., Lee, S. W., Lee, D. G., Hahm, K. S., Kim, K. K., Kim, H., Lee, Z. H. (2003). A hybrid peptide derived from cecropin-A and magainin-2 inhibits osteoclast differentiation. *Life Sciences*, *73*, 993–1005.
30. Matsuzaki, K., Murase, O., Fujii, N., Miyajima, K. (1996). An antimicrobial peptide, magainin 2, induced rapid flip-flop of phosopholipids couples with pore formation and peptide translocation. *Biochemistry*, *35*, 11361–11368.
31. Bechinger, B. (1997). Structure and functions of channel-forming peptides: Magainins, cecropins, melittin and alamethicin. *Journal of Membrane Biology*, *156*, 197–211.

32. Zasloff, M. (1987). Magainins, a class of antimicrobial peptides from Xenopus skin: Isolation, characterization of two active forms, and partial cDNA sequence of a precursor. *Proceedings of the National Academy of the Sciences, 84*, 5449–5453.
33. Tamba, Y., Yamazaki, M. (2005). Signal giant unilamellar vesicle method reveals effect of antimicrobial peptide magainin 2 on membrane permeability. *Biochemistry, 35*, 15823–15833.
34. Hirsh, D. J., Hammer, J., Maloy, W. L., Blazyk, J., Schaefer, J. (1996). Secondary structure and location of a magainin analogue in synthetic phospholid bilayers. *Biochemistry, 35*, 12733–12741.
35. Bechinger, B. (2005). Detergent-like properties of magainin antibiotic peptides: A 31P solid-state NMR spectroscopy study. *Biochemica et Biophysical Acta, 1712*, 101–108.
36. Munster, C., Spaar, A., Bechinger, B., Salditt, T. (2002). Magainin 2 in phospholipid bilayers: Peptide orientation and lipid chain ordering studied by X-ray diffraction. *Biochemistry, 1562*, 37–44.
37. Voet, D., Voet, J. G., Pratt, C. W. (2002). *Fundamentals of Biochemistry*, Wiley, New York, NY.
38. Jacobs, R. E., White, H. S. (1989). The nature of the hydrophobic binding of small peptides at the bilayer interface: Implications for the insertion of transbilayer helices. *Biochemistry, 28*, 3421–3437.
39. White, H. S., Wimley, W. C., Ladokhin, A. S., Hristova, K., (1998). Protein folding in membranes: Determining energetics of peptide-bilayer interactions. *Methods in Enzymology, 295*, 62–87.
40. Abrunhosa, F., Faria, S., Gomes, P., Tomaz, I., Pessoa, J. C., Andreu, D., Bastos, M. (2005). Interaction and lipid-induced conformation of two cecropin-melittin hybrid peptides depend on peptide and membrane composition. *Journal of Physical Chemistry B, 109*, 17311–17319.
41. Yang, L., Harroun, T. A., Weiss, T. M., Ding, L., Huang, H. W. (2001). Barrel-stave model or toroidal model? A case study on melittin pores. *Biophysical Journal, 81*, 1475–1485.
42. Allende, D., Simon, S. A., McIntosh, T. J., Melittin-induced bilayer leakage depends on lipid material properties: Evidence for toroidal pores. *Biophysical Journal* 2005, *88*. 1828–1837.
43. Matsuzaki, K., Nakayama, M., Fukui, M., Otaka, A., Funakoshi, S., Fujii, N., Bessho, K., Miyajima, K. (1993). Role of disulfide linkages in tachyplesin-lipid interactions. *Biochemistry, 32*, 11704–11710.
44. Katso, T., Nakao, S., Iwanaga, S. (1993). Mode of action of an antimicrobial peptide, tachyplesin I, on biomembranes. *Biological and Pharmaceutical Bulletin, 16*, 178–181.
45. Nakamura, T., Furunaka, H., Miyata, T., Tokunaga, F., Muta, T., Iwanaga, S., Niwa, M., Takao, T., Shimonishi, Y. (1988). Tachyplesin, a class of antimicrobial peptide from the meocytes of the horseshoe crab (tachypleus tridentatus). Isolation and chemical structure. *Journal of Biological Chemistry, 263*, 16709–16713.
46. Gururaj Rao, A. (1999). Conformation and antimicrobial activity of linear derivatives of tachyplesin lacking disulfide bonds. *Archives of Biochemistry, 361*, 127–134.
47. Nguyen, K., King, J. T., Chen, Z. (2010). Orientation determination of interfacial β-sheet structure. *Journal of Physical Chemistry B, 114*, 8291–8300.

48. Som, A., Yang, L., Wong, G. C. L., Tew, G. N. (2009). Divalent metal ion triggered activity of a synthetic antimicrobial in cardiolipin membranes. *Journal of the American Chemical Society*, *131*, 15102–15103.

49. Lienkamp, K., Kumar, K.-N., Som, A., Nusslein, K., Tew, G. N. (2009). "Double selective" antimicrobial polymers: How do they differentiate between bacteria? *Chemistry: A European Journal*, *15*, 11710–11714.

50. Colak, S., Nelson, C. F., Nusslein, K., Tew, G. N. (2009). Hydrophillic modifications of an amphiphillic polynorborene and the effects on its hemolytic and antibacterial activity. *Biomacromolecules*, *10*, 353–359.

51. Gabriel, G. J., Maegerlein, J. A., Nelson, C. F., Dabowski, J. M., Eren, T., Nusslein, K., Tew, G. N. (2009). Comparison of a facially amphiphilic versus segregated monomers in the design of antibacterial copolymers. *Chemistry: A European Journal*, *15*, 433–439.

52. Gabriel, G. J., Madkour, A. E., Dabowski, J. M., Nelson, C. F., Nusslein, K., Tew, G. N. (2008). Synthetic mimic of antimicrobial peptide with nonmembrane-disrupting antibacterial properties. *Biomacromolecules*, *9*, 2980–2983.

53. Slutsky, M. M., Phillip, J. S., Tew, G. N. (2008). Synthesis and characterization of amphiphilic o-phenylene ethynylene oligomers. *New Journal of Chemistry*, *32*, 670–675.

54. Even, M. A., Lee, S.-H., Wang, J., Chen, Z. (2006). Detection and analysis of trifluoromethyl groups at a surface by sum frequency generation vibrational spectroscopy. *Journal of Physical Chemistry B*, *110*, 26089–27097.

CHAPTER 17

GAS-BASED ANTIMICROBIALS IN ACTIVE PACKAGING

SIRIYUPA NETRAMAI,[1,2] MARIA RUBINO,[1] and LOONG-TAK LIM[3]
[1] School of Packaging, Michigan State University, East Lansing, Michigan
[2] Mahidol University, Karnjanaburi, Thailand
[3] Department of Food Science, University of Guelph, Guelph, Ontario, Canada

CONTENTS

17.1 Introduction . . . 459
17.2 Allyl Isothiocyanate . . . 460
  17.2.1 Food Packaging Applications . . . 461
  17.2.2 Mass Transfer of AITC . . . 464
17.3 Chlorine Dioxide . . . 469
  17.3.1 Biocide Performance . . . 469
  17.3.2 Sanitizing and Food Packaging Applications . . . 470
  17.3.3 Mass Transfer of ClO_2 in Selected Polymeric Packaging Materials . . . 470
  17.3.4 Effects of ClO_2 Gas on Properties of Polymeric Materials . . . 477
17.4 Future Trends . . . 482
References . . . 482

17.1 INTRODUCTION

The packaging system can be considered as a practical extension of the food production line to improve food safety and economically extend shelf life. Fresh produce, meats, and minimally processed as well as ready-to-eat foods

Antimicrobial Polymers, First Edition.
Edited by José M. Lagarón, María J. Ocio, and Amparo López-Rubio.

can benefit from packaging systems that provide additional antimicrobial protection [1–5].

The protective functions of packaging can be grouped into passive and active. When packaging is not part of the preparation process and preservation system [3, 6], it provides passive protection against mechanical forces and microbial contamination as well as heat and mass transfer. On the other hand, when packaging constitutes an integral part of food preservation processes, and is thus considered an indispensable element, it has an active role. Common examples of active systems include oxygen scavenging systems and antimicrobial packaging [3, 6]. Also, the addition of antimicrobial agents, such as organic acids, inorganic compounds, enzymes, and spices, to packaging materials for active packaging has been extensively studied in the past several decades [2, 3, 7].

For antimicrobial packaging, the disinfecting agents can be integrated into the packaging system by (1) adding sachets or pads that are carriers for the gas/vapor agents, (2) coating the agent onto the inner packaging surface, (3) embedding the agent into the polymer matrix, or (4) introducing the agent directly into the package headspace [2, 3, 7]. The antimicrobial agents, once generated and/or released, will inhibit and/or retard microbial growth within the package and on the food product surfaces, where the majority of contamination and spoilage occurs [2, 5, 8]. Examples of disinfecting agents that are incorporated into commercially available packaging systems are silver-substituted zeolite, triclosan, and ethanol vapor [2]. Other promising agents for use in vapor-phase decontamination within the package of the food products include allyl isothiocyanate and chlorine dioxide [2, 7, 9–13]. The characterization and applications of allyl isothiocyanate and chlorine dioxide are reviewed here.

Protection and containment are two of the most important roles of the package. The ability to protect and contain the product depends on the stability of the packaging material to environmental factors, such as temperature, humidity, and exposure to chemical compounds. Knowledge of the interaction and compatibility between packaging materials and gas-based antimicrobials is critical for the selection of materials for specific applications. This requires an understanding of the mass transfer properties of the antimicrobials in the packaging materials and the effects on packaging integrity and performance, as will be addressed in the following sections.

17.2 ALLYL ISOTHIOCYANATE

Allyl isothiocyanate (AITC) is a naturally occurring antimicrobial compound found in edible plants from the Cruciferae family (e.g., horseradish and mustard) and, to a lesser extent, in leafy vegetables of the Brassica genus (e.g., cabbage). The antimicrobial properties of this compound have been well demonstrated in the literature [14–21]. In the vapor phase, minimum inhibitory concentrations of AITC against bacteria, yeasts, and molds have been estimated at 34–110, 13–37, and 16–62 ng/mL, respectively [22]. Also, Tsunoda reported that the toxic limits of AITC against five fungi on wood ranged from 3.8 to 118 ppm [23]. In view of its antimicrobial potency and natural sources, AITC has gained considerable interest for use as an antimicrobial agent for food and nonfood applications.

The following sections provide a brief review of the literature on food packaging applications for AITC and discuss the compatibility of AITC for selected polymeric packaging materials.

17.2.1 Food Packaging Applications

Active packaging is defined as "packaging in which subsidiary constituents have been deliberately included in or on either the packaging material or the package headspace to enhance the performance of the package system" [24]. One form of active packaging involves the incorporation of an antimicrobial agent within the package structure, allowing for release into the product in a controlled manner to inhibit the proliferation of microorganisms during storage.

Sachet technology can be considered as one of the simplest methods to deliver AITC for active antimicrobial packages. AITC is impregnated in a solid support carrier (e.g., diatomaceous earth, silica gel, and plastic pellet) that is overwrapped with a sachet fabricated from semipermeable films, perforated barrier films, porous nonwovens, or papers. By strategically placing the sachet within a package, controlled release of the antimicrobial volatile can be achieved. For example, Sekiyama et al. [25] used diatomaceous earth as a solid support for AITC and studied the release behavior of AITC when the AITC-impregnated earth (~13% AITC by weight) was packaged in various materials (Figure 17.1). The researchers observed that the rate of AITC release from sachet could be adjusted by selecting different polymers (material type; multilayer versus monolayer), adopting different perforation sizes, and using optimal material thickness [25].

Another form of AITC-based commercial antimicrobial active packaging systems is pressure-sensitive labels (Wasaouro) from Mitsubishi-Kagaku Foods Corporation, Japan. In this system, AITC is impregnated in the adhesive layer of a label, which is protected by a barrier film (e.g., metalized polyester). During end use, the protective film is removed and the label is affixed to the exterior of a thermoplastic food package. As AITC permeates through the polymer, it is released gradually into the package headspace, thereby inhibiting microorganisms on food surfaces (Figure 17.2). Packages based on polyolefin polymers, e.g., poly(ethylene) (PE) and poly(propylene) (PP), are well suited for this application because they are highly permeable to AITC. For packaging structures that are less permeable to AITC, such as poly(ethylene terephthalate) (PET) and polystyrene (PS), the label must be attached to the interior surface of the package. In this case, the label is constructed of AITC-permeable film to allow for gradual release of AITC from the adhesive layer.

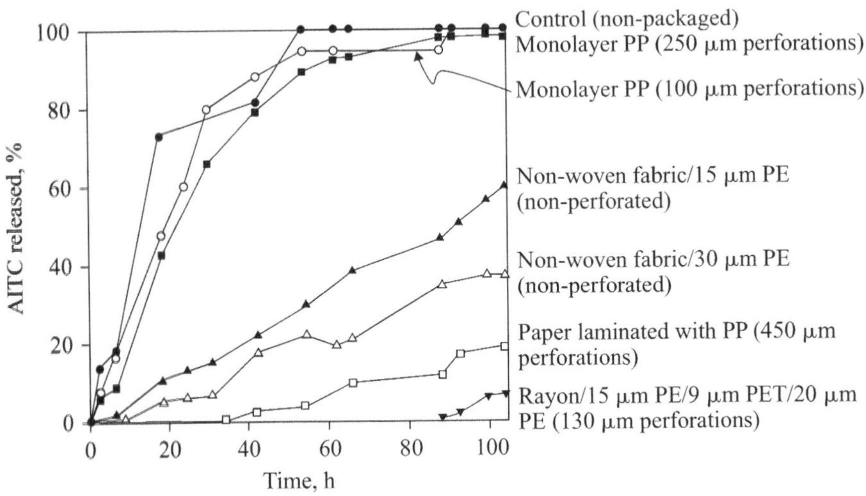

FIGURE 17.1 Percent AITC release as a function of time for 50 g of AITC-impregnated diatomaceous earth particles (100 parts earth to 15 parts AITC by weight) packaged in different materials. Percent of AITC release was determined gravimetrically. Adapted from Reference 25.

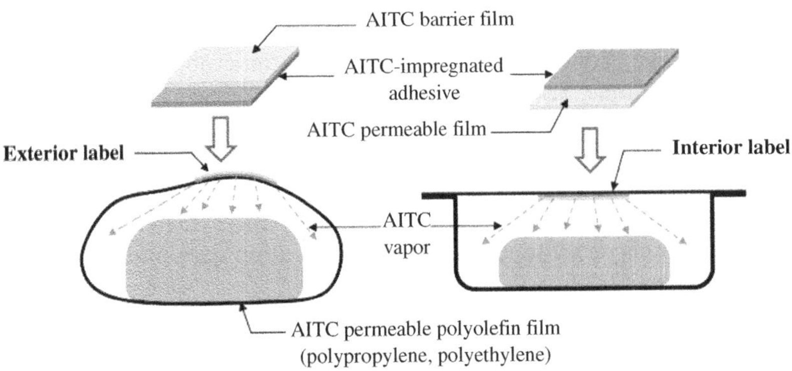

FIGURE 17.2 Controlled release of AITC vapor into food package headspace to inhibit the growth of microorganisms by using adhesive label as a carrier. Adapted from http://www.mfc.co.jp/wasaouro/e/products/labels.html.

Winther and Nielsen [26] investigated the efficacy of Wasaouro interior labels for shelf life extension of cheese packaged in thermoformed PET containers. The use of one AITC label (20 mm × 20 mm) extended the shelf life of the product up to 13 weeks, as compared with 4.5 weeks for the control samples packaged in the absence of the antimicrobial label. When two AITC labels were used, the shelf life was further extended to 28 weeks. The researchers reported that cheese stored for up to 12 weeks in the presence of the AITC label had an unacceptable mustard flavor because of extensive AITC absorption by the product. Interestingly, the off-flavor decreased to an acceptable level between 12 and 28 weeks, presumably attributed to reaction with nucleophiles (e.g., -SH and -OH groups) present in cheese. The study concluded that AITC-based active food packaging may be a good alternative to modified atmosphere packaging for cheese.

Recently, fibrous membranes prepared from electrospun fibers have been investigated as carriers for controlled release of AITC in active packaging applications. Unlike the typical melt spun fibers that are formed by mechanical extrusion, this method uses electrostatic force to draw polymer solution into fibers of diameters ranging from tens to hundreds of nanometers [27–29]. By virtue of their small diameter, electrospun fibers exhibit much larger surface-to-volume ratios as compared with conventional melt spun fibers and continuous films. This property enhances the surface activity of electrospun fibers, making them an ideal carrier for delivery of bioactive agents.

Vega-Lugo and Lim exploited the properties of electrospun fibers for controlled release of AITC [30]. In these studies, electrospun fibers were prepared from soy protein isolate/poly(ethylene oxide) (SPI/PEO) blends. Briefly, the spinning solution was made up of 15% SPI dissolved in 1% NaOH solution. PEO (0.6%) and Triton X-100 (1%) were added as processing aid and emulsifier, respectively. The solution was heated for 2 h at 60 °C to denature the protein, followed by the addition of β-cyclodextrin, and the mixture was stirred for another 30 min. After cooling to 24 ± 2 °C, AITC was added at a 0.4:2.4 AITC:β-cyclodextrin molar ratio and treated with a sonicator to form the AITC-cyclodextrin complex. The solution was then electrospun at 20 kV DC with a spinneret-collector distance of 26 cm. At 0% RH, the release of AITC from the SPI fibers was negligible. However, AITC release was triggered when the fibers were tested at 50% and 75% RH (Figure 17.3). Compared with the

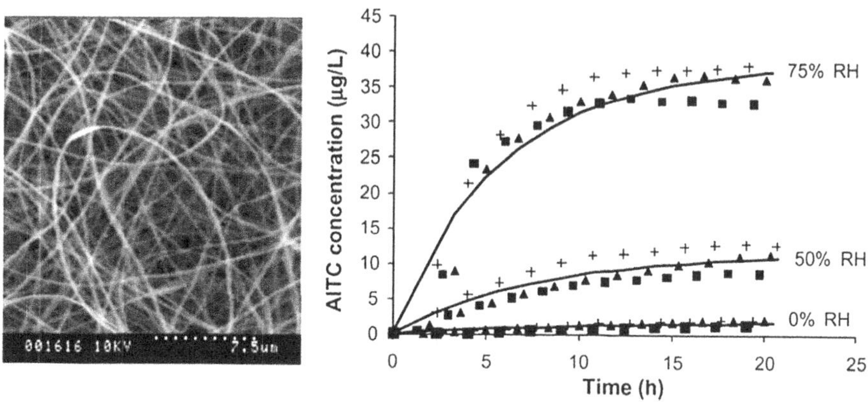

FIGURE 17.3 Interactive behavior of AITC-encapsulated SPI-PEO fibers toward relative humidity. The electron scanning micrograph illustrates the morphology of typical electrospun fibers at 3% AITC loading and 12% SPI concentration, with an average diameter of 281 ± 43 nm [31].

direct addition of AITC to the protein solution, the AITC-cyclodextrin approach resulted in greater loading of AITC ($\sim$70% higher) because of the reduced evaporative loss during spinning. The interactive behavior of these antimicrobial nanofibers may be promising in active packaging applications for foods to ensure that the release of AITC takes place only under elevated humidity conditions, such as within a food package. A similar concept may be applied for encapsulation of other antimicrobial agents for the shelf life extension of food products.

One of the potential shortfalls of AITC-based antimicrobial active packaging systems is that the food may absorb the AITC, resulting in off-flavors. Therefore, the incorporation of AITC in food packaging systems requires careful product selection. For instance, the garlic-like odor of AITC may be suitable for certain products such as raw beef, cured pork, sliced raw tuna, cheeses, egg sandwiches, noodles, pasta, and breads [21, 26, 32]. In certain jurisdictions, the migrated antimicrobial may be considered a food additive and, accordingly, the use of this type of active packaging may be subjected to regulatory additive limits.

17.2.2 Mass Transfer of AITC

17.2.2.1 Vapor Pressure Properties of AITC. Permeation of AITC in a polymer film is primarily driven by a concentration gradient that exists between the opposite surfaces of the film. Since the surface concentration of the permeant cannot be determined readily, vapor pressure is often used to characterize its mass transfer properties. The saturated vapor pressure of pure AITC, as a function of temperature, is summarized in Figure 17.4. The temperature dependence of AITC vapor pressure can be described reasonably well by the classical Clausius-Clapeyron equation, which gives an enthalpy of vaporization of about 48 kJ/mol (Figure 17.4) [33].

Because of its lachrymatory properties and strong odor, AITC is seldom used in its pure state. Instead it is often blended with a compatible liquid to depress its vapor pressure for various controlled-release applications. For instance, the vapor pressure of AITC can be effectively controlled by blending it with vegetable oil (Figure 17.5).

17.2.2.2 Mass Transfer in Selected Polymeric Packaging Materials. The permeability coefficient (P) is an important parameter for characterizing the permeation process. It is a normalized transmission flux that takes thickness and vapor pressure gradients into account, carrying a dimension of [quantity] · [thickness]/[time] · [area] · [pressure]. P is commonly determined using a permeability cell, comprising high (upstream) and low (downstream) concentration chambers separated by the test film. In a quasi-isostatic experimental setup, an excess amount of permeant liquid is placed in the hermetically sealed upstream chamber, whereas the downstream chamber is purged with a carrier gas (typically nitrogen) that is directed to a detector, such as a gas chromatograph, for quantifying the amount of permeant permeated. The permeation test may

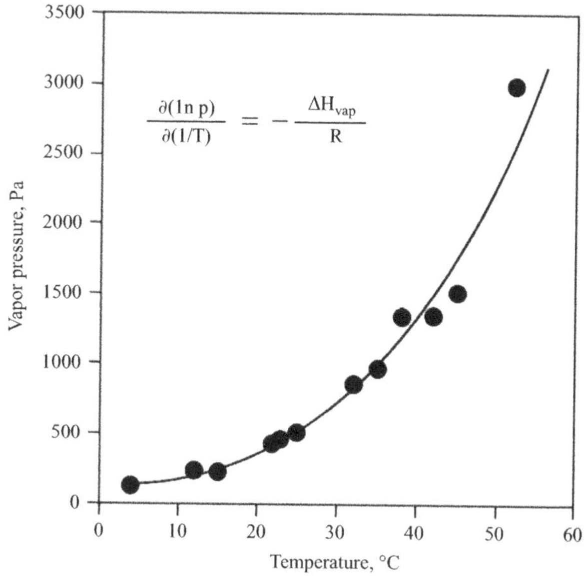

FIGURE 17.4 Vapor pressure of AITC as a function of temperature. Data are plotted based on References 25, 33, and 34. The solid curve is predicted from the Clausius-Clapeyron equation (insert) with an enthalpy of vaporization of 48 kJ/mol.

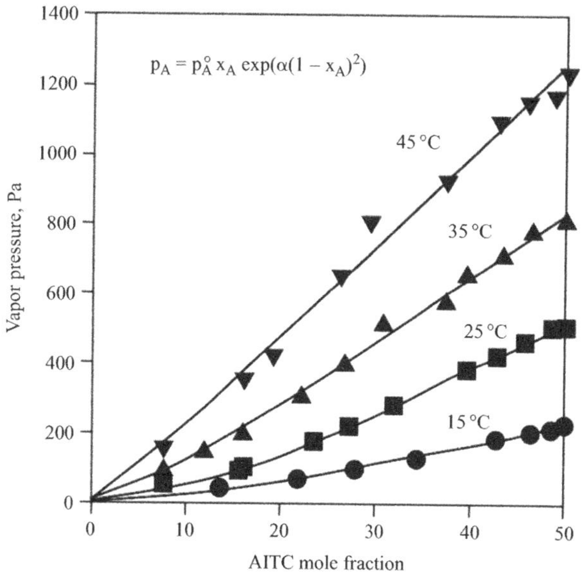

FIGURE 17.5 Depression of AITC vapor pressure using canola oil. The solid lines are predicted from the Margules equation (insert) where p_A° and x_A are temperature-dependent constants. α is related to the heat of mixing (ΔHmix): $\alpha = -\Delta Hmix/RT$, where R is a gas constant and T is the absolute temperature. Values of heat of mixing range from 3380 to 368 J/mol as temperature increases from 15 to 45 °C. Adapted from Reference 33.

also be conducted in an isostatic mode; in which case, the upstream chamber is continuously purged with a stream of gas carrying a constant concentration of the permeant to be tested. From the permeation curve (plot of amount of AITC permeated versus time), P can be determined from the slope, whereas diffusivity can be derived from the time lag (i.e., the intercept on the time axis when a straight line is extrapolated from the steady-state portion of the permeation curve) according to $D = L^2/6\tau$, where L is film thickness [35].

The mass transfer properties of AITC in poly(vinylidene chloride) and poly (vinyl chloride) (PVDC-PVC) copolymer film, which is not moisture sensitive, are illustrated in Figure 17.6. As shown, the P values of AITC in PVDC-PVC copolymer film increase exponentially with increasing AITC concentration. The concentration-dependent mass transfer phenomenon is indicative of a strong interaction between the organic vapor and the polymer. This interaction can lead to plasticization and swelling of the polymer matrix, enhancing polymer chain segmental movements, and thereby increasing the diffusivity and solubility of AITC in the polymer. It is noteworthy that at any given AITC vapor activity, the diffusion coefficient (D) correlates positively with temperature, as shown. This correlation is expected because of enhanced polymer chain motion at higher temperatures. In contrast, a negative relationship of the solubility

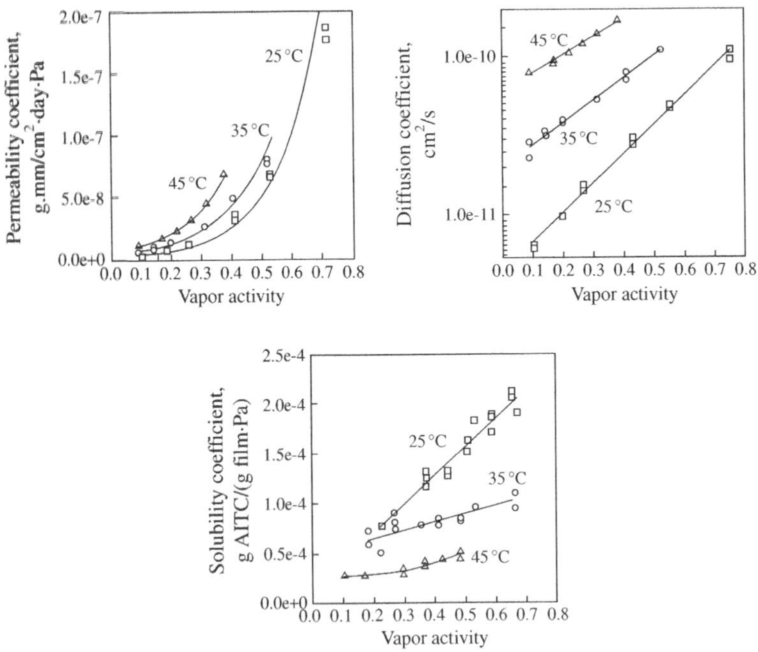

FIGURE 17.6 Permeability, diffusion, and solubility coefficients of AITC in PVDC-PVC copolymer film (10 m, Saran Plastic Wrap; SC Johnson, Racine, WI) as a function of AITC vapor pressure, at 25, 35, and 45 °C. Adapted from Reference 33.

coefficient (S) with temperature is evident, which can be attributed to the exothermic nature of the sorption process, possibly because of the substantial contribution from the heat of condensation. It is generally agreed that a relationship exists among the three mass transfer coefficients: $P = D \times S$. Accordingly, the positive correlation of overall permeability with temperature suggests that the diffusion phenomenon is more dominant than solubilization during the permeation of AITC in the PVDC-PVC copolymer.

Permeation of AITC in moisture-sensitive polymers, such as nylon and the ethylene vinyl alcohol (EVOH) copolymer, is markedly different. Under dry conditions, permeation of AITC through these polymers is minimal because of hydrogen bonds between the polymer chains, which hinder the penetration of the relatively hydrophobic AITC molecules. However, when a moisture-sensitive polymer is exposed to an elevated humidity environment, the barrier properties are weakened as the polymer absorbs moisture. One of the polymers that exhibits this moisture-dependent behavior is nylon 6,6.

As shown in Figure 17.7a, the time lag for AITC permeation decreases as relative humidity (RH) increases, indicating that the breakthrough of AITC in nylon 6,6 film occurs more quickly because of the increased diffusivity of the permeant. The increased permeability can be attributed to the glass transition temperature (T_g) depression effect of water upon the polyamide (Figure 17.8). As the RH increases, the moisture absorbed by the polymer causes it to undergo glass-to-rubber transition. The resulting swelling of the polymer matrix and increased polymer chain segmental movement are the main contributors to the increased diffusivity of AITC molecules. Figure 17.8 also highlights that when the polymer is dry (0% RH), the plasticization effect of AITC is minimal.

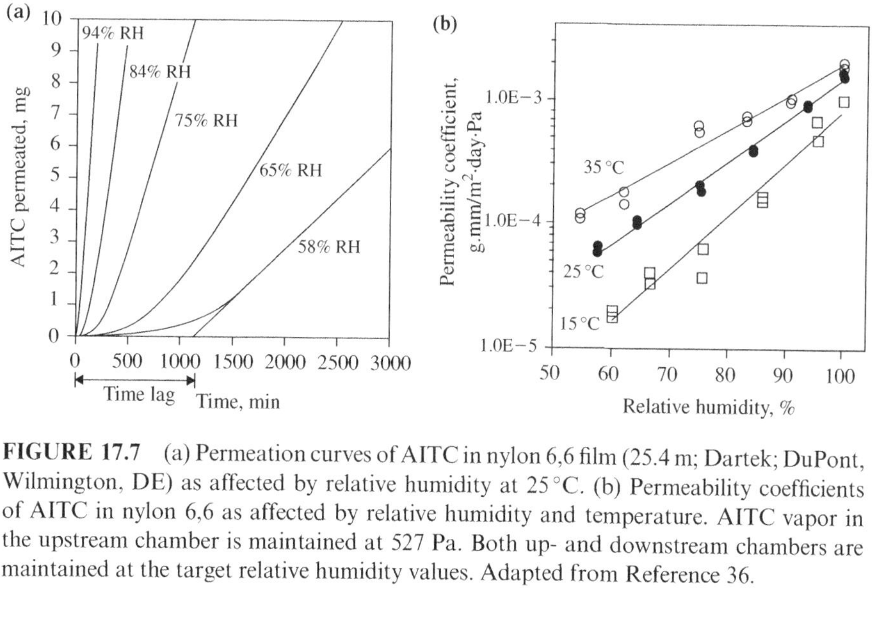

FIGURE 17.7 (a) Permeation curves of AITC in nylon 6,6 film (25.4 m; Dartek; DuPont, Wilmington, DE) as affected by relative humidity at 25 °C. (b) Permeability coefficients of AITC in nylon 6,6 as affected by relative humidity and temperature. AITC vapor in the upstream chamber is maintained at 527 Pa. Both up- and downstream chambers are maintained at the target relative humidity values. Adapted from Reference 36.

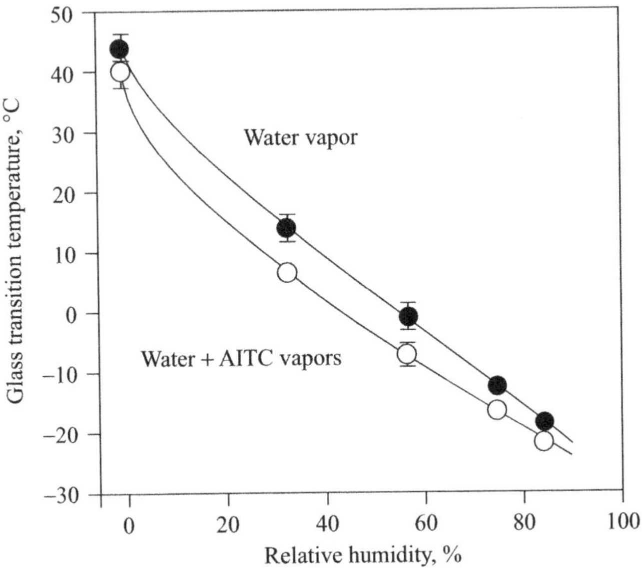

FIGURE 17.8 Glass transition temperature of nylon 6,6, film at different relative humidities when exposed to water vapor alone or water/AITC vapors at 25 °C. Adapted from Reference 36.

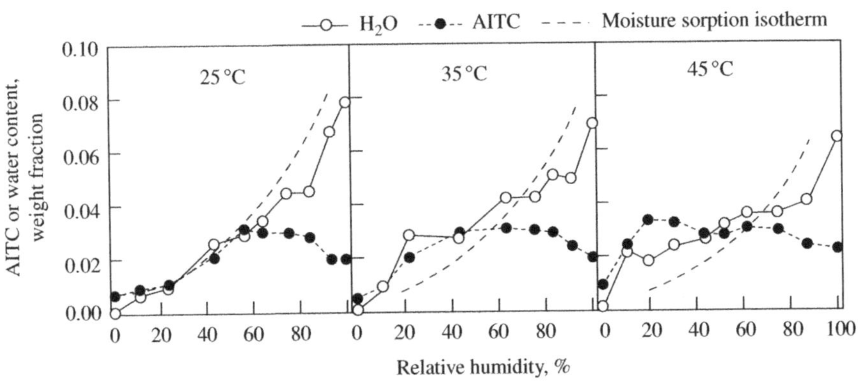

FIGURE 17.9 AITC and water sorption in nylon 6,6 film as affected by relative humidity and temperature. Saturated vapor pressures of AITC at 25, 35, and 45 °C are 527, 972, and 1726 Pa, respectively. Adapted from Reference 36.

However, as the water content of the polymer increases, the T_g depression effect of AITC becomes more prominent. This additional plasticization effect of AITC suggests that water may have facilitated the interaction between nylon 6,6 and AITC.

The evidence for this moisture-interactive behavior is provided in Figure 17.9. As shown, the equilibrium AITC contents are minimal at low RH but increase to about 3% as RH increases. The decreasing AITC solubility trend above 70% RH is likely as a result of competition between water and AITC for sorption sites. The formation of larger water clusters at high RH may have displaced some of the sorption sites for AITC. At the higher temperature (45 °C), the enhanced AITC sorption below intermediate RH can be attributed to the existence of nylon 6,6 in a rubbery state. The presence of AITC also appears to alter the sorption behavior of water. The higher water sorption observed for nylon films exposed to the AITC/water binary permeant as compared with water alone (i.e., moisture sorption isotherm) seems to suggest that the uptake of AITC also results in the exposure of additional active sites previously unavailable for water sorption when water is the only permeant present. The moisture-interactive behavior of nylon and other polymers may be useful for the controlled release of AITC and other antimicrobial agents.

17.3 CHLORINE DIOXIDE

Chlorine dioxide (ClO_2) is an oxidizing agent that has been used primarily as a bleaching agent in pulp and paper production. In the gas phase, ClO_2 has a greenish-yellow color and a pungent chlorine-like odor. The gas is highly soluble in water and remains a dissolved gas in solution. Because of its broad antimicrobial spectrum, ClO_2, in both liquid and gas forms, is also used in a wide range of sanitizing applications, particularly in wastewater treatment [12, 37]. The following sections provide a brief review of the performance of ClO_2 as an antimicrobial agent as well as a discussion of the effects of chlorine dioxide exposure on the properties and performance of packaging materials.

17.3.1 Biocide Performance

Detailed information on the chemistry, application, and performance of chlorine dioxide can be accessed in the guidance manual [37] issued by the U.S. Environmental Protection Agency (EPA) and is summarized only briefly here. Chlorine dioxide is recognized as a strong disinfectant against bacterial, viral, and protozoan pathogens [12, 37, 38]. The disinfecting efficacy of ClO_2 is generally reported to be equal to or higher than that of chlorine (Cl_2) on a mass-dose basis but less than that of ozone.

Chlorine dioxide disinfects by oxidation through the one-electron transfer mechanism in which it is usually reduced to chlorite (ClO_2^-). However, its primary mode of microbial inactivation is not yet well established and seems to depend on the type of target microorganism. Two possible primary microbial inactivation mechanisms have been reported: (1) interactions between ClO_2 and nucleic acid and/or peripheral structures, which lead to a disruption of protein synthesis [37, 39]; and (2) destruction of outer membrane proteins, which alters the permeability of the cell membrane [37, 40].

Inactivation efficiency of ClO_2 is not significantly affected by a change of pH [12, 37]; however, the disinfecting capacities of both ClO_2 and Cl_2 decrease as temperature decreases.

17.3.2 Sanitizing and Food Packaging Applications

As a registered bactericide, fungicide, algaecide, and biocide, ClO_2 is approved by the U.S. EPA for use in both liquid and gas forms as a sanitizing agent for food processing equipment, pharmaceutical and factory tools, as well as in potable and wastewater treatment [10, 11, 37].

Specifically, in food applications, ClO_2 is reported to be an effective disinfecting agent against many important spoilage and pathogenic microorganisms (Table 17.1), such as *Escherichia coli* O157:H7, *Listeria monocytogenes*, *Salmonella* spp., yeasts, and molds, which may reside on the surface of food products such as fresh and fresh-cut fruits and vegetables, as well as on fresh meats [4, 7, 67–69].

Although ClO_2 is often used as a disinfecting solution for fresh foods [10, 68], its gaseous form is gaining interest as a disinfectant for vapor-phase decontamination. Examples of ClO_2 gas use include cleaning food-contact surfaces [42]; treating fresh produce, such as green peppers, blueberries, and apples before packaging [9, 12, 70–72]; and sanitizing raw foods, such as chicken, directly within the package [1, 7, 13].

The introduction of ClO_2 gas within food packaging systems is often carried out by means of sachets, with either low- or fast-release characteristics [1, 73]. However, in 2001, the U.S. Food and Drug Administration (FDA) approved the incorporation of ClO_2 precursors within food packaging materials to be used as active packaging concepts for meats, poultry, and seafood [74].

Most studies regarding food packaging applications of ClO_2 gas focus on its sanitizing effect against particular microorganisms and/or for specific perishable foods [7, 9, 11, 42, 70]. Few studies have considered the effects of ClO_2 gas on the properties of the food packaging materials and systems [75, 76].

17.3.3 Mass Transfer of ClO_2 in Selected Polymeric Packaging Materials

To consider packaging as a strategy for the release and application of ClO_2 gas requires information on the mass transfer properties of ClO_2 as well as the interaction and compatibility between various polymeric materials and gaseous ClO_2 [77–81].

One of the most promising applications of ClO_2 is as an antimicrobial gas inside the packaging systems for a vapor-phase decontamination, to improve safety and extend the shelf-life of perishable food products [7, 9, 11, 70]. As a "device" to contain and deliver ClO_2 gas to food products, the mass transfer properties of ClO_2 through the package must be known because this will impact the selection of materials with an appropriate gas barrier.

TABLE 17.1 Summary of research studies on antimicrobial efficiencies of ClO_2 for food products and other surfaces

Microorganism	Food product/ Other surface	Treatment[1]	Result (log_{10} reduction)	Reference
Alicyclobacillus acidoterrestris (spore)	*Stainless steel*	50–200 ppm (l) for 1–2 min at 40–90 °C	≤2.17 log CFU/cm^2	[43]
Bacillus cereus	Apple	10–200 μm/mL (l) for 5 min	≤3.79 log CFU/fruit	[44]
	Stainless steel	100–200 μm/mL (l) for 5 min	≤3.71 log CFU/coupon	[45]
Bacillus cereus (spore)	*Stainless steel*	10–200 μm/mL (l) for 5 min	≤2.55 log CFU/coupon	[44]
Bacillus cereus (cell and spore)	*Stainless steel*	100–200 μm/mL (l) for 5 min	≤4.48 log CFU/coupon	[45]
		50–200 μm/mL (l) for 1–5 min	≤4.0 log CFU/coupon	[46]
Bacillus thuringiensis (spore)	Apple	10–200 μm/mL (l) for 5 min	≤4.25 log CFU/fruit	[44]
	Stainless steel	10–200 μm/mL (l) for 5 min	≤2.71 log CFU/coupon	[44]
Cryptosporidium parvum	Basil	4.1 mg/L (g) for 20 min	2.6 log CFU/g	[47]
	Lettuce	4.1 mg/L (g) for 20 min	3.3 log CFU/g	[47]
Encephalitozoon intestinalis (spore)	Basil	4.1 mg/L (g) for 20 min	3.6 log CFU/g	[47]
	Lettuce	4.1 mg/L (g) for 20 min	3.6 log CFU/g	[47]
Enterobacter sakazakii	Apple	10–100 μg/mL (l) for 1–5 min	≥5.46 log CFU/fruit	[48]
	Lettuce	10–100 μg/mL (l) for 1–5 min	≥4.81 log CFU/piece	[48]
	Tomato	10–100 μg/mL (l) for 1–5 min	≥5.31 log CFU/fruit	[48]
Escherichia coli	Apple	0.03–0.30 ppm (g) for 1–20 hours	≤5 log CFU/g	[9]
Escherichia coli O157:H7	Alfalfa seed	10–50 mg/L (l) for 3–10 min	≤1.22 log CFU/g	[49]
	Alfalfa sprout	50 mg/L (l) for 1–10 min	≤2.37 log CFU/g	[50]
		50 mg/L (l) + 0.5 g/100 mL fumaric acid for 1–10 min	≤3.96 log CFU/g	[50]
	Apple	5–40 ppm (l) for 3–10 min + 170-kHz Ultrasonication	≤3.9 log CFU/g	[51]
		3–5 ppm (l) for 5 min + storage at 4 °C	≤5 log CFU/g	[52]
	Apple (calyx, stem cavity and skin)	3–12 mg/L (g) for 10–30 min	≤8 log CFU/site	[53]
	Apple (fresh cut)	3–5 ppm (l) for 5 min + storage at 4 °C	≤5 log CFU/g	[52]

(*Continued*)

TABLE 17.1 (Continued)

Microorganism	Food product/ Other surface	Treatment[1]	Result (log_{10} reduction)	Reference
	Blueberries	4 mg/L (g) for 12 hours	4.5 log CFU/g	[54]
	Cabbage (fresh cut)	1.4–4.1 mg/L (g) for 6–21 min	≤3.13 log CFU/g	[52]
	Cantaloupe	3–5 ppm (l) for 5 min + storage at 4 °C	≤5 log CFU/g	[52]
	Carrot	5–20 mg/L (l) for 1–15 min	≤1.39 log CFU/g	[55]
	Carrot (fresh cut)	0.5–1.0 mg/L (g) for 5–15 min	≤3.08 log CFU/g	[55]
		1.4–4.1 mg/L (g) for 6–21 min	≤5.62 log CFU/g	[41]
	Green pepper	0.15–1.2 mg/L (g) for 30 min	≤7.3 log CFU/pepper	[56]
		5–40 ppm (l) for 3–10 min + 170-kHz Ultrasonication	≤2.5 log CFU/g	[51]
	Lettuce	3–5 ppm (l) for 5 min + storage at 4 °C	≤5 log CFU/g	[52]
		4.3–8.7 mg/L (g) for 30–180 min	≤6.9 log CFU/g	[57]
		5–50 ppm (l) for 10 min	≤1.44 log CFU/g	[58]
	Lettuce (fresh cut)	3–5 ppm (l) for 5 min + storage at 4 °C	≤5 log CFU/g	[52]
		1.4–4.1 mg/L (g) for 6–21 min	≤1.57 log CFU/g	[41]
		5–20 mg/L (l) for 1–15 min	≤0.90 log CFU/g	[55]
		0.5–1.0 mg/L (g) for 5–15 min	≤2.31 log CFU/g	[55]
	Potato	9 ppm (l) for 30–300 min with spray washing	≤1.4 log CFU/fruit	[52]
	Strawberries	3–5 ppm (l) for 5 min + storage at 4 °C	≤5 log CFU/g	[59]
		0.2–4.0 mg/L (g) for 15–30 min	≤5 log CFU/fruit	[60]
		0.6–3.0 mg/L (g) for 10 min	≥5 log CFU/fruit	[60]
		100–200 ppm (l) for 2 min	≤2 log CFU/fruit	[61]
Lactic acid bacteria	Rib eye steak	30–100 ppm (l) for 2 min + vacuum packaging for 4 weeks	≥1.0 log CFU/cm^2	[4]
Lactobacillus buchneri	*Stainless steel coated with epoxy*	2–14 mg/L (g) for 5–120 min	≤6 log CFU	[42]
Leuconostoc mesenteroides	*Stainless steel coated with epoxy*	8–10 mg/L (g) for 30 min	≤6 log CFU	[42]

Listeria monocytogenes	Alfalfa sprout	50 mg/L (l) for 1–10 min	≤2.36 log CFU/g	[50]
		50 mg/L (l) + 0.5 g/100 mL fumaric acid for 1–10 min	≤3.69 log CFU/g	[50]
	Apple	3–5 ppm (l) for 5 min + storage at 4 °C	≤5 log CFU/g	[52]
	Cantaloupe	3–5 ppm (l) for 5 min + storage at 4 °C	≤5 log CFU/g	[52]
	Carrot (fresh cut)	1.4–4.1 mg/L (g) for 10–29 min	≤5.88 log CFU/g	[41]
	Lettuce	3–5 ppm (l) for 5 min + storage at 4 °C	≤5 log CFU/g	[52]
		4.3–8.7 mg/L (g) for 30–180 min	≤5.4 log CFU/g	[57]
	Lettuce (fresh cut)	3–5 ppm (l) for 5 min + storage at 4 °C	≤5 log CFU/g	[52]
		5–50 ppm (l) for 10 min	≤1.20 log CFU/g	[58]
		1.4–4.1 mg/L (g) for 10–29 min	≤1.53 log CFU/g	[41]
		3–5 ppm (l) for 5 min + storage at 4 °C	≤5 log CFU/g	[52]
	Mungbean sprout	100 ppm (l) for 5 min + MAP in 7 days	≤2.08 log CFU/g	[5]
	Strawberries	0.2–4.0 mg/L (g) for 15–30 min or 0.6–3.0 mg/L for 10 min	≥5 log CFU/g	[60]
Pseudomonas	*Stainless steel*	100–200 μm/mL (l) for 5 min	≥6.15 log CFU/coupon	[45]
Salmonella spp.	Apple	5–40 ppm (l) for 3–10 min + Ultrasonication	≤4.3 log CFU/g	[51]
		1.4–4.1 mg/L (g) for 6–25 min	≤4.21 log CFU/g	[41]
	Bell pepper	100 mg (g) for 1 hour	≤5.97 log CFU/fruit	[62]
	Blueberries	4–8 mg/L (g) for 30–120 min	≤3.67 log CFU/g	[70]
		4 mg/L (g) for 12 hours	3.8 log CFU/g	[54]
	Cabbage (fresh cut)	1.4–4.1 mg/L (g) for 10–31 min	≤4.42 log CFU/g	[41
	Carrot (fresh cut)	1.4–4.1 mg/L (g) for 10–31 min	≤5.15 log CFU/g	[41]
	Lettuce	5–40 ppm (l) for 3–10 min + 170-kHz Ultrasonication	up to 3.5 log CFU/g	[51]
	Lettuce (fresh cut)	1.4–4.1 mg/L (g) for 10–31 min	up to 1.58 log CFU/g	[41]
	Onion	1.4–4.1 mg/L (g) for 5–20 min	up to 1.94 log CFU/g	[41]
	Peach	1.4–4.1 mg/L (g) for 5–20 min	up to 3.23 log CFU/g	[41]
	Raspberries	4–8 mg/L (g) for 30–120 min	up to 1.54 log CFU/g	[70]
	Strawberries	4–8 mg/L (g) for 30–120 min	up to 3.76 log CFU/g	[70]

(Continued)

TABLE 17.1 (Continued)

Microorganism	Food product/ Other surface	Treatment[1]	Result (log_{10} reduction)	Reference
	Tomato	1.4–4.1 mg/L (g) for 5–20 min	up to 4.33 log CFU/g	[41]
		5 ppm (l) + spray washing for 10–60 s	up to 5.6 log CFU/cm^2	[63]
Salmonella Montevideo	Strawberries	100–200 ppm (l) for 2 min	less than 2 log CFU/fruit	[61]
Salmonella Typhimurium	Alfalfa sprout	50 mg/L (l) for 1–10 min	≤2.23 log CFU/g	[50]
		50 mg/L (l) + 0.5 g/100 mL fumaric acid for 1–10 min	≤3.57 log CFU/g	[50]
	Chicken (breast)	2.25 mg (g) in 22 days and 6.6 mg (g) in 26 h	1 log CFU/breast less than package without ClO_2 after 15 days of storage	[1]
	Lettuce	4.3–8.7 mg/L (g) for 30–180 min	≤5.4 log CFU/g	[57]
		5–50 ppm (l) for 10 min	≤1.95 log CFU/g	[58]
	Mungbean sprout	100 ppm (l) for 5 min + MAP in 7 days	≤2.76 log CFU/g	[5]
	Potato	9 ppm (l) for 30–300 min with spray washing	≤1.9 log CFU/fruit	[59]
	Tomato	Up to 1 mg/kg fruit (g) for 2 h	≤7 log CFU/fruit	[64]
Total aerobic plate count	Crawfish	10–40 mg/L (l) for 2 min	≤4 log CFU	[65]
		high-pressure water spray at 600 psi for 30 s + 10–40 mg/L (l) for 2 min	≤4.5 log CFU	[65]
	Mungbean sprout	100 ppm (l) for 5 min + MAP in 7 days	≤1.35 log CFU/g	[5]
	Potato	9 ppm (l) for 30–300 min with spray washing	≤1.1 log CFU/fruit	[59]
	Rib eye steak	30–100 ppm (l) for 2 min + vacuum packaging for 4 weeks	≥1.0 log CFU/cm^2	[4]
	Shrimp	high-pressure water spray at 600 psi for 30 s + 10–40 mg/L (l) for 2 min	≤4 log CFU	[65]
Molds	Blueberries	4 mg/L (g) for 12 hours	3.0 log CFU/g	[54]

	Potato	9 ppm (l) for 30–300 min with spray washing	≤0.9 log CFU/fruit	[59]
	Stainless steel coated with epoxy	8–10 mg/L (g) for 10–30 min	≤4 log CFU	[42]
Yeasts	Blueberries	4 mg/L (g) for 12 hours	3.2 log CFU/g	[54]
	Potato	9 ppm (l) for 30–300 min with spray washing	≤1.1 log CFU/fruit	[59]
	Stainless steel coated with epoxy	8–10 mg/L (g) for 10–30 min	≤4 log CFU	[42]
Yeasts and Molds	Apple	1.4–4.1 mg/L (g) for 6–25 min	≤1.68 log CFU/g	[41]
	Blueberries (calyx, stem cavity and skin)	4–8 mg/L (g) for 30–120 min	≤2.78 log CFU/g	[70]
	Onion	1.4–4.1 mg/L (g) for 5–20 min	≤0.22 log CFU/g	[41]
	Peach	1.4–4.1 mg/L (g) for 5–20 min	≤2.65 log CFU/g	[41]
	Raspberries (calyx, stem cavity and skin)	4–8 mg/L (g) for 30–120 min	≤3.18 log CFU/g	[70]
	Strawberries (calyx, stem cavity and skin)	4–8 mg/L (g) for 30–120 min	≤4.16 log CFU/g	[70]
	Tomato	1.4–4.1 mg/L (g) for 6–25 min	≤1.16 log CFU/g	[41]
Penicillium expansum (spore)	*(Suspension)*	3 ppm (l) for 30–300 s	≤4 log CFU/mL	[66]
Poliovirus 1	Strawberries	100–200 ppm (l) for 2 min	≤2 log CFU/fruit	[61]

[1](g) = ClO_2 in a gas phase; (l) = ClO_2 in a liquid/solution form.

Plastic packaging materials, unlike glass, metals, and so on, are relatively permeable to small molecules such as gases, water vapor, organic vapors, and liquids. Plastic packaging materials provide a broad range of mass transfer characteristics, from high to low barrier values. The specific barrier requirements of the packaging system will depend on the product's characteristics and the intended end-use application [82, 83].

Ozen [75] studied the mass transfer of ClO_2 gas through polymeric materials using a quasi-isostatic method. The reported P values of ClO_2 gas, at 20 °C, for linear low-density poly(ethylene) (LLDPE), oriented poly(propylene) (OPP), and biaxially oriented nylon (BON) were 7.62×10^{-16}, 6.21×10^{-17}, and 2.34×10^{-17} kg $ClO_2 \cdot m \cdot m^{-2} \cdot s^{-1} \cdot Pa^{-1}$, respectively. According to the researcher, because steady state was reached before the detection system in use could respond, the assessment of the unsteady region for the calculation of the diffusion coefficient was not possible.

Mass transfer profiles of ClO_2 gas through 10 types of polymeric materials were obtained by Netramai et al. [79] using an isostatic method (Figure 17.10). The profiles represent cumulative permeated gaseous ClO_2 (kg) through each material versus time (s). The materials examined are typically used in packaging systems for perishable food products and various nonperishable and pharmaceutical goods. The corresponding P, D, and S values of the materials as determined at 23 °C are listed in Table 17.2.

Permeability values of gaseous ClO_2 through PET, nylon, biaxially oriented poly(propylene) (BOPP), poly(lactic acid) (PLA), and a multilayer structure of ethylene vinyl acetate (EVA) and EVOH (EVA/EVOH/EVA) were at least one order of magnitude lower than those of PE, PVC, and PS, indicating that the former materials are higher barriers to ClO_2 gas. The differences in the mass transfer properties of these materials could be attributed to the chemical composition of the material and the penetrant, the morphology of the polymer, the concentration of the penetrant, and the presence of co-permeants [84, 85].

A positive empirical correlation was found by the author [86] between the free volume and cohesive energy density ratio (FV/CED), expressed as m^3/kJ, and the P values of ClO_2 gas for polymeric films at 23 °C (Figure 17.11). The free volume (FV) represents the unoccupied spaces between molecules in a polymer structure, and the cohesive energy density (CED) considers the intermolecular forces within the polymer molecules [84, 85]. Polymers with high CED and low FV tend to have a higher barrier to a particular compound as compared with polymers with low CED and high FV. Both characteristics influence packing of polymeric chains that ultimately influences the material properties [85, 87].

The coefficients P, D, and S are temperature dependent and can be described by a Van't Hoff-Arrhenius equation [85]. Increasing the environmental temperature facilitates the diffusion of the penetrant molecules, resulting in increased mass transfer in the system [84, 85]. The activation energy (E_P) of ClO_2 for PET and PLA films is shown in Table 17.3. The higher E_P value of PLA, as compared with that of PET, indicates that the permeation of ClO_2 gas through PLA is less temperature dependent, which could be beneficial in a

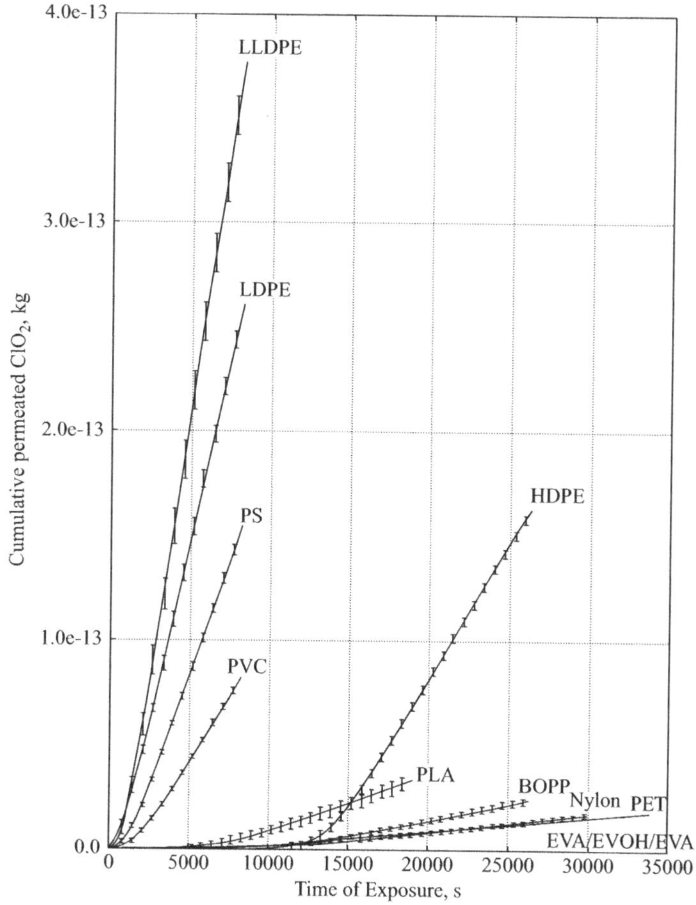

FIGURE 17.10 Mass transfer of 10 mg ClO_2/L ClO_2 gas (3600 ppmV) in polymeric materials at 23 °C. Adapted from Reference 79.

packaging system that is occasionally subjected to temperature fluctuations, such as during transportation or distribution.

17.3.4 Effects of ClO_2 Gas on Properties of Polymeric Materials

Including antimicrobial gas in the package headspace can be a potential complementary approach to improve fresh food safety. However, one important criterion to consider when the gas is a strong oxidizing agent, like chlorine dioxide, is the possibility of chemical changes in the packaging material itself. Such changes may impact the integrity and performance of the packaging material, resulting in a reduction of the packaged product shelf life [77, 78].

Changes in the mechanical, physical, and barrier properties of LDPE, LLDPE, OPP, BON, PS, and nylon after exposure to ClO_2 gas have been

TABLE 17.2 Permeability (P), diffusion (D), and solubility (S) coefficients of 10 mg ClO_2/L ClO_2 gas (3600 ppmV) in selected polymers at 23 °C

Polymer	$P \times 10^{-17} \left(\frac{KgClO_2 \cdot m}{m^2 \cdot s \cdot Pa}\right)$	$D \times 10^{-14} \left(\frac{m^2}{s}\right)$	$S \times 10^{-3} \left(\frac{Kg}{m^3 \cdot Pa}\right)$
HDPE	$24.1 \pm 0.42^{a,1}$	3.14 ± 0.06^{a}	7.68 ± 0.27^{a}
LDPE	66.0 ± 1.09^{b}	26.2 ± 5.11^{b}	$2.59 \pm 0.46^{b,d}$
LLDPE	96.8 ± 1.34^{c}	40.4 ± 2.40^{c}	2.40 ± 0.17^{b}
BOPP	3.04 ± 0.16^{d}	0.39 ± 0.02^{d}	7.84 ± 0.17^{a}
PS	41.8 ± 0.82^{e}	9.15 ± 0.47^{e}	4.57 ± 0.02^{c}
PVC	23.5 ± 0.46^{a}	7.65 ± 0.96^{e}	3.11 ± 0.38^{d}
PET	1.26 ± 0.03^{f}	0.32 ± 0.02^{d}	3.92 ± 0.19^{e}
PLA	5.40 ± 0.13^{g}	2.86 ± 0.18^{a}	1.90 ± 0.15^{f}
Nylon	1.80 ± 0.07^{h}	0.53 ± 0.02^{f}	3.43 ± 0.18^{d}
EVA/EVOH/EVA	Permeability is less than 0.07 $\frac{KgClO_2 \cdot m}{m^2 \cdot s \cdot Pa}$ (24 hours of exposure)		

[1] Within columns, means (±SD) sharing the same superscript letter are not significantly different ($P > 0.05$; $n = 4$).
Adapted from Reference 78.

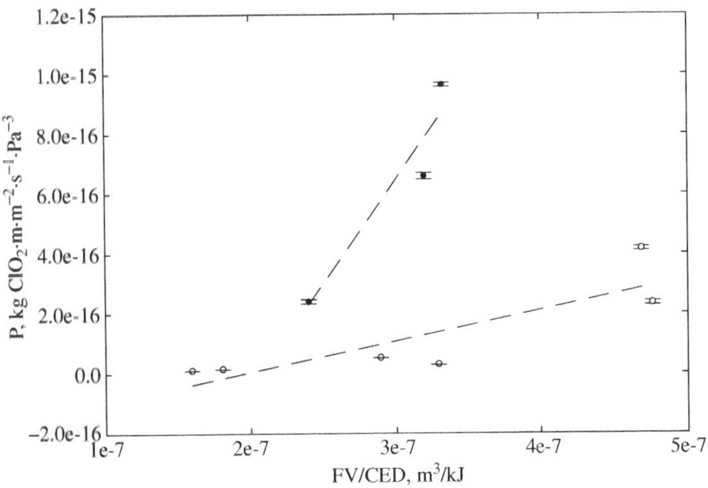

FIGURE 17.11 Correlation between the free volume and cohesive energy density ratio (FV/CED) and the permeability coefficient (P) of ClO_2 in polymeric films.

reported [75–77]. The degree of change varied depending on the gas concentration, relative humidity, and testing temperature [75]. Some chemical and barrier property changes are discussed below.

17.3.4.1 Chemical Changes in Polymers. Netramai et al. [77] reported several possible chemical changes, such as partial chlorination, main chain degradation, change in hydroxyl groups, and formation of carbonyl groups, as observed by comparing Fourier transform infrared (FTIR) spectra of polymers

TABLE 17.3 Permeability (P) coefficients at various temperatures and activation energy for permeation (E_p) of ClO_2 gas in PET and PLA films

	PET		PLA	
Temperature (°C)	$P \times 10^{-17}$ $\left(\frac{KgClO_2 \cdot m}{m^2 \cdot s \cdot Pa}\right)$	E_P (kJ/mol)	$P \times 10^{-17}$ $\left(\frac{KgClO_2 \cdot m}{m^2 \cdot s \cdot Pa}\right)$	E_P (kJ/mol)
23	$1.26 \pm 0.00^{a,1}$	51.05 ± 4.35	5.40 ± 0.13^{a}	129.03 ± 2.82
30	2.67 ± 0.04^{b}		21.5 ± 0.74^{b}	
40	3.98 ± 0.05^{c}		94.4 ± 0.52^{c}	

[1] Within columns, means (±SD) sharing the same superscript letter are not significantly different ($P > 0.05$; $n = 4$).
Adapted from Reference 78.

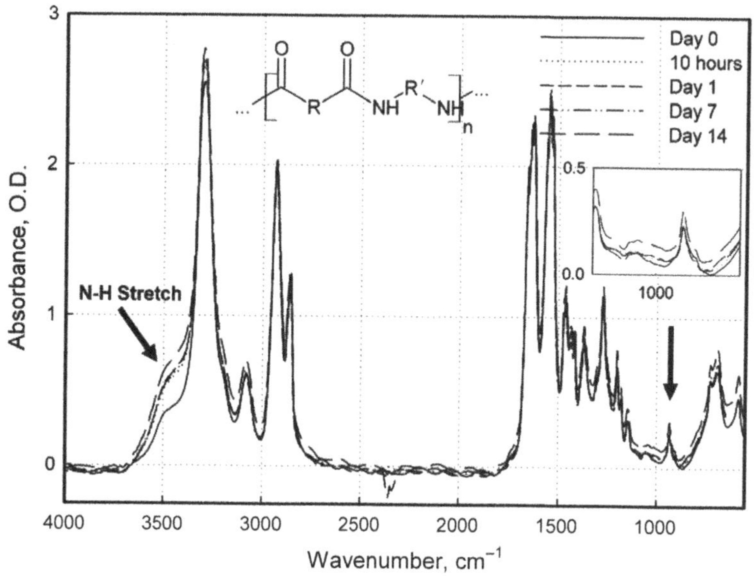

FIGURE 17.12 FT-IR spectra of nylon 6,6 obtained at various exposure time; (—) day 0; (······) 10 hours; (-----) day 1; (----) day 7; and (— · —) day 14. The arrow indicates stretching of N-H. Adapted from Reference 77.

obtained before and after ClO_2 exposure. For many polymers, exposure of ClO_2 gas at the level of 3600 ppm by volume (ppmV) for less than 24 h, which is within the upper range of ClO_2 gas used in the treatment of fresh produce, caused only transient changes in chemical characteristics. However, after exposure for more than 24 h, most of the chemical changes observed from the FTIR spectra tended to be permanent in many polymer samples, even after the removal of the samples from the testing environment. The exception to this behavior was nylon 6,6; permanent changes were observed in all nylon 6,6 samples exposed to ClO_2 gas, regardless of exposure time (Figure 17.12).

Furthermore, increasing the exposure time to ClO_2 gas increased the degree of change within the polymer matrix; however, as the exposure time lengthened, the rate of change decreased [77]. This behavior could be attributed to the reduction of available functional groups for the reaction because the oxidative degradation is typically a surface phenomenon [77, 88, 89].

Chemical changes found in nonpolar polymeric materials, such as polyolefins, after transient or permanent ClO_2 gas exposure, are partial chlorination and main chain degradation. These are the typical alterations reported when treating this type of material with oxidative agents [80, 90]. For polar polymeric materials, changes in the hydroxyl groups and C-N as well as the formation of carbonyl groups would be expected [77, 80].

17.3.4.2 Changes in Barrier Properties of Polymers.

The barrier properties of polymeric materials to vapors and gases are critical to the performance of packaging systems for many food applications. For example, understanding the film permeability to oxygen (O_2) and carbon dioxide (CO_2) is especially important in the packaging of fresh produce that continues to respire during the postharvest period [83, 91]. The concentrations of O_2 and CO_2 that accumulate in the package affect the produce respiration rate, which has a major impact on produce ripening and deterioration. For some modified atmosphere package (MAP) applications, the respiration rate of the product is reduced by starting off with a low O_2 level in the package headspace and selecting packaging material(s) with appropriate barrier properties to O_2 and CO_2 that create conditions to extend the shelf life of the packaged product. Once the material is selected for a particular commodity, it is critical that its permselectivity ratio (P_{CO_2}/P_{O_2}) is maintained throughout the shelf life of the product [83, 84, 91].

To utilize ClO_2 gas along with other typical headspace gases in a MAP system, the impact of ClO_2 gas on the barrier characteristics of polymeric materials to O_2, CO_2, and water vapor should be known. Netramai et al. [77] studied the effects of ClO_2 gas exposure on O_2, CO_2, and moisture permeability for several polymers. The changes in barrier properties, after the materials were exposed to 3600 ppmV of ClO_2 gas for 14 days, which is the general shelf life for many MAP products, are listed in Table 17.4.

After exposure to ClO_2 gas, the O_2 barrier of the nylon increased, which was in agreement with an earlier study on the effects of ozone treatment on biaxially oriented nylon [80]. In other polymer materials, the barrier properties to O_2 and CO_2 tended to decrease after ClO_2 exposure. Decreased barrier properties are likely the result of main chain scission, which would result in increased chain mobility and facilitate the transfer of gases through the polymer structure [81, 84, 88]. A comparison of the permselectivity ratios of the polymeric materials at day 0 and after 14 days of ClO_2 gas treatment, showed changes ranging up to 47%, with the most notable change occurring in the exposed HDPE film [77]. Table 17.4 also presents some results from that study for selected polymers.

TABLE 17.4 Changes in barrier properties of selected polymeric packaging materials exposed to gaseous ClO_2

Sample	Time of exposure (day)	$P_{H_2O} \times 10^{-18}$ $\left(\frac{Kg \cdot m}{m^2 \cdot s \cdot Pa}\right)$	$P_{CO_2} \times 10^{-18}$ $\left(\frac{Kg \cdot m}{m^2 \cdot s \cdot Pa}\right)$	$P_{O_2} \times 10^{-18}$ $\left(\frac{Kg \cdot m}{m^2 \cdot s \cdot Pa}\right)$	P_{CO_2}/P_{O_2}	% change
PE (HDPE)	0	0.49 ± 0.00^{a},[1]	58.28 ± 8.08^{a}	9.26 ± 1.24^{a}	6.30 ± 0.32^{a}	
	14	0.48 ± 0.00^{b}	77.56 ± 0.40^{b}	8.40 ± 0.29^{a}	9.25 ± 0.28^{b}	+46.77
PS	0	10.5 ± 0.74^{a}	111.07 ± 6.63^{a}	16.76 ± 0.57^{a}	6.58 ± 0.17^{a}	
	14	9.84 ± 1.06^{a}	123.21 ± 6.37^{b}	15.82 ± 0.64^{a}	7.77 ± 0.19^{b}	+17.95
BOPP	0	0.24 ± 0.00^{a}	5.30 ± 0.27^{a}	3.86 ± 0.20^{a}	4.75 ± 0.15^{a}	
	14	0.25 ± 1.01^{a}	5.38 ± 0.25^{a}	3.91 ± 0.18^{a}	4.91 ± 0.25^{a}	—
PLA	0	28.2 ± 1.18^{a}	37.32 ± 17.2^{a}	21.22 ± 3.25^{a}	1.73 ± 0.64^{a}	
	14	26.9 ± 1.56^{a}	43.39 ± 13.9^{a}	30.76 ± 12.34^{a}	1.41 ± 0.05^{a}	—
PET	0	1.64 ± 0.01^{a}	1.54 ± 0.04^{a}	0.22 ± 0.01^{a}	6.88 ± 0.13^{a}	
	14	1.94 ± 0.03^{b}	1.53 ± 0.04^{a}	0.26 ± 0.04^{a}	5.85 ± 0.65^{b}	−14.86
Nylon	0	n/a	1.23 ± 0.01^{a}	0.24 ± 0.01^{a}	5.08 ± 0.08^{a}	
	14	n/a	1.21 ± 0.03^{a}	0.22 ± 0.00^{b}	5.63 ± 0.06^{b}	+10.92
EVA/EVOH/EVA	0	0.40 ± 0.01^{a}	0.35 ± 0.04^{a}	0.10 ± 0.01^{a}	3.86 ± 0.04^{a}	
	14	0.41 ± 0.01^{a}	0.71 ± 0.04^{b}	0.13 ± 0.01^{b}	5.30 ± 0.24^{b}	+37.19

[1] Within columns, means (±SD) sharing the same superscript letter are not significantly different ($P > 0.05$; $n = 4$).
Adapted from Reference 92.

17.4 FUTURE TRENDS

Because of their busy lifestyle and increased awareness of the importance of health-promoting wholesome foods, consumers today prefer products that are easy to prepare, safe, and of high quality. These are some factors that drive the demand for the minimally processed food products (e.g., ready-to-eat fresh cut fruits and vegetables, fresh pastas, ready-to-bake bakery, etc.). This trend can present some challenges during the distribution because of the reduced stability of these products. Advanced packaging systems, such as those based on AITC and ClO_2 antimicrobials, are expected to experience significant growth because they are ideal vehicles for the delivery of antimicrobial agents to the surface, where postprocess contamination is most likely to occur. Foods are complex heterogeneous systems in which components may interact with the antimicrobials, which may present synergistic or antagonistic effects toward preserving the quality of the products. Moreover, the mechanical and thermal abuses likely to occur during distribution may also affect the efficacy of the controlled-release mechanisms. Therefore, further investigations involving the target products and using the conditions that are likely to occur in the distribution chain are important to demonstrate the robustness of the antimicrobial delivery system. From a consumer acceptance standpoint, the perception and willingness to accept new technologies, along with the approval by food and health legislative agencies, will ultimately determine the success or failure of active packaging technology.

Recent research in active packaging that need to be further developed include (1) incorporation of new organic or inorganic compounds into the packaging system; (2) combination of other processing or preservation means with antimicrobial packaging in hurdle technology; and (3) improvement of an active gas delivery system though packaging redesign [2, 3, 7, 91]. The packaging system needs to be designed in a manner that facilitates the uniform distribution of the antimicrobial gas within the package; this will maximize the contact between food products and the gases in an even fashion.

REFERENCES

1. Ellis, M., Cooksey, K., Dawson, P., Han, I., Vergano, P. (2006). Quality of fresh chicken breasts using a combination of modified atmosphere packaging and chlorine dioxide sachets. *Journal of Food Protection*, *69*, 1991–1996.
2. Appendini, P., Hotchkiss, J. H. (2002). Review of antimicrobial food packaging. *Innovative Food Science & Emerging Technologies*, *3*, 113–126.
3. Vermeiren, L., Devlieghere, F., Van Beest, M., De Krujif, N., Debevere, J. (1999). Developments in the active packaging of foods. *Trends in Food Science & Technology*, *10*, 77–86.
4. Unda, J., Molins, R. A., Zamojcin, C. A. (2006). Sanitization of fresh rib eye steaks with chlorine dioxide-generating binary systems. *Journal of Food Science*, *54*, 7–10.

5. Jin, H., Lee, S. Y. (2007). Combined effect of aqueous chlorine dioxide and modified atmosphere packaging on inhibiting *Salmonella Typhimurium* and *Listeria monocytogenes* in mungbean sprouts. *Journal of Food Science, 72*, M441–M445.
6. Multon, J. L. (1996). *The Role of Packaging in Preserving Foodstuffs*, Volome 1, 1st Edition. VCH Publishers, Inc., New York, NY.
7. Cooksey, K. (2005). Effectiveness of antimicrobial food packaging materials. *Food Additives & Contaminants: Part A, 22*, 980–987.
8. Ooraikul, B., Stiles, M. E. (1991). *Modified Atmosphere Packaging of Food*, 1st Edition, Ellis Horwood, New York, NY.
9. Sapers, G., Walker, P. N., Sites, J. E., Annous, B. A., Eblen, D. R. (2003). Vapor-phase decontamination of apples innoculated with *Escherichia coli*. *Journal of Food Science, 68*, 1003–1007.
10. Kim, J., Huang, T.-S., Marshall, M. R., Wei, C.-I. (1999). Chlorine dioxide treatment of seafoods to reduce bacterial loads. *Journal of Food Science, 64*, 1089–1093.
11. Kaczur, J., Cawlfield, D.W. (1992). Chlorine oxygen acids and salts, chlorous acid, chlorites, and chlorine dioxide. In *Kirk-Othmer Encyclopedia of Chemical Technology*. Ed.: Kroschwitz, J. I. Wiley, New York, NY.
12. Gómez-López, V., Rajkovic, A., Ragaert, P., Smigic, N., Devlieghere, F. (2009). Chlorine dioxide for minimally processed produce preservation: A review *Trends in Food Science & Technology, 20*, 17–26.
13. Novak, J., Demirci, A., Han, Y. (2008). Novel chemical processes: Ozone, supercritical CO2, electrolyzed oxidizing water, and chlorine dioxide gas. *Food Science and Technology International, 14*, 437–441.
14. Lin, C., Kim, J., Du, W. X., Wei, C. I. (2000). Bactericidal activity of isothiocyanate against pathogens on fresh produce. *Journal of Food Protection, 63*, 25–30.
15. Lin, C., Preston, J. F., Wei, C. I. (2000). Antibacterial mechanism of allyl isothiocyanate. *Journal of Food Protection, 63*, 727–734.
16. Delaquis, P., Sholberg, P. L. (1997). Antimicrobial activity of gaseous allyl isothiocyanate. *Journal of Food Protection, 60*, 943–947.
17. Delaquis, P., Mazza, G. (1995). Antimicrobial properties of isothiocyanates in food preservation. *Food Technology, 49*, 73–84.
18. Park, C., Taormina, P. J., Beuchat, L. R. (2000). Efficacy of allyl isothiocyanate in killing enterohemorrhagic Escherichia coli O157: H7 on alfalfa seeds. *International Journal of Food Microbiology 56*, 13–20.
19. Kim, Y. S., Ahn, E. S., Shin, D. H. (2002). Extension of shelf life by treatment with allyl isothiocyanate in combination with acetic acid on cooked rice. *Journal of Food Science, 67*, 274–279.
20. Nadarajah, D., Han, J. H., Holley, R. A. (2005). Inactivation of Escherichia coli O157: H7 in packaged ground beef by allyl isothiocyanate. *International Journal of Food Microbiology, 99*, 269–279.
21. Nielsen, P., Rios, R. (2000). Inhibition of fungal growth on bread by volatile components from spices and herbs, and the possible application in active packaging, with special emphasis on mustard essential oil. *International Journal of Food Microbiology, 60*, 219–229.
22. Isshiki, K., Tokuoka, K., Mori, R., Chiba, S. (1992). Preliminary examination of allyl isothiocyanate vapor for food preservation. *Bioscience, Biotechnology, and Biochemistry, 56*, 1476–1477.

23. Tsunoda, K. (1994). Effect of gaseous treatment with allyl isothiocyanate on the control of microbial growth on a wood substrate. *Journal of Antibacterial and Antifungal Agents*, *22*, 145–148.
24. Robertson, G. (2006). Active and intelligent packaging. In: *Food Packaging*. Ed.: Robertson, G. CRC Press, Boca Raton, FL.
25. Sekiyama, Y., Mizukami, Y., Takada, A. (1995). Corrosiveness of allyl isothiocyanate towards metals, rubbers and plastics and ability of allyl isothiocyanate vapor to permeate plastic films. *Journal of the Food Hygienic Society of Japan*, *36*, 375–382.
26. Winther, M., Nielsen, P. V. (2006). Active packaging of cheese with allyl isothiocyanate, an alternative to modified atmosphere packaging. *Journal of Food Protection*, *69*, 2430–2435.
27. Feng, J. J. (2002). The stretching of an electrified non-Newtonian jet – a model for electrospinning. *Physics of Fluid*, *14*, 3912–3926.
28. Carroll, C., Joo, Y. L. (2006). Electrospinning of viscoelastic Boger fluids: modeling and experments. *Physics of Fluid*, *18*, 1–14.
29. Reneker, D., Yarin, A. L., Fong, H., Koombhongse, S. (2000). Bending instability of electrically charged liquid jets of polymer solutions in electrospinning. *Journal of Applied Physics*, *87*, 4531–4547.
30. Vega-Lugo, A., Lim, L.-T. (2009). Controlled release of allyl isothiocyanate using soy protein and poly(lactic acid) electrospun fibers. *Food Research International*, *42*, 933–940.
31. Vega Lugo, A. C. (2008). Electrospinning of soy protein isolate and poly(lactic acid) ultrafine fibers for controlled release of allyl isothiocyante Ph.D. Dissertation.: University of Guelph, Guelph, ON, Canada.
32. Mari, M., Leoni, O., Lori, R., Cembali, T. (2002). Antifungal vapour-phase activity of allyl-isothiocyanate against Penicillium expansum on pears. *Plant Pathology*, *51*, 231–236.
33. Lim, L.-T., Tung, M. A. (1997). Vapor pressure of allyl isothiocyanate and its transport in PVDC/PVC copolymer packaging film. *Journal of Food Science*, *62*, 1061–1066.
34. Lewis, R. J. (1992). *Sax's Dangerous Properties of Industrial Materials*, Volume 2, 8th Edition. Van Nostrand Reinhold, New York, NY.
35. Robertson, G. (1993). *Food Packaging Principles and Practices*. Mercel Dekker, Inc., New York.
36. Lim, L. -T., Britt, I. J., Tung, M. A. (1998). Sorption and permeation of allyl isothiocyanate vapor in nylon 6,6 film as affected by relative humidity. *Journal of Plastic Film and Sheeting*, *14*, 207–225.
37. US Environmental Protection Agency. (1999). Chlorine dioxise. In: *Alternative Disinfectants and Oxidants Guidance Manual*. p. 4-1-4-41.
38. Erickson, M., Ortega, Y. R. (2006). Inactivation of protozoan parazites in food, water, and environmental systems. *Journal of Food Protection*, *69*, 2786–2808.
39. Bernarde, M., Snow, W. B., Olivieri, V. P. (1967). Kinetics and mechanism of bacterial disinfection by chlorine dioxide *Journal of Applied Microbiology*, *15*, 257–265.
40. Aieta, E., Berg, J. D. (1986). A review of chlorine dioxide in drinking water treatment. *Journal AWWA*, *78*, 62–72.

41. Sy, K., Murray, M. B., Harrison, M. D., Beuchat, L. R. (2005). Evaluation of gaseous chlorine dioxide as a sanitizer for killing Salmonella, Escherichia Coli O157: H7, listeria monocytogenes, and yeasts and molds on fresh and fresh-cut produce. *Journal of Food Protection*, *68*, 1176–1187.
42. Han, Y., Guentert, A. M., Smith, R. S., Linton, R. H., Nelson, P. E. (1999). Efficacy of chlorine dioxide gas as a sanitizer for tanks used for aseptic juice storage. *Food Microbiology*, *16*, 53–61.
43. Podolak, R., Elliott, P. H., Taylor, B. J., Khurana, A., Black D. G. (2009). Destruction of *Alicyclobacillus acidoterrestris* spores in apple juice on stainless steel surfaces by chemical disinfectants. *Journal of Food Protection*, *72*, 510–514.
44. Kreske, A., Ryu, J. H., Pettigrew, C. A., Beuchat, L. R. (2006). Lethality of chlorine, chlorine dioxide, and a commercial produce sanitizer to bacillus cereus and pseudomonas in a liquid detergent, on stainless steel, and in biofilm. *Journal of Food Protection*, *69*, 2621–2634.
45. Kreske, A., Ryu, J.-H., Beuchat, L. R. (2006). Evaluaion of chlorine, chlorine dioxide, and a peroxyacetic acid-based sanitizer for effectiveness in killing *Bacillus cereus* and *Bacillus thuringiensis* spores in suspensions, on the surface of stainless steel, and on apples. *Journal of Food Protection*, *69*, 1892–1903.
46. Ryu, J., Beuchat, L. R. (2005). Biofilm formation and sporulation by bacillus cereus on a stainless steel surface and subsequent resistance of vegetative cells and spores to chlorine, chlorine dioxide, and a peroxyacetic acid-based sanitizer. *Journal of Food Protection*, *68*, 2614–2622.
47. Ortega, Y., Mann, A., Torres, M. P., Cama, V. (2008). Efficacy of gaseous chlorine dioxide as a sanitizer against cryptosporidium parvum, cyclospora cayetanensis, and encephalitozoon intestinalis on produce. *Journal of Food Protection*, *71*, 2410–2414.
48. Kim, H., Ryu, J.-H., Beuchat, L. R. (2006). Survival of *Enterobacter Sakazakii* on fresh produce as affected by temperature, and effectiveness of sanitizers for its elimination. *International Journal of Food Microbiology*, *111*, 134–143.
49. Singh, N., Singh R. K., Bhunia, A. K. (2003). Sequential disinfection of Escherichia coli O157:H7 inoculated alfalfa seeds before and during sprouting using aqueous chlorine dioxide, ozonated water, and thyme essential oil. *Lebensmittel-Wissenschaft und-Technologie*, *36*, 235–243.
50. Kim, Y., Kim, M., Song, K. B. (2009). Combined treatment of fumaric acid with aqueous chlorine dioxide or UV-C irradiation to inactivate Escherichia coli O157: H7, Salmonella enterica serovar Typhimurium, and Listeria monocytogenes inoculated on alfalfa and clover sprouts. *Lebensmittel-Wissenschaft und-Technologie*, *42*, 1654–1658.
51. Huang, T.-S., Xu, C., Walker, K., West, P., Zhang, S., Weese, J. (2006). Decontamination efficacy of combined chlorine dioxide with ultrasonication on apples and lettuce. *Journal of Food Science*, *71*, M134–M139.
52. Rodgers, S., Cash, J. N., Siddiq, M., Ryser, E. T. (2004). A comparison of different chemical sanitizers for inactivating Escherichia coli O157:H7 and Listeria monocytogenes in solution and on apples, lettuce, strawberries and cantaloupe. *Journal of Food Protection*, *67*, 721–731.
53. Du, J., Han, Y., Linton, R. H. (2003). Efficacy of chlorine dioxide gas in reducing *Escherichia coli* O157:H7 on apple surfaces. *Food Microbiology*, *20*, 583–591.

54. Popa, I., Hanson, E. J., Todd, E. C., Schilder, A. C., Ryser, E. T. (2007). Efficacy of chlorine dioxide gas sachets for enhancing the microbiological quality and safety of blueberries. *Journal of Food Protection, 70*, 2084–2088.
55. Singh, N., Singh, R. K., Bhunia, A. K., Stroshine, R. L. (2002). Efficacy of chlorine dioxide, ozone, and thyme essential oil or a sequential washing in killing Escherichia coli O157:H7 on lettuce and baby carrots. *Lebensmittel-Wissenschaft und-Technologie, 35*, 720–729.
56. Han, Y., Linton, R. H., Nielsen, S. S., Nelson, P. E. (2000). Inactivation of *Escherichia coli* O157:H7 on surface-uninjured and -injured green pepper (*Capsicum annuum* L.) by chlorine dioxide gas as demonstrated by confocal laser scanning microscopy. *Food Microbiology, 17*, 643–655.
57. Lee, S., Costello, M., Kang, D. H. (2004). Efficacy of chlorine dioxide gas as a sanitizer of lettuce leaves. *Journal of Food Protection, 67*, 1371–1376.
58. Kim, Y., Lee, S. H., Park, J., Park, J., Chung, M., Kwon, K., Chung, K., Won, M., Song, K. B. (2008). Inactivation of *Escherichia coli* O157:H7, *Salmonella typhimurium*, and *Listeria monocytogenes* on stored iceberg lettuce by aqueous chlorine dioxide treatment. *Journal of Food Science, 73*, 18–22.
59. Park, E., Gray, P. M., Oh, S. W., Kronenberg, J., Kang, D. H. (2008). Efficacy of FIT produce wash and chlorine dioxide on pathogen control in fresh potatoes. *Journal of Food Science, 73*, M278–M282.
60. Han, Y., Selby, T. L., Schultze, K. K., Nelson, P. E., Linton, R. H. (2004). Decontamination of strawberries using batch and continuous chlorine dioxide gas treatments. *Journal of Food Protection, 67*, 2450–2455.
61. Lukasik, J., Bradley, M. L., Scott, T. M., Dea, M., Koo, A., Hsu, W. Y., Bartz, J. A., Farrah, S. R. (2003). Reduction of poliovirus 1, bacteriophages, *Salmonella montevideo*, and *Escherichia coli* O157:H7 on strawberries by physical and disinfectant washes. *Journal of Food Protection, 66*, 188–193.
62. Du, J., Han, Y., Linton, R. H. (2002). Inactivation by chlorine dioxide gas (ClO2) of Listeria monocytogenes spotted onto different apple surfaces. *Food Microbiology, 19*, 481–490.
63. Yuk, H.-G., Bartz, J. A., Schneider, K. R. (2006). The effectiveness of sanitizer treatments in inactivation of *Salmonella spp.* from bell pepper, cucumber, and strawberry. *Journal of Food Science, 71*, M95–M99.
64. Pao, S., Kelsey, D. F., Long, III, W. (2009). Spray washing of tomatoes with chlorine dioxide to minimize Salmonella on inoculated fruit surfaces and cross-contamination from revolving brushes. *Journal of Food Protection, 72*, 2448–2452.
65. Mahovic, M., Bartz, J. A., Schneider, K. R., Tenney, J. D. (2009). Chlorine dioxide gas from an aqueous solution: Reduction of *Salmonella* in wounds on tomato fruit and movement to sinks in a treatment chamber. *Journal of Food Protection, 72*, 952–958.
66. Andrews, L., Key, A. M., Martin, R. L., Grodner, R., Park, D. L. (2002). Chlorine dioxide wash of shrimp and crawfish an alternative to aqueous chlorine. *Food Microbiology, 19*, 261–267.
67. Taylor, J., Rogers, S. J., Holah, J. T. (1999). A comparison of the bactericidal efficacy of 18 disinfectants used in the food industry against *Escherichia coli* O157:H7 and *Pseudomonas aeruginosa* at 10 and 20 °C. *Journal of Applied Microbiology, 87*, 718–725.

68. Yuk, H. -G., Bartz, J. A., Schneider, K. R. (2005). Effectiveness of individual or combined sanitizer treatments for inactivating Salmonella spp. on smooth surface, stem scar, and wounds of tomatoes. *Journal of Food Science*, *70*, M409–M414.1
69. Ölmez, H., Kretzschmar, U. (2009). Potential alternative disinfection methods for organic fresh-cut industry for minimizing water consumption and environmental impact. *LWT - Food Science and Technology*, *42*, 686–693.
70. Sy, K., McWatters, K. H., Beuchat, L. R. (2005). Efficacy of gaseous chlorine dioxide as a sanitizer for killing salmonella, yeasts, and molds on blueberries, strawberries, and raspberries. *Journal of Food Protection*, *68*, 1165–1175.
71. Stopforth, J., O'Connor, R., Lopes, M., Kottapalli, B., Hill, W. E., Samadpour, M. (2007). Validation of individual and multiple-sequential interventions for reduction of microbial populations during processing of poultry carcasses and parts. *Journal of Food Protection*, *70*, 1393–1401.
72. Lin, W.-F., Huang, T.-S., Cornell, J. A., Lin, C. M., Wei, C.-I. (2006). Bactericidal activity of aqueous chlorine and chlorine dioxide solutions in a fish model system. *Journal of Food Science*, *61*, 1030–1034.
73. Shin, J. (2001). Evaluation of the effect of chlorine dioxide and allyl-isothiocyanate on the growth of salmonella typhimurium and listeria monocytogenes on fresh chicken breast and effect of chlorine dioxide exposure on the physical properties of plastic films. Ph.D. Dissertation. Michigan State University, East Lansing, MI.
74. FDA. Agency Response Letter GRAS Notice No. GRN 000062. Services, Ed.
75. Ozen, B. F. (2000). Effect of ozone and chlorine dioxide treatments used in perishable food applications on polymeric materials: changes in mechanical, thermal and mass transfer properties. Ph.D. Dissertation. Purdue University, West Lafayette, IN.
76. Shin, J., Ryser, E., Selke, S., Harte, B. (2006). Evaluation of the microbial inhibition performance of gas type antimicrobials (chlorine dioxide and allyl-isothiocyanate) with modified atmosphere packaging (MAP), and the effct of chlorine dioxide exposure on the physical properties of plastic films. *The 15th IAPRI World Conference on Packaging*. Tokyo, Japan: International Association of Packaging Research Institutes.
77. Netramai, S., Rubino, M., Auras, R., Annous, B. A. (2010). Effect of chlorine dioxide gas on physical, thermal, mechanical, and barrier properties of polymeric packaging materials. *Journal of Applied Polymer Science*, *115*, 1742–1750.
78. Ozen, B., Floros, J. D. (2001). Effects of emerging food processing techniques on the packaging materials. *Trends in Food Science & Technology*, *12*, 60–67.
79. Netramai, S., Rubino, M., Auras, R., Annous, B. A. (2009). Mass transfer study of chlorine dioxide gas through polymeric packaging materials. *Journal of Applied Polymer Science*, *114*, 2929–2936.
80. Ozen, B., Mauer, L. J., Floros, J. D. (2002). Effects of ozone exposure on the structural, mechanical and barrier properties of select plastic packaging films. *Packaging Technology and Science*, *15*, 301–311.
81. Kulshreshtha, A. K. (1992). 3: Chemical degradation. In: Handbook of Polymer Degradation. Eds.: Hamid, H., Amin, M. B., Maadhah, A. G. p. 55. Marcel Dekker, New York, NY.
82. Taub, I., Singh, R. P. (1988). *Food Storage Stability*. CRC Press LLC, Boca Raton, FL.

83. Sandhya (2010). Modified atmosphere packaging of fresh produce: Current status and future needs. *LWT - Food Science and Technology*, *43*, 381–392.

84. Selke, S., Culter, J. D., Hernandez, R. J. (2004). *Plastics Packaging: Properties, Processing, Applications, and Regulations*, 2nd Edition. Hanser Gardner Publications, Inc., Cincinnati, OH.

85. Van Krevelen, D. W. (1997). *Properties of Polymers*, 3rd Edition. Elsevier Science B.V., New York, NY.

86. Netramai, S., Correlation between free volume and cohesive energy density ratio and the permeability coefficient of chlorine dioxide gas for polymeric films. In: *The Use of Chlorine Dioxide Gas in Food Packaging Application*. Unpublished data School of Packaging Michigan State University, East Lansing, MI.

87. Alentiev, A., Yampolskii, Y. P. (2002). Meares equation and the role of cohesion energy density in diffusion in polymers. *Journal of Membrane Science*, *206*, 291–306.

88. Rodriguez, F., Cohen, C., Ober, C. K., Archer, L. A. (2003). *Principles of Polymer Systems*, 5th Edition. Taylor & Francis, New York, NY.

89. Rivaton, A., Gardette, J.-L. (1998). Photo-oxidation of aromatic polymers. *Die Angewandte Makromolekulare Chemie*, *261/262*, 173–188.

90. Walzak, M., Flynn, S., Foerch, R., Hill, J.M., Karbashewski, E., Lin, A., Strobel, M. (1995). UV and ozone treatment of polypropylene and poly(ethylene terephthalate). *Journal of Adhesion Science and Technology*, *9*, 1229–1248.

91. Martinez-Romero, D., Guillen, F., Castillo, S., Valero, D., Serrano, M. (2003). Modified atmosphere packaging maintains quality of table grapes. *Journal of Food Science*, *68*, 1838–1843.

92. Okull, D., Demirci, A., Rosenberger, D., LaBorde, L. F. (2006). Susceptibility of *Penicillium expansum* spores to sodium hypochlorite, electrolyzed oxidizing water, and chlorine dioxide solutions modified with nonionic surfactants. *Journal of Food Protection*, *69*, 1944–1948.

CHAPTER 18

CURRENT LEGISLATION IN ANTIMICROBIALS

JAVIER GÓMEZ
Treqtop Services, S.L., Valencia, Spain

CONTENTS

18.1 Introduction 490
18.2 European Union 493
 18.2.1 Regulatory Frame 493
 18.2.2 Biocidal Product Types and Their Descriptions on Directive 98/8 . 495
 18.2.3 Directive 98/8 Milestones. 497
 18.2.4 Directive 98/8. Actual Status 498
 18.2.5 Directive 98/8. Actual Implementation Guidelines 500
 18.2.6 Updated Information 501
 18.2.7 New Regulations Coming. 502
 18.2.8 Food Contact Applications 505
18.3 United States 507
 18.3.1 Regulatory Frame 507
 18.3.2 Definitions 508
 18.3.3 Types of Pesticides 508
 18.3.4 Exclusions 511
 18.3.5 EPA Registration Process 511
 18.3.6 EPA Registration Process for Antimicrobial Pesticides 512
 18.3.7 Claims, Labeling, and Labels 514
 18.3.8 Tests Requested 514
 18.3.9 Active and Inert Ingredients 520

Antimicrobial Polymers, First Edition.
Edited by José M. Lagarón, María J. Ocio, and Amparo López-Rubio.

18.3.10 Utilities. 520
18.3.11 Treated Article Exemptions 521
18.3.12 Food Contact Applications 521
18.4 Websites Listed 524

18.1 INTRODUCTION

Biocidal products contain or generate active substances and are used against harmful organisms such as pests and bacteria. They can be used to protect both human and animal health. They include household products such as disinfectants, rodenticides, repellents, and insecticides. Others are used in more industrial applications as wood and material preservatives, antifouling paints, and embalming products to avoid damage to natural or manufactured products.

Because of their intrinsic properties and uses, biocidal products may themselves pose health risks and be harmful to the environment. Therefore, it is vital to ensure that only biocidal products safe for use are placed on the market.

In most developed countries, all the chemical components of a biocide must be listed on the relevant national inventory of chemical substances before it can be supplied. See Table 18.1. In addition, the usage of biocides in precise applications such as for skin contact will always be subject to additional legislation that often requires a specific product registration.

Depending on the biocidal substance and on the intended application, the registration can be more or less complex in terms of the necessary documentation to submit and in terms of the time needed for the registration process.

In the United States, for example, biocides must be listed before on the national inventory of chemical substances, which is called the Toxic Substances Control Act (TSCA). Alongside the listing on the TSCA, there are regulations governing the use of biocides on both the federal (Environmental Protection Agency [EPA]) and the state level.

In Europe, the regulations governing the use of biocides, at the moment, are on a European Union (EU) level for the active substance but on a country level for registration of the biocide formulation intended to be marketed. For the rest of world, many countries have their own regulations governing the use of biocides; also for some countries, a recognized registration like EPA can be valid without the need for extra registration.

Figure 18.1 illustrates which countries regulate the use of biocides. For countries not listed, this should not be interpreted as meaning that sales are permitted without further regulatory action as some of those countries may require an import certificate of some sort or letter of no objection.

The approval by the regulatory body can take up to two years and requires suppliers to make a substantial commitment to support the registration process in terms of documentation regarding the physical, chemical, and toxicological properties of the biocide.

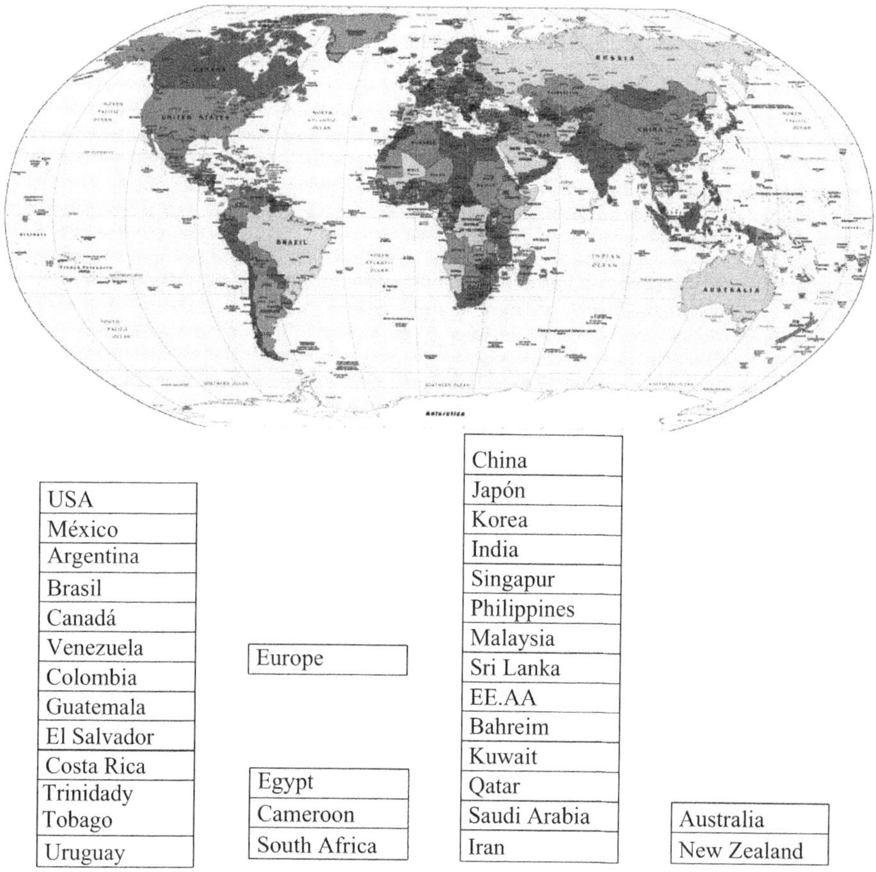

FIGURE 18.1 Countries regulating the use of biocides.

TABLE 18.1 Main inventories of chemical substances around the world

Country	Inventory
Australia	AICS
EU	ESIS
China	IECSC
Japan	ENCS
Korea	KECI
Philippines	PICCS
USA	TSCA
New Zealand	NZIoC

Table 18.1 illustrates some of the main inventories of chemical substances around the world that also have to be kept in mind before registration of biocidal products.

TABLE 18.2 Different terms often used when talking about biocides

Main legal denomination	Area	Other denominations	Legal definition	Remarks
Biocidal products	European Union	Antimicrobial	Active substances and preparations containing one or more active substances, put up in the form in which they are supplied to the user, intended to destroy, deter, render harmless, prevent the action of, or otherwise exert a controlling effect on any harmful organism by chemical or biological means.	It does not refer to plant protection products
Plant protection products	European Union	Pesticide (insecticide, herbicide, fungicide)	Active substances and preparations containing one or more active substances, put up in the form in which they are supplied to the user, intended to protect plants or plant products against all harmful organisms or prevent the action of such organisms; influence the life processes of plants, other than as a nutrient (e.g., growth regulators); preserve plant products; destroy undesired plants; or destroy parts of plants, check or prevent undesired growth of plants.	Only plant protection
Pesticides	USA / EPA	Insecticide, herbicide, fungicide	A pesticide is any substance or mixture of substances intended for preventing, destroying, repelling, or mitigating any pest. A pesticide is also any substance or mixture of substances intended for use as a plant regulator, defoliant, or desiccant.	Includes antimicrobials or biocides
Antimicrobial pesticides	USA		Antimicrobial pesticides are substances or mixtures of substances used to destroy or suppress the growth of harmful microorganisms whether bacteria, viruses, or fungi on inanimate objects and surfaces.	It is a subclass of pesticides

In this chapter, we will describe the legal requirements existing at the moment in the European Union and the United States.

Different terms are often used when talking about biocides. Some of them can lead to a misunderstanding, so it is convenient to establish some definitions in order to understand the different existing regulations and the scope of their application. These definitions are compiled in Table 18.2.

18.2 EUROPEAN UNION

18.2.1 Regulatory Frame

As stated, each European country currently regulates the use of biocides within their region, but this is changing with the European Biocidal Product Directive.

The European Biocidal Product Directive (Directive 98/8/EC of the European Parliament and of the Council of 16 February 1998 concerning the placing of biocidal products on the market) determines a step-wise assessment of existing biocidal active substances placed on the market before May 14, 2000. Only those active substances that conform to the requirements are included in the active substances lists of Annex I (authorized active substances) of the directive. The existing active substances will be evaluated up to the year 2013.

Biocidal products are assigned to 1 of 23 different product types (PTs), which are described in Annex V of the EU Directive. These PTs are divided into four main groups:

1. Disinfectants and general biocidal products (human hygiene, private and public health area, veterinary hygiene, food and feed area, and drinking water disinfectants).
2. Preservatives (e.g., in-can or film preservatives, wood preservatives, preservatives for fibers, leather, rubber, masonry, biocides for liquid-cooling and processing systems, slimicides, and metalworking-fluids biocides).
3. Pest control (rodenticides, avicides, molluscicides, piscicides, insecticides, and products against other arthropods, repellents, and attractants).
4. Other biocidal products (preservatives for food or feedstocks, antifouling products, embalming and taxidermist fluids, and control of other vertebrates).

On July 9, 2005, the assessment of active substances was allocated to competent authorities of the 25 EU member states. Annex V provides the details regarding the deadlines for submission of a full dossier to the EU member state. More information can be found at http://ec.europa.eu/environment/biocides/index.htm biocides.

Directive 98/8/EC, also now known as BPD (Biocides Product Directive), of the European Parliament and of the Council on the placing on the market of biocidal products was adopted in 1998. According to the Directive, member states had to transpose the rules before May 14, 2000 into national law.

The Commission adopted the original proposal for the Directive in 1993. Directive 91/414/EEC on plant protection products, adopted in 1991, served as a model for the new Directive.

The BPD aims to harmonize the European market for biocidal products and their active substances. At the same time, it tries to provide a high level of protection for humans, animals, and the environment.

The Directive will not apply to certain product types already covered by other Community legislation, such as food contact materials, plant protection products, medicines, and cosmetics.

The basic principles of the Directive are as follows:

- Active substances have to be assessed, and the decision on their inclusion into Annex I of the Directive shall be taken at the Community level.
- Comparative assessment will be made at the Community level when an active substance, although in principle acceptable, still causes concern. Inclusion to Annex I may be denied if there are less harmful, suitable substitutes available for the same purpose.
- Member states shall authorize the biocidal products in accordance with the rules and procedures set in Annex VI of the Directive. They can only authorize products that contain active substances included in Annex I.
- The producers and formulators responsible for the placing on the market the biocidal products and their active substances must apply for authorization and submit all necessary studies and other information needed for the assessments and the decision making.
- A biocidal product authorized in one member state shall be authorized upon application also in another member state unless there are specific grounds to derogate from this principle of mutual recognition.

The Directive established a period of 10 years starting with the year 2000 for the process of identifying existing substances, notifying the ones the market operators are interested in, and notifying and supporting them to be evaluated and approved by the EU.

The scope of the Directive is very wide, covering 23 different product types. These include disinfectants used in different areas, chemicals used for preservation of products and materials, nonagricultural pesticides, and antifouling products used on hulls of vessels.

The different annexes included in the Directive are as follows:

ANNEX I LIST OF ACTIVE SUBSTANCES WITH REQUIREMENTS AGREED AT COMMUNITY LEVEL FOR INCLUSION IN BIOCIDAL PRODUCTS

ANNEX IA LIST OF ACTIVE SUBSTANCES WITH REQUIREMENTS AGREED AT COMMUNITY LEVEL FOR INCLUSION IN LOW-RISK BIOCIDAL PRODUCTS

ANNEX IB LIST OF BASIC SUBSTANCES WITH REQUIREMENTS AGREED AT COMMUNITY LEVEL
ANNEX IIA COMMON CORE DATA SET FOR ACTIVE SUBSTANCES
ANNEX IIB COMMON CORE DATA SET FOR BIOCIDAL PRODUCTS
ANNEX IIIA ADDITIONAL DATA SET FOR ACTIVE SUBSTANCES
ANNEX IIIB ADDITIONAL DATA SET FOR BIOCIDAL PRODUCTS
ANNEX IVA DATA SET FOR ACTIVE SUBSTANCES
ANNEX IVB DATA SET FOR BIOCIDAL PRODUCTS
ANNEX V BIOCIDAL PRODUCT-TYPES AND THEIR DESCRIPTIONS
ANNEX VI COMMON PRINCIPLES FOR THE EVALUATION OF DOSSIERS FOR BIOCIDAL PRODUCTS

18.2.2 Biocidal Product Types and Their Descriptions on Directive 98/8

18.2.2.1 MAIN GROUP 1: Disinfectants and General Biocidal Products. These product types exclude cleaning products that are not intended to have a biocidal effect, including washing liquids, powders, and similar products.

- Product-type 1: Human hygiene biocidal products:
  Products in this group are biocidal products used for human hygiene purposes.
- Product-type 2: Private area and public health area disinfectants and other biocidal products:
  Products used for the disinfection of air, surfaces, materials, equipment, and furniture that are not used for direct food or feed contact in private, public, and industrial areas, including hospitals, as well as products used as algaecides.
  Usage areas include, *inter alia*, swimming pools, aquariums, bathing areas, and other waters; air-conditioning systems; walls and floors in health and other institutions; as well as chemical toilets, wastewater, hospital waste, soil, or other substrates (in playgrounds).
- Product-type 3: Veterinary hygiene biocidal products:
  Products in this group are biocidal products used for veterinary hygiene purposes including products used in areas in which animals are housed, kept, or transported.
- Product-type 4: Food and feed area disinfectants:
  Products used for the disinfection of equipment, containers, consumption utensils, surfaces, or pipework associated with the production, transport, storage, or consumption of food, feed, or drink (including drinking water) for humans and animals.
- Product-type 5: Drinking water disinfectants:
  Products used for the disinfection of drinking water (for both humans and animals).

18.2.2.2 MAIN GROUP 2: Preservatives.

- Product-type 6: In-can preservatives:

  Products used for the preservation of manufactured products, other than foodstuffs or feedingstuffs, in containers by the control of microbial deterioration to ensure their shelf life.
- Product-type 7: Film preservatives:

  Products used for the preservation of films or coatings by the control of microbial deterioration in order to protect the initial properties of the surface of materials or objects such as paints, plastics, sealants, wall adhesives, binders, papers, and art works.
- Product-type 8: Wood preservatives:

  Products used for the preservation of wood, from and including the saw-mill stage, or wood products by the control of wood-destroying or wood-disfiguring organisms.

  This product type includes both preventive and curative products.
- Product-type 9: Fiber, leather, rubber, and polymerized materials preservatives:

  Products used for the preservation of fibrous or polymerized materials, such as leather, rubber, or paper or textile products and rubber by the control of microbiological deterioration.
- Product-type 10: Masonry preservatives:

  Products used for preservation and remedial treatment of masonry or other construction materials other than wood by the control of microbiological and algal attack.
- Product-type 11: Preservatives for liquid-cooling and processing systems:

  Products used for the preservation of water or other liquids used in cooling and processing systems by the control of harmful organisms such as microbes, algae, and mussels.

  Products used for the preservation of drinking water are not included in this product type.
- Product-type 12: Slimicides:

  Products used for the prevention or control of slime growth on materials, equipment, and structures, used in industrial processes, e.g., on wood and paper pulp, porous sand strata in oil extraction.
- Product-type 13: Metalworking-fluid preservatives:

  Products used for the preservation of metalworking fluids by the control of microbial deterioration.

18.2.2.3 MAIN GROUP 3: Pest Control.

- Product-type 14: Rodenticides:

  Products used for the control of mice, rats, or other rodents.

- Product-type 15: Avicides:
  Products used for the control of birds.
- Product-type 16: Molluscicides:
  Products used for the control of mollusks.
- Product-type 17: Piscicides:
  Products used for the control of fish; these products exclude products for the treatment of fish diseases.
- Product-type 18: Insecticides, acaricides, and products to control other arthropods:
  Products used for the control of arthropods (e.g., insects, arachnids, and crustaceans).
- Product-type 19: Repellents and attractants:
  Products used to control harmful organisms (invertebrates such as fleas, and vertebrates such as birds), by repelling or attracting, including those that are used for human or veterinary hygiene either directly or indirectly.

18.2.2.4 MAIN GROUP 4: Other Biocidal Products.

- Product-type 20: Preservatives for food or feedstocks:
  Products used for the preservation of food or feedstocks by the control of harmful organisms.
- Product-type 21: Antifouling products:
  Products used to control the growth and settlement of fouling organisms (microbes and higher forms of plant or animal species) on vessels, aquaculture equipment, or other structures used in water.
- Product-type 22: Embalming and taxidermist fluids:
  Products used for the disinfection and preservation of human or animal corpses, or parts thereof.
- Product-type 23: Control of other vertebrates:
  Products used for the control of vermin.

18.2.3 Directive 98/8 Milestones.

Initially, the directive provided for a transitional period of 10 years (May 14, 2000–May 14, 2010), during which the biocides market would continue to be regulated by national rules. However, the deadline of May 14, 2010 has been extended.

It is interesting to understand the phases and timing that Directive 98/8 established for the Directive implementation.

The main steps and activities since publication have been as follows:

- An exhaustive list of identified existing active substances in the market before May 14, 2000 was compiled and published.
- A deadline was established for manufacturers and/or interested bodies to notify active substances (existing or new) and applications (PT as defined

on the Directive) that would be supported by them to be approved after evaluation by the EU.

- The exhaustive list of notified existing active substances and the product types was established, as well as the list of participants and of applicants having submitted a dossier. Supported substances were included in a provisional list as notified substances.
- Active substances identified as existing but not supported by anybody were dropped from the provisional list.
- Phase-out periods were established for substances that were only identified and notified substances for those product types were not notified.
- Regulators have started to make evaluations of substances and, if positive, including them in Annex I and Annex IA of the Directive 98/8/EC.
- Regulators have started to make evaluations of substances and, if negative, marking for them a decision of noninclusion into Annex I or Annex IA of Directive 98/8/EC (to be removed from the market within 12 months of entering as a result of such a decision).
- Prioritization of the product types in the review program has been done.
- Responsibility at a country level for evaluation of the submitted dossiers of the different notified active substances has been assigned (allocation of the Rapporteur member states for the substances).
- The details on the submission of the dossier, the evaluation, and the corresponding procedures have been defined.

18.2.4 Directive 98/8. Actual Status

The current progress rate of the review program has not permitted its completion by May 14, 2010 as planned. The reason for the delay is that before any review could start, it was necessary to establish an inventory of active substances used in biocidal products placed on the European market of biocidal products and to list the ones that the industry or specific member states were interested in having examined in view of their possible inclusion into Annex I or IA of the Directive (the Community positive list). This elaborated exercise has taken three full years to complete. It was only at the end of 2003 that the timetable, priorities, and list of Rapporteur member states for the review program could be set, while the first dossiers with studies for evaluation were not submitted before 2004. Overall, 964 active substances were identified, of which 468 were notified for evaluation.

Only 42 substances were included in Annex I and 2 substances were included in Annex IA as of August 2011. See Table 18.3 for the list of biocide substances included in Annex I or Annex IA of Directive 98/8/EC.

The experience so far with the review program indicates that the evaluation of a regular active substance dossier never takes less than three years, even under optimum conditions, and the average time required is in the order of four years.

TABLE 18.3 Biocide substances included in Annex I or Annex IA to Directive 98/8/EC (08/2011)

Annex I

Active substance	EC number	CAS number	Product type	Inclusion Directive
Acrolein	203-453-4	107-02-8	12	**2010/5/EU**
alphachloralose	240-016-7	15879-93-3	14	**2009/93/EC**
Aluminium phosphide releasing phosphine	244-088-0	20859-73-8	14	**2009/95/EC**
Aluminium phosphide releasing phosphine	244-088-0	20859-73-8	18	**2010/9/EU**
Bifenthrin	n/a	82657-04-3	8	**2011/10/EU**
Boric acid	233-139-2	10043-35-3	8	**2009/94/EC**
Boric oxide	215-125-8	1303-86-2	8	**2009/98/EC**
Brodifacoum	259-980-5	56073-10-0	14	**2010/10/EU**
Bromadiolone	249-205-9	28772-56-7	14	**2009/92/EC**
Carbon dioxide	204-696-9	124-38-9	14	**2008/75/EC**
Carbon dioxide	204-696-9	124-38-9	18	**2010/74/EU**
Chlorophacinone	223-003-0	3691-35-8	14	**2009/99/EC**
Clothianidin	433-460-1	210880-92-5	8	**2008/15/EC**
Coumatetralyl	227-424-0	5836-29-3	14	**2009/85/EC**
Dazomet	208-574-7	533-74-4	8	**2010/50/EU**
DEET	205-149-7	134-62-3	19	**2010/51/EU**
Dichlofluanid	214-118-7	1085-98-9	8	**2007/20/EC**
Difenacoum	259-978-4	56073-07-5	14	**2008/81/EC**
Difethialone	n/a	104653-34-1	14	**2007/69/EC**
Disodium octaborate tetrahydrate	234-541-0	12280-03-4	8	**2009/96/EC**
Disodium tetraborate	215-540-4	1330-43-4	8	**2009/91/EC**
Etofenprox	407-980-2	80844-07-1	8	**2008/16/EC**
Fenoxycarb	276-696-7	72490-01-8	8	**2011/12/EU**
Fenpropimorph	266-719-9	67564-91-4	8	**2009/86/EC**
Flocoumafen	421-960-0	90035-08-8	14	**2009/150/EC**
Indoxacarb	n/a	173584-44-6	18	**2009/87/EC**
IPBC	259-627-5	55406-53-6	8	**2008/79/EC**
K-HDO	n/a	66603-10-9	8	**2008/80/EC**
Magnesium phosphide releasing phosphine	235-023-7	12057-74-8	18	**2010/7/EU**
Metofluthrin	n/a	240494-70-6	18	**2010/71/EU**
Nitrogen	231-783-9	7727-37-9	18	**2009/89/EC**
Nonanoic acid	203-931-2	112-05-0	19	**2011/13/EU**
Propiconazole	262-104-4	60207-90-1	8	**2008/78/EC**
Spinosad	434-300-1	168316-95-8	18	**2010/72/EU**
Sulfuryl fluoride	220-281-5	2699-79-8	8	**2006/140/EC**
Sulfuryl fluoride	220-281-5	2699-79-8	18	**2009/84/EC**
Tebuconazole	403-640-2	107534-96-3	8	**2008/86/EC**

(*Continued*)

TABLE 18.3 (Continued)

Annex I

Active substance	EC number	CAS number	Product type	Inclusion Directive
Thiabendazole	205-725-8	148-79-8	8	**2008/85/EC**
Thiacloprid	n/a	111988-49-9	8	**2009/88/EC**
Thiamethoxam	428-650-4	153719-23-4	8	**2008/77/EC**
Tolylfluanid	211-986-9	731-27-1	8	**2009/151/EC**
Warfarin	201-377-6	81-81-2	14	**2010/11/EU**
Warfarin sodium	204-929-4	129-06-6	14	**2010/8/EU**
(Z,E)-tetradeca-9,12-dienyl acetate	n/a	30507-70-1	19	**2011/11/EU**

Annex IA

Active substance	EC number	CAS number	Product type	Inclusion Directive
Carbon dioxide	204-696-9	124-38-9	14	**2007/70/EC**
(Z,E)-tetradeca-9,12-dienyl acetate	n/a	30507-70-1	19	**2011/11/EU**

An extension of three years has been approved to end the review, with the new deadline set at May 14, 2013.

Existing active substances are being evaluated in the Review Program, according to Article 16 of the BPD. The Review Program was established via several Regulations. The latest Regulation is Regulation (EC) No 1451/2007, which repeals Regulation (EC) No 2032/2003, and was entered into force on December 31, 2007.

18.2.5 Directive 98/8. Actual Implementation Guidelines

18.2.5.1 Products Containing Existing Substances. The Directive provides a derogation allowing for the placing on the market of products containing existing substances, while they are being evaluated under the Review Program for existing active substances (the list of existing active substances is under evaluation).

Products containing existing active substances can be placed on the market in accordance with the rules and practices of member states until the inclusion of their active substance(s) in Annex I or Annex IA of the Directive.

After such inclusion decisions, products must be authorized in accordance with the terms of the Directive and applications must be submitted to the member states, where such authorizations are to be granted.

18.2.5.2 Products Containing New Active Substances. For products containing new active substances, an application for product authorization can be submitted in parallel with the application for the inclusion of the active substance in Annex I or Annex IA of the Directive.

In accordance with Article 15(2), member states may also grant provisional authorizations to products containing new active substances until an active substance is listed in Annex I or Annex IA. Such authorization may however only be granted after the application submitted for the active substance has been evaluated and pending a final decision to include the active substance in Annex I or Annex IA.

18.2.5.3 Registration Process Online. Applicants for product authorizations are advised to use the application forms generated by the Register for Biocidal Products (R4BP). This Register is available online at the following address: https://webgate.ec.europa.eu/env/r4bp.

18.2.6 Updated Information

The EU has developed a website where information is updated periodically. Please check the latest status of topics relating to the BPD on the following links:

http://ec.europa.eu/environment/biocides
http://ecb.jrc.ec.europa.eu/biocides/

Of special interest:

Manual of decisions for implementation of Directive 98/8/EC concerning the placing on the market of biocidal products containing questions and answers of general interest:

http://ec.europa.eu/environment/biocides/manual.htm
http://circa.europa.eu/Public/irc/env/bio_reports/library?l=/manual_decisions

To check substances withdrawed (noninclusion decisions):

http://ec.europa.eu/environment/biocides/withdrawals.htm
http://ec.europa.eu/environment/biocides/non_inclusions.htm
http://ec.europa.eu/environment/biocides/pdf/list_dates_product_phasing_out.pdf

Also for checking the updated list of substances assessed or under assessment and who is notifying, please refer to:

Searching by substance:

http://ec.europa.eu/environment/biocides/pdf/list_participants_applicants_subs.pdf

Searching by PT:

http://ec.europa.eu/environment/biocides/pdf/list_participants_applicants_prod.pdf

18.2.7 New Regulations Coming

On June 12, 2009, the European Commission adopted a proposal for a Regulation concerning the placing on the market and use of biocidal products (COM(2009)267). The proposed Regulation will repeal and replace the current Directive 98/8/EC concerning the placing of biocidal products on the market.

The objective of the proposal for a Regulation concerning the placing on the market and use of biocidal products (COM(2009)267) is to improve the functioning of the internal market in biocidal products while maintaining a high level of environmental and human health protection. The proposal will build on the principles laid down in Directive 98/8/EC, in particular the two-tier authorization process: first, the inclusion of the active substance in Annex I and, second, the authorization of the biocidal product. The proposed regulation is scheduled to enter into force on January 1, 2013.

Scope of this revision:

- Product authorization
- Data sharing
- Data requirements
- Fees

- What are the main differences between the current Directive and the new Regulation?

  The new Regulation increases the protection for human health and the environment, while being more efficient, notably through the active involvement of the European Chemicals Agency (ECHA). It will retain the two-step authorization process brought in by the current Directive, whereby active substances are first tested and approved and included in a Community list (known as Annex I), with subsequent authorization of a product containing the active substance.
- Scope

  The scope has been extended to cover articles and materials treated with biocidal products, including furniture and textiles. The Regulation will also apply to active substances generated *in situ*, and to biocidal products used in materials that come into contact with food. But other products that are sufficiently covered by existing legislation (including food and feed, food and feed additives, and processing aids) are excluded from the scope of the new regulation. Biocidal products approved under the International Convention for the Control and Management of Ships' Ballast Water and Sediments are considered as authorized.

- Product authorization and mutual recognition

  Under the current Directive, all biocidal products are authorized at the member state level. This will change for two types of biocidal products—biocidal products based on new active substances and low-risk biocidal products—which will have access to a Community authorization allowing them to be placed on the market throughout the Community. All other biocidal products will still be subject to national authorizations issued by member states.

  There will also be further changes to the rules on mutual recognition, the process whereby an authorization in one member state may be recognized by another member state. Under the Regulation, it will be possible to apply either for a mutual recognition of an existing authorization or for a mutual recognition that runs parallel to the first authorization process. The new Regulation will also clarify the conditions for obtaining a parallel trade permit.

- Data requirements

  Under the current Directive, the same set of data must be submitted for all biocidal products, not all of which is always necessary.

  Under the proposed Regulation, the data requirements will be more aligned with the actual needs of the evaluating authorities. It will be possible to waive requirements if data are not scientifically necessary, if they are technically impossible to supply, or if they are not relevant (there is no need for marine toxicity studies, for example, if a product is reserved for use on dry land). It will no longer be possible to repeat tests that have already been carried out on vertebrate animals, and information gained from such tests must be shared, in exchange for fair compensation.

- What active substances will be phased out?

  The proposed Regulation sets out "exclusion criteria" to prevent authorization of active substances with very poor hazard profiles, including substances that can cause cancer, mutations, reproductive problems, and hormonal imbalances.

  In the future, such substances will only be allowed for use provided that:

  1. Human exposure to the active substance in the biocidal product is negligible.
  2. The active substance is necessary to control a serious danger to public health such as an epidemic of particular insects.
  3. The negative effects of banning the active substance would be disproportionate to the impacts on human health or the environment, and no alternatives are available.

Biocidal products with problematic active substances will also be compared to ensure that only the products with the lower risk remain on the market.

- How will the rules on the inclusion of active substances change?

  There is a need to ensure equality of treatment for active substances evaluated before and after the entry into force of the new Regulation. It is therefore proposed that substances still being evaluated under the old rules on January 1, 2013 should continue to be evaluated under the same rules.

The changes introduced by the new Regulation, including the data requirements, data waiving, obligatory sharing of vertebrate animal data, and so forth will only apply to active substances whose evaluation starts after January 1, 2013.

The new proposal will not affect the rules on the examination of existing active substances (those on the market on May 14, 2000) laid down under Regulation (EC) No 1451/2007 (amendment of BPD).

- How will the rules on Community-wide authorization work?

  Community authorization will be available for two types of biocidal products: those based on new active substances and those based on low-risk biocidal products. Such authorizations will be granted by the European Commission and will allow products to be placed on the market across the EU without any need to apply for separate national authorizations or the mutual recognition process.

  Applications for a Community authorization will be submitted to the ECHA and to a competent authority of the applicant's choice. The competent authority will evaluate the application and send its conclusions to the ECHA for an opinion. The Commission will then decide whether a Community authorization can be granted and, if so, under which conditions.

  Community-wide authorizations should stimulate innovation and the development of new and improved products.

- What will be the impact on animals?

  The proposal strives to minimize animal testing as far as possible. Vertebrate tests may not be repeated, and a new obligation to share data involving vertebrate animal tests will come into force. This means that data owners will be obliged to share their data in exchange for fair compensation.

  The Regulation will also encourage the sharing of data from other types of animal tests, with a goal of reducing the overall costs and avoiding the duplication of tests.

- What will be the tasks of the ECHA under the new Regulation?

  The ECHA will coordinate the active substance evaluation for new and existing active substances, the reassessments of the already approved active substances in light of available new information, and some related evaluation work.

  The agency will also play a key role in the centralized authorization of products. The ECHA will be in charge of coordinating the technical and scientific work of this new centralized authorization process. In the event of any disputes over mutual recognition between the member states or the member states and applicants, the ECHA will provide the Commission with technical and scientific support.

The ECHA will intervene in cases of disagreements about data sharing from vertebrate animal tests. The ECHA will also be responsible for maintaining the relevant databases and other administrative and scientific tasks, such as coordinating the meeting of experts reviewing the draft risk assessments reports prepared by the member states in the context of the Review Program of the biocidal active substances.

- What are the costs and benefits of this proposal?

  Compared with the existing rules, the only change involving additional costs to the industry concerns the extension of the scope to treated articles and materials. These costs will mainly result from the inclusion of further active substances in Annex I and the compliance with the labeling obligations. But it should be noted that according to the impact assessment carried out to evaluate the proposal, the environmental and human health benefits of extending the scope to treated articles and materials will easily outweigh the costs.

  The other measures introduced by the proposal, such as improved authorization procedures, including the Community authorization, obligatory data sharing for vertebrate animal data, and streamlining the data requirements, will result in significant cost savings. The reduction of financial and administrative burden was also one of the main objectives of this proposal. However, this proposal shows that these savings can be achieved without compromising the high level of environmental and human health protection.

18.2.8 Food Contact Applications

Food contact materials are all materials and articles intended to come into contact with foodstuffs, including packaging materials but also cutlery, dishes, processing machines, containers, and so on. The term also includes materials and articles that are in contact with water intended for human consumption.

As mentioned, Directive 98/8 is in the process of being reviewed, including also food contact applications, but at the moment, biocides to be used in food contact applications are not falling within the scope of Directive 98/8/EC as they fall within the scope of Regulation (EC) No 1935/2004. This means that the use of biocidal substances in these materials has to comply with the rules laid down in this Regulation and its implementing measures and any relevant national rules that might exist in the member states.

The Framework Regulation (EC) 1935/2004 (L338/4) states that food contact materials shall be safe in the sense that they shall not transfer their components into the food in quantities that:

- Could endanger human health,
- Change the composition of the food in an unacceptable way, or
- Deteriorate the taste and odor of foodstuffs.

Regulation (EU) No 10/2011 of January 14, 2011 on plastic materials and articles intended to come into contact with food applies from May 1, 2011 and regulates plastic materials and articles intended to come into contact with foodstuffs and establishes a positive list with the substances that can be used in food contact plastic materials and articles. One of the new outcomes of this Regulation is the fact that nanomaterials will be assessed on a case-by-case basis.

A food contact materials database is managed by the EU and is available at https://webgate.ec.europa.eu/sanco_foods/main/?event=substances.search&substances.pagination=1.

One of the potential uses of biocides is in the field of food packaging materials. In this context, a food packaging material incorporating a biocide falls under the definition of an "active material or article."

In 2009, the European Union launched the Regulation (EC) No 450/2009, laying down specific rules for active and intelligent materials and articles intended to be used in contact with foodstuffs to be applied in addition to the general requirements established in Regulation (EC) No 1935/2004 for their safe use. The substance(s) responsible for the active and/or intelligent function of the material should be evaluated under the Regulation (EC) No 450/2009.

According to this regulation, "active materials and articles" means materials and articles that are intended to extend the shelf life or to maintain or improve the condition of packaged food; they are designed to deliberately incorporate components that would release or absorb substances into or from the packaged food or the environment surrounding the food;

A Community list of authorized substances, which can be used to manufacture an active or intelligent component of active and/or intelligent materials and articles, shall therefore be established after the European Food Safety Authority (EFSA) has performed a risk assessment and has issued an opinion on each substance. In some cases, restrictions may be proposed by the EFSA on a group of substances, especially when the active or intelligent function implies interactions between different substances.

Substances deliberately incorporated into active materials and articles to be released into the food or the environment surrounding the food do not need to be on the Community list. They shall be used in full compliance with the relevant Community and national provisions applicable to food, and they shall comply with the provisions of Regulation (EC) No 1935/2004 and its implementing measures. The same shall apply to substances that are incorporated into active materials and articles by techniques such as grafting and immobilization, in order to have a technological effect on the food. However, for these substances already approved in food legislation, their stability under the intended packaging manufacturing and processing conditions must be verified by the packaging manufacturer and a dossier for safety evaluation has to be submitted if chemical reaction, degradation, or decomposition of these substances is likely to occur.

The passive parts should be covered by the specific community or national legislation applicable to those materials. These guidelines do not apply to substances used behind a functional barrier as defined by Article 3 of Regulation (EC) No 450/2009. Substances behind such a barrier will not, by definition, migrate in amounts that could endanger human health or bring about unacceptable changes in the composition of the food or of its organoleptic properties. Consequently, these active and intelligent substances do not need a safety evaluation and are outside the scope of Regulation (EC) No 450/2009.

However, this functional barrier concept does not apply to substances in nanoparticulate form, which should be assessed on a case-by-case basis.

The EFSA released in July 2009 the document

"Guidelines on submission of a dossier for safety evaluation by the EFSA of active or intelligent substances present in active and intelligent materials and articles intended to come into contact with food."

The purpose of this document is to give guidance to applicants and other interested parties for the preparation and the submission of a dossier for the evaluation of the safety of active and/or intelligent substances responsible for active and/or intelligent functions of active and/or intelligent materials and articles intended to be used in contact with food. It gives guidance on the administrative and technical data required and on the format of a submission (hereinafter referred to as "dossier") for the evaluation by the EFSA.

18.3 UNITED STATES

18.3.1 Regulatory Frame

The U.S. Environmental Protection Agency (EPA) regulates the use of pesticides under the authority of two federal statutes: the Federal Insecticide, Fungicide, and Rodenticide Act (FIFRA) and the Federal Food, Drug, and Cosmetic Act (FFDCA).

FIFRA provides the basis for regulation, sale, distribution, and use of pesticides in the United States. FIFRA authorizes the EPA to review and register pesticides for specified uses. The EPA also has the authority to suspend or cancel the registration of a pesticide if subsequent information shows that continued use would pose unreasonable risks. Some key elements of FIFRA include:

- It is a product licensing statute; pesticide products must obtain an EPA registration before manufacture, transport, and sale.
- Registration is based on a risk/benefit standard.
- There is a strong authority to require data—authority to issue Data Call-ins.
- There is an ability to regulate pesticide use through labeling, packaging, composition, and disposal.

- Emergency exemption authority—permits approval of unregistered uses of registered products on a time-limited basis.
- The ability to suspend or cancel a product's registration: appeals process, adjudicatory functions, etc.

The EPA and the states (usually the state Department of Agriculture) register or license pesticides for use in the United States. In addition, anyone planning to import pesticides for use in the United States must notify the EPA. The EPA receives its authority to register pesticides under FIFRA.

In 2006, the EPA initiated a new program called registration review to reevaluate all pesticides on a regular cycle. The program's goal is to review each pesticide active ingredient every 15 years to make sure that, as the ability to assess risks to human health and the environment evolves and as policies and practices change, all pesticide products in the marketplace can still be used safely.

After a pesticide is registered by the EPA, it must be registered at the state level. States can register pesticides under specific state pesticide registration laws. A state may have more stringent requirements for registering pesticides for use in that state.

18.3.2 Definitions

A pesticide is any substance or mixture of substances intended for preventing, destroying, repelling, or mitigating any pest.

Although often misunderstood to refer only to insecticides, the term *pesticide* also applies to herbicides, fungicides, and various other substances used to control pests.

Under U.S. law, a pesticide is also any substance or mixture of substances intended for use as a plant regulator, defoliant, or desiccant.

Pests are living organisms that occur where they are not wanted or that cause damage to crops or humans or other animals. Examples include insects, mice, and other animals, unwanted plants (weeds), fungi, microorganisms such as bacteria and viruses, and prions.

18.3.3 Types of Pesticides

Pesticides are often referred to according to the type of pest they control. Another way to think about pesticides is to consider those that are chemical pesticides or are derived from a common source or production method. Other categories include biopesticides, antimicrobials, and pest control devices.

18.3.3.1 Chemical Pesticides. Some examples of chemically related pesticides follow:

Organophosphate Pesticides—These pesticides affect the nervous system by disrupting the enzyme that regulates acetylcholine, a neurotransmitter.

Most organophosphates are insecticides. They were developed during the early 19th century, but their effects on insects, which are similar to their effects on humans, were discovered in 1932. Some are very poisonous (they were used in World War II as nerve agents). However, they usually are not persistent in the environment.

Carbamate Pesticides—These pesticides affect the nervous system by disrupting an enzyme that regulates acetylcholine, a neurotransmitter. The enzyme effects are usually reversible. There are several subgroups within the carbamates.

Organochlorine Insecticides—These pesticides were commonly used in the past, but many have been removed from the market because of their health and environmental effects and their persistence (e.g., DDT and chlordane).

Pyrethroid Pesticides—These pesticides were developed as a synthetic version of the naturally occurring pesticide pyrethrin, which is found in chrysanthemums. They have been modified to increase their stability in the environment. Some synthetic pyrethroids are toxic to the nervous system.

18.3.3.2 Biopesticides. Biopesticides are certain types of pesticides derived from natural materials such as animals, plants, bacteria, and certain minerals. For example, canola oil and baking soda have pesticidal applications and are considered biopesticides. At the end of 2001, there were approximately 195 registered biopesticide active ingredients and 780 products. Biopesticides fall into three major classes:

a. **Microbial pesticides** contain a microorganism (e.g., a bacterium, fungus, or virus or protozoan) as the active ingredient. Microbial pesticides can control many different kinds of pests, although each separate active ingredient is relatively specific for its target pest[s]. For example, there are fungi that control certain weeds, and other fungi that kill specific insects.

   The most widely used microbial pesticides are subspecies and strains of *Bacillus thuringiensis*, or Bt. Each strain of this bacterium produces a different mixture of proteins and specifically kills one or a few related species of insect larvae. Although some Bts control moth larvae found on plants, other Bts are specific for larvae of flies and mosquitoes. The target insect species are determined by whether the particular Bt produces a protein that can bind to a larval gut receptor, thereby causing the insect larvae to starve.

b. **Plant-incorporated protectants (PIPs)** are pesticidal substances that plants produce from genetic material that has been added to the plant. For example, scientists can take the gene for the Bt pesticidal protein and introduce the gene into the plant's own genetic material. Then the plant, instead of the Bt bacterium, manufactures the substance that destroys the pest. The protein and its genetic material, but not the plant itself, are regulated by the EPA.

c. **Biochemical pesticides** are naturally occurring substances that control pests by nontoxic mechanisms. Conventional pesticides, by contrast, are generally synthetic materials that directly kill or inactivate the pest. Biochemical pesticides include substances, such as insect sex pheromones, that interfere with mating, as well as various scented plant extracts that attract insect pests to traps. Because it is sometimes difficult to determine whether a substance meets the criteria for classification as a biochemical pesticide, the EPA has established a special committee to make such decisions.

18.3.3.3 Pest Types. Pesticides that are related because they address the same type of pests include:

Algicides Control algae in lakes, canals, swimming pools, water tanks, and other sites.

Antifouling agents Kill or repel organisms that attach to underwater surfaces, such as boat bottoms.

Antimicrobials Kill microorganisms (such as bacteria and viruses).

Attractants Attract pests (for example, to lure an insect or rodent to a trap). (However, food is not considered a pesticide when used as an attractant.)

Biopesticides Biopesticides are certain types of pesticides derived from such natural materials as animals, plants, bacteria, and certain minerals.

Biocides Kill microorganisms.

Disinfectants and sanitizers Kill or inactivate disease-producing microorganisms on inanimate objects.

Fungicides Kill fungi (including blights, mildews, molds, and rusts).

Fumigants Produce gas or vapor intended to destroy pests in buildings or soil.

Herbicides Kill weeds and other plants that grow where they are not wanted.

Insecticides Kill insects and other arthropods.

Miticides (acaricides) Kill mites that feed on plants and animals.

Microbial pesticides Microorganisms that kill, inhibit, or out compete pests, including insects or other microorganisms.

Molluscicides Kill snails and slugs.

Nematicides Kill nematodes (microscopic, worm-like organisms that feed on plant roots).

Ovicides Kill eggs of insects and mites.

Pheromones Biochemicals used to disrupt the mating behavior of insects.

Repellents Repel pests, including insects (such as mosquitoes) and birds.

Rodenticides Control mice and other rodents.

The term *pesticide* also includes these substances:

Defoliants Cause leaves or other foliage to drop from a plant, usually to facilitate harvest.

Desiccants Promote drying of living tissues, such as unwanted plant tops.

Insect growth regulators Disrupt the molting, maturity from pupal stage to adult, or other life processes of insects.

Plant growth regulators Substances (excluding fertilizers or other plant nutrients) that alter the expected growth, flowering, or reproduction rate of plants.

18.3.3.4 Pest Control Devices. Regarding the pest control devices, the EPA also has a role in regulating them. More specifically, a "device" is any instrument or contrivance (other than a firearm) intended for trapping, destroying, repelling, or mitigating any pest. A mousetrap is an example of a device. Unlike pesticides, the EPA does not require devices to be registered with the Agency. Devices are subject to certain labeling, packaging, record keeping, and import/export requirements, however.

18.3.4 Exclusions

The U.S. definition of pesticides is broad, but it does have some exclusions:

- Drugs used to control diseases of humans or animals (such as livestock and pets) are not considered pesticides; such drugs are regulated by the Food and Drug Administration (FDA).
- Fertilizers, nutrients, and other substances used to promote plant survival and health are not considered plant growth regulators and thus are not pesticides.
- Biological control agents, except for certain microorganisms, are exempted from regulation by the EPA. (Biological control agents include beneficial predators such as birds or ladybugs that eat insect pests.)
- Products that contain certain low-risk ingredients, such as garlic and mint oil, have been exempted from federal registration requirements, although state regulatory requirements may still apply.

18.3.5 EPA Registration Process

Pesticide registration is the process through which the EPA examines the ingredients of a pesticide; the site or crop on which it is to be used; the amount, frequency, and timing of its use; and storage and disposal practices. The EPA evaluates the pesticide to ensure that it will not have unreasonable adverse effects on humans, the environment, and nontarget species. Pesticides must be registered or exempted by the EPA's Office of Pesticide Programs before they may be sold or distributed in the United States. Once registered, a pesticide may not legally be used unless the use is consistent with the approved directions for use on the pesticide's label or labeling.

The EPA has separate review processes for three categories of pesticides:

- Antimicrobials
- Biopesticides
- Conventional

The three processes share the same application materials, but there are differing data requirements and review policies that registrants must take into account in their submittal.

The process of registering a pesticide begins with submission to the EPA of an application package. The EPA's review of this application includes assessment of the hazards to human health and the environment that may be posed by the pesticide. Depending on the class of pesticide and the priority assigned to it, the review process can take from 120 days to several years.

18.3.6 EPA Registration Process for Antimicrobial Pesticides

Antimicrobial pesticides are substances or mixtures of substances used to destroy or suppress the growth of harmful microorganisms whether bacteria, viruses, or fungi on inanimate objects and surfaces. Antimicrobial products contain about 275 different active ingredients and are marketed in several formulations: sprays, liquids, concentrated powders, and gases.

More than 5000 antimicrobial products are currently registered with the EPA and sold in the marketplace. Nearly 60% of antimicrobial products are registered to control infectious microorganisms in hospitals and other health-care environments.

18.3.6.1 Major Uses. Antimicrobial pesticides have two major uses:

1. Disinfect, sanitize, reduce, or mitigate growth or development of microbiological organisms
2. Protect inanimate objects (for example, floors and walls), industrial processes or systems, surfaces, water, or other chemical substances from contamination, fouling, or deterioration caused by bacteria, viruses, fungi, protozoa, algae, or slime

These uses do not include certain pesticides intended for food use, but they do encompass pesticides with a wide array of other uses. For example, antimicrobial pesticides act as preserving agents in paints, metalworking fluids, wood supports, and many other products to prevent their deterioration.

18.3.6.2 Types of Antimicrobial Products. Antimicrobial products are divided into two categories based on the type of microbial pest against which the product works:

18.3.6.2.1 Nonpublic health products. These products are used to control the growth of:

- Algae
- Odor-causing bacteria
- Bacteria that cause spoilage, deterioration, or fouling of materials and microorganisms infectious only to animals. This general category includes products used in cooling towers, jet fuel, paints, and treatments for textile and paper products.

18.3.6.2.2 Public health products. They are intended to control microorganisms infectious to humans in any inanimate environment. The more commonly used public health antimicrobial products include the following:

- **Sterilizers (Sporicides):**

  Used to destroy or eliminate all forms of microbial life including fungi, viruses, all forms of bacteria, and their spores.

  Spores are considered to be the most difficult form of microorganism to destroy. Therefore, the EPA considers the term *sporicide* to be synonymous with *sterilizer*.

  Sterilization is critical to infection control and is widely used in hospitals on medical and surgical instruments and equipment. Types of sterilizers include:

  - steam under pressure (autoclaving)
  - dry heat ovens (used primarily for sterilization of medical instruments)
  - low-temperature gas (ethylene oxide) (used primarily for sterilization of medical instruments)
  - liquid chemical sterilants (used primarily for delicate instruments that cannot withstand high temperature and gases)
- **Disinfectants:**

  Used on hard inanimate surfaces and objects to destroy or irreversibly inactivate infectious fungi and bacteria but not necessarily their spores. Disinfectant products are divided into two major types:

  - Hospital-type disinfectants are the most critical to infection control and are used on medical and dental instruments, floors, walls, bed linens, toilet seats, and other surfaces.
  - General-use disinfectants are the major source of products used in households, swimming pools, and water purifiers.
- **Sanitizers:**

  Used to reduce, but not necessarily eliminate, microorganisms from the inanimate environment to levels considered safe as determined by public health codes or regulations. Sanitizers include:

  - Food contact products—These products are important because they are used on sites where consumable food products are placed and

TABLE 18.4 Overview of EPA approach on registration path depending on marketed claims

	EPA Registration	
	Antimicrobial Formulation	Treated Article
No claims	No	No
Claims - contains in preservative to inhibit odor causing bacteria - contains and antimicrobial preservative - *An article or substance treated with, or containing, a pesticide to protect the trented article or substance itself*	Yes	Exemption
Claim to control pest organisms that pose a threat to human health	Yes (+ efficay data)	Yes (+ efficay data)

stored. Sanitizing rinses for surfaces such as dishes and cooking utensils, equipment and utensils found in dairies, food-processing plants, eating and drinking establishments, and nonfood contact products

- Nonfood contact surface sanitizers include carpet sanitizers, air sanitizers, laundry additives, and in-tank toilet bowl sanitizers.

- **Antiseptics and Germicides:**
  Used to prevent infection and decay by inhibiting the growth of microorganisms. Because these products are used in or on living humans or animals, they are considered drugs and are thus approved and regulated by the FDA.

18.3.7 Claims, Labeling, and Labels

Pesticide product labels provide critical information about how to handle and use pesticide products safely. A critical aspect of registering a pesticide product is the approval of the product label. Depending on the claims stated by the pesticide label, the registration process can be different and the data needed for the dossier can be more or less complex. Table 18.4 gives an overview of the registration path depending on the claims.

18.3.8 Tests Requested

The Code of Federal Regulations 40CFR Part 161 covers the information regarding DATA REQUIREMENTS FOR REGISTRATION OF ANTIMICROBIAL PESTICIDES. Depending on final application, the requirements can be different. General requirements follow the pattern below:

Product Chemistry Data Requirements

§ 161.155 Product composition.
§ 161.160 Description of materials used to produce the product.
§ 161.162 Description of production process.
§ 161.165 Description of formulation process.
§ 161.167 Discussion of formation of impurities.
§ 161.170 Preliminary analysis.
§ 161.175 Certified limits.
§ 161.180 Enforcement analytical method.
§ 161.190 Physical and chemical characteristics.

Data Requirement

§ 161.240 Residue chemistry data requirements.
§ 161.290 Environmental fate data requirements.
§ 161.340 Toxicology data requirements.
§ 161.440 Spray drift data requirements.
§ 161.490 Wildlife and aquatic organisms data requirements.
§ 161.540 Plant protection data requirements.
§ 161.590 Non target insect data requirements.
§ 161.640 Product performance data requirements.

The following describes the reasons for each type of test and the kind of information the EPA obtains from the results of tests.

18.3.8.1 Product Performance. The requirements to develop data on product performance provide a mechanism to ensure that pesticide products will control the pests listed on the label and that unnecessary pesticide exposure to the environment will not occur as a result of the use of ineffective products. Specific performance standards are used to validate the efficacy data in the public health areas, including disinfectants used to control microorganisms infectious to humans in any area of the inanimate environment and those pesticides used to control vertebrates (such as rodents, birds, bats, and skunks) and invertebrates (ticks, mosquitoes, etc.) that may directly or indirectly transmit diseases to humans.

18.3.8.2 Data From Studies That Determine Hazard to Humans and Domestic Animals. The data required to assess hazards to humans and domestic animals are derived from a variety of acute, subchronic, and chronic toxicity tests and from tests to assess mutagenicity and pesticide metabolism.

18.3.8.2.1 Acute Studies. Determination of acute oral, dermal, and inhalation toxicity is usually the initial step in the assessment and evaluation of the toxic characteristics of a pesticide. These data provide information on health hazards likely to originate soon after and, as a result of, short-term exposure. Data from acute studies serve as a basis for classification and precautionary labeling. They also:

- provide information used in establishing the appropriate dose levels in subchronic and other studies;
- provide initial information on the mode of toxic action(s) of a substance;
- determine the need for child-resistant packaging; and
- determine the need to restrict use of the pesticide to trained applicators or in other ways to minimize human and environmental hazards.

Information derived from primary eye and primary dermal irritation studies serves to identify possible hazards from exposure of the eyes, associated mucous membranes, and skin.

18.3.8.2.2 Subchronic Studies. Subchronic tests provide information on health hazards that may originate from repeated exposures over a limited period of time. They provide information on target organs and accumulation potential. The resulting data are also useful in selecting dose levels for chronic studies and for establishing safety criteria for human exposure. These tests are not capable of detecting those effects that have a long latency period for expression (e.g., carcinogenicity).

18.3.8.2.3 Chronic Studies. Chronic toxicity (usually conducted by feeding the test substance to the test species) studies are intended to determine the effects of a substance in a mammalian species after prolonged and repeated exposure. Under the conditions of this test, effects that have a long latency period or are cumulative should be detected. The purpose of long-term carcinogenicity studies is to observe test animals over most of their life span for the development of neoplastic lesions during or after exposure to various doses of a test substance by an appropriate route of administration.

18.3.8.3 Data From Studies That Determine Hazard to Nontarget Organisms. The information required to assess hazards to nontarget organisms are derived from tests to determine pesticidal effects on birds, mammals, fish, terrestrial and aquatic invertebrates, and plants. These tests include short-term acute, subacute, reproduction, simulated field, and full field studies arranged in a hierarchical or tier system that progresses from the basic laboratory tests to the applied field tests. The results of each tier of tests must be evaluated to determine the potential of the pesticide to cause harmful effects and to determine whether further testing is required. A purpose common to all data requirements is to help determine the need for (and appropriate wording for) precautionary label statements to minimize the potential harm to nontarget organisms.

18.3.8.3.1 Acute and Subacute Studies. The short-term acute and subacute laboratory studies provide basic toxicity information that serves as a starting point for the hazard assessment. These data are used to:

- establish the acute toxicity levels of the active ingredient to the test organisms;
- compare toxicity information with measured or estimated pesticide residues in the environment in order to assess the potential effects on fish, wildlife, plants, and other nontarget organisms; and
- indicate whether further laboratory and/or field studies are needed.

18.3.8.3.2 Chronic and Field Studies. Additional studies (i.e., avian, fish, and invertebrate reproduction; life-cycle studies; and plant field studies) may be required when basic data and environmental conditions suggest possible problems. Data from these studies are used to:

- estimate the potential for chronic effects, taking into account the measured or estimated residues in the environment; and
- determine whether additional field or laboratory data are necessary to further evaluate hazards.

Simulated field and/or field data are used to examine the acute and chronic adverse effects on captive or monitored fish and wildlife populations under natural or near-natural environments. Such studies are required only when predictions as to possible adverse effects in less extensive studies cannot be made, or when the potential for harmful effects is high.

18.3.8.4 Postapplication Exposure Studies. The data required to assess hazards to employees resulting from reentry into areas treated with pesticides are derived from studies on toxicity, residue dissipation, and human exposure. Monitoring data generated during exposure studies are used to determine how much pesticide people may be exposed to after application and to establish how long workers must wait before reentering a treated area.

18.3.8.4.1 Applicator/User Exposure Studies. The EPA requires applicator/user exposure data for all pesticides to evaluate the potential risks to people applying the pesticide, i.e., those who may be exposed to higher concentrations of the pesticide through handling, including mixing or applying.

18.3.8.4.2 Pesticide Spray Drift Evaluation. The data required to evaluate pesticide spray drift are derived from studies on the range of droplet sizes and spray drift field evaluations. These data contribute to the development of the overall exposure estimate. Along with data on toxicity for humans, fish, wildlife, or plants, data on spray drift are used to assess the potential exposure of these organisms to pesticides. A purpose common to all these tests is to provide data to help determine the need (and appropriate wording) for precautionary labeling to minimize the potential harm to nontarget organisms.

18.3.8.4.3 Environmental Fate. The EPA uses the data generated by environmental fate studies to:

- assess the presence of widely distributed and persistent pesticides in the environment that may result in loss of usable land, surface water, ground water, and wildlife resources;
- assess the potential environmental exposure of other nontarget organisms, such as fish, wildlife, and plants, to pesticides; and
- help estimate expected environmental concentrations of pesticides in specific habitats where threatened or endangered species or other wildlife populations at risk are found.

18.3.8.4.4 Residue Chemistry. The EPA uses residue chemistry data to estimate the exposure of the general population to pesticide residues in food and for setting and enforcing tolerances for pesticide residues in food or feed. The Agency can estimate the amount and nature of residues likely to be present in food or animal feed because of a proposed pesticide usage by evaluating information on:

- the chemical identity and composition of the pesticide product;
- the amounts, frequency, and time of pesticide application; and
- test results on the amount of residues remaining on or in the treated food or feed.

Table 18.5 contains a list of test requested by EPA for a typical biocide used as preservative with indication of EPA guideline reference.

TABLE 18.5 Tests for a package submission for EPA registration for a biocide formulation

	Guideline Reference Number	Guideline Study Name
1	8301550	Product Identity and Composition
2	8301620	Beginning Materials and Manufacturing Process
3	8301670	Discussion of the Formation of Impurities
4	8301700	Preliminary Analysis
5	8301750	Certified Limits
6	8301800	Enforcement Analytical Method
7	8306302	Color
8	8306303	Physical State
9	8306304	Odor

	Guideline Reference Number	Guideline Study Name
10	8306314	Oxidation / Reduction
11	8306315	Flammability
12	8306316	Explodability
13	8306317	Storage Stability
14	8306320	Corrosion Characteristics
15	8307000	pH
16	8307300	Density
17	8701100	Acute Oral Toxicity—Rat
18	8701200	Acute Dermal Toxicity—Rabbit
19	8701300	Acute Inhalation Toxicity—Rat
20	8702400	Primary Eye Irritation—Rabbit
21	8702500	Primary Dermal Irritation—Rabbit
22	8702600	Dermal Sensitization—Guinea Pig
23	8703100	90-Day Oral Toxicity
24	8703150	90-Day Oral Toxicity—Nonrodent
25	8703250	90-Day Dermal Toxicity
26	8703700	Teratology—Rats
27	8703700	Prenatal Developmental Toxicity—Rabbit
28	8703800	Reproduction
29	8704100	Chronic toxicity
30	8704200	Oncogenicity
31	870.5000 series	Genotoxicity
32	8705100	Ames/Salmonella Mutagenicity
33	8705375	In vitro Mammalian Chromosome Aberration
34	8705915	In vivo Mammalian Chromosome Aberration
35	8706300	Neurotoxicity
36	8707485	Metabolism
37	8707800	Immunotoxicity
38	8502100	Acute Avian Oral Toxicity—Quail
39	8502200	Acute Avian Dietary Toxicity—Quail
40	8502100	Acute Avian Oral Toxicity—Quail
41	8502200	Acute Avian Dietary Toxicity—Quail
42	8501010	Freshwater Aquatic Invertebrate Toxicity—Daphnia
43	8501075	Freshwater Fish Toxicity—Rainbow Trout
44	8504400	Freshwater Plant (algae)
45	8352100	Hydrolysis
46		Particle Size Distribution

When substances in nanosize (at least one dimension between approximately 1 and 100 nm according to the EPA definition) are involved on the antimicrobial pesticide, the EPA can ask for additional requirements on a case-by-case analysis as nanomaterials are more carefully assessed by the EPA because of the potential toxicological risks resulting from nanoparticles.

18.3.9 Active and Inert Ingredients

Pesticide products contain both "active" and "inert" ingredients. The terms *active ingredient* and *inert ingredient* are defined by the federal law that governs pesticides (FIFRA). An active ingredient is one that prevents, destroys, repels, or mitigates a pest, or is a plant regulator, defoliant, desiccant, or nitrogen stabilizer. By law, the active ingredient must be identified by name on the pesticide product's label together with its percentage by weight.

All other ingredients in a pesticide product are called "inert ingredients." An inert ingredient means any substance (or group of similar substances) other than an active ingredient that is intentionally included in a pesticide product. Called "inerts" by the law, the name does not mean nontoxic.

Pesticide products often contain more than one inert ingredient. Inert ingredients play key roles in the effectiveness of pesticides.

- Pesticide Product Information System (PPIS)
  http://www.epa.gov/pesticides/PPISdata/index.html:
  The PPIS contains information concerning all pesticide products registered in the United States. It includes registrant name and address, chemical ingredients, toxicity category, product names, distributor brand names, site/pest uses, pesticidal type, formulation code, and registration status.
- National Pesticide Information Retrieval System (NPIRS)
  http://ppis.ceris.purdue.edu/:
  The NPIRS contains technical registration information on pesticide products.
- Minimum risk active substances
  http://www.epa.gov/opprd001/inerts/section25b_inerts.pdf:
  Substances permitted as active ingredients because of their low-risk profile.
- Inert Ingredients Permitted for Use in Nonfood Use Pesticide
  http://www.epa.gov/opprd001/inerts/inert_nonfooduse.pdf:
  The listed inert ingredients are permitted for use in nonfood use pesticide products.

18.3.10 Utilities

- Data Submitters List
  http://www.epa.gov/pesticides/DataSubmittersList/:
  This is a compilation of names and addresses of pesticide registrants who wish to be notified and offered compensation for use of their data. It was developed to assist pesticide applicants in fulfilling their obligation as required by Sections 3(c)(1)(f) and 3(c)(2)(D) of the FIFRA and 40 CFR Part 152 sub part E regarding ownership of data used to support registration.

It can be useful if you wish to avoid costing required tests that maybe have been already done by another company using same active ingredient.

- Pesticide Product Label System (PPLS)
  http://www.epa.gov/pesticides/pestlabels/:
  The PPLS is a collection of images of pesticide labels that have been approved by the Office of Pesticide Programs (OPP). The label images are indexed by EPA registration number and the date on which the label was initially registered or amended.

18.3.11 Treated Article Exemptions

Known as the "Treated Articles Exemption," section 40CFR 152.25(a) provides an exemption from all requirements of FIFRA for qualifying articles or substances treated with, or containing a pesticide, if:

1. the incorporated pesticide is registered for use in or on the article or substance; and
2. the sole purpose of the treatment is to protect the article or substance itself.

The exemption gives two examples of treatments that are intended to protect only the treated article or substance itself. In the first case, paint is being protected from deterioration of the paint film or coating. In the second case, wood is being protected from fungus or insect infestations that may originate on the surface of the wood. Pesticides used in this manner are generally classified as preservatives. Other pesticides are incorporated into treated articles because of their ability to inhibit the growth of microorganisms that may cause odors or to inhibit the growth of mold and mildew.

To qualify for the treated articles exemption, both conditions must be met. If both are not met, the article or substance does not qualify for the exemption and is subject to regulation under FIFRA.

18.3.12 Food Contact Applications

For any indirect or direct food contact application and before EPA registration, the antimicrobial pesticide and its intended application on the final article must be approved by the FDA.

- Determine the Regulatory Status of Components of a Food Contact Material
  The overall regulatory status of a food contact material is dictated by the regulatory status of each individual substance that comprises the article. The individual substance that is reasonably expected to migrate to food because of its intended use in the food contact material shall be covered by one of the following:

- A regulation listed in Title 21 Code of Federal Regulations
- A prior sanction letter
- Meeting the criteria for GRAS status (including but not limited to a GRAS regulation or GRAS notice)
- A Threshold of Regulation (TOR) exemption request
- An effective Food Contact Notification (FCN).

Be aware that Section 409(h)(2)(C) of the Federal Food, Drug, and Cosmetic Act (the Act) states that an FCN is effective for the manufacturer, the Food Contact Substance (FCS), and the conditions of use identified in the notification and not effective for a similar or identical substance produced or prepared by a manufacturer other than a manufacturer identified in the prior notification. FCNs are proprietary to the manufacturer for which the notification is effective; therefore, the FCS must be obtained from that manufacturer.

It is the responsibility of the manufacturer of an FCS to ensure that food contact materials comply with the specifications and limitations in all applicable authorizations. When reviewing your composite formulations to determine compliance, consider each authorization to be composed of three parts: the *identity* of the substance, *specifications* including purity or physical properties, and *limitations* on the conditions of use.

You may ask the manufacturer for a letter of guaranty to customers certifying that a particular product is acceptable for the intended food-contact use.

How do I determine the compliance of the components of my food contact article with the requirements of the Act?

There are a number of ways that a component of your food contact article may comply with the Act. The guides given in Figure 18.2 are designed to help you make that determination researching each component for category of authorization.

- **Consult 21 CFR 174-179 to see whether the use of the component is an appropriately regulated indirect additive**:
  - General Indirect Food Additives (21 CFR 174)
  - Adhesives and Components of Coatings (21 CFR 175)
  - Paper and Paperboard Components (21 CFR 176)
  - Polymers (21 CFR 177)
  - Adjuvants, Production Aids, and Sanitizers (21 CFR 178)
  - Irradiation in the Production, Processing and Handling of Food (21 CFR 179)

The requirement for premarket approval in Section 409 of the Act in 1958 resulted in the development of a petition process by which a person could request approval of a food additive for an intended use. The approval resulted in a regulation listed in 21 CFR. Components of a food packaging material used in compliance with a regulation in 21 CFR (174-179) need no further FDA

Consult 21 CFR 174-179 to see if the use of the component is an appropriately regulated indirect additive.	Consult 21 CFR 182-186 to see if the use of the component is Generally Recognized as Safe (GRAS)* or consult the list of GRAS Notices
Consult 21 CFR 181 to see if the use of the component is Prior Sanctioned*	Consult the listing of Threshold of Regulation Exemptions
Consult the listing of Effective Food Contact Substance Notifications	

If any component of my food contact article is not covered by any of the categories above, what options do I have?

Submit a Threshold of Regulation Exemption Request	Or ↔	Submit a Food Contact Substance Notification.	Or ↔	Satisfy the criteria for GRAS status

FIGURE 18.2 Basics for determining the regulatory status of components of a food contact material.
Source: FDA Web.

review. Most of the regulated indirect food additives can be found in CFSAN's "Indirect Additive" Database.

- **Consult 21 CFR 182-186 to see whether the use of the component is listed as Generally Recognized As Safe (GRAS).**
  - Substances GRAS in food (21 CFR 182)
  - Substances affirmed as GRAS in food (21 CFR 184)
  - Substances affirmed as GRAS for use in food packaging (21 CFR 186)

Not all substances that are GRAS are listed in the FDA's regulations. The FDA has instituted a procedure whereby someone may inform the FDA of their own GRAS determination. A list of these GRAS notices, with the FDA's response letter to the notifier, is also available at "Summary of all GRAS Notices."

- **Consult 21 CFR 181 to see whether the use of the component is listed as Prior Sanctioned.***

  Prior Sanctioned substances are those substances whose use in contact with food is the subject of a letter issued by the FDA or USDA before 1958 offering no objection to a specific use of a specific substance.

- **Consult the listing of Threshold of Regulation Exemptions to check whether the component is exempted from a petition or an FCN as a food additive because it becomes a component of food at levels that are below the threshold of regulation.**
  A substance used in a food contact article may be exempted by the FDA from the need of an FCN or a petition (regulation) as a food additive if the use in question has been shown to result in a very low concentration (0.5 ppb). For details, see "Submitting Requests under 21 CFR 170.39 Threshold of Regulation for Substances Used in Food Contact Articles."
- **Consult the listing of effective Food Contact Substance Notifications.**
  The listing of effective food contact substance notifications, the regulation, guidance documents, and additional information regarding the notification program are listed on the Food Contact Substance web page. However, you should be aware that FCNs are proprietary and users must be able to trace the substance they use back to the manufacturer for which the notification is effective.
- **Submit a Food Contact Formulation (FCF) Notification.**
  Occasionally, individuals may wish to verify compliance of the components of a particular food contact material. In such instances they may submit a notification for a food contact substance formulation. The purpose of a formulation notification is to verify that components of a food contact material may legally be used and not to authorize a new food contact substance. Because these notifications are for the purpose of regulatory status and not safety assessments, notifications for formulations do not require resubmission of the information supporting the safety of the intended use of each food contact substance in the formulation. A notifier for a formulation need only submits a completed FDA Form 3479 and any additional documentation required to establish that each of the components of the formulation is authorized for its intended use. In cases where the basis for compliance of an individual FCS in a formulation is an effective notification, a notifier of the formulation should establish that he can rely on the notification cited for the intended use of the FCS in the formulation.

18.4 WEBSITES LISTED

http://ec.europa.eu/environment/biocides/index.htm
https://webgate.ec.europa.eu/env/r4bp
http://ec.europa.eu/environment/biocides
http://ecb.jrc.ec.europa.eu/biocides/
http://circa.europa.eu/Public/irc/env/bio_reports/library?l=/manual_decisions
http://ec.europa.eu/environment/biocides/withdrawals.htm

http://ec.europa.eu/environment/biocides/non_inclusions.htm

http://ec.europa.eu/environment/biocides/pdf/list_dates_product_phasing_out.pdf

http://ec.europa.eu/environment/biocides/pdf/list_participants_applicants_subs.pdf

http://ec.europa.eu/environment/biocides/pdf/list_participants_applicants_prod.pdf

https://webgate.ec.europa.eu/sanco_foods/main/?event = substances.search&substances.pagination = 1

http://www.epa.gov/pesticides/PPISdata/index.html

http://ppis.ceris.purdue.edu/

http://www.epa.gov/opprd001/inerts/section25b_inerts.pdf

http://www.epa.gov/opprd001/inerts/inert_nonfooduse.pdf

http://www.epa.gov/pesticides/DataSubmittersList/

http://www.epa.gov/pesticides/pestlabels/

CHAPTER 19

HUMAN SAFETY AND ENVIRONMENTAL CONCERNS ASSOCIATED WITH THE USE OF BIOCIDES

ORLA CONDELL,[1] CAROL IVERSEN,[2] KAREN POWER,[1] and SÉAMUS FANNING[1]
[1] University College Dublin, Belfield, Dublin, Ireland
[2] Nestlé Research Center, Lausanne, Switzerland

CONTENTS

19.1 Risks to Health Associated with the General Use of Biocides 528
19.2 Human Safety Concerns Arising from Biocide Misuse 530
19.2.1 Tolerance to Biocides 530
19.2.2 Categories of Biocide Tolerance 531
19.2.3 Mechanisms of Biocide Tolerance 533
19.2.4 Misuse of Biocides and Cross-Resistance with Clinically Important Antibiotics 537
19.2.5 Biocide Cross-Resistance to Preservatives 539
19.2.6 Biocides May Select for Strains Displaying Enhanced Virulence or Resistance to Stresses 540
19.2.7 Implications for and Prevention of a Biocide-Induced Enhanced Tolerance 541
19.2.8 Human Health Impacts of Biocide Use 542
19.3 Environmental Impacts of Biocide Use 544
19.3.1 Biocide Release, Exposure, and Accumulation in the Environment 544

Antimicrobial Polymers, First Edition.
Edited by José M. Lagarón, María J. Ocio, and Amparo López-Rubio.

19.3.2 Toxicity of Biocides and the Food Chain 545
19.3.3 Breakdown of Biocides in the Environment May Lead to the Formation of Compounds with Increased Toxicity 546
19.3.4 Interactions of Multiple Biocides May Lead to a Synergistic Toxicity 547
19.3.5 Reducing the Impact of Biocides on the Environment 547
19.4 Biocides and Food Packaging 548
19.4.1 Antimicrobial Packaging—An Overview 548
19.4.2 Examples of Antimicrobial Packaging 548
19.4.3 Biopolymers 549
19.4.4 Antimicrobial Agents Used in Food Packaging 550
19.4.5 Challenges to the Use of Antimicrobials in Food Packaging 552
19.4.6 Conclusions Regarding Biocides and Food Packaging 554
19.5 Summary and Future Trends 554
References 555

19.1 RISKS TO HEALTH ASSOCIATED WITH THE GENERAL USE OF BIOCIDES

Over the past decade, there has been an increase in the use of various biocides in industry, in the clinical environment, and in the home. The range of commercially available biocide-based products has also increased substantially. This increase in product range is correlated with a shift in use from controlled and specific applications, typical of the clinical setting, to the more casual and unreserved use in the modern household [1].

In the food industry, an increase in the scale of processed food allied to consumer demands for healthy nutritious and minimally processed food devoid of additives, such as chemical preservatives and other antimicrobial agents, has had an important impact on the volume of biocide(s) used in this environment. In attempts to ensure food safety and improve hygiene measures, the food industry has increased its use of biocides and chemical-based disinfection to control the microbial ecology of the production environment [2]. Similar trends have also been observed in the home wherein consumers are encouraged through advertisements to use these compounds to eliminate microbial contamination on food preparation surfaces and elsewhere [3]. This marketing strategy is contributing to an increased liberal use of biocides by consumers aimed at the decontamination of *at-risk* areas in the home [4]. The latter strategies trade on a lack of consumer understanding of the issues concerned and the fear associated with a family member falling ill.

In the clinical setting, biocides have been used for decades. Chemical disinfectants are routinely used, following carefully developed application protocols to control microorganisms in the hospital environment, on medical devices, and various surfaces including the skin and mucosa.

Increased exposure of microorganisms to humans and the environment alike to biocides is not without risk. Concern has been growing over the potential for the overuse of biocides to select for bacteria and other microbes displaying an increased tolerance to these chemical agents. Bacteria with the ability to survive treatment with different classes of biocides have emerged and the possibility of associated cross-resistance to valuable clinical antibiotics is an important public health concern [2, 3]. Recently, attention has focused on the link between biocide resistance in spoilage microorganisms and cross-resistance to preservatives designed to control their growth [2]. The impact of biocide use in industrial settings is affecting beneficial bacteria, which are also known as probiotics, thereby limiting their desirable effects [5]. Additionally, the potential for biocides to select for more virulent strains of bacteria has recently been highlighted [6, 7]. The application of a biocide imposes a selective pressure that drives the adaptation of microorganisms to modify their genetic makeup enhancing their survival. These altered genetic and physiological traits may lead to the emergence of more virulent bacterial strains. Additionally, the selective pressure imposed after the use of biocides may result in the emergence of bacteria with the capacity to evade various processing stresses, such as pH, heat, and others, that are designed to eliminate them from the food production line.

Humans encounter biocides on a regular basis, although the specific exposure levels remain to be determined. Individuals can be exposed to biocides after dermal contact and through inhalation routes [8] arising from the use of a range of personal care products, cleaning and sanitizing agents, as well as textiles, children's toys, cutting boards, toothbrush handles, and even bowling ball finger inserts [9]. Reports have also highlighted the increasing exposure of individuals to biocides *via* the oral route from residues that remain in food and beverages after the use of biocides during production [8]. Concerns have been raised in regard to the toxic [10] estrogenic [11], allergenic [12], and in some cases even carcinogenic effects [13] these exposures may have on public health.

Several studies have outlined the widespread exposure of the environment, as well as aquatic, terrestrial, and oceanic life forms, to biocides from various sources. Evidence has shown that many biocides accumulate in the environment because of their recalcitrant nature, or bioaccumulate to levels where they can have a detrimental effect on indigenous environmental biospecies [14]. Additionally little information is available describing the catabolic fate of biocides after their release. Although toxicity studies are often carried out on the chemical component(s) themselves, this testing frequently takes into account only the acute effect of the biocide(s) and not the chronic long-term effect(s) after release into the environment. Similarly, the toxicity of the breakdown products once the original biocide is degraded are not investigated, and toxicity testing does not take into account the potential synergistic negative effect of multiple biocides and associated breakdown products in the environment, something that is near impossible to study in the laboratory setting. Nonetheless, the latter is gaining in recognition as being harmful to the complex interaction and balance between organisms and resources in the environment.

19.2 HUMAN SAFETY CONCERNS ARISING FROM BIOCIDE MISUSE

19.2.1 Tolerance to Biocides

Biocides play an essential and effective role in limiting the spread of infectious agents and disease. However, consideration must be given to the ever-increasing use of these agents and the associated risk to human health. The current scientific literature continues to report on the links between the overuse and abuse of biocides in the clinical environment and the subsequent selection of bacterial isolates displaying an increased tolerance to these agents. Specifically, evidence points to a relationship between tolerance to biocides and the emergence of bacteria resistant to clinically important antimicrobial compounds. This feature contributes to the increased risk of therapeutic failure, rendering certain classes of antimicrobial compound essentially useless. Undoubtedly, this has the potential to pose a substantial threat to public health.

Tolerance to biocides has been documented for most classes of agent. It can also be selected for *in vitro* in the laboratory. Unlike antimicrobial compounds, biocides do not have a distinct bacterial cell target. Consequently, biocidal tolerance is mediated by mechanisms that are less well defined compared with those involved in antibiotic resistance. Antibiotic resistance often emerges after one or more mutations in a distinct antibiotic target gene. In contrast, tolerance to a biocide, reflecting the multiplicity of their action, often involves much broader cellular changes, such as an upregulated efflux pump activity or alterations in cell wall permeability [15].

Biocides are typically applied at concentrations far exceeding those used for antimicrobial compounds without the same toxicity concerns [3]. At high concentration biocides cause cell death from chemical and physical interactions (Figure 19.1). Oxygen-releasing biocides facilitate the indiscriminate oxidation of a range of membrane and cytoplasmic thiol- and other chemical groups leading to the inhibition of multiple key enzymes and modification of structural proteins [3, 16].

It has been proposed that when used incorrectly, the biocide loses its spectrum of activity. Examples of incorrect use include the use of the biocide at a concentration below that which is recommended, the application of the biocide in the presence of organic matter (which can interfere with and disrupt the biocidal activity), or if an initial high concentration of biocide is applied but allowed to remain in the environment where the concentration reduces overtime. Even when applied at sufficiently high concentrations and used under optimal conditions, in an environment a gradient of biocide concentration will exist ranging from a high level, to a sublethal concentration through to no active agent [3].

A sublethal concentration of biocide can select the most tolerant bacteria within a bacterial population by creating a selective pressure favoring those cells in the population that overexpress a biocide-degrading enzyme. Isothiazolone at high concentrations interacts nonspecifically with thiol groups, but at low

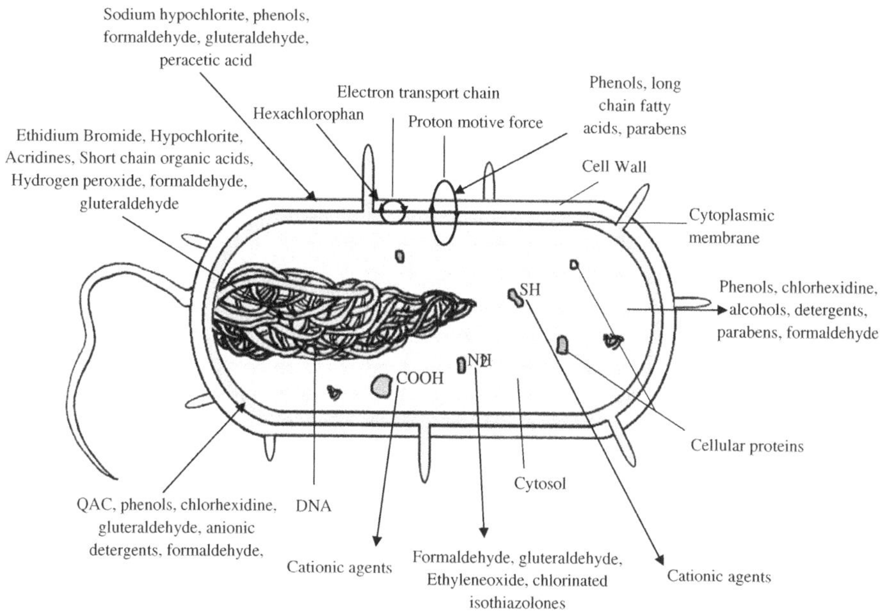

FIGURE 19.1 Sites of action of biocides on the typical bacterial cell.

concentrations it acts specifically on various respiratory enzymes [3]. Thus, at these concentrations and in this case, tolerance could develop through the mutation of a distinct target [1]. Theoretically, sublethal concentrations of biocides could also promote the spread of mobile biocide resistance determinants among surviving cells. Although to date, little evidence of this possibility has been reported [3].

Biocide tolerance can be divided into several categories (Table 19.1). First, an intrinsic tolerance can be described, wherein an entire species of bacteria demonstrates a reduced susceptibility to a specific biocide, or class of biocide (Figure 19.2). Second, extrinsic resistance can also occur, wherein a bacterial isolate alone demonstrates reduced susceptibility to a biocide where other members of that species are completely susceptible to the same agent. The third group can be described by a phenotypic population or adaptive tolerance.

19.2.2 Categories of Biocide Tolerance

Intrinsic tolerance to biocides has been attributed to several features within bacteria designed to reduce their intracellular concentration [17], thereby preventing the biocide from exerting its effect. These intrinsic resistance mechanisms involve features of the bacterial cell that can act either to decrease or to limit the permeability of the cell to one or more biocides, or that can efflux the biocidal agent from the cell after entry [18–20].

TABLE 19.1 Tolerance features and associated mechanisms of biocide tolerance

Tolerance	Resistance mechanism	Examples
Intrinsic	Permeability	Tolerance to multiple agents due to Mycoylachyarabinogalactan in *Mycobabacterium*
	Efflux	QAC, phenols, and chlorhexidine tolerance in gram-negative organisms for example because of MexAB-OrpM efflux system in *Pseudomonas aeruginosa* and the AcrAB-TolC efflux system in *Escherichia coli* In gram-positive for example because of the QuaA-D, Smr efflux systems in *Staphylococcus aureus*.
Extrinsic	Permeability	Phenol, chlorhexidine and cetrimide tolerance in *Escherichia coli* due to an RP1 plasmid, reducing the level of porin protein expression and altering the LPS composition
	Efflux	*qac*A/B, efflux genes, in *Staphylococcus* species associated with a tolerance to antiseptics
	Target alteration	Triclosan tolerance associated with mutation in the *fabI* gene and *fabI* homologues in *Escherichia coli*, *Staphylococcus aureus*, and *Mycobacterium smegmatis*
	Degradation	Formaldehyde tolerance due to plasmid encoded formaldehyde dehydrogenase found in *Escherichia coli* and *Serratia* Chlorhexidine, phenylethanol and QAC degradation in species of *Pseudomonas*
Adaptive	Biofilm	Increased triclosan tolerance in *Salmonella* and *Proteus*, cells present as biofilms Increased tolerance to peroxides in *Listeria* compounds
	Stress response	Tolerance to pine oil, triclosan and multiple antibiotics through the induction of the MDR phenotype in *Pseudomonas aeruginosa*

In contrast to the preceding discussion, extrinsic tolerance develops after mutation in part of the bacterial chromosome, upregulation of specific bacterial genes, or through the acquisition of extrachromosomal elements such as plasmids or transposons that encode resistance determinants. Extrinsic tolerance generally occurs as a result of one or more of four recognized mechanisms. These tolerance mechanisms include target site modification, cell permeability alteration, upregulation of membrane-associated efflux pumps, and target inactivation [21].

Population or phenotypic tolerance is an increased tolerance to biocides because of a phenotypic trait or pattern of gene expression. For example, when

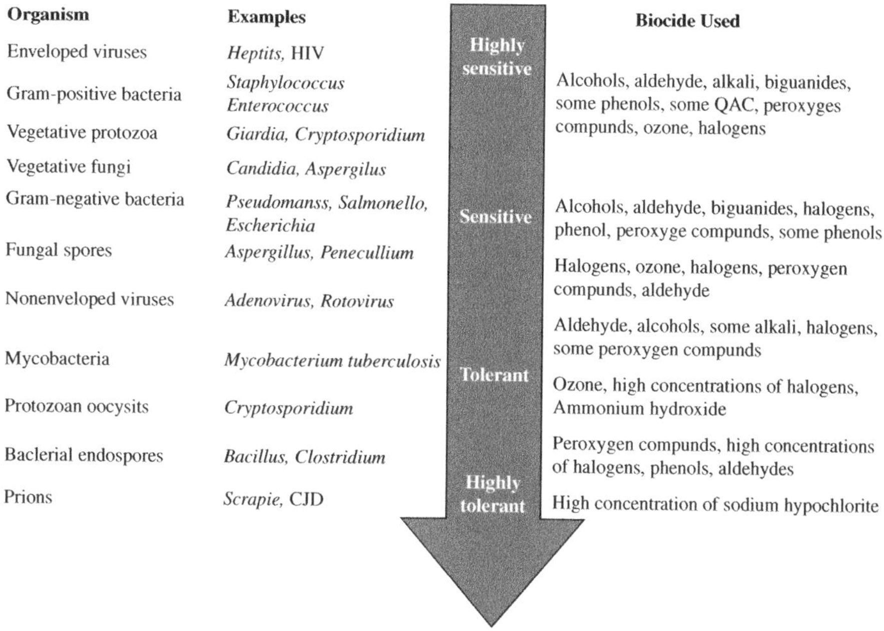

FIGURE 19.2 A schematic illustration showing the relative susceptibility of different microorganisms to biocides.

bacteria form part of a biofilm or in a surface-attached state, the bacterial population demonstrated a higher resistance collectively compared with that demonstrated by an individual bacterial cell in its planktonic state [3, 19]. Similarly, if bacteria are preexposed to a sublethal concentration of a biocide these organisms may induce a stress response, altering patterns of gene expression that then result in a modified phenotype [19, 22]. Some bacteria may become persistors, or dormant cells, displaying a higher tolerance to the adverse physical environment or chemical compounds present therein, without any mutation or genetic exchange.

19.2.3 Mechanisms of Biocide Tolerance

19.2.3.1 Modifications in Cell Permeability. Certain bacterial traits, including intrinsic, extrinsic, or phenotypic mechanisms, which lead to a decreased cell permeability, have been linked to a tolerance to biocides and antimicrobial compounds alike.

Some gram-negative bacteria, in particular enteric *Enterobacteriaceae* and *Pseudomonas* species, display a higher intrinsic resistance to biocides than gram-positive organisms. This increased resistance can be attributed to a reduction in the permeability toward the active compound. Unlike gram-positive bacteria, gram-negative organisms have a multilayered membrane structure containing an outer membrane. This outer membrane acts as a barrier and provides for an increased tolerance to biocides. Components of the outer membrane, and indeed the structure of the outer membrane itself, have been found to prevent, or limit, the penetration of biocides [17, 19].

The lipid content of the outer membrane, particularly the lipopolysaccharide (LPS) layer, which is a unique component of the outer membrane in some gram-negative bacteria, forms a rigid cell barrier that limits or slows the diffusion of hydrophilic molecules into the cell [17]. The role of LPS as a permeability barrier in gram-negative bacteria has been extensively reported.

Water-filled outer membrane protein channels, or porins, allow for the selective entry of molecules into cells. Porins facilitate the entry of small hydrophilic molecules while restricting the influx of hydrophobic and large hydrophilic molecules [15, 17]. The outer membrane proteins of *Escherichia coli*, such as OmpF, are examples of such membrane porins [17]. Other membrane characteristics such as surface hydrophobicity, for example, through certain LPS modifications have also been shown to influence the permeability of disinfectants into the cell [17, 19].

Gram-positive endospore-forming bacteria also display an intrinsic resistance. Endospores are the most resilient cell from known. These organisms are resistant to most biocides, such as alcohols, biguanides, esters, phenols, quaternary ammonium compounds (QACs), organic acids, and organomercurials, which are usually lethal to vegetative cells. The spore structure provides a robust physical barrier to many biocides, rendering them ineffective as sporocidal agents [23].

Extrinsic resistance often involves alterations in membrane proteins, lipid content, membrane ultrastructure, surface hydrophobicity, phospholipids, fatty acid composition, and electrical charge properties [21]. For example, studies showed that resistance to QACs was induced in a staphylococcus strain after the a modification of a single amino acid in a membrane permease. In *E. coli*, the presence of the RP1 plasmid reduced the level of porin protein expression and altered the LPS composition. These cellular modifications were associated with an increased tolerance to phenol, chlorhexidine, and cetrimide [24].

19.2.3.2 Efflux Pump Activity. Bacteria demonstrate intrinsic or extrinsic tolerance to biocides because of a mechanism called efflux [15, 17]. Biocides that can penetrate into the bacteria cell are actively extruded or pumped from the cell. In this example, biocides do not reach a sufficiently high intracellular concentration within the cell to be bactericidal. The efflux of biocidal compounds is mediated by one or more membrane associated transport systems.

Efflux pumps are membrane-associated proteins that extrude a range of structurally dissimilar toxic compounds from the cytoplasm of the cell. Five classes of efflux pumps are recognized and found widespread among bacteria. These include the small multidrug resistance (SMR) family [now part of the drug/metabolite transporter super-family], the major facilitator super-family

(MFS), the ATP-binding cassette (ABC) family; the resistance-nodulation-division family, and the multidrug and toxic drug extrusion super-family [25].

In gram-positive bacteria, efflux determinants are often located on plasmids, where biocide tolerance will develop after the acquisition of a plasmid encoding these genes [21]. Genes encoding SMR and MFS families of efflux pumps are frequently located on these plasmids. The *qac*A/B-encoding efflux pumps have been found in *Staphylococcus aureus* and other *Staphylococcus* species displaying resistance to antiseptics [26].

Gram-negative bacteria such as *E. coli* possess chromosomally encoded broad-spectrum efflux pumps that can actively expel a wide range of compounds, including biocides from the cell. Expression of these efflux pumps is induced by exposure to a wide range of chemical agents [17].

An example of such an efflux system includes the AcrAB-TolC, AcrEF-TolC, and EmrE found in *E. coli* and *Salmonella* [15]. AcrAB has been found to contribute to resistance to QAC compounds, chlorhexidine and triclosan as well as possessing the ability to pump tetracycline, ciprofloxacin, fluoroquinolones, β-lactams, novobiocin, ethidium bromide, acriflavine, phenylethylalcohol, sodium dodecyl sulfate, and deoxycholate from the bacterial cell [24].

Despite their ability to export biocides, chromosomally encoded efflux pumps are not known to promote acquired tolerance to biocides. However, recent studies showed the overexpression of such efflux pumps has been associated with an increase in tolerance to biocides. For example, a constitutive *mar* mutant of *Pseudomonas aeruginosa* hyperexpresses efflux pumps including the MexCD and MexJK systems. These can extrude many biocides such as triclosan and antimicrobial compounds including fluroquinolones [3].

Efflux genes associated with biocide resistance have also been found associated with mobile genetic elements in other gram-negative bacteria. In *Acinetobacter* species, *E. coli, Salmonella* species, *Pseudomomas* species, and others, transporters such as QacE and QacE 1 have been found located on plasmids and on the chromosome [27].

19.2.3.3 Target Alteration. For any biocides with a specific target or site of action, a mutation in this target could potentially alter the susceptibility of that cell to the biocide. Because of the target multiplicity of the antimicrobial action of most biocides, this form of resistance is rare [15]. However, it has been reported for triclosan, which is a phenolic biocide common in hand soap, shampoos, and other cleaning products [28]. Triclosan inhibits FabI, an enzyme that functions as an enol acyl-carrier protein (ACP) reductase, involved in fatty acids biosynthesis [15].

It has been demonstrated *in vitro* that exposing *E. coli* cells to subinhibitory levels of triclosan can lead to the selection of cells with mutations mapping *to fabI* or involving the deletion of the *fabI* gene [3]. Similarly, mutation(s) in the *inh*A gene in *S. aureus* (the gene encoding the FabI homolog) have been shown to render this bacterium resistant to triclosan, although it is thought that other factors also play a role [15].

19.2.3.4 Enzymatic Inactivation. The enzymatic inactivation of biocides can also confer a cell with a biocide-tolerant phenotype. This mechanism is less common than the enzymatic inactivation of antibiotics [3]. Biocide inactivation has been reported for numerous heavy metals used as biocides; for example, mercury resistance has been attributed to the plasmid-encoded *mer* genes [21]. The inactivation of aldehydes in some species of *Pseudomonas* has also been reported, as has inactivation of chlorhexidine [3, 21], phenylethanol, and QAC compounds. Tolerance from the degradation of compounds including formaldehyde have also been reported for *Pseudomonas putida* [3] and, more recently, was linked to a plasmid encoded formaldehyde dehydrogenase identified in *E. coli* and *Serratia* species [21].

19.2.3.5 Biofilms. During the last few years, population-mediated or phenotypic tolerance to biocides has been recognized as an important mechanism of resistance *in vivo*. Most bacteria *in vivo* are not present as free-living planktonic cells but rather as surface-attached communities called biofilms. Populations of bacteria in biofilms display orders of magnitude higher biocide resistance compared with their planktonic counterparts [19, 29]. The bacterial communities consist of an organized three-dimensional structure of multiple bacterial species embedded in an exopolysaccharide matrix, or glycocalyx [30]. Biofilms containing *Salmonella* and *Proteus* display an increased resistance to triclosan [19, 31, 32]. Similarly, *Listeria*, which is present in a biofilm display increased resistance to peroxides [33], and *Cronobacter* (*Enterobacter sakazakii*) in a biofilm shows enhanced resistance to quaternary ammonium compounds [19, 34].

Increased tolerance to biocides in a biofilm has been attributed to multiple factors. First, the extracellular matrix acts a physical barrier, limiting diffusion of the biocide, to prevent it from contacting and, therefore, entering the bacterial cells. The glycocalyx also contains trapped extracellular enzymes, which are concentrated within its matrix. If a species can produce an inactivating enzyme to a particular biocide is present in a biofilm, it will confer a resistance on the entire community [30].

Additionally, the resistance to biocides in biofilms can be attributed to an alteration in the physiological state of the bacteria within the biofilm [17]. Bacteria present in biofilms have a slower growth rate because of the limited availability of nutrients. Thus, bacterial cells existing in such a state may be metabolically inactive or dormant [19, 35]. Slow-growing cells are generally more resistant to biocidal compounds than rapidly growing cells. Similarly, the biofilm-related phenotype is associated with an alteration in outer membrane proteins that reduce the permeability of the cells as well as an upregulation in efflux, which also reduce the biological effect of the biocide [17].

19.2.3.6 Stress Adaptation. Stress adaptation is a form of phenotypic tolerance regarded as transient and dependent on the prevailing physiochemical conditions the bacterium is encountering at that time. It has been suggested that the general stress response of a cell to conditions (such as the depletion of nutrients, the transition into stationary phase, or exposure to sublethal concentration of a toxic chemical) is associated with an increased tolerance to various physical and chemical agents.

The general bacterial stress response, or stress adaptation, is associated with an altered pattern of gene expression including a reorganization of metabolic pathways. This is a nonspecific, or general response, where a response to one stress can confer cross-resistance to other stresses. As an example, certain multigene pathways may be overexpressed including *soxRS*, *oxyR*, *mar*, and production of ppGpp is increased in response to a stress [19, 36, 37]. These pathways lead to the expression of genes that aid in the survival of the cell under adverse conditions. For example, the stress response is associated with induction of the multidrug resistance (MDR) phenotype. This phenotype confers a resistance to multiple chemically and structurally unrelated compounds and has been found to cause an increased tolerance to triclosan and other natural toxic compounds such as pine oil [38].

19.2.4 Misuse of Biocides and Cross-Resistance with Clinically Important Antibiotics

Much concern has been raised in the literature over the potential for the overuse and abuse of biocides to select for antibiotic-resistant strains. It has been proposed that cross-resistance to antimicrobial compounds, after exposure and adaptation to a biocide could occur in one of four situations, as follows: first, if a biocide and the antibiotic act on the same cellular target; second, if a biocide and the antibiotic have the same transport mechanism; third, if a biocide and antibiotic can be accommodated by the same resistance mechanism [3]; and finally, if the biocide and antibiotic resistance determinant are carried on the same mobile genetic element (Table 19.2) [21].

19.2.4.1 Common Targets Between Biocides and Antibiotics. Several bacterial targets of biocides are also the targets of clinically important antibiotics. It has been reported that mutations that develop after the exposure

TABLE 19.2 Mechanisms of bacterial tolerance to biocide, potential for transfer, and cross-resistance with other antimicrobials

Mechanism	Class of resistance	Antibiotic cross-resistance	Biocide cross-resistance	Transferable
Efflux	Intrinsic/extrinsic	Yes	Yes	Yes
Impermeability	Intrinsic/extrinsic	Yes	Yes	
Target alteration	Intrinsic/extrinsic	Yes	No	
Degradation	Extrinsic	No	No	Yes
Stress response	Adaptive		Yes	
Biofilm formation	Adaptive	Yes	Yes	No

to biocides can lead to the development of cross-resistance to antimicrobial compounds. Examples of these include the front-line antibiotic used in the treatment of *Mycobacterium tuberculosis*, isoniazid. This drug targets enoyl reductase, which is also a target for the biocide triclosan. Triclosan resistance can arise because of a mutation in *inhA*, a *fabI* homolog, which provides the bacteria with a cross-resistance to isoniazid [21]. Previously it was reported that deletion of *inhI*, increases the MIC of triclosan by 4- to 6.3-fold and similarly increases the MIC of isoniazid 1.2- to 8.5-fold. FabI is also a target for the diazaborine family of antimicrobials and the biocide hexachlorophenol in *E. coli*, and it is possible that cross-resistance could be induced with these compounds in a similar manner. This commonality of targets is not confined to *fabI* and can be found with the subinhibitory concentrations of many biocides. Chlorhexidine and gentamicin have a nucleic acid binding site in common, and a resistance determinant conferring reduced susceptibility to both has been identified [3, 39].

19.2.4.2 Common Transport Mechanism. "Self-promotion" is defined as the transport of an antimicrobial agent into the cell after the interaction of the agent with the cell surface, in particular with cell envelope associated cations. This interaction results in the destabilization of these divalent cations. The destabilization induces a reorganization of the lipopolysaccharide and the formation of permeable pores, facilitating the entry of the agent into the cell. [3, 40]

Self-promotion transport is shared between polymeric biguanide biocides and aminoglycoside antibiotic classes of a compound. Changes in the cell envelope that result in a reduced binding of cations to the cell envelope can lead to a reduction in the susceptibility to both classes of agent. This was found to be the case with a polymyxin-resistant *P. aeruginosa*, which also displayed a reduced susceptibility to quaternary ammonium compounds [3].

19.2.4.3 Common Resistance Mechanisms. Some determinants that cause an increase in biocide tolerance are nonspecific and may be mediated through a decrease in cellular permeability or an increase in efflux. If the use of a biocide selects for bacteria displaying these traits, they may also have a reduced susceptibility to antimicrobial compounds that can be accommodated by the same resistance mechanism. An increase in efflux pump activity selected for by exposure to biocides can result in a concurrent increase in the resistance to antibiotics. The exposure of a susceptible population of a *mexAB* deleted *P. aeruginosa* to triclosan, selected for strains that were multidrug resistant because of the hyperexpression of the *mexCD* efflux system [39]. Similarly, strains that constitutively express the Mar regulon demonstrate a change in their metabolic pathways and overexpress up to 60 genes [4], including *acrAB* [39]. This modification can confer a resistance to an array of chemical agents including triclosan, chloramphenicol, tetracycline, and pine oil [4].

If a biocide and an antimicrobial compound have a common pathway of transport into the cell, then a mutation that renders the cell less permeable to that biocide can also confer a cross-resistance phenotype to that antibiotic or class of antibiotic. For example, the mechanism by which aminoglycosides and many biocides, such as polymeric biguanide, enter the cell is common. Changes that affect this transport mechanism, such as a decrease in the envelope-associated cations, could decrease the susceptibility to biocides and antibiotics alike. In *P. aeruginosa*, a cross-resistance to polymixin and QACs was identified [3].

19.2.4.4 Mobile Genetic Elements. If a mobile genetic element encodes genes for both biocide and antibiotic resistance determinants the long-term exposure to subinhibitory levels of that biocide could select for strains containing these elements. The disinfectant resistance gene *qacC*, an efflux pump [41], has been found together with genes encoding resistance to kanamycin, gentamicin, and trimethoprim on a conjugative plasmid [2]. A gentamicin resistance strain of MRSA was found to be carrying a multidrug-resistance plasmid, pSAJ1, which contained resistance determinants for multiple antimicrobial agents such as ethidium bromide, benzalkonium chloride, chlorhexidine and amionoglycosides [4, 42].

19.2.5 Biocide Cross-Resistance to Preservatives

Many biocides used as disinfectants are also used as preservatives in the clinical, pharmaceutical, food, and cosmetics industries. QAC, chlorhexidine, and triclosan are all used in the food industry as biocides and as preservatives. Cross-resistance can occur when chemicals used as biocides are also employed as preservatives. Alternatively, biocide cross-resistance with preservatives not used as disinfectants can occur when the preservatives and disinfectants have a similar mechanism of action and/or tolerance [43]. However, preservatives are typically used at concentrations below those used as sanitizers and disinfectants; therefore, bacterial tolerance may pose more of a threat to the failure of preservatives than disinfectants [44].

Electrophilic-based preservatives have a similar mechanism of action to the peroxide disinfectants. Preservatives such as formalydehyde and chloromethylisothiazolone, as well as disinfectants like hydrogen peroxide and hypochlorous acid, exert their bacteriocidal activity on the cell through the release of free radicals. In the latter example, a resistance mechanism mediated through the *oxyR* radical defense system, following the expression of the *oxyR* regulon, can be induced through exposure to peroxide disinfectants that can result in cross-resistance between these two classes of chemical agents. Cross-resistance has also been reported between formalydehyde and hydrogen peroxide, chloromethylisothiazolone and hypochlorous acid, as well as chloromethylisothiazolone and hydrogen peroxide [43, 45].

Some preservatives can, like antibiotics, be accommodated by broad-spectrum resistance mechanisms that also mediate tolerance to biocides. An example of this would be the multidrug efflux pump NorA in *S. aureas*, which can efflux preservatives berberine, palmatime, and disinfectants chlorhexidine and benzalkonium chloride [46]. In this case, an increased tolerance to the biocides chlorhexidine and benzalkonium resulting from efflux could lead to a cross-resistance to preservatives.

The extensive use of biocides as preservatives in the cosmetics industry, for example the use of parabens, formaldehyde, and phenoxy-ethanol, resulted in the development of resistant bacteria [19, 47]. In response, more preservative chemicals and mixtures of preservative chemicals are added to cosmetic products, resulting in the increased exposure of humans to these agents. Although these bacterial isolates of relevance in this case are usually considered as spoilage microorganisms, pathogenic organisms like *P. aeruginosa* have been cultured from cosmetics. In the latter example, this isolate was resistant to multiple preservatives and to clinically important antibiotics [19, 48].

19.2.6 Biocides May Select for Strains Displaying Enhanced Virulence or Resistance to Stresses

19.2.6.1 Enhanced Virulence. It has been proposed that factors conferring biocide tolerance may also promote an increase in virulence among some strains. It has been shown that the growth of *Porphyromonas gingivalis* in the presence of low levels of triclosan selects for strains with an increased resistance to defensins. In addition, in the same strain, triclosan altered the expression of 13 genes associated with virulence, 9 were upregulated, and 4 were downregulated. This finding may suggest that biocide use can alter the pattern of virulence gene expression and, in this example, can affect the progression of periodontal infections caused by this bacterium [49]. In a similar manner, it was reported that in a collection of enterotoxigenic *E. coli* (ETEC) increased tolerance to chlorhexidine correlated with four virulence factor genotypes. These authors suggested that the use of chlorhexidine was selecting for ETEC [7]. An extensive study carried out using 11 disinfectants, of 4 classes, commonly used in the food industry examined the effect of sublethal concentration of biocides on the expression of virulence genes in the food-borne pathogen *Listeria monocytogenes*. Interestingly, the results showed that the effect on the transcriptional pattern of virulence genes was dependent on the class of disinfectants. Peroxy- and chlorine-based compounds decreased the level of expression of all virulence genes examined, whereas QAC induced the expression of virulence genes [50].

19.2.6.2 Enhanced Tolerance to Food-Processing-Related Stress. The misuse and exposure of bacteria to biocides can induce a stress response in exposed cells. This stress response may also drive these bacteria toward the development of a higher tolerance to commonly used food-processing methods designed to inactivate them. In the food-processing industry, heat and a low pH are often used as a control measure to ensure the safety of the final food product.

It is known that a sublethal concentration of biocide induces several stress response pathways including SoxRS and the SOS pathways. These pathways induce a combined nonspecific and specific stress response. They exert a level of control over genes coding for proteins designed to protect the cell from other stresses including heat, desiccation, and extreme pH. In fact, it was shown that exposure of *E. coli* to the biocide polyhexamethylene biguanide induces the heat shock response (σH), this leads to the upregulation of many stress genes that protect the cell from heat-related damage. In this example, a 7.4-fold upregulation of a chaperone heat shock protein 70 was recorded. The exposure of *E. coli* to this biocide also induced changes in global metabolism, a transcriptional change, that led to the upregulation of genes associated with alkali, acid, and osmotic stress [51].

19.2.7 Implications for and Prevention of a Biocide-Induced Enhanced Tolerance

The development of biocide tolerance has important implications for public health. This tolerance would undoubtedly compromise the important role of disinfectants in controlling the spread of infectious disease and reducing the contamination and spoilage of food, pharmaceuticals, and medicines.

There is a possibility that the overuse and abuse of biocides could continue to have implications in the future, particularly with respect to the emergence of antibiotic resistance. Sublethal exposure of a bacterium can lead to the alteration of specific biocide targets. If these targets were the sole targets for clinically important antimicrobial compounds, then the overuse of biocide could be reflected in the emergence of resistance to antibiotics and subsequent treatment failure. This scenario might therefore not be detected as a failure of the biocide [24]. Overuse of biocides could also lead to the development of a pool of environmental biocide tolerant bacteria that contain resistance-encoding mobile genetic elements that could transmit this genotype to other bacteria including pathogens [2]. Additionally, constant low levels of biocide could select for the persistence of antibiotic-resistant strains in the absence of any antibiotic(s) selection [24]. Biocide overuse could exacerbate the future discovery potential for new antibiotics, especially those sharing the same bacterial cell target [24]. The fatty acid biosynthesis pathway was thought to be an excellent candidate to study for developing new antimicrobials [9]; however, the overuse of triclosan is compromising this as a potential direction for the future [3, 24]

Biocide tolerance can be avoided by following suitable guidelines and using an effective disinfectant and cleaning plan [2]. To avoid the selection of strains possessing an intrinsic resistance (most problems caused by the failure of disinfectants are caused by the persistence of intrinsically resistant strains), it is important to choose a biocide with a spectrum of efficacy that is appropriate to the needs of the area to be sanitized. It has been proposed also that biocide rotation could be an effective way of preventing the development of resistance, but only when used correctly. Disinfectants from different classes of biocides with different mechanisms of action should be rotated. The rotation of biocides could be beneficial when combined with an effective surveillance plan, where the prevalence of persistent strains can be detected and the disinfectant protocol modified accordingly, either by increasing the concentration of biocide currently being used or by changing to a new class of disinfectant [52]. It is important also to ensure that the presence of organic matter does not interfere with the activity of the biocide, that the biocide is stored and applied at the correct temperature and pH and to ensure that residual biocide is not left remaining after the completion of the cleaning routine as might arise after improper rinsing and drying [2].

Additionally, to prevent the development of biofilms along with surface dried bacteria, cleaning frequencies and regimes should be designed carefully, monitored, and modified appropriately. A cleaning plan that ensures a sufficient surface disinfectant contact time, using biocides in a concentration range sufficient to kill cells present in a biofilm, and avoids the dilution of this applied disinfectant with rinsing water or any remaining product should prevent the persistence of adapted strains [52].

19.2.8 Human Health Impacts of Biocide Use

Given the widespread use of biocides and the ongoing exposure of humans to biocides whether it is from cleaning agents, clothing, kitchen ware, cosmetics or food packaging, concern has arisen over the potential toxic effects of these biocides on humans (Figure 19.3).

Biocides, like other chemicals used in the past, such as bisphenol A and p-nonylphenol, may possess an estrogenic activity and/or they may be endocrine-disrupting chemicals. The latter may have serious consequences for human health including being involved in the development of hormone-dependent cancers capable of disrupting the endocrine development of an exposed fetus. In a study by Nakama et al. [11], 12 of 20 organic biocides were shown to possess potential oestrogenic activity, showing an affinity for the human estrogen receptor.

With the level of biocides used in the food and beverage manufacturing industry, there is concern that these biocides could be toxic and ultimately find their way into the final product. For example, ortho-phenylphenol (OPP) is used as a preservative in textiles, cosmetics, and leather among other goods and is also used as a surface disinfectant in hospitals, industry, and the household. Yet OPP is a toxic and potentially carcinogenic compound. Canned soft drinks in the United States and Germany were tested for the presence of this biocide, and 49 of 55 cans tested contained OPP [13]. A similar study surveyed canned beer and detected OPP in samples analyzed from more than 27 countries at

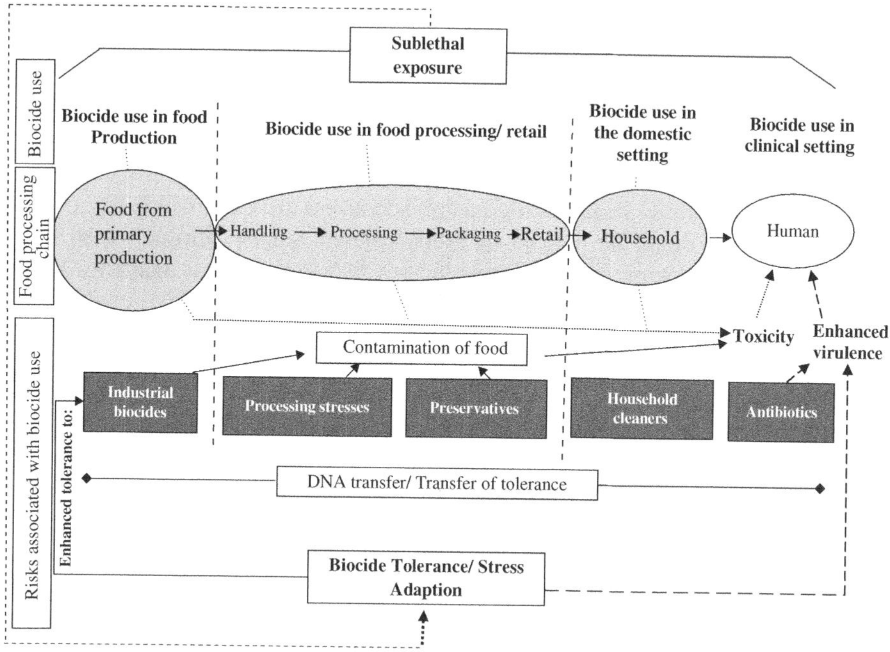

FIGURE 19.3 Risks associated with the overuse of biocides in the food industry. Sublethal exposure of bacteria to biocides along the food-processing chain may lead to the development of biocide tolerance. Biocide tolerance may result in an enhanced tolerance to industrial biocides, processing stresses, preservatives, household cleaners, and antibiotics, as well as an enhanced virulence. This tolerance may be transmittable and may transfer to biocide sensitive bacteria. Biocides may be associated with toxicity to humans.

levels ranging from >0.1 to 40 μg/L [53]. This finding may have serious consequences for human health, as OPP was shown in animal experiments to have a low acute toxicity, leading to bladder cancer in male rats [13]. OPP is also an androgen receptor antagonist and is known to interfere with testosterone receptor binding [53]. This chemical was recognized by the Environmental Protection Agency and was included on the list of chemicals considered likely to be carcinogenic. OPP was included on the European Union prioritization list as being endocrine damaging [54].

Biocides may leach and persist in the environment, and potentially they can enter the food chain. The biocide pentachlorophenol, a compound commonly used as a herbicide/pesticide [55], was one of the most widely used disinfectants in the past. Use of this agent was banned in 1987. However, an environmental modeling study carried out in 1989 showed that it entered the food chain after partition into the soil (perhaps from its use as a wood preservative). This study claimed that humans in the United States were exposed to on average 16 μg of the agent each day [10], and a similar study in the United Kingdom 5 years after the ban estimated a total absorption of 5.7 μg per day [56]. These concentrations are above the Maximum Contaminant Level Goal, the level of a contaminant in drinking water below which there is no known or expected risk to health [57]. At these concentrations, this chemical could potentially cause adverse effects resulting in liver and kidney damage.

Similarly, the allergenic impact of biocide use on human health has been known for some time. Biocides used as preservatives are responsible for a large proportion of contact dermatitis. A study carried out by Schnuch et al. [12] examined the allergenic response to biocides by patch testing against a large cohort of patients with contact dermatitis. More than 28,000 people were studied, using the patch test, for an allergic response against a standard series of preservatives; more than 11,000 individuals were tested with an additional preservative series whereas another 1,700 were evaluated with an industrial biocide tray. After a dermatological diagnosis, sensitization rates in more than 1% of people tested were recorded for the standard panel of preservatives. Many commonly used biocides such as benzalkonium chloride, bromopol, chloracetamide, and alkylaminobenzoate were shown to be important allergens with sensitization rates of 1.8%, 1.1%, 1.4%, and 2.5% respectively. Schnuch et al. [12] suggested that every preservative could be considered as a potential allergen and thus of concern from an occupational and public health perspective.

19.3 ENVIRONMENTAL IMPACTS OF BIOCIDE USE

Large quantities of biocides are used and ultimately released into the environment. Although environmental degradation of biocides has been documented and can occur, it is not a quick process and rarely occurs under optimal conditions, often being incomplete [1]. Thus, biocides tend to accumulate in the environment, where they may be degraded over time or they may persist. This buildup of biocides in the environment can have several negative effects. These effects include causing the death and mutilation of endogenous organisms, disruption of food chain interactions, and selection for tolerant microorganisms. Furthermore, it is difficult to monitor or predict the impact of these biocides or their degraded products in the environment; thus, the fate of biocides released into the environment is largely unknown.

19.3.1 Biocide Release, Exposure, and Accumulation in the Environment

Studies have been undertaken to determine the levels of certain accumulating biocides after their release into the environment, both in aquatic and terrestrial settings. Triclosan is commonly used as a model compound. An estimated 350 tons of triclosan are used in Europe annually, yet little data exist on the behavior of this compound once it is released into the environment. Because of its many uses, it is not unreasonable to suggest that substantial concentrations have already been released into the environment. Triclosan can be degraded by several different mechanisms, including photolytic degradation, chemical degradation (through the Fenton reaction), and microbial degradation both in soil and water. This chemical can be removed during the water-treatment processes, as it is commonly used in water treatment plants. It is estimated that 79% of triclosan is degraded by biological means; another 15% is absorbed by sludge, leaving 6% being partitioned to receiving surface water in such treatment plants [28]. Nonetheless, its accumulation, persistence in the environment, and exposure to endogenous environmental species is almost ubiquitous.

Singer et al. [28] reported that levels of triclosan in waste water, streams, sea water, and sediment ranged from 0.07 to 14,000 μg/L, 50 to 2300 ng/L , 50 to 150 ng/L, and 1 to 35μg/kg respectively. Interestingly, for triclosan, a no observed effect concentration (NOEC) of 500 ng/L and a predicted no effect concentration (PNEC) of 50 ng/L for the algae species *Scenedesmus subspicatus* along with a NOEC of 34 μg/L for rainbow trout has been calculated [58]. Therefore, although triclosan can be degraded in the environment, these latter values suggest that it persists and is accumulating to levels well above its PNEC. This presents an important environmental challenge that needs to be addressed.

19.3.2 Toxicity of Biocides and the Food Chain

Several studies have been carried out recently in order to assess the potential eco-toxicity of biocidal compounds. Tributytin (TBT) is a good example wherein an adverse effect can be attributed to a single compound in the environment. TBT was widely used as a biocide, beginning between the years 1959 and 1961. By 1985, TBT was used in 20% to 30% of shipping vessels worldwide. It is an organotin, which is an organic industrial biocide that was commonly added to antifouling paints. This chemical was also used as a fungicide, bactericide, insecticide, and as a preservative of wood, leather, electrical equipment, and paper, among others [14]. Thus, TBT was released into the environment at high levels by various means, including through leaching from boats. Although TBT degrades quickly in water, usually in a matter of days or weeks, this is not the case for TBT in sediment, wherein it persists [59]. Its half-life is estimated to be between 6 to 9 months, and the breakdown product of this compound also displays a high degree of toxicity [60]. The first reports of TBT-based toxicity arose from commercially cultivated oysters, *Crussostreu gigus*; the oysters decreased in abundance and exhibited shell malformations. TBT had multiple adverse effects on many aquatic animals, such as the development of imposex in aquatic creatures including sea snails *Anassa obsoleta* and dog whelk *Nucelio lapilli*. The toxicity effects of TBT were observed at a concentration as low as parts per billion or trillion [61]. The compound was found at levels in the environment significantly higher than this and was particularly high in estuaries, in coastal regions, and in semiclosed marinas [60].

One main problem regarding the toxicity of biocides in the environment is the phenomenon of bioaccumulation. Although biocides may be released at levels below those considered toxic, the accumulation of a biocide over time may result in subsequent toxic effects. In the case of TBT, a target concentration of 20 ng/L was considered to be safe and was used as a standard by many countries. However, this concentration did not account for the fact that TBT could bioaccumulate, thereby increasing in concentration over time, in the food chain to toxic levels. Some organisms such as fish, crabs, mollusks, and microorganisms can bioconcentrate TBT by three or more orders of magnitude greater than that to which they were originally exposed [14]. TBT is toxic [60] to organisms that would normally be sensitive to lower concentrations of the chemical in the environment and chronically toxic to the organisms exposed to higher concentrations through bioaccumulation in the food chain [60].

19.3.3 Breakdown of Biocides in the Environment May Lead to the Formation of Compounds with Increased Toxicity

Natural breakdown products of commonly used biocides may display a greater toxicity compared with the original biocides themselves, which causes an environmental problem because the chemical structure of the breakdown product may be unknown, and thus, its environmental impact is difficult to predict. A study reported by Fernández-Alba et al. [60] examined the effects of biocides and their breakdown products on an aquatic ecosystem, which included the bacterium *Vibrio fischeri*, the algae *Selenastrum capricornotum*, and the planktonic crustacean *Daphnia magna*. This study used the biocides Irgarol 1051 and diuron, both of which are classified as herbicides. Photodegradation, biodegradation, and chemical hydrolysis of Igarol 1051 occurring in the environment all yield the *n*-dealkylated principle degradation product 2-methylthio-4-tert-butylamino-6-amino-s-triazine. Although this degradation product was equally effective against *D. magna* (an organism widely used for ecotoxicity assays because of its central role in the food chain and its high sensitivity to chemical agents), the breakdown product was even more bactericidal. Similarly, the breakdown product of diuron, 1-(3,4-dichlorophenyl), exhibited 215-fold higher toxicity against *D. magna*. Both of these breakdown products when assessed in biocidal assays had a higher nonspecific toxic action compared with the original parent compound [60].

Studies have also shown that photolysis is an important breakdown route for triclosan. Photolysis can create toxic breakdown products such as dioxins, an example of which is the major breakdown product 2,8-dichlorodibenzo-p-dioxin [58, 62]. Dioxins are known to be extremely toxic chemicals, and in humans, they can act as carcinogens [63]. The U.S. Food and Drugs Administration considers that long-term exposure to dioxin can result in negative reproductive and developmental effects [64]. Additionally, in an aquatic environment, in the presence of low concentrations of free chlorine, degradation of triclosan can lead to the formation of toxic chlorophenols such as 2,4-dichlorophenol and two-, tetra-, and penta-chlorinated hydroxylated diphenyl ether [65].

Consequently, it is important that when assessing the potential toxicity of a compound, its degradation product(s) and potential toxicity must also be considered [60].

19.3.4 Interactions of Multiple Biocides May Lead to a Synergistic Toxicity

Recently, it has been recognized that the combined synergistic effects of multiple biocides is an important area when considering the ecotoxicity of such compounds. Mixtures of biocides may have a greater toxicity than each individual component biocide. Given the high level of biocides released into the environment, it is inevitable that for any given environment, particularly water bodies, a mixture of biocides and biocidal degradation products is almost always present. Data derived from models that describe toxicity interactions suggest that for certain biocides or their breakdown product(s), which act at a common site or on a common metabolic process or pathway, their toxicity will be additive. For example, a binary mixture of the antifoulants 2-thiocyanomethylthiobenzothiazole and Irgarol 1051 was found to be 2,500 times more toxic to the test bacteria *V. fisheri* and the test crustacean *D. magna*. Thus, interactions of biocides producing a toxic effect need to be assessed when studying the environmental impact of a biocide [60].

19.3.5 Reducing the Impact of Biocides on the Environment

It is clear from the current scientific literature that biocide toxicity testing is not ideal. Most adverse affects on an organism are not evident based on current screening methods [59]. As it stands, degradation and good monitoring data are rarely available for biocides currently in use [66]. This problem originates from the fact that contemporary testing protocols are based on acute toxicity tests, and these do not take into account the long-term chronic effect(s) the product may have on the environment [59].

To reduce the adverse effect of biocides on the environment, toxicity testing should include assays to evaluate acute and chronic toxicity, and the environment profile of the biocide, along with its ability to partition in sediment [67]. The potential for synergistic toxic interactions with existing biocides, the potential to accumulate in the environment as well as the development of analytical tests to monitor the fate and toxicity of the biocide in the environment should also be assessed [59]. Ideally, these protocols should assess the availability and toxicity of a compound to relevant indicator species [61, 67].

The complexity of the environment, its interactions, and fluxes demand an integrated approach to risk assessment of biocide toxicity allied to predictive modeling. Finally, a standardized method of testing needs to be developed to assure a level of consistency between agencies and companies carrying out such testing [61].

19.4 BIOCIDES AND FOOD PACKAGING

Over recent years, there has been a drive to develop new antimicrobials for use in the food industry with a wide spectrum of activity and low toxicity. There has been a lot of interest throughout the food industry with regard to the exciting new area of biopolymer-based antimicrobial packaging. Antimicrobial packaging can be defined as any packaging technique that contains antimicrobials or a packaging technique that is used to control the growth of microorganisms in the food product. These new techniques satisfy the ever increasing demand expressed by modern consumers for minimally processed food, free from preservatives and nutrient-depleting harsh preservative techniques. Biopolymer packaging is defined as any packaging where the raw materials are obtained from marine and agricultural sources [68]. This new emerging industry satisfies consumer's demands for biodegradable and environmentally friendly products; it moves away from the traditional heavy petroleum-based approach toward a more green and renewable ideal. This in effect moves packaging, from what was originally considered as the provision of an inert barrier, to so-called *active packaging*, bringing with it a new dimension to the packaging industry, focused on the safety and stability of the food we eat. However, this new development is not without its problems, risks, and safety concerns.

19.4.1 Antimicrobial Packaging—An Overview

Food spoilage accounts for major losses to the food industry. It generally occurs through the metabolic activity of microorganisms found on the surface of the food [68, 69]. The use of antimicrobials in food packaging may delay or prevent the growth of specific microorganisms that contribute to food spoilage, thereby extending shelf life. Antimicrobial packaging has the added advantage of improving the safety of food by impeding the growth of any pathogens that may also be present in the food, albeit at lower numbers. Additionally, antimicrobial packaging would have the potential to impart a self-sterilizing ability to foods.

Antimicrobials can be used in various ways. The active agent can be incorporated into the food itself or it can be incorporated into the food packaging material [69]. There are two approaches to develop antimicrobial packaging: Packaging where the antimicrobial agent(s) are designed to migrate onto the surface of the food and those approaches in which the antimicrobial agent exerts its effect without migrating from the packaging [68].

19.4.2 Examples of Antimicrobial Packaging

Several types of antimicrobial packaging show promise for use as food preservatives. The addition of sachets containing volatile antimicrobial agents can be placed into the packaging to prevent spoilage. Alternatively, antimicrobials can be incorporated directly into a polymer coating, absorbed onto polymers, covalently linked to polymers, or polymers themselves can be used as packaging that display an inherent antimicrobial activity [69].

Antimicrobial polymer-based active packaging can be used in several ways. The package can consist of a polymer and an antimicrobial agent where the agent is incorporated directly into the polymer. This form of packaging is commonly used in the pharmaceutical industry, household goods, textiles, and certain biomedical devices. The antimicrobial agent or combination of agents, in this case, is incorporated with the molten polymer, if thermostable, or by solvent compounding, if thermolabile. The antimicrobial compound incorporated into the polymer may either be volatile, later diffusing out and coming into contact with the surface of the food, or the active compound may be non-volatile, in which case the food packaging material must be in intimate contact with the food to achieve the desired effect [69].

In another iteration of the process, antimicrobials can be coated onto the polymer after synthesis. This is often the case with antimicrobial compounds that cannot withstand the heat of polymer processing. Films can be treated or manipulated to increase the adsorption of the antimicrobial compound to the polymer [69].

Antimicrobials may also be immobilized onto films. This method can be used if there are functional groups on the polymer that can react with the antimicrobial molecule. This form of packaging has its advantages, as the ionic bonding of the antimicrobial agent to the polymer will allow for a slow, steady diffusion of the agent from the polymer leading to a slow release onto the food [69].

Finally, polymers with an inherent antimicrobial activity can be used as active polymer packaging. Cationic polymers such as chitosan display an intrinsic antimicrobial activity; in this example, membrane disruption and cellular leakage results from the interaction of the films amine groups with the negatively charged cell surface of the microorganism [69].

19.4.3 Biopolymers

The use of biopolymers is growing in popularity across the food industry. Biopolymer-based food packaging, active packaging in particular, is biodegradable, can be cost-effective, and has the potential to offer continued protection once the package is opened [68]. When choosing the class and type of active packaging, several factors must be considered.

Several classes of biopolymers are available; these include protein-, lipid-, polysaccharide-, and composite-based biopolymers [68]. Protein-based biopolymers are commonly derived from milk, gelatin, soybean, wheat, and corn. They are permeable to water, but in dry conditions, they can act as a good diffusion barrier to CO_2 and O_2. These biopolymers are, however, appropriate for use in the packaging of fruits and vegetables. In this case, they impart a delayed ripening while retarding microbial respiration. Their poor water vapor barrier action can also serve as an advantage; it prevents the condensation of water, which is a common source of food spoilage and microbial contamination. An example of a promising protein-based biopolymer is corn zein. Corn zein originates from the prolamin portion of corn proteins. This biopolymer is used in the packaging of sweets, nuts, and pharmaceuticals. Like most protein-based films, corn zein is a good barrier to O_2. Park et al. [70] reported that corn zein is an appropriate packaging material suitable for tomatoes, both delaying color change and ripening during storage. Many antimicrobials can be incorporated into corn zein by heat press, such as nicin, lysozyme, lauric acid, and EDTA. Hoffman *et al.*, [71] showed that corn zein polymers containing EDTA, lauric acid, and nicin were effective at reducing *L. monocytogenes* and *Salmonella* Enteritidis by 4 logs after 48 hours [68].

Lipid-based biopolymers are composed of natural waxes, surfactants, and acetylated monoglycerides. These lipid-based biopolymers possess the useful property of being water vapor impermeable. The functional properties of any given lipid-based biopolymer is dependent on the properties of the lipid molecules themselves and includes factors such as the degree of saturation, chain length, structure, physical state, dimension of crystal, and distribution of the lipids in the film. Waxes are an example of a nonpolar class of lipids commonly used in packaging and include bees wax, paraffin, and rice bran. There are, however, disadvantages to these lipid-based biopolymers. Unlike protein- and polysaccharide-based biopolymers, lipid biopolymers provide a good water barrier; however, they have a tendency to crack and confer a waxy taste to the food [68].

Polysaccharide-based biopolymers present effective barriers to CO_2 and O_2 but not to water vapor and, therefore, they are a good material for packaging fruits and vegetables. There are many different examples of polysaccharide-based biopolymers such as starch, cellulose, agar, chitin, carageenan, and alginate. Chitin is a natural polysaccharide synthesized by many crustaceans and fungi. It has many beneficial properties as a packaging material. It can maintain the quality and integrity of fruit during low-temperature storage and prolong the postharvest life of the fruit it surrounds. Chitin films can inhibit the growth of common food spoilage organisms *Penicillin notatum* and *Rhodotorula rubra*. The incorporation of acids such as propionic and acetic acid into chitin can produce an effective antimicrobial film. Chitin lactate and glutamate can inhibit important food pathogens such as *E. coli* and *S. aureus*, and *Saccharomyces cerevisea* [68].

Another class of biopolymers is termed composite polymers, which combine the advantages of different classes of biopolymers. Composite biopolymers are composed of multilayer films, bilayers, or emulsions; an example of which would include a mixture of fatty acids, proteins, or polysaccharides, such as would be found in a combination of starch and gelatin [68]. Similarly, a film composed of soy protein and alginate showed a greater stability when compared with films formed from soy alone [68].

19.4.4 Antimicrobial Agents Used in Food Packaging

Several classes of antimicrobial agents are used in food packaging. These are chosen based on their stability. They are used in combination with a biopolymer as a preservative. However, no antimicrobial agents used in the antimicrobial packaging industry are effective against all microorganisms, and it has been proposed that combinations of antimicrobial agents with different antimicrobial efficacy should be used [68]. Several classes of antimicrobial agents can be used in active packaging: These include weak acids, organic oils, bacteriocins, enzymes, and fungicides, among others.

19.4.4.1 Weak Acids. Weak acids are the most common form of preservatives used in the food industry, which include citric, acetic, benzoic, glutamic, malic, lactic, propionic, and sorbic acids, among others. Weak acids such as these inhibit growth of bacteria and fungi. Some including sorbic acid inhibit the germination of bacterial spores. The effectiveness of weak acid preservatives was demonstrated by Ghosh et al. [72]; a wrapper with sorbic acid incorporated as an antifungal compound was used to wrap bread followed by heating for 30–60 min at 95°C to 100°C. The bread remained fungus-free and edible for up to 3 months thereafter. However, there are some disadvantages to the use of acid-based preservatives; acids can have an organoleptic effect on the food they surround. Nevertheless, there are ways to combat this effect by using antioxidants incorporated into the wrapper and odor absorbers inside the bread packaging. Weak acids have also been used in a biopolymer-based system, anhydride sorbic salts have been incorporated into starch-glycerol films, and acetic acid and propionic acid have been incorporated into chitin-based films [68].

19.4.4.2 Enzymes. Enzymes can be incorporated into various packaging matrices as preservative-based antimicrobial compounds. These can be used if they are nontoxic to humans. An example of an enzyme possessing these characteristics is lysozyme. It displays a broad spectrum of antimicrobial activity against both gram-positive and gram-negative organisms because of its peptidoglycan-cleaving activity. Lysozyme is a single peptide enzyme that cleaves the beta 1-4 glycoside linkage between the *N*-acetylmuramic acid and *N*-acetyl glucosamine subunits that make up peptideglycan layer, which is a major constituent in the cell wall of gram-positive and gram-negative bacteria. Lyzozyme causes instability, malformations, and ultimately lysis of the cell. This enzyme can be incorporated into different biopolymers, such as cellulose triacetate. Enzymes such as the glucose oxidase and naringinase have been used as preservatives of fresh fish and grapefruit, respectively [68]. Other enzymes exhibiting antimicrobial properties include lactoperoxidase and lactoferrin; these may be of use in future packaging applications. Antimicrobial peptides such as defensins, cecropins, and magainins may also be useful [69].

19.4.4.3 Bacteriocins. Bacteriocins have attracted attention as a potential strategy for use in the preservation of food. These antimicrobial peptides are synthesized by a range of different bacteria. Nicin produced by *Lactobacillus lactis* is one example of a bacteriocin showing promise for use in the food-packaging industry. This bacterial species has a Generally Regarded as Safe status and is therefore approved for use in the food industry. Nisin has been used as a preservative for over 50 years and along with pediocin and lacticin has been tested as a potential antimicrobial agent for incorporation into biopolymers. Pediocin incorporated into a cellulose polymer was found to inhibit the growth of *L. monocytogenes* when stored for a period of 12 weeks at 4 °C. Similarly when nicin and lacticin were incorporated into a bioactive package design, these successfully extended the shelf life of sliced ham and cheese at refrigeration temperatures by reducing cell counts of lactic acid bacteria. Nicin applied to a nylon membrane was found to be effective at inhibiting *Salmonella* Typhimurium when applied to chicken skin [68].

19.4.4.4 Essential Oils. Like bacteriocins, natural compounds such as essential oils have the potential to be used in the food industry as antimicrobial agents. Essential oils are derived from plant extracts. Thymol, carvacol, and thyme essential oils are potent antimicrobial agents. It is thought that essential oils possess antimicrobial properties because of the high concentration of phenolic compounds they contain. Although their mode of action is not yet fully elucidated, it has been suggested that their antimicrobial activity stems from the partition of the phenolic compounds into the inner membranes, which leads to increased permeability of the cell membrane to protons and potassium [73]. A disadvantage associated with the use of essential oils is that they have a strong taste and can affect the sensory quality if directly incorporated into the food. However, they display potential for use in active packaging as part of a hurdle technology strategy. Essential oils incorporated into whey protein biopolymers have been found to be effective in the controlling the growth of *L. monocytogenes* and *E. coli* O15:H7 [74].

19.4.4.5 Other Antimicrobial Agents. The cationic polymer chitosan has also been evaluated as a potential antimicrobial compound for use in the food industry. Charged amines on chitosan interact with the negative charge on the cell membrane surface causing cell adhesion and leading to the leakage of the bacterial intracellular constituents. Chitosan also possesses antifungal activity and has been used in the coating of fruit and vegetables, protecting them from fungal damage and degradation [68].

Other compounds suggested for the use in the food industry to prevent microbial spoilage include grape seed extracts, which are a natural agent containing numerous antimicrobial compounds, and allylisothiocyanate, which is a natural constituent of mustard oil as well as garlic, clove, and cinnamon.

19.4.5 Challenges to the Use of Antimicrobials in Food Packaging

The incorporation of an antimicrobial compound may affect the physical properties of the packaging films [75], such as optical and mechanical properties, tensile strength, seal strength [69], diffusion coefficients [68], and barrier properties may decrease after the addition of the antimicrobial compound [69]. The addition of benzoic anhydride, ethyl paraben, and propyl paraben to a biopolymer resulted in a reduction in its tensile strength [75]. Additionally, heat sealing, printing, and adhesion properties of the polymer may be altered. Inclusion of acetic and propionic acids to chitosan-based films resulted in the reduction in the diffusion coefficient of the polymer, as the temperature decreased from 24°C to 4°C [68]. These changes in properties are polymer-antimicrobial agent specific.

Similarly, the heat, pressure, shearing force, or other processing conditions in the manufacturing of active packaging may affect the stability of the antimicrobial agent. Postmanufacturing processing such as printing, drying, lamination, and chemical exposure to adhesives and solvents may have an impact on the activity of the antimicrobial agent(s). There is also the risk that volatile antimicrobial agents may be lost during periods of storage. The food itself may also have an adverse effect on the preservative activity of the active packaging. Food properties such as a low water activity (a_w) may affect the stability of the antimicrobial agents. Diffusion of potassium sorbate through polysaccharide films was increased in food with a_w. External factors may also affect the stability and efficacy of active packaging, such as storage temperature [75].

The diffusion rate of the antimicrobial agent through the film must remain sufficiently high for the entire duration of the storage period to remain effective [75]. The failure to do so will expose any microflora in the food to a sublethal concentration of the antimicrobial agent, facilitating the survival of the organism. Similarly, if any unexpected changes in the property of the film or antimicrobial agents emerge, as a result of, for example, any of the situations discussed previously, the microflora may again be exposed to sublethal concentrations. This has the potential to select for microorganisms that become tolerant. Focused studies to assess whether antimicrobial resistance to these preservatives could develop are required.

Finally, regulatory issues limiting the use and future development of active packaging need to be addressed. Compounds coming into contact with food need to be on a list of approved chemicals and a migration limit of 60 mg/kg is required for approval. This migration limit is not compatible with the goals of active packaging. With current regulations, the applications and potential uses of active packaging is restricted [75]. Regulations need to be updated to take into account the novel area of active packaging before their widespread use can become a reality.

Some concern has been expressed about the use of biocides and other chemicals in active polymer packaging and in traditional food packaging in general. These chemicals have the potential to migrate unintentionally from packaging materials into the food with which they are in contact. There is a risk that human exposure may be toxic or have other disruptive ill-health effects.

Nonylphenol, a compound contained in food packaging has been found to migrate from the packaging into food. Specifically, it was shown to migrate from milk containers into milk [53]. Evidence has been found that phthalates, which are common additives of polyethylene terephthalate (PET), have been found to migrate from PET packaging into soft drinks and yogurt drinks. This migration was dependent on multiple factors such as time and temperature, and could reach levels as high as 2.4 mg/L. Phthalate can have negative side effects on a developing fetus [53].

Additionally, it has been suggested that combinations of certain materials or the interaction between materials and their breakdown products contained in food packaging may present a risk to human health. Although cytological testing must be carried out on all materials used in food packaging, these tests are often carried out in an isolated manner. The toxicological characteristics of the starting material used to make such packaging should be examined and the final product is deemed safe only if all components are found to be nontoxic. This testing strategy fails to take into account any compounds that may form unsuspectingly from the starting materials after the manufacturing of the final packaging. This form of toxicological testing does not evaluate the interaction between various compounds; it does not assess the potential toxicological synergy between various elements of the final food packaging material. Fauris et al. [76] carried out cytotoxicological testing and examined the inhibition of RNA synthesis in HeLa cells after exposure to 6 papers and 15 boards used in food packaging. This study reported that cytotoxicity observed was not caused by any one individual compound but rather by combinations of compounds, a significant proportion of the toxicity was attributed to 3-chloro- 4(dichlor-omethyl)-5-hydroxy-2(5H)-furanone (MX), which forms from chemical reactions with chlorine and lignin [76].

Although the preceding examples are not associated with biopolymers, the principles are the same. Risk assessment of the potential for these compounds to leech into food and any risk they may pose to humans, along with their breakdown products and synergistic interactions needs to be thoroughly evaluated to assure they are safe as a method of food preservation.

19.4.6 Conclusions Regarding Biocides and Food Packaging

The ultimate aim of antimicrobial packaging would be to create a food-packaging matrix that does not pose any risk to human health and neither disrupts the sensory nor chemical composition of the food it contains but yet it must have broad and stable antimicrobial properties.

Antimicrobial polymers are a novel development in the field of food safety, not only for food packaging, but also if proven successful, for other uses such as self-sterilizing surfaces for food processing and preparation, for gloves, garments, and personal protection clothing [69]. Antimicrobial polymers can be expected to play a significant role in the food industry of the future, both in the protection against pathogens and in the extension of food shelf life.

19.5 SUMMARY AND FUTURE TRENDS

Although the widely available use of biocides has undoubtedly brought many advantages to the clinical, industrial, and domestic settings, they are not without their risks. Biocides are by their nature toxic and are of concern from a public health and environmental perspective. It has been demonstrated that the misuse of biocides can lead to the development of biocide tolerant bacteria, potentially reducing the antibacterial efficacy of these compounds. Similarly, biocide misuse may lead to the selection of bacteria displaying a reduced susceptibility to important clinical antibiotics, preservatives, processing techniques and furthermore, may select for strains that display an enhanced virulence. Studies focusing on an improved understanding of the mechanisms involved in biocide tolerance and research into the epidemiology of any emerging biocide tolerant strains will be of increasing importance in the future.

Control and monitoring protocols must be established to ensure that biocide tolerance does not become widespread in the same manner and with the same devastating consequences as antibiotic resistance.

Efforts must be made in the future to develop and implement appropriate toxicological testing, both acute and chronic testing alongside methods with other studies to improve the understanding of the fate of biocide compounds in use, their potential breakdown routes and products. This testing is important in ensuring a limited impact of any biocide(s) on the environment and in eliminating or reducing any potential threats they pose to human health.

Many new and exciting developments involve antimicrobial polymers, including the use of antimicrobial-based food packaging that are being developed. However, it is important to remember that biocides, such as biopolymer-based active packaging, do not make up for a lack of hygiene, failings in good manufacturing practice, or poor-quality raw materials. Rather, biocides constitute an important element as part of an integrated hurdle technology.

REFERENCES

1. Marshall, B. M., McMurray, L. M. (2005). Biocides and resistance. In: *Frontiers in Antimicrobial Resistance: A Tribute to Stuart B. Levy*. Eds.: White, D. G., Alekshun, M. N., McDermott, P.F., pp. 174, 190. ASM Press, Washington, D.C.
2. Langsrud, S., Sidhu, M., Heir, E., Holck, A. (2003). Bacterial disinfectant resistance—a challenge for the food industry. *International Biodeterioration & Biodegradation*, *51*, 283–290.
3. Gilbert, P., McBain, A. J. (2003). Potential impact of increased use of biocides in consumer products on prevalence of antibiotic resistance. *Clinical Microbiology Reviews*, *16*, 189–208.
4. Fraise, A. P. (2002). Biocide abuse and antimicrobial resistance–a cause for concern? *The Journal of Antimicrobial Chemotherapy*, *49*, 11–2.
5. Lebert, I., Leroy, S., Talon, R. (2007). Effect of industrial and natural biocides on spoilage, pathogenic and technological strains grown in biofilm. *Food Microbiology*, *24*, 281–287.
6. Aarestrup, F. M., Hasman, H. (2004). Susceptibility of different bacterial species isolated from food animals to copper sulphate, zinc chloride and antimicrobial substances used for disinfection. *Veterinary Microbiology*, *100*, 83–89.

7. Beier, R. C., Bischoff, K. M., Ziprin, R. L., Poole, T. L., Nisbet, D. J. (2005). Chlorhexidine susceptibility, virulence factors, and antibiotic resistance of beta-hemolytic Escherichia coli isolated from neonatal swine with diarrhea. *Bulletin of Environmental Contamination and Toxicology*, *75*, 835–844.
8. Van Hemmen, J. J. (1999). Biocides steering group on human exposure assessment: A preliminary report. *Toxicology Letters*, *107*, 95–102.
9. Schweizer, H. P. (2001). Triclosan: A widely used biocide and its link to antibiotics. *FEMS Microbiology Letters*, *202*, 1–7.
10. Hattemer-Frey, H. A., Travis, C. C. (1989). Pentachlorophenol; environmental partitioning and human exposure. *Archives of Environmental Contamination and Toxicology*, *18*, 482–489.
11. Nakama, A., Funasaka, K., Shimizu, M. (2007). Evaluation of estrogenic activity of organic biocides using ER-binding and YES assay. *Food and Chemical Toxicology*, *45*, 1558–1564.
12. Schnuch, A., Geier, J., Uter, W., Frosch, P. J. (1998). Patch testing with preservatives, antimicrobials and industrial biocides. Results from a multicentre study. *The British Journal of Dermatology*, *138*, 467–476.
13. Coelhan, M., Yu, J. T., Roberts, A. L. (2009). Presence of the biocide ortho-phenylphenol in canned soft drinks in the United States and Germany. *Food Chemistry*, *112*, 515–519.
14. Clark, E. A., Sterritt, R. M., Lester, J. N. (1988). The fate of tributyltin in the aquatic environment. *Environmental Science & Technology*, *22*, 600–604.
15. Poole, K. (2002). Mechanisms of bacterial biocide and antibiotic resistance. *Symposium Series (Society for Applied Microbiology)*, *92*, 55S–64S.
16. Lambert, P. (2004). Mechanism of action of biocides. In: *Russell, Hugo & Ayliffe's Principles and Practice of Disinfection, Preservation and Sterilization*. 4th ed, Eds.: Fraise, F., Lambert, P., Maillare, J.-Y. pp. 139, 153. Blackwell Publishing, Oxford, UK.
17. Stickler, D. (2004). Intrinsic resistance of gram-negitive bacteria. In: *Russell, Hugo & Ayliffe's Principles and Practice of Disinfection, Preservation and Sterilization*, 4th ed, Eds.: Fraise, F., Lambert, P., Maillare, J.-Y., pp. 154, 169. Blackwell Publishing, Oxford, UK.
18. Levy, S. B. (2002). Active efflux, a common mechanism for biocide and antibiotic resistance. *Symposium Series (Society for Applied Microbiology)*, *92*, 65S–71S.
19. SCENIHR. (2009). Scientific Committee on Emerging and Newly Identified Health Risks SCENIHR Assessment of the Antibiotic Resistance Effects of Biocides. http://ec.europa.eu/health/ph_risk/committees/04_scenihr/docs/scenihr_o_021.pdf. accessed April 28, 2011.
20. Lambert, P. A. (2002). Cellular impermeability and uptake of biocides and antibiotics in gram-positive bacteria and mycobacteria. *Symposium series (Society for Applied Microbiology)*, *92*, 46S–54S.
21. Poole, K. (2004). Acquired resistance. In: *Russell, Hugo & Ayliffe's Principles and Practice of Disinfection, Preservation and Sterilization*. Eds.: Fraise, F., Lambert, P., Maillare, J.-Y. 4th ed, pp. 170, 183. Blackwell Publishing, Oxford, U.K.
22. Abdel, M. S., Winder, C., Buultjens, T. (2002). Antimicrobial susceptibility changes and T-OMP shifts in pyrithione-passaged planktonic cultures of Pseudomonas aeruginosa PAO1. *Journal of Applied Microbiology*, *92*, 729–736.

23. Lambert, P. (2004). Resistance of bacterial spores to chemical agents. In: *Russell, Hugo & Ayliffe's Principles and Practice of Disinfection, Preservation and Sterilization.* Eds.: Fraise, F., Lambert, P., Maillare, J.-Y., 4th ed, pp. 184, 190. Blackwell Publishing, Oxford, U.K.

24. McBain, A. J., Gilbert, P. (2001). Biocide tolerance and the harbingers of doom. *International Biodeterioration & Biodegradation, 47*, 55–61.

25. Brown, M. H., Paulsen, I. T., Skurray, R. A. (1999). The multidrug efflux protein NorM is a prototype of a new family of transporters. *Molecular Microbiology, 31*, 394–395.

26. Wang, J. I., Wang, J. T., Sheng, W. H., Chen, Y. C., Chang, S. C. (2010). Nosocomial methicillin-resistant Staphylococcus aureus (MRSA) bacteremia in Taiwan: Mortality analyses and the impact of vancomycin, MIC = 2 mg / L, by the broth microdilution method. *BMC Infectious Diseases, 10*, 1–7.

27. Piddock, L. J. V. (2006). Clinically relevant chromosomally encoded multidrug resistance efflux pumps in bacteria. *Clinical Microbiology Reviews, 19*, 382–402.

28. Singer, H., Müller, S., Tixier, C., Pillonel, L. (2002). Triclosan: Occurrence and fate of a widely used biocide in the aquatic environment: Field measurements in wastewater treatment plants, surface waters, and lake sediments. *Environmental Science & Technology, 36*, 4998–5004.

29. Gilbert, P., Allison, D. G., McBain, A. J. (2002). Biofilms in vitro and in vivo: Do singular mechanisms imply cross-resistance? *Symposium Series (Society for Applied Microbiology), 92*, 98S–110S.

30. Gilbert, P., Rickard, A., McBain, A. (2004). Biofilms and antimicrobial resistance. In: *Russell, Hugo & Ayliffe's Principles and Practice of Disinfection, Preservation and Sterilization.* Eds.: Fraise, F., Lambert, P., Maillare, J.-Y., 4th ed, pp. 128, 138. Blackwell Publishing, Oxford, U.K.

31. Stickler, D. J., Jones, G. L. (2008). Reduced susceptibility of Proteus mirabilis to triclosan. *Antimicrobial Agents and Chemotherapy, 52*, 991–994.

32. Tabak, M., Scher, K., Hartog, E., Romling, U., Matthews, K. R., Chikindas, M. L., Yaron, S. (2007). Effect of triclosan on Salmonella Typhimurium at different growth stages and in biofilms. *FEMS Microbiology Letters, 267*, 200–6.

33. Pan, Y., Breidt, F., Kathariou, S. (2006). Resistance of Listeria monocytogenes biofilms to sanitizing agents in a simulated food processing environment. *Applied and Environmental Microbiology, 72*, 7711–7717.

34. Kim, H., Ryu, J.-H., Beuchat, L. R. (2007). Effectiveness of disinfectants in killing Enterobacter sakazakii in suspension, dried on the surface of stainless steel, and in a biofilm. *Applied and Environmental Microbiology, 73*, 1256–1265.

35. Lewis, K. I. M. (2001). Minireview riddle of biofilm resistance. *Antimicrobial Agents and Chemotherapy, 45*, 999–1007.

36. Dukan, S. A. M., Monod, I. J. (1996). Hypochlorous acid stress in Escherichia coli: Resistance, DNA damage, and comparison with hydrogen peroxide stress. *Microbiology, 178*, 6145–6150.

37. Greenway, D. L., England, R. R. (1999). The intrinsic resistance of Escherichia coli to various antimicrobial agents requires ppGpp and sigma s. *Letters in Applied Microbiology, 29*, 323–326.

38. Moken, M. C., McMurry, L. M., Levy, S. B. (1997). Selection of multiple-antibiotic-resistant (mar) mutants of Escherichia coli by using the disinfectant pine oil: Roles of the mar and acrAB loci. *Antimicrobial Agents and Chemotherapy*, *41*, 2770–2772.
39. Gilbert, P. (2002). Biocide abuse and antimicrobial resistance: Being clear about the issues. *Journal of Antimicrobial Chemotherapy*, *50*, 137–139.
40. Denyer, S. P., Maillard, J. Y. (2002). Cellular impermeability and uptake of biocides and antibiotics in gram-negative bacteria. *Symposium series (Society for Applied Microbiology)*, *92*, 35S–45S.
41. Levy, S. B. (1992). Minireview active efflux mechanisms for antimicrobial resistance. *Microbiology*, *36*, 695–703.
42. Yamamoto, T., Tamura, Y., Yokota, T. (1988). Antiseptic and antibiotic resistance plasmid in staphylococcus aureus that possesses ability to confer chlorhexidine and acrinol resistance. *Microbiology*, *32*, 932–935.
43. Chapman, J. (2003). Disinfectant resistance mechanisms, cross-resistance, and co-resistance. *International Biodeterioration & Biodegradation*, *51*, 271–276.
44. Russell, A. D. (2004). Bacterial adaptation and resistance to antiseptics, disinfectants and preservatives is not a new phenomenon. *The Journal of Hospital Infection*, *57*, 97–104.
45. Chapman, J. (1998). Characterizing bacterial resistance to preservatives and disinfectants. *International Biodeterioration & Biodegradation*, *41*, 241–245.
46. Brul, S., Coote, P. (1999). Preservative agents in foods. mode of action and microbial resistance mechanisms. *International Journal of Food Microbiology*, *50*, 1–17.
47. Flores, M., Morillo, M., Crespo, M. (1997). Deterioration of raw materials and cosmetic products by preservative resistant micro-organisms. *International Biodeterioration and Biodegradation*, *40*, 157–156.
48. Scully, B., Parry, M., Neu, H., Mandell, W. (1986). Oral ciprofloxacin therapy of infections due to Pseudomonas aeruginosa. *The Lancet*, *327*, 819–822.
49. Kujala, N. (2011). Biocide-induced antimicrobial peptide reisitance in Porphyromonas gingivalis. http://dspace.nitle.org/handle/10090/4371. accessed April 15, 2011.
50. Kastbjerg, V. G., Larsen, M. H., Gram, L., Ingmer, H. (2010). Influence of sublethal concentrations of common disinfectants on expression of virulence genes in Listeria monocytogenes. *Applied and Environmental Microbiology*, *76*, 303–309.
51. Allen, M. J., White, G. F., Morby, A. P. (2006). The response of Escherichia coli to exposure to the biocide polyhexamethylene biguanide. *Microbiology*, *152*, 989–1000.
52. Meyer, B. (2006). Does microbial resistance to biocides create a hazard to food hygiene? *International journal of food microbiology*, *112*, 275–9.
53. Muncke, J. (2009). Environmental hormones in food packaging: Migration into food and the environment. http://www.apikyan.com/PDFFiles/Hormones_en.pdf. accessed April 15, 2011.
54. PAN Pesticide Database. Toxicity information for ortho-phenylphenol. http://www.pesticideinfo.org/Detail_Chemical.jsp?Rec_Id = PC33048#Toxicityaccessed April 15, 2011.
55. United States Environmental Protection Agency. Basic information about pentachlorophenol in drinking water. http://water.epa.gov/drink/contaminants/basicinformation/pentachlorophenol.com, accessed April 15, 2011.

56. Wild, S. R., Jones, K. C. (1992). Pentachlorophenol in the UK environment: II: Human exposure and an assessment of pathways. *Chemosphere, 24*, 847–855.

57. United States Environmental Protection Agency. National primary drinking water regulations. http://water.epa.gov/drink/contaminants/index.cfm#Disinfectants, accessed April 15, 2011.

58. Aranami, K., Readman, J. W. (2007). Photolytic degradation of triclosan in freshwater and seawater. *Chemosphere, 66*, 1052–1056.

59. Evans, S. (2000). The TBT ban: Out of the frying pan into the fire? *Marine Pollution Bulletin, 40*, 204–211.

60. Fernández-Alba, A. (2002). Toxicity evaluation of single and mixed antifouling biocides measured with acute toxicity bioassays. *Analytica Chimica Acta, 456*, 303–312.

61. Huggert, R. J., Unger, M. A., Seligman, P. F., Valkirs, A. O. (1992). The marine biocide tributyltin. *Environmental Science and Technology, 26*, 232–237.

62. Buth, J. M., Steen, P. O., Sueper, C., Blumentritt, D., Vikesland, P. J., Arnold, W. A., McNeill, K. (2010). Dioxin photoproducts of triclosan and its chlorinated derivatives in sediment cores. *Environmental Science & Technology, 44*, 4545–51.

63. Harper, J. W. (2007). Chemical biology: A degrading solution to pollution. *Nature, 446*, 499–500.

64. Betts, K. (2011). River sediment contains dioxins linked to triclosan. http://pubs.acs.org/cen/news/88/i21/8821news8.html, accessed April 15, 2011.

65. Canosa, P., Morales, S., Rodríguez, I., Rubí, E., Cela, R., Gómez, M. (2005). Aquatic degradation of triclosan and formation of toxic chlorophenols in presence of low concentrations of free chlorine. *Analytical and Bioanalytical Chemistry, 383*, 1119–1126.

66. Weyers, A., Sokull-Klüttgen, B., Knacker, T., Martin, S., and Van Gestel CA. (2004). Use of terrestrial model ecosystem data in environmental risk assessment for industrial chemicals, biocides and plant protection products in the EU. *Ecotoxicology, 13*, 163–76.

67. Thomas, K., Fileman, T., Readman, J., Waldock, M. (2001). Antifouling paint booster biocides in the UK coastal environment and potential risks of biological effects. *Marine Pollution Bulletin, 42*, 677–688.

68. Chinnan, M., Cha, D. (2004). Biopolymer-based antimicrobial packaging: A review. *Critical Reviews in Food Science and Nutrition, 44*, 223–237.

69. Appendini, P., Hotchkiss, J. (2002). Review of antimicrobial food packaging. *Innovative Food Science & Emerging Technologies, 3*, 113–126.

70. Park, H., Chinnan, M., Shewfelt, R. (1994). Edible corn-zein film coatings to extend storage life of tomatoes. *Journal of Food Processing and Preservatives, 18*, 317–331.

71. Hoffman, K., Han, I., Dawson, P. (2001). Antimicrobial effects of corn zein films impregnated with nisin, lauric acid, and EDTA. *Journal of Food Protection, 64*, 885–889.

72. Ghosh, K., Srivatsava, A., Nirmala, N., Sharma, T. (1977). Development and application of fungistatic wrappers in food preservation. Part II. Wrappers made by coating process. *Journal of Food Science and Technology, 14*, 261–264.

73. Burt, S. (2004). Essential oils: their antibacterial properties and potential applications in foods- a rewiew. *International Journal of Food Microbiology*, *3*, 232–235.
74. Seydim, A., Sarikus, G. (2006). Antimicrobial activity of whey protein based edible films incorporated with oregano, rosemary and garlic essential oils. *Food Research International*, *39*, 639–644.
75. Suppakul, P., Miltz, J., Sonneveld, K., Bigger, S. W. (2003). Active packaging technologies with an emphasis on antimicrobial concise reviews in food science. *Journal of Food Science*, *68*, 408–420.
76. Fauris, C., Lundström, H., Vilaginès, R. (1998). Cytotoxicological safety assessment of papers and boards used for food packaging. *Food Additives and Contaminants*, *15*, 716–728.

INDEX

Absorption, distribution, metabolism, and excretion (ADME), of nanoparticles, 341
Absorption packaging system, 105
Acid antibiotics, 409
Acid-base interactions, 393, 394
Acid-resistant microorganisms, 6–7
Acinetobacter, 535
Acquired bacterial resistance, to biocides, 35–37
Acquired biocide resistance, mechanisms of, 35–36
Acquired resistance, 26–27, 37
Acrylic acid, as a coupling agent, 358
Actinomycetes, 161–162
Activation energy, of chlorine dioxide, 476
Activator molecules, 160
Active agents
  controlled release of, 177–184
  diffusion-limited release of, 178
  release from matrix system, 181
  in swelling-controlled systems, 181
Active antimicrobial packaging, performance of, 129
Active bacteriocins, designing films and surfaces containing, 126–143
Active coatings, 172, 401
Active compounds
  in biobased films, 184
  release kinetics of, 184
Active eco-friendly packaging solutions, 185
Active ingredients
  FIFRA regulations related to, 520
  steps in releasing, 178
Active localized drug administration, 401–402
Active metals, nanoscaling, 317
Active nanobiocomposite systems, 106
Active packaging, 104–105, 128, 167, 209, 461
  coating procedure and, 144–145
  future trends in, 147, 482
  gas-based antimicrobials in, 459–488
  regulatory issues related to, 553
Active packaging material, based on zein-thymol films, 106
Active packaging solution, 174
Active packaging storage, antimicrobial activity and, 131
Active packaging strategies, 12
Active packaging technologies, regulatory constraints on, 185
Active plastic film, development of, 144
Active principle(s)
  concentration of, 179
  mass balance of, 179, 180
  release from microencapsulated drugs, 178–179
Active systems, research activity aimed at, 160
Active technologies, combination of, 111
Active wound dressing material, 280
Acute hazard studies, 515–516
  for nontarget organisms, 517
Adhesins, 411
Adhesion-mediated infections, 386
Adhesive interactions, modeling, 391–392
ADP database, 228
Advanced packaging systems, 482

Antimicrobial Polymers, First Edition.
Edited by José M. Lagarón, María J. Ocio, and Amparo López-Rubio.

Aerobacter aerogenes, 54
Agar diffusion method, 9
Agar-well diffusion method, 125
Ag-based zeolites, 13
Ag^+ ions, inorganic materials for supporting, 320. *See also* Silver ions
Agricultural applications
of antimicrobial peptides, 247–248
chitosan in, 73–74
Agriculture, antibiotic resistance in, 33
Agrobacterium tumefaciens nuclear transformation methods, 233
agr quorum system, 384
Ag/SiO_2 coating solutions, 320
Ahmed, Nibras A. A. M., ix, 23
AITC blends, 464. *See also* Allyl isothiocyanate (AITC)
AITC controlled release
electrospun fibers for, 463–464
fibrous membranes for, 463
AITC-cyclodextrin approach, 463–464
AITC-encapsulated SPI-PEO fibers, 463
AITC labels, shelf life and, 462–463
AITC permeation, in moisture-sensitive polymers, 467
AITC permeation curves, 467
AITC P values, in PVDC-PVC copolymer film, 466
AITC release percent, 462
AITC vapor, 462
AITC vapor pressure, 465
Alf-AFP alfalfa defensin, 232
Alfalfa antifungal peptide, 248
Algicides, 510
Alginate-based edible coatings, 175
Alginate-entrapped naringinase, 173
Alicyclobacillus acidoterrestris, 173
Aliphatic polyester functionalization, CuAAC reaction in, 61–62
Aliphatic polyester synthesis, 52, 53
Aliphatic polyesters
degradation of, 61–62
grafting ammonium salts onto, 62
grafting antimicrobial agents onto, 60
imparting antibacterial behavior to, 62
N-Alkyl-pyridinium, 404
Alkyne coupling, 61
Alkynes, synthesis of, 66
Allaker, R. P., ix, 327
All-D antimicrobial peptides, 242
Allergenic impact, of biocide use, 544
Allyl acetylacetone, as a coupling agent, 358
Allyl isothiocyanate (AITC), 460–469. *See also* AITC entries
antimicrobial potency of, 461
food packaging applications of, 461–464
mass transfer of, 464–469
mass transfer properties of, 466
permeability, diffusion, and solubility coefficients, 466
vapor pressure properties of, 464
α-helical AMPs, 239. *See also* Antimicrobial peptides (AMPs)
α-helical peptides, 436
Alternaria solani, 232
Amide I bands, of peptides, 443
Amide I signal, 437, 438
Amino-acid-enriched cationic peptides, 234
Ammonium salts, grafting, 65–66
AMP activity. *See also* Antimicrobial peptides (AMPs)
imbalance in, 246
requirements for, 233–244
AMP biosynthesis
in plant expression systems, 232–233
in prokaryotic expression systems, 229–231
in yeast expression systems, 231–232
AMP cell-penetrating properties, 205
AMP databases, 228–229
AMP defensin, 232
AMP genes, in plants and animals, 247–248
Amphipathicity
of antimicrobial peptides, 235
forms of, 239
quantitative measure of, 238
Amphipathic secondary structures, 240
Amphiphilicity, 211
Amphiphilic peptide dendrimers, 251
AMP mechanisms, 203, 204
AMP mimics, 451–452, 453
AMP monomers, covalent multimerization of, 202
AMP production, improving, 230
AMP–target interactions, 235
AMP toxicity, preventing, 229
AMSdb, 228
Anassa obsoleta, 545
Anatase nanoparticles, 353
Anatase phase, 352

Anatase-TiO_2, nanostructured, 352–353
Anatase TiO_2 nanoparticles, modification of, 362–363
Anatase-TiO_2 system photoactivity, optimizing, 354
Animal defenses, 232
Animals, AMP genes in, 247–248. *See also* Farm animals
Anionic antimicrobial peptides, 234
Anionic peptide fragments, 234
Anionic peptides, cysteine-containing and disulfide-bond-forming, 234
"Anomalous diffusion," 182, 183, 184
Anthocyanin extracts, spray-dried, 177
Antiadhesive PEMs, 398. *See also* Polyelectrolyte multilayers (PEMs)
Antiadhesive strategies, 409
Antibacterial activity, of silver nanoparticles, 292
Antibacterial aliphatic polyesters, by click chemistry, 60–65
Antibacterial copolymers, 62
Antibacterial diblock copolymer synthesis, 56
Antibacterial layers, multifaceted/multifunctional, 407
Antibacterial molecules, application of, 24
Antibacterial PEMs, 407. *See also* Polyelectrolyte multilayers (PEMs)
Antibacterial polymer synthesis, 54
  by quaternization, 55
Antibacterial polypropylene, by click chemistry, 58–60
Antibacterial properties, photocatalytic, 368–369
Antibacterial surface design, strategies for, 391
Antibiofilm surfaces, developing, 206–207
Antibiotic agents
  antimicrobial peptides as, 244
  release from neutral polymer surfaces, 401–402
Antibiotic delivery vehicles, 271
Antibiotic-free strategies, 248
Antibiotic loading, 274
Antibiotic loading dosage, varying, 272
Antibiotic pollution, environmental, 32
Antibiotic polymer affinity, 272
Antibiotic products, demand for, 232
Antibiotic release, prolonged, 402
Antibiotic release profiles, 272–273
Antibiotic resistance, 2
Antibiotic revolution, 24
Antibiotics
  biocide cross-resistance with, 537–539
  cross-resistance to, 37–38
  efficacy of, 443
  innate resistance to, 26
  releasable, 400
  resistance to, 25
  subinhibitory concentrations of, 27, 30–31
  vs. biocides, 39
Antibody-based therapies, 410
Antifouling agents, 510
Antifungal activity, of natural antimicrobial peptides, 197
Antifungal agents, metals as, 328
Anti-inflammatory properties, of silver nanoparticles, 296
Antilisterial bacteriocins, 142
Antimicrobial action
  of chitosan, 77
  of quaternary ammonium salts, 52
Antimicrobial activity, 317–318
  of antimicrobial peptides, 203, 244–245
  of encapsulated lemon grass essential oil, 176
  extrusion process and, 130–131
  against Gram-negative bacteria, 142
  measurement of, 131
  methods to determine, 9–10
  peptide cationicity and, 236
  relationship to nanoparticle size, 331
  residual, 131–133
  solvent casting and, 131
Antimicrobial (AM) additives, 324–325
  naturally derived, 101–102
Antimicrobial agent release kinetic, 140
Antimicrobial agents, 317–318
  adding to packaging systems, 105
  diffusion rate of, 553
  effectiveness of, 86
  essential oils as, 552
  in food packaging, 550–552
  for food preservation, 8–9
  grafting onto aliphatic polyesters, 60
  nanometals as, 328–334
  nanoparticulate metal oxides as, 335–336
  resistance to, 25
Antimicrobial biodegradable films, 105
Antimicrobial bone interfaces, 276–279
Antimicrobial cathelicidins, 200

Antimicrobial coatings, 549
inactivation of, 339
Antimicrobial compounds, biocide cross-resistance to, 537
Antimicrobial effects, product acceptability and, 172
Antimicrobial efficacy, fat and, 172
Antimicrobial electrospun compositions, postsurgical infections and, 278–279
Antimicrobial enzymes
in food packaging, 184
in plastics, 159–194
Antimicrobial films
forming, 16
water-resistant chitosan-based, 79
Antimicrobial food contact surfaces, 142
Antimicrobial food packaging
bacteriocins in, 128
basil extract in, 103
Antimicrobial functionality development, electrospinning in, 276–282
Antimicrobial implant/tissue interfaces, 388–396
Antimicrobial interfaces, tissue-implant, 379–428
Antimicrobial lipopeptides, 202
Antimicrobial materials, for food packaging, 366. *See also* Antimicrobials; Antimicrobial substances
Antimicrobial nanofibers
layer deposition of, 280–281
in water microfiltration, 281–282
Antimicrobial nanolayered clays, functioning of, 319
Antimicrobial networks, synthesis of, 64–65
Antimicrobial packaging, 12, 105, 128, 249, 548
action of, 129
bacteriocin-based, 130–141
challenges to the use of, 552–554
chitosan-based, 81–88
disinfecting agents in, 460
examples of, 548–549
for extending food shelf life, 145–146
future perspectives on, 90–91
Antimicrobial packaging materials, concerns related to, 319
Antimicrobial packaging polymers, 1–22
Antimicrobial packaging technologies, 12–13
Antimicrobial peptide defensin, 399–400
Antimicrobial peptide encapsulation, by liposomes or microparticles, 209–210
Antimicrobial peptides (AMPs), 195–225, 228–229. *See also* AMP entries; Natural antimicrobial peptides; Novel antimicrobial peptides; Recombinant antimicrobial peptides
action stages of, 203
activity spectra of, 196, 237, 245
advantages of, 196–197
anticancer activity of, 247
antimicrobial and antiviral activity of, 196
antimicrobial properties of, 236
attachment to materials, 206
bacterial resistance and, 443–444
binding to microbial membranes, 236
characteristics of, 233
classes of, 233–234
databases including, 228–229
by design, 201–203
future trends for, 211
hybrid, 231
hydrophobic and hydrophilic stereogeometries in, 238
hydrophobic residues in, 238
immunomodulatory properties of, 245–246
impact on membranes, 241
incorporating into polymers, 206–209
intein-mediated expression of, 230
interaction with membrane bilayers, 430
large-scale production of, 229
limitations of and future trends in, 250–251
mechanisms of action of, 239–240
mechanistic classes of, 241
membrane-disruptive, 240
mode of action of, 203–205
molecular structures of, 199
nanostructure-forming, 210–211
non-bacteriocin, 209
novel/specific applications of, 243–250
potential of, 195, 196
production of, 205–206
properties of, 196–197
secondary structures of, 240
self-assembling nanostructure properties of, 210–211
structural classes of, 239
structural properties of, 235

successful delivery of, 209
therapeutic applications of, 244–247
Antimicrobial peptide sources, 201
Antimicrobial pesticides
data requirements for registering, 514–519
EPA registration process for, 512–514
major uses of, 512
Antimicrobial phyllosilicates, 324–325
Antimicrobial plant peptides (phytAMP) database, 228
Antimicrobial plastic packaging, 318
Antimicrobial plastics, viii
based on metal-containing nanolayered clays, 317–326
production of, 15
Antimicrobial polymer-based active packaging, 549
Antimicrobial polymers, 33, 267–270
determining the effectiveness of, 83
development of, 250
Antimicrobial polymeric systems, release kinetics of, 318
Antimicrobial products, types of, 512–514
Antimicrobial proteins, disadvantages of, 172
Antimicrobial-releasing materials, 41
Antimicrobials, 510. *See also* Antimicrobial materials; Antimicrobial substances
chitosan and chitosan blends as, 71–99
immobilized onto films, 549
interactions with cell membranes, 429–457
interest in, 2
legislation related to, 489–525
nanometals as, 327–350
in polymers, 12–16
Antimicrobial silver
advantages of, 298
applications of, 288–289
efficacy of, 289, 290, 293, 304
mechanisms of, 289–297
optimizing, 292–293
stability *vs.* efficiency of, 293–295
Antimicrobial solubility, 318–319
Antimicrobial substances, 128. *See also* Antimicrobial materials; Antimicrobials
adding to nanocomposite packaging, 106
Antimicrobial surfaces, 33
polymer-based, 396–404
Antimicrobial technologies, for industrial applications, vii
Antioxidant activity, chitosan in, 75
Antioxidant film, rosemary extract in, 175
Antioxidant protection, of food, 175
Antioxidants, synthetic *vs.* natural, 166
Antipathogenic drugs, 409
Antiseptics, 514
Applicator/user exposure studies, 517
Arminin 1a, 245
Aseptic plastics, 16
Aspergillus niger, 162
Asymmetric coatings, on drug tablets, 171
Asymmetric-membrane capsules, 171
Atom transfer radical polymerization (ATRP), 363
ATP-binding cassette (ABC) family, 535
ATP synthesis, 242
Attenuated total reflection Fourier transformed infrared (ATR-FTIR) spectroscopy, 77, 82, 83
Attractants, 510
Autocatalytic electroless chemical plating (AECP), 301
Autochthonous microorganisms, monitoring the growth of, 146
Avery, Christopher W., ix, 429
Azide/alkyne cycloaddition reaction
"click" copper-mediated, 58
metal-free, 66
Azides, 66
Azido-containing polymers, synthesis of, 66

Bac-7, 237
Bacillus, in food-borne outbreaks, 4
Bacillus cereus, 84
chitosan temperature and, 85
Bacillus coagulans, 143
Bacillus pumilis, 102
Bacillus saccharolyticus, 102
Bacillus stearothermophilus, 102
Bacillus subtilis, 54, 102
silver nanoparticle action on, 333
Bacillus thuringiensis, 102, 509
Bactenecin, 202
Bacteremia, 381
Bacteria. *See also* Microbes; Microorganisms
acquired resistance mechanisms in, 27–28

Bacteria (*Continued*)
bacteriocin-producing, 118
biosorption of metals on, 300–301
multi-drug resistance in, 28
ponericin G1 activity against, 207
quorum sensing, 384
selective toxicity to, 335
silver-resistant, 37
silver-treated, 290–291
Bacterial adhesion, 382–385
mechanisms of, 389–390
prediction of, 396
preventing, 391
steps in, 389
surface characteristics influencing, 391–396
surface roughness and, 396
varying reports on, 391
Bacterial adhesion model, 389
Bacterial attachment
barring, 409
transient to permanent, 383
Bacterial colonization, 383
in implants, 401–402
protecting against, 339
of surgical implants, 380
Bacterial cultures, spontaneous mutation in, 27
Bacterial food-borne illness, 6
Bacterial infections, 51
host-pathogen interaction in, 410
Bacterial membranes
differentiating from erythrocyte membranes, 435–436
metal nanoparticle interaction with, 329
Bacterial mutation, 25
Bacterial pathogens
transmembrane potentials of, 236
Bacterial resistance, 23–33, 443
to biocides, 35–37
consequences of, 31–33
mechanisms of, 25–31
against metal nanoparticles, 331
rise in, 24
Bacterial targets, biocide and antibiotic, 537–538
Bacteria-membrane disruption, 399
Bacteria populations, slowing the growth of, 146
Bacteria type, influence on chitosan biocide properties, 84–85
Bactericidal activity, of rmHD5, 245
Bacteriocin, migration from packaging material, 134–141. *See also* Bacteriocins
Bacteriocin activity, expression of, 125
Bacteriocin-based antimicrobial packaging, factors in producing, 130–141
Bacteriocin-containing antimicrobial films, effectiveness of, 145
Bacteriocin-containing films, *in vitro* antimicrobial activity of, 142
Bacteriocin database (BACTIBASE), 228
Bacteriocin detection, new and rapid methods of, 125
Bacteriocin diffusion coefficient, 140
Bacteriocin-EDTA coupling, 143. *See also* EDTA (ethylenediaminetetraacetic acid)
Bacteriocin genome web-base (BAGEL), 229
Bacteriocin inactivation, 130–131
Bacteriocin incorporation, by casting methods, 131
Bacteriocinogenic strains, 124
Bacteriocin-producer cells, incorporating into plastic materials, 147
Bacteriocin-producing bacteria, detecting antimicrobial activity of, 124–125
Bacteriocin-producing species, 118
Bacteriocin-producing strains, steps in searching for, 124
Bacteriocin release, film thickness and, 140
Bacteriocin release kinetics, evaluating, 140
Bacteriocins, 34, 36, 200, 209, 228, 551–552. *See also* Active bacteriocins
antimicrobial action of, 122, 131
combined with natural food preservatives, 147
controlled release of, 147
defined, 118
for developing antimicrobial food packaging materials, 147
diffusion coefficient of, 134
effect on mechanical properties, 141
in food packaging applications, 143–147
generalities related to, 117–119
incorporation into food, 119
nontoxicity of, 125–126
in plastics, 117–158
production of, 124

references related to, 135–139
use in generating active plastic materials, 128
Bacteriostatic agents, 8
Bactiblock Company, 14, 321
Basil essential oils, 103. *See also Ocimum basilicum*
in food preservation, 174
Basil extracts, blending LDPE with, 103
Bellomo, F., ix, 379
Beta defensins, 246
β-sheet structure, 440–443
Binding properties, of gold, 333
Bioactive food packaging, applications of, 167, 168
Bioactive glass ceramic (BGC) biomaterials, 89
Bioactive packaging, 111. *See also* Bioactive food packaging
Bioactive packaging materials, electrospun nanofibers in, 280–281
Bioactive peptides, encapsulation in liposomes, 210
Biobased-derived plastics, 11
Biobased films, active compounds in, 184
Biobased materials, 128
Biobased polymer films, 142
Biocatalyst-based technology, 174
Biocatalysts, 162
disadvantages of, 162
incorporating, 184
Biochemical pesticides, 510
Biocidal action, inactivation of, 40
Biocidal activity, of silver nanoparticles, 332
Biocidal oxide, 355
Biocidal polymers, sustainability of, 40–41
Biocidal products, 493
under Directive 98/8 (EU), 497
safety of, 490
Biocidal product types, under Directive 98/8 (EU), 495–497
Biocidal properties, titanium dioxide nanocomposites with, 366–369
Biocidal surfaces, nonleaching, 399
Biocidal tests, standardizing, 304
Biocide, breakdown in the environment, 546–547. *See also* Biocides
Biocide–antibiotic resistance mechanisms, 37–38
Biocide compound leaching, 391
Biocide concentrations, 530–531
Biocide cross-resistance, to preservatives, 539–540
Biocide disinfectants, 24
Biocide drugs, in electrospun nanofiber mats, 271
Biocide efficiency, of thymol, 101–104
Biocide formulations, EPA registration tests for, 518–519
Biocide-induced enhanced tolerance, 541–542
Biocide leaching PEMs, 400–401. *See also* Polyelectrolyte multilayers (PEMs)
Biocide-loaded nanofibers, 270–276
Biocide misuse, 537–539
safety concerns related to, 530–544
Biocide nanofibers, in plastic matrices, 281
Biocide performance, of chlorine dioxide, 469–470
Biocide plastics, 33–41
challenges of, 34–41
efficacy of, 38
efficiency of, 40
role in microbial resistance, 41
sustainability of, 40–41
Biocide polymeric fiber mats, 267–276
Biocide properties, of chitosan films, 77–78, 83
Biocide regulation, in the European Union, 490, 493–507
Biocide regulations, federal and state, 490
Biocides, 51–52, 510. *See also* Liquid biocides
acquired bacterial resistance to, 35–37
applications of, vii
countries regulating the use of, 491
cross-resistance with antibiotics, 537–539
embedded, 270–271
exposure to, 529
in the food chain, 543–544
food packaging and, 548–554
future trends in, 554–555
health risks associated with, 528–529
incorporation into polymers and surfaces, 38
microorganism susceptibility to, 533
migration of, 553–554
multiple, 547
naturally derived, viii
nonleaching, 412
overuse of, 541
in plastics, 12–13, 25

Biocides (*Continued*)
potential overuse of, 528–529
reducing environmental impact of, 547–548
release, exposure, and accumulation of, 544–545
resistance to, 34–37
risks related to, 555
role in limiting infectious agents/disease, 530
safety and environmental concerns related to, 527–560
selection for virulent or resistant strains, 540–541
toxicity of, 545–546
use in food contact applications, 505–507
various terms related to, 492
vs. antibiotics, 39
Biocides Product Directive (BPD), 493. *See also* European Biocidal Product Directive
Biocide tolerance, 530–531
avoiding, 541–542
categories of, 531–533
features of, 532
increase in, 538
mechanisms of, 532, 533–537
Biocide use
allergenic impact of, 544
environmental impacts of, 544–548
human health impacts of, 542–544
Biocompatibility, nanoparticle, 343–344
Biocorrosion, 33
Biodegradable antibacterial polymers, 66
Biodegradable materials, 11, 128
Biodegradable matrices, 167
Biodegradable polymer blends, 272
Biodegradable polymers, 61–62, 278, 402
Biodegradable scaffold, 277
Bioengineering, active systems intended for, 160
Biofilm, transition to, 384. *See also* Biofilms
Biofilm-controlling materials, 388–396
nanometals as, 336–337
Biofilm development, preventing, 542
Biofilm formation, 36, 38, 384
preventing, 208, 412
processes governing, 382–385
proteins involved in, 410
resistance to, 368
Biofilm growth, preventing, 207
Biofilm infections, 34
Biofilm matrix, 30
Biofilm penetration, 40
Biofilm phenotype, 31
Biofilm promotion signals, 384
Biofilm resistance, mechanisms of, 29–30
Biofilm resistance mechanisms, on implants, 388
Biofilms
antimicrobial characteristics of, 30–31
architectural structure of, 385
biocide tolerance and, 536
on implant surfaces, 380
light-activated killing of, 334
physiological resistance and, 35
role in biofouling and biocorrosion, 33
types of, 383
Biofouling, 33
Biological molecules, interactions with membrane bilayer, 430
Biomacromolecule films, 161
Biomass-derived materials, 127
Biomass resources, materials extracted from, 11
Biomaterial design, developments in, 408–409
Biomaterials
bacterial adhesion to, 385–386
for enzyme immobilization, 170
foreign-body reaction to, 386–387
implantable, 397
infection involving, 339
in medicine, 31–32
peptides as, 210
Biomaterial surface, killing bacteria with, 391
Biomedical applications
chitosan-based systems in, 88–90
chitosan in, 73, 77
Biomedical coating applications, 207
Biomedical implants, increase in, 380
Biomembrane destabilization, 211
Biomimetic peptides, 249
Biomimetics, 277
Biopesticides, 509–510
Bioplastics
as biocide matrices, viii
development of, 147
in packaging, 11–12
used in food packaging, 127
widespread use of, 160

Biopolymer film properties, modifying, 79–80
Biopolymer matrices, chitosan in, 79
Biopolymers, 249, 267
  in the food industry, 549–550
  silk-based, 208
Biopreservatives, 13
  importance of, 101–102
Biosensors
  chitosan in, 74
  for pathogen detection, 208–209
Biotechnological processes, enzymes in, 161
Bispecific monoclonal antibody complex, 410–411
Biteral, 274
Black carrot anthocyanins, spray drying of, 177
Blowing-assisted electrospinning, 264
BM0 peptide, 198, 201
Bone interfaces, antimicrobial, 276–279
Bottom-up technique, for silver particle production, 299
Bound antimicrobial agents, 34
Brochothrix thermosphacta, controlling the growth of, 145–146
Brochothrix thermosphacta ATCC 11509, antimicrobial activity against, 144
Brucella, in food-borne outbreaks, 4
"Burst" effect, 339
Busolo, Maria A., ix, 317

Campylobacter, in food-borne outbreaks, 4
Campylobacter infections, in the European Union, 3
Campylobacter jejuni, 2, 6, 102
  effect of essential oils on, 101
Cancer, antimicrobial peptides and, 247
Cancer diagnosis, chitosan-based systems in, 90
Cancer therapy, silver nanoparticles and, 296–297
Candida albicans, 208
Candida albicans reduction, 342
Candida glabrata, 380
Candida krusei, 380
Candida lambica, 88
Candida parapsilosis, 380
Candida spp., 380
Candida tropicalis, 380
Capping agents, 293, 300, 344
γ-Caprolactone, synthesis of, 63
Carbamate pesticides, 509
Carbohydrates, microencapsulation using, 176
Carbonated hydroxyapatite coatings, 409
Carbon–carbon multiple bond additions, 57
Carbonyl chemistry, non-aldol, 57
Cardamom oleoresin, microencapsulation of, 177
Cardiovascular implants, mortality associated with, 380
Carnosic acid, 175
"Case II diffusion," 182
Catalytic remnants, 66
Cathelicidin LL-37, 230
Cathelicidins, 245–246
Catheter-related infections, 334
Cation exchange capacity (CEC) methodology, 320
Cation exchange method, 320
Cationic amphipathic peptides, 203
Cationic antimicrobial peptides, cytotoxic activity of, 243
Cationic biocides, 52
Cationic contact-killing surfaces, 398–399
Cationic host-defense peptides, antimicrobial activity of, 240
Cationicity, 211
Cationic particles, toxicity of, 344
Cationic peptide fragments, 234, 237
Cationic peptides
  amino-acid-enriched, 234
  cysteine-containing and disulfide-bond-forming, 234
Cationic-salt-functionalized monomers, synthesis and polymerization of, 52–54
Cationic surfactants, 342
Cecropins, 197, 198, 232, 247
Cell death
  link to membrane-disruptive events, 242
  mechanism leading to, 356
Cell envelope permeability, decreased, 26
Cell membranes, interactions with antimicrobials, 429–457
Cell parameters, effect on chitosan properties, 83–85
Cell-penetrating peptides, 205, 241–242
Cell permeability modifications, biocide tolerance and, 533–534

Cellulose-based film, containing pediocin and EDTA, 146
Cellulose triacetate (CTA) films, 170
Cellulosic derivatives, use in enzyme immobilization, 163
Cell wall hydrophilicity, 395
Cerrada, María L., ix, 351
Cetyl trimethyl ammonium bromide (CTAB), 300
CF/enzyme fibers, 163
CF/GOX/HRP electrospun fibers, 163
"Challenge test," 144
Characterization techniques, for embedded particles, 302–303
Chemical antimicrobial additives, 13
Chemical antimicrobial agents, 8–9
Chemical changes
  in packaging materials, 477–480
  in polymers, 478–480
Chemically synthetic antimicrobials, 9
Chemical pesticides, 508–509
Chemical preservatives, 9
Chemical substances, national inventories of, 490, 491
Chemokine induction, 243
Chen, Zhan, ix, 429
Chitin/nanoBGC, 89
Chitooligosaccharides (COS), 81
Chitosan
  antimicrobial activity of, 72
  as an antimicrobial agent, 71–99, 552
  antimicrobial capacity of, 77, 78, 87–88
  biocide capacity of, 15
  biomedical applications of, 77
  effect of cell parameters on properties of, 83–85
  effect on real food systems, 87–88
  general properties of, 72–75
  in food applications, 76
  limitations of, 75–77
  as a natural food preservative, 87–88
  potential of, 72
  production of, 76
  properties of, 268
  uses of, 72–75
Chitosan-based antimicrobial packaging, 81–88
Chitosan-based fibers, electrospun, 268
Chitosan-based packaging, storage temperature of, 86
Chitosan-based systems, in biomedical applications, 88–90
Chitosan biocide properties
  effect of cell parameters on, 83–85
  effect of chitosan formulation on, 81–83
  effect of culture conditions on, 85–86
Chitosan biopolymers, 11
Chitosan blends, 79
  as antimicrobials, 71–99
Chitosan coatings, 399
  oleoresin-enriched, 175
Chitosan degradation, 78
Chitosan-dried films, 73
Chitosan electrospinning, 265
Chitosan encapsulated mannosylated zinc sulphide, 90
Chitosan films, 76
  antimicrobial capacity of, 83
  antimicrobial performance of, 82
  biocide properties of, 77–78
  effect of film-forming and storage conditions on, 86–87
  garlic oil in, 176
  preserving biocide capacity of, 86–87
  water-resistant, 78–81
Chitosan formulation, effect of, 81–83
Chitosan-gelatin blends, 87
Chitosan matrices
  clay additives to, 81
  as food contact materials, 82–83
  thermal treatments of, 86–87
Chitosan matrix thickness, influence on biocide properties, 83
Chitosan membranes, 269
Chitosan/methyl cellulose films, vanillin incorporated into, 175
Chitosan/MMTNa hybrid materials, 80
Chitosan molecular weight, influence on antimicrobial properties, 81
Chitosan/nanoBGC composite scaffolds, 89
Chitosan nanobiocomposites, tensile strength of, 80
Chitosan nanofibers
  bacterial inhibition by, 269
  electrospun, 266
Chitosan/PET matrices, 269
Chitosan/PLA blends, chitosan porous matrices from, 269
Chitosan–polymer blends, 269
Chitosan powder, 73
Chitosan/PVA composites, 269
Chitosan-SiO_2 hybrid gel, preparing, 165–166

Chitosan solutions, biocide properties of, 81
Chitosonium acetate, 77–78
Chitosonium acetate films, biocide properties of, 82, 86
Chitosonium acetate solutions, biocide properties of, 83
χ_{ZZZ}/χ_{YYZ} ratio, 440
Chlorine dioxide, 469–481
  biocide performance of, 469–470
  chemical changes related to, 478–480
  mass transfer of, 470–477
  microbial inactivation mechanisms of, 469
  polymer barrier property changes related to, 480
  sanitizing/food packaging applications of, 469, 470
Chlorine dioxide antimicrobial efficiencies, research studies on, 471–475
Chlorine dioxide gas
  effects on polymeric materials properties, 477–481
  mass transfer profiles of, 476
  mass transfer through polymeric materials, 476
Chronic hazard studies, 516
  for nontarget organisms, 517
ChronoFlex AR, 163. *See also* CF entries
Cinnamon essential oils, in alginate-based edible coatings, 175
Claims, pesticide-related, 514
Class I LAB bacteriocins, 120, 121–122. *See also* LAB bacteriocins; Lactic acid bacteria (LAB)
Class II LAB bacteriocins, 120, 123–124
Class IIa LAB bacteriocins, 123
Class IIb LAB bacteriocins, 123
Class III LAB bacteriocins, 120
Class IV LAB bacteriocins, 120
Class V LAB bacteriocins, 120
Clausius-Clapeyron equation, 464
Clay concentration, thymol release as a function of, 107–111
Clay-containing nanocomposite system, *vs.* PCL nanocomposite system, 107
"Click chemistry"
  antibacterial aliphatic polyesters by, 60–65
  antibacterial polypropylene by, 58–60
  to derived antimicrobial polymers, 51–70
"Click" copper-mediated azide/alkyne cycloaddition reaction, 58. *See also* CuAAC reaction
"Clicked" antibacterial polymers, CuAAC reaction for synthesizing, 66
Click reactions, 57
  metal-free, 66
Click-reaction synthesis, of polymers bearing pendant cationic salts, 57
Clinical resistance, 31–33
Clostridium, in food-borne outbreaks, 4
Clostridium botulinum, 8
Clostridium perfringens, 87
Clostridium tyrobutyricum, 161
Clothing, with antimicrobial properties, 281
CM4, 230
Coagulase-negative staphylococci, 381
Coatings
  antimicrobial, 549
  biodegradable, 402
  chitosan in, 73–74
  Cu-CFx, 340
  obtention/incorporation of silver into, 297–302
  passive and active, 401
Coaxial electrospinning, 262–263, 266, 273, 279
Code of Federal Regulations 40CFR Part161, pesticide tests requested by, 514–519
C=O groups, orientation analysis of, 451
Colony-forming units (CFUs), 324
Combinatorial peptide libraries, 202
Common acquired resistance, 37
Common innate resistance, 37
Communicable diseases, dissemination of, 24
Complex materials, classification of, 356–357
Composite biopolymers, 550
Composite films, 357
Composites, high-performance, 337
Compounds
  broad-spectrum activity of, 444
  synthetic antimicrobial, 443–452
Concentration-dependent mass transfer phenomenon, 466
Condell, Orla, ix, 527
Condensed RNA vaccines, 411

"Conditioning film," 389
Conditioning layers, 38, 40
Consumer products, nanosilver in, 289, 306–307
Contact-active PEMs, 398–400. *See also* Polyelectrolyte multilayers (PEMs)
  preparation of, 399
Contact-resistant PEM, 407. *See also* Polyelectrolyte multilayers (PEMs)
Contamination
  epidemiological investigations of, 7
  primary, 2–7
Conte, Amalia, ix, 159
Contiguous infections, 381
Controlled drug release, chitosan in, 73
Controlled release
  of bacteriocins, 147
  modeling and delivery systems for, 177–184
Controlled-release packaging, 105
Controlled-release packaging materials, smart blending for, 170
Controlled-release systems, developing, 178
Control release plastic, 318
Cooking, inadequate, 7
Copolyesters, with nonquaternized amines, 62
Copolymers, antibacterial, 62
Copper. *See also* CuAAC reaction
  as an antimicrobial agent, 332–333
  antiseptic properties of, 328
Copper-containing coatings, 340
Copper-fluoropolymer (Cu-CFx) nanocomposite films, 340
Copper nanoparticles, 332
  bactericidal effect of, 333
  embedded, 340
Copper oxide
  as an antimicrobial agent, 335
  toxicity of, 344
Copper oxide nanoparticles, 335
Core dissolution, 179
Corn zein, 143, 550
Corn zein films, 176
Cosmetics, chitosan in, 73
Cotter classification scheme, for LAB bacteriocins, 120–121
Covalent attachment technique, for enzyme immobilization, 169
Crassostrea gigas, 545
Critical dilution assay, 125
Crohn's colitis, 246
Cronobacter, 536
Crop contamination, 3
Cross-contamination, 7
Cross-resistance, to antibiotics, 37–38
Cryptosporidium, in food-borne outbreaks, 4
CuAAC reaction, 58–59. *See also* "Click" copper-mediated azide/alkyne cycloaddition reaction
  in aliphatic polyester functionalization, 61–62
  coupling reaction during, 59
  CuI as catalyst of, 62, 65
  drawbacks of, 66
  for grafting antimicrobial agents, 60
CuAAC reaction sequence, reversing, 62
CuI, as a CuAAC reaction catalyst, 62, 65
Culture conditions, effect on chitosan antimicrobial effectiveness, 85–86
Culture pH, effect on chitosan biocide properties, 85
Cyclic lipopeptides, 201
Cyclic *N*-halamines, 250
Cycloaddition, of unsaturated species, 57
Cycloaddition reaction, "click" copper-mediated azide/alkyne, 58
Cysteine-rich β-sheet AMPs, 239. *See also* Antimicrobial peptides (AMPs)
Cytotoxic activity, against cancer cells, 243
Cytotoxicity, of antimicrobial peptides, 196

D4E1 peptide, 198, 201
Daphnia magna, 546, 547
Data Submitters List, 520–521
d-DPPG/d-DPPG bilayer, 436
d-DPPG/DPPG asymmetric bilayer, 439
Defensins, 200
  antimicrobial peptide, 399–400
Defensins knowledgebase, 229
Deferred antagonism test, 124, 125
Defoliants, 511
Degree of deacetylation (DD), 82, 269
Delivery systems, swelling-controlled, 181–184
Del Nobile, Matteo Alessandro, ix, 159
Denatured enzymes, 161
Dendrimeric antimicrobial peptides, 202
Dendrimeric peptides, 250
Dental polymeric materials, silver nanoparticles in, 337–339

Depectinization, 174
Depolymerized chitosan, 73
Derived antimicrobial polymers, "click chemistry" to, 51–70
Derjaguin, Landau, Verwey, Overbeek (DLVO) theory, 383, 392–393, 395. *See also* Extended DLVO theory
Dermaseptic S1, 208
Dermaseptin K4K20S4, 249
Dermaseptin peptides, 245
Dermcidin-1L, 230
Desiccants, 511
"Desorption coefficient," 134
Desorption modeling, 140
Device-centered infections, antibiotic therapies related to, 382
Device infections, development of, 412
Dietary habits, food-borne illnesses and, 2
Diffusion coefficient, 466
Diffusion-controlled systems, types of, 178
Diffusion-limited release, 178–181
Diffusion rate, of antimicrobial agents, 553
Dioxins, 546
Directive 98/8 (EU). *See* European Biocidal Product Directive
Direct polymer grafting reaction, 362
Disinfectants, 493, 510, 513
  under Directive 98/8 (EU), 495
  rotating, 542
Disinfecting agents, in antimicrobial packaging, 460
Disinfection, by titanium dioxide, 354–356
Disk assay, 9
Disulfide bonds, 440–441, 442–443
Disulfide bridge-containing small proteins, 200
Disulfide bridges, 239
Dithiothreitol (DTT), 442–443
Divalent cations, destabilization of, 538
DMA/THF cosolvent, 270
DMF, 275
DNA-damaging effects, 344
DNA recombinant techniques, for antimicrobial peptides, 229
DNA vaccine, 411
Docking molecules, 122
Dodecapeptide bactenacin, 197, 198
Donati, I., ix, 379
Doped nanoparticles, 368
Dose-dumping, 272
DPPG bilayers, 435
DPPG/d-DPPG bilayer, 444
Drug delivery
  gold nanoparticles and, 333–334
  silver nanoparticles and, 296–297
Drug-releasing tunability, 272
Drugs
  multitarget, 196, 203
  release rate of, 171
DTT-treated tachyplesin I, 442–443
Dual ion-beam sputtering technique, 340
Dye-leakage tests, 433
  on oligomers, 448–450

E14LKK, 207
Eco-friendly packaging solutions, 185
Edible bioplastics, 128
Edible coatings, 172–173
  alginate-based, 175
  lemon grass, oregano oil, and vanillin in, 175
Edible films
  incorporating essential oils into, 174
  mechanical properties of, 141
Edible polymers, bacteriocins in, 147
Edible soy protein films, 143
EDTA (ethylenediaminetetraacetic acid)
  cellulose-based film containing pediocin and, 146
  effect on mechanical properties, 141
  incorporated into zein films, 171
EDTA-bacteriocin coupling, 143
Efflux genes, 535
Efflux pump activity, biocide tolerance and, 534–535
Efflux pumps, 26, 27, 35
Elastin-like proteins (ELPs), 231
Electrospinning, 163, 171. *See also* Electrospun entries
  in antimicrobial functionality development, 276–282
  of chitosan, 268–269
  in creating nanofibrous membranes, 277
  fundamentals of, 261–264
  innovations in, 262–264
  novel antimicrobials obtained by, 261–285
Electrospinning apparatus, 262, 263
Electrospinning process, 264
Electrospinning technology, applications of, 277
Electrospraying, 266

Electrospun biocide-loaded nanofibers, 271
Electrospun fiber-based membranes, 276
Electrospun fibers. *See also* Electrospun nanofibers
  for AITC controlled release, 463–464
  medical applications of, 276
  smart textiles based on, 281
Electrospun materials, 263–264
Electrospun nanofibers
  antimicrobial applications of, 280–282
  binding of silver nanoparticles on, 274–275
Electrospun polymers, physical and morphological properties of, 264–266
Electrospun porous membranes, as wound dressings, 279–280
Electrospun resorbable nanofibers, in tissue/organ repair, 278
Electrospun surface structure, controlling, 265–266
Embedded silver particles, thermal properties of, 303
Emulsification, chitosan in, 75
Encapsulation method, of enzyme immobilization, 169
Engineered antimicrobial phyllosilicates, 324–325
Engineered polymers, 267
Engineering approaches, for treating implant-related infections, 405–412
Enhanced tolerance, biocide-induced, 541–542
Enterobacteriaceae, 533
Enterobacter sakazakii, 536
Enterococcus casseliflavus, 142
Enterococcus faecalis, 366
Enterococcus faecium CTC492, 145
Enterotoxigenic *Escherichia coli* (ETEC), 540
Entrapped enzymes, applications to foodstuffs, 172–174
Environment
  biocide breakdown in, 546–547
  exposure to biocides, 529
Environmental antibiotic pollution, 32
Environmental concerns, biocide-related, 527–560
Environmental conditioning layers, 38
Environmental fate data, 518
Environmental impacts, of biocide use, 544–548
Environmental Protection Agency (EPA), 507–508. *See also* EPA entries
Enzymatic browning control, chitosan in, 74
Enzymatic inactivation, biocide tolerance and, 536
Enzyme activity, 160–161
Enzyme binding, mineral supports for, 169–170
Enzyme entrapment methods, 170–171
Enzyme immobilization. *See also* Enzyme nanoscale immobilization
  benefits of, 168–169
  cellulosic derivatives in, 163
  chitosan in, 74
  improving, 169
  techniques for, 169
Enzyme layer deposition technique, 171
Enzyme nanoscale immobilization, 171–172
Enzyme-polymer matrices, 407–408
Enzyme release model, 184
Enzyme reuse, 168, 169
Enzymes, 160–167
  advantages and limitations of, 160–162
  in biotechnological processes, 161
  chemical binding of, 167–170
  depectinization and, 174
  efficacy of, 172
  as food additives, 162
  history of, 160
  immobilization of, 162–163
  in packaging materials, 174
  physical retention of, 170–171
  as preservatives, 551
  sources of, 161–162
  structure of, 161
  use in juice industry, 173
EPA guidance manual, 469. *See also* Environmental Protection Agency (EPA)
EPA Office of Pesticide Programs, 511
EPA registration process, 511–512
  for antimicrobial pesticides, 512–514
EPA registration review program, 508
EPA registration tests, for biocide formulations, 518–519
Erythrocyte membranes, differentiating from bacterial membranes, 435–436
Escherichia coli, 2, 5–6, 54, 102, 143, 163, 291, 534, 535, 536, 541, 550, 552
  basil antimicrobial agents and, 103

in food-borne outbreaks, 4
inhibition via liposomal nisin, 210
keratin/chitosan film reduction of, 79
oregano and garlic oil against, 174
silver nanoparticle action on, 333
silver-nanoparticle-treated, 291
spread of resistance genes in, 32
vanillin inhibitory effect on, 175
Escherichia coli cells, susceptibility to chitosan, 84
Escherichia coli expression system, use in producing recombinant AMPs, 229
Escherichia coli O157:H7, 146
Escherichia coli outbreaks, in the European Union, 3
Essential oil delivery, via microencapsulation processes, 176
Essential oils, 166. *See also* Basil essential oils; Cinnamon essential oils; Lemon grass essential oil; Oregano essential oil; Plant essential oils; Thyme essential oil
antifungal properties of, 102
antimicrobial activity of, 102, 104
as antimicrobial agents, 552
antimicrobial effect of, 101–102
bactericidal activity of, 102
incorporating into food, 174
in plastic systems, 15
Ethylene-vinyl alcohol (EVOH) copolymer, 366–369
EU antimicrobial regulations, 14. *See also* European Union (EU)
Eukaryotic posttranslational modifications, 232
Eulerian approach, 184
Europe, regulation of silver in, 305
European Biocidal Product Directive, 493–494
Annexes included in, 494–495
basic principles of, 494
biocidal product types under, 495–497
biocide substances in Annexes 1 or 1A of, 499–500
food contact applications under, 505–507
forthcoming regulations under, 502–505
implementation guidelines under, 500–501
milestones under, 497–498
status of, 498–500
updated information related to, 501–502
website, 501–502
European Chemicals Agency (ECHA), 502, 504–505
European Commission (EC), 502, 505, 506, 507
European Commission regulations, 2, 502, 504, 505, 506, 507
European Economic Community food additive list, 122, 126
European Food Safety Authority (EFSA), 2, 305, 324, 506, 507
European Union (EU)
biocide regulation in, 490, 493–507
food-borne outbreaks in, 3–5
regulatory frame of, 493–495
EVOH/TiO_2 nanocomposites, 366
biocidal characteristics of, 367–368
Exopolysaccharides (EPS), 384–385, 388
Experimental composite adhesives (ECAs), 339
Expression vectors, commercial, 230
Extended DLVO theory, 395. *See also* Derjaguin, Landau, Verwey, Overbeek (DLVO) theory
Extracellular matrix (ECM), 277
Extrinsic biocide tolerance, 532
Extrinsic resistance, 534
Extrusion process, antimicrobial activity and, 130–131

FabI enzyme, 535, 538
Fanning, Séamus, ix, 527
Farm animals, microbial hazards of, 3
Fat, antimicrobial efficacy and, 172
Fat-trapping agents, chitosan in, 73
FDA antimicrobial regulations, 14. *See also* Food and Drug Administration (FDA)
Fecal contamination, 6
Federal biocide regulations, 490
Federal Food, Drug, and Cosmetics Act (FFDCA), 507
Federal Insecticide, Fungicide, and Rodenticide Act (FIFRA), 507–508
Fernández-García, Marcos, ix, 351
Fernández-García, Marta, ix, 351
Fernandez-Saiz, Patricia, ix, 71
Fiber formation technology, 261–262
Fibers, electrospun, 171
Fibrous membranes, for AITC controlled release, 463

Fickian diffusion, 182
Fick's second law, 181, 182
Field studies, for nontarget organisms, 517
Film coatings, AMP-releasing, 207
Film contact method, 133
Film-forming conditions, effect on chitosan films, 86–87
Film-released ponericin, 207–208
Films
  containing active bacteriocins, 126–143
  designing, 167–177
  synthesizing with embedded metal nanoparticles, 340
  testing antimicrobial effects of, 173
  water vapor transmission rate of, 324
Film thickness, bacteriocin release and, 140
Finite elements method, 184
"First principles" model, 178–179
Fish preservation, chitosan in, 87
Flavivirus, in food-borne outbreaks, 4
Fleming, Alexander, 24
Flexible packaging industry, 129
Flip-flop, of lipids, 436, 450
Fluorescence viable staining method, 131–133
Food. *See also* Fresh foods
  antioxidant protection of, 175
  bacteriocin use in, 125–126
  incorporating essential oils into, 174
  microbial contamination of, vii
  pathogens in, 2
  prolonging the shelf life of, 172–173
Food and Drug Administration (FDA), 2. *See also* FDA antimicrobial regulations
  approval of food contact applications by, 521–524
  medical device approval by, 277–278
Food antimicrobials, 8–9
Food applications
  chitosan in, 76
  of liposomes, 210
Food-borne diseases
  increase in, 2
  prevention and control of, 7–8
Food-borne outbreaks
  causative agents responsible for, 4
  in the European Union, 3–5
Food-borne pathogenic bacteria, effect of plant extracts on, 101–102
Food-borne pathogens, public health and, 2
Food chain, biocides in, 545–546
Food coatings, antimicrobial, 249
Food components, enzyme efficacy and, 172
Food contact applications
  biocides used in, 505–507
  FDA approval of, 521–524
Food-contact-approved kitchen products, 14
Food-contact-approved nanotechnology systems, 13–14
Food-contact-approved organoclays, 106
Food Contact Formulation (FCF) Notification, 524
Food Contact Notification (FCN), 522
Food Contact Substance (FCS), 522
Food contamination, 2
"Food-grade" organisms, 118
Food handling, inadequate, 7
Food industry
  antibiotic resistance in, 33
  biocide use in, 528
  biopolymers in, 549–550
Food industry applications, antimicrobial peptides in, 248–249
Food packaging
  antimicrobial agents in, 550–552
  antimicrobial enzymes in, 184
  antimicrobial materials for, 366
  antimicrobial peptide-based materials in, 209
  biocides and, 548–554
  consumer safety and, 12
  controlled-release materials for, 171
  microbial target of, 133–134
  plastics and bioplastics in, 10–12, 127
  preservation role in, 167
Food packaging applications
  for AITC, 461–464
  bacteriocins in, 143–147
  of chlorine dioxide, 470
Food packaging materials, interactions among, 554
Food packaging systems, chitosan-based, 78–79
Food pathogens, inhibiting, 8
Food preservation
  active packaging in, 209
  antimicrobial agents for, 8–9
  technologies for, 117–118

Food preservatives
  as antimicrobial substances, 14
  LAB bacteriocins as, 119
Food processing, inadequate, 7
Food-processing-related stress, enhanced tolerance to, 540–541
Food products, chlorine dioxide antimicrobial efficiencies for, 471–475
Food-related bacteria, types of, 133–134
Food safety
  active packaging and, 104–105
  prevention technologies for, 7
Food spoilage, slowing of, 145–146
Food spoilage microorganisms, chitosan effectiveness against, 87–88
Foodstuffs
  applications of entrapped enzymes to, 172–174
  silver nitrate in, 305
Food systems
  effect of chitosan on, 87–88
  use of bacteriocins in, 126
Food transformation processes, enzymes in, 162
Foreign-body reaction, 386–387
Fourier transformed infrared (FTIR) analysis, 303
Fourier transform infrared (FTIR) spectra, 478–479
Free halogens, 408
Free nanoparticles, risk posed by, 341
Free-radical formation, antimicrobial activity and, 332
Free radical suspension polymerization, 358
Free volume/cohesive energy density (FV/CED) ratio, 476
Freeze-drying technique, for microencapsulated fish oil, 177
Fresh foods, disinfecting, 470
Fruit juice processing, chitosan in, 74–75
Fruits, contamination of, 3
Fumigants, 510
Functionalized aliphatic polyesters, synthesized by ring-opening polymerization, 64
Functionalized metal nanoparticles, 328–329
Functional nanoadditives, 106
Functional polymers, interaction with inorganic component, 363
Functional protein extracts, for controlling lysozyme release, 171
Fungi, effect of silver nanoparticles on, 292
Fungicides, 510
Fungistatic agents, 8
Fusion proteins, 411

García-Arévalo, Carmen, ix, 227
Garlic oil
  in chitosan films, 176
  in protein-based films, 174
Gas-based antimicrobials, in active packaging, 459–488
Gelatin fiber mats, 276
Gel-phase lipids, 436
Gel-phase models, 436
General biocidal products, under Directive 98/8 (EU), 495
Generally recognized as safe (GRAS) designation, 118
Gene transfer mechanism, 28
Gentamicin, encapsulation of, 273
Gentamicin resistance, in MRSA, 539
Germicides, 514
Germ theory, 24
Giardia, in food-borne outbreaks, 4
Girotti, Alessandra, x, 227
Glass-to-rubbery transition process, 182, 467
Glass transition temperature, 467, 468
Gliadin/chitosonium acetate, *Staphylococcus aureus* growth and, 79
Gliadin-chitosonium acetate blends, 78
Globalization, infection outbreaks and, vii
Glucose oxidase (GOX), 162–166
  in biocatalyst-based technology, 174
  covalently immobilized onto biorientated polypropylene films, 165
  immobilization by entrapment, 164
  immobilization in conductive matrices, 164
  immobilization of, 169
  use in the food industry, 162
Glutaraldehyde, PVOH cross-linked with, 173–174
Glutathione-S-transferase, 230
Gold
  as an antibacterial treatment, 328
  as an antimicrobial agent, 333–334
Gold nanoparticles, 333
  for delivering antimicrobials, 333–334

Gómez, Javier, x, 489
Gramicidin A, 207
Gramicidins, 201
Gram-negative bacteria, 35
  antimicrobial activity against, 134, 142
  biocides and, 535
  protection for, 173
Gram-negative microaerophiles, 6
Gram-positive bacteria, 84–85. *See also* Lactic acid bacteria (LAB)
  antifungal activity against, 102
  antimicrobial activity against, 133
  biocides and, 535
  polymer activity against, 54
"Green" synthesis, 300
Green tea leaf extracts, 166
Growth inhibition halos, 140
Growth inhibition studies, 333
Gum arabic, in encapsulation, 177

Hazard Analysis Critical Control Point (HACCP) process, 7
Hazard studies, data from, 515–516
hBD-2, 245, 246
HD5, 232
Health impacts, of biocide use, 542–544
Health products
  nonpublic, 513
  public, 513–514
Health risks, biocide-associated, 528–529
Hematogenous infections, 381, 386
Heng classification scheme, for LAB bacteriocins, 121
Herbicides, 510
Herbs, extracts from, 166
Heterologous proteins, production of, 231–232
High-molecular-weight chitosan, 77
Hinnavin II/α-melanocyte-stimulating hormone, 230
His-rich AMP histatin, 242. *See also* Antimicrobial peptides (AMPs)
Histamine, in food-borne outbreaks, 4
Histatin 5, 205, 208, 242
Histatins, 200
"Hole-poking" mechanism, 403
Horizontal gene transfer, 27, 28
Horseradish peroxidase (HRP), 163
Hospital-related infections, 31
Host-defense peptides (HDPs), 228. *See also* Antimicrobial peptides (AMPs)
Host immunity, AMP activation of, 243
HPMC films, nisin-containing, 142. *See also* Hydroxypropyl methylcellulose (HPMC)
Human b-defensins 4 (HbD4), 231
Human/domestic animal hazard studies, data from, 515–516
Human health
  food packaging risk related to, 554
  silver nanoparticles and, 296
Human health impacts, of biocide use, 542–544
Human safety concerns, biocide-related, 527–560
"Hurdle technology," 8
Hybrid antimicrobial peptides, 231
Hybrid materials
  manufacturing, 358–359
  producing, 363
Hybrid recombinant proteins, 229
Hydrogen-bonded multilayer films, 406
Hydrophilic polymers, 397
Hydrophobic interactions, 235
Hydrophobic moment, 238
Hydroxypropyl methylcellulose (HPMC), 134. *See also* HPMC films
Hygiene, insufficient, 7

IDR-1, 243
IL-10, 243
Immobilization, effect on enzyme activity, 169
Immobilization matrices, 164
Immobilization packaging system, 105
Immobilization systems, 318
Immune-compromised patients, 25, 31
Immune response
  AMP modulation of, 243
  implant-related, 387
  innate, 227–228
Immunity, innate and adaptive, 411
Immunoincompetent zone, 387
Immunomodulatory activities, 228, 243
Immunomodulatory properties, of antimicrobial peptides, 245–246
Immunotherapeutic approaches, for inhibiting bacterial virulence, 410
Immunotherapy, 402
Implantable biomaterials, 397
Implantable devices, infection and, 339
Implant applications, materials for, 208
Implant-associated biofilms, preventing, 401

Implant-associated infections, 380–382
  early, 386
  engineering approaches for treating, 405–412
  pathogenesis of, 385–387
  proof of, 381
Implants
  bacterial colonization in, 401–402
  biofilm resistance mechanisms on, 388
  compromised, 386
  surgical removal of, 381–382
Implant/tissue interfaces, antimicrobial, 388–396
Impregnated materials, limitations of, 402
Incorporation biocides, 33
Incubation temperature, effect on chitosan biocide properties, 85–86
Indian Collection of antimicrobial peptides (CAMP) database, 228
Indolicidin, 200, 242
Industrial membrane bioreactors, chitosan in, 74
Inert ingredients, FIFRA regulations related to, 520
Infection(s)
  antimicrobial peptide defense against, 196
  bacterial, 51
  biofilm, 34
  biofilms as hazards for, 29
  biomaterial-associated, 32
  catheter-related, 334
  food-related, 2
  immune system defense against, 228
  implant-related, 380–382
  implantable devices and, 339
  multi-drug-resistant, 244
  nosocomial, 31
  postsurgical, 278–279
  Pseudomonas aeruginosa, 26
  reducing the spread of, 24–25
  untreatable, 41
Infection immunity, engineering, 411
Infection outbreaks, vii
Infrared (IR) spectroscopy, 107
Inhibitor molecules, 160
Innate biocide resistance, 34–35
  mechanisms of, 35
Innate immune defense, roles of antimicrobial peptides in, 228
Innate immune response, 227–228
Innate resistance, 37
  to antibiotics, 26
Insect growth regulators, 511
Insecticides, 510
In situ emulsion polymerization, 361
In situ suspension polymerization, 359
Intein protein purification strategy, 231
Intoxications, food-related, 2
Intrinsic biocide tolerance, 531
Intrinsic resistance, 26
Invertase, immobilization by entrapment, 164
In vitro antimicrobial activity, of bacteriocin-containing films, 142
Irradiation methods, nanoparticle tailoring via, 299–300
Isoniazid, 538
Isotactic polypropylene (iPP), as a matrix, 368
Isothiazolone, 530–531
Isotopically asymmetric bilayers, 436
Iversen, Carol, x, 527

Jérôme, C., x, 51
Juice industry, use of enzymes in, 173

K4K20-S4, 209
Keratin/chitosan films, 79
Klaenhammer classification scheme, for LAB bacteriocins, 120
Klebsiella, in food-borne outbreaks, 4
Kubacka, Anna, x, 351

LAB bacteriocins, 118–119. *See also* Lactic acid bacteria (LAB)
  advantages and limitations of, 125–126
  as food preservatives, 119
  classes of, 120–124
  defined, 119
  inhibitory capacity of, 118–119
  sources of, 124–125
  types and characteristics of, 119–124
Labeling, pesticide-related, 514
Labiatae oils, 166
Lactase, immobilization of, 169
Lactic acid bacteria (LAB), 118. *See also* LAB bacteriocins
  inhibitory compounds produced by, 118
Lactobacillus curvatus, 142
Lactobacillus curvatus 32Y, 144
Lactobacillus lactis, 551

Lactobacillus plantarum, 143
oregano and garlic oil against, 174
vanillin inhibitory effect on, 175
Lactococcin K, 230
Lactococcus lactis, 118, 248
Lactoferrampins, 201
Lactoferricins, 201, 245
Lagarón, José María, x, 1, 317
Lagrangian approach, 184
Langmuir-Blodgett-Schaeffer method, 435
Lantibiotics, 120, 200, 248
binding of, 122
Layer-by-layer (LbL) assembly, 397
Layer-by-layer polyelectrolyte deposition, 207–208
Layered silicates, food-contact-approved, 106
Layers, synthesizing with embedded metal nanoparticles, 340
Leachable antimicrobial peptides, 400–401
Leachable antimicrobial agents, 400
Leaching antimicrobials, 41
Lecomte, P., x, 51
Legionella pneumophila, 291
Legislation, antimicrobial-related, 489–525
Legislation-related websites, 524–525
Leishmania, 208–209
Lemon grass
in alginate-based edible coatings, 175
in edible coatings, 175
Lemon grass essential oil, microencapsulation of, 176
Lewis acid-base interactions, 395–396
LFB15, 230
LfcinB15-W4,10, 230
Lifshitz-van der Walls forces, 392, 393, 394
"Lift-off" PEMs, 405, 406–407. *See also* Polyelectrolyte multilayers (PEMs)
Light-activated killing, 334
Lim, Loong-Tak, x, 459
Linalool, release of, 318
Linear antimicrobial peptides, production of, 206
Linear cationic α-helical peptides, 234
Linear peptides, 200
Lipid II, 203–205
Lipid-based biopolymers, 550
Lipid bilayer destabilization, 211
Lipid bilayers, 240
solid-supported, 433–436
Lipid flip-flop, 436, 450
Lipid molecules, names and abbreviations for, 435
Lipid oxidation, chitosan and, 75
Lipids, transmembrane movement of, 450. *See also* Lipid II
Lipopeptides, 206
antimicrobial, 202
cyclic, 201
synthetic, 202
Lipopolysaccharide layer, 534
Lipopolysaccharides (LPSs), 228, 291
Liposomes, antimicrobial peptide encapsulation by, 209–210
Liquid biocides, efficacy of, 38–40
Lister, Joseph, 24
Listeria, 536
in food-borne outbreaks, 4
Listeria-active peptides, 123
Listeria innocua, 209
antimicrobial activity against, 175
basil antimicrobial agents and, 103
as a microbial target, 142
Listeria monocytogenes, 2, 5, 8, 102, 209, 540, 550, 552
antimicrobial packaging for controlling, 145
biofilm formation by, 36
chitosan-coated plastic films and, 87
chitosan temperature and, 85
growth inhibition halos of, 140
as a microbial target, 142
oregano and garlic oil against, 174
Listeria monocytogenes ATCC 19115, 144
Listeria monocytogenes mutants, 36
Listeria outbreaks, in the European Union, 3
LIVE/DEAD bacterial viability kit, 145
LL-37, 198, 200, 208, 245–246
LLDPE films, 142. *See also* Low-density polyethylene films
Local drug release strategies, 401
López-Rubio, Amparo, x, 1
Low-density polyethylene (LDPE), 127, 142
Low-density polyethylene films, 103. *See also* LLDPE films
nisin-coated, 146
Low-molecular-weight biocides, dispersion of, 52
Low-molecular-weight chitosan, 72
Luciferase-loaded polyvinyl alcohol nanofibers, 172

LuxR-LuxI quorum system, 384
Lycopene microcapsules, 177
Lysine dendrimers, 250–251
Lysozyme, 161, 551
  antimicrobial spectrum of, 172
  controlled release of, 170–171
  as a food preservative, 172
  for limiting bacterial growth, 173
  immobilization of, 169
  immobilized on PVOH, 173
  partially purified, 170–171
Lysozyme entrapment, 170–171
Lysozyme films, 161
Lysozyme-nisin-EDTA treatment, 173. *See also* EDTA (ethylenediaminetetraacetic acid)
Lysozyme release kinetic, 183–184

MAC activity, determining, 433. *See also* Membrane-active compounds (MACs)
Magainin I, 207
Magainin 2, 436–439
  multiple-AMP experiments with, 437–438
  orientation of, 453
Magainins, 197, 198, 247
Magneporthe grisea, 232
Magnetron sputtering, 301
Major facilitator super-family (MFS), 535
Malic acid, in alginate-based edible coatings, 175
Manzanares, Paloma, x, 195
Marcos, Jose F., x, 195
Marsich, E., x, 379
Martín, Laura, x, 227
Martínez-Abad, Antonio, x, 287
Mass diffusion, 182
Mass transfer
  of chlorine dioxide, 470–477
  effect of temperature on, 476
  in polymeric packaging materials, 464–469
Mass transfer characteristics, of plastic packaging materials, 476
Mass transfer coefficients, 180, 467
Mass transport, 182
Mastromatteo, Marcella, x, 159
Mastromatteo, Marianna, x, 159
Materials
  biofilm-controlling, 388–396
  electrospun, 263–264
Material surfaces
  effects on bacterial adhesion, 390
  modifying, 409
Mathematical modeling, role in polymeric network design, 178
Matrix system, 178
  release of active agent from, 181
Mauriello, Gianluigi, x, 117
Maximum Contaminant Level Goal, 544
Meat
  challenge tests on, 144
  contaminated, 6
Meat/meat product oxidation, chitosan and, 75
Meat spoilage Gram-positive bacteria, nisin-EDTA solution for, 146
Meat-spoiling agent, 145
Medical applications
  of antimicrobial silver, 288–289
  chitosan membranes for, 269
  of liposomes, 210
Medical devices
  FDA-approved, 277–278
  nanosilver in, 305–306
Medical fields, antimicrobials and, viii
Medicine, biomaterials in, 31–32
MEH-PPV conjugated polymer, 365
Melittin, 439–440
  χ_{ZZZ}/χ_{YYZ} ratio for, 440
  orientation distribution of, 440, 441
Melittin-bilayer interaction kinetics, 439–440
Melt electrospinning, 264
Membrane-active compounds (MACs), 430. *See also* MAC activity
  synthesized, 436
Membrane activity, of MACs, 433
Membrane bilayers, antimicrobial peptide interaction with, 430
Membrane-disruptive antimicrobial peptides, 240, 241
Membrane permeability, reduced, 29
Membrane permeabilization, 241
Membrane perturbation, 241
Membranes, electrospun fiber-based, 276
Membrane translocation, 242
Membrane types, modeling structural differences between, 435–436
Mercury, as a medicinal agent, 328
Mersacidin, 200
Mesoporous materials, 208
Mesoporous silica systems, 208

Messenger RNA (mRNA) translation, 205. *See also* mRNA vaccine
Metabolic pathways, enzymes and, 160
Metal-containing nanolayered clays, antimicrobial plastics based on, 317–326
Metallic antimicrobial compounds, release of, 400
Metallic (metal) nanoparticles
  biocidal effectiveness of, 328–329
  susceptibility to, 341
Metal oxide nanoparticles, toxicity of, 344
Metal oxides, nanoparticulate, 335–336
Metals. *See also* Nanometals; Nanoparticulate metals
  in nanoparticulate form, 328
  nanoscaling, 317
Methicillin-resistant staphylococci, 381
Methicillin-resistant *Staphylococcus aureus* (MRSA), 25, 244, 330, 334, 335
  gentamicin resistance strain of, 539
Microbes. *See also* Bacteria; Microorganism entries
  resilience of, 24
  role in disease, 24
  silver inactivation of, 275
Microbial adhesion, 392
Microbial adhesion to hydrocarbons (MATH) experiments, 394
Microbial adhesion to solvents (MATS) approach, 394
Microbial biofilm growth, preventing, 207
Microbial cell surface hydrophobicity, 393–394
Microbial food-borne disease, increase of, 3
Microbial membrane, penetrating/permeabilizing/disrupting, 241
Microbial pathogen infection, antimicrobial peptide defense against, 196
Microbial pesticides, 509, 510
Microbial targets, 133–134, 142
Microbial viability, peptide effect on, 241
Microbiological tests, 173
Micrococcus glutamicus, 102
Micrococcus luteus, 142–143, 171
Micrococcus luteus ATCC 10240, 146
Microencapsulated drugs, release of active principles from, 178–179
Microencapsulated fish oil, freeze-drying technique for, 177
Microencapsulated systems, natural extracts in, 176–177
Microencapsulation, spray-dry, 177
Microenvironments, alteration of, 31
Microorganisms. *See also* Bacteria; Microbes
  biofilm-related, 380
  killing with light, 334
  natural extracts from, 167
  production of linear AMPs in, 206
  silver ion action on, 332
Microorganism survival curves, 369
Microparticles, antimicrobial peptide encapsulation by, 209–210
Microstructuring techniques, 405
Milk proteins, for natural flavoring encapsulation, 176–177
Minimal inhibitory concentration (MIC) values, 104
Minimum bactericidal concentrations (MBCs), 330, 333
Minimum inhibitory concentrations (MICs), 133, 235, 242, 318, 333, 444, 447, 450
Minimum risk active substances, 520
Miticides, 510
Mitochondrial function, silver nanoparticles and, 296
Mj-AMP1 jalapa defensin, 232. *See also* Antimicrobial peptides (AMPs)
Mobile genetic elements, 539
Modified atmosphere package (MAP) applications, 480
Moisture-interactive behavior, 468–469
Moisture-sensitive polymers, 467
Molluscicides, 510
Monodisperse nanocomposite particles, 359
Monomers, cationic-salt-functionalized, 52–54
Monoterpene phenols, 103
Montmorillonite (MMT) clay, 79–80. *See also* Na-MMT
mRNA vaccine, 411. *See also* Messenger RNA (mRNA) translation
MRSA outbreaks, 25. *See also* Methicillin-resistant *Staphylococcus aureus* (MRSA)
MSI-99, 248
Mueller Hinton broth (MHB), 324

Multi-drug resistance (MDR), 25, 28–29
Multidrug resistance (MDR) phenotype, 537
Multidrug tolerance (MDT), 388
Multilayer film, 171
Multiple biocide interactions, 547
Multiresistant pathogenic bacteria, 250
Multitarget drugs, 196
  antimicrobial peptides as, 203
Mutations, resistant, 27
Mycobacterium tuberculosis, 538

nAg-loaded electrospun fiber mats, 275–276. *See also* Nanosilver entries; Silver entries; Silver nanoparticles (nAg)
N-alkyl-pyridinium, 404
Na-MMT, silver-exchanged, 320. *See also* Montmorillonite (MMT) clay; Sodium–silver ion exchange
Nanoadditive incorporation, 111
Nanoaggregation, 211
NanoBGC, 89
Nanobiocomposites, 106, 323
  thermal transitions of, 80
Nanoclays, for solubility and diffusion control, 106–111
Nanocomposite films, 79–80
  chitosan-based, 81
Nanocomposite organic-inorganic materials, interest in, 356–357
Nanocomposite polymeric electrolytes, 358
Nanocomposites, 106
  thymol in, 101–116
  titanium dioxide-polymeric, 356–365
Nanocomposite systems, PCL *vs.* clay-containing, 107
Nanocomposite technology, value of, 323
Nanocrystalline silver dressings, 306
Nanofiber electrospinning, research on, 282
Nanofibers
  biocide-loaded, 270–276
  production of, 262
Nanofibrous mesh, 263–264
Nanohybrid thin films, transparent, 357
Nanolayered clays
  metal-containing, 317–326
  silver-containing, 320–324
Nanomaterial–biosystem interface, 340–341
  biophysicochemical influences on, 343
Nanomaterials, photoactive, 365
Nanometals. *See also* Nanoparticulate metals
  as antimicrobials, 327–350
  biofilm control and, 336–337
  combining with polymers, 339–340
  polymeric films incorporating, 340–341
  toxicity issues related to, 341–345
Nanoparticle activity, importance of shape to, 332
Nanoparticle aggregation, preventing, 344
Nanoparticle biocompatibility, 343–344
Nanoparticle coating, effect on rheological properties, 365
Nanoparticle incorporation, 331
  scientific articles related to, 307, 308
Nanoparticle morphology, 353–354
  of silver, 293
Nanoparticle preparations, antimicrobial potential of, 330
Nanoparticles
  adhesion of, 302
  sedimentation stability of, 359
  surfactant-coated, 342
  toxico-kinetic properties of, 341
Nanoparticulate cytotoxicity, mechanisms of, 344
Nanoparticulate gold, photodynamic therapy and, 334. *See also* Gold nanoparticles
Nanoparticulate metal oxides, as antimicrobial agents, 335–336
Nanoparticulate metals
  potential substrates for, 331
  toxic properties of, 329–330
Nanoscale fibers, electrospinning in producing, 267
Nano silicon dioxide, 329–330
Nanosilver
  in consumer products, 306–307
  incorporating, 301–302
  mastering the size and shape of, 299–301
  in medical devices, 305–306
Nanosilver-based plastic technologies, 287–316
Nanosilver technologies, future trends in, 307–308
Nanosize substances, in antimicrobial pesticides, 519
Nanostructured fiber mats, biocide properties of, 15

Nanostructure-forming antimicrobial peptides, 210–211
Nanotechnology
  potential of, 328
  safe use of, 341–342
Nano titanium dioxide, 329–330
Naringinase, 173
Natamycin, 143
National Centre for Epidemiology (Spain), 5
National inventories, of chemical substances, 490, 491
National Pesticide Information Retrieval System (NPIRS), 520
Natural active compounds, disadvantages of, 160
Natural antimicrobial additives, 13. *See also* Naturally derived antimicrobial additives
Natural antimicrobial agents, release of, 318. *See also* Natural antimicrobials
Natural antimicrobial peptides, 197–201
  amino acid sequences of, 198
  antifungal activity of, 197
  biophysical properties of, 197
  peptide analogs of, 201–202
Natural antimicrobials, 8, 9. *See also* Natural antimicrobial agents
Natural antimicrobial technologies, 319
Natural antioxidants, 166
Natural biocides, 160
Natural extracts, 160–167
  designing, 167–177
  effect on food-borne pathogenic bacteria, 101–102
  from microorganisms, 167
  physical retention of, 174–176
  in plastics, 159–194
  types and sources of, 166–167
Natural flavorings, encapsulation of, 176–17
Natural food preservatives, combined with bacteriocins, 147
Naturally derived antimicrobial additives, 101–102. *See also* Natural antimicrobial additives
Naturally occurring antimicrobials, 9
Naturally organic matter (NOM), 304
  silver toxicity and, 297
Nematicides, 510
Net positive charge, of antimicrobial peptides, 235–236
Netramai, Siriyupa, x, 459
Networks, synthesis of, 64–65
Neutral polymer surfaces, antibiotic agent release from, 401–402
Nisaplin, 122, 145
  effect on mechanical properties, 141
Nisaplin ND, 122
Nisin, 120, 122–123, 200, 551–552
  activity spectrum of, 134
  adsorption to packaging materials, 209
  antimicrobial activity spectrum of, 123
  back-absorption of, 141
  encapsulation in liposomes, 210
  in food industry applications, 248–249
  resistance to, 36
  use in food, 125–126
Nisin-activated film, 140
Nisin activity, lipid II and, 203–205
Nisin/chitosan-coated paperboard, 146
Nisin-coated films, 209
Nisin-coated paperboard, 146
Nisin-containing cellulose film, 145
Nisin desorption, 134–140
Nisin desorption kinetics, 140
Nisin-EDTA solution, 146. *See also* EDTA (ethylenediaminetetraacetic acid)
Nisin-impregnated films, 145
Nisin variants, 122
Nitrate, in silver nanoparticle preparation, 299
Nitric oxide, release from polymer coatings, 409–410
Nitroxide-mediated radical polymerization (NMP), 362
Non-aldol carbonyl chemistry, 57
Non-food-use pesticides, inert ingredients permitted for, 520
Non-lantibiotics, 120
Nonleaching biocidal surfaces, 399
Nonleaching biocide plastics, 41
Non-membrane-disruptive antimicrobial peptides, 241
Non-natural AMP modifications, 202–203. *See also* Antimicrobial peptides (AMPs)
Nonpublic health products, 513
Nonquaternized amines, copolyesters with, 62
Nonribosomal peptide synthetases (NRPSs), 200–201

Nonstereospecific membranolytic mechanisms, 237
Nontarget organism hazard studies, data from, 516–517
Nonylphenol, 553
No observed effect concentration (NOEC), 545
Nosocomial infections, 24, 31
  implant-related, 380
Nosocomial pathogens, 26
Novel antimicrobial peptides, recombinant routes for generating, 229–233
Novel antimicrobials, obtained by electrospinning, 261–285
Novel antimicrobial technologies, viii
Novel carrier molecules, 231
Novel porous gels, preparing, 165–166
Nucelio lapilli, 545
Nucleic acid binding, 205
Nucleophilic substitution reactions, 57

Ocimum basilicum, 102, 174. *See also* Basil essential oils
Office of Pesticide Programs (OPP), 521
O-H stretching mode, 437
Oleoresin-enriched chitosan coating, 175
Oligomer 1, 443–446, 447, 450–451
  structure of, 445
Oligomer 2, 446–452
Oligomer 3, 446–452
Oligomer 4, 446–448
Oligomer concentration, effect on lipid bilayer, 447
Oligomer interaction, differences in, 450
One-dimensional mass diffusion equation, 181
One-pot procedure, 66
o-phenylenediamine (OPDA), 360–361
Opsonization, 410
Optical parametric/amplification and difference frequency generation (OPG/OPA/DFG) system, 433
Oregano, in protein-based films, 174
Oregano essential oil, 104
  in edible coatings, 175
Organic conditioning layer, 38
Organic solvents, influence on chitosan film properties, 82
Organochlorine insecticides, 509
Organomercurial compounds, 328
Organomodified nanolayered clay, silver-containing, 323
Organophosphate pesticides, 508–509
Orientation analysis, 438
  of C=O groups, 451
Orientation distribution, 440, 441, 444
Orthodontic adhesives, silver nanoparticles in, 337–339
Ortho-phenylphenol (OPP), 542–543
Osseo-integration, 278
Osteoconductive interfaces, 278
Ovicides, 510
Oxide-surface doping process, 368
Oxidizing agents, 469. *See also* Chlorine dioxide
Oxygen-scavenger packaging solution, 174

Packaging. *See also* Active packaging; Advanced packaging systems; Antimicrobial food packaging; Antimicrobial packaging entries; Bioactive packaging entries; Controlled-release packaging entries; Food packaging entries
  consumer safety and, 12
  plastics and bioplastics in, 10–12
  protective functions of, 460
Packaging material(s). *See also* Plastic packaging materials; Polymeric packaging materials
  bacteriocin migration from, 134–141
  chemical changes in, 477–480
  enzymes in, 174
  improved functionality of, 126
  traditional, 167
Packaging solutions, active eco-friendly, 185
Packaging systems, 459–460
  with antimicrobial agent activity, 105
PAF26 peptide, 198, 201
Palmarosa, in alginate-based edible coatings, 175
PANI/TiO_2 nanocomposites, 360
Paoletti, S., x, 379
Passive coatings, 401
Pasteur, Louis, 24
Pathogenic bacteria, 133
Pathogens
  biosensors for detecting, 208–209
  multidrug-resistant, 24
Pathogen-specific immune responses, 228
PCL electrospun mats, 274. *See also* Poly (γ-caprolactone) (PCL)

PCL/layered organoclay nanocomposite, 106
PCL-nanoclay systems, release of thymol from, 107
PCL nanocomposites, 318
PCL nanocomposite system, *vs.* clay-containing nanocomposite system, 107
PCL nanofibers, 276
PCU (polycarbonate urethane) nanofibers, 270
PDMS/PPy copolymer matrix, 164–165. *See also* Polydimethylsiloxane (PDMS); Polypropylene (PP)
PE + nisin + EDTA polymer film, 145. *See also* EDTA (ethylenediaminetetraacetic acid); Polyethylene (PE)
PE + PEO + nisin polymer film, 145. *See also* Poly(ethylene oxide) (PEO)
Pea defensin PSD1, 205
Peanut RNase AhPR-10, 205
Pectinases, 173
Pectin esterase (PE), 174
Pectin lyase, 174
Pediocin, 552
  cellulose-based film containing EDTA and, 146
Pediocin-like bacteriocins, 123
Pediocin PA-1, 123
Pendant amines, quaternization of, 62
Pendant ammonium salts, 63, 64–65
Pendant-azide-bearing aliphatic polyesters, grafting alkynes onto, 61
Pendant-cationic-salt-bearing polymers, click-reaction synthesis of, 57
Pendant functional groups, tailoring, 61
Pendant pyridines, quaternization of, 55
Penetration theory, 179–180
Penicillium notatum, 550
Penicillium spp., 162
PEO-PPO-PEO, 359. *See also* Poly (ethylene oxide) (PEO)
Peptaibol database, 229
Peptaibols, 250
Peptide cyclation, 202
Peptide design, properties modulated by, 201
Peptide efficacy, problems affecting, 211
Peptide folding, 241
Peptide half-life, 250
Peptide libraries, combinatorial, 202
Peptide-polymerization approach, 231
Peptides
  antimicrobial, 195–225
  in clinical/preclinical trials, 244
  nanostructure-forming, 210–211
  secondary structures of, 238
  synthetic *vs.* biotechnological approaches to producing, 206
Peptide structure, independent molecular determinants of, 238
Peptide synthesis services, 206
Peptidoglycan precursor, 203
Peptidomimetics, 202–203
Peptoids, 202
Peri-implant colonization, 387
Peri-implant infections, 382–388
Perinerin, 232
Perioperative infections, 381
Permanent bacterial attachment, 383
Permeability barrier, 35
Permeability cell, 464
Permeability coefficient (P), 464–467, 478–479
Permeability values, of gaseous chlorine dioxide, 476
Permeabilization, 203, 205
Permeation test, 464–466
Persister cells, 31, 388
Pest control, 493
  under Directive 98/8 (EU), 496–497
Pest control devices, 511
Pesticide labels, 514
Pesticide Product Information System (PPIS), 520
Pesticide Product Label System (PPLS), 521
Pesticide regulation, in the United States, 507–524
Pesticides
  defined, 508
  exclusions from, 511
  pest-type-related, 510–511
  types of, 508–511
Pesticide spray drift evaluation, 517–518
Pest-type-related pesticides, 510–511
pET-31b(+) vector, 230
pET32a(+) plasmid, 230
Phagocytosis, frustrated, 387
Pharmaceutical applications, chitosan in, 73
Pharmaceutical field, active systems intended for, 160

Phenolic acids, 175
Phenolic biocides, low-level resistance to, 35
Phenolic compounds, in spices, 174
Phenotypic biocide tolerance, 532–533
o-Phenylenediamine (OPDA), 360–361
Pheromones, 510
Phosphatidylcholine, 236
Phosphatidylcholine nanovesicles, antimicrobial activity of, 147
Phosphatidylglycerol (PG), 236
Phosphonium salts, 51, 52
grafting, 65–66
Photoactive nanomaterials, 365
Photocatalysis, 336, 352
Photocatalysts, characteristics of, 352
Photocatalytic disinfection, 354–356
Photocatalytic polymerization, 360
Photodynamic therapy (PDT), nanoparticulate gold and, 334
Photoinduced copolymerization, 368–369
Photoinduced processes, 352
Photokilling mechanisms, cell death and, 356
Photokilling performance, experimental factors related to, 354–355
Phyllosilicates, engineered antimicrobial, 324–325
Physical vapor deposition (PVD), 301
Physiological resistance, 29–31
biofilms and, 35
Pichia pastoris, 231–232
PLA film, 143. *See also* Poly(lactic acid) (PLA)
PLA nanobiocomposites, 323
PLA nanofibers, silver-loaded, 275
Plant antimicrobial peptides, 228
"Plant bioreactors," 232
Plant bioreactor technology, 232–233
Plant crop diseases, 232
Plant essential oils, 166
efficacy of, 167
Plant expression systems, AMP biosynthesis in, 232–233
Plant extracts
antimicrobial activity of, 102
combining bacteriocins with, 147
in plastic systems, 15
Plant growth regulators, 511
Plant-incorporated protectives (PIPs), 509
Plant pathogens, protection against, 232
Plants
AMP genes in, 247–248
as a source of recombinant proteins, 232
Plant-tissue defense processes, chitosan-activated, 72
Plasma immersion ion implantation (PIII), 301
Plasmid-mediated gene-transfer technique, 248
Plasmodium falciparum, 245
Plastic films
antimicrobial effectiveness of, 144
chitosan-coated, 87
mechanical properties of, 141
Plastic materials
bacteriocin inclusion in, 129
references related to, 135–139
Plastic packaging materials, 126–127
mass transfer characteristics of, 476
permeability of, 476
Plastics
antimicrobial enzymes and natural extracts in, 159–194
bacteriocins in, 117–158
biocidal action in, 34
biocide, 33–41
as biocide matrices, viii
in packaging, 10–12
used in food packaging, 127
widespread use of, 160
Plastic technologies
interactive, 12
silver- and nanosilver-based, 287–316
titanium dioxide-based, 351–377
Plectasin, 200
PLGA nanoparticles, 334
PMBTA, 164
PMBTA/PPy copolymer, 164
PMMA acrylic resin, silver particles in, 338. *See also* Poly(methyl methacrylate) (PMMA)
PMMA hybrid films, synthesis of, 364
PMMA nanofiber, 337
PMMA/TiO_2 nanocomposite materials, 359
preparation of, 359–360
polymeric, 358
PMMA/TiO_2 nanocomposite particles, transparent, 362
Polar angle, 238
Poly[2-(*tert*-butylamino)ethyl methacrylate] (PTBAEMA)
synthesis, 58–60

Poly(3-vinylpyridine) (P3VP)-modified nanoparticles, 362
Poly(4-vinyl-*N*-alkylpyridinium bromide), surfaces functionalized with, 404
Polyacrylamide, bifunctional, 410
Polyacrylonitriles (PAN), thermal annealing of silver nanoparticles in, 301–302
Polyamide-6 (PA-6) microfiltration membrane matrix, 165
Poly(amide-imide)/TiO_2 (PAI/TiO_2) composite films, 357
Polyamide nanofibers, in water filtration, 282
Polyaniline (PANI), grafting on SAM-coated TiO_2 nanoparticles, 360
Polycaprolactone (PCL) copolyester, 60
Polycaprolactones (PCLs), 11. *See also* PCL entries; Poly(γ-caprolactone) (PCL)
  antimicrobial films based on, 105
  thermodynamic miscibility of, 62
Polycondensation, 52, 54, 64
Polydimethylsiloxane (PDMS), 208
Polyelectrolyte multilayer films, 400
Polyelectrolyte multilayers (PEMs), 250, 397–404
  responsive, 405
  types of, 398–401
Polyelectrolytes, layer-by-layer deposition of, 207–208
Poly(γ-caprolactone) (PCL), 271
Polyester films, surface antibacterial activity of, 52
Polyethersulfone/TiO_2 membranes, 364
Polyethylene (PE), 127
Poly(ethylene-co-vinyl acetate) (PEVA), 272
Poly(ethylene glycol) (PEG), 300, 398
Polyethylene-oriented polyamide (PE-OPA) plastic films, 144–145
Poly(ethylene oxide) (PEO), 65. *See also* PEO-PPO-PEO
  in chitosan electrospinning, 269
Polyethylene terephthalate (PET), 553–554
Polygalacturonase (PG), 174
Polyhydroxyalcanoates (PHAs), 11
Poly(lactic acid) (PLA), 271. *See also* PLA entries
Poly(lactic acid) (PLA)/CS nanoparticles, 89
Poly(lactide-co-glycolide) acid (PLAGA), 271, 272
Polylactide (PLA) films, 323
Poly(L-glutamic acid) (PLGa), 399
Poly(L-lysine) (PLL), 398
Polymer-based antimicrobial surfaces, 396–404
Polymer blends
  characteristics of, 264
  chitosan in, 269
Polymer chains, functionalization of, 55–57
Polymer/clay nanocomposites, 79–80
Polymer coatings
  active principle within, 179–180
  nitric oxide release from, 409–410
Polymeric biocides, 267
Polymeric films, incorporating nanometals, 340–341
Polymeric material barrier properties, changes in, 480
Polymeric materials, attaching antimicrobial peptides to, 207
Polymeric materials properties, chlorine dioxide effects on, 477–481
Polymeric-matrix nanocomposites, properties of, 337–341
Polymeric network design, mathematical modeling in, 178
Polymeric packaging materials
  changes in barrier properties of, 481
  mass transfer in, 464–469
  mass transfer of chlorine dioxide in, 470–477
Polymeric surfaces, binding of antimicrobials to, 129
Polymeric ultrafine fibers, 171
Polymer impregnation, advantages of, 340
Polymerization, of cationic-salt-functionalized monomers, 52–54
Polymer matrices
  incorporating antimicrobials into, 249
  obtention/incorporation of silver into, 297–302
Polymer-matrix nanocomposites, methods for formulating, 339
Polymer nanocomposites, 328
Polymer nanofibers, introducing silver nanoparticles into, 275
Polymer relaxation, 183
Polymers. *See also* Electrospun polymers
  antimicrobials in, 12–16

biodegradable, 11, 278
changes in barrier properties of, 480
chemical changes in, 478–480
combining nanometals with, 339–340
extracted from biomass, 12
functionalization of, 128
incorporating antimicrobial peptides into, 206–209
incorporating biocides into, 36, 38
incorporating nanosilver into, 301–302
with inherent antimicrobial activity, 549
lysozyme retention in, 170
molecular weight of, 264
nanoparticle-containing, 334
producing fibers from, 261–262
silver-based, 302–304
sterile-surface, 402–404
swelling of, 181–182
Polymer scaffolds, requirements for designing, 89
Polymer surface modifications, 79
Polymer surfaces, antimicrobial immobilization on, 270
Poly(methyl methacrylate) (PMMA), 208, 331. *See also* PMMA entries
silver-containing, 337
titanium dioxide nanoparticles encapsulated in, 358
Polymorphonucleated (PMN) cells, 387
Polymyxins, 201, 202
Polyolefin polymers, packages based on, 461
Polyphenols, 166
Polypropylene (PP), 127. *See also* PP-*g*-PTBAEMA graft copolymer
Polypropylene–titania nanocomposites, 365
Polysaccharide-based biopolymers, 550
Poly(sodium 4-styrene sulfonate) (SPS), 405
Polystyrene (PS)
grafting of, 361–362
surface modification of, 249
synthesis of, 53
Poly(styrene-*alt*-maleic anhydride) (PSMA), 363
Polyurethane-based leukocyte-inspired biocidal materials, 407–408
Polyurethane cationomers (PUCs), 270
Polyurethane/titania hybrid materials, 363–364
Polyvinylacetate (PVAc), 357
Poly(vinyl alcohol) (PVA), 300. *See also* PVA fibers
Poly(vinyl alcohol) nanofibers, 269
Polyvinyl-alcohols (PVOHs), 11
cross-linked with glutaraldehyde, 173–174
lysozyme immobilization on, 170
lysozyme immobilized on, 173
Poly(vinyl chloride) (PVC), 127. *See also* PVC nanocomposite films
hydrophilicity of, 395
Polyvinylidene copolymer films, 209
Poly(vinylidene fluoride-co-hexafluoropropene) (PVdF-HFP) mat, 279. *See also* TFA/HFP cosolvent
Poly(vinyl pyrrolidone) (PVP), 275, 300
Ponericin G1, 207
POPC bilayer, 438
POPG bilayer, 438
Population biocide tolerance, 532–533
Porcine antimicrobial peptides, 248
Porins, 534
Porphyromonas gingivalis, 540
Postapplication exposure studies, data from, 517–519
Postmanufacturing processing, impact on antimicrobial agent activity, 553
Postsurgical adhesion, preventing, 280
Postsurgical infections, preventing, 278–279
Poultry, contaminated, 6
Power, Karen, x, 527
PP-*g*-PTBAEMA graft copolymer, 60. *See also* Polypropylene entries
Precious metals, antimicrobial effects of, 328
Preservation role, in food packaging, 167
Preservation technologies, vii
Preservatives, 493
as antimicrobial substances, 14
applications of, 542
bacteriocins as, 551–552
biocide cross-resistance to, 539–540
chemical, 9
under Directive 98/8 (EU), 496
enzymes as, 551
weak acids as, 551
Pressure-sensitive labels, 461–463
Prevention technologies, importance to food safety, 7
Primary contamination, causes of, 2–7

Pro/Arg-rich antimicrobial peptides, 242
Process parameters, in electrospinning, 264
Product performance requirements, 515
Product types (PTs), groups of, 493
Prokaryotic expression systems, AMP biosynthesis in, 229–231
Proline-novispirin G10, 247
Pro-rich indolicidin, 239
Prostheses, complications related to, 380
Proteinaceous substances, inhibition from, 125
Protein-based biopolymers, 549–550
Protein-based films, oregano and garlic oil in, 174
Protein-expression kits, 232
Protein interactions, 430
Proteolysis resistance, 250
Proteus, 536
Pseudomonas, 87, 533
Pseudomonas aeruginosa, 26, 54, 291, 330, 333, 366, 535, 538, 539, 540
Pseudomonas putida, 536
PS-modified nanoparticles, 362. *See also* Polystyrene entries
PS/TiO_2 nanocomposite membranes, 364–365
pTYB11 vector, 230
Public health
  crop contamination and, 3
  food-borne pathogens and, 2
Public health products, 513–514
Pulse-filtered cathodic vacuum arc deposition, 301
PVA fibers, electrospun, 275. *See also* Poly (vinyl alcohol) (PVA)
PVC nanocomposite films, 365. *See also* Poly(vinyl chloride) (PVC)
Pyrethroid pesticides, 509

Quantitative quaternization reactions, 56–57
Quaternary ammonium compounds (QACs), 399, 403, 534
  incorporating into PEMs, 400
Quaternary ammonium salts, 51–52
Quaternary phosphonium salts, 404
Quaternization, of prefunctionalized polymer chains, 55
Quaternization reactions, efficiency of, 56–57
Quorum sensing, 384

Radical polymerization, 52–54
Rapid biodegradable nanofibers, 271–272
Reactive oxygen species (ROS), 290, 329, 334
Ready-to-eat (RTE) products, pathogens, 5
Real food system experiments, importance of, 144
Real food systems, effect of chitosan on, 87–88
Recombinant antimicrobial peptides, 227–260
  expression and purification of, 231
Recombinant dermcidin-1L (rDCD-1L), 237
Recombinant DNA techniques/ technologies, 248, 249, 250
Recombinant fusion proteins, 229–230
Recombinantly produced antimicrobial peptides (RADP) database, 228
Recombinant mature human α-defensin 5 (rmHD5), 245
Recombinant proteins, production of, 232
Recombinants routes, for generating novel antimicrobial peptides, 229–233
Refrigeration, inadequate, 7
Register for Biocidal Products (R4BP), 501
Regulation-related websites, 524–525
Regulatory issues, silver-related, 305
Relative humidity (RH), 467–469
Relaxation phenomenon, 182
Release kinetics, 140, 304
Release packaging system, 105
  nanoclays in, 106
Release systems, 318
Ren, G. G., x, 327
Repeat sequence protein polymer technology, 249
Repellants, 510
Research, in active packaging, 482
Research studies, on chlorine dioxide antimicrobial efficiencies, 471–475
Reservoir system, 178
Residual bacteriocin activity, evaluation of, 131–133
Residue chemistry data, 518–519
Resistance
  to antibiotics, 25
  antibiotic use and, 27
  to biocides, 34–37
  mechanisms of, 25–31

mobile genetic elements of, 28–29
physiological, 29–31
spread of, 27–28
to surface-incorporated biocides, 36–37
Resistance genes, spread of, 35
Resistance mechanisms
between biocides and antibiotics, 37–38
common, 538–539
Resistant mutations, 27
Resistant strains, biocide selection for, 540–541
Responsive-PEM technology, 405. *See also* Polyelectrolyte multilayers (PEMs)
Restaurants, food poisoning in, 5–6
Rhizopus nigrans, 102
Rhodotorula rubra, 550
Ring-opening polymerization, 64
Ring-opening polymerization sequence, reversing, 62
Riva, R., xi, 51
RNase activity, 205
RNA vaccines, condensed, 411
Rodenticides, 510
Rodríguez-Cabello, J. Carlos, xi, 227
Rosemary extract, in antioxidant film, 175
Rubino, Maria, xi, 459
Rutile phase, 352

Saccharomyces cerevisiae, 550
basil antimicrobial agents and, 103
inhibition of, 340
vanillin inhibitory effect on, 175
Sachet technology, 461
Sacrificial-anode electrochemical synthesis, 340
Safety
of biocidal products, 490
of food packaging, 12
Safety concerns
biocide-related, 527–560
related to biocide misuse, 530–544
Sakacin A-containing pullulan film, 145
Salmonella, 535, 536, 550, 552
in food-borne outbreaks, 4
Salmonella enterica, 102
chitosan temperature and, 85
effect of essential oils on, 101
Salmonella enteritidis, 143
oregano and garlic oil against, 174
reductions in, 175
Salmonella outbreaks, in the European Union, 3
Salmonella spp., 2, 5, 84, 323
Salmonella typhimurium, 143
Salmonella weltevreden, 102
Sanchez-Garcia, Maria Dolores, xi, 101
Sanitizers, 510, 513–514
Sanitizing applications, of chlorine dioxide, 469, 470
Santos, Mercedes, xi, 227
Sarcina lutea, 102
Scenedesmus subspicatus, 545
Seafood product oxidation, chitosan and, 75
Seamless cloning technique, 231
Selective pressure, 27
Selenastrum capricornotum, 546
Self-assembled monolayer (SAM)-coated TiO_2 nanoparticles, 360
Self-assembled nanostructures, 210–211
Self-assembly method/technique, 299, 301
Self-biomineralization, cationic-AMP-catalyzed, 211
Self-promoted uptake, 236
Self-promotion transport, 538
Semmelweis, Ignaz, 24
Serratia, 536
SFG-active vibrational modes, 432. *See also* Sum-frequency generation (SFG) vibrational spectroscopy
SFG bilayer structures, 434
SFG experimental setup, 432, 433
SFG experimentation with tachyplesin I, 441–442
SFG experiments, 431
SFG hyperpolarizability, 432
SFG laser system, 433
SFG signal, 431
SFG signal intensity, 437
SFG spectra, 436, 447, 449–450
SFG studies, lipid bilayers in, 431
Shai-Matsuzaki-Huang (SMH) model, 241
Shake flask method, 62, 65
Shelf life, prolonging, 172–173
Shigella, in food-borne outbreaks, 4
Short antimicrobial peptides, 197, 198
Short peptides, production of, 205–206
Shrimp antimicrobial peptide penaeidin database (PenBase), 229
Side chain functional groups, of oligomers, 452

Silicon dioxide (SiO_2), 329–330
Silicone, nanoparticulate-silver-impregnated, 339
Silicon surfaces, cellular adhesion and, 142
Silk-based biopolymers, 208
Silk fibroin films, 208
Silver. *See also* Ag entries; Antimicrobial silver; Nanosilver entries
  advantages and limitations of using, 293–297
  antibacterial and antifungal nanofibers based on, 274–276
  as an antimicrobial agent, 288, 331–332
  antimicrobial properties of, 328
  bacterial adhesion of, 303
  bacterial susceptibility to, 291–292
  benefits and toxicity of, 295–297
  as a biocidal leaching agent, 400
  environmental effects of, 297
  interaction with sulphur groups, 293
  nanoscaling, 317
  obtention and incorporation of, 297–302
  properties of, 289
  recommended dosages of, 322
  regulatory issues related to, 305
  WAXS diffractograms of, 322, 323
Silver antimicrobial agents, resistance to, 36–37. *See also* Silver-based antimicrobial agents
Silver-based antimicrobial additives, 319
Silver-based antimicrobial agents, biofilm prevention and, 336–337
Silver-based antimicrobial systems, widespread use of, 289
Silver-based plastic technologies, 287–316
Silver-based polymers
  characterization of, 302–303
  properties of, 302–304
  release issues related to, 303–304
Silver-based technologies, future trends in, 307–308
Silver bromide precipitation, 331
Silver cations, bacterial viability and, 290
Silver chloride, antimicrobial activity of, 295
Silver-containing coatings, 340
Silver-containing nanolayed clays, 320–324
Silver/copper nanoparticle combination, 333
Silver ions, 332. *See also* Ag^+ ions
  in biomedical applications, 331–332
  skin discoloration and, 295
Silver-loaded montmorillonite, 320
Silver nanoparticle images, 294
Silver nanoparticles (nAg), 275, 319–320, 331. *See also* nAg-loaded electrospun fiber mats
  approaches to producing, 299
  benefits and toxicity of, 295–296
  biological reduction of, 300–301
  chemical reduction of, 300
  in dental polymeric materials, 337–339
  effects on viruses and fungi, 292–293
  mechanism of action of, 292
  physical reduction of, 299–300
  preparing, 299–301
  properties of, 289
  size and shape dependence of, 293
  superiority of, 337
Silver nitrate, 288
Silver particle production, at the nanoscale, 298–299
Silver redox potential, 295
Silver release, measuring, 304
Silver-TiO_2 combination, 302
Silver zeolites, 306–307, 331
  ion-exchange-induced release mechanism in, 297–298
Silver-zirconium phosphate nanoparticles, 276
Simultaneous antagonism, 124
Skin infections, 245–246
Small multidrug resistance (SMR), 534–535
Small-size nanoparticles, in the biomedical field, 331
SMAP-29, 230
Smart blending, for controlled-release packaging materials, 170
Smart textiles, 281
Society of Industrial Technology for Antimicrobial Articles, 133
Sodium–silver ion exchange, 297–298
Sol–gel encapsulation method, 169
Solid-supported lipid bilayers, 433–436
Solution parameters, in the electrospinning process, 264
Solvent-casting methods, 131
Sonochemical process, 302
Sorbate, release from tapioca starch edible film, 182

Soy-based films, 145
Soy protein films, edible, 143
Spectral regions, SFG monitoring of, 435
Spelt bran, 176
Sphingomyelin, 236
Spices
  extracts from, 166
  phenolic compounds in, 174
Spoilage bacteria, 133
Spoilage index, 146
Spoilage phenomenon, validating packaging for controlling, 146–147
Spontaneous mutation, 27
Spores, microbial tests on, 173
Sporicides, 513
"Spot-on-lawn" approach, 124
Spray drying method, 177
"Stand-up" orientation, 451
Staphylococcus, in food-borne outbreaks, 4
Staphylococcus aureus, 2, 5–6, 163, 209, 380, 535, 540, 550
  basil antimicrobial agents and, 103
  chitosan films and, 86
  chitosan temperature and, 83, 84
  effect of essential oils on, 102
  as a microbial target, 142
  oregano and garlic oil against, 174
  silver nanoparticle action on, 333
  surface antibacterial activity against, 52
Staphylococcus aureus outbreaks, in the European Union, 3
Staphylococcus epidermidis, 32, 395
Staphylococcus epidermidis reduction, 341
Starting culture, effect on chitosan biocide properties, 85
State biocide regulations, 490
Stereospecific-dependent uptake, 237
Sterile-surface material, design of, 402–403
Sterile-surface polymers, 402–404
Sterilization guidelines, 34
Sterilizers, 513
Stochastic diffusion, 182
Storage conditions, effect on chitosan films, 86–87
Streptococcus, in food-borne outbreaks, 4
Streptomyces natalensis, 143
Stress adaptation, biocide tolerance and, 536–537
Stress resistance, 36
Subacute studies, for nontarget organisms, 517
Subchronic hazard studies, 516
Subinhibitory biocide concentrations, 37
Subinhibitory concentrations, of biocides, 35
Substratum hydrophobicity, adhesion and, 395
Sum-frequency generation (SFG) vibrational spectroscopy, 430–433. *See also* SFG entries
  for an α-helical peptide, 438
  capabilities of, 432–433
  for molecular interaction studies, 453
  for surface/interface studies, 452–453
  uses of, 453
Sun protection factor indexes, 359
"Super case II transport," 182
Surface antimicrobials, 34
Surface biocontamination, 31
Surface charge, importance of, 342
Surface coating, with antimicrobial materials, 207
Surface-containing active enzymes, designing, 167–177
Surface contamination, reductions in, 34
Surface-enhanced Raman scattering (SERS), 292
Surface-incorporated biocides, resistance to, 36–37
Surface layer, characteristics of, 342
Surface modification, 397
Surface plasmon resonance, 303
Surface roughness, bacterial adhesion and, 396
Surfaces
  adhesion-impeding, 394
  chlorine dioxide antimicrobial efficiencies for, 471–475
  containing active bacteriocins, 126–143
  incorporation of biocides into, 38
Surfactants, antibiotic release profile of, 272
Sustainable matrices, 167
Swelling-controlled delivery systems, 181–184
Swelling polymer, lysozyme release from, 184
Synergistic toxicity, 547
Synthetic additives, carcinogenic nature of, 8–9
Synthetic analogs, of antimicrobial peptides, 201–202

Synthetic antibiotic peptides database (SAPD), 229
Synthetic antibiotics, 24
Synthetic antimicrobial compounds, 443–452
as AMP mimics, 451–452
Synthetic antimicrobial peptides, 207
Synthetic antioxidants, 166
Synthetic copolymers, 128
Synthetic lipopeptides, 202
Synthetic peptide libraries, 202
Synthetic peptides, 201–203
non-natural modifications of, 202
Synthetic polymeric matrices, 271
Synthetic polymers, 128
in enzyme immobilization, 163

Tachyplesin I, 440–443
membrane interaction of, 453
Target alteration, biocide tolerance and, 535
Target microorganisms, references related to, 135–139
Temperature
effect on chitosan biocide properties, 85–86
effect on mass transfer, 476
photocatalytic disinfection and, 355
Tensile strength, of plastic films, 141
Terminal methyl groups, orientation of, 432
Tetraoctylammonium (TAO) salts, 340
Textiles
antimicrobial electrospun nanofibers in, 281
antimicrobial peptides in, 249
enzymatic processes of, 161
silver content in, 305
TFA/HFP cosolvent, 269. *See also* Poly (vinylidene fluoride-co-hexafluoropropene) (PVdF-HFP) mat; Trifluoroacetic acid (TFA)
Therapeutic applications, of antimicrobial peptides, 244–247
Thermal annealing, of silver nanoparticles, 301–302
Thermal gravimetrical analysis (TGA), 303
Thermal plasma technology, 330
Thermodynamic model, 392
Thickness, of plastic films, 141
Thin films, contact-killing, 399
Thiol-containing ligands, silver efficacy and, 293
Thymbra spicata, 166–167
Thyme essential oil, 104
Thymol
antifungal activity of, 102
as an antimicrobial agent, 104
in nanocomposites, 101–116
release from PCL-nanoclay systems, 107
release of, 318
understanding the biocide efficiency of, 101–104
Thymol-based films, generating by casting methods, 104–106
Thymol-containing nanobiocomposites, 106
Thymol diffusion coefficient, 107, 108–109, 111, 112–113, 318
Thymol extracts, 103–104
Thymol release
ATR-FTIR spectra of, 108
controlling, 176
as a function of clay concentration, 107–111
from zein-based films, 110–111
Thymol release kinetics, 106
Thymol release modeling, 107–111
Thymol solubility coefficient, 109, 112–113
TiO_2 nanocomposites, preparation by melting process, 365. *See also* Titania entries; Titanium dioxide entries
TiO_2 nanoparticles
influence on biokilling potential, 366–367
modification of, 361–362
surface-modified, 358
TiO_2/PMMA composite cross-linked microspheres, 359
TiO_2-polymer nanocomposites, antimicrobial performance of, 368
Tissue engineering, 276–277
chitosan-based systems in, 89
chitosan in, 73, 77
Tissue-implant antimicrobial interfaces, 379–428
Tissue/organ repair, electrospun resorbable nanofibers in, 278
Titania hybrid nanocomposite films, in polyimide (PI) matrix, 363. *See also* TiO_2 entries; Titanium dioxide entries

Titania nanocomposites, applications of, 357–365
Titania nanoparticles, commercial, 359
Titania performance tests, 355
Titania photokilling performance, 355
Titanium butoxide, 358
Titanium dioxide (TiO_2), 329–330, 352–354. *See also* TiO_2 entries; Titania entries
  as an antimicrobial agent, 336
  crystallographic phases of, 352
  disinfection by, 354–356
  in PMMA-based gel polymeric electrolytes, 358
Titanium dioxide-based plastic technologies, 351–377
Titanium dioxide nanocomposites, 356–365
  with biocidal properties, 366–369
Titanium dioxide nanoparticles, 341–342
Titanium isopropoxide, polymerization in PVAc, 357
Tollens method/process, 300, 302
Top-down technique, for silver particle production, 299
Toroidal pores, 450–451
Torres-Giner, Sergio, xi, 261
Toxicity
  of biocides, 545–546
  synergistic, 547
Toxicity issues, nanometal-related, 341–345
Toxicological testing, 555
Toxic properties, of nanoparticulate metals, 329–330
Toxic Substances Control Act (TSCA), 490
Toxins, in food-borne outbreaks, 4
Transgenic plants, AMP-expressing, 232
Transient bacterial attachment, 383
Travan, A., xi, 379
Treated Articles Exemption, 521
Tributyltin (TBT), 545–546
Trichinella, in food-borne outbreaks, 4
Triclosan, 544–545
  resistance to, 538
Trifluoroacetic acid (TFA), 268. *See also* TFA/HFP cosolvent
Trp-rich tritrpticin, 239
Trx gene, 230
Tryptone soy agar (TSA), 324
Tryptone soy broth (TSB), 323, 324
Tryptophan-rich antimicrobial peptides, 200
Turco, G., xi, 379
Type A lantibiotics, 120, 121, 122
Type B lantibiotics, 120, 121, 122

Ulcerative colitis, 246
Ultrafiltration, 173, 364–365
United States. *See also* Federal entries; National entries
  pesticide regulation in, 507–524
  regulatory frame in, 507–508
Unsaturated species, cycloaddition of, 57
Useful bacteria, 133

Vancomycin, 25
Vancomycin-resistant enterococci (VRRE), 244
Vanillin
  in edible coatings, 175
  inhibitory activity of, 175
Vapor-phase decontamination, 470
Vapor transmission, in plastic films, 141
Vargas-Reus, M. A., xi, 327
Vegetables, contamination of, 3
Vertebrate species, antimicrobial peptides in, 228
Vertical gene transfer, 28
Verticillium dahlia, 232, 248
Veterinary medicine, antibiotic resistance in, 33
Viable spores, microbial tests on, 173
Vibrio fischeri, 546, 547
Vibrio parahaemolyticus, effect of essential oils on, 102
Villani, Francesco, xi, 117
Virulent strains, biocide selection for, 540–541
Viruses
  effect of silver nanoparticles on, 292–293
  in food-borne outbreaks, 4
  inactivating, 330
Vitamin D, 246
Volatile substances, as packaging material antimicrobial agents, 14

Wastewater treatment, chitosan in, 72–73
Water microfiltration, antimicrobial nanofibers in, 281–282
Water purification, chitosan in, 72–73

Water-resistant chitosan-based films, designing, 78–81
Water uptake kinetic, 184
Water vapor transmission rate (WVTR), 324
WAXS diffractograms, of silver, 322, 323
WAXS experiments, 323
Weak acids, as preservatives, 551
Websites
  European Biocidal Product Directive, 501–502
  legislation-related, 524–525
Wide-angle X-ray scattering (WAXS), 323. *See also* WAXS entries
World Health Organization (WHO), 2
"Wormhole" pores, 241
Wound dressings
  chitosan-based systems in, 89
  electrospun porous membranes as, 279–280
Wound healing, antimicrobial peptides and, 246–247

X-ray diffraction (XRD) patterns, 302–303
X-ray photoelectron spectroscopy (XPS), 360

Yeast expression systems, AMP biosynthesis in, 231–232
Yersinia, in food-borne outbreaks, 4

Zapata, María José Ocio, xi, 1
Zein-based films, 173
  EDTA incorporated into, 171
Zein-based mono- and multi-layer films, 176
Zein biopolymers, 11
Zein/chitosan blends, 269, 270
Zein-thymol active packaging materials, 106
Zeolites, 297–298
  silver-substituted, 306
Zinc oxide, 329
  as an antimicrobial agent, 335–336
Zinc oxide nanoparticles, 336
Zinc salts, antiseptic properties of, 328
Zones of inhibition, 125
Zoonotic bacteria, 2

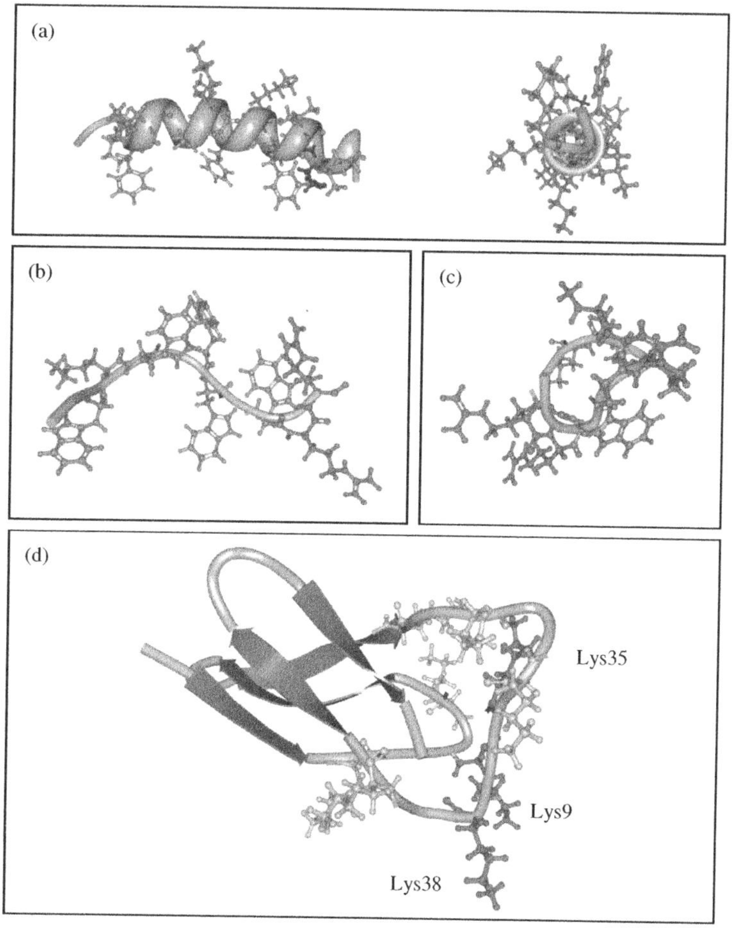

FIGURE 8.1 Selected molecular structures of antimicrobial peptides determined by solution nuclear magnetic resonance methods. Peptide 3D structures and identifications (PDB ID) were obtained from Protein Data Bank (PDB, www.rcsb.org/pdb/) and visualized with PDB ProteinWorkshop v3.9 viewer [36]. **(a)** α-helix folding of the 23-residue peptide magainin 2 in dodecylphosphocholine micelles (PDB ID: 2MAG) [31]. Side view (left) and top view from the N-terminus (right) are shown. The side chains of hydrophobic isoleucine, leucine and phenylalanine residues are shown in green, whereas cationic arginine and lysine residues are in red, to visualize the separation and amphipathicity of the peptide. **(b)** Structure of the 13-residue indolicidin bound to dodecylphosphocholine micelles (PDB ID: 1G89) [28]. **(c)** Structure of the cyclic synthetic derivative of bovine lactoferrin LfcinB4-14Disu in SDS micelles (PDB ID: 1Y58) [29]. In **(b)** and **(c)**, residue coloring as in **(a)**. **(d)** The structure of the antifungal protein from *Penicillium chrysogenum* [37]. The five antiparallel β-strands forming two orthogonal β-sheets are shown in blue at the left side. The cationic side chains exposed in the two largest loops at the right are shown in light red, including the three lysine residues (labeled in dark red) that were demonstrated to enhance toxicity by site-directed mutagenesis.

Printed and bound by CPI Group (UK) Ltd, Croydon, CR0 4YY
18/09/2026
14964970-0003